Colliders and Neutrinos

The Window into
Physics beyond the
Standard Model

Colliders and Neutrinos

The Window into Physics beyond the Standard Model

Editors

Sally Dawson
Brookhaven National Laboratory, USA

Rabindra N. Mohapatra
University of Maryland, USA

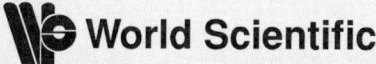

NEW JERSEY · LONDON · SINGAPORE · BEIJING · SHANGHAI · HONG KONG · TAIPEI · CHENNAI

Published by

World Scientific Publishing Co. Pte. Ltd.

5 Toh Tuck Link, Singapore 596224

USA office: 27 Warren Street, Suite 401-402, Hackensack, NJ 07601

UK office: 57 Shelton Street, Covent Garden, London WC2H 9HE

British Library Cataloguing-in-Publication Data
A catalogue record for this book is available from the British Library.

Images on cover design are from www.interactions.org and
http://cdsweb.cern.ch/collection/Photos?ln=en

COLLIDERS AND NEUTRINOS
The Window into Physics beyond the Standard Model

Copyright © 2008 by World Scientific Publishing Co. Pte. Ltd.

All rights reserved. This book, or parts thereof, may not be reproduced in any form or by any means, electronic or mechanical, including photocopying, recording or any information storage and retrieval system now known or to be invented, without written permission from the Publisher.

For photocopying of material in this volume, please pay a copying fee through the Copyright Clearance Center, Inc., 222 Rosewood Drive, Danvers, MA 01923, USA. In this case permission to photocopy is not required from the publisher.

ISBN-13 978-981-281-925-3
ISBN-10 981-281-925-8

Printed in Singapore by B & JO Enterprise

Preface

With many experimental undertakings in particle physics around the corner such as the Large Hadron Collider at CERN, many neutrino experiments are under way and in the horizon, with aspirations for perhaps an International Linear Collider in the not too distant future. The field of particle physics is poised to take a giant leap into unravelling the unknown world of new particles and forces in the coming decades, and build on its success of the past four decades. Combined with the spectacular developments in the field of cosmology, which has perhaps already given us the standard model of the universe and begging for new ideas from particle theory for a deeper understanding of observations, the promise of major breakthroughs and deep insights have filled the air. Many exciting ideas such as supersymmetry, extra dimensions and grand unification are reaching a stage of maturity waiting to be tested. We may also learn about the true nature of the dark constituent of the universe, as well as about the happenings at the early moments of the Big Bang embodied in the ideas of inflation. These discoveries may also provide a better understanding of the formation of structure and evolution of stars and galaxies.

In order to prepare for this new era, the TASI summer school has always been structured to bring to Ph. D. students in the US and abroad the latest ideas and information in a cogent and pedagogical manner, so as to build the intellectual base for tackling the new theoretical challenges that will emerge and are already emerging. The 2006 TASI school was charged with bringing the new phenomenological, cosmological and model building frontier to the students and researchers of tomorrow. With this in mind, we decided to focus on two main themes: Colliders and Neutrinos at the frontier of Physics and inviting experts in the related fields to lecture at the school.

Acknowledgments

We are grateful to all the speakers for taking time out of their busy schedule to prepare lectures and interact with students and contribute to the success of the school. We are especially grateful to those who submitted the written version of the lectures that can be an invaluable educational resource for students both now and in future. We thank Susan Spika and Elizabeth Price for efficient secretarial assistance before and during TASI, Abdul Bachri and Sogee Spinner for organizing the student seminars, Erin DePree and Nicholas Setzer for designing and distributing the TASI06 tee-shirts and Thomas Degrand for organizing the hikes. We thank the National Science Foundation, the Department of Energy, and the University of Colorado for financial and material support. We are very grateful to Prof. K. T. Mahanthappa for superb organization and for being a wonderful host.

Sally Dawson
Rabindra N. Mohapatra

Contents

Preface	v
Extra Dimensions *K. Agashe*	1
The International Linear Collider *M. Battaglia*	49
Astrophysical Aspects of Neutrinos *J. F. Beacom*	101
Leptogenesis *M.-C. Chen*	123
Neutrino Experiments *J. M. Conrad*	177
String Theory, String Model-Building, and String Phenomenology — A Practical Introduction *K. R. Dienes*	255
Theoretical Aspects of Neutrino Masses and Mixings *R. N. Mohapatra*	379
Searching for the Higgs Boson *D. Rainwater*	435
Z' Phenomenology and the LHC *T. G. Rizzo*	537
Neutrinoless Double Beta Decay *P. Vogel*	577
Supersymmetry in Elementary Particle Physics *M. E. Peskin*	609

Chapter 1

Extra Dimensions

Kaustubh Agashe

*Department of Physics, Syracuse University,
Syracuse, NY 13244, USA**

We begin with a discussion of a model with a *flat* extra dimension which addresses the flavor hierarchy of the Standard Model (SM) using profiles for the SM fermions in the extra dimension. We then show how flavor violation and contributions to the electroweak precision tests can be suppressed [even with $O(\text{TeV})$ mass scale for the new particles] in this framework by suitable modifications to the basic model. Finally, we briefly discuss a model with a *warped* extra dimension in which all the SM fields propagate and we sketch how this model "mimics" the earlier model in a flat extra dimension. In this process, we outline a "complete" model addressing the Planck-weak as well as the flavor hierarchy problems of the SM.

1.1. Introduction

Extra dimensions is a vast subject so that it is difficult to give a complete review in 5 lectures. The reader is referred to excellent lectures on this subject already available such as references [1–4] among others. Similarly, the list of references given here is incomplete and the reader is referred to the other lectures for more references.

We begin with some (no doubt this is an incomplete list) motivations for studying models with extra dimensions:

(i) Extra dimensional models can address or solve many of the problems of the Standard Model (SM): for example, the various hi-

---
*After August 1, 2007:
Maryland Center for Fundamental Physics,
Department of Physics,
University of Maryland,
College Park, MD 20742, USA

erarchies unexplained in the SM – that between the Planck and electroweak scales [often called the "(big) hierarchy problem"] and also among the quark and lepton masses and mixing angles (often called the flavor hierarchy). We will show how both these problems are solved using extra dimensions in these lectures.

Extra dimensional models can also provide particle physics candidates for the dark matter of the universe (such a particle is absent in the SM). We will *not* address this point in these lectures.

(ii) Extra dimensions seem to occur in (and in fact are a necessary ingredient of) String Theory, the only known, complete theory of quantum gravity (see K. Dienes' lectures at this and earlier summer schools).

(iii) Although we will *not* refer to this point again, it turns out [5] that, under certain circumstances, extra dimensional theories can be a (weakly coupled) "dual" description of strongly coupled four-dimensional ($4D$) theories as per the correspondence between $5D$ anti-de Sitter (AdS) spaces and $4D$ conformal field theories (CFT's) [6].

The goal of these lectures is a discussion of the theory and phenomenology of some types of extra dimensional models, especially their applications to solving some of the problems of the SM of particle physics. The main concept to be gleaned from these lectures is that

- extra dimensions appear as a tower of particles (or modes) from the $4D$ point of view (a la the standard problem of a particle in $1D$ box studied in quantum mechanics).

The lightest mode (which is often massless and hence is called the zero-mode) is identified with the observed or the SM particles. Whereas, the heavier ones are called Kaluza-Klein (KK) modes and appear as new particles (beyond the SM). It is these particles which play a crucial role in solving problems of the SM, for example they could be candidates for dark matter of the universe or these particles can cut-off the quadratically divergent quantum corrections to the Higgs mass. These particles also give rise to a variety of signals in high-energy collider (i.e., via their on-shell or real production) and in low-energy experiments (via their off-shell or virtual effects). This is especially true if the masses of these KK modes are around the TeV scale, as would be the case if the extra dimension is relevant to explaining the Planck-weak hierarchy.

Here is a rough outline of the lectures. In lecture 1, we begin with the basics of KK decomposition in *flat* spacetime with one extra dimension compactified on a circle. We will show how obtaining chiral fermions requires an *orbifold* compactification instead of a circle. In lecture 2, we will consider a simple solution to the flavor hierarchy using the profiles of the SM fermions in the extra dimension. However, we will see that such a scenario results in too large contributions to flavor changing neutral current (FCNC) processes (which are ruled out by experimental data) if the KK scale is around the TeV scale – this is often called a flavor *problem*. Then, in lecture 3, we will consider a solution to this flavor problem based on the idea of large kinetic terms (for $5D$ fields) localized on a "brane". Another kind of measurement of properties of the SM particles (not involving flavor violation), called Electroweak Precision Tests, will be also be studied in this lecture, including the problem of large contributions to one such observable called the T (or ρ) parameter. In lecture 4, we will solve this problem of the T parameter by implementing a "custodial isospin" symmetry in the extra dimension. We will then briefly discuss some collider phenomenology of such models and some questions which are unanswered in these models. Finally, we will briefly study models based on *warped* spacetime in lecture 5, indicating how such models "mimic" the models in *flat* spacetime (with large brane kinetic terms) studied in the previous lectures. We will sketch how some of the open questions mentioned in lecture 4 can be addressed in the warped setting, resulting in a "complete" model.

1.2. Lecture 1

1.2.1. *Basics of Kaluza-Klein Decomposition*

Consider the following $5D$ action for a (real) scalar field (here and henceforth, the coordinates x^μ will denote the usual $4D$ and the coordinate y will denote the extra dimension):

$$S_{5D} = \int d^4x \int dy \left[\left(\partial^M \Phi \right) \left(\partial_M \Phi \right) - M^2 \Phi \Phi \right] \tag{1.1}$$

Since gravitational law falls off as $1/r^2$ and not $1/r^3$ at long distances, it is clear that we must compactify the extra dimension. Suppose we compactify the extra dimension on a circle (S^1), i.e., with y unrestricted ($-\infty < y < \infty$), but with y identified with $y + 2\pi R$.[a] We

---

[a]Equivalently, we can restrict the range of y: $0 \leq y \leq 2\pi R$, imposing the condition that $y = 0$ same as $y = 2\pi R$.

impose periodic boundary conditions on the fields as well, i.e., we require $\Phi(y = 2\pi R) = \Phi(y)$. Then, we can (Fourier) expand the $5D$ scalar field as follows:

$$\Phi = \frac{1}{\sqrt{2\pi R}} \sum_{n=-\infty}^{n=+\infty} \phi^{(n)}(x) e^{iny/R} \qquad (1.2)$$

where the coefficient in front has been chosen for proper normalization.

Substituting this expansion into S_{5D} and using the orthonormality of profiles of the Fourier modes in the extra dimension (i.e., $e^{iny/R}$) to integrate over the extra dimension, we obtain the following $4D$ action:

$$S_{4D} = \int d^4 x \sum_n \left[\left(\partial_\mu \phi^{(n)} \right) \left(\partial^\mu \phi^{(n)} \right) - \left(M^2 + \frac{n^2}{R^2} \right) \phi^{(n)} \phi^{(n)} \right] \qquad (1.3)$$

This implies that from the $4D$ point of view the $5D$ scalar field appears as an (infinite) tower of $4D$ fields which are called the Kaluza-Klein (KK) modes: $\phi^{(n)}$ with mass^2, $m_n^2 = M^2 + n^2/R^2$ (note that the n^2/R^2 contribution to the KK masses arises from ∂_5 acting on the profiles) [see Fig. 1.1(a)].

The lightest or zero-mode ($n = 0$) has mass M (strictly speaking it is massless only for $M = 0$). The non-zero KK modes start at $\sim 1/R$ (for the case $M \ll 1/R$) which is often called the *compactification scale*. We can easily generalize to the case of δ extra dimensions, each of which

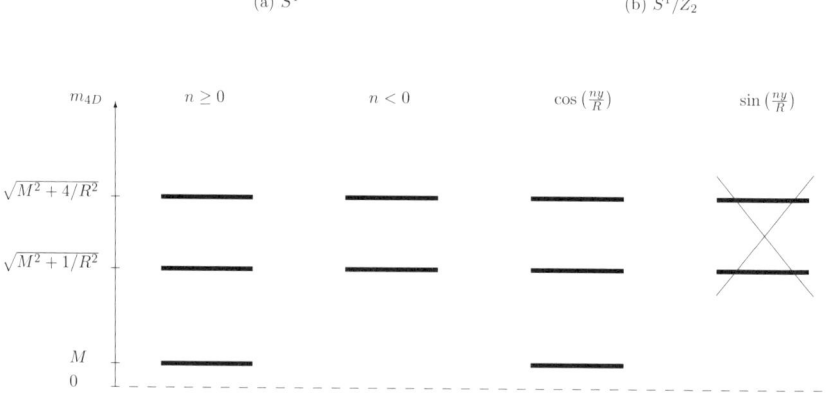

Fig. 1.1. KK decomposition of a $5D$ scalar on a circle (a) and an orbifold (b), choosing even parity.

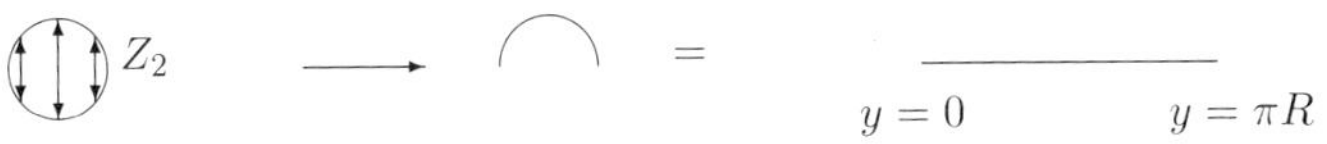

Fig. 1.2. Going from a circle to an orbifold using Z_2 symmetry

is compactified on a circle of same radius to obtain the spectrum: $m_n^2 = M^2 + \sum_{i=1}^{\delta} n_i^2/R^2$. However, in these lectures, *we will restrict to only one extra dimension*.

Thus, we see that the signature of an extra dimension from the 4D point of view is the appearance of infinite tower of KK modes: to repeat, the lightest (zero)-modes is identified with the SM particle and the heavier ones (KK modes) appear as new particles beyond the SM.

1.2.2. *Orbifold*

Mathematically speaking, a circle is a (smooth) manifold since it has no special points. We can "mod out" this smooth manifold by a discrete symmetry to obtain an "orbifold". Specifically, we impose the discrete (Z_2) identification: $y \leftrightarrow -y$ in addition to $y \equiv y + 2\pi R$. Thus, the physical or fundamental domain extends only from $y = 0$ to $y = \pi R^b$ – this compactification is denoted by S^1/Z_2: see Fig. 1.2.

The endpoints of the orbifold ($y = 0, \pi R$) do *not* transform under Z_2 and hence are called *fixed* points of the orbifold. Also, note that the end points of this extra dimension are not identified with each other either by the periodicity condition $y \equiv y + 2\pi R$ (unlike the endpoints $y = 0, 2\pi R$ on S^1) or by the Z_2 symmetry.

Let us consider how the KK decomposition is modified in going from a circle to an orbifold. We can rewrite the earlier KK decomposition in terms of functions which are even and odd under $y \to -y$:

$$\Phi(x,y) = \frac{1}{\sqrt{2\pi R}} \phi^{(0)} + \sum_{n=1}^{\infty} \frac{1}{\sqrt{\pi R}} \left[\phi_+^{(n)} \cos \frac{ny}{R} + \phi_-^{(n)} \sin \frac{ny}{R} \right] \quad (1.4)$$

with the identification $\phi_\pm^{(n>0)} \equiv 1(i)/\sqrt{2} \left(\phi^{(n)} \pm \phi^{(-n)} \right)$.

---
[b]Equivalently, we can still pretend that it extends from $y = 0$ to $y = 2\pi R$ as before, but with the region $y = \pi R$ to $y = 2\pi R$ *not* being independent of the region $y = 0$ to $y = \pi R$.

We must require the physics, i.e., S_{5D}, to be invariant under $y \to -y$. For this purpose, we assign an (intrinsic) parity transformation to Φ:

$$\Phi(x, -y) = P\Phi(x, y) \qquad (1.5)$$

with $P = \pm 1$, i.e., Φ being even or odd. This assignment sets $\phi_-^{(n>0)} = 0$ for $P = +1$ and $\phi_+^{(n)} = 0$ [including $\phi^{(0)}$] for $P = -1$ see Fig. 1.1(b).

Thus, a summary of orbifold compactification is that[c]: (i) it reduces the number of modes by a factor of 2 and (ii) it removes or projects out the *zero*-mode for the case of the $5D$ field being *odd* under the parity.

1.2.3. *Fermions on a Circle: Chirality Problem*

One possible representation of the $5D$ Clifford algebra for fermions:

$$\{\Gamma_M, \Gamma_N\} = 2\eta_{MN} \qquad (1.6)$$

is provided by the usual Dirac (4×4) matrices

$$\Gamma_\mu = \gamma_\mu, \quad \Gamma_5 = -i\gamma_5 \qquad (1.7)$$

Thus, we see that the smallest (irreducible) representation for $5D$ fermions has 4 (complex) components (cf. 2-component complex or Weyl spinor in $4D$, where the 2×2 Pauli matrices form a representation of Clifford algebra).

Consider the following $5D$ action for fermions

$$S_{5D} = \bar\Psi \left(i\partial_M \Gamma^M - M \right) \Psi \qquad (1.8)$$

When the extra dimension is compactified on a circle, we can plug in the usual decomposition $\Psi_{\alpha=1-4} = \sum_n \psi_\alpha^{(n)} e^{iny/R}$ to find the $4D$ action:

$$S_{4D} = \sum_n \bar\psi^{(n)} \left(i\gamma_\mu \partial^\mu - M - in/R \right) \psi^{(n)} \qquad (1.9)$$

Thus, we obtain a tower of Dirac (4-component) spinors from the $4D$ point of view: $m_n^2 = M^2 + n^2/R^2$: see Fig. 1.3(a).

---
[c]We will see later how an orbifold is "useful" in the case of $5D$ fermion/gauge fields because of these properties.

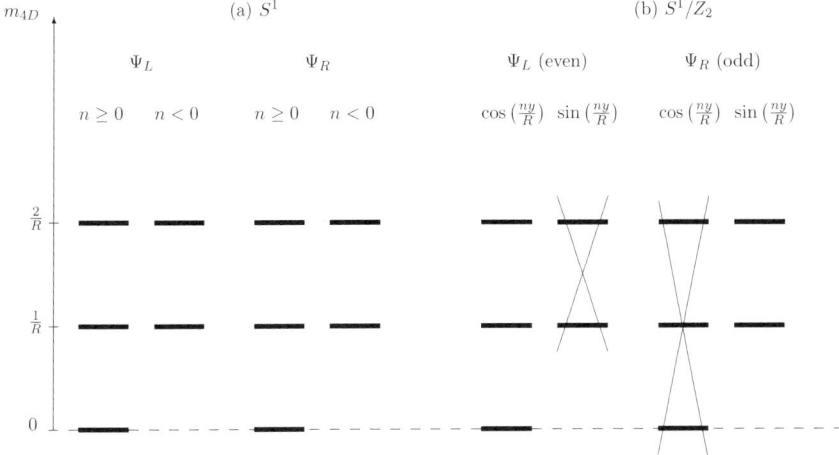

Fig. 1.3. KK decomposition for a 5D fermion on a circle (a) and an orbifold (b) with even parity for Ψ_L.

Consider the case $M = 0$. We see that there are *non*-chiral massless (or zero) modes: explicitly, in the Weyl representation of Dirac matrices, i.e.,

$$\gamma_\mu = \begin{pmatrix} 0 & \sigma_\mu \\ \sigma_\mu & 0 \end{pmatrix} \tag{1.10}$$

$$\gamma_5 = \begin{pmatrix} 1 & 0 \\ 0 & -1 \end{pmatrix} \tag{1.11}$$

$$\sigma_\mu = (\sigma_{i=1..3}, \mathbf{1}), \tag{1.12}$$

$\psi^{(0)}_{\alpha=1-4}$ decomposes as $\sim \left[\psi^{(0)}_L(\alpha = 1, 2), \psi^{(0)}_R(\alpha = 3, 4)\right]$, where L (R) refers to left (right) chirality (or helicity) under the 4D Lorentz transformation. The problem is that if the 5D fermion transforms under some 5D gauge symmetry, then the L and R (massless) chiralities (zero-modes) transform identically under this gauge symmetry. Hence, such a scenario cannot correspond to the SM, where the fermions are known to be chiral, i.e., the left-handed (LH) and right-handed (RH) ones transform as doublets and singlets, respectively under the $SU(2)_{weak}$ gauge symmetry.

1.2.4. *Fermion Chirality from Orbifold*

We can obtain chiral fermions by compactifying the 5D theory on an orbifold instead of a circle as follows. Suppose we choose Ψ_L to be even under the Z_2 parity. Then, Ψ_R must be odd since the $5D$ action contains the term $\bar{\Psi}\Gamma^5\partial_5\Psi \ni \Psi_L^\dagger \partial_5 \Psi_R$, which must be even so that the $5D$ action is Z_2-invariant (note that ∂_5 is odd under parity).

We obtain the following decomposition:

$$\Psi_{L\ (R)} \sim \sum_n \psi^{(n)}_{L\ (R)} \cos\frac{ny}{R}\ (\sin\frac{ny}{R}) \tag{1.13}$$

Thus, (for case of the $5D$ mass, $M = 0^{\text{d}}$) we get a massless zero-mode only for Ψ_L (even field): see Fig. 1.3(b). Of course, we could have chosen Ψ_R to be even instead to obtain a RH zero-mode.

1.3. Lecture 2

1.3.1. Zero-Mode Fermion Profiles

We see that the massless (chiral) mode on an orbifold has a *flat* profile [see Eq. (1.13)]. So, if all the SM fermions have $M = 0$, then the extra dimension does not provide any resolution of the flavor hierarchy, i.e., we need to put hierarchies in $5D$ Yukawa couplings (similar to the situation in the SM) in order to obtain hierarchies in the $4D$ Yukawa couplings.

We must then consider modifying the profiles of the fermion zero-modes in order to solve the flavor hierarchy problem using the extra dimension. We can try adding a bare mass term: $\bar{\Psi}\Psi = \Psi_L^\dagger \Psi_R + h.c.$, but such a mass term breaks the Z_2 symmetry (again since $\Psi_{L,R}$ transform oppositely under the parity). The solution to this problem [7] is to couple the $5D$ fermion to a Z_2-odd scalar with the following $5D$ Lagrangian:

$$\mathcal{L}_{5D} = \bar{\Psi}\left(i\partial_M \Gamma^M - h\Phi\right)\Psi \\ + \left(\partial_M \Phi\right)^2 - \lambda\left(\Phi^2 - V^2\right)^2 \tag{1.14}$$

The point is that the potential $V(\Phi) = \lambda\left(\Phi^2 - V^2\right)^2$ forces a vacuum expectation value (vev) for Φ which is a constant in y in-between the endpoints of the extra dimension (often called the "bulk"). However, such a vev tends to "clash" with $\Phi = 0$ at the endpoints (as required by the scalar being odd under the Z_2 parity). As a result, we obtain a (approximately) "kink-anti-kink" profile for the scalar vev (see references [7] for more details) as in Fig. 1.4. Such a profile for the scalar vev is equivalent to adding a Z_2-*odd* $5D$ mass for the fermion. The point is that with such a scalar vev

---
[d] We will see in the next section that only a "special" form of mass term is allowed on an orbifold.

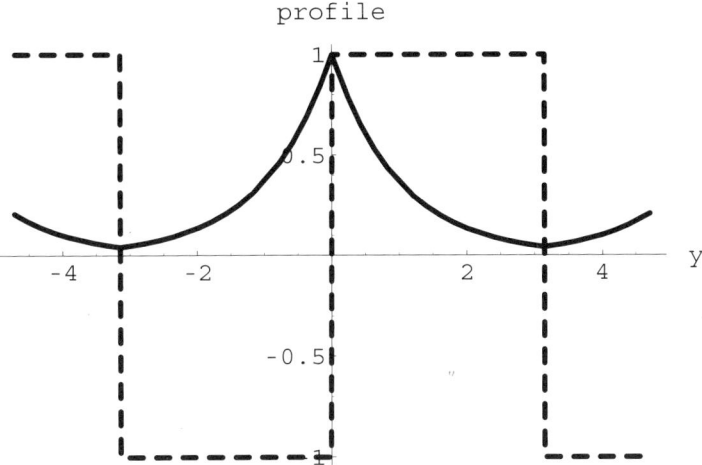

Fig. 1.4. Profile of odd mass term (dashed line) and fermion zero-mode (solid line). Here and henceforth, we set radius of extra dimension, $R = 1$ in all figures.

we have a *spontaneous* breaking of the Z_2 symmetry – recall that it is this Z_2 symmetry which prevented us from writing such a mass term to begin with, i.e., a *bare* mass term would correspond to an *explicit* breaking of this symmetry.

Let us then consider how the KK decomposition is modified in the presence of such an (odd) bulk fermion mass term. The $5D$ action is

$$S_{5D} = \bar{\Psi}\left[i\partial_M \Gamma^M + M\epsilon(y)\right]\Psi \qquad (1.15)$$

where $\epsilon(y) = +1(-1)$ for $\pi R > y > 0 (-\pi R < y < 0)$. It is easy to see that the eigenmodes are no longer *single* sin or cos, but instead are linear *combinations* of these basis functions. Hence, we have to work harder to obtain the eigenmodes.

1.3.2. General Procedure for KK Reduction

We will now take a slight detour to discuss the procedure to obtain the KK decomposition for a general $5D$ action and return to apply this procedure to the above $5D$ fermion case.

For simplicity, consider a $5D$ scalar field decomposed into modes as follows: $\Phi(x, y) = \sum_n \phi^{(n)}(x) f_n(y)$. Plug this expansion into the simple

5D action:

$$S_{5D} = \int d^4x \int dy \left[(\partial^M \Phi)(\partial_M \Phi) - M^2 \Phi \Phi \right] \quad (1.16)$$

We *require* that, *after* integrating over the extra dimension, we get

$$S_{4D} = \int d^4x \sum_n \left[\left(\partial_\mu \phi^{(n)}\right)\left(\partial^\mu \phi^{(n)}\right) - \left(M^2 + \frac{n^2}{R^2}\right) \phi^{(n)} \phi^{(n)} \right] \quad (1.17)$$

so that we can interpret $\phi^{(n)}$'s as particles (KK modes) from the 4D point of view.

This requirement gives us the following two equations: matching kinetic terms in S_{4D} of Eq. (1.17) to the ∂_μ (or 4D) part of the kinetic term obtained from S_{5D} gives us the following:

(i) orthonormality condition

$$\int dy\, f_n^*(y) f_n(y) = 1 \quad (1.18)$$

whereas matching the mass terms in S_{4D} of Eq. (1.17) to the 5D mass term (M) and the action of ∂_5 on the profiles in S_{5D} gives us the

(ii) differential equation:

$$\partial_y^2 f_n(y) - M^2 f_n^2(y) = -m_n^2 f_n^2(y) \quad (1.19)$$

Thus the KK decomposition reduces to an eigenvalue problem, solving which gives us the KK masses (eigenvalues) m_n and their profiles $f_n(y)$ (eigenfunctions). This is very reminiscent of solving the problem of Schroedinger equation for a particle in a 1D box in quantum mechanics.

For the above simple case of a 5D scalar with a bulk mass, we get the following solutions to the differential equation [i.e., Eq. (1.19)]: $f_n(y) \sim e^{\pm i\sqrt{m_n^2 - M^2}\, y}$ for $m_n^2 \geq M^2$. In addition, the periodicity condition, i.e., $f_n(y) = f_n(y + 2\pi R)$ requires $\sqrt{m_n^2 - M^2} = n^2/R^2$ so that $m_n^2 = M^2 + n^2/R^2$ (as before). The reader should think about the possibility $m_n^2 < M^2$ (where we get exponentially rising or decaying profiles) to show that we cannot satisfy the continuity of derivative at $y = 0, \pi R$ in this case and hence we cannot have such solutions for a scalar.

The above procedure can be generalized to more complicated 5D actions and for other spin fields.

1.3.3. *Solution to Flavor Puzzle*

Next, we return to the problem of the KK decomposition of a $5D$ fermion with the (odd) mass term and with $\Psi_{L\ (R)}$ being even (odd) under Z_2 parity. As outlined above, we plug $\Psi_{L,R} = \psi^{(n)}(x) f_{L,R\ n}(y)$ into S_{5D} to obtain the differential equations:

$$\left[-\partial_5 + M\epsilon(y)\right] f_{L\ n} = m_n f_R \tag{1.20}$$

$$\left[\partial_5 + M\epsilon(y)\right] f_{R\ n} = m_n f_L \tag{1.21}$$

Note that (as mentioned before) cos or sin are solutions only for $M = 0$, but not for $M \neq 0$ [On a circle, the mass term M has no $\epsilon(y)$ so that $f_{L,R\ n} \sim e^{iny/R}$ are indeed solutions.].

It is easy to solve for the zero-mode profile ($m_n = 0$) even for $M \neq 0$ (the $m_n \neq 0$ case is difficult to solve due to the two differential equations being coupled):

$$\begin{aligned} f_{L\ 0}(y) &= N e^{My} \quad (0 \leq y \leq \pi R) \\ &= N e^{-My} \quad (0 \geq y \geq -\pi R) \end{aligned} \tag{1.22}$$

(N is a normalization factor: see exercise 1 in appendix).

Note that for RH modes, $f_{R\ 0} \sim e^{\mp My}$ solves the eigenvalue equation, but it clashes with vanishing of $f_{R\ 0}(y)$ at $y = 0, \pi R$ as required by Ψ_R being odd under Z_2 parity. Thus, as expected from the parity choice, there is no RH zero-mode. Note that there is a discontinuity in the derivative of $f_{L\ 0}$ at $y = 0, \pi R$ (Fig. 1.4), which precisely matches the $\epsilon(y)$ term (cf. scalar case earlier where such profiles cannot satisfy the requirement of continuity of derivative at the fixed points). The point is that $M \neq 0$ still gives a *massless* fermion mode (unlike for a scalar).

We will now see how the flavor hierarchy can be accounted for without any large hierarchies in the $5D$ theory: see exercise 1 and Fig. 1.5. For simplicity, suppose the SM Higgs field is localized at $y = \pi R$ (each end of the extra dimension is often called a "brane", motivated by String Theory) and add the following coupling of $5D$ fermions to it:

$$S_{5D} \ni \int d^4x dy \delta(y - \pi R) H \Psi_L \Psi'_R \lambda_{5D} \tag{1.23}$$

where Ψ and Ψ' are two *different* $5D$ fermion fields which are $SU(2)_L$ doublets and singlets with M, M' being their $5D$ masses, respectively. Note that Ψ_L and Ψ'_R are chosen to be even under Z_2 so that they give the LH and RH zero-modes, respectively. Since Ψ_R and Ψ'_L vanish at the $y = \pi R$

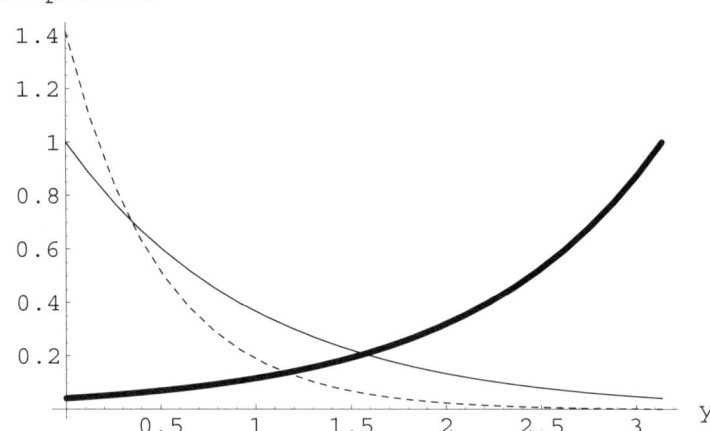

Fig. 1.5. Profiles for down (dashed line: $5D$ mass, $M = -2$), strange (thin solid line: $M = -1$) and top quarks (thick solid line: $M = +1$). The SM Higgs is localized on the $y = \pi R$ brane.

brane, they do not couple to the Higgs as seen in Eq. (1.23). Plugging in the zero-mode profiles, we obtain the effective $4D$ Yukawa coupling, i.e., $\lambda_{4D} H \psi_L^{(0)} \psi_R'^{(0)}$:

$$\lambda_{4D} \approx \lambda_{5D} \times f_{L\,0}(\pi R) f_{R\,0}(\pi R)$$
$$\propto \lambda_{5D} e^{(M-M')} \quad (1.24)$$

Let us consider the hierarchy between the down (d) and strange (s) quark masses for example. For simplicity, we set λ_{5D} to be the same for d, s and also $M = -M'$ for each quark to obtain (up to small dependence of normalization on M's)

$$\frac{m_d}{m_s} \sim e^{2\Delta M \pi R}$$
$$\sim 1/100 \text{ which is the required, i.e., experimental value} \quad (1.25)$$

so that $\Delta M \equiv M_d - M_s \sim -2$ [for example, $M_d = -3$, $M_s = -1$] in units of $1/(\pi R)$ suffices to obtain the hierarchy in $4D$ masses (or Yukawa couplings).

The crucial point is that we did not invoke any large hierarchies in the $5D$ or fundamental parameters (M or λ_{5D}), but we can still obtain large hierarchies in the $4D$ Yukawa couplings.

1.3.4. *Intermediate Summary: Basic Concepts*

Before moving on, let us summarize:

(i) A $5D$ field appears as a tower of KK modes from $4D$ point of view, with each mode having a profile in the extra dimension.
(ii) The profiles and the KK masses are obtained by solving an eigenvalue problem (or wave equations in $5D$ space-time).
(iii) The coupling of particles (i.e., zero and KK modes) is proportional to the overlap of their profiles in the extra dimension.

1.3.5. *Gauge Field on a Circle*

Next, we consider $5D$ gauge fields with the following $5D$ action[e]:

$$S_{5D} = \int d^4x dy \frac{1}{4} \mathcal{F}_{MN} \mathcal{F}^{MN} \tag{1.26}$$

$$= \int d^4x dy \frac{1}{4} \left(\mathcal{F}_{\mu\nu} \mathcal{F}^{\mu\nu} + \mathcal{F}_{\mu 5} \mathcal{F}^{\mu 5} \right) \tag{1.27}$$

with

$$\mathcal{A}_M = \mathcal{A}_\mu + \mathcal{A}_5 \tag{1.28}$$

As usual, the KK decomposition is achieved by plugging in the expansion $\mathcal{A}_{\mu,\,5} = \sum_n A^{(n)}_{\mu,\,5} f_{\mu,\,5\,n}(y)$ into S_{5D}. It is easy to see that this procedure is similar to that for a $5D$ scalar, up to the presence of Lorentz index and gauge fixing. It is straightforward to include the Lorentz index in the KK decomposition, but there are subtleties with gauge fixing – we will not go into details of the latter issue in these lectures (for a discussion of this issue, see, for example, 1st reference in [3]).

The end result is that, on a circle, both $\mathcal{A}_\mu$ and $\mathcal{A}_5$ components have zero-modes – the former is a vector, whereas the latter is a scalar from the $4D$ point of view: see Fig. 1.6(a).

Thus, we encounter a *unification of spins* in the sense that massless $4D$ scalars can be obtained from $5D$ gauge fields. If the $4D$ scalar $A_5^{(0)}$ remains massless, then it will result in an extra long range force which would be ruled out by experiments. However, this scalar does acquire a mass from

---
[e]Once the SM fermions propagate in the extra dimension, we can show that the SM gauge fields also have to do the same to preserve gauge invariance.

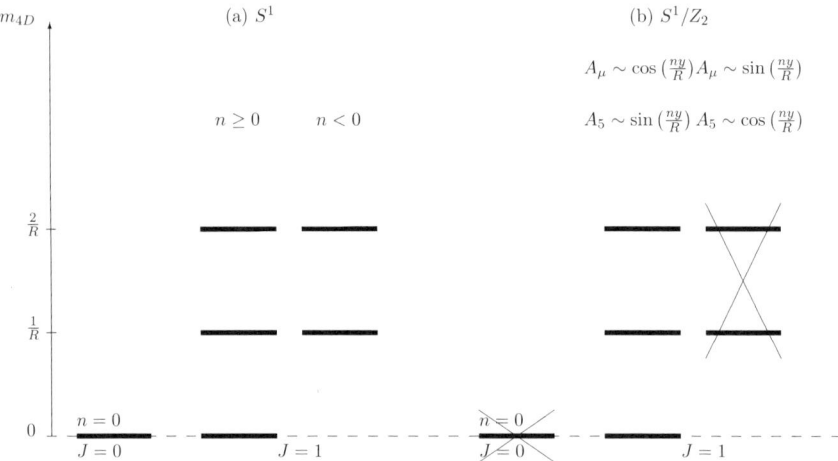

Fig. 1.6. KK decomposition for a 5D gauge field on a circle (a) and on a orbifold (b) with choice of even parity for $\mathcal{A}_\mu$.

loop corrections (see lecture 5) so that such a light scalar (almost zero-mode) might not be a *robust* problem (unlike the chirality problem with fermions on a circle).

1.3.6. *Gauge Field on an Orbifold*

In any case, it is possible to get rid of the $\mathcal{A}_5$ zero-mode using orbifold compactification as follows. Notice that for

$$\mathcal{F}_{\mu 5} = \partial_\mu \mathcal{A}_5 - \partial_5 \mathcal{A}_\mu \quad (1.29)$$

to have a well-defined Z_2 parity, we have two choices:

(i) $\mathcal{A}_\mu$ is even – it has a zero-mode which is identified with the SM gauge boson – which implies that $\mathcal{A}_5$ is odd and so does not have a zero-mode [see Fig. 1.6(b)] or

(ii) $\mathcal{A}_\mu$ is odd (no zero-mode gauge boson) so that $\mathcal{A}_5$ is even and has a zero-mode.

As we will see later, the $\mathcal{A}_5$ zero-mode in case (ii) can play the role of SM Higgs, but for now, we will make the choice (i), i.e., $\mathcal{A}_{\mu\,(5)}$ is even (odd) so that we do have a zero-mode (i.e., SM) gauge boson.

Hence, we obtain the following KK decomposition for this gauge field on an orbifold [Fig. 1.6(b)]:

$$f_{\mu\,0} = \frac{1}{\sqrt{2\pi R}} \quad \text{(i.e., a flat profile)} \tag{1.30}$$

$$f_{\mu\,n}(y) = \frac{1}{\sqrt{\pi R}} \cos ny/R \tag{1.31}$$

$$f_{5\,n}(y) = \frac{1}{\sqrt{\pi R}} \sin ny/R \tag{1.32}$$

We have normalized the modes over $-\pi R \leq y \leq +\pi R$, even though the physical domain is from $y = 0$ to $y = \pi R$. We can show that $A_\mu^{(n \neq 0)}$ "eats" $A_5^{(n)}$ to form a massive spin-1 gauge boson from the following mass terms

$$\mathcal{F}_{\mu 5}^2 \ni \partial_\mu A_5 \partial_5 \mathcal{A}^\mu \tag{1.33}$$

$$\sim \sum_n A_\mu^{(n)} \partial^\mu A_5^{(n)} \partial_y f_{\mu\,n}(y) \tag{1.34}$$

These mass terms mixing $A_\mu^{(n)}$ and $A_5^{(n)}$ are similar to the ones in the SM: $W_\mu \partial^\mu H \langle H \rangle$ (which indicate that the longitudinal polarization of W is the unphysical component of Higgs, i.e., the equivalence theorem).

1.3.7. *Couplings of Gauge Modes*

We now calculate the couplings of the various gauge modes to the matter particles (in this case fermions) based on their profiles. We can show that the coupling of zero-mode is the same to all fermion modes (whether zero or KK):

$$\int d^4x dy \bar{\Psi} \Gamma^M \left(\partial_M + g_5 \mathcal{A}_M \right) \Psi \ni \sum_n \bar{\psi}_L^{(n)} \gamma^\mu \psi_L^{(n)} \times \int dy f_{L\,n}^2(y)$$

$$\left(\partial_\mu + A_\mu^{(0)} \frac{g_5}{\sqrt{2\pi R}} \right) \tag{1.35}$$

$$= \bar{\psi}_L^{(n)} \gamma^\mu \psi_L^{(n)} \left(\partial_\mu + g_4 A_\mu^{(0)} \right)$$

$$\text{(for all } n\text{)} \tag{1.36}$$

with

$$g_4 \text{ (or } g_{SM}) = \frac{g_5}{\sqrt{2\pi R}} \tag{1.37}$$

The point is that the profile of the gauge zero-mode is flat so that the overlap integrals appearing in the kinetic term for fermion mode and in the coupling to gauge zero-mode are identical. This universality of the zero-mode gauge coupling is actually guaranteed by $4D$ gauge invariance.

However, the couplings of zero-mode fermions to gauge KK modes (coming from the overlap of profiles) are *non*-universal, i.e., these couplings depend on the $5D$ fermion mass (see Fig. 1.7):

$$g(n, M) = g_5 \int dy \left(N e^{-My} \right)^2 \times f_{\mu\, n}(y) \qquad (1.38)$$

$$\equiv g_4 \times a(n, M) \qquad (1.39)$$

where a is an $O(1)$ quantity (see exercise 1). The reason is that the gauge KK profile is not flat (unlike for zero-mode) or equivalently there is no analog of $4D$ gauge invariance for the massive (KK) gauge modes.

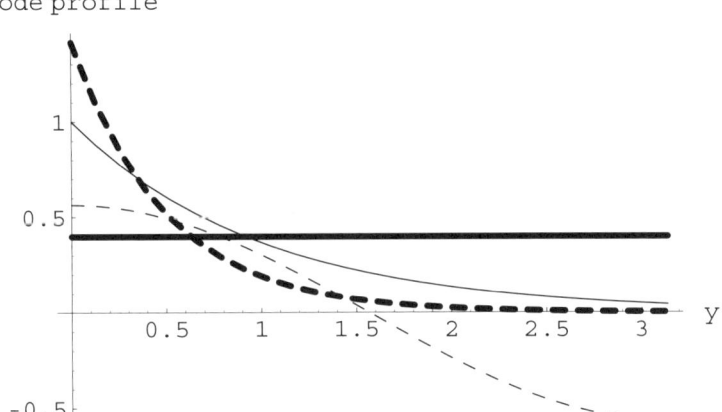

Fig. 1.7. Profiles for down (thick dashed line) and strange (thin solid line) quarks and the gauge zero-mode (thick solid line) and 1st KK mode (thin dashed line). The SM Higgs is localized on the $y = \pi R$ brane.

1.3.8. *Flavor Problem from Gauge KK Modes*

Such non-universal couplings of gauge KK modes to fermion zero-modes results in flavor violation as follows [8]. The point is that the couplings of the gauge KK modes to zero-mode fermions are flavor *diagonal*, but non-universal in the interaction (or weak) basis:

$$g_4 \left(\bar{d}_{L\,\text{weak}}\ \bar{s}_{L\,\text{weak}} \right) \begin{pmatrix} a_d & 0 \\ 0 & a_s \end{pmatrix} \gamma^\mu A^{(n)}_\mu \begin{pmatrix} d_{L\,\text{weak}} \\ s_{L\,\text{weak}} \end{pmatrix} \qquad (1.40)$$

which results in the appearance of flavor violating couplings *after* a unitary rotation to the mass basis:

$$...g_4 D_L^\dagger \text{diag}(a_d, a_s) D_L ... \to g_4 (a_s - a_d)(D_L)_{12} \times$$
$$\bar{d}_{L\,\text{mass}} \gamma^\mu A_\mu^{(n)} s_{L\,\text{mass}} \quad (1.41)$$

where D_L is the unitary transformation to go from the interaction (or weak) basis to the mass basis (for left-handed down-type quarks).

Hence, we obtain a contribution to, for example, $K - \bar{K}$ mixing amplitude:

$$\mathcal{M}_{KK} \sim \frac{g_4^2}{M_{KK}^2} (a_s - a_d)^2 (D_L)_{12}^2 \quad (1.42)$$

The SM contribution to $K - \bar{K}$ mixing amplitude has a suppression mechanism (see below):

$$\mathcal{M}_{SM} \sim \frac{g_4^4}{16\pi^2} \frac{m_c^2}{M_W^4} (V_{us} V_{ud})^2 \quad (1.43)$$

where $V_{us,\,ud}$ are the Cabibbo-Kobayashi-Maskawa (CKM) mixing angles. Since the data agrees with the SM prediction, we must require the KK contribution to be smaller than the SM one and hence we can set a bound on the KK mass. Using

$$(a_s - a_d) \sim O(1/10) \quad (1.44)$$

(see exercise 1), i.e., the fact that the couplings of gauge KK modes to down and strange quarks are $O(1)$ different, we get

$$M_{KK} \gtrsim 20 \text{ TeV} \quad (1.45)$$

assuming that the the D_L mixing angles are of order the CKM mixing angles. Such a large KK mass scale could result in a tension with a solution to the Planck-weak hierarchy problem: we would like the KK scale to be $\sim$ TeV for this purpose (we will see later how the KK mass scale is related to the EW scale).

For completeness, we briefly review FCNC's in the SM below. We begin with the transformation of quarks from weak to mass basis. The Yukawa couplings of the SM fermions to the Higgs (or the mass terms) are 3×3 complex matrices (denoted by M_d in the down quark sector) in the generation space. Such matrices can be diagonalized by *bi*-unitary transformations, $D_{L,R}$. For simplicity, consider the 2 generation case (this analysis can be easily generalized to the case of 3 generations), where this transformation can be explicitly written as

$$\begin{pmatrix} \bar{d}_{L\,\text{weak}} & \bar{s}_{L\,\text{weak}} \end{pmatrix} (M_d)_{2\times 2} \begin{pmatrix} d_{R\,\text{weak}} \\ s_{R\,\text{weak}} \end{pmatrix}$$
$$= \begin{pmatrix} \bar{d}_{L\,\text{mass}} & \bar{s}_{L\,\text{mass}} \end{pmatrix} M_d^{\text{diag.}} \begin{pmatrix} d_{R\,\text{mass}} \\ s_{R\,\text{mass}} \end{pmatrix} \quad (1.46)$$

where

$$\begin{pmatrix} d_{L,R\,\text{weak}} \\ s_{L,R\,\text{weak}} \end{pmatrix} = D_{L,R} \begin{pmatrix} d_{L,R\,\text{mass}} \\ s_{L,R\,\text{mass}} \end{pmatrix} \quad (1.47)$$

$$M_d^{\text{diag.}} \equiv D_L^\dagger M_d D_R$$
$$= \begin{pmatrix} m_d & 0 \\ 0 & m_s \end{pmatrix} \quad (1.48)$$

There are no tree-level FCNC in the SM since the gluon, γ and Z vertices preserve flavor in spite of the above transformations. Of course, the reason is that the couplings of gluon, γ and Z in the weak (or interaction) basis are universal. Explicitly,

$$g_Z \left(-\frac{1}{2} + \frac{1}{3}\sin^2\theta_W\right) \begin{pmatrix} \bar{d}_{L\,\text{weak}} & \bar{s}_{L\,\text{weak}} \end{pmatrix} Z_\mu \gamma^\mu \begin{pmatrix} 1 & 0 \\ 0 & 1 \end{pmatrix} \begin{pmatrix} d_{L\,\text{weak}} \\ s_{L\,\text{weak}} \end{pmatrix}$$
$$= \ldots \begin{pmatrix} \bar{d}_{L\,\text{mass}} & \bar{s}_{L\,\text{mass}} \end{pmatrix} Z_\mu \gamma^\mu D_L^\dagger \begin{pmatrix} 1 & 0 \\ 0 & 1 \end{pmatrix} D_L \begin{pmatrix} d_{L\,\text{mass}} \\ s_{L\,\text{mass}} \end{pmatrix}$$
$$= \ldots \sum_{i=d,s} \bar{d}_{L\,\text{mass}}^i Z_\mu \gamma^\mu d_{L\,\text{mass}}^i \quad (1.49)$$

as compared to Eqs. (1.40) and (1.41).

However, the charged current (W) couplings *are* non-diagonal in the mass basis:

$$\frac{g}{\sqrt{2}} \begin{pmatrix} \bar{u}_{L\,\text{weak}} & \bar{c}_{L\,\text{weak}} \end{pmatrix} W_\mu \gamma^\mu \begin{pmatrix} 1 & 0 \\ 0 & 1 \end{pmatrix} \begin{pmatrix} d_{L\,\text{weak}} \\ s_{L\,\text{weak}} \end{pmatrix}$$
$$= \ldots \begin{pmatrix} \bar{u}_{L\,\text{mass}} & \bar{c}_{L\,\text{mass}} \end{pmatrix} W_\mu \gamma^\mu U_L^\dagger \begin{pmatrix} 1 & 0 \\ 0 & 1 \end{pmatrix} D_L \begin{pmatrix} d_{L\,\text{mass}} \\ s_{L\,\text{mass}} \end{pmatrix}$$
$$= \ldots \sum_{i=u,c\;j=d,s} \bar{u}_{L\,\text{mass}}^i W^\mu \gamma_\mu V_{CKM\;ij} d_{L\,\text{mass}}^j \quad (1.50)$$

where the CKM matrix
$$V_{CKM} \equiv U_L^\dagger D_L$$
$$\neq 1 \tag{1.51}$$

since the transformations in the up and down sectors are, in general, not related. Hence, the charged currents do convert up-type quark of one generation to a down-type quark of a *different* generation. So, we can use the charged current interactions more than once, i.e., in loop diagrams, to change one down-type quark to another down-type quark, for example, to obtain a $\Delta S = 2$ process via a box diagram.

Naively, we can estimate the size of this box diagram

$$\mathcal{M}_{SM} \sim g_2^4 \int \frac{d^4k}{(2\pi)^4} V^*_{CKM\ is} V^*_{CKM\ js} V_{CKM\ id} V_{CKM\ jd} \frac{1}{\slashed{k}-m_i} \frac{1}{\slashed{k}-m_j} \frac{1}{k^2 - M_W^2}$$
$$\sim g_2^4 \left(V_{us} V_{cd}\right)^2 \frac{1}{16\pi^2 M_W^2} \tag{1.52}$$

(neglecting $m_{i,j}$ in the up quark propagators: more on this assumption below) which turns out to be too large compared to the experimental value!

However, this is where the Glashow-Iliopoulos-Maiani (or GIM) mechanism comes in. Using the unitarity of the CKM matrix,

$$\sum_i V^\dagger_{si} V_{id} = 0, \tag{1.53}$$

we find that $\mathcal{M}_{SM}$ vanishes if $m_i = m_j$, in particular if we neglect the quark masses as we did above. Hence, the amplitude must be proportional to the non-degeneracy of the up-type quark masses, i.e., for the two generation case we find that

$$\mathcal{M}_{SM} \sim \frac{g_2^4}{16\pi^2} \left(V_{us} V_{cd}\right)^2 \frac{m_c^2 - m_u^2}{M_W^4} \tag{1.54}$$

which was used earlier in Eq. (1.43). The point is that we get an extra suppression of $\sim m_c^2/M_W^2 \sim 10^{-4}$ compared to the naive estimate in 2nd line of Eq. (1.52).

1.4. Lecture 3

As we saw in the previous lecture, the extra dimensional model which addresses the flavor hierarchy does *not* have analog of the GIM suppression in the gauge KK contribution to flavor violation. The reason is that the couplings of the strange and down quarks to the gauge KK modes, denoted by $a_{s,d}$ (in units of g_4), are $O(1)$, and different.

In order to solve this problem, we would like to modify the gauge KK profile, for example, a more favorable picture would be as in Fig. 1.8, where gauge KK modes are localized near the $y = \pi R$ brane whereas light fermions are localized near the $y = 0$ brane as usual. The point is that in this case couplings of fermions to the gauge KK modes (even though still non-universal) are $\ll 1$ (in units of g_4) so that the FCNC's are suppressed. So, the question is how to modify KK decomposition in general and, in particular, how to obtain the profiles as in Fig. 1.8.

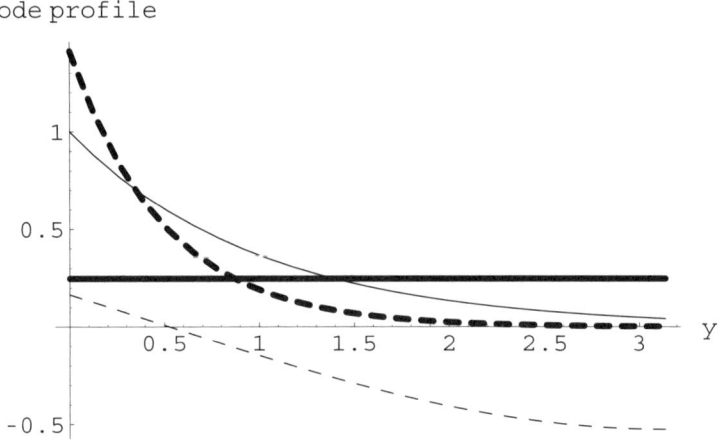

Fig. 1.8. Same as Fig. 1.7, but with brane kinetic term, $r/R = 10$, for gauge fields on $y = 0$ brane.

1.4.1. Brane Kinetic Terms

We consider a modification to the extra dimensional model by adding interactions for the 5D gauge fields which are localized at the fixed points (branes). The point is that such interactions are allowed for an orbifold, but not on a circle, where there are no such "special" points in the extra dimension. In fact, consistency of the model at the quantum level requires the presence of such terms since such terms are generated by loops even if they are absent at tree-level [9].

Specifically, we study the Lagrangian:

$$\mathcal{L}_{5D} = -\frac{1}{4}\Big[\mathcal{F}_{MN}\mathcal{F}^{MN} + \delta(y)r\mathcal{F}_{\mu\nu}\mathcal{F}^{\mu\nu}\Big]$$
$$+ \bar{\Psi}\left(\partial_M + g_5\mathcal{A}_M\right)\Gamma^M\Psi \quad (1.55)$$

Simple dimensional analysis gives $\left[\mathcal{A}_M\right] = 3/2$, $[\Psi] = 2$, $[g_5] = -1/2$ (here $[...]$ denotes mass dimension) so that the brane kinetic term has mass dimension -1 (i.e., it has dimension of a length) and is therefore denoted by r.

It is sometimes convenient to use a different normalization for $\mathcal{A}_M$: $\mathcal{A}_M \to \hat{\mathcal{A}}_M/g_5$ in terms of which the action is:

$$\mathcal{L}_{5D} = -\frac{1}{4}\Big[\frac{1}{g_5^2}\hat{\mathcal{F}}_{MN}\hat{\mathcal{F}}^{MN} + \delta(y)\frac{r}{g_5^2}\hat{\mathcal{F}}_{\mu\nu}\hat{\mathcal{F}}^{\mu\nu}\Big]$$
$$+ \bar{\Psi}\left(\partial_M + \hat{\mathcal{A}}_M\right)\Gamma^M\Psi \quad (1.56)$$

With this normalization, we have $\left[\hat{\mathcal{A}}_M\right] = 1$ (as in $4D$) so that the brane kinetic term is dimensionless: we can then define a brane-localized "coupling" as $1/g_{\text{brane}}^2 \equiv r/g_5^2$.

We will now study how the KK decomposition is modified in the presence of these brane kinetic terms. Consider the case of a scalar field for simplicity (the gauge case which we are really interested in is similar). Here, we will only give a summary: for details, see exercise 2 and reference [10] for example.

Following the procedure outlined in lecture 2, we find that the orthonormality condition is modified (relative to the case of no brane terms):

$$\int dy f_n^*(y)f_m(y)\Big[1 + r\delta(y)\Big] = \delta_{mn} \quad (1.57)$$

and the profiles and mass eigenvalues are given by solving the differential equation:

$$\Big[\partial_y^2 + m_n^2 + r\delta(y)m_n^2\Big]f_n(y) = 0 \quad (1.58)$$

The solutions $f_n(y)$ of this equation are linear combination of sin and cos, in particular, a *different* one for $y = 0$ to $y = \pi R$ and $y = -\pi R$ to $y = 0$.

In addition, in order to solve for the coefficients of sin, cos in these linear combinations, we must impose conditions such as continuity of $f_n(y)$ at $y = 0$, periodicity of $f_n(y)$ and matching the discontinuity in derivative of $f_n(y)$ to $\delta(y)$ in Eq. (1.57).

1.4.2. *Couplings of gauge modes*

It turns out that the zero-mode of the gauge field continues to have a flat profile: only its normalization affected by brane term such that

$$g_4 = \frac{g_5}{\sqrt{r + 2\pi R}} \qquad (1.59)$$

For large brane kinetic terms,

$$g_4 \approx \frac{g_5}{\sqrt{r}} \qquad (1.60)$$

Let us now consider couplings of gauge KK modes to particles localized on the branes in the limit of large brane terms. We find that

(i) the coupling of gauge KK mode to a particle (say light SM fermion) localized at $y = 0$ is *suppressed* (compared to zero-mode): $g_5 \times f_n(0) \sim g_4/\sqrt{r/R}$.

(ii) Whereas, the coupling to particles (such as the Higgs) localized at $y = \pi R$ is *enhanced* compared to the zero-mode (or SM) gauge coupling : $g_5 \times f_n(\pi R) \sim g_4 \times \sqrt{r/R}$

The intuitive understanding is that large brane kinetic terms "repel" gauge KK mode from that brane (see Fig. 1.8).

1.4.3. *Solution to Flavor* Problem

In reality, the light SM fermions are not exactly localized at the $y = 0$ brane, but we find a similar suppression in their coupling to gauge KK mode for the actual profiles of the light fermions which are exponentials *peaked* at $y = 0$. Hence, based on the rough size of the coupling mentioned in point (i) above, we can show that FCNC's from exchange of gauge KK modes are suppressed by a factor of r/R relative to the case of without brane kinetic terms, i.e., large brane kinetic terms provide an *analog* of GIM suppression in the SM.

One might wonder if we are introducing a new hierarchy since we need $r/R \gg 1$. However, that's not really the case since a mild hierarchy of $O(10)$ is enough. In fact, we will see in lecture 5 how we can *effectively* obtain the same effect as that of such large brane kinetic terms in a *warped* extra dimension without introducing *any* brane terms and therefore any hierarchy in the 5D theory at all.

1.4.4. *Electroweak Precision Tests*

Having seen how to suppress contributions of the gauge KK modes to FCNC's, we will now consider their contributions to flavor-*preserving* observables called electroweak precision tests (EWPT). There are 3 such effects which we discuss in turn.

1.4.4.1. *4-fermion operators*

Tree-level exchange of gauge KK modes also generates flavor-preserving 4-fermion operators, Fig. 1.9. We can compare these effects to SM (i.e., zero-mode) Z exchange which has coefficient $\sim g_Z^2/m_Z^2$ and use the fact that the experimental data on these operators agrees with the SM prediction at the $\sim 0.1\%$ level. For $r = 0$ (no brane term), we found that gauge KK coupling $\approx \sqrt{2}g_4$ for fermions localized at $y = 0$ (recall that light fermions are localized near $y = 0$) so that we obtain a limit of $m_{KK} \gtrsim$ a few TeV. However, for large brane kinetic terms, the gauge KK couplings and hence the coefficients of these operators are further suppressed by a factor of $\sim r/R$ so that $m_{KK} \sim$ TeV is *easily* allowed by the data.

The other 2 effects originate from the mixing of zero and KK modes for W, Z via the Higgs vev which we now discuss. The gauge group in the bulk is $SU(2)_L \times U(1)_Y$. We first perform the KK decomposition (i.e., obtain zero and KK modes) for $W_{i=1,2,3}$ and B (hypercharge) setting $v = 0$. At this level, there is no kinetic or mass mixing between these modes.

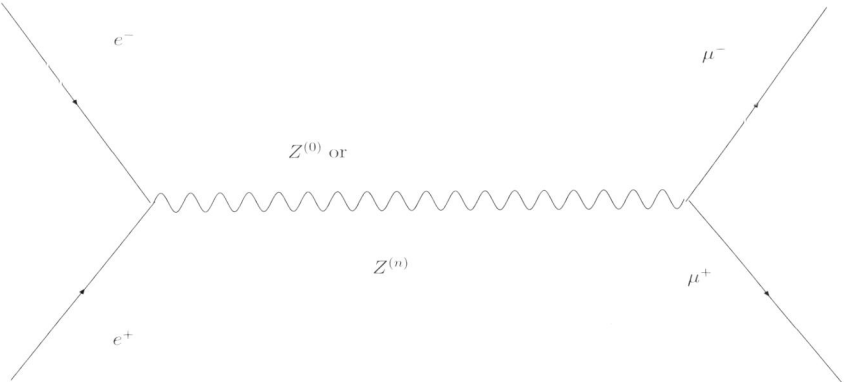

Fig. 1.9. 4-fermion operators generated by exchange of zero and KK modes of Z.

Next, we turn on the Higgs vev. For $v \neq 0$, we obtain masses for zero-modes of B and W_i and mass *mixing* between W_3 and B zero-modes (as in the SM). We define photon and Z zero-modes, $Z_\mu^{(0)}$ and $A_\mu^{(0)}$, to be combinations of $W_3^{(0)}$ and $B^{(0)}$ such that the *zero*-mode mass mixing is diagonalized (as in the SM). We first define the zero-mode gauge couplings (we neglect the brane terms for simplicity here, but it is straightforward to include them): $g_{W^{(0)}} = g_{5\,2}/\sqrt{2\pi R}$, $g_{Z^{(0)}} = g_{5\,Z}/\sqrt{2\pi R}$, where $(g_{5\,Z}^2 = g_{5\,2}^2 + g_5'^2)$. The weak mixing angle between $W_3^{(0)}$ and $B^{(0)}$, i.e., $\sin^2\theta_W$ is the ratio of these zero-mode gauge couplings.

It turns out to be convenient to define the KK modes, $Z^{(n)}$ and $A^{(n)}$ ($n \neq 0$), using *same* (0-mode) mixing angles. The reason is that with this definition, the KK photon modes $A_\mu^{(n)}$ do not couple to Higgs (just like zero-mode) and hence decouple from the other modes.

However, the crucial point is that the $W^\pm$ zero mode mixes with the KK modes of $W^\pm$ via mass terms coming from the Higgs vev localized at $y = \pi R$ (similarly for Z). Therefore, the mass eigenstates, i.e., SM $W^\pm$ and Z, are *admixtures* of zero and KK modes. To understand this effect, we can diagonalize the 2×2 mass matrix (for zero and 1st KK mode) for simplicity (see exercise 3).

1.4.4.2. *Shift in coupling of SM fermions to Z*

The above zero-KK mode mixing for W, Z induced by Higgs vev results in a shift in the coupling of SM W, Z to a fermion localized at $y = 0$ from the pure zero-mode coupling, i.e., SM Z has a (small) KK Z component so that $g_Z = g_{Z^{(0)}} + \delta g_Z$. We can estimate this effect via mass insertion diagrams as in Fig. 1.10 which are valid for $v \times$ couplings $\ll m_{KK}$ to find $\delta g_Z/g_{Z^{(0)}} \sim g_{Z^{(0)}}^2 v^2/m_{KK}^2$: see exercise 3 for a more accurate calculation. Note that there is *no* enhancement in δg_Z for large brane kinetic terms ($r/R \gg 1$). The point is that the enhancement in the coupling (relative to the zero-mode coupling) at the Higgs-KK Z vertex cancels the suppression in the coupling at the fermion-KK Z vertex (cf. the effect on the W, Z masses below). Just like the case of 4-fermion operators, the measured couplings of SM fermions to Z agree with the SM prediction at the $\sim 0.1\%$ level so that we obtain a limit of $m_{KK} \gtrsim$ a few TeV.

1.4.4.3. *Shift in ratio of W and Z masses or ρ parameter*

The mixing of zero and KK W modes induced by the Higgs vev also results in a shift in SM W mass from the pure zero-mode mass (a similar effect

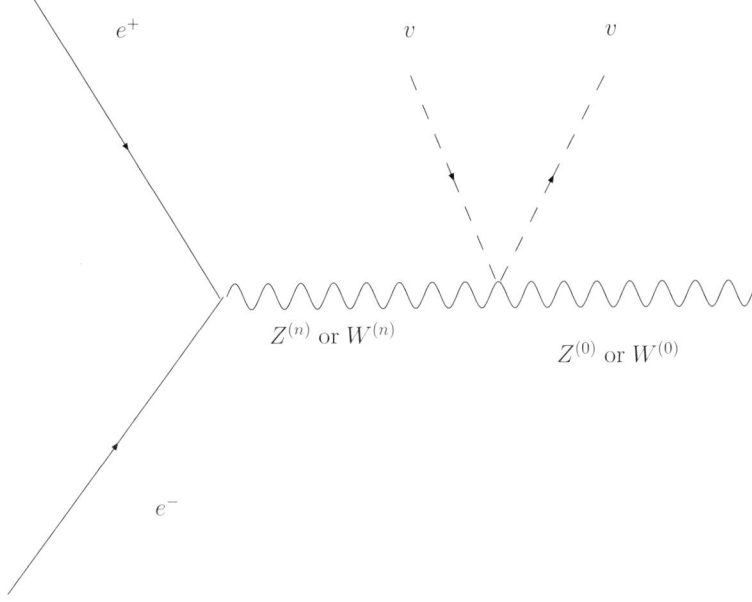

Fig. 1.10. Shift in the coupling of a SM fermion to SM Z from the zero-mode gauge coupling due to the mixing of zero and KK modes of Z.

also happens for SM Z) as in Fig. 1.11:

$$M_W^2 = M_{W^{(0)}}^2 + \delta M_W^2, \text{ where} \tag{1.61}$$

$$M_{W^{(0)}}^2 = \frac{1}{4} g_{W^{(0)}}^2 v^2 \tag{1.62}$$

$$\delta M_W^2 \sim g_{W^{(0)}}^4 \frac{v^4}{m_{KK}^2} \frac{r}{R} \tag{1.63}$$

This effect, in turn, shifts the ρ parameter defined as

$$\rho = \frac{M_W^2}{M_Z^2} \times \frac{g_Z^2}{g_2^2} \tag{1.64}$$

The point is that $\rho = 1$ in the SM (at the tree-level) and $\Delta\rho_{expt.} \equiv \rho_{expt.} - 1 \sim 10^{-3}$. Actually, there is a subtlety in this definition for the 5D model due to the fact that the couplings of the Z boson to the SM fermions are also modified from the pure zero-mode Z coupling: $g_Z = g_{Z^{(0)}} + \delta g_Z$. However, as we discussed earlier, $\delta g_{Z,\,W}$ are not enhanced by $r/R \gg 1$ so that we

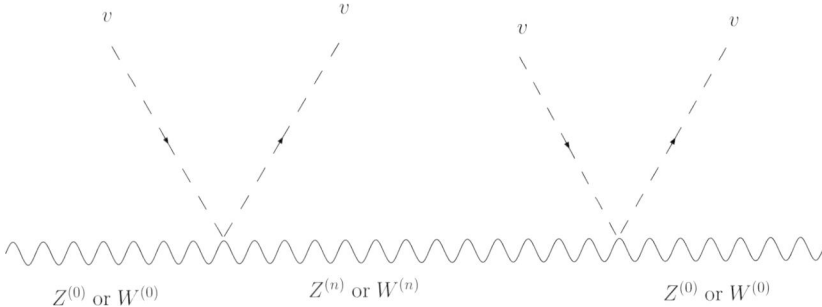

Fig. 1.11. Shift in the masses of SM W, Z from the zero-mode masses due to the mixing of zero and KK modes.

can set $g_Z \approx g_{Z^{(0)}}$ in $\Delta \rho$ to find

$$\delta \rho \equiv \rho - 1 \sim \left(g_{Z^{(0)}}^2 - g_{W^{(0)}}^2\right) \frac{v^2}{m_{KK}^2} \times \frac{r}{R} \qquad (1.65)$$

The crucial point is that $\Delta \rho$ is *enhanced* by the presence of large brane kinetic terms such that we must require $m_{KK} \gtrsim 10$ TeV for $r/R \sim 10$ (as needed to solve the flavor problem).

1.5. Lecture 4

In this lecture, we will show how to solve the problem of large corrections to the ρ parameter discussed in lecture 3. For this purpose, we have to introduce a "custodial isospin" symmetry in the extra dimension. We will then discuss some signals of this extra dimensional scenario.

1.5.1. *Custodial Isospin in SM*

We will first review why $\rho = 1$ in the SM at the tree-level. The starting point is that the Higgs potential, $V(|H|)$ in the SM with the *complex* doublet Higgs written as

$$H = (h_1, h_2, h_3, h_4) \qquad (1.66)$$

has a global $SO(4)$ symmetry (corresponding to rotations among the 4 *real* fields, h_i). Moreover, $SO(4)$ is isomorphic to $SU(2) \times SU(2)$ – one of these $SU(2)$'s in fact corresponds to the usual gauged $SU(2)_L$ group and the other one is usually denoted by $SU(2)_R$. The crucial point is that the global symmetry of the Higgs potential is enhanced compared to the gauged $SU(2)_L$ symmetry. The Higgs vev:

$$\langle H \rangle = (0, 0, 0, v) \tag{1.67}$$

breaks the global $SO(4)$ symmetry of the Higgs sector (in isolation) to $SO(3)$ – the gauged $SU(2)_L$ symmetry is broken in this process so that the W_i^L gauge bosons acquire masses. The unbroken $SO(3)$ symmetry (which is global) is isomorphic to an $SU(2)$ – clearly this unbroken $SU(2)$ is the *diagonal* subgroup of the 2 original $SU(2)$'s and is often called *custodial isospin*. It is this remnant symmetry which enforces equal masses for $W^L_{i=1,2,3}$.

Of course, W_3^L only mixes with B (there is no mixing for $W_L^\pm$). This mixing results in the neutral mass, $M_Z^2 = 1/4\ v^2 \left(g_2^2 + g'^{\,2} \right)$, not being equal to the charged mass, $M_W^2 = 1/4\ v^2 g_2^2$. That is the reason why there is a factor of g_Z^2/g_2^2 in the definition $\rho = M_W^2/M_Z^2\ g_Z^2/g_2^2$: this factor takes the "violation of custodial symmetry" due to the gauging of hypercharge into account.

1.5.2. *Custodial Isospin* Violation *in* 5D

Based on the above discussion, the sizable $\Delta\rho$ in the $5D$ model signals violation of custodial isospin symmetry somewhere in the $5D$ theory. First we begin with identifying the precise origin of custodial isospin *violation* and then we will come up with a solution to this problem. As we saw in lecture 3, $\Delta\rho$ from gauge KK modes $\propto \left(g^2_{Z^{(0)}} - g^2_{W^{(0)}} \right) \sim g^2_{B^{(0)}}$ just as in the SM. So, the origin of large $\Delta\rho$ or custodial isospin violation seems to be similar to that in the SM, i.e., it is due to gauging of hypercharge and the resulting mixing of W_3 with B. However, the point is that there are *additional* mixing effects (compared to the SM) in the $5D$ model due to the presence of KK modes (the mixing of zero-modes amongst each other is same as in the SM) In particular, $W_{L\,3}^{(0)} - B^{(n)}$ mixing occurs only in neutral sector and has no charged counterpart, whereas $W_L^{(0)} - W_L^{(n)}$ mixing is symmetric between charged and neutral sectors.

The origin of this dichotomy between charged and neutral sectors is the fact that the symmetry gauged in $5D$ is same as in the SM, i.e., $SU(2)_L \times U(1)_Y$, so that we have KK modes only for $W_L^{3,\,\pm}$ and B: there are no charged partners for the B KK modes. This new effect (the custodial isospin violation due to B KK modes) is not taken into account by the factor of g_Z^2/g_2^2 in the definition of ρ – the point is that this factor only accounts for the mixing only amongst *zero*-modes, i.e., the $W_{L\,3}^{(0)} - B^{(0)}$

mixing. To repeat, $W_{L\,3}^{(0)} - W_{L\,3}^{(n)}$ mixing *does* have a counterpart in the charged sector. Moreover, $W_{L\,3}^{(0)} - B^{(n)}$ mass term $\sim g_{W^{(0)}} g_5' \times f_n(\pi R) v^2 \sim g_{W^{(0)}} g_{B^{(0)}} v^2 \sqrt{r/R}$ so that this effect is *enhanced* for large brane terms!

1.5.3. *Custodial Isospin* Symmetry *in 5D*

It is clear that we need *extra* charged KK modes to partner $B^{(n)}$ if we wish to suppress $\Delta \rho$. We can achieve this goal by promoting the hypercharge gauge boson to be a triplet. Hence, we can restore custodial isospin symmetry in the $5D$ model by enlarging the $5D$ *gauge* symmetry to $SU(2)_L \times SU(2)_R$ [11]. It turns out that we need something like $SU(2)_L \times SU(2)_R \times U(1)_{B-L}$ to obtain the correct fermion hypercharges as follows. Hypercharge is identified with a subgroup of $U(1)_R$ and $U(1)_{B-L}$: $Y = T_{3R} + (B - L)/2$, with $T_{3R} = \pm 1/2$ for $(u, d)_R$ and $(\nu, e)_R$ and $B - L = 1/3, -1$ for q, l (it is easy to check that this reproduces the SM hypercharges). Note that we still have extra neutral KK modes from $U(1)_{B-L}$ (which have no charged counterpart), but these KK modes do not couple to Higgs since the Higgs has $B - L$ charge of zero: only KK $W_{L,R}^{3,\pm}$ couple to Higgs such that the KK exchanges which give the shifts in masses respect custodial isospin (i.e., they are the same in the charged and the neutral channels).

Of course, we must break $SU(2)_R \times U(1)_{B-L}$ down to $U(1)_Y$, i.e., we must require that there are no zero-modes for $W_R^\pm$ and the extra $U(1)$ which is the combination of $U(1)_R$ and $U(1)_{B-L}$ orthogonal to $U(1)_Y$. However, this breaking must (approximately) preserve degeneracy for (at least the lighter) $W_R^\pm$ and W_R^3 modes such that $\Delta\rho$ continues to be (at least approximately) protected. It is clear that for this purpose we require degeneracy in both the mass of these modes and their coupling to the Higgs. This might seem to be challenging at first, but note that, for large brane kinetic terms ($r/R \gg 1$), KK modes are localized near $y = \pi R$. Therefore, if we break custodial isospin on the $y = 0$ brane, then the degeneracy between W_R^3 and $W_R^\pm$ is not significantly affected by this breaking. Specifically, we write down a *large* mass term for $W_R^\pm$ and the extra $U(1)$ at $y = 0$ which can originate from a localized scalar vev (different from the SM Higgs). We can show that this is equivalent to requiring vanishing of these gauge fields at $y = 0$ (odd or Dirichlet boundary condition: section 3.3 of reference [2]). This illustrates the general idea that breaking a $5D$ gauge symmetry by a large mass term localized on a brane is equivalent to breaking by boundary condition.

1.5.4. *Signals*

Let us consider some of the signals of this extra-dimensional set-up. A quick glance at Fig. 1.8 tells us that the coupling of gauge KK modes to top quark is enhanced compared to the SM couplings, whereas the couplings to the light SM fermions are suppressed (all based on the profiles for these modes).

We begin with real production of gauge KK modes, for example, the KK gluon. Due to the $\sim$ TeV mass for these particles, it is clear that we have to consider such a process at the Large Hadron Collider (LHC). Based on the above couplings, we typically find a broad resonance decaying into top pairs making it a challenge to distinguish the signal from SM background. It turns out that due to a constraint from a shift in the $Z \to \bar{b}b$ coupling,[f] we cannot localize b_L and hence its partner t_L too close to the Higgs brane, forcing us to localize t_R near the Higgs brane in order to obtain the large top mass. Hence the KK gluon dominantly decays to RH top quark. We can use this fact (and noting that the SM $t\bar{t}$ production is approximately same for LH and RH top quarks) for the purpose of signal versus background discrimination [12]. It is easy to distinguish this signal for the extra dimension from SUSY: there is no missing energy (at least in this process) and top quark is treated as "special" in the sense that it has a larger coupling (than the other SM fermions) to the new particles, namely KK modes, unlike in SUSY.

We can also consider *virtual* exchange of gauge KK modes.

(i) In analogy with the shift in the coupling of SM fermion to the Z that we considered earlier, we see that $\bar{t}tZ$ is shifted compared to the SM prediction (or compared to $\bar{u}uZ$ and $\bar{c}cZ$) since top quark (up quark) is localized near $y = \pi R$ ($y = 0$) brane. Such an effect can be easily measured at the International Linear Collider (ILC) [13].

(ii) From the above discussion, it is clear that the couplings of the top and charm quarks to the KK Z are diagonal, but not universal in the weak or interaction basis. Once we rotate to the mass basis, there is a flavor violating coupling to KK Z to the top and the charm quark. In turn, this effect induces a flavor violating coupling

---

[f]This shift in the coupling originates from diagrams similar to the ones we considered earlier for the shift in coupling of SM fermion to the Z: see Fig. 1.10. Such shifts are enhanced if SM fermion is localized near $y = \pi R$ brane, where gauge KK mode is peaked.

of the *SM Z* to the top and charm quarks (via mixing of KK and zero-mode Z), resulting in a flavor violating decay of the top quark: $t \to cZ$. Such decays can be probed at the LHC [14].

1.5.5. *Summary of Model and Unanswered Questions*

So, far we have considered a model with the SM gauge and fermion propagating in the bulk of a *flat* extra dimension, with the Higgs localized on or near one of the branes. The other SM particles (gauge bosons and fermions) are identified with zero-modes of the corresponding $5D$ fields.

We have seen that a solution to the flavor hierarchy of the SM is possible using profiles for the SM fermions (again, these are the zero-modes of the $5D$ fields) in the extra dimension; in particular, top and bottom quarks can be localized near the Higgs brane, whereas the 1st and 2nd generation (or light) fermions can be localized near the other brane. Moreover, the resulting flavor problem due to non-universal couplings of gauge KK modes to the SM fermions (for a few TeV KK scale) can be ameliorated with large brane kinetic terms for $5D$ gauge fields on *non*-Higgs brane (i.e., where the light fermions are localized).

We also studied constraints from electroweak precision tests on this set-up and found that these constraints can also be satisfied for $m_{KK} \sim$ TeV, provided there is a custodial isospin symmetry in the bulk to protect the observable related to the ratio of W/Z masses (the ρ parameter).

This set-up still leaves some questions unanswered:

(i) We have assumed so far that $m_{KK} \sim$ TeV, but why is it $\ll M_{Pl}$?
(ii) Is there a mild hierarchy problem associated with having large brane kinetic terms? Moreover, it seems a bit arbitrary that such terms appear only at $y = 0$ brane (where light SM fermions are localized) and not at $y = \pi R$.

We will see in the next lecture that both these questions can be answered by using a *warped* geometry (instead of flat extra dimension).

Furthermore,

(iii) Why does Higgs have a negative (mass)^2 or why does electroweak symmetry breaking (EWSB) occur? What sets this mass scale? Specifically, can the hierarchy $m_H \ll M_{Pl}$ be due to some dynamics giving $m_H \sim m_{KK}$ which, in turn, is $\sim$ TeV?
(iv) Why is the Higgs localized on or near one of the branes?

These questions will be answered by a *combination* of Higgs being A_5, i.e., the 5th component of bulk gauge field and warped geometry.

1.6. Lecture 5

In this lecture, we will be brief: for details and a more complete set of references, see the excellent set of lectures by Sundrum [3].

1.6.1. *Warped Extra Dimension (RS1)*

We begin with a review of the original Randall-Sundrum model (RS1) [15]: see Fig. 1.12. It consists of an extra-dimensional interval ($y = 0$ to πR as before), but with the gravitational action containing a bulk cosmological constant (CC) and brane tensions (localized or 4D CC's):

$$S_{5D} = \int d^4x dy \sqrt{-\det G} \left(M_5^3 \mathcal{R}_5 - \Lambda \right)$$

$$S_{brane\,1,\,2} = \int d^4x \sqrt{-\det g_{1,\,2}} T_{1,\,2} \quad (1.68)$$

with $g_{\mu\nu\,1}(x) = G_{\mu\nu}(x, y=0)$ and $g_{\mu\nu\,2}(x) = G_{\mu\nu}(x, y=\pi R)$, where $g_{\mu\nu}$'s are the induced metrics on the branes and G_{MN} is the bulk metric. Also, M_5 is the 5D Planck scale and $\mathcal{R}_5$ is the 5D Ricci scalar.

With the following two fine-tunings:

$$T_1 = -T_2 = 24k M_5^3, \quad (1.69)$$

where the (curvature) scale, k is defined using $\Lambda = 24k^2 M_5^3$, we obtain a flat (or Minkowski), but y-dependent 4D metric as a solution of the 5D Einstein's equations:

$$(ds)^2 = e^{-2ky} \eta_{\mu\nu} (dx)^\mu (dx)^\nu + (dy)^2 \quad (1.70)$$

Thus, the geometry is that of a slice of anti-de Sitter space in 5D (AdS$_5$). The y-dependent coefficient of the 4D metric, i.e., e^{-ky} is called the "warp factor".

4D gravity: The 4D graviton (which is the zero-mode of the 5D gravitational) corresponds to fluctuations around the flat spacetime background, i.e., $g_{\mu\nu}^{(0)}(x) \approx \eta_{\mu\nu} + h_{\mu\nu}^{(0)}(x)$. As usual, we plug this fluctuation into the 5D action and integrate over the extra dimensional coordinate to find an effective 4D action for $g_{\mu\nu}^{(0)}(x)$:

$$S_{4D} = \frac{M_5^3}{k} \left(1 - e^{-2k\pi R} \right) \int d^4x \sqrt{-\det g^{(0)}} \mathcal{R}_4[g^{(0)}] \quad (1.71)$$

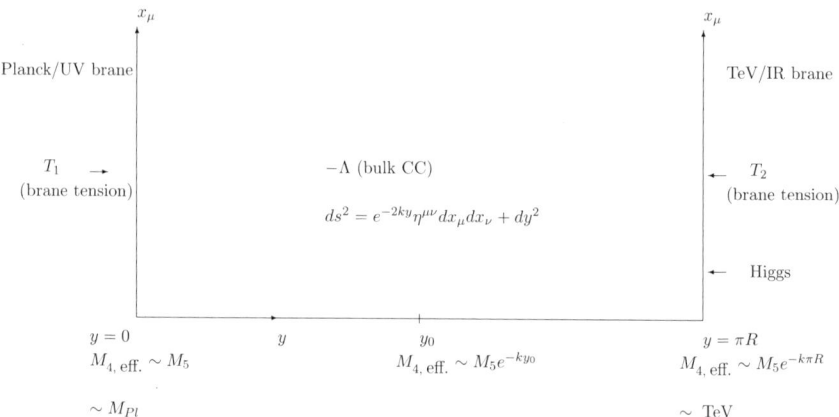

Fig. 1.12. The Randall-Sundrum (RS1) model.

from which we can deduce the $4D$ Planck scale:

$$M_{Pl}^2 = \frac{M_5^3}{k}\left(1 - e^{-2k\pi R}\right)$$
$$\approx \frac{M_5^3}{k} \text{ for } kR \gg 1 \qquad (1.72)$$

We choose $k \lesssim M_5$ so that the higher curvature terms in the $5D$ action are small and hence can be neglected. Thus, we get the following order of magnitudes for the various mass scales:

$$k \lesssim M_5 \lesssim M_{Pl} \sim 10^{18} \text{ GeV} \qquad (1.73)$$

It turns out that the $4D$ graviton is (automatically) localized near $y = 0$ (which is hence called the Planck or UV brane) - that is why the $4D$ Planck scale is finite even if we go to the decompactified limit of $R \to \infty$ in Eq. (1.72). Specifically, its profile is $\sim e^{-2ky}$.

1.6.2. *Solution to Planck-Weak Hierarchy*

The motivation for the RS1 model is to solve the Planck-weak hierarchy problem. Let us now see how this model achieves it. Assume that a $4D$ Higgs field is localized on the $y = \pi R$ brane which is hence called the TeV or IR brane:

$$S_{\text{Higgs}} = \int d^4x \sqrt{-\det g_2} \Big[g^{\mu\nu}_{\text{ind.}} \, \partial_\mu H \partial_\nu H \\ - \lambda \left(|H|^2 - v_0^2 \right)^2 \Big] \quad (1.74)$$

where the natural size for v_0 is the 5D gravity or fundamental scale (M_5). Using the metric induced on the TeV brane, $g_{\mu\nu\,2} = G_{\mu\nu}(y = \pi R) = g^{(0)}_{\mu\nu} e^{-2k\pi R}$, the action for the Higgs field becomes

$$S_{\text{Higgs}} = \int d^4x \sqrt{-\det g^{(0)}} \Big[e^{-2k\pi R} g^{(0)\,\mu\nu} \partial_\mu H \partial_\nu H \\ - e^{-4k\pi R} \lambda \left(|H|^2 - v_0^2 \right)^2 \Big] \quad (1.75)$$

Now comes the crucial point: we must rescale the Higgs field to go to canonical normalization, $H \equiv \hat{H} e^{k\pi R}$, which results in

$$S_{\text{Higgs}} = \int d^4x \sqrt{\det g^{(0)}} \Big[g^{(0)\,\mu\nu} \partial_\mu \hat{H} \partial_\nu \hat{H} \\ - \lambda \left(|\hat{H}|^2 - v_0^2 e^{-2k\pi R} \right)^2 \Big] \quad (1.76)$$

Note that the Higgs mass is "warped-down" to $\sim$ TeV from the 5D (or the 4D) Planck scale if we have the following *modest* hierarchy between the radius (or the proper distance) of the extra dimension and the AdS curvature scale.

$$k\pi R \sim \log (M_{Pl}/\text{TeV}) \\ \sim 30 \text{ or} \\ R \sim \frac{10}{k} \quad (1.77)$$

Moreover, the quartic coupling is unchanged and hence the Higgs vev (or weak scale) is also at the TeV scale, assuming $\lambda \sim O(1)$.

Note that the radius of the extra dimension is not a fundamental or 5D parameter, rather it is determined by the dynamics of the theory. Hence, in order to have complete solution to the hierarchy problem (without any hidden fine-tuning), we must show that the radius can be stabilized at the required size without further (large) fine-tuning of parameters of the 5D theory. In fact, stabilization of such a radius can be achieved using a bulk scalar (Goldberger-Wise mechanism) [16], provided we invoke a *mild* hierarchy $M^2/k^2 \sim O(1/10)$, where M is the 5D mass of the scalar.

Thus, we see that the Planck-weak hierarchy can be obtained from $O(10)$ hierarchy in the fundamental or $5D$ theory! In general, a large ("exponential") hierarchy for the $4D$ mass scales can be obtained from a small hierarchy in the $5D$ parameters.

The central feature of a warped extra dimension is that the *effective $4D$ mass scale depends on position* in the extra dimension. In order to have a more intuitive understanding of this feature, consider the position $y \sim y_0$ where the metric is:

$$(ds)^2_{y \sim y_0} \sim e^{-2ky_0} \eta_{\mu\nu}(dx)^\mu(dx)^\nu + (dy)^2 \tag{1.78}$$

In terms of the rescaled coordinate and mass scale: $\hat{x} \equiv e^{-ky_0}x$, $\hat{m}_{4D} \equiv e^{ky_0}m_{4D}$, we get

$$(ds)^2_{y \sim y_0} \sim \eta_{\mu\nu}(d\hat{x})^\mu(d\hat{x})^\nu + (dy)^2 \tag{1.79}$$

The advantage of the new coordinates $\hat{x}$ is that we have a "flat" metric in terms of it so that we expect $\hat{m}_{4D} \sim m_{5D}$ (such a relationship is valid in the absence of warping). Converting back to original mass scales, we find $m_{4D} \sim e^{-ky_0} m_{5D}$, i.e., $4D$ mass scales are warped compared to $5D$ mass scales. *An analogy with the expanding Universe is useful:* just as $3D$ space expands with time, in the warped extra dimension, the $4D$ space-*time* "expands" (or contracts) with motion along the 5th dimension.

1.6.3. *Summary of RS1*

The preceding discussion leads us to the "master equation" for a warped extra dimension:

$$M_{4,\text{ eff.}}(y) \sim M_5 \times e^{-ky}$$

relating the effective $4D$ mass scales on the left-hand side (LHS) of the above equation to the fundamental or $5D$ mass scale on the right-hand side (RHS) by the warp factor. Applying it to the case of the $4D$ graviton localized at $y \sim 0$, we get

$$M_{Pl} \sim M_5 \tag{1.80}$$

so that we must choose the $5D$ Planck scale to be

$$M_5 \sim 10^{18} \text{GeV} \tag{1.81}$$

Whereas, the Higgs sector is localized at $y \sim \pi R$ so that

$$M_{\text{weak}} \sim M_5 \times e^{-k\pi R} \tag{1.82}$$

so that

$$M_{\text{weak}} \sim \text{TeV} \qquad (1.83)$$

provided we have a mild hierarchy

$$k\pi R \sim \log(M_{Pl}/\text{TeV})$$
$$\sim 30 \qquad (1.84)$$

1.6.4. *Similarity with* Flat *TeV-Size Extra Dimension with Large Brane Terms*

In the original RS1, it was assumed that the entire SM, i.e., including fermion and gauge fields, is localized on the TeV brane. However, it was subsequently realized that, in oder to solve the Planck-weak hierarchy problem, only the SM Higgs boson has to be localized on or near the TeV brane – the masses of non-Higgs fields, i.e., fermions and gauge bosons, are protected by gauge and chiral symmetries, respectively.

So, we are led to consider RS1 with the SM gauge [17] and fermion fields [18] propagating in the bulk (with the Higgs still being on or near the TeV brane). It turns out that the profiles for the SM fermions in the bulk can address the flavor hierarchy just as in the case of flat extra dimension. Moreover, solving the wave equation in curved spacetime, we can show [17–19] that all KK modes are localized near the IR brane (that too automatically, i.e., with*out* actual brane terms) and the KK masses are given by $m_{KK} \sim ke^{-k\pi R}$ and *not* $1/R$ [note that, based on Eqs. (1.73) and (1.77) $1/R$ is of the size of the 4D Planck scale!]. Hence, we find $m_{KK} \sim$ TeV given the choice of parameters to solve the Planck-weak hierarchy problem! A very rough intuition for localization of KK modes near the TeV brane is that the KK modes can minimize their mass by "living" near IR brane, where all mass scales are warped down. In this sense, the warped extra dimension "mimics" large brane kinetic terms of flat geometry – recall that the large brane kinetic terms in a flat extra dimension result in a similar localization of KK modes. In addition, the hierarchy $m_{KK} \ll M_{Pl}$ is explained by the warped geometry. This addresses the 1st and 2nd questions outlined at the end of the previous lecture.

Because of this localization of KK modes near the TeV brane, we find that the solution to the flavor problem and the discussion of the electroweak precision tests (including custodial isospin) goes through (roughly) as in the case of a flat extra dimension.

1.6.5. Unification of Spins: Higgs as A_5

We now return to the other (3rd and 4th) questions asked at the end of the previous lecture, namely, what sets the scale of EWSB or Higgs mass and why is Higgs localized on the TeV brane?

We will show in this and the next subsection that obtaining the SM Higgs as the 5th component of $5D$ gauge field (or A_5) can resolve the 3rd question and then outline in the final subsection how combining the idea of Higgs as A_5 with the warped geometry answers the 4th question, resulting in a "complete" model.

As a warm-up for the idea of Higgs as A_5 (see the review [20] for references), consider an $SU(2)$ gauge theory in an extra dimension which is compactified on a circle (S^1). As we saw earlier, for $n \neq 0$, the $A_\mu^{(n)}$ modes "eat" $A_5^{(n)}$ modes to form massive spin-1 states. Moreover, there is a (massless) zero-mode A_5, which is in adjoint representation of $SU(2)$, i.e., it is charged under the $SU(2)$ gauge symmetry. We can introduce a $SU(2)$ doublet fermion in the bulk which will acquire a Yukawa coupling $\sim g$ to the A_5 zero-mode from the interaction $\bar{\Psi}_L A_5 \Psi_R$ coming from the $5D$ covariant derivative. Hence, this scenario is often called "Gauge-Yukawa unification".

Note that this scalar has no potential at the tree-level since it is part of a $5D$ gauge field. We will now discuss the potential for A_5 zero-mode induced by loop effects to find that it is *finite*. Naively, the scalar (mass)^2 gets quadratically divergent loop corrections: $m^2_{A_5^{(0)}} \sim g_4^2/\left(16\pi^2\right) \Lambda_{UV}^2$. However, from the $5D$ point of view, it is clear that $5D$ gauge invariance protects the A_5 scalar mass from receiving divergent loop corrections (there is no counter-term to absorb such divergences and so these must be absent). The reader is referred to the 1st reference in [3] for a detailed calculation of $m^2_{A_5^{(0)}}$ coming from a fermion loop for the simpler case of a $U(1)$ gauge field in the bulk. The summary is that loop contributions to $m^2_{A_5^{(0)}}$ are "cut-off" by R^{-1}:

$$m^2_{A_5^{(0)}} \sim \frac{g_4^2}{16\pi^2} R^{-2} \tag{1.85}$$

Intuitively, the understanding is that A_5 behaves as a "regular" scalar till $E \sim R^{-1}$: see Fig. 1.13(a). Beyond these energies, the quantum corrections "realize" that A_5 is part of a $5D$ gauge field. Therefore, the loop contributions from $E \gtrsim R^{-1}$ are highly suppressed, in particular, there is no divergence. Thinking in terms of KK modes, there is a cancellation in the loop diagram among the different modes. We can then ask: what

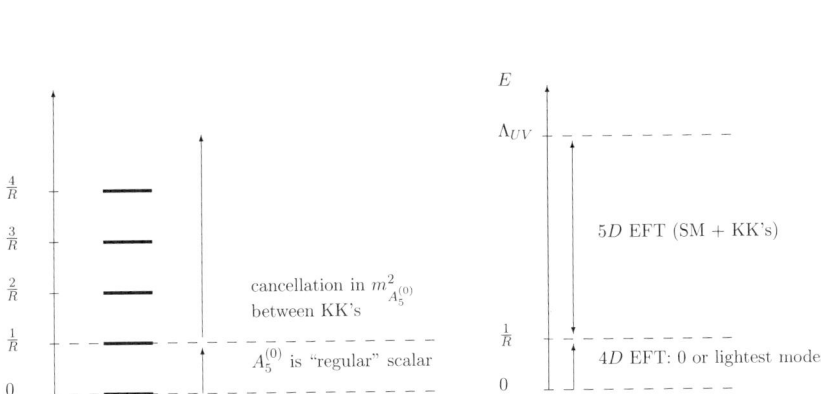

Fig. 1.13. Contributions to mass of A_5 (a) and various energy scales in the $5D$ model (b).

did we gain relative to a "regular" scalar (which is not an A_5 zero-mode, but is localized on a brane or originates in a $5D$ scalar field)? To answer this question, we need to know what is Λ_{UV}, the scale which cuts of the divergence in the case of a regular scalar. The $4D$ SM (without gravity) is renormalizable so that the cut-off is the Planck scale (where quantum gravity becomes important). However, the $5D$ *gauge theory*, even without gravity, *is non-renormalizable* and therefore must be defined with a cut-off (which is not related to the Planck scale): see Fig. 1.13(b). The reason is that the $5D$ gauge coupling constant is dimensionful so that the $5D$ loop expansion grows with energy: $g_5^2 E / \left(16\pi^2\right)$. Since we cannot extrapolate the $5D$ gauge theory beyond the energy scale where the loop expansion parameter becomes ~ 1, we must introduce a cut-off at this scale:

$$\Lambda_{UV} \sim \frac{16\pi^2}{g_5^2}$$
$$\sim \frac{16\pi}{g_4^2} R^{-1} \quad (1.86)$$

where we have set the brane terms to be small so that $g_4 \sim g_5/sqrtR$. Note that this cut-off is not much larger than the compactification scale since $g_4 \sim 1$ in the SM. Thus, we find that $m^2_{A_5^{(0)}}$ is suppressed relative to the mass^2 in the case of a regular scalar by $\sim (\Lambda_{UV} R)^2 \sim \left(16\pi/g_4^2\right)^2$: we *do* gain by going to A_5.

Next, we discuss how to use A_5 for *radiative symmetry breaking (often called Hosotani mechanism)* [21]. Continuing with the case of $SU(2)$ on S^1, we see that a vev for the A_5 zero-mode, $\langle A_5^{(0)} \rangle$ can break $SU(2)$ gauge symmetry to a $U(1)$ gauge symmetry. The point is that fermion loops typically give $m^2_{A_5^{(0)}} < 0$, whereas gauge loops are of opposite sign. However, the fermion contributions can win if the number of fermion degrees of freedom is larger than that of gauge bosons.

Thus, we have a "cartoon" of the SM in the following sense. We can identify the $SU(2)$ gauge group that we considered above with the SM W's. We will then get $M_{W^\pm} \sim R^{-1}$ (coming from $\langle A_5 \rangle$), whereas W_3 [corresponding to the unbroken $U(1)$ gauge symmetry] remains massless (it is the "photon"). Finally, the $\bar{\Psi}_L A_5 \Psi_R$ coupling mentioned above gives a fermion mass $M_{\psi^{(0)}} \sim R^{-1} \sim M_W$ which is roughly correct for top quark (since $m_t \sim M_W$).

Of course, this model is far from being realistic:

(i). We must require $1/R \gg 100$ GeV since we have not seen any KK modes in experiments so far which have probed energy scales up to $\sim$ TeV (either directly in the highest energy colliders or indirectly via virtual effects of new particles). To satisfy this constraint, we can fine-tune the fermion versus the gauge loop contributions to A_5 mass such that $M_{W^\pm}$ or $\langle A_5 \rangle \sim 100$ GeV $\ll R^{-1}$.

(ii). More importantly, we do not have fermion chirality on a circle.

1.6.6. *Towards Realistic Higgs as A_5: Chirality and Enlarging the Gauge Group*

As we saw earlier, we can obtain chiral fermions by going to an orbifold: S^1/Z_2. However, if we require A_μ of $SU(2)$ to be even under Z_2 (such that we get a corresponding zero-mode, i.e., a massless $4D$ gauge boson), then the A_5's are necessarily odd. Thus, we lose the scalar zero-mode. In any case, the scalar was in the adjoint representation of $SU(2)$, whereas we need a doublet for EW symmetry breaking.

The trick is to enlarge the gauge group to $SU(3)$ and to break it down to $SU(2) \times U(1)$ by boundary condition as follows. Choose the following parities under Z_2 for the fundamental representation

$$\begin{pmatrix} \\ 3 \\ \end{pmatrix} \to P \begin{pmatrix} \\ 3 \\ \end{pmatrix}, \text{ where}$$

$$P = \begin{pmatrix} + & & \\ & + & \\ & & - \end{pmatrix} \quad (1.87)$$

Given this parity choice, can derive the transformation of any other representation under Z_2. For example, consider fields in the adjoint representation, Φ_a ($a = 1...8$), written as a 3×3 matrix, $\Phi_a T^a$, where T^a's are generators of the fundamental representation. This matrix transforms as

$$\begin{pmatrix} \\ 8 \\ \end{pmatrix} \to P^\dagger \begin{pmatrix} \\ 8 \\ \end{pmatrix} P \sim \begin{pmatrix} + & + & - \\ + & + & - \\ - & - & + \end{pmatrix} \quad (1.88)$$

This implies that if the A_μ's belonging to $SU(2) \times U(1)$ are chosen to be even (and hence have a zero-mode), then the A_μ's of the coset group $SU(3)/[SU(2) \times U(1)]$ are odd (i.e., do not have a zero-mode). This choice of parities thus achieves the desired breaking pattern $SU(3) \to SU(2) \times U(1)$. Moreover, the A_5's of $SU(3)/[SU(2) \times U(1)]$ are even, giving us a scalar zero-mode which is a doublet of the unbroken $SU(2)$ group as desired.

Furthermore, just like in the case of the breaking $SU(2) \to U(1)$ discussed earlier, the breaking of $SU(2) \times U(1)$ can be achieved by vev of A_5 which is generated by loop corrections. Moreover, due to usage of fundamental representation for this radiative symmetry breaking, the rank of the gauge group is also broken, i.e., we have an unbroken $U(1)$ symmetry.

A 5D fermion which is a triplet of $SU(3)$ gives zero-modes for LH $SU(2)$ doublet *and* RH singlet:

$$\Psi_L = \begin{pmatrix} \Psi_L^D \ + \\ \Psi_L^S \ - \end{pmatrix}$$

$$\Psi_R = \begin{pmatrix} \Psi_R^D \ - \\ \Psi_R^S \ + \end{pmatrix} \quad (1.89)$$

where D and S denote $SU(2)$ doublet and singlet, respectively – recall that the parities of the RH and LH fields must be opposite. Moreover, the Yukawa coupling for the zero-mode fermions comes from the interaction $\bar{\Psi}_L^D A_5 \Psi_R^S$. Thus, we are getting closer to the SM!

1.6.7. *Realistic Higgs as* A_5 *in* **Warped** *Extra Dimension*

When we construct the previous model in a warped extra dimension, it turns out that the $A_5^{(0)}$ is automatically localized near the TeV brane [22] – recall that in order to solve the hierarchy problem, we would like the Higgs to be localized precisely there. Thus, A_5 zero-mode is an excellent candidate for SM Higgs!

As "finishing touches", we can add an extra $U(1)$ to obtain the correct hypercharges for the fermions and similarly a custodial isospin symmetry to protect the ρ parameter [23]. Also, it turns out that the Ψ_L^D and Ψ_R^S have (effectively) "opposite" sign of $5D$ mass, M (recall that this mass is not coming from $\langle A_5^{(0)} \rangle$) in the sense that if the LH zero-mode is localized near $y = 0$, then the RH zero-mode must be near $y = \pi R$ (or vice versa): see exercise 1. To relax this constraint, i.e., to obtain more freedom in localization of LH versus RH zero-modes, we can instead obtain LH and RH SM fermions as zero-modes of different bulk multiplets. However, then the question arises: since A_5 only couples fermions within the same fermionic multiplet, how do we obtain Yukawa couplings? The solution is to mix fermionic multiplets by adding mass terms localized at the endpoints of the extra dimension.

1.6.8. *Epilogue*

Due to lack of time, we have not considered other extra dimensional models with connections to the weak scale (and gravitational aspects of extra dimensional models in general). Here, we give a summary of the essential features of these other models: for details, see the references below and other lectures [1–4]. Arkani-Hamed, Dimopoulos and Dvali (ADD) proposed a scenario where only gravity propagates in extra dimensions, with all the SM fields localized on a brane [24]. The idea is that the fundamental or higher-dimensional gravity scale is $\sim$ TeV (and not the $4D$ Planck scale), while the weakness of gravity (or largeness of $4D$ or observed Planck scale) is accounted for by diluting the strength of gravity using extra dimensions which are much larger in size than the fundamental length scale, i.e., $R \gg 1/$ TeV. The crucial point is that the gravitational force law has been tested only for distances larger than $O(100)\,\mu m$ so that such very large extra dimensions could be consistent with current experiments. Only the graviton has KK modes in this framework, that too very light, resulting in interesting phenomenology both from real and virtual production of these KK modes. These KK modes couple with the usual $4D$ gravitational strength, but their large multiplicity can compensate for this very weak coupling.

At the other extreme is the model called Universal Extra Dimensions (UED) [25]. This scenario has a flat extra dimension(s) in which *all* the SM fields (including Higgs) propagate. The $5D$ fields have no brane localized interactions at the tree-level: of course, loops will generate small brane terms. Moreover, there are no $5D$ masses for fermions and Higgs so that profiles for all zero-modes (including all fermions, gauge fields and Higgs) are flat. Hence, we do not have a solution to the flavor hierarchy of the SM unlike in the scenario considered in these lectures. The motivation for UED is more phenomenological: there is a remnant of extra dimensional momentum or KK number conservation (dubbed KK parity) which forbids a coupling of a *single* lightest (level-1 and in general, odd level) KK mode to SM particles. Such a coupling *is* allowed for level-2 (and in general, even level) KK modes, but it is still suppressed by the small (loop-induced size) of brane kinetic terms.[g] Hence, the contributions from KK exchange to precision tests are suppressed (in particular, tree-level exchange of odd level modes is forbidden), *easily* allowing KK mass scale *below* a TeV for level-1 and even level-2 modes (cf. the lower limit of *a few* TeV in the scenario studied in these lectures). Thus, KK modes can be more easily produced at colliders (even though it is clear that the odd level KK modes have to be pair produced). Moreover, the lightest KK particle (LKP) is stable and can be a good dark matter candidate [26].

Finally, we mention the $5D$ Higgsless models [27], where EW symmetry itself is broken by boundary conditions like the breaking of $5D$ custodial isospin symmetry mentioned in lecture 4 [or the breaking $SU(3) \rightarrow SU(2) \times U(1)$ considered in lecture 5 in order to obtain Higgs as A_5]. The idea is that there is no light Higgs in the spectrum in order to unitarize WW scattering, which is instead accomplished by exchange of gauge KK modes. These KK modes then must have mass $\lesssim 1$ TeV. It turns out that due to such a low KK scale, the simplest such models are severely constrained by precision tests, but it is possible to avoid some of these constraints by suitable model-building.

---

[g]This coupling does not preserve KK number conservation or extra dimensional translation invariance and hence must arise from interactions localized on the branes which violate these symmetries.

Acknowledgments

KA was supported in part by the U. S. DOE under Contract no. DE-FG-02-85ER 40231 and would like to thank organizers of the summer school for the invitation to give lectures and all the students at the summer school for a wonderful and stimulating experience.

Appendix A. Excercises

A.1. Exercise 1

A.1.1. *Zero-Mode Fermion and 4D Yukawa Coupling*

Show that the normalized profile for LH zero-mode fermion (i.e., choosing Ψ_L to be even) is (lecture 2):

$$f_{L\,0}(y) = \sqrt{\frac{M}{e^{2M\pi R} - 1}} e^{My} \quad (0 \leq y \leq \pi R)$$

$$= \sqrt{\frac{M}{e^{2M\pi R} - 1}} e^{-My} \quad (0 \geq y \geq -\pi R) \quad (A.1)$$

where the normalization is over $0 \leq y \leq 2\pi R$ (even though the physical domain is from $y = 0$ to $y = \pi R$). Similarly, if we choose Ψ_R to be even instead of Ψ_L, then the RH zero-mode profile is

$$f_{R\,0}(y) = \sqrt{\frac{-M}{e^{-2M\pi R} - 1}} e^{-My} \quad (0 \leq y \leq \pi R)$$

$$= \sqrt{\frac{-M}{e^{-2M\pi R} - 1}} e^{+My} \quad (0 \geq y \geq -\pi R) \quad (A.2)$$

Note the opposite sign of M in the LH versus RH zero-mode profiles [following from Eqs. (1.20) and 1.21)]. Assuming that the SM Higgs field is localized at $y = \pi R$, we see that we need $M < 0 \, (> 0)$ for LH (RH) fermion to obtain small fermion wavefunction at the location of the Higgs and hence small 4D Yukawa couplings for light fermions (1st and 2nd generations). So, we can neglect $e^{\pm M\pi R}$ compared to 1 wherever appropriate.

The zero-mode (4D or SM) Yukawa coupling in terms of the 5D Yukawa coupling: $\int dy d^4 x \delta(y) \lambda_{5D} H \Psi_L \Psi'_R$ [where Ψ_L is $SU(2)_L$ doublet and Ψ'_R is $SU(2)_L$ singlet] is:

$$\lambda_{4D} \approx \lambda_{5D} M e^{2M\pi R} \quad (A.3)$$

and the 4D mass of fermion is

$$m \approx \lambda_{4D} v, \quad (A.4)$$

where, for simplicity, we assume equal size of 5D masses, i.e., $M = -M'$, for doublet and singlet fermions.

A.1.2. Coupling of Zero-mode Fermion to Gauge KK mode: No Brane Kinetic Terms

The profile for n^{th} gauge KK mode ($m_n = n/R$) is:

$$f_n(y) = \frac{1}{\sqrt{\pi R}} \cos(m_n y) \tag{A.5}$$

Calculate the coupling of zero-mode fermion to gauge KK modes in terms of the coupling of zero-mode gauge field (i.e., SM gauge coupling), $g_4 \equiv g_5/\sqrt{2\pi R}$:

$$g(n, M) = g_4 a(n, M) \tag{A.6}$$

You should obtain:

$$a(n, M) \approx \sqrt{2} \frac{4M^2}{4M^2 + (n/R)^2} \tag{A.7}$$

Use $m_{d,s} = 1$ MeV, 100 MeV and the Higgs vev $v \approx 100$ GeV. Assume, for simplicity, that $\lambda_{5D} M = 1$ for both s, d — otherwise, we have to solve a transcendental equation to obtain M (given the 4D Yukawa coupling). Calculate the 5D masses $M_{s,d}$ and show that $a(1, M_s) - a(1, M_d) \approx 0.1$. Compare $K - \bar{K}$ mixing from KK Z exchange as in lecture 2

$$-\frac{g_Z^2}{m_{KK}^2} \left[a(1, M_s) - a(1, M_d) \right]^2 \text{(mixing angle)}^2 \tag{A.8}$$

to the SM amplitude

$$\frac{g_2^4}{16\pi^2} \frac{m_c^2}{M_W^4} \text{(mixing angle)}^2 \tag{A.9}$$

to obtain bound on m_{KK} of ≈ 20 TeV, using $g_Z \approx 0.75$ and $g_2 \approx 0.65$ for the SM Z and $SU(2)_L$ gauge couplings.

It turns out that another observable called ϵ_K (which is the imaginary or CP-violating part of the above $K - \bar{K}$ mixing amplitude) gives a stronger bound on KK mass scale of ~ 100 TeV.

A.2. Exercise 2

A.2.1. *General Brane Kinetic Terms*

The Lagrangian is

$$\mathcal{L}_{5D} \ni -\frac{1}{4}\mathcal{F}_{MN}\mathcal{F}^{MN} - \frac{1}{4}\delta(y)r\mathcal{F}_{\mu\nu}\mathcal{F}^{\mu\nu} \tag{A.10}$$

where r has dimension of length.

Go through the derivation outlined in lecture 3, i.e., f_n satisfies the orthonormality condition:

$$\int dy f_n^*(y) f_m(y) \left[1 + r\delta(y)\right] = \delta_{mn} \tag{A.11}$$

and the differential equation:

$$\left[\partial_y^2 + m_n^2 \left(1 + r\delta(y)\right)\right] f_n(y) = 0 \tag{A.12}$$

The solution is

$$\begin{aligned} f_n(y) &= a_n \cos(m_n y) + b_n \sin(m_n y) \text{ for } y \geq 0 \\ &= \tilde{a}_n \cos(m_n y) + \tilde{b}_n \sin(m_n y) \text{ for } y \leq 0 \end{aligned} \tag{A.13}$$

Use the following 4 conditions to obtain relations between coefficients a, b's and to solve for m_n: (i) continuity at $y = 0$, (ii) discontinuity in derivative matches brane term, (iii) f_n is even and (iv) periodicity of f_n. In particular, condition (iv) is satisfied by repeating (or copying) f_n between $-\pi R$ and πR to between πR and $3\pi R$ and so on. However, continuity of f_n at $y = \pi R$ has to be imposed and similarly that of derivative of f_n (assuming no brane kinetic term at $y = \pi R$).

You should find

$$\begin{aligned} a_n &= \tilde{a}_n \\ \frac{b_n}{a_n} &= -\frac{rm_n}{2} \\ b_n &= -\tilde{b}_n \\ \frac{b_n}{a_n} &= \tan(m_n \pi R) \end{aligned} \tag{A.14}$$

so that eigenvalues are solutions to

$$\tan(m_n \pi R) = -\frac{rm_n}{2} \tag{A.15}$$

Finally, calculate

$$\frac{1}{a_n^2} = \pi R \left(1 + \frac{1}{4} r^2 m_n^2 + \frac{r}{2\pi R} \right) \quad (A.16)$$

from normalization.

A.2.2. Large *Brane Kinetic Terms*

Verify approximate results shown in lecture 3 for large brane kinetic terms, $r/R \gg 1$, namely,

(i) $m_n \approx (n + 1/2)/R$,

(ii) $1/g_4^2 \approx r/g_5^2$

and for lightest KK modes (small n)

(iii) coupling of a fermion localized at $y = 0$ to gauge KK mode $\sim g_4/\sqrt{r/R}$

(iv) coupling of gauge KK mode to a fermion/Higgs field localized on $y = \pi R$ brane $\sim g_4 \sqrt{r/R}$.

We can generalize these couplings of gauge KK mode to the case of a zero-mode fermion with a profile in the bulk – it's just that we have to do an overlap integral as in problem 2 of exercise 1. Calculate the new $a(1, M_s) - a(1, M_d)$. For $r/R \gg 1$, show that it is smaller than before (i.e., without brane terms) so that $K - \bar{K}$ mixing is suppressed and a lower KK mass scale is allowed.

A.3. Exercise 3

As discussed in lecture 3, the zero and KK modes of Z are defined by setting the Higgs vev to zero. However, due to non-zero Higgs vev, the zero and KK modes of Z mix via mass terms – kinetic terms are still diagonal. The $Z^{(0)}$-$Z^{(1)}$ (i.e., 1st KK mode of Z) mass matrix is:

$$\mathcal{L}_{mass} \ni \begin{pmatrix} Z^{(0)}_\mu & Z^{(1)}_\mu \end{pmatrix} \begin{pmatrix} m^2 & \Delta m^2 \\ \Delta m^2 & M^2 \end{pmatrix} \begin{pmatrix} Z^{\mu\,(0)} \\ Z^{\mu\,(1)} \end{pmatrix} \quad (A.17)$$

where $m^2 = 1/4\, g_{Z^{(0)}}^2 v^2$, mixing term $\Delta m^2 = 1/4\, g_{Z^{(0)}} g_{5\,Z} f_1(\pi R)\, v^2$ and $M^2 = m_{KK}^2 + 1/4\, g_{5\,Z}^2 f_1^2(\pi R)\, v^2$. Here, $f_1(\pi R)$ is wavefunction of $Z^{(1)}$ evaluated at the Higgs brane ($y = \pi R$). Also, $g_{Z^{(0)}} = g_{5\,Z}/\sqrt{2\pi R + r}$ denotes the coupling of $Z^{(0)}$ (where r is the brane kinetic term at $y = 0$) and $g_{5\,Z} = \sqrt{g_{5\,2}^2 + g_5'^2}$ denotes the 5D coupling of Z, with $g_{5\,2}$ and g_5' being the 5D gauge couplings of $SU(2)$ and $U(1)_Y$, respectively (assume, for simplicity, the same brane kinetic term r for all gauge fields).

Diagonalize this mass matrix, assuming $v^2/m_{KK}^2 \times$ gauge couplings $\ll 1$ where appropriate, i.e., determine

(i) the unitary transformation to go from $\left(Z^{(0)} Z^{(1)}\right)$ to physical basis and

(ii) the eigenvalues of the mass matrix.

There are 2 effects of this diagonalization.

A.3.1. *Shift in Coupling of a Fermion to Z*

Given couplings of a fermion to $Z^{(0)}$ and $Z^{(1)}$ (KK basis)

$$\mathcal{L}_{coupling} \ni \bar{\psi}\gamma^\mu (g, G) \begin{pmatrix} Z_\mu^{(0)} \\ Z_\mu^{(1)} \end{pmatrix} \psi \quad (A.18)$$

use the above unitary transformation to calculate the couplings to the fermion in the physical basis, denoted by Z_{light} (which is SM Z) and Z_{heavy}.

Specifically, calculate the coupling of a fermion localized at $y = 0$ to the SM Z using $g = g_{Z^{(0)}}$ and $G = g_{5\,Z} f_1(0)$ in the above equation, where $f_1(0)$ is wavefunction of $Z^{(1)}$ evaluated at the fermion brane ($y = 0$).

Verify that the shift in the coupling of this fermion to the SM Z from the zero-mode Z coupling (i.e., $g_{Z^{(0)}}$) is as shown in lecture 3: $\delta g_Z/g_{Z^{(0)}} \sim g_{Z^{(0)}}^2 v^2/m_{KK}^2$, in particular, that there is *no* enhancement for large brane kinetic terms, $r/R \gg 1$.

A.3.2. *Shift in Z mass*

The lighter eigenvalue of mass matrix is the SM Z mass. Verify that the shift in the SM Z mass from the purely zero-mode mass, i.e., $1/4 g_{Z^{(0)}}^2 v^2$, is as shown in lecture 3, in particular, that there *is* an enhancement in this shift due to $r/R \gg 1$ (when the shift is expressed in terms of $g_{Z^{(0)}}$).

References

[1] K. Dienes, this summer school proceedings and http://scipp.ucsc.edu/haber/tasi_proceedings/dienes.ps.
[2] C. Csaki, arXiv:hep-ph/0404096.
[3] R. Sundrum, arXiv:hep-th/0508134; http://www-conf.slac.stanford.edu/ssi/2005/lec_notes/ Sundrum1/default.htm
 (+ "...Sundrum2..." and "...Sundrum3...")
[4] G. D. Kribs, arXiv:hep-ph/0605325.
[5] N. Arkani-Hamed, M. Porrati and L. Randall, JHEP **0108**, 017 (2001) [arXiv:hep-th/0012148]; R. Rattazzi and A. Zaffaroni, JHEP **0104**, 021 (2001) [arXiv:hep-th/0012248].
[6] J. M. Maldacena, Adv. Theor. Math. Phys. **2**, 231 (1998) [Int. J. Theor. Phys. **38**, 1113 (1999)] [arXiv:hep-th/9711200]; S. S. Gubser, I. R. Klebanov and A. M. Polyakov, Phys. Lett. B **428**, 105 (1998) [arXiv:hep-th/9802109]; E. Witten, Adv. Theor. Math. Phys. **2**, 253 (1998) [arXiv:hep-th/9802150].
[7] H. Georgi, A. K. Grant and G. Hailu, Phys. Rev. D **63**, 064027 (2001) [arXiv:hep-ph/0007350]. For further discussion, see, for example, D. E. Kaplan and T. M. P. Tait, JHEP **0111**, 051 (2001) [arXiv:hep-ph/0110126].
[8] A. Delgado, A. Pomarol and M. Quiros, JHEP **0001**, 030 (2000) [arXiv:hep-ph/9911252].
[9] H. Georgi, A. K. Grant and G. Hailu, Phys. Lett. B **506**, 207 (2001) [arXiv:hep-ph/0012379]. For further discussion, see, for example, H. C. Cheng, K. T. Matchev and M. Schmaltz, Phys. Rev. D **66**, 036005 (2002) [arXiv:hep-ph/0204342].
[10] M. Carena, T. M. P. Tait and C. E. M. Wagner, Acta Phys. Polon. B **33**, 2355 (2002) [arXiv:hep-ph/0207056].
[11] K. Agashe, A. Delgado, M. J. May and R. Sundrum, JHEP **0308**, 050 (2003) [arXiv:hep-ph/0308036].
[12] K. Agashe, A. Belyaev, T. Krupovnickas, G. Perez and J. Virzi, arXiv:hep-ph/0612015; B. Lillie, L. Randall and L. T. Wang, arXiv:hep-ph/0701166.
[13] See, for example, F. del Aguila and J. Santiago, Phys. Lett. B **493**, 175 (2000) [arXiv:hep-ph/0008143] and section 4.1 of A. Juste et al., arXiv:hep-ph/0601112.
[14] K. Agashe, G. Perez and A. Soni, Phys. Rev. D **75**, 015002 (2007) [arXiv:hep-ph/0606293].
[15] L. Randall and R. Sundrum, Phys. Rev. Lett. **83**, 3370 (1999) [arXiv:hep-ph/9905221].
[16] W. D. Goldberger and M. B. Wise, Phys. Rev. Lett. **83**, 4922 (1999) [arXiv:hep-ph/9907447].
[17] H. Davoudiasl, J. L. Hewett and T. G. Rizzo, Phys. Lett. B **473**, 43 (2000) [arXiv:hep-ph/9911262]; A. Pomarol, Phys. Lett. B **486**, 153 (2000) [arXiv:hep-ph/9911294]; S. Chang, J. Hisano, H. Nakano, N. Okada and M. Yamaguchi, Phys. Rev. D **62**, 084025 (2000) [arXiv:hep-ph/9912498].
[18] Y. Grossman and M. Neubert, Phys. Lett. B **474**, 361 (2000) [arXiv:hep-

ph/9912408]; T. Gherghetta and A. Pomarol, Nucl. Phys. B **586**, 141 (2000) [arXiv:hep-ph/0003129].

[19] L. Randall and R. Sundrum, Phys. Rev. Lett. **83**, 4690 (1999) [arXiv:hep-th/9906064].

[20] M. Quiros, arXiv:hep-ph/0302189.

[21] Y. Hosotani, Phys. Lett. B **126**, 309 (1983); Phys. Lett. B **129**, 193 (1983) and Annals Phys. **190**, 233 (1989).

[22] R. Contino, Y. Nomura and A. Pomarol, Nucl. Phys. B **671**, 148 (2003) [arXiv:hep-ph/0306259];

[23] K. Agashe, R. Contino and A. Pomarol, Nucl. Phys. B **719**, 165 (2005) [arXiv:hep-ph/0412089].

[24] N. Arkani-Hamed, S. Dimopoulos and G. R. Dvali, Phys. Lett. B **429**, 263 (1998) [arXiv:hep-ph/9803315]; I. Antoniadis, N. Arkani-Hamed, S. Dimopoulos and G. R. Dvali, Phys. Lett. B **436**, 257 (1998) [arXiv:hep-ph/9804398]; N. Arkani-Hamed, S. Dimopoulos and G. R. Dvali, Phys. Rev. D **59**, 086004 (1999) [arXiv:hep-ph/9807344].

[25] T. Appelquist, H. C. Cheng and B. A. Dobrescu, Phys. Rev. D **64**, 035002 (2001) [arXiv:hep-ph/0012100].

[26] G. Servant and T. M. P. Tait, Nucl. Phys. B **650**, 391 (2003) [arXiv:hep-ph/0206071] and New J. Phys. **4**, 99 (2002) [arXiv:hep-ph/0209262]; H. C. Cheng, J. L. Feng and K. T. Matchev, Phys. Rev. Lett. **89** (2002) 211301 [arXiv:hep-ph/0207125].

[27] C. Csaki, C. Grojean, H. Murayama, L. Pilo and J. Terning, Phys. Rev. D **69**, 055006 (2004) [arXiv:hep-ph/0305237]; C. Csaki, C. Grojean, L. Pilo and J. Terning, Phys. Rev. Lett. **92**, 101802 (2004) [arXiv:hep-ph/0308038]; for a review, see C. Csaki, J. Hubisz and P. Meade, arXiv:hep-ph/0510275.

Chapter 2

The International Linear Collider

Marco Battaglia

*Department of Physics, University of California at Berkeley and
Lawrence Berkeley National Laboratory
Berkeley, CA 94720, USA
MBattaglia@lbl.gov*

The International Linear Collider (ILC) is the next large scale project in accelerator particle physics. Colliding electrons with positrons at energies from 0.3 TeV up to about 1 TeV, the ILC is expected to provide the accuracy needed to complement the LHC data and extend the sensitivity to new phenomena at the high energy frontier and answer some of the fundamental questions in particle physics and in its relation to Cosmology. This paper reviews some highlights of the ILC physics program and of the major challenges for the accelerator and detector design.

2.1. Introduction

Accelerator particle physics is completing a successful cycle of precision tests of the Standard Model of electro-weak interactions (SM). After the discovery of the W and Z bosons at the $Sp\bar{p}S$ hadron collider at CERN, the concurrent operation of hadron and e^+e^- colliders has provided a large set of precision data and new observations. Two e^+e^- colliders, the SLAC Linear Collider (SLC) at the Stanford Linear Accelerator Center (SLAC) and the Large Electron Positron (LEP) collider at the European Organization for Nuclear Research (CERN), operated throughout the 1990's and enabled the study of the properties of the Z boson in great detail. Operation at LEP up to 209 GeV, the highest collision energy ever achieved in electron-positron collisions, provided detailed information on the properties of W bosons and the strongest lower bounds on the mass of the Higgs boson and of several supersymmetric particles. The collision of point-like, elementary particles at a well-defined and tunable energy offers advantages for precision measurements, as those conducted at LEP and SLC, over proton colliders. On the other hand experiments at hadron machines, such as the Tevatron $p\bar{p}$ collider at Fermilab, have enjoyed higher constituent energies. The CDF and D0 experiments eventually observed the direct production of top quarks, whose mass had been predicted on the basis of precision data obtained at LEP and SLC.

While we await the commissioning and operation of the LHC pp collider at CERN, the next stage in experimentation at lepton colliders is actively under study. For more than two decades, studies for a high-luminosity accelerator, able to collide electrons with positrons at energies of the order of 1 TeV, are being carried out world-wide.

2.2. The path towards the ILC

The concept of an e^+e^- linear collider dates back to a paper by Maury Tigner[1] published in 1965, when the physics potential of e^+e^- collisions had not yet been appreciated in full. This seminal paper envisaged collisions at 3-4 GeV with a luminosity competitive with that of the SPEAR ring at SLAC, i.e. 3×10^{30} cm^{-2} s^{-1}. *A possible scheme to obtain e^-e^- and e^+e^- collisions at energies of hundreds of GeV* is the title of a paper[2] by Ugo Amaldi published a decade later in 1976, which sketches the linear collider concept with a design close to that now developed for the ILC. The

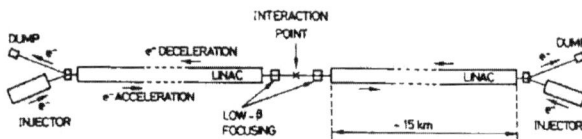

Fig. 2.1. The linear collider layout as sketched in 1975 in one of the figures of Ref. 2. The paper discussed the possibility to achieve e^-e^- and e^+e^- collisions at 0.3 TeV using superconducting linacs with a gradient of 10 MV/m.

parameters for a linear collider, clearly recognised as the successors of e^+e^- storage rings on the way to high energies, were discussed by Burt Richter at the IEEE conference in San Francisco in 1979[3] and soon after came the proposal for the *Single Pass Collider Project* which would become SLC at SLAC.

From 1985, the CERN Long Range Planning Committee considered an e^+e^- linear collider, based on the CLIC[4] design, able to deliver collisions at 2 TeV with 10^{33} cm^{-2} s^{-1} luminosity, *vis-a-vis* a hadron collider, with proton-proton collisions at 16 TeV and luminosity of 1.4×10^{33} cm^{-2} s^{-1}, as a candidate for the new CERN project after LEP. That review process eventually led to the decision to build the LHC, but it marked an important step to establish the potential of a high energy e^+e^- collider. It is important to note that it was through the contributions of several theorists, including John Ellis, Michael Peskin, Gordon Kane and others, that the requirements in terms of energy and luminosity for a linear collider became clearer in the mid 1980's.[5] The SLC project gave an important proof of principle for a high energy linear collider and the experience gained has shaped the subsequent designs in quite a significant way.

After a decade marked by important progress in the R&D of the basic components and the setup of advanced test facilities, designs of four different concepts emerged: TESLA, based on superconducting RF cavities, the NLC/JLC-X, based on high frequency (11.4 GHz) room-temperature copper cavities, JLC-C, based on lower frequency (5.7 GHz) conventional cavities and CLIC, a multi-TeV collider based on a different beam acceleration technique, the two-beam scheme with transfer structures operating at 30 GHz. Accelerator R&D had reached the maturity to assess the technical feasibility of a linear collider project and take an informed choice of the most advantageous RF technology. The designs were considered by the International Linear Collider Technical Review Committee (ILC-TRC), originally formed in 1994 and re-convened by the International Committee for Future Accelerators (ICFA) in 2001 under the chairmanship of Greg A. Loew. The ILC-TRC assessed their status using common criteria, identified outstanding items needing R&D effort and suggested areas of collaboration. The TRC report was released in February 2003[6] and the committee found that there were *no insurmountable show-stoppers to build TESLA, NLC/JLC-X or JLC-C in the next few years and CLIC in a more distant future, given enough resources*. Nonetheless, significant R&D remained to be done. At this stage, it became clear that, to make further progress, the international effort towards a linear collider should be focused on a single design. ICFA gave mandate to an International Technology Recommendation Panel (ITRP), chaired by Barry Barish, to make a definite recommendation for a RF technology that would be the basis of a global project. In August 2004 the ITRP made the recommendation in favour of superconducting RF cavities.[7] The technology choice, which was promptly accepted by all laboratories and groups involved in the R&D process, is regarded as a major step towards the realization of the linear collider project. Soon after it, a truly world-wide, centrally managed design effort, the Global Design Effort (GDE),[8] a team of more than 60 persons, started, with the aim to produce an ILC Reference Design Report by beginning of 2007 and an ILC Technical Design Report by end of 2008. The GDE responsibility now covers the detailed design concept, performance assessments, reliable international costing, industrialization plan, siting analysis, as well as detector concepts and scope. A further important step has been achieved with release of the Reference Design Report in February 2007.[9] This report includes a preliminary value estimate of the cost for the ILC in its present design and at the present level of engineering and industrialisation. The value estimate is structured in three parts: 1.78 Billion ILC Value Units for site-related costs, such as those of tunneling in a specific region, 4.87 Billion ILC Value Units for the value of the high technology and conventional components and 13,000 person-years for the required supporting manpower. For this estimate the conversion factor is 1 ILC Value Unit = 1 US Dollar = 0.83 Euro = 117 Yen. This estimate, which is comparable to the LHC cost, when the pre-existing facilities, such as the LEP tunnel, are included, provides guidance for optimisation of both the design and the R&D to be done during the engineering phase, due to start in Fall 2007.

Technical progress was paralleled by increasing support for the ILC in the scientific community. At the 2001 APS workshop *The Future of Physics* held in Snowmass, CO, a consensus emerged for the ILC as the right project for the next large scale facility in particle physics. This consensus resonated and expanded in a number of statements by highly influential scientific advisory panels world-wide. The ILC role in the future of scientific research was recognised by the OECD Consultative Group on High Energy Physics,[10] while the DOE Office of Science ranked the ILC as its top mid-term project. More recently the EPP 2010 panel of the US National Academy of Sciences, in a report titled *Elementary Particle Physics in the 21st Century* has endorsed the ILC as the next major experimental facility to be built and its role in elucidating the physics at the high energy frontier, independently from the LHC findings.[11] Nowadays, the ILC is broadly regarded as the highest priority for a future large facility in particle physics, needed to extend and complement the LHC discoveries with the accuracy which is crucial to understand the nature of New Physics, test fundamental properties at the high energy scale and establish their relation to other fields in physical sciences, such as Cosmology. A matching program of physics studies and detector R&D efforts has been in place for the past decade and it is now developing new, accurate and cost effective detector designs from proof of concepts towards that stage of engineering readiness, needed for being adopted in the ILC experiments.

2.3. ILC Accelerator Parameters

2.3.1. *ILC Energy*

The first question which emerges in defining the ILC parameters is the required centre-of-mass energy $\sqrt{s}$. It is here where we most need physics guidance to define the next thresholds at, and beyond, the electro-weak scale. The only threshold which, at present, is well defined numerically is that of top-quark pair production at $\sqrt{s} \simeq 350$ GeV. Beyond it, there is a strong prejudice, supported by precision electro-weak and other data, that the Higgs boson should be light and new physics thresholds may exist between the electro-weak scale and approximately 1 TeV. If indeed the SM Higgs boson exists and the electro-weak data is not affected by new physics, its mass M_H is expected to be below 200 GeV as discussed in section 2.4.1. Taking into account that the Higgs main production process is in association with a Z^0 boson, the maximum of the $e^+e^- \to H^0 Z^0$ cross section varies from $\sqrt{s} = 240$ GeV to 350 GeV for 120 GeV $< M_H <$ 200 GeV. On the other hand, we know that the current SM needs to be extended by some New Physics. Models of electroweak symmetry breaking contain new particles in the energy domain below 1 TeV. More specifically, if Supersymmetry exists and it is responsible for the dark matter observed in the Universe, we expect that a significant fraction of the supersymmetric spectrum would be accessible at $\sqrt{s} = 0.5$-1.0 TeV. In particular, the ILC should be able to study in detail those particles determining the dark matter relic density in the Universe by operating at energies not exceeding 1 TeV, as discussed in section 2.4.2. Another useful perspective on the ILC energy is an analysis of the mass scale sensitivity for new physics vs. the $\sqrt{s}$ energy for lepton and hadron colliders in view of their synergy. The study of electro-weak processes at the highest available energy offers a window on mass scales well beyond its kinematic reach. A comparison of the mass-scale sensitivity for various new physics scenarios as a function of the centre-of-mass energy for e^+e^- and pp collisions is given in section 2.4.3. These and similar considerations, emerged in the course of the world-wide studies on physics at the ILC, motivate the choice of $\sqrt{s} = 0.5$ TeV as the reference energy parameter, but requiring the ILC to be able to operate, with substantial luminosity, at 0.3 TeV as well and to be upgradable up to approximately 1 TeV.

It is useful to consider these energies in an historical perspective. In 1954 Enrico Fermi gave a talk at the American Physical Society, of which he was chair, titled *What can we learn with high energy accelerators ?*. In that talk Fermi considered a proton accelerator with a radius equal to that of Earth and 2 T bending magnets, thus reaching a beam energy of 5×10^{15} eV.[12] Stanley Livingstone, who had built with Ernest O. Lawrence the first circular accelerator at Berkeley in 1930, had formulated an empirical linear scaling law for the available centre-of-mass energy vs. the construction year and cost. Using Livingstone curve, Fermi predicted that such an accelerator could be built in 1994 at a cost of 170 billion $. We have learned that, not only such accelerator could not be built, but accelerator physics has irrevocably fallen off the Livingstone curve, even in its revised version, which includes data up to the 1980's. As horizons expanded, each step has involved more and more technical challenges and has required more resources. The future promises to be along this same path. This underlines the need of coherent and responsible long term planning while sustaining a rich R&D program in both accelerator and detector techniques.

The accelerator envisaged by Enrico Fermi was a circular machine, as the almost totality of machines operating at the high energy frontier still are. Now, as it is well known, charged particles undergoing a centripetal acceleration $a = v^2/R$ radiate at rate $P = \frac{1}{6\pi\epsilon_0}\frac{e^2 a^2}{c^3}\gamma^4$. If the radius R is kept constant, the energy loss is the above rate P times $t = 2\pi R/v$, the time spent in the bending section of the accelerator. The energy loss for electrons is $W = 8.85 \times 10^{-5}\frac{E^4(\text{GeV}^4)}{R(\text{km})}$ MeV per turn while for protons is $W = 7.8 \times 10^{-3}\frac{E^4(\text{TeV}^4)}{R(\text{km})}$ keV per turn. Since the energy transferred per turn by the RF cavities to the beam is constant, $G \times 2R \times F$, where G is the cavity gradient and F the tunnel fill factor, for each value of the accelerator ring radius R there exists a maximum energy E_{max} beyond which the energy loss exceeds the energy transferred. In practice, before this value of E_{max} is reached, the real energy limit is set by the power dumped by the beam as synchrotron radiation. To make a quantitative example, in the case of the LEP ring, with a radius $R = 4.3$ km, a beam of energy $E_{beam} = 250$ GeV, would lose 80 GeV/turn. Gunther Voss is thought to be the author of a plot comparing the guessed cost of a storage ring and a linear collider as a function of the e^+e^- centre-of-mass energy. A $\sqrt{s} = 500$ GeV storage ring, which would have costed an estimated 14 billions CHF in 1970's is aptly labelled as the Crazytron.[13] LEP filled the last window of opportunity for a storage ring at the high energy frontier. Beyond LEP-2 energies the design must be a linear collider, where no bending is applied to the accelerated particles. Still the accelerator length is limited by a number of constraints which include costs, alignment and siting. Therefore, technology still defines the maximum reachable energy at the ILC.

The ILC design is based on superconducting (s.c.) radio-frequency (RF) cavities. While s.c. cavities had been considered already in the 1960's, it was Ugo Amaldi to first propose a fully s.c. linear collider in 1975.[2] By the early

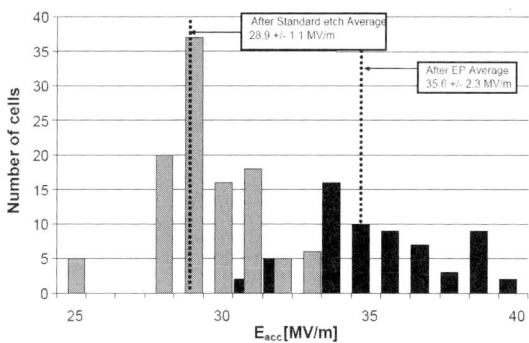

Fig. 2.2. Distributions of gradients measured for pure niobium, nine-cell cavities. After electro-polishing an average gradient in excess to 35 MV/m has been obtained.

1990's, s.c. cavities equipped already one accelerator, TRISTAN at KEK in Japan, while two further projects were in progress, CEBAF at Cornell and the LEP-2 upgrade at CERN. LEP-2 employed a total of 288 s.c. RF cavities, providing an average gradient of 7.2 MV/m. It was the visionary effort of Bjorn Wijk to promote, from 1990, the TESLA collaboration, with the aim to develop s.c. RF cavities pushing the gradient higher by a factor of five and the production costs down by a factor of four, thus reducing the cost per MV by a factor of twenty. Such reduction in cost was absolutely necessary to make a high energy collider, based on s.c. cavities, feasible. Within less than a decade 1.3 GHz, pure niobium cavities achieved gradients in excess to 35 MV/m. This opened the way to their application to a e^+e^- linear collider, able to reach centre-of-mass energies of the order of 1 TeV, as presented in detail in the TESLA proposal published in 2001[14] and recommended for the ILC by the ITRP in 2004.[7]

Today, the ILC baseline design aims at matching technical feasibility to cost optimisation. One of the major goals of the current effort in the ILC design is to understand enough about its costs to provide a reliable

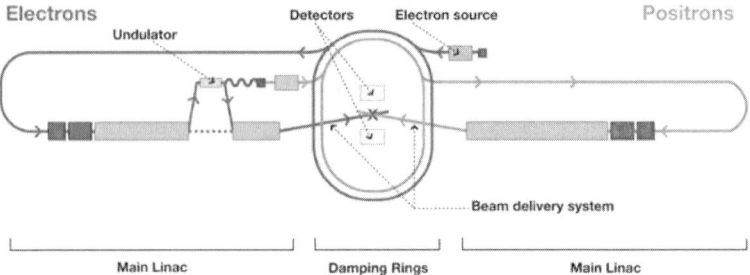

Fig. 2.3. Schematic layout of the International Linear Collider. This diagram reflects the recommendations of the Baseline Configuration Document, a report published in December 2005 that outlines the general design of the machine. (Credit ILC Global Design Effort)

indication of the scale of funding required to carry out the ILC project. Preparing a reliable cost estimate for a project to be carried out as a truly world-wide effort at the stage of a conceptual design that still lacks much of the detailed engineering designs as well as agreements for responsibility and cost sharing between the partners and a precise industrialisation plan is a great challenge. Still having good cost information as soon as possible, to initiate negotiations with the funding agencies is of great importance. An interesting example of the details entering in this process is the optimisation of the cost vs. cavity gradient for a 0.5 TeV collider. The site length scales inversely with the gradient G while the cost of the cryogenics scales as G^2/Q_0 resulting in a minimum cost for a gradient of 40 MW/m, corresponding to a tunnel length of 40 km, and a fractional cost increase of 10 % for gradients of 25 MV/m or 57 MV/m. The chosen gradient of 35 MV/m, which is matched by the average performance of the most recent prototypes after electro-polishing, gives a total tunnel length of 44 km with a cost increment from the minimum of just 1 %.

Beyond 1 TeV, the extension of conventional RF technology is more speculative. In order to attain collisions at energies in excess of about 1 TeV, with high luminosity, significantly higher gradients are necessary. As the gradient of s.c. cavities is limited below ~ 50 MV/m, other avenues should be explored. The CLIC technology,[15] currently being developed at CERN and elsewhere, may offer gradients of the order of 150 MV/m,[16] allowing collision energies in the range 3-5 TeV with a luminosity of 10^{35} cm^{-2} s^{-1}, which would support a compelling physics

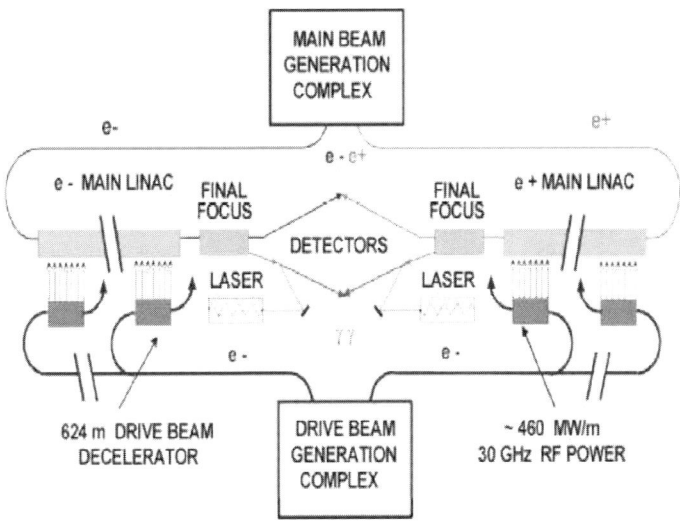

Fig. 2.4. Schematics of the overall layout of the CLIC complex for e^+e^- collisions at $\sqrt{s} = 3$ TeV. (from Ref. 17)

program.[17] While RF cavities are limited to accelerating fields of order of 100-200 MV/m, or below, laser-wakefield accelerators are capable, in principle, of producing fields of 10-100 GV/m. Recently a 1 GeV e^- beam has been accelerated over just 3.3 cm using a 40 TW peak-power laser pulse,[18] thus opening a possible path towards ultra-high energies in e^+e^- collisions in some more distant future.

2.3.2. ILC Luminosity

The choice of a linear collider, rather than a circular storage ring, while solving the problem of the maximum reachable energy, introduces the challenge of achieving collisions with the required luminosity. The luminosity, $\mathcal{L}$, defined as the proportionality factor between the number of events produced and the process cross section σ, has requirements which depend on the typical values of s-channel cross sections and so scale as $1/s$. First luminosity requirements were already outlined in the 1980s[19,20] as $\mathcal{L} \simeq \frac{2E_{beam}}{\text{TeV}} \times 10^{33}$ cm^{-2} s^{-1}, based on the estimated discovery potential. But in the present vision of the ILC role in probing the high energy frontier new requirements must be considered. One example is the precision study of electro-weak processes to look for deviations from the SM predictions, due to effect of new physics at high scales. The $e^+e^- \to b\bar{b}$ cross section at 1 TeV is just 96 fb, so this would corresponds to less than 10^3 events per year at 10^{33} cm^{-2} s^{-1}, which is certainly insufficient for the kind of precision measurements which we expect from the ILC. Another example is offered by one of the reactions most unique to the ILC: the double-Higgs production $e^+e^- \to HHZ$ sensitive to the Higgs self-coupling, which has a cross section of order of only 0.2 fb at 0.5 TeV. Therefore a luminosity of 10^{34} cm^{-2} s^{-1} or more is required as baseline parameter.

The luminosity can be expressed as a function of the accelerator parameters as:

$$\mathcal{L} = f_{rep} n_b \frac{N^2}{4\pi\sigma_x \sigma_y}. \tag{2.1}$$

Now, since in a linear machine the beams are collided only once and then dumped, the collision frequency, f_{rep}, is small and high luminosity should be achieved by increasing the number of particles in a bunch N, the number of bunches n_b and decreasing the transverse beam size σ. Viable values for N are limited by wake-field effects and the ILC parameters have the same number of electrons in a bunch as LEP had, though it aims at a luminosity three orders of magnitude higher. Therefore, the increase must come from a larger number of bunches and a smaller transverse beam size. The generation of beams of small transverse size, their preservation during acceleration and their focusing to spots of nanometer size at the interaction region presents powerful challenges which the ILC design must solve. A small beam size also induces beam-beam interactions. On one hand the beam self-focusing, due to the electrostatic attraction of particles of opposite charges enhances the luminosity. But beam-beam interactions also result in an increase of beamstrahlung with a larger energy spread of the colliding particles, a degraded luminosity spectrum and higher backgrounds. Beamstrahlung is energy loss due to particle radiation triggered by the trajectory bending in the interactions with the charged particles in the incoming bunch.[21] The mean beamstrahlung energy loss, which has to be minimised, is given by:

$$\delta_{BS} \simeq 0.86 \frac{er_e^3}{2m_0 c^2} \frac{E_{cm}}{\sigma_z} \frac{N_b^2}{(\sigma_x + \sigma_y)^2}. \tag{2.2}$$

Since the luminosity scales as $\frac{1}{\sigma_x \sigma_y}$, while the beamstrahlung energy loss scales as $\frac{1}{\sigma_x + \sigma_y}$, it is advantageous to choose a large beam aspect ratio, with the vertical beam size much smaller than the horizontal component. The parameter optimisation for luminosity can be further understood by expressing the luminosity in terms of beam power $P = f_{rep}NE_{cm} = \eta P_{AC}$ and beamstrahlung energy loss as:

$$\mathcal{L} \propto \frac{\eta P_{AC}}{E_{cm}} \sqrt{\frac{\delta_{BS}}{\epsilon_y}} H_D \qquad (2.3)$$

which highlights the dependence on the cavity efficiency η and the total power P_{AC}. The H_D term is the pinch enhancement factor, that accounts for the bunch attraction in the collisions of oppositely charged beams. In summary, since the amount of available power is necessarily limited, the main handles on luminosity are η and ϵ_y. The efficiency for transferring power from the plug to the beam is naturally higher for s.c. than for conventional copper cavities, so more relaxed collision parameters can be adopted for a s.c. linear collider delivering the same luminosity. The main beam parameters for the ILC baseline design are given in Table 2.1.

Table 2.1. ILC baseline design beam parameters.

Parameter	$\sqrt{s}$ 0.5 TeV	$\sqrt{s}$ 1.0 TeV
Luminosity L (10^{34} cm^{-2}s^{-1})	2.0	2.8
Frequency (Hz)	5.0	5.0
Nb. of particles (10^{10})	2.0	2.0
Nb. of bunches N_b	2820	2820
Bunch spacing (ns)	308	308
Vertical beam size σ_y (nm)	5.7	3.5
Beamstrahlung Parameter δ_{BS}	0.022	0.050
H_D	1.7	1.5

2.4. ILC Physics Highlights

The ILC physics program, as we can anticipate it at present, is broad and diverse, compelling and challenging. The ILC is being designed for operation at 0.5 TeV with the potential to span the largest range of collision energies, from the Z^0 peak at 0.091 TeV up to 1 TeV, collide electrons with positrons, but optionally also electrons with electrons, photons with photons and photons with electrons, and combine various polarization states of the electron and positron beams. Various reports discussing the linear collider physics case, including results of detailed physics studies, have been published in the last few years.[5,17,22–26] Here, I shall focus on three of the main ILC physics themes: the detailed study of the Higgs boson profile, the determination of neutralino dark matter density in the Universe from accelerator data, and the sensitivity to new phenomena beyond the ILC kinematic reach, through the analysis of two-fermion production, at the highest $\sqrt{s}$ energy. Results discussed in the following have been obtained mostly using realistic, yet parametric simulation of the detector response. Only few analyses have been carried out which include the full set of physics and machine-induced backgrounds on fully simulated and reconstructed events. With the progress of the activities of detector concepts and the definition of well-defined benchmark processes, this is becoming one of the priorities for the continuation of physics and detector studies.

2.4.1. *The Higgs Profile at the ILC*

Explaining the origin of mass is one of the great scientific quests of our time. The SM addresses this question by the Higgs mechanism.[27] The first direct manifestation of the Higgs mechanism through the Higgs sector will be the existence of at least one Higgs boson. The observation of a new spin-0 particle would represent a first sign that the Higgs mechanism of mass generation is indeed realised in Nature. This has motivated a large experimental effort, from LEP-2 to the Tevatron and, soon, the LHC, actively backed-up by new and more accurate theoretical predictions. After a Higgs discovery, which we anticipate will be possible at the LHC, full validation of the Higgs mechanism can only be established by an accurate study of the Higgs boson production and decay properties. It is here where the ILC potential in precision physics will be crucial for the validation of the Higgs mechanism, through a detailed study of the Higgs profile.[31]

The details of this study depend on the Higgs boson mass, M_H. In the SM, $M_H = \sqrt{2\lambda}v$ where the Higgs field expectation value v is determined as $(\sqrt{2}G_F)^{-1/2} \approx 246$ GeV, while the Higgs self-coupling λ is not specified, leaving the mass as a free parameter. However, we have strong indications that M_H must be light. The Higgs self-coupling behaviour at high energies,[28] the Higgs field contribution to precision electro-weak data[30] and the results of direct searches at LEP-2[29] at $\sqrt{s} \geq 206$ GeV, all point towards a light Higgs boson. In particular, the study of precision electro-weak data, which are sensitive to the Higgs mass logarithmic contribution to radiative corrections, is based on several independent observables, including masses (m_{top}, M_W, M_Z), lepton and quark asymmetries at the Z^0 pole, Z^0 line-shape and partial decay widths. The fit to eighteen observables results in a 95% C.L. upper limit for the Higgs mass of 166 GeV, which becomes 199 GeV when the lower limit from the direct searches at LEP-2, $M_H > 114.4$ GeV, is included. As a result, current data indicates that the Higgs boson mass should be in the range 114 GeV $< M_H <$ 199 GeV. It is encouraging to observe that if the same fit is repeated, but excluding this time m_{top} or M_W, the results for their values, 178^{+12}_{-9} GeV and 80.361±0.020 GeV respectively, are in very good agreement with the those obtained the direct determinations, $m_{top} = 171.4 \pm 2.1$ GeV and $M_W = 80.392 \pm 0.029$ GeV.

At the ILC the Higgs boson can be observed in the Higgs-strahlung production process $e^+e^- \to HZ$ with $Z \to \ell^+\ell^-$, independent of its decay mode, by the distinctive peak in the di-lepton recoil mass distribution. A data set of 500 fb^{-1} at $\sqrt{s} = 350$ GeV, corresponding to four years of ILC running, provides a sample of 3500-2200 Higgs particles produced in the di-lepton HZ channel, for $M_H = 120$-200 GeV. Taking into account the SM backgrounds, dominated by $e^+e^- \to Z^0Z^0$ and W^+W^- production, the Higgs boson observability is guaranteed up to its production kinematical limit, independent of its decays. This sets the ILC aside from the LHC, since the ILC sensitivity to the Higgs boson does not depend on its detailed properties.

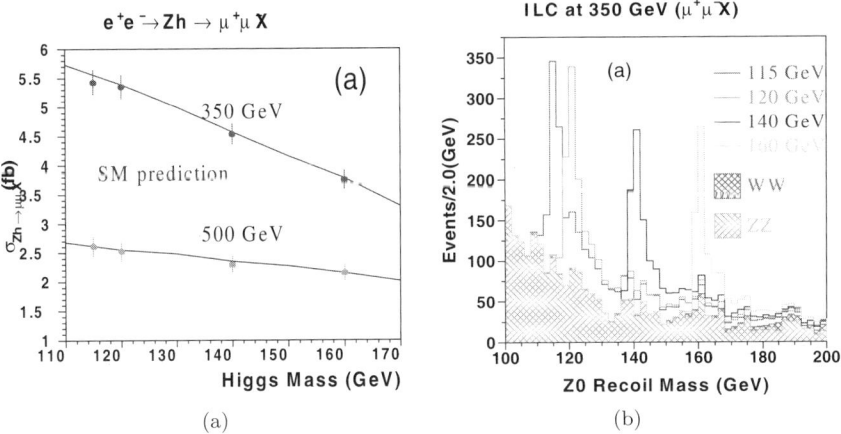

Fig. 2.5. The Higgs-strahlung process at the ILC. (a) $e^+e^- \to HZ$ cross section vs. M_H for $\sqrt{s} = 0.35$ TeV and 0.5 TeV, (b) reconstructed $\mu^+\mu^-$ recoil mass for various values of the Higgs boson mass (Credit: ALCPG Study Group).

After observation of a new particle with properties compatible with those of the Higgs boson, a significant experimental and theoretical effort will be needed to verify that this is indeed the boson of the scalar field responsible for the electro-weak symmetry breaking and the generation of mass. Outlining the Higgs boson profile, through the determination of its mass, width, quantum numbers, couplings to gauge bosons and fermions and the reconstruction of the Higgs potential, stands as a most challenging, yet compelling, physics program. The ILC, with its large data sets at different centre-of-mass energies and beam polarisation conditions, the high resolution detectors providing unprecedented accuracy on the reconstruction of the event properties and the use of advanced analysis techniques, developed from those successfully adopted at LEP and SLC, promises to promote Higgs physics into the domain of precision measurements. Since the Higgs mass M_H is not predicted by theory, it is of great interest to measure it precisely. Once this mass, and thus λ, is fixed, the profile of the Higgs particle is uniquely determined in the SM. In most scenarios we expect the LHC to determine the Higgs mass with a good accuracy. At the ILC, this measurement can be refined by exploiting the kinematical characteristics of the Higgs-strahlung production process $e^+e^- \to Z^* \to H^0 Z^0$ where the Z^0 can be reconstructed in both its leptonic and hadronic decay modes. The $\ell^+\ell^-$ recoil mass for leptonic Z^0 decays yields an accuracy of 110 MeV for 500 fb^{-1} of data, without any requirement on the nature of the Higgs decays. Further improvement can be obtained by explicitly selecting $H \to b\bar{b}$ (WW) for $M_H \leq (>)$ 140 GeV. Here a kinematical 5-C fit, imposing energy and momentum conservation and the mass of a jet pair to correspond to M_Z, achieves an accuracy of 40 to 90 MeV for 120$< M_H <$ 180 GeV.[32]

The total decay width of the Higgs boson is predicted to be too narrow to be resolved experimentally for Higgs boson masses below the ZZ threshold. On the contrary, above $\simeq$ 200 GeV, the total width can be measured directly from the reconstructed width of the recoil mass peak, as discussed below. For the lower mass range, indirect methods must be applied. In general, the total width is given by $\Gamma_{tot} = \Gamma_X/\mathrm{BR}(H \to X)$. Whenever Γ_X can be determined independently of the corresponding branching fraction, a measurement of Γ_{tot} can be carried out. The most convenient choice is the extraction of Γ_H from the measurements of the WW fusion cross section and the $H \to WW^*$ decay branching fraction . A relative precision of 6% to 13% on the width of the Higgs boson can be obtained at the ILC with this technique, for masses between 120 GeV and 160 GeV. The spin, parity

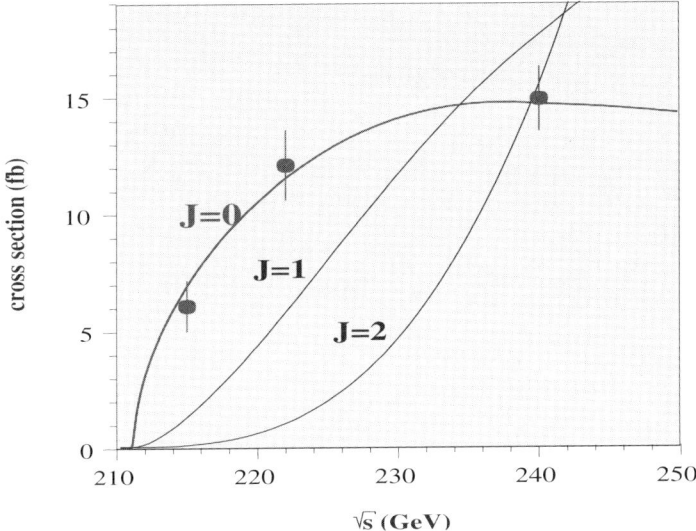

Fig. 2.6. Determination of the Higgs boson spin from a scan of the $e^+e^- \to HZ$ cross section at threshold at the ILC (from Ref. 23).

and charge-conjugation quantum numbers J^{PC} of Higgs bosons can be determined at the ILC in a model-independent way. Already the observation of either $\gamma\gamma \to H$ production or $H \to \gamma\gamma$ decay sets $J \neq 1$ and $C = +$. The angular dependence $\frac{d\sigma_{ZH}}{d\theta} \propto \sin^2\theta$ and the rise of the Higgs-strahlung cross section:

$$\sigma_{ZH} \propto \beta \sim \sqrt{s - (M_H + M_Z)^2} \qquad (2.4)$$

allows to determine $J^P = 0^+$ and distinguish the SM Higgs from a CP-odd 0^{-+} state A^0, or a CP-violating mixture of the two.[33,34] But where the ILC has a most unique potential is in verifying that the Higgs boson does its job of providing gauge bosons, quarks and leptons with their masses. This requires to precisely test the relation $g_{HXX} \propto m_X$ between the Yukawa couplings, g_{HXX}, and the corresponding particle masses, m_X. In fact, the SM Higgs couplings to fermion pairs $g_{Hff} = m_f/v$ are fully determined by the fermion mass m_f. The corresponding decay partial widths only depend on these couplings and on the Higgs boson mass, QCD corrections do not represent a significant source of uncertainty.[35] Therefore, their accurate determination will represent a comprehensive test of the Higgs mechanism of mass generation.[36] Further, observing deviations of the measured values

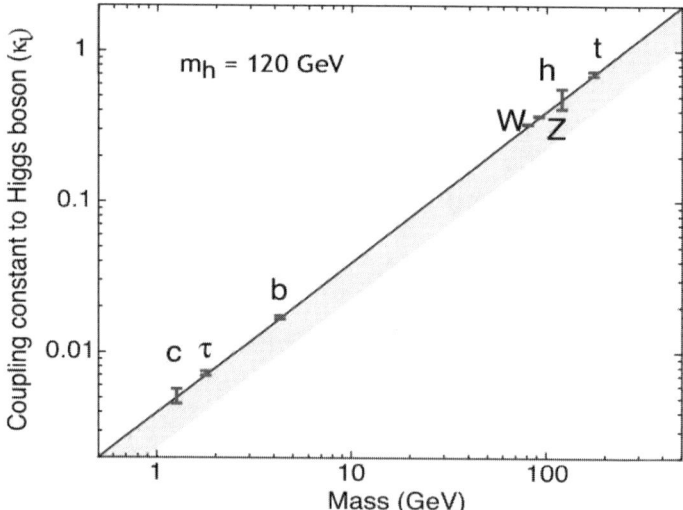

Fig. 2.7. Particle couplings to the Higgs field, for a 120 GeV boson, as a function of the particle masses. The error bars show the expected ILC accuracy in determining each of the couplings. The dark line is the SM prediction, while the shaded gray area shows the range of predictions from theories of new physics beyond the SM with extra dimensions (Credit: ACFA ILC Study Group).

from the SM predictions will probe the structure of the Higgs sector and may reveal a non-minimal implementation of the Higgs model or the effect of new physics inducing a shift of the Higgs couplings.[37-39] The accuracy of these measurements relies on the performances of jet flavour tagging and thus mostly on the Vertex Tracker, making this analysis an important benchmark for optimising the detector design. It is important to ensure that the ILC sensitivity extends over a wide range of Higgs boson masses and that a significant accuracy is achieved for most particle species. Here, the ILC adds the precision which establishes the key elements of the Higgs mechanism. It is important to point out that these tests are becoming more stringent now that the B-factories have greatly improved the determination of the b- and c-quark masses. When one of these studies was first presented in 1999,[40] the b quark mass was known to ± 0.11 GeV and the charm mass to ± 0.13 GeV, with the expectation that e^+e^- B-factory and LHC data could reduce these uncertainties by a factor of two by the time the ILC data would be analysed. Today, the analysis of a fraction of the BaBar data[41] has already brought these uncertainties down to 0.07 GeV for

m_b and, more importantly, 0.09 GeV for m_c, using the spectral moments technique in semi-leptonic B decays, which had been pioneered on CLEO[42] and DELPHI data.[43] Extrapolating to the anticipated total statistics to be collected at PEP-II and KEKB, we can now confidently expect that the b quark mass should be known to better than ± 0.05 GeV and the charm mass to better than ± 0.06 GeV. This translates into less than ± 0.4 % and ± 6.5 % relative uncertainty in computing the Higgs SM couplings to b and c quarks, respectively, and motivates enhanced experimental precision in the determination of these couplings at the ILC. Detailed simulation shows that these accuracy can be matched by the ILC.[44,47]

While much of the emphasis on the ILC capabilities in the study of the Higgs profile is for a light Higgs scenario, preferred by the current electro-weak data and richer in decay modes, the ILC has also the potential of precisely mapping out the Higgs boson properties for heavier masses. If the Higgs boson turns out to weigh of order 200 GeV, the 95% C.L. upper limit indicated by electro-weak fits, or even heavier, the analysis of the recoil mass in $e^+e^- \to HZ$ at $\sqrt{s} = 0.5$ TeV allows to precisely determine M_H, Γ_H and the Higgs-strahlung cross section. Even for $M_H = 240$ GeV, the mass can be determined to a 10^{-3} accuracy and, more importantly, the total width measured about 10% accuracy. Decays of Higgs bosons produced in $e^+e^- \to H\nu\bar{\nu}$ give access to the Higgs couplings. The importance of the WW-

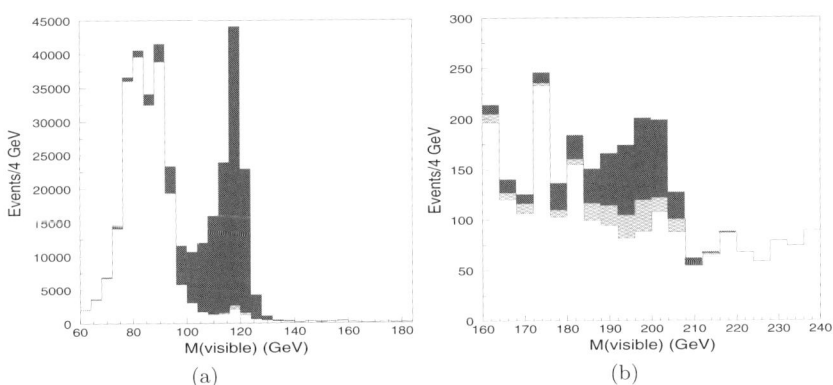

Fig. 2.8. $H \to b\bar{b}$ signal after full event selection at the ILC for (a) $M_H = 120$ GeV and (b) $M_H = 200$ GeV (from Ref. 47).

fusion process $e^+e^- \to H^0\nu\bar{\nu}$ to probe rare Higgs decays at higher energies, emerged in the physics study for a multi-TeV linear collider.[45] Since this cross section increases as $log \frac{s}{M_H^2}$, it becomes dominant around $\sqrt{s} = 1$ TeV. Detailed studies have been performed and show that 1 ab^{-1} of data at $\sqrt{s} = 1$ TeV, corresponding to three to four years of ILC running, can significantly improve the determination of the Higgs couplings, especially for the larger values of M_H.[46,47] WW and ZZ couplings can be determined with relative accuracies of 3 % and 5 % respectively, while the coupling to $b\bar{b}$ pairs, a rare decay with a branching fraction of just 2×10^{-3} at such large masses, can be determined to 4 % to 14 % for 180 GeV $< M_H <$ 220 GeV. This measurement is of great importance, since it would offer the only opportunity to learn about the fermion couplings of such an heavy Higgs boson, and it is unique to a linear collider.

A most distinctive feature of the Higgs mechanism is the shape of the Higgs potential:

$$V(\Phi) = -\frac{\mu^2}{2}\Phi^2 + \frac{\lambda}{4}\Phi^4 \qquad (2.5)$$

with $v = \sqrt{\frac{\mu^2}{\lambda}}$. In the SM, the triple Higgs coupling, $g_{HHH} = 3\lambda v$, is related to the Higgs mass, M_H, through the relation

$$g_{HHH} = \frac{3}{2}\frac{M_H^2}{v}. \qquad (2.6)$$

By determining g_{HHH}, the above relation can be tested. The ILC has access to the triple Higgs coupling through the double Higgs production processes $e^+e^- \to HHZ$ and $e^+e^- \to HH\nu\nu$.[48] Deviations from the SM relation for the strength of the Higgs self-coupling arise in models with an extended Higgs sector.[49] The extraction of g_{HHH} is made difficult by their tiny cross sections and by the dilution effect, due to diagrams leading to the same double Higgs final states, but not sensitive to the triple Higgs vertex. This makes the determination of g_{HHH} a genuine experimental *tour de force*. Other modes, such as $e^+e^- \to HHb\bar{b}$, have also been recently proposed[50] but signal yields are too small to provide any precise data. Operating at $\sqrt{s} = 0.5$ TeV the ILC can measure the HHZ production cross section to about 15% accuracy, if the Higgs boson mass is 120 GeV, corresponding to a fractional accuracy on g_{HHH} of 23%.[51] Improvements can be obtained first by introducing observables sensitive to the presence of the triple Higgs vertex and then by performing the analysis at higher energies where the $HH\nu\bar{\nu}$ channel contributes.[52] In the HHZ process events from diagrams containing the HHH vertex exhibit a lower invariant mass of the HH system compared to double-Higgsstrahlung events. When

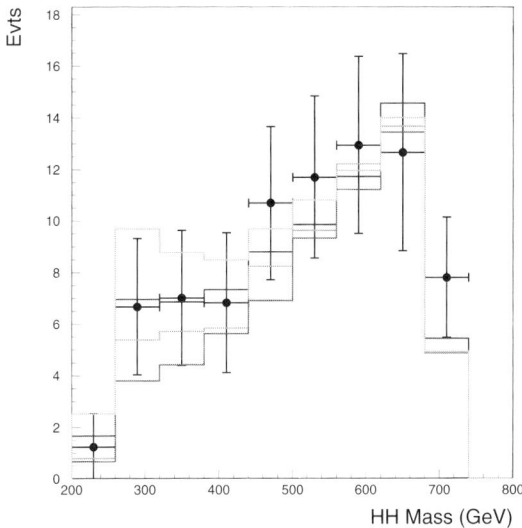

Fig. 2.9. Invariant mass of the HH system in $e^+e^- \to HHZ$ events reconstructed with 1 ab^{-1} of data at 0.8 TeV. The histograms show the predicted distribution for various values of g_{HHH} demonstrating that the low mass region is sensitive to the contribution of the triple Higgs vertex.

the M_{HH} spectrum is fitted, a relative statistical accuracy of ± 0.20 can be obtained with 1 ab^{-1} at $\sqrt{s} = 0.5$ TeV. The availability of beam polarization increases the HHZ cross section by a factor of two and that for $HH\nu\bar{\nu}$ by a factor of four, thus offering a further possible significant improvement to the final accuracy. The ILC and, possibly, a multi-TeV e^+e^- collider represent a unique opportunity for carrying out this fundamental measurement. In fact, preliminary studies show that, the analysis of double Higgs production at the LHC is only possible after a luminosity upgrade and, even then, beyond the observation of double Higgs production, it would provide only a very limited information on the triple-Higgs coupling.[53,54]

2.4.2. Understanding Dark Matter at the ILC

The search for new physics beyond the Standard Model has a central role in the science program of future colliders. It is instructive to contrast the LHC and the ILC in terms of their potential in such searches. Running

Table 2.2. Summary of the accuracies on the determination of the Higgs boson profile at the ILC. Results are given for a 350-500 GeV ILC with $\mathcal{L}$=0.5 ab^{-1}. Further improvements, expected from a 1 TeV ILC are also shown for some of the measurements.

	M_H (GeV)	$\delta(X)/X$ ILC-500 | ILC-1000 0.5 ab^{-1} | 1 ab^{-1}
$\delta M_H/M$	120-180	(3-5) $\times 10^{-4}$
$\delta \Gamma_{tot}/\Gamma$	120-200	0.03- -- | 0.03 - 0.05
$\delta g_{HWW}/g$	120-240	0.01-0.03 | 0.01 - 0.01
$\delta g_{HZZ}/g$	120-240	0.01-0.05
$\delta g_{Htt}/g$	120-200	0.02- -- | 0.06 - 0.13
$\delta g_{Hbb}/g$	120-200	0.01-0.06 | 0.01 - 0.05
$\delta g_{Hcc}/g$	120-140	0.06-0.12
$\delta g_{H\tau\tau}/g$	120-140	0.03-0.05
$\delta g_{H\mu\mu}/g$	120-140	0.15 | 0.04-0.06
CP test	120	0.03
$\delta g_{HHH}/g$	120	0.20 | 0.1

at $\sqrt{s} \leq 1$ TeV the ILC might appear to be limited in reach, somewhere within the energy domain being probed by the Tevatron and that to be accessed by LHC. And yet its potential for fully understanding the new physics, which the LHC might have manifested, and for probing the high energy frontier beyond the boundaries explored in hadron collisions is of paramount importance. There are several examples of how the ILC will be essential for understanding new physics. They address scenarios where signals of physics beyond the SM, as observed at the LHC, may be insufficient to decide on the nature of the new phenomena. One such example, which has been studied in some details, is the case of Supersymmetry and Universal Extra Dimensions (UED), two very different models of new physics leading to the very same experimental signature: fermion pairs plus missing energy. Here, the limited analytical power of the LHC may leave us undecided,[56,57] while a single spin measurement performed at the ILC precisely identifies the nature of the observed particles.[58] But the ILC capability to fully understand the implications of new physics, through fundamental measurements performed with high accuracy, is manifested also in scenarios where the LHC could observe a significant fraction of the new particle spectrum. An especially compelling example, which can be studied quantitatively, is offered by Supersymmetry in relation to Dark Matter (DM). Dark Matter has been established as a major component of the Universe. We know from several independent observations, including the cosmic microwave background (CMB), supernovas (SNs) and galaxy clusters, that DM is responsible for approximately 20 % of the energy density of the universe. Yet, none of the SM particles can be responsible for it and the observation of DM is likely the first direct signal of new physics beyond the SM. Several particles and objects have been nominated as candidates for DM. They span a wide range of masses, from 10^{-5} eV, in the case of axions, to 10^{-5} solar masses, for primordial black holes. Cosmology tells us that a significant fraction of the Universe mass consists of DM, but does not provide clues on its nature. Particle physics tells us that New Physics must exist at, or just beyond, the EW scale and new symmetries may result in new, stable particles. Establishing the inter-relations between physics at the microscopic scale and phenomena at cosmological scale will represent a major theme for physics in the next decades. The ILC will be able to play a key role in elucidating these inter-relations. Out of these many possibilities, there is a class of models which is especially attractive since its existence is independently motivated and DM, at about the observed density, arises naturally. These are extensions of the SM, which include an extra symmetry protecting the lightest particle in the new sector from decaying into ordinary SM states. The lightest particle becomes stable and can be chosen to be neutral. Such a particle is called a weakly interacting massive particle (WIMP) and arises in Supersymmetry with conserved R-parity (SUSY) but also in Extra Dimensions with KK-parity (UED).[66] Current cosmological data, mostly through the WMAP satellite measurements of the CMB, determine the DM density in the Universe with a 6 % relative accuracy.[59] By the next decade, the PLANCK satellite will push this uncertainty to ~ 1 %, or below.[60] Additional astrophysical data manifest a possible evidence of DM annihilation. The EGRET data show excess of γ emission in the inner galaxy, which has been interpreted as due to DM[61] and the WMAP data itself may show a signal of synchrotron emission in the Galactic center.[62] These data, if confirmed, may be used to further constrain the DM properties. Ground-based DM searches are also approaching the stage where their sensitivity is at the level predicted by Supersymmetry for some combinations of parameters.[63] The next decades promise to be a time when accelerator experiments will provide new breakthroughs and highly accurate data to gain new insights, not only on fundamental questions in particle physics, but also in cosmology, when studied alongside the observations from satellites and other experiments. The questions on the nature and the origin of DM offer a prime example of the synergies of new experiments at hadron and lepton colliders, at satellites and ground-based DM experiments.

It is essential to study, in well defined, yet general enough, models, which are the properties of the new physics sector, such as masses and couplings, most important to determine the resulting relic density of the DM particles. Models exist which allow to link the microscopic particle properties to the present DM density in the Universe, with mild assumptions. If DM consists of WIMPs, they are abundantly produced in the very early Universe when $T \simeq (t(\sec))^{-1/2} > 100$ GeV and their interaction cross section is large enough that they were in thermal equilibrium for some period in the early universe. The DM relic density can be determined by solving the Boltzmann equation governing the evolution of their phase space number density.[64] It can be shown that, by taking the WMAP result for the DM relic density in units of the Universe critical density, $\Omega_{DM} h^2$, the thermal averaged DM annihilation cross section times the co-moving velocity, $<\sigma v>$, should be $\simeq 0.9$. From this result, the mass of the DM candidate can be estimated as:

$$M_{DM} = \sqrt{\frac{\pi \alpha^2}{8 <\sigma v>}} \simeq 100 \text{ GeV}. \qquad (2.7)$$

A particle with mass $M = \mathcal{O}(100$ GeV$)$ and weak cross section would naturally give the measured DM density. It is quite suggestive that new physics, responsible for the breaking of electro-weak symmetry, also introduce a WIMP of about that mass. In fact, in essentially every model of electroweak symmetry breaking, it is possible to add a discrete symmetry that makes the lightest new particle stable. Often, this discrete symmetry is required for other reasons. For example, in Supersymmetry, the conserved R parity is needed to eliminate rapid proton decay. In other cases, such as models with TeV-scale extra dimensions, the discrete symmetry is a natural consequence of the underlying geometry.

Data on DM density already set rather stringent constraints on the parameters of Supersymmetry, if the lightest neutralino χ_1^0 is indeed responsible for saturating the amount of DM observed in the Universe. It is useful to discuss the different scenarios, where neutralino DM density is compatible with the WMAP result, in terms of parameter choices in the context of the constrained MSSM (cMSSM), to understand how the measurements that the ILC provides can establish the relation between new physics and

DM. The cMSSM reduces the number of free parameters to just five: the common scalar mass, m_0, the common gaugino mass, $m_{1/2}$, the ratio of the vacuum expectation values of the two Higgs fields, $\tan\beta$, the sign of the Higgsino mass parameter, μ, and the common trilinear coupling, A_0. It is a remarkable feature of this model that, as these parameters, defined at the unification scale, are evolved down to lower energies, the electroweak symmetry is broken spontaneously and masses for the $W^\pm$ and Z^0 bosons generated automatically. As this model is simple and defined by a small number of parameters, it is well suited for phenomenological studies. The cosmologically interesting regions in the m_0 - $m_{1/2}$ parameter plane are shown in Figure 2.10. As we move away from the bulk region, at small values of m_0 and $m_{1/2}$, which is already severely constrained by LEP-2 data, the masses of supersymmetric particles increase and so does the dark matter density. It is therefore necessary to have an annihilation process, which could efficiently remove neutralinos in the early universe, to restore the DM density to the value measured by WMAP. Different processes define three main regions: i) the focus point region, where the χ^0_1 contains an admixture of the supersymmetric partner of a neutral Higgs boson and annihilates to W^+W^- and Z^0Z^0, ii) the co-annihilation region, where the lightest slepton has a mass very close to $M_{\chi^0_1}$, iii) the A annihilation funnel, where $M(\chi^0_1)$ is approximately half that of the heavy A^0 Higgs boson, providing efficient s-channel annihilation, $\chi\chi \to A$. In each of these regions, researchers at the ILC will be confronted with several different measurements and significantly different event signatures.

Table 2.3. cMSSM parameters of benchmark points.

Point	m_0	$m_{1/2}$	$\tan\beta$	A_0	$Sgn(\mu)$	$M(t)$
LCC1	100	250	10	-100	+	178
LCC2	3280	300	10	0	+	175
LCC3	210	360	40	0	+	178
LCC4	380	420	53	0	+	178

It is interesting to observe that the DM constraint, reduces the dimensionality of the cMSSM plane, by one unit, since the allowed regions are tiny lines in the m_0 - $m_{1/2}$ plane, evolve with $\tan\beta$ and depend only very weakly on A_0.[65] Representative benchmark points have been defined and their parameters are summarised in Table 2.3. Even though these points have been defined in a specific supersymmetric model, their phenomenol-

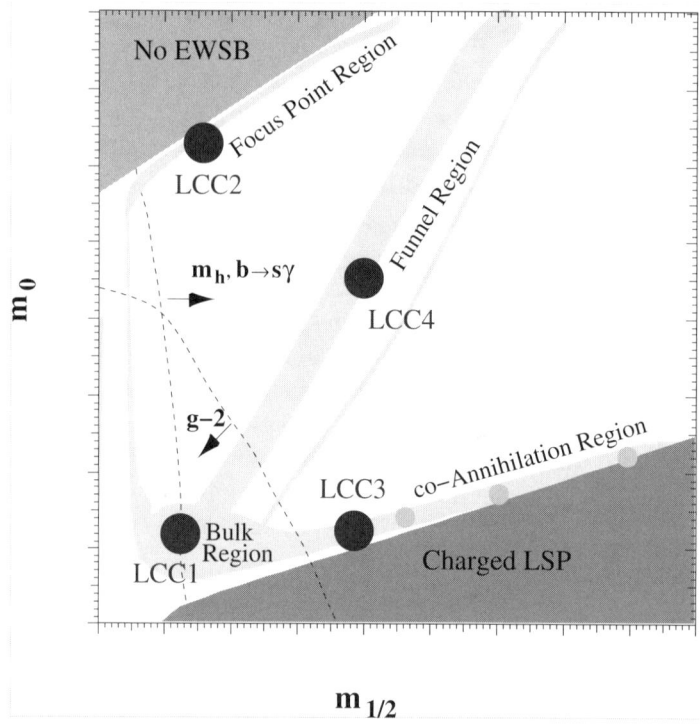

Fig. 2.10. The DM-favoured regions in the m_0 - $m_{1/2}$ plane of the cMSSM and existing constraints. The precise locations of these regions vary with the $\tan\beta$ parameter and therefore the axis are given without units. The indicative locations of the four benchmark points adopted, are also given. Lower limits on the Higgs boson mass and, in a portion of the parameter space, the measurement of the $b \to s\gamma$ decay branching fraction, exclude the region at low values of $m_{1/2}$. A discrepancy of the measured anomalous magnetic moment of the muon value with the SM prediction would favour the region on the left of the curve labeled $g-2$.

ogy is common to the more general supersymmetric solutions and we shall soon discuss the extension of results derived in this constrained model to the general MSSM. There are several features which are common to all these regions. First, the relic density depends on the mass of the lightest neutralino and of few additional particles, close in mass to it. The heavier part of the SUSY spectrum decouples from the value of $\Omega_\chi h^2$. This is of particular importance for the ILC. Running at $\sqrt{s} \leq 1$ TeV, the ILC will not be able to study supersymmetric particles exceeding $\simeq$450-490 GeV, in particular scalar quarks and heavy Higgs bosons in some regions of the parameter phase space. But, independently of the LHC results, the ILC will either observe and measure these particles if they may be relevant to determine the relic DM density, or it will set bounds that ensure their decoupling. A second important observation is that $\Omega_\chi h^2$ typically depends on SUSY parameters which can be fixed by accurate measurements of particle masses, particle mass splittings, decay branching fractions and production cross sections. In some instances the availability of polarised beams is advantageous. The LHC can often make precise measurements of some particles, but it is difficult for the LHC experiments to assemble the complete set of parameters needed to reconstruct annihilation cross section. It is also typical of supersymmetry spectra to contain light particles that may be very difficult to observe in the hadron collider environment. The ILC, in contrast, provides just the right setting to obtain both types of measurements. Again, it is not necessary for the ILC to match the energy of the LHC, only that it provides enough energy to see the lightest charged particles of the new sector.

Rather detailed ILC analyses of the relevant channels for each benchmark point have been performed,[67-70] based on parametric simulation, which includes realistic detector performances and effects of the ILC beam characteristics. It has been assumed that the ILC will be able to provide collisions at centre-of-mass energies from 0.3 TeV to 0.5 TeV with an integrated luminosity of 500 fb^{-1} in a first phase of operation and then its collision energy can raised to 1 TeV to provide an additional data set of 1 ab^{-1}, corresponding to an additional three to four years of running. Results are summarised in terms of the estimated accuracies on masses and mass differences in Table 2.4.

In order to estimate the implications of these ILC measurements on the estimation of neutralino dark DM density $\Omega_\chi h^2$, broad scans of the multi-parameter supersymmetric phase space need to be performed. For each benchmark point, the soft parameters (masses and couplings) at the electroweak scale can be computed with the full 2-loop renormalization group equations and threshold corrections using `Isajet 7.69`.[71] Supersymmetric loop corrections to the Yukawa couplings can also be included. The electroweak-scale MSSM parameters are extracted from the high scale cMSSM parameters. The dark matter density $\Omega_\chi h^2$ can be estimated using the `DarkSUSY`[72] and `Micromegas`[73] programs. These programs use the same `Isajet` code to determine the particle spectrum and couplings, including the running Yukawa couplings, and compute the thermally averaged cross section for neutralino annihilation, including co-annihilation and solve the equation describing the evolution of the number density for the DM candi-

Table 2.4. Summary of the accuracies (in GeV) on the main mass determinations by the ILC at 0.5 TeV for the four benchmark points. Results in [] brackets also include ILC data at 1 TeV.

Observable	LCC1	LCC2	LCC3	LCC4
$\delta M(\tilde{\chi}_1^0)$	± 0.05	± 1.0	± 0.1	[± 1.4]
$\delta M(\tilde{e}_R)$	± 0.05	-	[± 1.0]	[± 0.6]
$\delta M(\tilde{\tau}_1)$	± 0.3	-	± 0.5	± 0.9
$\delta M(\tilde{\tau}_2)$	± 1.1	-	-	-
$\delta(M(\mu_R) - M(\tilde{\chi}_1^0))$	± 0.2		[±0.2]	± 0.6
$\delta(M(\tilde{\tau}_1) - M(\tilde{\chi}_1^0))$	0.3	-	± 1.0	± 1.0
$\delta(M(\tilde{\tau}_2) - M(\tilde{\chi}_1^0))$	± 1.1		[± 3.0]	
$\delta(M(\tilde{\chi}_2^0) - M(\tilde{\chi}_1^0))$	± 0.07	± 0.3	± 0.6	[± 1.8]
$\delta(M(\tilde{\chi}_3^0) - M(\tilde{\chi}_1^0))$	± 4.0	± 0.2	[± 2.0]	[± 2.0]
$\delta(M(\tilde{\chi}_1^+) - M(\tilde{\chi}_1^0))$	± 0.6	± 0.25	[± 0.7]	± 2.0
$\delta(M(\tilde{\chi}_2^+) - M(\tilde{\chi}_1^+))$	[± 3.0]	-	[± 2.0]	± 2.0
$\delta M(A^0)$	[± 1.5]	-	[± 0.8]	[± 0.8]
$\delta\Gamma(A^0)$		-	[± 1.2]	[± 1.2]

date. While the assumptions of the cMSSM are quite helpful for defining a set of benchmark points, the cMSSM is not representative of the generic MSSM, since it implies several mass relations, and its assumptions have no strong physics justification. Therefore, in studying the accuracy on $\Omega_\chi h^2$, the full set of MSSM parameters must be scanned in an uncorrelated way and the mass spectrum evaluated for each parameter set. A detailed study has recently been performed.[74] I summarise here some of the findings, Table 2.5 gives results for the neutralino relic density estimates in MSSM for the LHC, the ILC at 0.5 TeV and the ILC at 1 TeV.

The LCC1 point is in the bulk region and the model contains light sleptons, with masses just above that of the lightest neutralino. The most

Table 2.5. Summary of the relative accuracy $\frac{\delta\Omega_\chi h^2}{\Omega_\chi h^2}$ for the four benchmark points obtained with full SUSY scans.

Benchmark Point	Ωh^2	LHC	ILC 0.5 TeV	ILC 1.0 TeV
LCC1	0.192	0.072	0.018	0.024
LCC2	0.109	0.820	0.140	0.076
LCC3	0.101	1.670	0.500	0.180
LCC4	0.114	4.050	0.850	0.190

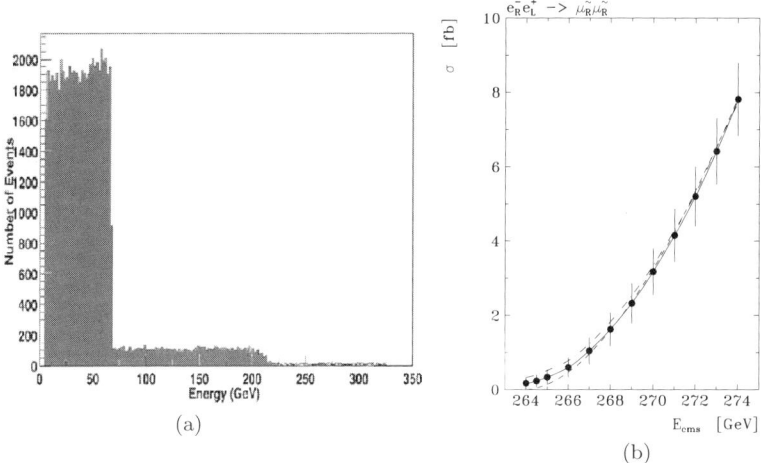

Fig. 2.11. Mass reconstruction at ILC: (a) momentum endpoint in $\tilde{\mu} \to \mu\chi_1^0$ (from Ref. 76) and (b) threshold scan for $e^+e^- \to \tilde{\mu}^+\tilde{\mu}^-$ (from Ref. 23)

important annihilation reactions are those with t-channel slepton exchange. At the LHC, many of the SUSY spectrum parameters can be determined from kinematic constraints. At the ILC masses can be determined both by the two-body decay kinematics of the pair-produced SUSY particles and by dedicated threshold scans. Let us consider the two body decay of a scalar quark $\tilde{q} \to q\chi_1^0$. If the scalar quarks are pair produced $e^+e^- \to \tilde{q}\tilde{q}$, $E_{\tilde{q}} = E_{beam}$ and the χ_1^0 escapes undetected, only the q (and the $\bar{q}$) are observed in the detector. In a 1994 paper, J. Feng and D. Finnell[75] pointed out that the minimum and maximum energy of production for the quark can be related to the mass difference between the scalar quark $\tilde{q}$ and the χ_1^0:

$$E_{max,\,min} = \frac{E_{beam}}{2}\left(1 \pm \sqrt{1 - \frac{m_{\tilde{q}}^2}{E_{beam}^2}}\right)\left(1 - \frac{m_\chi^2}{m_{\tilde{q}}^2}\right). \quad (2.8)$$

The method can also be extended to slepton decays $\tilde{\ell} \to \ell\chi_1^0$, which share the same topology, and allows to determine slepton mass once that of the neutralino is known or determine a relation between the masses and get $m_{\chi_1^0}$ if that of the slepton can be independently measured. The measurement requires a precise determination of the endpoint energies of the lepton momentum spectrum, E_{min} and E_{max}. It can be shown that accuracy is limited by beamstrahlung, affecting the knowledge of E_{beam} in the equation above, more than by the finite momentum resolution, $\delta p/p$ of the detector. The ILC has a second, and even more precise, method for mass measurements. The possibility to precisely tune the collision energy allows to perform scans of the onset of the cross section for a specific SUSY particle pair production process. The particle mass and width can be extracted from a fit to the signal event yield as function of $\sqrt{s}$. The accuracy depends rather weakly on the number of points, N, adopted in the scan and it appears that concentrating the total luminosity at two or three different energies close to the threshold is optimal.[77,78] The mass accuracy, δm can be parametrised as:

$$\delta m \simeq \Delta E \frac{1 + 0.36/\sqrt{N}}{\sqrt{18NL\sigma}} \qquad (2.9)$$

for S-wave processes, where the cross section rises as β and as

$$\delta m \simeq \Delta E \frac{1}{N^{1/4}} \frac{1 + 0.38/\sqrt{N}}{\sqrt{2.6NL\sigma}} \qquad (2.10)$$

for P-wave processes, where the cross section rises as β^3. The combination of these measurements allows the ILC to determine the χ_1^0 mass to ±0.05 GeV, which is two orders of magnitude better than the anticipated LHC accuracy, while the mass difference between the $\tilde{\tau}_1$ and the χ_1^0 can be measured to ±0.3 GeV, which is more than a factor ten better. Extension of ILC operation to 1 TeV gives access to the $e^+e^- \to H^0 A^0$ process. As a result of the precision of these measurements, the ILC data at 0.5 TeV will allow to predict the neutralino relic density to ±2 % and the addition of 1.0 TeV data will improve it to ±0.25 %. It is suggestive that this accuracy is comparable, or better, than that expected by the improved CMB survey by the PLANCK mission. For comparison, the LHC data should provide a ±7 % accuracy. This already a remarkable result, due the fact that, a large number of measurements will be available at the LHC and SUSY decay chains can be reconstructed. Still, the overall mass scale remains uncertain at the LHC. The direct mass measurements on the ILC data remove this uncertainty.

The LCC1 point is characterised by the relatively low SUSY mass scale, most of the particles can be observed at the LHC and their masses accurately measured at the ILC. However, in more general scenarios, the information available from both collider will be more limited. This is the case at

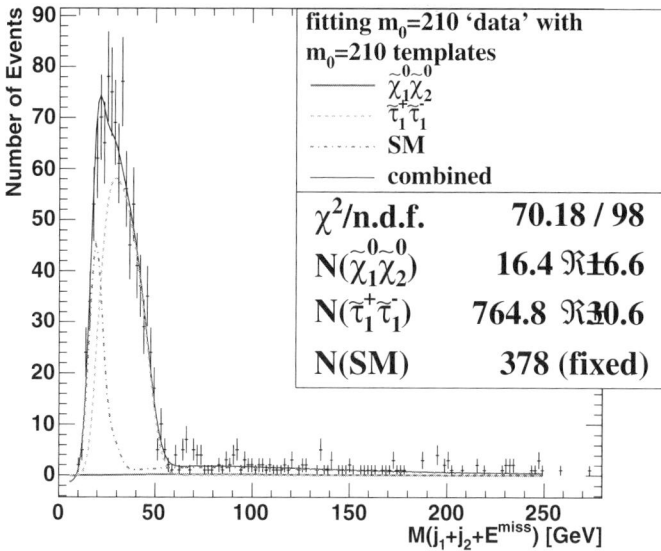

Fig. 2.12. DM-motivated SUSY $\tilde{\tau}$ reconstruction at ILC: determination of the stau-neutralino mass difference from a reconstruction of $e^+e^- \to \tilde{\tau}_1\tilde{\tau}_1$ at 0.5 TeV for LCC3 (from Ref. 69).

the LCC2 point, located in the focus point region, where masses of scalar quarks, sleptons and heavy Higgs bosons are very large, typically beyond the ILC but also the LHC reach, while gauginos masses are of the order of few hundreds GeV, thus within the kinematical domain of the ILC. In this specific scenario, the LHC will observe the SUSY process $\tilde{g} \to q\bar{q}\chi$ and the subsequent neutralino and chargino decays. Still the neutralino relic density can only be constrained within $\pm 40\%$ and the hypothesis $\Omega_\chi h^2 = 0$, namely that the neutralino does not contribute to the observed dark matter density in the universe, cannot be ruled out, based only on LHC data. At a 0.5 TeV collider, the main SUSY reactions are $e^+e^- \to \chi_1^+\chi_1^-$ and $e^+e^- \to \chi_2^0\chi_3^0$. Operation at 1 TeV gives access also to $e^+e^- \to \chi_2^+\chi_2^-$ and $e^+e^- \to \chi_3^0\chi_4^0$. Not only the gaugino mass splittings but also the polarised neutralino and chargino production cross section can be accurately determined at the ILC.[68] These measurements fix the gaugino-Higgsino mixing angles, which play a major role in determining the neutralino relic density. The decoupling of the heavier, inaccessible part of the SUSY spectrum, can be insured with the data at the highest energy. The combined ILC

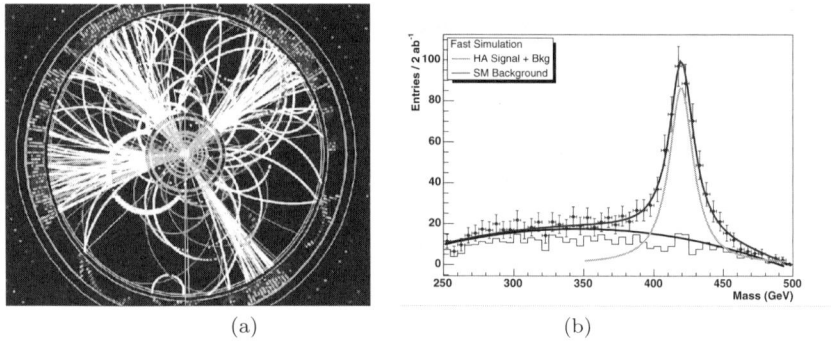

Fig. 2.13. DM-motivated SUSY Higgs reconstruction at ILC: (a) an event $e^+e^- \to A^0 H^0 \to b\bar{b}b\bar{b}$ at 1 TeV in the LDC detector and (b) di-jet invariant mass spectrum for $e^+e^- \to A^0 H^0 \to b\bar{b}b\bar{b}$ at 1 TeV for LCC4 (from Ref. 70).

data at 0.5 TeV and 1 TeV provide an estimate of the neutralino relic density to ± 8 % accuracy, which matches the current WMAP precision. The characteristics featured by the LCC2 point persist, while the SUSY masses increase, provided the gaugino-Higgsino mixing angle remains large enough. This DM-motivated region extends to SUSY masses which eventually exceed the LHC reach, highlighting an intriguing region of parameters where the ILC can still observe sizable production of supersymmetric particle, compatible with dark matter data, while the LHC may report no signals of New Physics.[82]

Instead, the last two points considered, LCC3 and LCC4, are representative of those regions where the neutralino relic density is determined by accidental relationships between particle masses. Other such regions may also be motivated by baryogenesis constraints.[83] The determination of the neutralino relic density, in such scenarios, depends crucially on the precision of spectroscopic measurements, due to the large sensitivity on masses and couplings. The conclusions of the current studies are that the LHC data do not provide quantitative constraints. On the contrary, the ILC can obtain interesting precision, especially when high energy data is available.

The LCC3 point is in the so-called $\tilde{\tau}$ co-annihilation region. Here, the mass difference between the lightest neutralino, χ_1^0, and the lightest scalar tau, $\tilde{\tau}_1$, is small enough that $\tilde{\tau}_1 \chi_1^0 \to \tau\gamma$ can effectively remove neutralinos in the early universe. The relative density of $\tilde{\tau}$ particles to neutralinos scales as $e^{-\frac{m_{\tilde{\tau}} - m_\chi}{m_\chi}}$, so this scenario tightly constrain the $m_{\tilde{\tau}} - m_\chi$ mass difference. Here, the precise mass determinations characteristic of LCC1 will not be

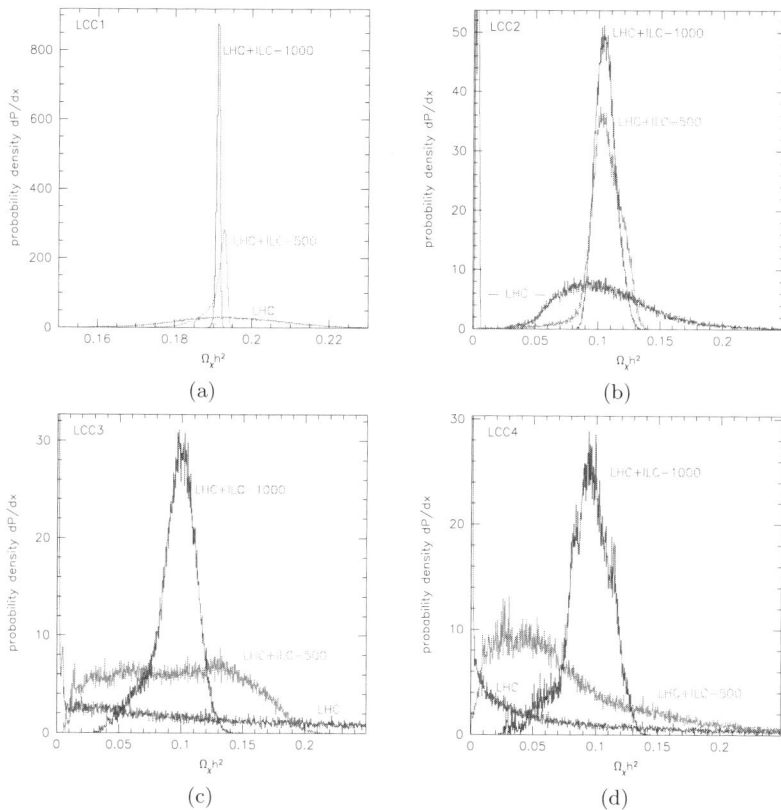

Fig. 2.14. Relic DM density determination based on simulation from LHC, ILC at 0.5 TeV and ILC at 1.0 TeV for the four SUSY benchmark points studied: a) LCC1, b) LCC2, c) LCC3 and d) LCC4. The plots show the probability density functions of the Ωh^2 values corresponding to MSSM points compatible with the accelerator data (from Ref. 74).

available: at 0.5 TeV, the ILC will observe a single final state, $\tau^+\tau^- + E_{missing}$, from the two accessible SUSY processes,[69] $e^+e^- \to \tilde{\tau}_1\tilde{\tau}_1$, $\tilde{\tau} \to \tau\chi_1^0$ and $e^+e^- \to \chi_1^0\chi_2^0$, $\chi_2^0 \to \chi_1^0\tilde{\tau} \to \chi_1^0\chi_1^0\tau\tau$. The signal topology consists of two τ-jets and missing energy. Background processes, such as $e^+e^- \to ZZ$ can be suppressed using cuts on event shape variables. The mass splitting can be determined by a study of the distribution of the invariant mass of the system made by the two τ-jets and the missing energy vector, $M_{j_1 j_2 E_{missing}}$. In this variable, the remaining SM background is confined to low values and the shape and upper endpoint of the $\tilde{\tau}_1\tilde{\tau}_1$ contribution depends on the stau-neutralino mass difference, $\Delta M = M_{\tilde{\tau}_1} - M_{\chi_1^0}$. Templates functions can be generated for different values of ΔM and the mass difference is extracted by a χ^2 fit of these templates to the "data". As the ΔM value decreases, the energy available to the τ leptons decreases. Since τ decays involve neutrinos, additional energy is lost from detection. When the $\tau\tau$ system becomes soft, the four fermion background process $ee \to ee\tau\tau$, the so-called $\gamma\gamma$ background which has cross sections at the nb level, makes its detection increasingly difficult. What makes possible to reject these $\gamma\gamma$ events is the presence of the two energetic primary electrons at small angle w.r.t. the beamline.[79] This is a significant challenge for low angle calorimetry, since the electron has to be detected in an hostile environment populated by a large number of other electrons, of lower energy, arising from pairs created during the bunch collision.[80,81] A detailed study,[69] performed for a statistics of 500 fb^{-1}, shows that values of ΔM as small as 5 GeV can be measured at the ILC, provided the primary electrons can be vetoed down to 17 mrad. In the specific case of the LCC3 point, where the mass splitting, ΔM, is 10.8 GeV, an accuracy of 1 GeV can be achieved. Heavier gauginos, as well as the A^0 boson, become accessible operating the ILC at 1 TeV. These data constrain both the mixing angles and $\tan\beta$. As a result the neutralino relic density can be estimated with an 18 % accuracy. Finally, the LCC4 point, chosen in the A funnel, has the DM density controlled by the $\chi\chi \to A$ process. This point is rather instructive in terms of the discovery-driven evolution of a possible experimental program at the ILC. The ILC can obtain the neutralino and $\tilde{\tau}$ masses at 0.5 TeV, following the same technique as for LCC3. We would also expect LHC experiments to have observed the A^0 boson, but it is unlikely M_A could be determined accurately in pp collisions, since the available observation mode is the decay in τ lepton pairs. At this stage, it would be apparent that the mass relation between the neutralino mass, accurately measured by the ILC at 0.5 TeV, and the A boson mass, from the LHC data, is compatible with $M_A \simeq 2M_\chi$, as required for the s-channel annihilation process to be effective. Three more measurements have to be performed at the ILC: the A^0 mass, M_A, and width Γ_A and the μ parameter, which is accessible through the mass splitting between heavier neutralinos, χ_3^0, χ_4^0 and the lighter χ_1^0, χ_2^0. All these measurements are available by operating the ILC at 1 TeV. M_A and Γ_A can be determined by studying the A^0 production in association with a H^0 boson, in the reaction $e^+e^- \to A^0 H^0 \to b\bar{b}b\bar{b}$. This process results in spectacular events with four b jets, emitted almost symmetrically, due to low energy carried by the heavy Higgs bosons (see Figure 2.13a). The cross section, for the parameters of LCC4 corresponding to $M_A = 419$ GeV, is just 0.9 fb highlighting the need of large luminosity at the highest energy. Jet flavour tagging and event shape analysis significantly reduces the major multi-jet backgrounds, such as WW, ZZ and $t\bar{t}$. The SM $b\bar{b}b\bar{b}$ electro-weak background has a cross section of $\sim$3 fb, but since it includes Z^0 or h^0 as intermediate states it can be efficiently removed by event shape and mass cuts. After event selection, the A^0 mass and width must be reconstructed from the measured di-jet invariant masses. This is achieved by pairing jets in the way that minimises the resulting di-jet mass difference, since the masses of the A and H bosons are expected to be degenerate within a few GeV, and the di-jet masses are computed by imposing constraints on energy and momentum conservation to improve the achievable resolution and gain sensitivity to the boson natural width (see Figure 2.13b). The result is a determination of the A mass to 0.2 % and of its width to $\simeq$15 % if a sample of 2 ab^{-1} of data can be collected. The full set of ILC data provides a neutralino relic density evaluation with 19 % relative accuracy. The full details of how these numbers were obtained can be found in Ref.[74].

SUSY offers a compelling example for investigating the complementarity in the search and discovery of new particles and in the study of their properties at the LHC and ILC. The connection to cosmology, through the study of dark matter brings precise requirements in terms of accuracy and completeness of the anticipated measurements and puts emphasis on scenarios at the edges of the parameter phase space. The interplay of satellite, ground-based and collider experiments in cosmology and particle physics will be unique and it will lead us to learn more about the structure of our Galaxy and of the Universe as well as of the underlying fundamental laws of the elementary particles. This quest will represent an major effort for science in the next several decades. The scenarios discussed above highlight the essential role of the ILC in this context. It will testing whether the particles observed at accelerators are responsible for making up a sizeable fraction of the mass of the Universe, through precision spectroscopic measurements. The data obtained at the ILC will effectively remove most particle physics uncertainties and become a solid ground for studying dark matter in our galaxy through direct and indirect detection experiments.[84]

2.4.3. *Indirect Sensitivity to New Physics at the ILC*

Beyond Supersymmetry there is a wide range of physics scenarios invoking new phenomena at, and beyond, the TeV scale. These may explain the origin of electro-weak symmetry breaking, if there is no light elementary Higgs boson, stabilise the SM, if SUSY is not realised in nature, or embed the SM in a theory of grand unification. The ILC, operating at high energy, represents an ideal laboratory for studying this New Physics in ways that are complementary to the LHC.[85,86] Not only it may directly produce some of the new particles predicted by these theories, the ILC also retains an indirect sensitivity, through precision measurements of virtual corrections to electro-weak observables, when the new particle masses exceed the available centre-of-mass energy.

One of the simplest of such SM extensions consists of the introduction of an additional $U(1)$ gauge symmetry, as predicted in some grand unified theories.[87,88] The extra Z' boson, associated to the symmetry, naturally mixes with the SM Z^0. The mixing angle is already strongly constrained, by precision electroweak data, and can be of the order of few mrad at most, while direct searches at Tevatron for a new Z' boson set a lower limit on its mass around 800 GeV, which may reach 1 TeV by the time the LHC will start searching for such a state. The search for an extended gauge sector offers an interesting framework for studying the ILC sensitivity to scales beyond those directly accessible. It also raises the issue of the discrimination between different models, once a signal would be detected. The main classes of models with additional Z' bosons include E_6 inspired models and left-right models (LR). In the E_6 models, the Z' fermion couplings depend on the angle, θ_6, defining the embedding of the extra $U(1)$ in the E_6 group. At the ILC, the indirect sensitivity to the mass of the new boson, $M_{Z'}$, can be parametrised in terms of the available integrated luminosity, $\mathcal{L}$, and centre-of-mass energy, $\sqrt{s}$. A scaling law for large values of $M_{Z'}$ can be obtained by considering the effect of the $Z' - \gamma$ interference in the two fermion production cross section $\sigma(e^+e^- \to f\bar{f})$ ($\sigma_{f\bar{f}}$ in the following). For $s << M_{Z'}^2$, and assuming the uncertainties $\delta\sigma$ to be statistically dominated, we obtain the following scaling for the difference between the SM cross section and that in presence of the Z', in units of the statistical accuracy:

$$\frac{|\sigma_{f\bar{f}}^{SM} - \sigma_{f\bar{f}}^{SM+Z'}|}{\delta\sigma} \propto \frac{1}{M_{Z'}^2}\sqrt{s\mathcal{L}} \qquad (2.11)$$

from which we can derive that the indirect sensitivity to the Z' mass scales with the square of the centre-of-mass energy and the luminosity as:

$$M_{Z'} \propto (s\mathcal{L})^{1/4}. \qquad (2.12)$$

In a full analysis, the observables sensitive to new physics contribution in two-fermion production are the cross section $\sigma_{f\bar{f}}$, the forward-backward asymmetries $A_{FB}^{f\bar{f}}$ and the left-right asymmetries $A_{LR}^{f\bar{f}}$. The ILC gives us the possibility to study a large number of reactions, $e_R^+e_L^-$, $e_R^+e_R^-$ → ($u\bar{u} + d\bar{d}$), $s\bar{s}$, $c\bar{c}$, $b\bar{b}$, $t\bar{t}$, e^+e^-, $\mu^+\mu^-$, $\tau^+\tau^-$ with final states of well defined flavour and, in several cases, helicity. In order to achieve this, jet flavour tagging is essential to separate b quarks from lighter quarks and c quarks from both b and light quarks. Jet-charge and vertex-charge reconstruction allows then to tell the quark from the antiquark produced in the same event.[89,90] Similarly to LEP and SLC analyses, the forward-backward asymmetry can be obtained from a fit to the flow of the jet charge Q^{jet}, defined as $Q^{jet} = \frac{\sum_i q_i |p_i T|^k}{\sum_i |p_i T|^k}$, where q_i is the particle charge, p_i its momentum, T the jet thrust axis and the sum is extended to all the particles in a given jet. Another possible technique uses the charge of secondary particles to determine the vertex charge and thus the quark charge. The application of this technique to the ILC has been studied in some details in relation to the optimisation of the Vertex Tracker.[91] At ILC energies, the $e^+e^- \to f\bar{f}$ cross sections are significantly reduced, compared to those at LEP and SLC: at 1 TeV the cross section $\sigma(e^+e^- \to b\bar{b})$ is only 100 fb, so high luminosity is essential and new experimental issues emerge. At 1 TeV, the ILC beamstrahlung parameter doubles compared to 0.5 TeV, beam-beam effects becoming important, and the primary e^+e^- collision is accompanied by $\gamma\gamma \to$ hadrons interactions.[92] Being mostly confined in the forward regions, this background may reduce the polar angle acceptance for quark flavour tagging and dilute the jet charge separation using jet charge techniques. The statistical accuracy for the determination of $\sigma_{f\bar{f}}$, $A_{FB}^{f\bar{f}}$ and $A_{LR}^{f\bar{f}}$ has been studied, for $\mu^+\mu^-$ and $b\bar{b}$, taking the ILC parameters at $\sqrt{s} =$ 1 TeV. The additional particles from the $\gamma\gamma$ background cause a broadening of the Q^{jet} distribution and thus a dilution of the quark charge separation. Detailed full simulation and reconstruction is needed to fully understand these effects. Despite these backgrounds, the anticipated experimental accuracy in the determination of the electro-weak observables in two-fermion processes at 1 TeV is of the order of a few percent, confirming the ILC role as the precision machine. Several scenarios of new physics have been investigated.[93,94] The analysis of the cross section and asymmetries at 1 TeV would reveal the existence of an additional Z' boson up to $\simeq$ 6-15 TeV, depending on its couplings. As a comparison the LHC direct sensitivity extends up to approximately 4-5 TeV. The ILC indirect sensitivity also extends to different models on new physics, such as 5-dimensional exten-

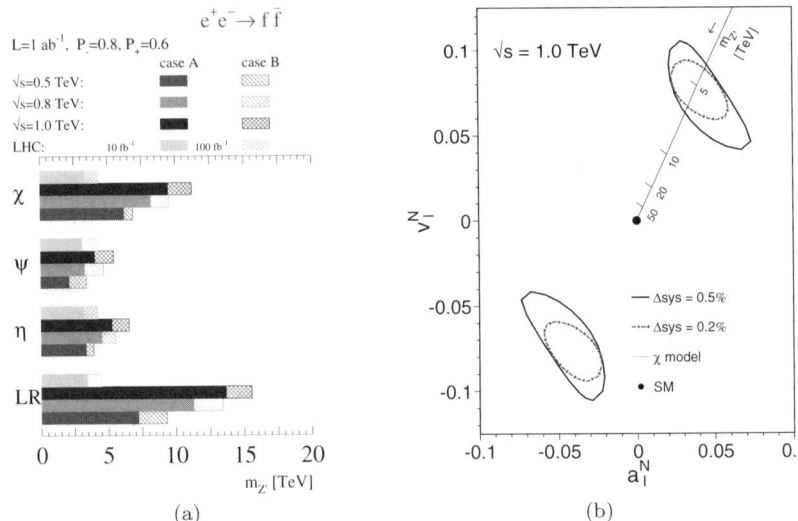

Fig. 2.15. Indirect sensitivity to Z' bosons at ILC: (a) mass sensitivity to different Z' models for 1 ab^{-1} of data at different centre-of-mass energies compared to that of LHC and (b) accuracy on leptonic couplings for a 5 TeV Z' boson (from Ref. 23)

sion of the SM with fermions on the boundary for a compactification where scales up to about 30 TeV can be explored. Finally, fermion compositeness or the exchange of very heavy new particles can be described in terms of effective four-fermion contact interactions.[96] The interaction depends on a scale $\Lambda = M_X/g$, where M_X is the mass of the new particle and g the coupling. Limits to this scale Λ can be set up to $\simeq 100$ TeV, which shows that the ILC sensitivity to new phenomena can exceed its centre-of-mass energy by a significant factor. In order to maximise this indirect sensitivity to new physics, the precision of the SM predictions should match the experimental accuracy. Now, at TeV energies, well above the electroweak scale, the ILC will face the effects of large non-perturbative corrections. Large logarithms $\propto \alpha^n \ log^{2n}(M^2/s)$ arise from the exchange of collinear, soft gauge bosons and are known as Sudakov logarithms.[95] At 1 TeV the logarithmically enhanced W corrections to $\sigma_{b\bar{b}}$, of the form $\alpha \ log^2(M_W^2/s)$ and $\alpha \ log(M_W^2/s)$ amount to 19% and -4% respectively. The effect of these large logarithmic corrections has been studied in some details.[17,97] It will be essential to promote a program of studies to reduce these theoretical uncertainties, to fully exploit the ILC potential in these studies.

2.4.4. Run Plan Scenario

One of the points of strength of the ILC is in its remarkable flexibility of running conditions. Not only the centre-of-mass energy can be changed over approximately an order of magnitude, but the beam particle and their polarization state can be varied to suit the need of the physics processes under study. At the same time, the ILC program is most diversified and data taken at the same centre-of-mass energy may be used for very different analyses, such as precise top mass determination, Higgs boson studies and reconstruction of SUSY decays. This has raised concerns whether the claimed ILC accuracies can be all achieved with a finite amount of data. A dedicated study was performed in 2001, under the guidance of Paul Grannis, taking two physics scenarios with Supersymmetry realised at relatively low mass, one being the LCC1 benchmark point, rich in pair-produced particles and requiring detailed threshold scans.[98] The study assumes a realistic profile for the delivered luminosity, which increases from 10 fb^{-1} in the first year to 200 fb^{-1} in the fifth year and 250 fb^{-1} afterward, for a total integrated equivalent luminosity $\int \mathcal{L} = 1$ ab^{-1}. The proposed run plan starts at the assumed maximum energy of 0.5 TeV for a first determination of the sparticle masses through the end-point study and then scans the relevant thresholds, including $t\bar{t}$ in short runs with tuned polarization states. A summary is given in Table 2.6. This plan devotes approximately

Table 2.6. ILC Run plan scenario for LCC1.

Beams	$\sqrt{s}$ (TeV)	Pol.	$\int \mathcal{L}$ (fb^{-1})	Comments
e^+e^-	0.500	L/R	335	Sit at max. energy for sparticle endpoint measurements
e^+e^-	0.270	L/R	100	Scan $\chi_1^0\chi_2^0$ (R pol.) and $\tilde{\tau}_1\tilde{\tau}_1$ (L pol.)
e^+e^-	0.285	R	50	Scan $\tilde{\mu}_R\tilde{\mu}_R$
e^+e^-	0.350	L/R	40	Scan $t\bar{t}$, $\tilde{e}_R\tilde{e}_L$ (L& R pol.), $\chi_1^+\chi_1^-$ (L pol.)
e^+e^-	0.410	L/R	100	Scan $\tilde{\tau}_2\tilde{\tau}_2$
e^-e^-	0.285	RR	10	Scan for $\tilde{e}_R$ mass

two third of the total luminosity at, or near, the maximum energy, so the program will be sensitive to unexpected new phenomena at high energy, while providing accurate measurements of masses through dedicated scans.

2.5. Sensors and Detectors for the ILC

The development of the ILC accelerator components and the definition of its physics case has been paralleled by a continuing effort in detector design and sensor R&D. This effort is motivated by the need to design and construct detectors which match the ILC promise to provide extremely accurate measurements over a broad range of collision energies and event topologies. It is important to stress that, despite more than a decade of detector R&D for the LHC experiments, much still needs to be done to obtain sensors matching the ILC requirements. While the focus of the LHC-motivated R&D has been on sensor radiation hardness and high trigger rate, the ILC, with its more benign background conditions and lower interaction cross sections, admits sensors of new technology which, in turn, have better granularity, smaller thickness and much improved resolution. Sensor R&D and detector design are being carried out world-wide and are starting deploying prototype detector modules on test beamlines.

2.5.1. *Detector Concepts*

The conceptual design effort for an optimal detector for the ILC interaction region has probed a wide spectrum of options which span from a spherical detector structure to improved versions of more orthodox barrel-shaped detectors. These studies have been influenced by the experience with SLD at the SLC, ALEPH, DELPHI and OPAL at LEP, but also with ATLAS and CMS at the LHC. The emphasis on accurate reconstruction of the particle flow in hadronic events and thus of the energy of partons is common to all designs. The main tracker technology drives the detector designs presently being studied. Four detector concepts have emerged, named GLD, LDC, SiD and 4th Concept.[99] A large volume, 3D continuous tracking volume in a Time Projection Chamber is the centerpiece of the GLD, the LDC and the so-called 4th Concept designs. The TPC is followed by an highly segmented electro-magnetic calorimeter for which these three concepts are contemplating different technologies A discrete tracker made of layers of high precision Silicon microstrip detectors, and a larger solenoidal field, which allows to reduce the radius, and thus the size, of the calorimeter is being studied in the context of the SiD design. Dedicated detector design studies are being carried out internationally[100,101] to optimise, through physics benchmarks,[102] the integrated detector concepts. Such design activities provide a bridge from physics studies to the assessment of priorities

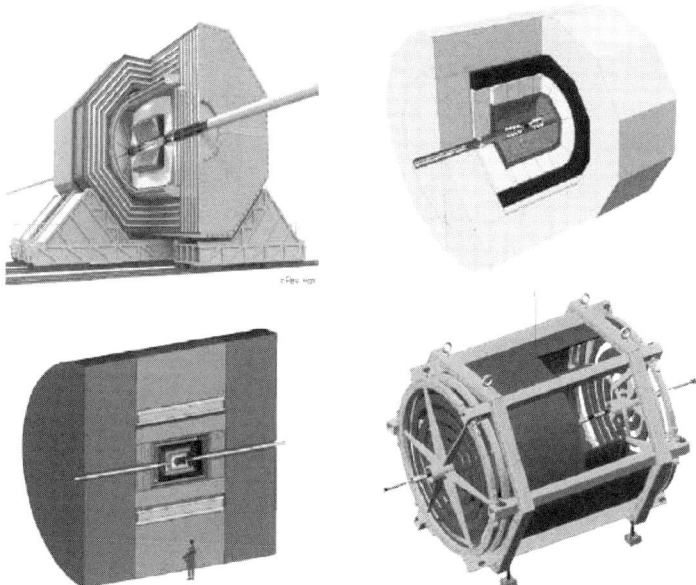

Fig. 2.16. View of the four ILC detector concepts presently being studied: GLD (upper left), LDC (upper right), SiD (lower left) and 4th Concept (lower right).

in detector R&D and are evolving towards the completion of engineered design reports at the end of this decade, synchronously with that foreseen for the ILC accelerator.

2.5.2. Vertexing and Tracking

The vertex and main tracker detectors must provide jet flavour identification and track momentum determination with the accuracy which makes the ILC such a unique facility for particle physics. The resolution in extrapolating charged particle trajectories to their production point, the so-called impact parameter, is dictated by the need to distinguish Higgs boson decays to $c\bar{c}$ from those to $b\bar{b}$ pairs, but also $\tau^+\tau^-$ and gluon pairs, as discussed in section 2.4.1. In addition, vertex charge measurements put emphasis on precise extrapolation of particle tracks down to very low momenta. Tagging of events with multiple b jets, such as $e^+e^- \to H^0 A^0 \to b\bar{b}b\bar{b}$, discussed in section 2.4.2, underscores the need of high tagging efficiency, ϵ_b, since the overall efficiency scales as ϵ_b^N, where N is the number of jets to be tagged. This is best achieved by analysing the secondary vertex structures in hadronic jets. A B meson, from a Higgs boson produced at 0.5 TeV, has an average energy of $x_B \sqrt{s}/4 \simeq 100$ GeV, where $x_B \simeq 0.7$ represents the average b fragmentation function, or a γ value of $\simeq 70$. Since $c\tau \simeq 500$ μm, the average decay distance $\beta\gamma c\tau$ is 3.5 mm and the average impact parameter, $\beta\gamma c\tau \sin\theta$, is 0.5 mm. In comparison, a D meson from a $H \to c\bar{c}$ decay has a decay length of 1.3 mm. More importantly, the average charged decay multiplicity for a B meson is 5.1, while for a D meson is 2.7. Turning these numbers into performance requirements sets the target accuracy for the asymptotic term a and the multiple scattering term b defining the track extrapolation resolution in the formula

$$\sigma_{\text{extrapolation}} = a \oplus \frac{b}{p_t} \quad (2.13)$$

The ILC target values are compared to those achieved by the DELPHI experiment at LEP, those expected for ATLAS at the LHC and the best performance ever achieved at a collider experiment, that of SLD, in Table 2.7. This comparison shows that the improvements required for ILC

Table 2.7. Values for the asymptotic term a and multiple scattering term b defining the track extrapolation resolution required for the ILC compared to those obtained by other collider experiments.

Experiment	a (μm)	b (μm/GeV)
ILC	5	10
DELPHI	28	65
ATLAS	15	75
SLD	8	33

on state-of-the-art technology is a factor 2-5 on asymptotic resolution and another factor 3-7 on the multiple scattering term.

At the ILC, particle tracks in highly collimated jets contribute a local track density on the innermost layer of 0.2-1.0 hits mm^{-2} at 0.5 TeV, to reach 0.4-1.5 hits mm^{-2} at 1.0 TeV. Machine-induced backgrounds, mostly pairs, add about 3-4 hits mm^{-2}, assuming that the detector integrates 80 consecutive bunch crossings in a train. These values are comparable to, or even exceed, those expected on the innermost layer of the LHC detectors: 0.03 hits mm^{-2} for proton collisions in ATLAS and 0.9 hits mm^{-2} for heavy ion collisions in ALICE. Occupancy and point resolution set the pixel size to 20x20 μm^2 or less. The impact parameter accuracy sets the layer material budget to $\leq 0.15\%$ X_0/layer. This motivates the development of thin monolithic pixel sensors. Charge coupled devices (CCD) have been a prototype architecture after the success of the SLD VXD3.[103] However, to match the ILC requirements in terms of radiation hardness and read-out speed significant R&D is needed. New technologies, such as CMOS active pixels,[104] SOI[105] and DEPFET[106] sensors, are emerging as promising, competitive alternatives, supported by an intensive sensor R&D effort promoted for the ILC.[107]

The process $e^+e^- \to H^0 Z^0$, $H^0 \to X$, $Z^0 \to \ell^+\ell^-$ gives access to Higgs production, irrespective of the Higgs decay properties. Lepton momenta must be measured very accurately for the recoil mass resolution to be limited by the irreducible smearing due to beamstrahlung. Since the centre-of-mass energy $\sqrt{s} = E_H + E_Z$ is known and the total momentum $p_H + p_Z = 0$, the Higgs mass, M_H can be written as:

$$M_H^2 = E_H^2 - p_H^2 = (\sqrt{s} - E_Z)^2 - p_Z^2 = s + E_Z^2 - 2\sqrt{s}E_Z - p_Z^2 = s - 2\sqrt{s}E_Z + M_Z^2 \tag{2.14}$$

In the decay $Z^0 \to \mu^+\mu^-$, $E_Z = E_{\mu^+} + E_{\mu^-}$ so that the resolution on M_H depends on that on the muon momentum. In quantitative terms the resolution required is

$$\delta p/p^2 < 2 \times 10^{-5} \tag{2.15}$$

A comparison with the performance of trackers at LEP and LHC is given in Table 2.8. The ability to tag Higgs bosons, independent on their decay mode

Table 2.8. Values for the momentum resolution $\delta p/p^2$ for the main tracker and the full tracking system at ILC, LEP and LHC. These values do not include the vertex constraint.

Experiment	Main Tracker Only	Full Tracker
ILC	1.5×10^{-4}	5×10^{-5}
ALEPH	1.2×10^{-3}	5×10^{-4}
ATLAS	–	2×10^{-4}

is central to the ILC program in Higgs physics. A degraded momentum resolution would correspond to larger background, mostly from $e^+e^- \to$

ZZ^*, being accepted in the Higgs signal sample. This degrades the accuracy on the determination of the Higgs couplings both in terms of statistical and systematic uncertainties. The particle momentum is measured through its bending radius R in the solenoidal magnetic field, B. The error on the curvature, $k = 1/R$, for a particle track of high momentum, measured at N equidistant points with an accuracy, σ, over a length L, applying the constraint that it does originate at the primary vertex (as for the leptons from the Z^0 in the Higgstrahlung reaction) is given by:[108]

$$\delta k = \frac{\sigma}{L^2}\sqrt{\frac{320}{N+4}} \qquad (2.16)$$

This shows that the same momentum resolution can be achieved either by a large number of measurements, each of moderate accuracy, as in the case of a continuous gaseous tracker, or by a small number of points measured with high accuracy, as in the case of a discrete Si tracker. Continuous tracking capability over a large area, with timing information and specific ionization measurement, and its robust performance make the Time Projection Chamber an attractive option for precision tracking at the ILC. The introduction of Micro Pattern Gaseous Detectors[109,110] (MPGD) offers significant improvements in terms of reduced $E \times B$, larger gains, ion suppression and faster, narrower signals providing better space resolution. Improving on the space resolution requires an optimal sampling of the collected charge, while the high solenoidal magnetic field reduces the diffusion effects. Several paths are presently being explored with small size prototypes operated on beamlines and in large magnetic fields.[111,112]

A multi-layered Si strip detector tracker in an high B field may offer a competitive $\delta p/p^2$ resolution with reduced material budget and afford a smaller radius ECAL, thus reducing the overall detector cost. This is the main rationale promoting the development of an all-Si concept for the main tracker, which follows the spirit of the design of the CMS detector at LHC. Dedicated conceptual design and module R&D is being carried out as a world-wide program.[113] There is also considerable R&D required for the engineering of detector ladders, addressing such issues as mechanical stability and integration of cooling and electrical services. These modules may also be considered as supplemental tracking devices in a TPC-based design to provide extra space points, with high resolution, and in endcap tracking planes. Assessing the required detector performance involves realistic simulation and reconstruction code accounting for inefficiencies, noise, overlaps and backgrounds.

2.5.3. *Calorimetry*

The ILC physics program requires precise measurements of multi-jet hadronic events, in particular di-jet invariant masses to identify W, Z and Higgs bosons, through their hadronic decays. An especially demanding reaction is $e^+e^- \to Z^0 H^0 H^0$, which provides access to the triple Higgs coupling as discussed in section 2.4.1. The large background from $e^+e^- \to Z^0 Z^0 Z^0$ can be reduced only by an efficient H^0/Z^0 separation, based on their masses. This impacts the parton energy resolution through the measurement of hadronic jets. Detailed simulation[51] shows that a jet energy resolution $\frac{\sigma_{E_{jet}}}{E_{jet}} \simeq \frac{0.30}{\sqrt{E}}$ is required, in order to achieve an interesting resolution on the g_{HHH} coupling. The analysis of other processes, such as $e^+e^- \to W^+W^-\nu\bar{\nu}$ and Higgs hadronic decays, leads to similar conclusions.[115] In the case of the determination of $H^0 \to W^+W^-$ branching fractions, the statistical accuracy degrades by 22 % when changing the jet energy resolution from $\frac{0.30}{\sqrt{E}}$ to $\frac{0.60}{\sqrt{E}}$. Such performance is unprecedented

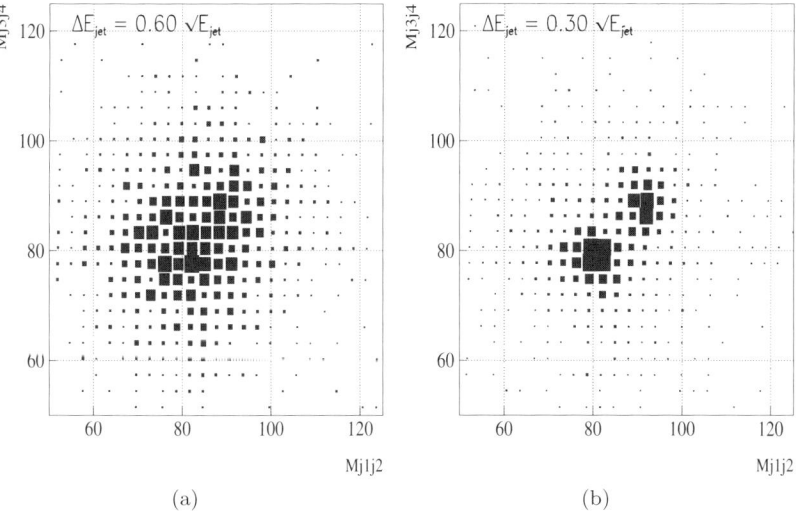

Fig. 2.17. $W^\pm$ and Z^0 gauge boson pair production separation at the ILC: invariant mass of the first di-jet pair vs. that of the second for a sample of WW and ZZ for two different assumptions on the jet energy resolution (a) $\frac{0.30}{\sqrt{E}}$ and (b) $\frac{0.60}{\sqrt{E}}$ (from Ref. 114).

and requires the development of an advanced calorimeter design as well as new reconstruction strategies. The most promising approach is based on the *particle flow algorithm* (PFA). The energy of each particle in an hadronic jet is determined based on the information of the detector which can measure it to the best accuracy. In the case of charged particles, this is achieved by measuring the particle bending in the solenoidal field with the main tracker. Electromagnetic neutrals (γ and π^0) are measured in the electromagnetic calorimeter and hadronic neutrals (K_L^0, n) in the hadronic calorimeter. The jet energy is then obtained by summing these energies:

$$E_{jet} = E_{charged} + E_{em\ neutral} + E_{had\ neutral} \qquad (2.17)$$

each being measured in a specialised detector. The resolution is given by:

$$\sigma^2_{E_{jet}} = \sigma^2_{charged} + \sigma^2_{em\ neutral} + \sigma^2_{had\ neutral} + \sigma^2_{confusion}. \qquad (2.18)$$

Assuming the anticipated momentum resolution, $\sigma_E \simeq 0.11/\sqrt{E}$ for the e.m. calorimeter, $\sigma_E \simeq 0.40/\sqrt{E}$ for the hadronic calorimeter and the fractions of charged, e.m. neutral and hadronic neutral energy in an hadronic jet we get:

$$\sigma^2_{charged} \simeq (0.02\text{GeV})^2 \frac{1}{10} \sum \frac{E^4_{charged}}{(10\text{GeV})^4} \qquad (2.19)$$

$$\sigma^2_{em\ neutral} \simeq (0.6\text{GeV})^2 \frac{E_{jet}}{100\text{GeV}} \qquad (2.20)$$

$$\sigma^2_{had\ neutral} \simeq (1.3\text{GeV})^2 \frac{E_{jet}}{100\text{GeV}} \qquad (2.21)$$

In case of perfect energy-particle association this would correspond to a jet resolution $\simeq 0.14/\sqrt{E}$. But a major source of resolution loss turns out to be the confusion term, $\sigma_{confusion}$, which originates from inefficiencies, double-counting and fakes, which need to be minimised by an efficient pattern recognition. This strategy was pioneered by the ALEPH experiment at LEP, where a resolution $\simeq 0.60/\sqrt{E}$ was obtained, starting from the stochastic resolutions of $\sigma_E \simeq 0.18/\sqrt{E}$ for the e.m. calorimeter, and $\sigma_E \simeq 0.85/\sqrt{E}$ for the hadronic calorimeter.[116] At hadron colliders, the possible improvement from using tracking information together with calorimetric measurements is limited, due to underlying events and the shower core size. On the contrary, at the ILC these limitations can be overcome, by developing an imaging calorimeter, where spatial resolution becomes as important as energy resolution. The minimisation of the confusion rate can then be obtained by choosing a large solenoidal field, B, and calorimeter radius, R, to increase the separation between charged and neutral particles

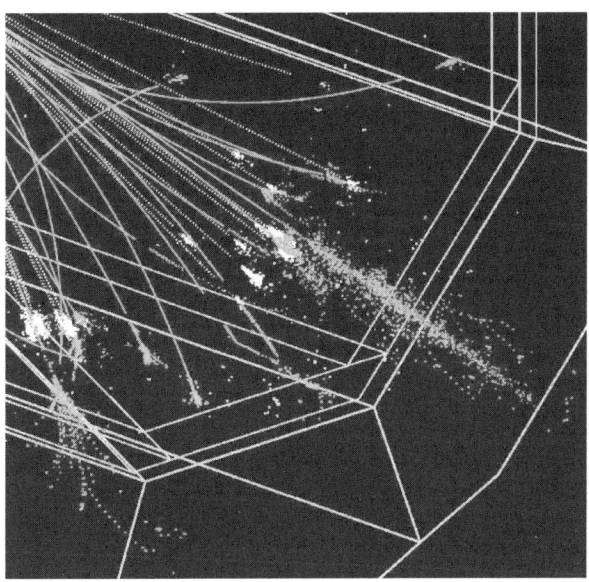

Fig. 2.18. Visualisation of the imaging calorimeter for the ILC: simulated response of a SiW calorimeter to a jet from $e^+e^- \to W^+W^- \to jets$ at $\sqrt{s}$=0.8 TeV.

in dense jets, a small Moliere radius, R_M, for the e.m. calorimeter, to reduce the transverse shower spread and small cells, R_{pixel}, with large longitudinal segmentation. The distance between a neutral and a charged particle, of transverse momentum p_t, at the entrance of the e.m. calorimeter located at a radius R is given by $0.15BR^2/p_t$, where B is the solenoidal magnetic field. A useful figure of merit of the detector in terms of the particle flow reconstruction capability is then offered by:

$$\frac{BR^2}{R_M^2 R_{pixel}^2} \qquad (2.22)$$

which is a measure of the particle separation capability. The value of BR^2 is limited to about 60 Tm^2 by the mechanical stability. An optimal material in terms of Moliere radius is Tungsten, with $R_M = 9$ mm. In four-jet events at $\sqrt{s}$=0.8 TeV, there are on average 28 GeV per di-jet carried by photons, which are deposited within 2.5 cm from a charged particle at the e.m. calorimeter radius. With pixel cells of order of 1×1 cm^2 to ensure sufficient transverse segmentation and 30 to 40 layers in depth, the e.m. calorimeter would consists of up to 30 M channels and 3000 m^2 of active Si. Due to the large amount of channels and the wish to use an absorber with the smallest possible Moliere radius, the e.m. calorimeter is the main cost-driver of the ILC detector and its optimisation in terms of performance and cost requires a significant R&D effort. A Silicon-Tungsten calorimeter (SiW) was first proposed in the framework of the TESLA study[114,117] and it is currently being pursued by large R&D collaborations in both Europe and the US. Alternative technologies are also being studied by the GLD and the 4th Concept. This R&D program involves design, prototyping and tests with high energy particle beams and it is being carried out world-wide,[118–120] supported by efforts on detailed simulation and reconstruction.

2.6. Epilogue

The ILC promises to complement and expand the probe into the TeV scale beyond the LHC capabilities, matching and improving its energy reach while adding precision. Its physics program will address many of the fundamental questions of today's physics from the origin of mass, to the nature of Dark Matter. After more than two decades of intense R&D carried out world-wide, the e^+e^- linear collider, with centre-of-mass energies up to 1 TeV, has become technically feasible and a costed reference design is now available. Detectors matching the precision requirements of its anticipated physics program are being developed in an intense R&D effort carried out world-wide. Now, theoretical predictions matching the anticipated experimental accuracies are crucially needed, as well as further clues on what physics scenarios could be unveiled by signals that the LHC may soon be observing. These will contribute to further define the physics landscape for the ILC. A TeV-scale electron-positron linear collider is an essential component of the research program that will provide in the next decades new insights into the structure of space, time, matter and energy. Thanks to the efforts of many groups from laboratories and universities around the world, the technology for achieving this goal is now in hand, and the prospects for the ILC success are extraordinarily bright.

Acknowledgments

I am grateful to the TASI organisers, in particular to Sally Dawson and Rabindra N. Mohapatra, for their invitation and the excellent organization. I am indebted to many colleagues who have shared with me both the excitement of the ILC physics studies and detector R&D, over many years, as well as many of the results included in this article. I would like to mention here Ugo Amaldi, Timothy Barklow, Genevieve Belanger, Devis Contarato, Stefania De Curtis, Jean-Pierre Delahaye, Albert De Roeck, Klaus Desch, Daniele Dominici, John Ellis, JoAnne Hewett, Konstantin Matchev, Michael Peskin and Tom Rizzo. I am also grateful to Barry Barish, JoAnne Hewett, Mark Oreglia and Michael Peskin for reviewing the manuscript and their suggestions.

This work was supported in part by the Director, Office of Science, of the U.S. Department of Energy under Contract No.DE-AC02-05CH11231.

References

1. M. Tigner, *Nuovo Cim.*, **37** 1228 (1965).
2. U. Amaldi, *Phys. Lett.* **B61** 313 (1976).
3. B. Richter, IEEE Trans. Nucl. Sci. **26**, 4261 (1979).
4. W. Schnell, *A Two Stage Rf Linear Collider Using A Superconducting Drive Linac*, CERN-LEP-RF/86-06 (1976).
5. C. r. Ahn et al., *Opportunities and Requirements for Experimentation at a Very High-Eenergy e^+e^- Collider*, SLAC-0329 (1988).
6. G. Loew (editor), *International Linear Collider Technical Review Committee: Second Report*, SLAC-R-606 (2003).
7. http://www.ligo.caltech.edu/~skammer/ITRP_Home.html
8. http://www.linearcollider.org/
9. http://media.linearcollider.org/rdr_draft_v1.pdf
10. Report of the OECD Consultative Group on High-Energy Physics, June 2002 (http://www.oecd.org/dataoecd/2/32/1944269.pdf)
11. http://www7.nationalacademies.org/bpa/EPP2010.html
12. L. Maiani, prepared for the *9th International Symposium on Neutrino Telescopes*, Venice, Italy, 6-9 March 2001.
13. D. Treille, *Nucl. Phys. Proc. Suppl.* **109B**, 1 (2002).
14. R. Brinkmann, K. Flottmann, J. Rossbach, P. Schmueser, N. Walker and H. Weise (editors), *TESLA: The superconducting electron positron linear collider with an integrated X-ray laser laboratory. Technical design report.*, DESY-01-011B (2001).
15. R. W. Assmann et al., *A 3-TeV e^+e^- linear collider based on CLIC technology*, CERN-2000-008 (2000).
16. W. Wuensch, *Progress in Understanding the High-Gradient Limitations of Accelerating Structures*, CLIC-Note-706 (2007).
17. M. Battaglia, A. De Roeck, J. Ellis and D. Schulte (editors), *Physics at the CLIC multi-TeV linear collider: Report of the CLIC Physics Working Group*, CERN-2004-005 (2004) and arXiv:hep-ph/0412251.
18. W.P. Leemans et al., *Nature Physics* **2** 696 (2006).
19. B. Richter, SLAC-PUB-2854 (1981)
20. U. Amaldi, *Summary talk given at Workshop on Physics at Future Accelerators, La Thuile, Italy, Jan 7-13, 1987*, CERN-EP/87-95 (1987).

21. R. J. Noble, Nucl. Instrum. Meth. A **256**, 427 (1987).
22. H. Murayama and M. E. Peskin, Ann. Rev. Nucl. Part. Sci. **46**, 533 (1996) [arXiv:hep-ex/9606003].
23. J. A. Aguilar-Saavedra et al. [ECFA/DESY LC Physics Working Group], *TESLA Technical Design Report Part III: Physics at an e+e- Linear Collider*, DESY-2001-011C (2001) and arXiv:hep-ph/0106315.
24. T. Abe et al. [American Linear Collider Working Group], *Linear collider physics resource book for Snowmass 2001*, SLAC-R-570 (2001).
25. K. Abe et al. [ACFA Linear Collider Working Group], *Particle physics experiments at JLC*, KEK-REPORT-2001-11 (2001) and arXiv:hep-ph/0109166.
26. S. Dawson and M. Oreglia, Ann. Rev. Nucl. Part. Sci. **54**, 269 (2004) [arXiv:hep-ph/0403015].
27. P.W. Higgs, *Phys. Rev. Lett.* **12** 132 (1964); idem, *Phys. Rev.* **145** 1156 (1966); F. Englert and R. Brout, *Phys. Rev. Lett.* **13** 321 (1964); G.S. Guralnik, C.R. Hagen and T.W. Kibble, *Phys. Rev. Lett.* **13** 585 (1964).
28. A. Hasenfratz et al., *Phys. Lett.* **B199** 531 (1987); M. Lüscher and P. Weisz, *Phys. Lett.* **B212** 472 (1988); M. Göckeler et al., *Nucl. Phys.* **B404** 517 (1993).
29. R. Barate et al. [LEP Working Group for Higgs boson searches], *Phys. Lett.* B **565**, 61 (2003) [arXiv:hep-ex/0306033].
30. LEP Electroweak Working Group, Report CERN-PH-EP-2006 (2006), arXiv:hep-ex/0612034 and subsequent updates available at http://lepewwg.web.cern.ch/LEPEWWG/.
31. S. Heinemeyer et al., arXiv:hep-ph/0511332.
32. P. Garcia-Abia, W. Lohmann and A. Raspereza, Note LC-PHSM-2000-062 (2000).
33. D. J. Miller, S. Y. Choi, B. Eberle, M. M. Muhlleitner and P. M. Zerwas, Phys. Lett. B **505**, 149 (2001) [arXiv:hep-ph/0102023].
34. M. Schumacher, Note LC-PHSM-2001-003 (2001).
35. A. Djouadi, M. Spira and P. M. Zerwas, Z. Phys. C **70**, 427 (1996) [arXiv:hep-ph/9511344].
36. M. D. Hildreth, T. L. Barklow and D. L. Burke, Phys. Rev. D **49**, 3441 (1994).
37. M. Carena, H. E. Haber, H. E. Logan and S. Mrenna, Phys. Rev. D **65**, 055005 (2002) [Erratum-ibid. D **65**, 099902 (2002)] [arXiv:hep-ph/0106116].
38. K. Desch, E. Gross, S. Heinemeyer, G. Weiglein and L. Zivkovic, JHEP **0409**, 062 (2004) [arXiv:hep-ph/0406322].
39. M. Battaglia, D. Dominici, J. F. Gunion and J. D. Wells, arXiv:hep-ph/0402062.
40. M. Battaglia, arXiv:hep-ph/9910271.
41. B. Aubert et al. [BABAR Collaboration], *Phys. Rev. Lett.* **93**, 011803 (2004) [arXiv:hep-ex/0404017].
42. C. W. Bauer, Z. Ligeti, M. Luke and A. V. Manohar, *Phys. Rev.* **D67**, 054012 (2003) [arXiv:hep-ph/0210027].
43. M. Battaglia et al., *Phys. Lett.* **B556**, 41 (2003) [arXiv:hep-ph/0210319].

44. T. Kuhl, prepared for the *International Conference on Linear Colliders (LCWS 04)*, Paris, France, 19-24 April 2004.
45. M. Battaglia and A. De Roeck, arXiv:hep-ph/0211207.
46. M. Battaglia, arXiv:hep-ph/0211461.
47. T. L. Barklow, arXiv:hep-ph/0312268.
48. A. Djouadi, W. Kilian, M. Muhlleitner and P. M. Zerwas, *Eur. Phys. J.* **C10** (1999) 27 [arXiv:hep-ph/9903229].
49. S. Kanemura, Y. Okada, E. Senaha and C. P. Yuan, Phys. Rev. D **70**, 115002 (2004) [arXiv:hep-ph/0408364].
50. A. Gutierrez-Rodriguez, M. A. Hernandez-Ruiz and O. A. Sampayo, arXiv:hep-ph/0601238.
51. C. Castanier, P. Gay, P. Lutz and J. Orloff, arXiv:hep-ex/0101028.
52. M. Battaglia, E. Boos and W. M. Yao, in *Proc. of the APS/DPF/DPB Summer Study on the Future of Particle Physics (Snowmass 2001)* ed. N. Graf, E3016, [arXiv:hep-ph/0111276].
53. U. Baur, T. Plehn and D. L. Rainwater, *Phys. Rev.* **D67**, 033003 (2003) [arXiv:hep-ph/0211224].
54. U. Baur, T. Plehn and D. L. Rainwater, *Phys. Rev.* **D69**, 053004 (2004) [arXiv:hep-ph/0310056].
55. T. L. Barklow, arXiv:hep-ph/0411221.
56. A. Datta, K. Kong and K. T. Matchev, Phys. Rev. D **72**, 096006 (2005) [Erratum-ibid. D **72**, 119901 (2005)] [arXiv:hep-ph/0509246].
57. J. M. Smillie and B. R. Webber, JHEP **0510**, 069 (2005) [arXiv:hep-ph/0507170].
58. M. Battaglia, A. Datta, A. De Roeck, K. Kong and K. T. Matchev, JHEP **0507**, 033 (2005) [arXiv:hep-ph/0502041].
59. D. N. Spergel et al. [WMAP Collaboration], Astrophys. J. Suppl. **148**, 175 (2003) [arXiv:astro-ph/0302209]
60. J. R. Bond, G. Efstathiou and M. Tegmark, Mon. Not. Roy. Astron. Soc. **291**, L33 (1997) [arXiv:astro-ph/9702100]
61. W. de Boer, C. Sander, V. Zhukov, A. V. Gladyshev and D. I. Kazakov, Astron. Astrophys. **444**, 51 (2005) [arXiv:astro-ph/0508617].
62. D. P. Finkbeiner, arXiv:astro-ph/0409027.
63. D. S. Akerib *et al.* [CDMS Collaboration], Phys. Rev. Lett. **96**, 011302 (2006) [arXiv:astro-ph/0509259].
64. R. J. Scherrer and M. S. Turner, Phys. Rev. D **33**, 1585 (1986) [Erratum-ibid. D **34**, 3263 (1986)].
65. M. Battaglia, A. De Roeck, J. R. Ellis, F. Gianotti, K. A. Olive and L. Pape, Eur. Phys. J. C **33**, 273 (2004) [arXiv:hep-ph/0306219].
66. K. Kong and K. T. Matchev, JHEP **0601**, 038 (2006) [arXiv:hep-ph/0509119].
67. G. Weiglein *et al.* [LHC/LC Study Group], arXiv:hep-ph/0410364.
68. R. Gray *et al.*, arXiv:hep-ex/0507008.
69. V. Khotilovich, R. Arnowitt, B. Dutta and T. Kamon, Phys. Lett. B **618**, 182 (2005) [arXiv:hep-ph/0503165].
70. M. Battaglia, arXiv:hep-ph/0410123.

71. F. E. Paige, S. D. Protopescu, H. Baer and X. Tata, arXiv:hep-ph/0312045.
72. P. Gondolo, J. Edsjo, P. Ullio, L. Bergstrom, M. Schelke and E. A. Baltz, JCAP **0407**, 008 (2004) [arXiv:astro-ph/0406204].
73. G. Belanger, F. Boudjema, A. Pukhov and A. Semenov, arXiv:hep-ph/0607059.
74. E. A. Baltz, M. Battaglia, M. E. Peskin and T. Wizansky, Phys. Rev. D **74**, 103521 (2006) [arXiv:hep-ph/0602187].
75. J. L. Feng and D. E. Finnell, Phys. Rev. D **49**, 2369 (1994) [arXiv:hep-ph/9310211].
76. G. A. Moortgat-Pick et al., arXiv:hep-ph/0507011, based on work of U. Nauenberg et al..
77. G. A. Blair, in Proc. of the APS/DPF/DPB Summer Study on the Future of Particle Physics (Snowmass 2001) ed. N. Graf, E3019.
78. H. U. Martyn and G. A. Blair, Note LC-TH-2000-023.
79. P. Bambade, M. Berggren, F. Richard and Z. Zhang, arXiv:hep-ph/0406010.
80. P. Chen and V. I. Telnov, Phys. Rev. Lett. **63**, 1796 (1989).
81. T. Tauchi, K. Yokoya and P. Chen, Part. Accel. **41**, 29 (1993).
82. H. Baer, A. Belyaev, T. Krupovnickas and X. Tata, JHEP **0402**, 007 (2004) [arXiv:hep-ph/0311351].
83. C. Balazs, M. Carena and C. E. M. Wagner, Phys. Rev. D **70**, 015007 (2004) [arXiv:hep-ph/0403224].
84. J. L. Feng, in Proc. of the *2005 Int. Linear Collider Workshop (LCWS 2005)*, Stanford, California, 18-22 Mar 2005, pp 0013 and [arXiv:hep-ph/0509309].
85. M. Battaglia et al., in *Physics and Experiments with Future Linear e^+e^- Colliders*, (A. Para and H.E. Fisk editors), AIP Conference Proceedings, New York, 2001, 607 [arXix:hep-ph/0101114].
86. D. Dominici, arXiv:hep-ph/0110084.
87. J. L. Hewett, arXiv:hep-ph/9308321.
88. T. G. Rizzo, arXiv:hep-ph/0610104.
89. K. Ackerstaff et al. [OPAL Collaboration], Z. Phys. C **75**, 385 (1997).
90. K. Abe et al. [SLD Collaboration], Phys. Rev. Lett. **94**, 091801 (2005) [arXiv:hep-ex/0410042].
91. S. Hillert [LCFI Collaboration], *In the Proceedings of 2005 International Linear Collider Workshop (LCWS 2005), Stanford, California, 18-22 Mar 2005, pp 0313*.
92. P. Chen, T. L. Barklow and M. E. Peskin, Phys. Rev. D **49**, 3209 (1994) [arXiv:hep-ph/9305247].
93. S. Riemann, arXiv:hep-ph/9710564.
94. M. Battaglia, S. De Curtis, D. Dominici and S. Riemann, in *Proc. of the APS/DPF/DPB Summer Study on the Future of Particle Physics (Snowmass 2001)* ed. N. Graf, E3020, [arXiv:hep-ph/0112270].
95. M. Melles, Phys. Rept. **375**, 219 (2003) [arXiv:hep-ph/0104232].
96. E. Eichten, K. D. Lane and M. E. Peskin, Phys. Rev. Lett. **50**, 811 (1983).
97. P. Ciafaloni and D. Comelli, Phys. Lett. B **476**, 49 (2000) [arXiv:hep-ph/9910278].

98. M. Battaglia et al., in *Proc. of the APS/DPF/DPB Summer Study on the Future of Particle Physics (Snowmass 2001)* ed. N. Graf, E3006, [arXiv:hep-ph/0201177].
99. http://physics.uoregon.edu/~lc/wwstudy/concepts/
100. T. Behnke, *In the Proceedings of 2005 International Linear Collider Workshop (LCWS 2005), Stanford, California, 18-22 Mar 2005, pp 0006.*
101. K. Abe et al. [GLD Concept Study Group], arXiv:physics/0607154.
102. M. Battaglia, T. Barklow, M. Peskin, Y. Okada, S. Yamashita and P. Zerwas, *In the Proceedings of 2005 International Linear Collider Workshop (LCWS 2005), Stanford, California, 18-22 Mar 2005, pp 1602* [arXiv:hep-ex/0603010].
103. T. Abe [SLD Collaboration], Nucl. Instrum. Meth. A **447** (2000) 90 [arXiv:hep-ex/9909048].
104. R. Turchetta et al., Nucl. Instrum. Meth. A **458** (2001) 677.
105. J. Marczewski et al., Nucl. Instrum. Meth. A **549** (2005) 112.
106. R. H. Richter et al., Nucl. Instrum. Meth. A **511** (2003) 250.
107. M. Battaglia, Nucl. Instrum. Meth. A **530**, 33 (2004) [arXiv:physics/0312039].
108. W.-M. Yao et al, J. Phys. G **33**, 1 (2006)
109. Y. Giomataris, P. Rebourgeard, J. P. Robert and G. Charpak, Nucl. Instrum. Meth. A **376**, 29 (1996).
110. F. Sauli, Nucl. Instrum. Meth. A **386**, 531 (1997).
111. S. Kappler et al., IEEE Trans. Nucl. Sci. **51**, 1039 (2004).
112. P. Colas et al., Nucl. Instrum. Meth. A **535**, 506 (2004).
113. J. Kroseberg et al., arXiv:physics/0511039.
114. T. Behnke, S. Bertolucci, R. D. Heuer and R. Settles, *TESLA Technical design report. Pt. 4: A detector for TESLA* DESY-01-011 (2001).
115. J. C. Brient and H. Videau, in *Proc. of the APS/DPF/DPB Summer Study on the Future of Particle Physics (Snowmass 2001)* ed. N. Graf, E3047, [arXiv:hep-ex/0202004].
116. D. Buskulic et al. [ALEPH Collaboration], Nucl. Instrum. Meth. A **360**, 481 (1995).
117. H. Videau, *Prepared for 5th International Linear Collider Workshop (LCWS 2000), Fermilab, Batavia, Illinois, 24-28 Oct 2000*
118. D. Strom et al., IEEE Trans. Nucl. Sci. **52**, 868 (2005).
119. G. Mavromanolakis, *In the Proceedings of 2005 International Linear Collider Workshop (LCWS 2005), Stanford, California, 18-22 Mar 2005, pp 0906* [arXiv:physics/0510181].
120. D. Strom et al., *In the Proceedings of 2005 International Linear Collider Workshop (LCWS 2005), Stanford, California, 18-22 Mar 2005, pp 0908.*

Chapter 3

Astrophysical Aspects of Neutrinos

John F. Beacom

Center for Cosmology and Astro-Particle Physics,
Departments of Physics and Astronomy,
191 W. Woodruff Ave., Columbus, OH 43210, USA

Neutrino astronomy is on the verge of discovering new sources, and this will lead to important advances in astrophysics, cosmology, particle physics, and nuclear physics. This paper is meant for non-experts, so that they might understand the basic issues in this field.

3.1. General Introduction

It has long been appreciated that neutrino astronomy would have unique advantages. The principal one, due to the weak interactions of neutrinos, is that they would be able to penetrate even great column densities of matter. This could be in dense sources themselves, like stars, supernovae, or active galactic nuclei. It could also be across the universe itself. Of course, the small interaction cross section is also the curse of neutrino astronomy, and to date, only two extraterrestrial sources have been observed: the Sun, and Supernova 1987A. That's it.

However, a new generation of detectors is coming online, and their capabilities are significantly better than anything built before. Additionally, a great deal of theoretical effort, taking advantage of the very rapid increases in the quality and quantity of astrophysical data, has refined estimates of predicted fluxes. The basic message is that the detector capabilities appear to have nearly met the theoretical predictions, and that the next decade should see several exciting first discoveries.

For these two talks, I was asked to introduce the topics of supernova neutrinos and high-energy neutrinos. See the other talks in this volume for more about these and related topics. To increase the probability of this paper being read, I have condensed the material covered in my computer presentation, focusing on the basic framework instead of the details. In the actual lectures, I made extensive use of the blackboard, and of interaction with the students through questions from them (and to them). It isn't possible to represent that here. I thank the students for their active participation, and hope that they've all solved the suggested problems!

3.2. PART ONE: Supernova Neutrinos

3.2.1. *Preamble*

Over the centuries, supernovae, which appear as bright stars and then disappear within a few months, have amazed and confused us. We're still amazed, and as Fermi said, we're still confused, just on a higher level. The historical observations of supernovae were of rare objects in our own Milky Way Galaxy (here and elsewhere, "Galaxy" is used for the Milky Way, and "galaxy" for the generic case). Now that we know their distances, we know that supernovae are extremely luminous in the optical, in fact comparable to the starlight from the whole host galaxy. But that's not the half of it, literally. If you had neutrino-detecting eyes, you'd see the neutrino burst from a single core-collapse supernova outshine the steady-state neutrino emission from all the stars in a galaxy (the analog of solar neutrinos) by a factor more like 10^{15} (that's a lot!). This is what enabled the detection of about 20 neutrinos from Supernova (SN) 1987A, despite its great distance.

A good general rule in decoding physical processes is "Follow the energy," much like "Follow the money" for understanding certain human endeavors. For core-collapse supernovae, this means the neutrinos, while for thermonuclear supernovae, this means the gamma rays. These are the direct messengers that reveal the details of the explosions. In the following, I'll discuss this in more detail, mostly focusing on the "observational" perspective, since it's easy to be convinced that observing these direct messengers is important, while hard to think of how to actually do it. As I will emphasize, this is much more than just astronomy for its own sake: these data play a crucial role in testing the properties of neutrinos, and more generally, in probing light degrees of freedom beyond the Standard Model.

3.2.2. *Introduction*

Stars form from the collapse and fragmentation of gas clouds, and empirically, the stellar Initial Mass Function is something like $dn/dm \sim m^{-2.35}$, where $m = M_{\text{star}}/M_{\text{sun}}$, as first pointed out by Salpeter in 1955, and refined by many authors since. You'll notice that this distribution is not renormalizable, but don't start worrying about dimensional regularization – a simple cutoff near $m = 0.1$ is enough for our purposes. What is the fate of these stars? There are two interesting broad categories. The "types" are observational distinctions, based on spectral lines, but the divisions below are based on the physical mechanisms.

- **Thermonuclear (Type Ia) supernovae**
  These have progenitors with $m \sim 3\text{--}8$, and live for $\sim$ Gyr. The interesting case is when the progenitor has ended its nuclear fusion processes at the stage of being a carbon/oxygen white dwarf, while it has a binary companion that donates mass through accretion. Once the mass of the progenitor grows above the Chandrasekhar mass of $m = 1.4$, this carbon and oxygen will explosively burn all the way up to elements near iron, generating a tremendous amount of energy. The most important isotope produced is ^{56}Ni, which decays to ^{56}Co with $\tau = 9$ days, which then decays to stable ^{56}Fe with $\tau = 110$ days. These decays produce MeV gamma rays and positrons that power the optical light curve. Indeed, a plot of luminosity versus time directly shows the two exponential components.

- **Core-collapse (Type II/Ib/Ic) supernovae**
  These have progenitors with $m \sim 8\text{--}40$, and live for less than ~ 0.1 Gyr. Importantly, the dynamics depend only on single stars, and not whether they happen to be in binaries or not. As you know, the source of stellar energy is nuclear fusion reactions, which burn light elements into progressively heavier ones, until elements near iron are reached, and the reactions stop being exothermic. Until that point, as each nuclear fuel is exhausted, the star contracts until the core is hot and dense enough to ignite the next one (remember, these reactions are suppressed by the Coulomb barrier). The cutoff of $m \sim 8$ denotes the requirement of being able to burn all the way up to iron. So what happens at that point? Once there is a $m \sim 1.4$ iron core, it is no longer generating nuclear energy, but it could support itself by electron degeneracy pressure, except for the fact that the massive envelope of the star is weighing down on it. As discussed below, this leads to the collapse of the core and the formation of a hot and dense proto-neutron star, which cools primarily by neutrino emission over a timescale of seconds.

In both cases, a tremendous amount of energy is released in a time that is very short compared to the lifetimes of stars, the resulting optical displays are crudely similar, and shell remnants are left behind. For thermonuclear supernovae, the source of the energy is nuclear fusion reactions, primarily revealed by the gamma rays from nuclear decays. The neutrino emission is subdominant, and no compact remnant is left behind. For core-collapse supernovae (often referred to as type-II supernovae, even when this is inclusive of types Ib and Ic as well), the source of the energy is gravitational, and is primarily revealed by the neutrinos emitted from the newly-formed neutron star (which may ultimately become a black hole). There is also gamma-ray emission, but it is subdominant compared to the neutrinos. Finally, one interesting fact is that for both categories of supernova, the explosion energy is about 10^{51} erg, known as 1 "f.o.e." (fifty-one erg) or 1 "Bethe." Note that this is about $10^{-3} Mc^2$ for 1 solar mass of material.

The neutrino and gamma-ray emissions from supernovae could in principle be detected from individual objects, or as diffuse glows from all past supernovae. Although low-mass stars are much more common than high-mass stars, type Ia supernovae are more rare than core-collapse supernovae by a factor of several, due to the requirement of being in a suitable binary. Before we get into details, here's where things stand on observations of the direct messengers.

- **Gamma rays from thermonuclear supernovae**
  These have never been robustly detected from individual objects, though in a few cases the COMPTEL instrument set interesting limits. While a diffuse background of gamma rays is seen in the MeV range (and beyond), it is now thought that supernovae do not contribute significantly, making it more of a mystery what does.

- **Neutrinos from core-collapse supernovae**
  These have been seen just once, from SN 1987A, but only with about 20 events. No diffuse background of neutrinos has been seen yet, placing interestingly tight limits on the contribution from supernovae.

For particle physicists, the primary interest is on two points. If neutrinos have unexpected properties, or if there are new light particles that effectively carry away energy, then the neutrino emission per supernova could be altered. If there are processes in the universe that produce MeV gamma rays, directly or after redshifting, e.g., dark matter decay, then these may explain the observed gamma-ray background.

Now let's turn to the basics of the neutrino emission from core-collapse supernovae. The gravitational binding energy release can be simply estimated. The gravitational self-energy of a constant-density sphere is $(3/5)G_N M^2/R$, and so

$$\Delta E \simeq \frac{3}{5}\frac{G_N M_{NS}^2}{R_{NS}} - \frac{3}{5}\frac{G_N M_{NS}^2}{R_{core}} \simeq 3 \times 10^{53} \text{ erg} \simeq 2 \times 10^{59} \text{ MeV},$$

using the observed facts that neutron stars have masses of about $m = 1.4$ and radii of about 10 km. Note that the second term in the difference is negligible. This is a tremendous amount of energy, trapped inside a very dense object, and so no particles can escape and carry away energy except neutrinos. In fact, even the neutrinos must diffuse out, as the density is high enough to counteract the smallness of their interaction cross sections.

The core collapses until it reaches near-nuclear densities, at which point it cannot proceed further, and hitting this wall creates an outgoing shock. If successful, the shock will propagate though the envelope of the star, lifting it off and creating the optical supernova. If not, it will stall, and then the inflow of further material will lead to black hole formation and no optical supernova.

The neutrinos are emitted from the core, within seconds of the collapse, and carry nearly the full binding energy release noted above. It takes perhaps hours or days for the shock to break through the envelope and begin the optical supernova, which is then bright for months. Importantly, the neutrinos are received *before* the light. It's not that they are tachyons, but rather just that they were emitted first. The kinetic energy of the supernova ejecta is only $\sim 1\%$ of the total energy, and the energy in the optical emission is even less. The neutrinos are the most interesting, since they carry most of the energy, are emitted in the shortest and earliest time, and come from the densest regions. Other than gravitational waves, which have yet to be observed, only neutrinos can reveal the inner dynamics of the core collapse process.

As noted, the neutrinos diffuse through the proto-neutron star, meaning that they leave on a longer timescale and with lower energies than they would if it were less dense. It is typically assumed that the neutrino emission per flavor (all six, counting neutrinos and antineutrinos) is comparable. That is, each takes about 1/6 of the binding energy, and has thermal spectral with average energies of 10–20 MeV. There is a vast literature about

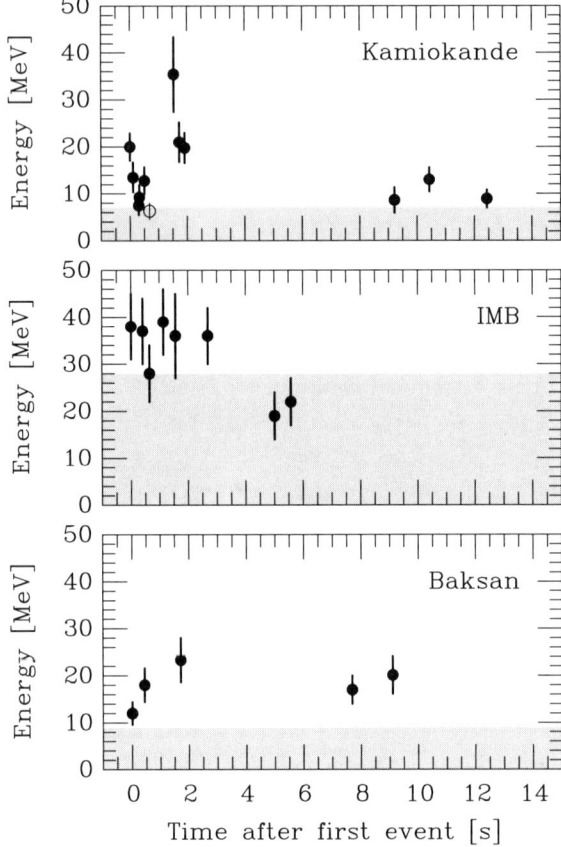

Fig. 3.1. Scatterplots of the neutrino events associated with SN1987A, as seen in the Kamiokande [2], IMB [3], and Baksan [4] detectors. The shaded regions indicate the nominal detector energy thresholds. Figure taken from Ref. [5].

the differences between flavors, and using this to test neutrino mixing, but this is beyond our scope.

The SN 1987A data are shown in Fig. 3.1. These are consistent with mostly being signal events due to inverse beta decay on free protons, $\bar{\nu}_e + p \rightarrow e^+ + n$. This reaction channel is special due to its large cross section, and the fact that the outgoing positron carries nearly the full antineutrino energy. The other flavors are much harder to detect. The first thing to notice is that the duration of the burst was about 10 seconds. The second is that the typical energies were low tens of MeV. (This is complicated somewhat by the nontrivial response function of the detectors, especially IMB, which was only effective at the highest energies.) At zeroth order, the Kamiokande and IMB data are consistent with each other and theoretical expectations. The Baksan data are quite puzzling, as this detector was about ten times smaller than Kamiokande, and thus they should have seen ~ 1 event; probably detector backgrounds were present.

The most important message is that these data are consistent with the picture of slow diffusion out of a very hot and dense object, i.e., with the birth of a neutron star, as is suggested also by the total energetics, assuming a comparable neutrino emission per flavor. You can easily estimate the number of detected events yourself, using the total energy noted above, the inverse beta cross section, and the distance of 50 kpc. (Interestingly, there is still no good astronomical evidence for such a compact object in the SN 1987A remnant.) This kind of basic confirmation of the explosion mechanism is what can do with such a small number of neutrino events.

How can we gather more supernova neutrinos? There are three possibilities. First, **Milky Way** objects, with $D \simeq 10$ kpc. Taking into account the fact that we have much larger detectors now, and assuming a typical distance in the Milky Way, we expect about 10^4 detected events in Super-Kamiokande. Unfortunately, the frequency is probably only 2 or 3 times per century, but we might get lucky. It will be very obvious if it happens. Second, **Nearby** objects with $D \sim 10$ Mpc or less. For these, one would need a much larger detector, at the 1 Mton scale, and the number of detected events per supenova is ~ 1. On the other hand, the frequency is about once per year. To reduce backgrounds, this would require a coincident detection of say two or more neutrinos, or one neutrino and the optical signal. Third, **Distant** objects from redshifts $z \sim 1\text{--}2$ or less. As a crude guide to how this works, imagine a supernova at a distance such that the expected number of detected events in Super-Kamiokande is 10^{-6}. Almost all of the time, nothing happens, but for one supernova in a million, one neutrino will be detected. This seems crazy until you realize that the supernova rate of the universe is a few per second. Putting this together more carefully leads to an expectation of several detected supernova events per year in Super-Kamiokande (these will be uncorrelated with the optical supernovae, due to the nearly isotropic nature of the detection cross section). A strong rejection of detector backgrounds is required to make this work.

Of these three detection modes, I'll focus on the last, as it is the least familiar.

3.2.3. Supernovae in the Milky Way

At present, the flagship supernova neutrino detector is Super-Kamiokande, which is located in a deep mine in Japan. It is the largest detector with the ability to separate individual supernova neutrino events from detector backgrounds. Its huge fiducial volume contains 22.5 kton of ultrapure water. Relativistic charged particles in a material emit optical Čerenkov radiation, which is detected by photomultiplier tubes around the periphery.

With $\sim 10^4$ events detected for a Milky Way supernova, the Super-Kamiokande data could be used to map out the details of the neutrino spectrum and luminosity profile. Additionally, other neutrino detection reactions, for which the yields are at the 1–10% level in comparison to inverse beta decay, would become important, revealing more about the flavors besides $\bar{\nu}_e$. The aspects of detecting a Milky Way supernova are very interesting, and have been extensively discussed elsewhere.

3.2.4. Supernovae in Nearby Galaxies

If Super-Kamiokande can detect 10^4 events at a supernova distance of 10 kpc, then it can expect to detect 1 event for a supernova distance of 1 Mpc, somewhat larger than the distance to the M31 (Andromeda) and M33 (Triangulum) galaxies. Unfortunately, a single event isn't exciting by itself, and anyway, these galaxies appear to have even lower supernova rates than the Milky Way. Still, it makes one wonder about greater distances. The number of galaxies in each new radial shell in distance increases like D^2, while the flux of each falls like $1/D^2$. As mentioned, one can beat even small Poisson expectations with enough tries, so this is intriguing.

An estimate based on the known nearby galaxies shows that the supernova rate with 10 Mpc should be about one per year, and this is shown in Fig. 3.2. In fact, the observed rates in the past few years have been even higher. A detailed calculation shows that a larger detector than Super-Kamiokande, something more on the 1 Mton scale, could detect about one supernova neutrino per year. (Such detectors are being considered for proton decay studies and as targets for long-baseline neutrino beams.) That seems like a small rate, but bear in mind that in the twenty years since SN 1987A, exactly zero supernova neutrinos have been (identifiably) detected. To reduce backgrounds, these nearby supernovae would need a coincidence of at least two neutrinos or one neutrino and the optical signal. Perhaps most importantly, the detection of even a single neutrino would fix the start time of the collapse to about ten seconds, compared to the precision

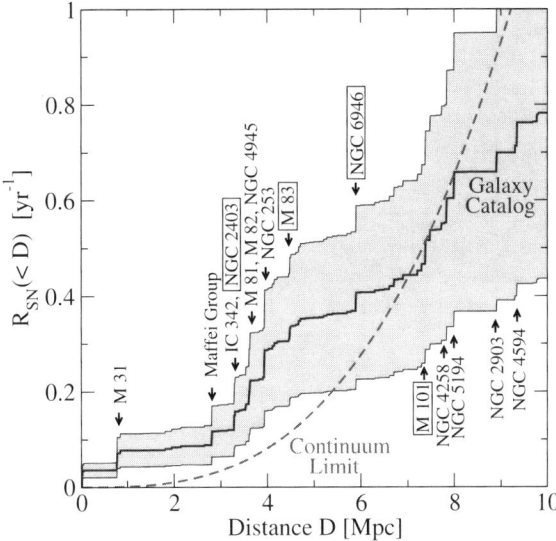

Fig. 3.2. The predicted cumulative supernova rate for nearby galaxies is shown by the blue line, and its uncertainty by the grey band (together denoted as "Galaxy Catalog"). The redshift $z = 0$ limit of the cosmic supernova rate is also shown ("Continuum Limit"). The observed local supernova rate in recent years has been higher than either prediction. Figure taken from Ref. [6].

of about one day that might be determined from the optical signal. This would be very useful for refining the window in which to look for a faint gravitational wave signal.

Related to this is an effort called NO SWEAT (Neutrino-Oriented Supernova Whole-Earth Telescope), led by Avishay Gal-Yam, to use a network of telescopes worldwide to find all supernovae in nearby galaxies.

3.2.5. DSNB: First Good Limit

The star formation rate was larger in the past, and in particular, was about 10 times larger at redshift $z \simeq 1$ than it is today. Since the lifetimes of massive stars are short, the core-collapse supernova rate should closely follow the evolution of the star formation rate, up to a constant factor. This gives more weight to distant supernovae than if the rate were constant. On the other hand, for supernova beyond $z \sim 1$, the neutrinos are so redshifted that their detection probabilities are too low (at lower energy, the detection cross section goes down while the detector background rates increase).

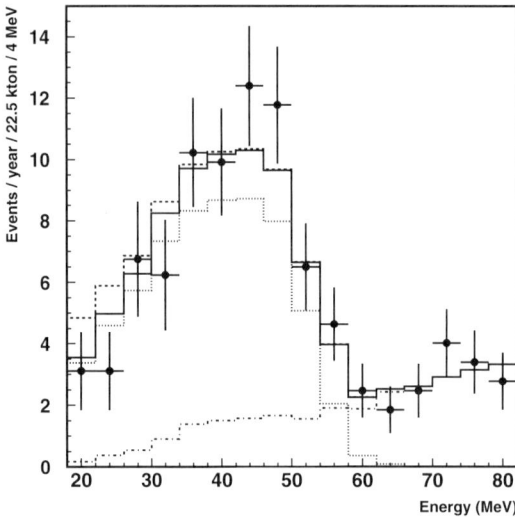

Fig. 3.3. The event spectrum measured in Super-Kamiokande is denoted by the points with error bars. The solid line indicates the expected total detector background rate (the dotted component is due to muon neutrinos, and the dot-dashed component is due to electron neutrinos). The dashed line above the solid line indicates how large of an excess due to DSNB events could be present, given the statistical uncertainties. Figure taken from Ref. [7].

Integrating the neutrino emission per supernova with the evolving supernova rate, and taking into account the cosmological factors, the accumulated spectrum of all past supernovae can be calculated. This is known as the Diffuse Supernova Neutrino Background (DSNB), or sometimes as relic supernova neutrinos (which is a confusing and deprecated term, i.e., these have nothing to do with the 2 K relic background of neutrinos that decoupled just before big-bang nucleosynthesis).

In 2000, a paper by Kaplinghat, Steigman, and Walker calculated the largest plausible DSNB flux, and found it to be 2.2 cm^{-2} s^{-1} for electron antineutrinos above 19.3 MeV. This was about 100 times smaller than the existing limit from Kamiokande, so the prospects for detection didn't look great. Other calculations with reasonable inputs (by modern standards) gave results that were a few to several times smaller.

In 2003, the Super-Kamiokande collaboration published a limit that was 1.2 in the above units. This was a milestone, because it showed for the first time that there was hope of reaching the range in which a detection might be made. Still, as shown in Fig. 3.3, there are large detector backgrounds that make it difficult to identify the DSNB signal. Note that for a background-limited search, like this one, to improve the signal sensitivity by a factor of 3 takes a factor 9 more statistics. Since this figure was based on 4 years of data, this would take a long time to collect (comparable to the wait for a Galactic supernova!).

3.2.6. *DSNB: Detection with Gadolinium*

In order to make progress, it is necessary to find a way to eliminate or at least severely reduce the detector background. Mark Vagins (a member of the Super-Kamiokande collaboration) and I decided to put our heads together to find a way to isolate the DSNB signal. This resulted in a 2004 article in Physical Review Letters, though we were forced to remove the code name of the project, "GADZOOKS!," from the title and text (but see the arXiv version). Recall that the detection reaction is $\bar{\nu}_e + p \to e^+ + n$, and that at present, only the positron is detected. We realized that the key was to detect the neutron in time and space coincidence with the positron. This is an old idea, and was used by Reines and Cowan in the first detection of neutrinos (antineutrinos from a nuclear reactor).

Saying that we had to detect the neutron was the easy part. It was more challenging to find a way to do this in a water-based detector, where normally the neutrons capture on free protons. That produces a 2.2 MeV gamma ray that Compton scatters electrons, but they are too low in energy to be detectable. We pointed out that the required neutron tagging might be possible by using a 0.2% admixture of dissolved gadolinium trichloride ($GdCl_3$). Gadolinium has a huge neutron capture cross section, and produces an 8 MeV gamma-ray cascade that reconstructs as an equivalent single electron of about 5 MeV, which is readily detectable.

The really hard part was in establishing that this technique might be possible in practice, which involved raising and answering many difficult technical questions. (Among them, finding a suitable water-soluble compound of gadolinium.) Somewhat to our surprise, we found no obvious obstacles. Mark Vagins has been leading a detailed research and development effort, and so far, the prospects look very good.

In Fig. 3.4, the spectra expected in Super-Kamiokande if gadolinium is added are shown. The atmospheric neutrino backgrounds mentioned above are reduced by a factor of about 5. Additionally, backgrounds at lower energies are severely reduced, allowing the use of a much lower threshold

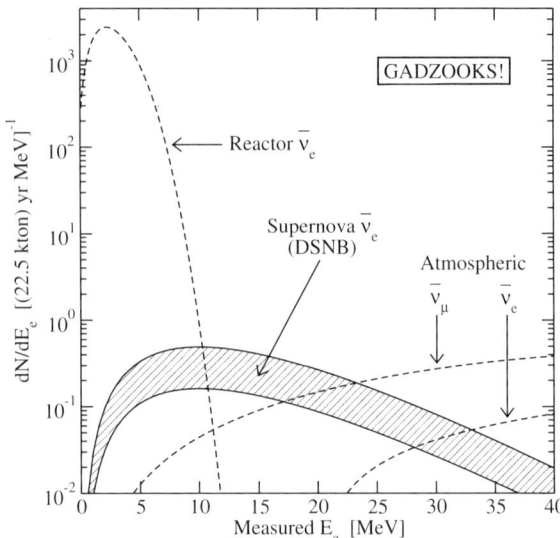

Fig. 3.4. The DSNB signal and detector backgrounds expected in Super-Kamiokande if gadolinium is added. Figure taken from Ref. [8].

energy. At moderate energies, it should be possible to cleanly identify DSNB signal events.

3.2.7. *DSNB: Astrophysical Impact*

Now let's return to the predicted DSNB spectrum. If either the assumed star formation rate or the neutrino emission per supernova were too large, then the predicted DSNB flux would already be ruled out the the Super-Kamiokande data.

Even since the time of the Super-Kamiokande limit, the astrophysical data have improved substantially. Andrew Hopkins and I synthesized a wide variety of data to constrain the star formation and supernova rate histories. An example fit is shown in Fig. 3.5. The uncertainty band is much more narrow now than it was just a few years ago. The normalization of the cosmic star formation rate depends on dust corrections. If the true star formation rate were even somewhat larger than determined here, then the DSNB neutrino flux would be too large relative to the Super-Kamiokande limit. The only way out would be to require a substantially lower neutrino emission per supernova.

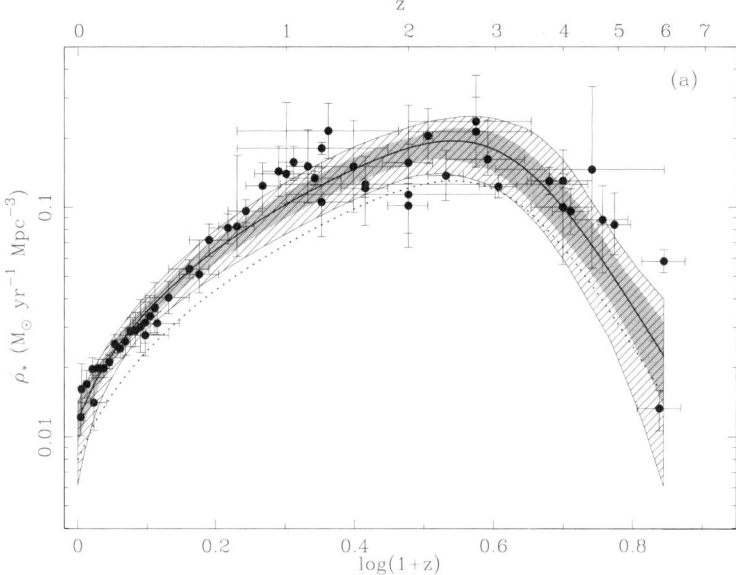

Fig. 3.5. The star formation rate history, with selected data shown by points and the fit and uncertainty shown by the bands. Figure taken from Ref. [9].

The corresponding calculated supernova rates are in good agreement with data. As an interesting aside, it was shown that the diffuse gamma-ray background from type Ia supernovae is too small to account for the observed data in the MeV range. That is particularly significant because many limits on exotic particle physics depend on just this energy range.

3.2.8. *Back to the Scene of the Crime: SN 1987A*

If we now know the star formation history, then the only remaining unknown is the neutrino emission per supernova. Hasan Yüksel, Shin'ichiro Ando, and I considered how well the Super-Kamiokande data already restrict the neutrino emission per supernova. The emission models are usually parametrized in terms of the time-integrated luminosity (or portion of the binding energy release) and the average energy per neutrino (related to the temperature of the spectrum). I mentioned above that the Kamiokande and IMB data on the emission from SN 1987A were mostly consistent. In fact, when fitted with thermal spectra, there are some discrepancies.

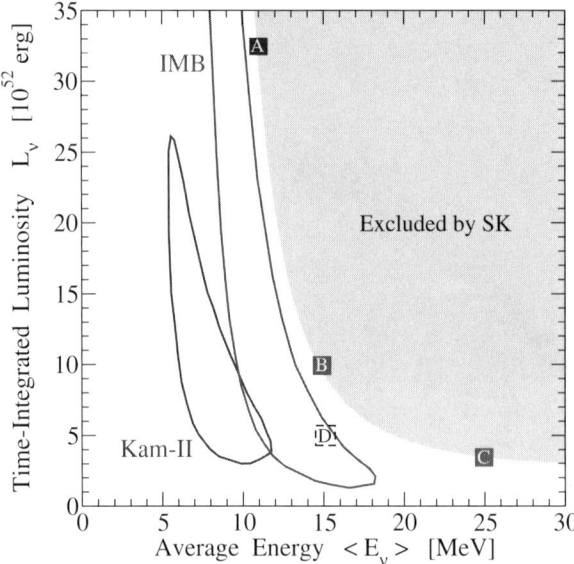

Fig. 3.6. The contours labeled Kam-II and IMB are the allowed regions from the SN 1987A data, assuming a thermal spectrum [10]. The shaded region is what is already excluded by the non-observation of a DSNB signal in Super-Kamiokande. Figure taken from Ref. [11].

In Fig. 3.6, I show that the DSNB data are probing neutrino emission parameters only slightly larger than those deduced from the SN 1987A data. With reduced detector backgrounds, the DSNB spectrum would be a new way to measure the neutrino emission per supernova.

3.2.9. *Conclusions*

Why is understanding supernovae interesting and important? For particle physics, it is to test the properties of neutrinos, and to search for new low-mass particles that cool the proto-neutron star. For nuclear physics, it is to constrain the neutron star equation of state and to shed light on the formation of the elements. For astrophysics, it is to understand the stellar life and death cycles and to understand the supernova mechanisms. For cosmology, it is to better understand the details of whether type Ia supernovae are standard candles, and to probe the origins of the gamma-ray and neutrino backgrounds. With more data, we can't lose.

3.3. PART TWO: High-Energy Neutrinos

3.3.1. *Introduction*

Now that we've covered the specific example of supernova neutrinos, let's step back and comment on the general status and outlook in neutrino astrophysics.

Unique among the Standard Model fermions, neutrinos are neutral, and more generally, have only weak interactions. This makes them potentially sensitive to even very feeble postulated new interactions. While the discovery of neutrino mass and mixing was "new physics" beyond the minimal Standard Model, the discovery of any new interactions would be a much more radical step, as it would require new particles as well.

This is one reason that we're interested in neutrinos. The other, already discussed, is that they will be especially powerful probes of astrophysical objects, once these neutrinos are detected. Already with the neutrinos from the Sun and SN 1987A, the scientific return was very rich: not only confirmation of the physics of their interiors, but also a crucial piece in the discovery of neutrino mass and mixing. Ray Davis and Masatoshi Koshiba shared in the 2002 Nobel Prize for this work, and their citation reads, "...for pioneering contributions to astrophysics, in particular for the detection of cosmic neutrinos...."

The general achievements in neutrino physics in just the recent past might be summarized as follows. The **cosmological** results are the consistency of big-bang nucleosynthesis yields with three flavors of neutrinos, and the exclusion of neutrinos as the (hot) dark matter. In both cases, these facts have been established independently in the laboratory as well. The **astrophysical** results are the discovery of neutrinos from SN 1987A and the solution of the solar neutrino problem. The **fundamental** results are the discovery of neutrino mass and mixing, and the clear exclusion of a huge range of formerly allowed models of exotic neutrino properties.

One of the lessons from this list is that we need data from new sources to make new discoveries, and that those discoveries may have a broader impact than initially thought. Astrophysical sources reach extremes of density, distance, and energy, and this will allow unprecedented tests of neutrino properties, for example.

We can identify three frontiers where new sources will likely be discovered soon. By the rough energy scale of the neutrinos, we might call these the MeV (10^{-6} TeV) scale, the TeV scale, and the EeV (10^6 TeV) scale. At the MeV scale, the focus is on the **Visible Universe**, i.e., stars and supernovae, and Super-Kamiokande is the main detector. At the TeV scale, the focus is on the **Nonthermal Universe**, i.e., jets powered by black holes, and the primary detector is AMANDA, which is being succeeded by IceCube. At the EeV scale, the focus is on the **Extreme Universe**, i.e., at the energy frontier of the highest-energy cosmic rays, and one of the key detectors is ANITA.

Why do we think that high-energy neutrinos even exist? First, because cosmic rays (probably mostly protons) are observed at energies as high as 10^{20} eV, and they are increasingly abundant down to at least the GeV range. Something is accelerating these cosmic rays, and it is very likely that these sources also produce neutrinos. Second, because extragalactic gamma-ray sources have been observed with energies up to about 10 TeV (and galactic sources up to about 100 TeV). Again, something is producing these particles, and in large fluxes, and it is likely that neutrinos are also produced.

So then why do we need neutrinos? The problem with cosmic rays is that they are easily deflected by magnetic fields, and so only their isotropic flux has been observed, making the identification of their sources very difficult. The problem with photons is that they are easily attenuated: a TeV gamma ray colliding with an eV starlight photon is able to produce an electron-positron pair. Thus at high energies, only nearby objects can be seen.

High-energy neutrinos can be made through either proton-proton or proton-photon collisions, depending on energies. In either case, pions are readily produced, and typically comparable numbers of neutral and charged pions are made. Neutral pions decay as $\pi^0 \to \gamma+\gamma$, and charged pions decay as $\pi^+ \to \mu^+ + \nu_\mu$, followed by $\mu^+ \to e^+ + \nu_e + \bar{\nu}_\mu$ (with obvious changes for the charge conjugate). This is the **hadronic** mechanism for producing gamma rays and neutrinos. There is also a **leptonic** mechanism, based on the inverse Compton scattering reaction $e^- + \gamma \to \gamma + e^-$, where fast electrons collide with low-energy photons and promote them to high-energy gamma rays. Note that the leptonic process produces no neutrinos. It is a major mystery whether the observed high energy gamma-ray sources are powered by the hadronic or leptonic mechanism. This is a key to uncovering the sources of the cosmic rays.

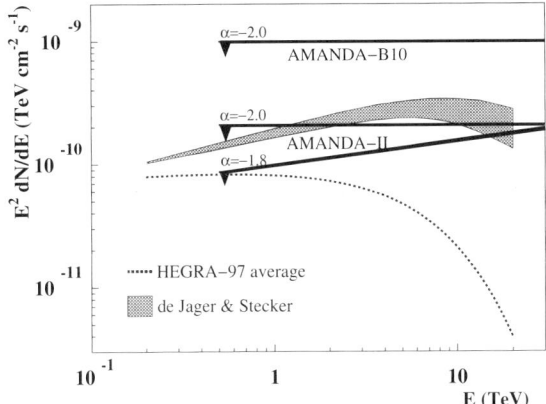

Fig. 3.7. The dotted line is based on gamma-ray observations of the nearby AGN Markarian 501 by the HEGRA experiment, and the shaded band is a calculation that removes the assumed affects of attenuation en route. The labeled solid lines indicate AMANDA limits on the neutrino flux. This object flares, and the gamma ray and neutrino data are not contemporaneous. Figure taken from Ref. [12].

3.3.2. Sources and Detection at ~ 1 TeV

At the simplest level, hadronic sources produce nearly equal fluxes of gamma rays and neutrinos (the corrections due to multiplicities, decay energies, and neutrino mixing can be easily taken into account). Therefore, the observed gamma-ray spectrum of an object like an AGN is a strong predictor of the neutrino spectrum, if the source is hadronic (if it is leptonic, then the neutrino flux will be zero). Any attenuation of the gamma-ray spectrum en route would mean that the neutrino flux would be even larger. An example is illustrated in Fig. 3.7, where it is shown that the neutrino detectors are now approaching the required level of flux sensitivity.

For hadronic sources, the initial neutrino flavor ratios (adding neutrinos and antineutrinos) are $\phi_e : \phi_\mu : \phi_\tau = 1 : 2 : 0$, following simply from the pion and muon decay chains. After vacuum neutrino mixing en route, these will become $\phi_e : \phi_\mu : \phi_\tau = 1 : 1 : 1$.

Of all flavors, the muon neutrinos are the easiest to detect and identify. Through charged-current deep inelastic scattering reactions, these produce muons that carry most of the neutrino energy, and which have only a very small deflection from the neutrino direction. Muons and other charged particles produce optical Čerenkov radiation in the detector, which is registered by photomultiplier tubes throughout the volume. Muons produce

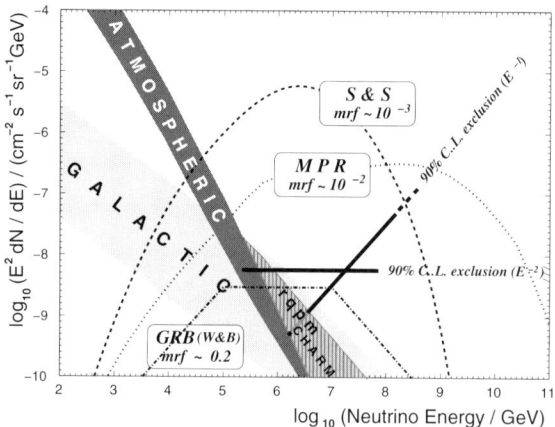

Fig. 3.8. The sensitivity of IceCube is marked with heavy solid lines, as labeled. The broken lines indicate various astrophysical diffuse flux models. The shaded regions indicate the atmospheric neutrino, prompt/charm component thereof, and Galactic neutrino backgrounds. Figure taken from Ref. [13].

spectacular long tracks that can range through the kilometer of the detector and beyond. The detection of electron and tau neutrinos is interesting and important too, but beyond our scope here.

To screen out enormous backgrounds from downgoing atmospheric muons, these detectors only look for upgoing events. Since muons cannot pass through Earth, these muons must have been created just below the detector by upgoing neutrinos. Even after this, there are backgrounds due to atmospheric neutrinos, themselves produced on the other side of Earth, and thus hardly extraterrestrial.

An astrophysical point source can be identified as an excess in a given direction, whereas the atmospheric neutrino background is smoothly varying. Transient point sources are even easier to recognize. On the other hand, diffuse astrophysical neutrino fluxes are quite hard to separate from the atmospheric neutrino background. The principal technique is that the former are believed to have spectra close to E^{-2}, while the latter is closer to E^{-3}, and steeper at higher energies. Thus at high energies the astrophysical diffuse fluxes should emerge as dominant. Once cannot go too high in energy – the event rates get too low, and Earth becomes opaque to neutrinos at around 100 TeV. An example of the diffuse flux sensitivity of IceCube is shown in Fig. 3.8.

3.3.3. Testing Neutrino Properties

As an example of a novel neutrino property that could be tested once astrophysical sources are observed, consider neutrino decay. Why should neutrinos decay? Other than the fact that there is no interaction that can cause fast neutrino decay, why shouldn't they decay? The other massive fermions all decay into the lowest-mass generation in their family. (Neutrinos can too, via the weak interaction, but it is exceedingly slow.) We'll consider simply neutrino disappearance, i.e., that the other particle in the decay of one neutrino mass eigenstate to another is too weakly interacting to be detected. It is quite hard to test for the effects of such decays.

Decay will deplete the original flux as

$$\exp\left(-t/\tau_{lab}\right) = \exp\left(-\frac{L}{E} \times \frac{m}{\tau}\right),$$

where L is distance, E the energy, m the mass, and τ the proper lifetime. For the Sun, the τ/m scale that can be probed is up to about 10^{-4} s/eV. On the other hand, for distant astrophysical sources of TeV neutrinos, L/E may be such that τ/m up to about 10^{+4} s/eV is relevant!

How can we tell if decay has occurred, if the neutrino fluxes are uncertain? As mentioned, the flavor ratios after vacuum oscillations are expected to be $\phi_e : \phi_\mu : \phi_\tau = 1 : 1 : 1$. However, it is among the mass eigenstates, not the flavor eigenstates, where decays take place. Suppose that the heaviest two mass eigenstates have decayed, leaving only the lightest mass eigenstate. What is its flavor composition? In the normal hierarchy, it has flavor ratios $\phi_e : \phi_\mu : \phi_\tau \sim 5 : 1 : 1$, whereas in the inverted hierarchy, they are $\sim 0 : 1 : 1$. In either case, they are quite distinct from the no-decay case, and the flavor identification capabilities of IceCube should be able to distinguish these possibilities.

3.3.4. Sources and Detection at $\sim 10^6$ TeV

Cosmic rays have been observed at energies above 10^{20} eV, and there are no good answers as to what astrophysical accelerators may have produced them. However, this becomes even more puzzling when it is noted that the universe should be opaque to protons above about 3×10^{19} eV traveling over more than 100 Mpc. There are no obvious sources within that distance.

The process by which protons are attenuated is $p + \gamma \rightarrow p + \pi^0, n + \pi^+$, where both final states are possible, and the target photon is from the cosmic microwave background. As with the hadronic processes discussed

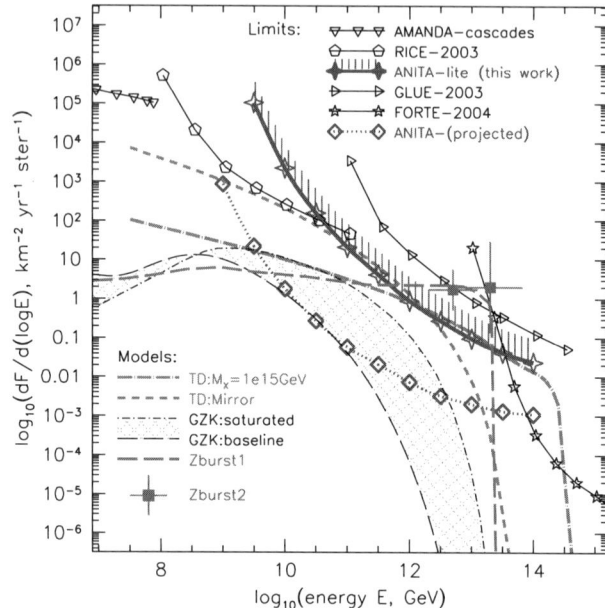

Fig. 3.9. Real and projected neutrino flux sensitivities of various experiments (lines with points), along with various models (as labeled). Figure taken from Ref. [14].

above, the neutral pion decays produce gamma rays and the charged pion decays produce neutrinos. The gamma rays are themselves attenuated, but the neutrino flux builds up when integrating over sources everywhere in the universe. Since the attenuation process for the protons is called the GZK process (Greisen-Zatsepin-Kuzmin), these are called GZK neutrinos. Typical energies are in the EeV range, and an isotropic diffuse flux is expected.

New experiments are being deployed to search for the GZK neutrino flux, as shown in Fig. 3.9. Unlike IceCube, which is based on optical Čerenkov radiation, ANITA and other experiments are based on radio Čerenkov radiation that is emitted coherently from the whole shower initiated by a neutrino in the ice or other transparent medium. ANITA is using the Antarctic ice cap as the detector, and is observing it with radio antennas mounted on a balloon. So far, detector backgrounds appear to be negligible, meaning that it should be straightforward to improve the signal sensitivity with more exposure.

In Fig. 3.10, I show the results of a very recent calculation of the expected GZK neutrino fluxes.

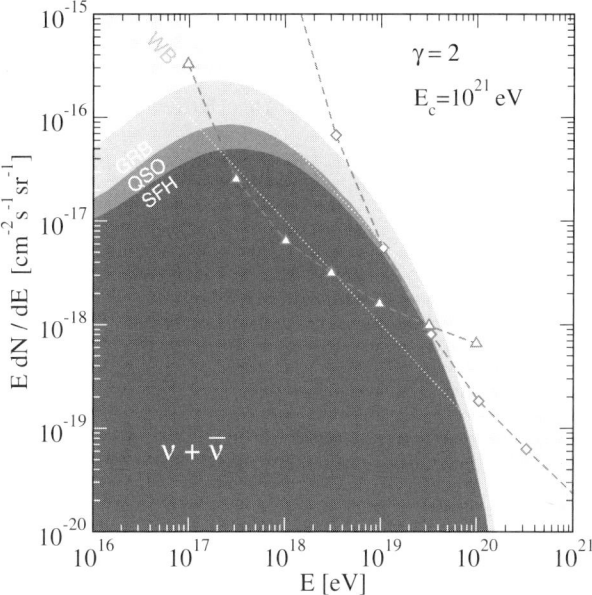

Fig. 3.10. Predicted GZK neutrino fluxes, assuming that ultrahigh energy cosmic rays are produced in gamma-ray bursts, and according to how the latter rate evolves with redshift (i.e., following the star formation rate alone, or rising like that of the quasars, or depending on both the star formation rate and the evolving local metallicity. The "WB" band is the Waxman-Bahcall bound. The curves with points are projected sensitivities for ANITA (upper) and ARIANNA (lower). Figure taken from Ref. [15].

Interestingly, when adjusted for the neutrino-quark center of mass energy, the detection reactions are probing above the TeV scale, opening the prospect of sensitivity to new physics in the detection alone.

3.3.5. *Conclusions*

So far, zero high-energy astrophysical neutrinos have been detected. However, the near-term prospects are very good, and are strongly motivated by measured data on high-energy protons and photons. Still, this will not be easy, and large detectors with strong background rejection will be needed. If successful, these experiments will make important astrophysical discoveries, e.g., whether gamma-ray sources are based on the hadronic or leptonic mechanisms, the origins of cosmic rays at all energies, etc. We might even learn something new about neutrinos in the process!

Acknowledgments

JFB was supported by National Science Foundation CAREER grant PHY-0547102, and by CCAPP at the Ohio State University.

References

[1] **Disclaimer:** I have been very light on referencing, in fact only noting the sources of the figures shown, to make the paper more readable.
[2] K. Hirata *et al.*, Phys. Rev. Lett. **58**, 1490 (1987).
[3] R. M. Bionta *et al.*, Phys. Rev. Lett. **58**, 1494 (1987).
[4] E. N. Alekseev, L. N. Alekseeva, I. V. Krivosheina and V. I. Volchenko, Phys. Lett. B **205**, 209 (1988).
[5] G. G. Raffelt, Ann. Rev. Nucl. Part. Sci. **49**, 163 (1999) [arXiv:hep-ph/9903472].
[6] S. Ando, J. F. Beacom and H. Yuksel, Phys. Rev. Lett. **95**, 171101 (2005) [arXiv:astro-ph/0503321].
[7] M. Malek *et al.*, Phys. Rev. Lett. **90**, 061101 (2003) [arXiv:hep-ex/0209028].
[8] J. F. Beacom and M. R. Vagins, Phys. Rev. Lett. **93**, 171101 (2004) [arXiv:hep-ph/0309300].
[9] A. M. Hopkins and J. F. Beacom, Astrophys. J. **651**, 142 (2006) [arXiv:astro-ph/0601463].
[10] B. Jegerlehner, F. Neubig and G. Raffelt, Phys. Rev. D **54**, 1194 (1996) [arXiv:astro-ph/9601111]; A. Mirizzi and G. G. Raffelt, Phys. Rev. D **72**, 063001 (2005) [arXiv:astro-ph/0508612].
[11] H. Yuksel, S. Ando and J. F. Beacom, Phys. Rev. C **74**, 015803 (2006) [arXiv:astro-ph/0509297].
[12] J. Ahrens *et al.*, Phys. Rev. Lett. **92**, 071102 (2004) [arXiv:astro-ph/0309585].
[13] J. Ahrens *et al.*, Astropart. Phys. **20**, 507 (2004) [arXiv:astro-ph/0305196].
[14] S. W. Barwick *et al.*, Phys. Rev. Lett. **96**, 171101 (2006) [arXiv:astro-ph/0512265].
[15] H. Yuksel and M. D. Kistler, Phys. Rev. D **75**, 083004 (2007) [arXiv:astro-ph/0610481].

Chapter 4

Leptogenesis

Mu-Chun Chen

Theoretical Physics Department, Fermi National Accelerator Laboratory
Batavia, IL 60510-0500, U.S.A.
and
Department of Physics & Astronomy, University of California
Irvine, CA 92697-4575, U.S.A.[*]
muchunc@uci.edu

The origin of the asymmetry between matter and anti-matter of the Universe has been one of the great challenges in particle physics and cosmology. Leptogenesis as a mechanism for generating the cosmological baryon asymmetry of the Universe has gained significant interests ever since the advent of the evidence of non-zero neutrino masses. In these lectures presented at TASI 2006, I review various realizations of leptogenesis and allude to recent developments in this subject.

4.1. Introduction

The understanding of the origin of the cosmological baryon asymmetry has been a challenge for both particle physics and cosmology. In an expanding Universe, which leads to departure from thermal equilibrium, a baryon asymmetry can be generated dynamically by charge-conjugation (C), charge-parity (CP) and baryon (B) number violating interactions among quarks and leptons. Possible realizations of these conditions have been studied for decades, starting with detailed investigation in the context of grand unified theories. The recent advent of the evidence of non-zero neutrino masses has led to a significant amount of work in leptogenesis. This subject is of special interests because the baryon asymmetry in this scenario is in principle entirely determined by the properties of the neutrinos.

---

[*]Address after January 1, 2007.

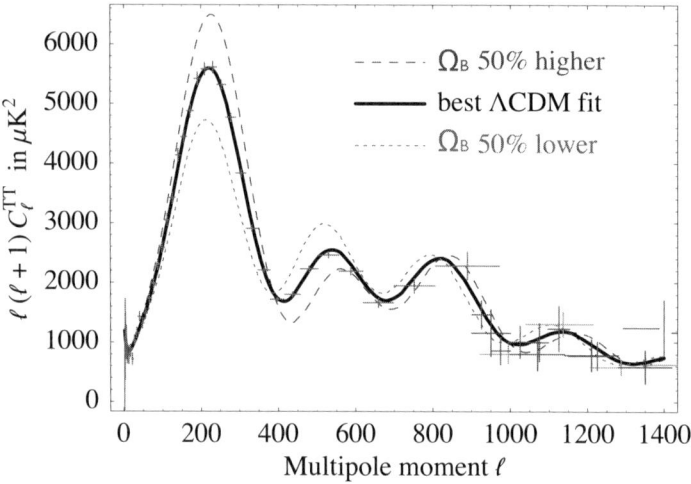

Fig. 4.1. The power spectrum anisotropies defined in Eqs. 4.2 and 4.3 as a function of the multiple moment, l. Figure taken from Ref. [2].

In these lectures, I discuss some basic ingredients of leptogenesis as well as recent developments in this subject.

These lectures are organized as follows: In Sec. 4.1, I review the basic ingredients needed for the generation of baryon asymmetry and describe various mechanisms for baryogenesis and the problems in these mechanisms. In Sec. 4.2, I introduce the standard leptogenesis and Dirac leptogenesis as well as the problem of gravitino over-production that exists in these standard scenarios when supersymmetry is incorporated. This is followed by Sec. 4.3, in which several alternative mechanisms that have been invented to alleviate the gravitino over-production problem are discussed. Section 4.4 focuses on the subject of connecting leptogenesis with low energy leptonic CP violating processes. Section 4.5 concludes these lectures with discussions on the recent developments. For exiting reviews on the subject of leptogenesis and on baryogenesis in general, see *e.g.* Refs. [1–4] and [5–7].

4.1.1. *Evidence of Baryon Number Asymmetry*

One of the main successes of the standard early Universe cosmology is the predictions for the abundances of the light elements, D, ^{3}He, ^{4}He and ^{7}Li. (For a review, see, Ref. [8]. See also Scott Dodelson's lectures.) Agreement between theory and observation is obtained for a certain range of parameter, η_B, which is the ratio of the baryon number density, n_B, to photon density, n_γ,

$$\eta_B^{\text{BBN}} = \frac{n_B}{n_\gamma} = (2.6 - 6.2) \times 10^{-10} \,. \tag{4.1}$$

The Cosmic Microwave Background (CMB) is not a perfectly isotropic radiation bath. These small temperature anisotropies are usually analyzed by decomposing the signal into spherical harmonics, in terms of the spherical polar angles θ and ϕ on the sky, as

$$\frac{\Delta T}{T} = \sum_{l,m} a_{lm} Y_{lm}(\theta, \phi) \,, \tag{4.2}$$

where a_{lm} are the expansion coefficients. The CMB power spectrum is defined by

$$C_l = \langle |a_{lm}|^2 \rangle \,, \tag{4.3}$$

and it is conventional to plot the quantity $l(l+1)C_l$ against l. The CMB measurements indicate that the temperature of the Universe at present is $T_{now} \sim 3°K$. Due to the Bose-Einstein statistics, the number density of the photon, n_γ, scales as T^3. Together, these give a photon number density at present to be roughly $400/\text{cm}^3$. It is more difficult to count the baryon number density, because only some fraction of the baryons form stars and other luminous objects. There are two indirect probes that point to the same baryon density. The measurement of CMB anisotropies probe the acoustic oscillations of the baryon/photon fluid, which happened around photon last scattering. Figure 4.1 illustrates how the amount of anisotropies depends on n_B/n_γ. The baryon number density, $n_B \sim 1/\text{m}^3$, is obtained from the anisotropic in CMB, which indicates the baryon density Ω_B to be 0.044. Another indirect probe is the Big Bang Nucleosynthesis (BBN), whose predictions depend on n_B/n_γ through the processes shown in Fig. 4.2. It is measured independently from the primordial nucleosynthesis of the light elements. The value for n_B/n_γ deduced from primordial Deuterium abundance agrees with that obtained by WMAP [9]. For ^{4}He and ^{7}Li, there are nevertheless discrepancies which may be due to the under-estimated errors. Combining WMAP measurement and the Deuterium abundance gives,

$$\frac{n_B}{n_\gamma} \equiv \eta_B = (6.1 \pm 0.3) \times 10^{-10} \,. \tag{4.4}$$

126 M.-C. Chen

4.1.2. *Sakharov's Conditions*

A matter-anti-matter asymmetry can be dynamically generated in an expanding Universe if the particle interactions and the cosmological evolution satisfy the three Sakharov's conditions [10]: (*i*) baryon number violation; (*ii*) C and CP violation; (*iii*) departure from thermal equilibrium.

4.1.2.1. *Baryon Number Violation*

As we start from a baryon symmetric Universe ($B = 0$), to evolve to a Universe where $B \neq 0$, baryon number violation is necessary. Baryon number violation occurs naturally in Grand Unified Theories (GUT), because quarks and leptons are unified in the same irreducible representations. It is thus possible to have gauge bosons and scalars mediating interactions among fermions having different baryon numbers. In the SM, on the other hand, the baryon number and the lepton number are accidental symmetries. It is thus not possible to violate these symmetries at the tree level. t'Hooft realized that [11] the non-perturbative instanton effects may give rise to processes that violate $(B + L)$, but conserve $(B - L)$. Classically, B and L are conserved,

$$B = \int d^3 x J_0^B(x), \quad L = \int d^3 x J_0^L(x) , \qquad (4.5)$$

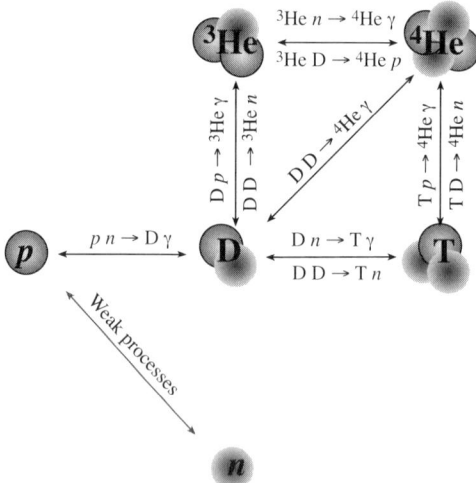

Fig. 4.2. Main reactions that determine the primordial abundances of the light elements. Figure taken from Ref. [2].

Table 4.1. Standard model fermions and their B and L charges.

	$q_L = \begin{pmatrix} u \\ d \end{pmatrix}_L$	u_L^c	d_L^c	$\ell_L = \begin{pmatrix} \nu \\ e \end{pmatrix}_L$	e_L^c
B	1/3	-1/3	-1/3	0	0
L	0	0	0	1	-1

where the currents associated with B and L are given by,

$$J_\mu^B = \frac{1}{3} \sum_i \left(\overline{q}_{L_i} \gamma_\mu q_{L_i} - \overline{u}_{L_i}^c \gamma_\mu u_{L_i}^c - \overline{d}_{L_i}^c \gamma_\mu d_{L_i}^c \right), \tag{4.6}$$

$$J_\mu^L = \sum_i \left(\overline{\ell}_{L_i} \gamma_\mu \ell_{L_i} - \overline{e}_{L_i}^c \gamma_\mu e_{L_i}^c \right). \tag{4.7}$$

Here q_L refers to the $SU(2)_L$ doublet quarks, while u_L and d_L refer to the $SU(2)_L$ singlet quarks. Similarly, ℓ_L refers to the $SU(2)_L$ lepton doublets and e_L refers to the $SU(2)_L$ charged lepton singlets. The B and L numbers of these fermions are summarized in Table 4.1. The subscript i is the generation index. Even though B and L are individually conserved at the tree level, the Adler-Bell-Jackiw (ABJ) triangular anomalies [12] nevertheless do not vanish, and thus B and L are anomalous [13] at the quantum level through the interactions with the electroweak gauge fields in the triangle diagrams (see, for example Ref. [14] for details). In other words, the divergences of the currents associated with B and L do not vanish at the quantum level, and they are given by

$$\partial_\mu J_B^\mu = \partial_\mu J_L^\mu = \frac{N_f}{32\pi^2} \left(g^2 W_{\mu\nu}^p \widetilde{W}^{p\mu\nu} - g'^2 B_{\mu\nu} \widetilde{B}^{\mu\nu} \right), \tag{4.8}$$

where $W_{\mu\nu}$ and $B_{\mu\nu}$ are the $SU(2)_L$ and $U(1)_Y$ field strengths,

$$W_{\mu\nu}^p = \partial_\mu W_\nu^p - \partial_\nu W_\mu^p \tag{4.9}$$

$$B_{\mu\nu} = \partial_\mu B_\nu - \partial_\nu B_\mu, \tag{4.10}$$

respectively, with corresponding gauge coupling constants being g and g', and N_f is the number of fermion generations. As $\partial^\mu (J_\mu^B - J_\mu^L) = 0$, $(B-L)$ is conserved. However, $(B+L)$ is violated with the divergence of the current given by,

$$\partial^\mu (J_\mu^B + J_\mu^L) = 2N_F \partial_\mu K^\mu, \tag{4.11}$$

where

$$K^\mu = -\frac{g^2}{32\pi^2} 2\epsilon^{\mu\nu\rho\sigma} W_\nu^p (\partial_\rho W_\sigma^p + \frac{g}{3}\epsilon^{pqr} W_\rho^q W_\sigma^r) \quad (4.12)$$
$$+ \frac{g'^2}{32\pi^2} \epsilon^{\mu\nu\rho\sigma} B_\nu B_{\rho\sigma} .$$

This violation is due to the vacuum structure of non-abelian gauge theories. Change in B and L numbers are related to change in topological charges,

$$B(t_f) - B(t_i) = \int_{t_i}^{t_f} dt \int d^3x \, \partial^\mu J_\mu^B \quad (4.13)$$
$$= N_f [N_{cs}(t_f) - N_{cs}(t_i)] ,$$

where the topological charge of the gauge field (*i.e.* the Chern-Simons number) N_{cs} is given by,

$$N_{cs}(t) = \frac{g^3}{96\pi^2} \int d^3x \epsilon_{ijk} \epsilon^{IJK} W^{Ii} W^{Jj} W^{Kk} . \quad (4.14)$$

There are therefore infinitely many degenerate ground states with $\Delta N_{cs} = \pm 1, \pm 2, \ldots$, separated by a potential barrier, as depicted by Fig. 4.3. In semi-classical approximation, the probability of tunneling be-

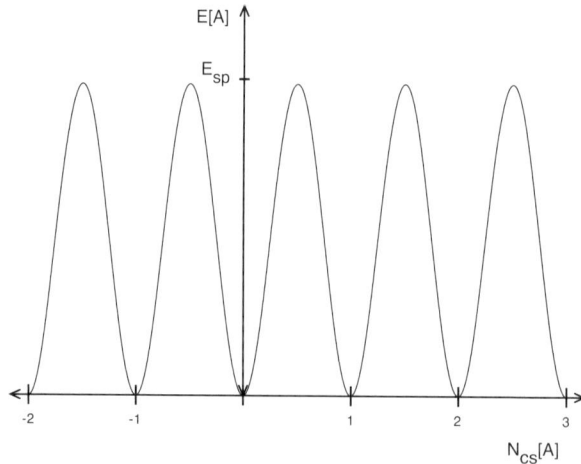

Fig. 4.3. The energy dependence of the gauge configurations A as a function of the Chern-Simons number, $N_{cs}[A]$. Sphalerons correspond to the saddle points, *i.e.* maxima of the potential.

tween neighboring vacua is determined by the instanton configurations. In

SM, as there are three generations of fermions, $\Delta B = \Delta L = N_f \Delta N_{cs} = \pm 3n$, with n being an positive integer. In other words, the vacuum to vacuum transition changes ΔB and ΔL by multiples of 3 units. As a result, the $SU(2)$ instantons lead to the following effective operator at the lowest order,

$$\mathcal{O}_{B+L} = \prod_{i=1,2,3} (q_{L_i} q_{L_i} q_{L_i} \ell_{L_i}), \tag{4.15}$$

which gives 12 fermion interactions, such as,

$$\bar{u} + \bar{d} + \bar{c} \to d + 2s + 2b + t + \nu_e + \nu_\mu + \nu_\tau. \tag{4.16}$$

At zero temperature, the transition rate is given by, $\Gamma \sim e^{-S_{int}} = e^{-4\pi/\alpha} = \mathcal{O}(10^{-165})$ [11]. The resulting transition rate is exponentially suppressed and thus it is negligible. In thermal bath, however, things can be quite different. It was pointed out by Kuzmin, Rubakov and Shaposhnikov [15] that, in thermal bath, the transitions between different gauge vacua can be made not by tunneling but through thermal fluctuations over the barrier. When temperatures are larger than the height of the barrier, the suppression due to the Boltzmann factor disappear completely, and thus the $(B+L)$ violating processes can occur at a significant rate and they can be in equilibrium in the expanding Universe. The transition rate at finite temperature in the electroweak theory is determined by the sphaleron configurations [16], which are static configurations that correspond to unstable solutions to the equations of motion. In other words, the sphaleron configurations are saddle points of the field energy of the gauge-Higgs system, as depicted in Fig. 4.3. They possess Chern-Simons number equal to 1/2 and have energy

$$E_{sp}(T) \simeq \frac{8\pi}{g} \langle H(T) \rangle, \tag{4.17}$$

which is proportional to the Higgs vacuum expectation value (vev), $\langle H(T) \rangle$, at finite temperature T. Below the electroweak phase transition temperature, $T < T_{EW}$, (i.e. in the Higgs phase), the transition rate per unit volume is [17]

$$\frac{\Gamma_{B+L}}{V} = k \frac{M_W^7}{(\alpha T)^3} e^{-\beta E_{ph}(T)} \sim e^{\frac{-M_W}{\alpha k T}}, \tag{4.18}$$

where M_W is the mass of the W gauge boson and k is the Boltzmann constant. The transition rate is thus still very suppressed. This result can be extrapolated to high temperature symmetric phase. It was found that, in the symmetric phase, $T \geq T_{EW}$, the transition rate is [18]

$$\frac{\Gamma_{B+L}}{V} \sim \alpha^5 \ln \alpha^{-1} T^4 , \qquad (4.19)$$

where α is the fine-structure constant. Thus for $T > T_{EW}$, baryon number violating processes can be unsuppressed and profuse.

4.1.2.2. C and CP Violation

To illustrate the point that both C and CP violation are necessary in order to have baryogenesis, consider the case [19] in which superheavy X boson have baryon number violating interactions as summarized in Table 4.2. The

Table 4.2. Baryon number violating decays of the superheavy X boson in the toy model.

process	branching fraction	ΔB
$X \to qq$	α	2/3
$X \to \bar{q}\ell$	$1-\alpha$	-1/3
$\overline{X} \to \bar{q}\bar{q}$	$\bar{\alpha}$	-2/3
$\overline{X} \to q\ell$	$1-\bar{\alpha}$	1/3

baryon numbers produced by the decays of X and $\overline{X}$ are,

$$B_X = \alpha \left(\frac{2}{3}\right) + (1-\alpha)\left(-\frac{1}{3}\right) = \alpha - \frac{1}{3} , \qquad (4.20)$$

$$B_{\overline{X}} = \bar{\alpha}\left(-\frac{2}{3}\right) + (1-\bar{\alpha})\left(\frac{1}{3}\right) = -\left(\bar{\alpha} - \frac{1}{3}\right) , \qquad (4.21)$$

respectively. The net baryon number produced by the decays of the X, $\overline{X}$ pair is then,

$$\epsilon \equiv B_X + B_{\overline{X}} = (\alpha - \bar{\alpha}) . \qquad (4.22)$$

If C or CP is conserved, $\alpha = \bar{\alpha}$, it then leads to vanishing total baryon number, $\epsilon = 0$.

To be more concrete, consider a toy model [19] which consists of four fermions, $f_{1,...4}$, and two heavy scalar fields, X and Y. The interactions among these fields are described by the following Lagrangian,

$$\mathcal{L} = g_1 X f_2^\dagger f_1 + g_2 X f_4^\dagger f_3 + g_3 Y f_1^\dagger f_3 + g_4 Y f_2^\dagger f_4 + h.c. , \qquad (4.23)$$

where $g_{1,...,4}$ are the coupling constants. The Lagrangian $\mathcal{L}$ leads to the following decay processes,

$$X \to \overline{f}_1 + f_2, \; \overline{f}_3 + f_4 , \quad (4.24)$$

$$Y \to \overline{f}_3 + f_1, \; \overline{f}_4 + f_2 , \quad (4.25)$$

and the tree level diagrams of these decay processes are shown in Fig. 4.4. At the tree level, the decay rate of $X \to \overline{f}_1 + f_2$ is,

$$\Gamma(X \to \overline{f}_1 + f_2) = |g_1|^2 I_X , \quad (4.26)$$

where I_X is the phase space factor. For the conjugate process $\overline{X} \to f_1 + \overline{f}_2$, the decay rate is,

$$\Gamma(\overline{X} \to f_1 + \overline{f}_2) = |g_1^*|^2 I_{\overline{X}} . \quad (4.27)$$

As the phase space factors I_X and $I_{\overline{X}}$ are equal, no asymmetry can be generated at the tree level.

At the one-loop level, there are additional diagrams, as shown in Fig. 4.5, that have to be taken into account. Including these one-loop contributions, the decay rates for $X \to \overline{f}_1 + f_2$ and $\overline{X} \to f_1 + \overline{f}_2$ become,

$$\Gamma(X \to \overline{f}_1 + f_2) = g_1 g_2^* g_3 g_4^* I_{XY} + c.c. , \quad (4.28)$$

$$\Gamma(\overline{X} \to f_1 + \overline{f}_2) = g_1^* g_2 g_3^* g_4 I_{XY} + c.c. , \quad (4.29)$$

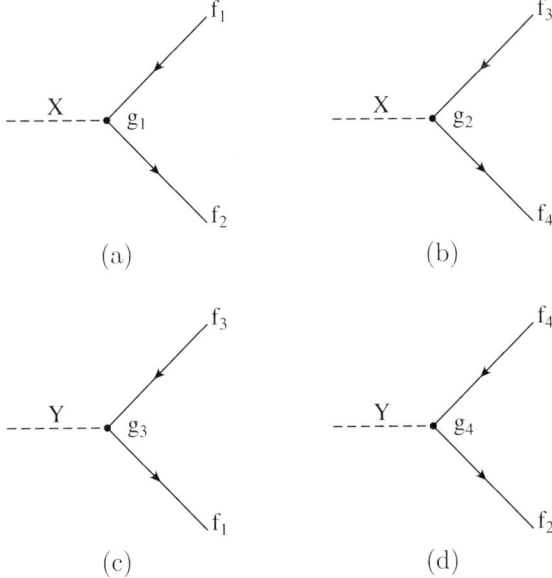

Fig. 4.4. Tree level diagrams for the decays of the heavy scalar fields, X and Y.

where c.c. stands for complex conjugation. Now I_{XY} includes both the phase space factors as well as kinematic factors arising from integrating over the internal loop momentum due to the exchange of J in I decay. If fermions $f_{1,...4}$ are allowed to propagate on-shell, then the factor I_{XY} is complex. Therefore,

$$\Gamma(X \to \overline{f}_1 + f_2) - \Gamma(\overline{X} \to f_1 + \overline{f}_2) = 4\,\text{Im}(I_{XY})\text{Im}(g_1^* g_2 g_3^* g_4)\,. \quad (4.30)$$

Similarly, for the decay mode, $X \to \overline{f}_3 + f_4$, we have,

$$\Gamma(X \to \overline{f}_3 f_4) - \Gamma(\overline{X} \to f_3 + \overline{f}_4) = -4\,\text{Im}(I_{XY})\text{Im}(g_1^* g_2 g_3^* g_4)\,. \quad (4.31)$$

Note that, in addition to the one-loop diagrams shown in Fig. 4.5, there are also diagrams that involve the same boson as the decaying one. However, contributions to the asymmetry from these diagrams vanish as the interference term in this case is proportional to $Im(g_i g_i^* g_i g_i^*) = 0$. The total baryon number asymmetry due to X decays is thus given by,

$$\epsilon_X = \frac{(B_1 - B_2)\Delta\Gamma(X \to \overline{f}_1 + f_2) + (B_4 - B_3)\Delta\Gamma(X \to \overline{f}_3 + f_4)}{\Gamma_X}, \quad (4.32)$$

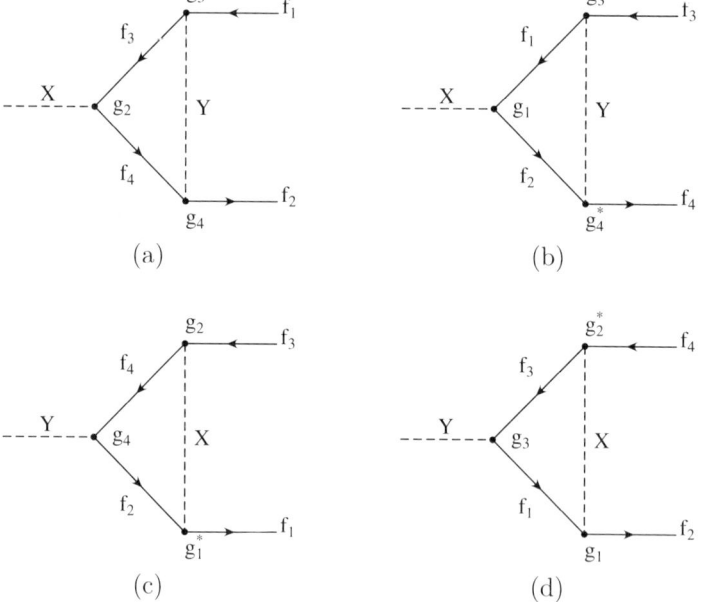

Fig. 4.5. One loop diagrams for the decays of the heavy scalar fields, X and Y, that contribute to the asymmetry.

where

$$\Delta\Gamma(X \to \overline{f}_1 + f_2) = \Gamma(X \to \overline{f}_1 + f_2) - \Gamma(\overline{X} \to f_1 + \overline{f}_2), \quad (4.33)$$
$$\Delta\Gamma(X \to \overline{f}_3 + f_4) = \Gamma(X \to \overline{f}_3 + f_4) - \Gamma(\overline{X} \to f_3 + \overline{f}_4). \quad (4.34)$$

Similar expression can be derived for the Y decays. The total asymmetries due to the decays of the superheavy bosons, X and Y, are then given, respectively, by

$$\epsilon_X = \frac{4}{\Gamma_X}\text{Im}(I_{XY})\text{Im}(g_1^* g_2 g_3^* g_4)[(B_4 - B_3) - (B_2 - B_1)], \quad (4.35)$$
$$\epsilon_Y = \frac{4}{\Gamma_Y}\text{Im}(I'_{XY})\text{Im}(g_1^* g_2 g_3^* g_4)[(B_2 - B_4) - (B_1 - B_3)]. \quad (4.36)$$

By inspecting Eq. 4.35 and 4.36, it is clear that the following three conditions must be satisfied to have a non-zero total asymmetry, $\epsilon = \epsilon_X + \epsilon_Y$:

- The presence of the two baryon number violating bosons, each of which has to have mass greater than the sum of the masses of the fermions in the internal loop;
- The coupling constants have to be complex. The C and CP violation then arise from the interference between the tree level and one-loop diagrams. In general, the asymmetry generated is proportional to $\epsilon \sim \alpha^n$, with n being the number of loops in the lowest order diagram that give non-zero asymmetry and $\alpha \sim g^2/4\pi$;
- The heavy particles X and Y must have non-degenerate masses. Otherwise, $\epsilon_X = -\epsilon_Y$, which leads to vanishing total asymmetry, ϵ.

4.1.2.3. Departure from Thermal Equilibrium

The baryon number B is odd under the C and CP transformations. Using this property of B together with the requirement that the Hamiltonian, H, commutes with CPT, the third condition can be seen by calculating the average of B in equilibrium at temperature $T = 1/\beta$,

$$<B>_T = \text{Tr}[e^{-\beta H} B] = \text{Tr}[(CPT)(CPT)^{-1} e^{-\beta H} B)] \quad (4.37)$$
$$= \text{Tr}[e^{-\beta H}(CPT)^{-1} B (CPT)] = -\text{Tr}[e^{-\beta H} B].$$

In equilibrium, the average $<B>_T$ thus vanishes, and there is no generation of net baryon number. Different mechanisms for baryogenesis differ in the way the departure from thermal equilibrium is realized. There are three possible ways to achieve departure from thermal equilibrium that have been utilized in baryogenesis mechanisms:

- Out-of-equilibrium decay of heavy particles: GUT Baryogenesis, Leptogenesis;
- EW phase transition: EW Baryogenesis;
- Dynamics of topological defects.

In leptogenesis, the departure from thermal equilibrium is achieved through the out-of-equilibrium decays of heavy particles in an expanding Universe. If the decay rate Γ_X of some superheavy particles X with mass M_X at the time when they become non-relativistic (i.e. $T \sim M_X$) is much smaller than the expansion rate of the Universe, the X particles cannot decay on the time scale of the expansion. The X particles will then remain their initial thermal abundance, $n_X = n_{\overline{X}} \sim n_\gamma \sim T^3$, for $T \lesssim M_X$. In other words, at some temperature $T > M_X$, the superheavy particles X are so weakly interacting that they cannot catch up with the expansion of the Universe. Hence they decouple from the thermal bath while still being relativistic. At the time of the decoupling, $n_X \sim n_{\overline{X}} \sim T^3$. Therefore, they populate the Universe at $T \simeq M_X$ with abundance much larger than their abundance in equilibrium. Recall that in equilibrium,

$$n_X = n_{\overline{X}} \simeq n_\gamma \quad \text{for} \quad T \gtrsim M_X, \tag{4.38}$$

$$n_X = n_{\overline{X}} \simeq (M_X T)^{3/2} e^{-M_X/T} \ll n_\gamma \quad \text{for} \quad T \lesssim M_X. \tag{4.39}$$

This over-abundance at temperature below M_X, as shown in Fig. 4.6, is the departure from thermal equilibrium needed to produce a final non-vanishing baryon asymmetry, when the heavy states, X, undergo B and CP violating decays. The scale of rates of these decay processes involving X and $\overline{X}$ relative to the expansion rate of the Universe is determined by M_X,

$$\frac{\Gamma}{H} \propto \frac{1}{M_X}. \tag{4.40}$$

The out-of-equilibrium condition, $\Gamma < H$, thus requires very heavy states: for gauge bosons, $M_X \gtrsim (10^{15-16})$ GeV; for scalars, $M_X \gtrsim (10^{10-16})$ GeV, assuming these heavy particles decay through renormalizable operators. Precise computation of the abundance is carried out by solving the Boltzmann equations (more details in Sec. 4.2.1.2).

4.1.3. Relating Baryon and Lepton Asymmetries

One more ingredient that is needed for leptogenesis is to relate lepton number asymmetry to the baryon number asymmetry, at the high temperature, symmetric phase of the SM [1]. In a weakly coupled plasma with temperature T and volume V, a chemical potential μ_i can be assigned to each of the quark, lepton and Higgs fields, i. There are therefore $5N_f + 1$ chemical potentials in the SM with one Higgs doublet and N_f generations of fermions. The corresponding partition function is given by,

$$Z(\mu, T, V) = \text{Tr}[e^{-\beta(H - \sum_i \mu_i Q_i)}] \qquad (4.41)$$

where $\beta = 1/T$, H is the Hamiltonian and Q_i is the charge operator for the corresponding field. The asymmetry in particle and antiparticle number densities is given by the derivative of the thermal-dynamical potential,

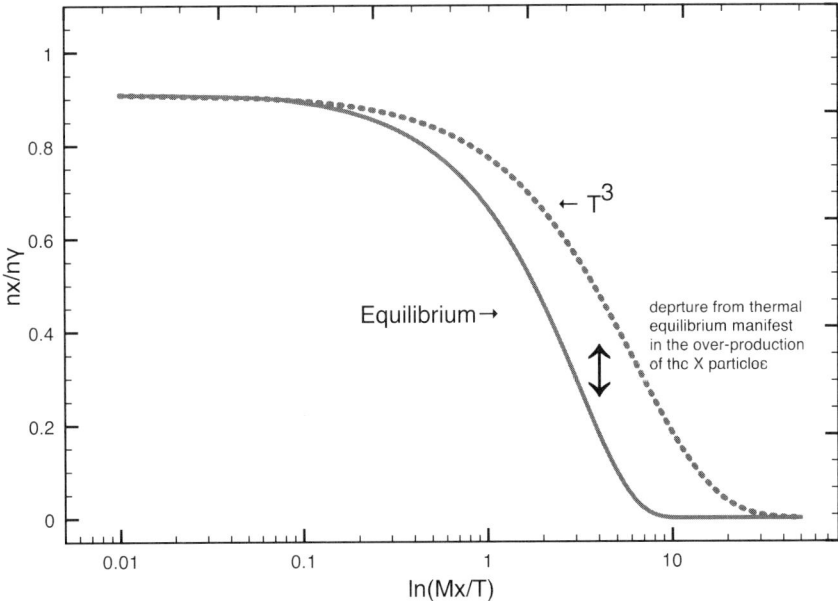

Fig. 4.6. The distribution of the X particles in thermal equilibrium (blue curve) follows Eq. 4.38 and 4.39. When departure from the thermal equilibrium occurs, the distribution of the X particles remains the same as the thermal distribution (red dashed curve).

$\Omega(\mu, T)$, as

$$n_i - \overline{n}_i = -\frac{\partial \Omega(\mu, T)}{\partial \mu_i}, \qquad (4.42)$$

where $\Omega(\mu, T)$ is defined as,

$$\Omega(\mu, T) = -\frac{T}{V} \ln Z(\mu, T, V). \qquad (4.43)$$

For a non-interacting gas of massless particles, assuming $\beta \mu_i \ll 1$,

$$n_i - \overline{n}_i = \frac{1}{6} g T^3 \begin{cases} \beta\mu_i + \mathcal{O}((\beta\mu_i)^3), & \text{fermions} \\ 2\beta\mu_i + \mathcal{O}((\beta\mu_i)^3), & \text{bosons} \end{cases} \qquad (4.44)$$

In the high temperature plasma, quarks, leptons and Higgs interact via the guage and Yukawa couplings. In addition, there are non-perturbative sphaleron processes. All these processes give rise to constraints among various chemical potentials in thermal equilibrium. These include [1]:

(1) The effective 12-fermion interactions $\mathcal{O}_{B+L}$ induced by the sphalerons give rise to the following relation,

$$\sum_i (3\mu_{q_i} + \mu_{\ell_i}) = 0. \qquad (4.45)$$

(2) The SU(3) QCD instanton processes lead to interactions between LH and RH quarks. These interactions are described by the operator, $\prod_i (q_{L_i} q_{L_i} u^c_{R_i} d^c_{R_i})$. When in equilibrium, they lead to,

$$\sum_i (2\mu_{q_i} - \mu_{u_i} - \mu_{d_i}) = 0. \qquad (4.46)$$

(3) Total hypercharge of the plasma has to vanish at all temperatures. This gives,

$$\sum_i \left(\mu_{q_i} + 2\mu_{u_i} - \mu_{d_i} - \mu_{\ell_i} - \mu_{e_i} + \frac{2}{N_f} \mu_H \right) = 0. \qquad (4.47)$$

(4) The Yukawa interactions yield the following relations among chemical potential of the LH and RH fermions,

$$\mu_{q_i} - \mu_H - \mu_{d_j} = 0, \qquad (4.48)$$
$$\mu_{q_i} + \mu_H - \mu_{u_j} = 0, \qquad (4.49)$$
$$\mu_{\ell_i} - \mu_H - \mu_{e_j} = 0. \qquad (4.50)$$

From Eq. (4.44), the baryon number density $n_B = \frac{1}{6}gBT^2$ and lepton number density $n_L = \frac{1}{6}gL_iT^2$, where L_i is the individual lepton flavor number with $i = (e, \mu, \tau)$, can be expanded in terms of the chemical potentials. Hence

$$B = \sum_i (2\mu_{q_i} + \mu_{u_i} + \mu_{d_i}) \tag{4.51}$$

$$L = \sum_i L_i, \quad L_i = 2\mu_{\ell_i} + \mu_{e_i} . \tag{4.52}$$

Consider the case where all Yukawa interactions are in equilibrium. The asymmetry $(L_i - B/N_f)$ is then preserved. If we further assume equilibrium among different generations, $\mu_{\ell_i} \equiv \mu_\ell$ and $\mu_{q_i} \equiv \mu_q$, together with the sphaleron and hypercharge constraints, all the chemical potentials can then be expressed in terms of μ_ℓ,

$$\mu_e = \frac{2N_f + 3}{6N_f + 3}\mu_\ell, \quad \mu_d = -\frac{6N_f + 1}{6N_f + 3}\mu_\ell, \quad \mu_u = \frac{2N_f - 1}{6N_f + 3}\mu_\ell$$
$$\mu_q = -\frac{1}{3}\mu_\ell, \quad \mu_H = \frac{4N_f}{6N_f + 3}\mu_\ell . \tag{4.53}$$

The corresponding B and L asymmetries are

$$B = -\frac{4}{3}N_f \mu_\ell , \tag{4.54}$$

$$L = \frac{14N_f^2 + 9N_f}{6N_f + 3}\mu_\ell . \tag{4.55}$$

Thus B, L and $B - L$ are related by:

$$B = c_s(B - L), \quad L = (c_s - 1)(B - L) , \tag{4.56}$$

where

$$c_s = \frac{8N_f + 4}{22N_f + 13} . \tag{4.57}$$

For models with N_H Higgses, the parameter c_s is given by,

$$c_s = \frac{8N_f + 4N_H}{22N_f + 13N_H} . \tag{4.58}$$

For $T = 100$ GeV $\sim 10^{12}$ GeV, which is of interest of baryogenesis, gauge interactions are in equilibrium. Nevertheless, the Yukawa interactions are in equilibrium only in a more restricted temperature range. But these effects are generally small, and thus will be neglected in these lectures. These effects have been investigated recently; they will be discussed in Sec. 4.5.

4.1.4. *Mechanisms for Baryogenesis and Their Problems*

There have been many mechanisms for baryogenesis proposed. Each has attractive and problematic aspects, which we discuss below.

4.1.4.1. *GUT Baryongenesis*

The GUT baryogenesis was the first implementation of Sakharov's B-number generation idea. The B-number violation is an unavoidable consequence in grand unified models, as quarks and leptons are unified in the same representation of a single group. Furthermore, sufficient amount of CP violation can be incorporated naturally in GUT models, as there exist many possible complex phases, in addition to those that are present in the SM. The relevant time scales of the decays of heavy gauge bosons or scalars are slow, compared to the expansion rate of the Universe at early epoch of the cosmic evolution. The decays of these heavy particles are thus inherently out-of-equilibrium.

Even though GUT models naturally encompass all three Sakharov's conditions, there are also challenges these models face. First of all, to generate sufficient baryon number asymmetry requires high reheating temperature. This in turn leads to dangerous production of relic particles, such as gravitinos (see Sec. 4.2.3). As the relevant physics scale $M_{GUT} \sim 10^{16}$ GeV is far above the electroweak scale, it is also very hard to test GUT models experimentally using colliders. The electroweak theory ensures that there are copious B-violating processes between the GUT scale and the electroweak scale. These sphaleron processes violate $B+L$, but conserve $B-L$. Therefore, unless a GUT mechanism generates an excess of $B-L$, any baryon asymmetry produced will be equilibrated to zero by the sphaleron effects. As $U(1)_{B-L}$ is a gauged subgroup of $SO(10)$, GUT models based on $SO(10)$ are especially attractive for baryogenesis.

4.1.4.2. *EW Baryogenesis*

In electroweak baryogenesis, the departure from thermal equilibrium is provided by strong first order phase transition. The nice feature of this mechanism is that it can be probed in collider experiments. On the other hand, the allowed parameter space is very small. It requires more CP violation than what is provided in the SM. Even though there are additional sources of CP violation in MSSM, the requirement of strong first order phase transition translates into a stringent bound on the Higgs mass, $m_H \lesssim 120$ GeV.

To obtain a Higgs mass of this order, the stop mass needs to be smaller than, or of the order of, the top quark mass, which implies fine-tuning in the model parameters.

4.1.4.3. Affleck-Dine Baryogensis

The Affleck-Dine baryogenesis [20] involves cosmological evolution of scalar fields which carry B charges. It is most naturally implemented in SUSY theories. Nevertheless, this mechanism faces the same challenges as in GUT baryogenesis and in EW baryogenesis.

4.1.5. *Sources of CP Violation*

In the SM, C is maximally broken, since only LH electron couples to the $SU(2)_L$ gauge fields. Furthermore, CP is not an exact symmetry in weak interaction, as observed in the Kaon and B-meson systems. The charged current in the weak interaction basis is given by,

$$\mathcal{L}_W = \frac{g}{\sqrt{2}} \overline{U}_L \gamma^\mu D_L W_\mu + h.c. \;, \tag{4.59}$$

where $U_L = (u, c, t)_L$ and $D_L = (d, s, b)_L$. Quark mass matrices can be diagonalized by bi-unitary transformations,

$$\text{diag}(m_u, m_c, m_t) = V_L^u M^u V_R^u \;, \tag{4.60}$$

$$\text{diag}(m_d, m_s, m_d) = V_L^d M^d V_R^d \;. \tag{4.61}$$

Thus the charged current interaction in the mass eigenstates reads,

$$\mathcal{L}_W = \frac{g}{\sqrt{2}} \overline{U}'_L U_{CKM} \gamma^\mu D'_L W_\mu + h.c. \;, \tag{4.62}$$

where $U'_L = V_L^u U_L$ and $D'_L = V_L^d D_L$ are the mass eigenstates, and $U_{CKM} \equiv V_L^u (V_L^d)^\dagger$ is the CKM matrix. For three families of fermions, the unitary matrix K can be parameterized by three angles and six phases. Out of these six phases, five of them can be reabsorbed by redefining the wave functions of the quarks. There is hence only one physical phase in the CKM matrix. This is the only source of CP violation in the SM. It turns out that this particular source is not strong enough to accommodate the observed matter-antimatter asymmetry. The relevant effects can be parameterized by [21],

$$B \simeq \frac{\alpha_w^4 T^3}{s} \delta_{CP} \simeq 10^{-8} \delta_{CP} \;, \tag{4.63}$$

where δ_{CP} is the suppression factor due to CP violation in the SM. Since CP violation vanishes when any two of the quarks with equal charge have degenerate masses, a naive estimate gives the effects of CP violation of the size

$$A_{CP} = (m_t^2 - m_c^2)(m_c^2 - m_u^2)(m_u^2 - m_t^2) \quad (4.64)$$
$$\cdot (m_b^2 - m_s^2)(m_s^2 - m_d^2)(m_d^2 - m_b^2) \cdot J \,.$$

Here the proportionality constant J is the usual Jarlskog invariant, which is a parameterization independent measure of CP violation in the quark sector. Together with the fact that A_{CP} is of mass (thus temperature) dimension 12, this leads to the following value for δ_{CP}, which is a dimensionless quantity,

$$\delta_{CP} \simeq \frac{A_{CP}}{T_C^{12}} \simeq 10^{-20} \,, \quad (4.65)$$

and T_C is the temperature of the electroweak phase transition. The baryon number asymmetry due to the phase in the CKM matrix is therefore of the order of $B \sim 10^{-28}$, which is too small to account for the observed $B \sim 10^{-10}$.

In MSSM, there are new sources of CP violation due to the presence of the soft SUSY breaking sector. The superpotential of the MSSM is given by,

$$W = \mu \hat{H}_1 \hat{H}_2 + h^u \hat{H}_2 \hat{Q} \hat{u}^c + h^d \hat{H}_1 \hat{Q} \hat{d}^c + h^e \hat{H}_1 \hat{L} \hat{e}^c \,. \quad (4.66)$$

The soft SUSY breaking sector has the following parameters:

- tri-linear couplings: $\Gamma^u H_2 \widetilde{Q} \widetilde{c}^c + \Gamma^d H_1 \widetilde{Q} \widetilde{d}^c + \Gamma^e H_1 \widetilde{L} \widetilde{e}^c + h.c.$, where $\Gamma^{(u,d,e)} \equiv A^{(u,d,e)} \cdot h^{(u,d,e)}$;
- bi-linear coupling in the Higgs sector: $\mu B H_1 H_2$;
- gaugino masses: M_i for $i = 1, 2, 3$ (one for each gauge group);
- soft scalar masses: $\widetilde{m}_f$.

In the constrained MSSM (CMSSM) model with mSUGRA boundary conditions at the GUT scale, a universal value is assumed for the tri-linear coupling constants, $A^{(u,d,e)} = A$. Similarly, the gaugino masses and scalar masses are universal, $M_i = M$, and $\widetilde{m}_f = \widetilde{m}$. Two phases may be removed by redefining the phase of $\hat{H}_2$ such that the phase of μ is opposite to the phase of B. As a result, the product μB is real. Furthermore, the phase of M can be removed by R-symmetry transformation. This then modifies the tri-linear couplings by an additional factor of $e^{-\phi_M}$, while other coupling constants are invariant under the R-symmetry transformation. There are thus two physical phases remain,

$$\phi_A = \text{Arg}(AM), \quad \phi_\mu = -\text{Arg}(B) . \tag{4.67}$$

These phases are relevant in soft leptogenesis, which is discussed in Sec. 4.3.2.

If the neutrinos are massive, the leptonic charged current interaction in the mass eigenstates of the leptons is given by,

$$\mathcal{L}_W = \frac{g}{\sqrt{2}} \bar{\nu}'_L U^\dagger_{MNS} \gamma^\mu \ell'_L W_\mu + h.c. , \tag{4.68}$$

where $U_{MNS} = (V^\nu_L)^\dagger V^e_L$. (For a review on physics of the massive neutrinos, see, e.g. Refs. [22] and [23]. See also Rabi Mohapatra's lectures.) The matrices V^ν_L and V^e_L diagonalize the effective neutrino mass matrix and the charged lepton mass matrix, respectively. If neutrinos are Majorana particles, which is the case if small neutrino mass is explained by the seesaw mechanism [24], the Majorana condition then forbids the phase redefinition of ν_R. Unlike in the CKM matrix, in this case only three of the six complex phases can be absorbed, and there are thus two additional physical phases in the lepton sector if neutrinos are Majorana fermions. And due to this reason, CP violation can occur in the lepton sector with only two families. (Recall that in the quark sector, CP violation can occur only when the number of famalies is at least three). The MNS matrix can be parameterized as a CKM-like matrix and a diagonal phase matrix,

$$U_{MNS} = \begin{pmatrix} c_{12}c_{13} & s_{12}c_{13} & s_{13}e^{-i\delta} \\ -s_{12}c_{23} - c_{12}s_{23}s_{13}e^{i\delta} & c_{12}c_{23} - s_{12}s_{23}s_{13}e^{i\delta} & s_{23}c_{13} \\ s_{12}s_{23} - c_{12}c_{23}s_{13}e^{i\delta} & -c_{12}s_{23} - s_{12}c_{23}s_{13}e^{i\delta} & c_{23}c_{13} \end{pmatrix}$$
$$\cdot \begin{pmatrix} 1 & & \\ & e^{i\alpha_{21}/2} & \\ & & e^{i\alpha_{31}/2} \end{pmatrix} . \tag{4.69}$$

The Dirac phase δ affects neutrino oscillation (see Boris Kayser's lectures),

$$P(\nu_\alpha \to \nu_\beta) = \delta_{\alpha\beta} - 4 \sum_{i>j} \text{Re}(U_{\alpha i} U_{\beta j} U^*_{\alpha j} U^*_{\beta i}) \sin^2\left(\Delta m^2_{ij} \frac{L}{4E}\right) \tag{4.70}$$
$$+ 2 \sum_{i>j} J^{\text{lep}}_{\text{CP}} \sin^2\left(\Delta m^2_{ij} \frac{L}{4E}\right)$$

where the parameterization invariant CP violation measure, the leptonic Jarlskog invariant $J_{\text{CP}}^{\text{lep}}$, is given by,

$$J_{\text{CP}}^{\text{lep}} = -\frac{Im(H_{12}H_{23}H_{31})}{\Delta m_{21}^2 \Delta m_{32}^2 \Delta m_{31}^2}, \quad H \equiv (M_\nu^{eff})(M_\nu^{eff})^\dagger \,. \tag{4.71}$$

The two Majorana phases, α_{21} and α_{31}, affect neutrino double decay (see Petr Vogel's lectures). Their dependence in the neutrinoless double beta decay matrix element is,

$$|\langle m_{ee}\rangle|^2 = m_1^2 |U_{e1}|^4 + m_2^2 |U_{e2}|^4 + m_3^2 |U_{e3}|^4 \tag{4.72}$$
$$+ 2m_1 m_2 |U_{e1}|^2 |U_{e2}|^2 \cos \alpha_{21}$$
$$+ 2m_1 m_3 |U_{e1}|^2 |U_{e3}|^2 \cos \alpha_{31}$$
$$+ 2m_2 m_3 |U_{e2}|^2 |U_{e3}|^2 \cos(\alpha_{31} - \alpha_{21}) \,.$$

The Lagrangian at high energy that describe the lepton sector of the SM in the presence of the right-handed neurinos, ν_{R_i}, is given by,

$$\mathcal{L} = \overline{\ell}_{L_i} i\gamma^\mu \partial_\mu \ell_{L_i} + \overline{e}_{R_i} i\gamma^\mu \partial_\mu e_{R_i} + \overline{N}_{R_i} i\gamma^\mu \partial_\mu N_{R_i} \tag{4.73}$$
$$+ f_{ij} \overline{e}_{R_i} \ell_{L_j} H^\dagger + h_{ij} \overline{N}_{R_i} \ell_{L_j} H - \frac{1}{2} M_{ij} N_{R_i} N_{R_j} + h.c. \,.$$

Without loose of generality, in the basis where f_{ij} and M_{ij} are diagonal, the Yukawa matrix h_{ij} is in general a complex matrix. For 3 families, h has nine phases. Out of these nine phases, three can be absorbed into wave functions of ℓ_{L_i}. Therefore, there are six physical phases remain. Furthermore, a real h_{ij} can be diagonalized by a bi-unitary transformation, which is defined in terms of six mixing angles. After integrating out the heavy Majorana neutrinos, the effective Lagrangian that describes the neutrino sector below the seesaw scale is,

$$\mathcal{L}_{eff} = \overline{\ell}_{L_i} i\gamma^\mu \partial_\mu \ell_{L_i} + \overline{e}_{R_i} i\gamma^\mu \partial_\mu e_{R_i} + f_{ii} \overline{e}_{R_i} \ell_{L_i} H^\dagger \tag{4.74}$$
$$+ \frac{1}{2} \sum_k h_{ik}^T h_{kj} \ell_{L_i} \ell_{L_j} \frac{H^2}{M_k} + h.c. \,.$$

This leads to an effective neutrino Majorana mass matrix whose parameters can be measured at the oscillation experiments. As Majorana mass matrix is symmetric, for three families, it has six independent complex elements and thus six complex phases. Out of these six phases, three of them can be absorbed into the wave functions of the charged leptons. Hence at low energy, there are only three physical phases and three mixing angles in the lepton sector. Going from high energy to low energy, the numbers of mixing angles and phases are thus reduced by half. Due to the presence of the additional mixing angles and complex phases in the heavy neutrino sector, it is generally not possible to connect leptogenesis with low energy CP violation. However, in some specific models, such connection can be established. This will be discussed in more details in Sec. 4.4.

4.2. Standard Leptogenesis

4.2.1. *Standard Leptogenesis (Majorana Neutrinos)*

As mentioned in the previous section, baryon number violation arises naturally in many grand unified theories. In the GUT baryogenesis, the asymmetry is generated through the decays of heavy gauge bosons (denoted by "V" in the following) or leptoquarks (denoted by "S" in the following), which are particles that carry both B and L numbers. In GUTs based on $SU(5)$, the heavy gauge bosons or heavy leptoquarks have the following B-non-conserving decays,

$$V \to \bar{\ell}_L u_R^c, \quad B = -1/3, \quad B - L = 2/3 \qquad (4.75)$$

$$V \to q_L d_R^c, \quad B = 2/3, \quad B - L = 2/3 \qquad (4.76)$$

$$S \to \bar{\ell}_L \bar{q}_L, \quad B = -1/3, \quad B - L = 2/3 \qquad (4.77)$$

$$S \to q_L q_L, \quad B = 2/3, \quad B - L = 2/3. \qquad (4.78)$$

Since $(B - L)$ is conserved, *i.e.* the heavy particles V and S both carry $(B-L)$ charges $2/3$, no $(B-L)$ can be generated dynamically. In addition, due to the sphaleron processes, $\langle B \rangle = \langle B - L \rangle = 0$. In $SO(10)$, $(B - L)$ is spontaneously broken, as it is a gauged subgroup of $SO(10)$. Heavy particles X with $M_X < M_{B-L}$ can then generate a $(B - L)$ asymmetry through their decays. Nevertheless, for $M_X \sim M_{GUT} \sim 10^{15}$ GeV, the CP asymmetry is highly suppressed. Furthermore, one also has to worry about the large reheating temperature $T_{RH} \sim M_{GUT}$ after the inflation, the realization of thermal equilibrium, and in supersymmetric case, the gravitino problem. These difficulties in GUT baryogenesis had led to a lot of interests in EW baryogenesis, which also has its own disadvantages as discussed in Sec. 4.1.4.

The recent advent of the evidence of neutrino masses from various neutrino oscillation experiments opens up a new possibility of generating the asymmetry through the decay of the heavy neutrinos [25]. A particular attractive framework in which small neutrino masses can naturally arise is GUT based on $SO(10)$ (for a review, see, *i.e.* Ref. [22]). $SO(10)$ GUT models accommodate the existence of RH neutrinos,

$$\psi(16) = (q_L,\ u_R^c,\ e_R^c,\ d_R^c,\ \ell_L,\ \nu_R^c)\,, \qquad (4.79)$$

which is unified along with the fifteen known fermions of each family into a single 16-dimensional spinor representation. For hierarchical fermion masses, one easily has

$$M_N \ll M_{B-L} \sim M_{GUT}\,, \qquad (4.80)$$

where $N = \nu_R + \nu_R^c$ is a Majorana fermion. The decays of the right-handed neutrino,

$$N \to \ell H, \quad N \to \bar\ell \overline{H}\,, \qquad (4.81)$$

where H is the $SU(2)$ Higgs doublet, can lead to a lepton number asymmetry. After the sphaleron processes, the lepton number asymmetry is then converted into a baryon number asymmetry.

The most general Lagrangian involving charged leptons and neutrinos is given by,

$$\mathcal{L}_Y = f_{ij}\bar e_{R_i}\ell_{L_j}H^\dagger + h_{ij}\bar\nu_{R_i}\ell_{L_j}H - \frac{1}{2}(M_R)_{ij}\bar\nu_{R_i}^c\nu_{R_j} + h.c.\,. \qquad (4.82)$$

As the RH neutrinos are singlets under the SM gauge group, Majorana masses for the RH neutrinos are allowed by the gauge invariance. Upon the electroweak symmetry breaking, the SM Higgs doublet gets a VEV, $\langle H \rangle = v$, and the charged leptons and the neutrino Dirac masses, which are much smaller than the RH neutrino Majorana masses, are generated,

$$m_\ell = fv, \quad m_D = hv \ll M_R\,. \qquad (4.83)$$

The neutrino sector is therefore described by a 2×2 seesaw matrix as,

$$\begin{pmatrix} 0 & m_D \\ m_D^T & M_R \end{pmatrix}. \qquad (4.84)$$

Diagonalizing this 2×2 seesaw matrix, the light and heavy neutrino mass eigenstates are obtained as,

$$\nu \simeq V_\nu^T \nu_L + V_\nu^* \nu_L^c, \quad N \simeq \nu_R + \nu_R^c \qquad (4.85)$$

with corresponding masses

$$m_\nu \simeq -V_\nu^T m_D^T \frac{1}{M_R} m_D V_\nu, \quad m_N \simeq M_R\,. \qquad (4.86)$$

Here the unitary matrix V_ν is the diagonalization matrix of the neutrino Dirac matrix.

At temperature $T < M_R$, RH neutrinos can generate a lepton number asymmetry by means of out-of-equilibrium decays. The sphaleron processes then convert ΔL into ΔB.

4.2.1.1. *The Asymmetry*

At the tree level, the ith RH neutrino decays into the Higgs doublet and the charged lepton doublet of α flavor, $N_i \to H + \ell_\alpha$, where $\alpha = (e, \mu, \tau)$. The total width of this decay is,

$$\Gamma_{D_i} = \sum_\alpha [\Gamma(N_i \to H + \ell_\alpha) + \Gamma(N_i \to \overline{H} + \overline{\ell}_\alpha)] \qquad (4.87)$$

$$= \frac{1}{8\pi}(hh^\dagger)_{ii} M_i \ .$$

Suppose that the lepton number violating interactions of the lightest right-handed neutrino, N_1, wash out any lepton number asymmetry generated in the decay of $N_{2,3}$ at temperatures $T \gg M_1$. (For effects due to the decays of $N_{2,3}$, see Ref. [26].) In this case with N_1 decay dominating, the final asymmetry only depends on the dynamics of N_1. The out-of-equilibrium condition requires that the total width for N_1 decay, Γ_{D_1}, to be smaller compared to the expansion rate of the Universe at temperature $T = M_1$,

$$\Gamma_{D_1} < H \Big|_{T=M_1} . \qquad (4.88)$$

That is, the heavy neutrinos are not able to follow the rapid change of the equilibrium particle distribution, once the temperature dropped below the mass M_1. Eventually, heavy neutrinos will decay, and a lepton asymmetry is generated due to the CP asymmetry that arises through the interference of the tree level and one-loop diagrams, as shown in Fig. 4.7,

$$\epsilon_1 = \frac{\sum_\alpha [\Gamma(N_1 \to \ell_\alpha H) - \Gamma(N_1 \to \overline{\ell}_\alpha \overline{H})]}{\sum_\alpha [\Gamma(N_1 \to \ell_\alpha H) + \Gamma(N_1 \to \overline{\ell}_\alpha \overline{H})]} \qquad (4.89)$$

$$\simeq \frac{1}{8\pi} \frac{1}{(h_\nu h_\nu)_{11}} \sum_{i=2,3} \mathrm{Im}\{(h_\nu h_\nu^\dagger)_{1i}^2\} \cdot \left[f\left(\frac{M_i^2}{M_1^2}\right) + g\left(\frac{M_i^2}{M_1^2}\right) \right] .$$

In Fig. 4.7, the diagram (b) is the one-lop vertex correction, which gives the term, $f(x)$, in Eq. 4.89 after carrying out the loop integration,

$$f(x) = \sqrt{x}\left[1 - (1+x)\ln\left(\frac{1+x}{x}\right)\right] . \qquad (4.90)$$

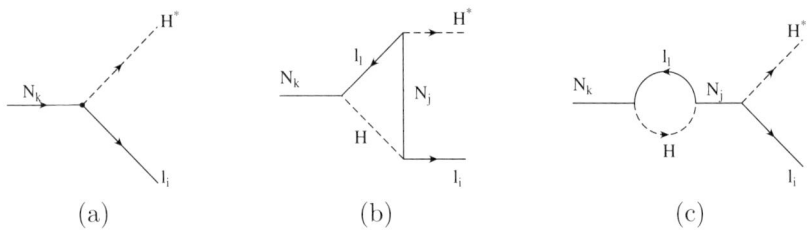

Fig. 4.7. Diagrams in SM with RH neutrinos that contribute to the lepton number asymmetry through the decays of the RH neutrinos. The asymmetry is generated due to the interference of the tree-level diagram (a) and the one-loop vertex correction (b) and self-energy (c) diagrams.

Diagram (c) is the one-loop self-energy. For $|M_i - M_1| \gg |\Gamma_i - \Gamma_1|$, the self-energy diagram gives the term

$$g(x) = \frac{\sqrt{x}}{1-x}, \qquad (4.91)$$

in Eq. 4.89. For hierarchical RH neutrino masses, $M_1 \ll M_2, M_3$, the asymmetry is then given by,

$$\epsilon_1 \simeq -\frac{3}{8\pi} \frac{1}{(h_\nu h_\nu^\dagger)_{11}} \sum_{i=2,3} \text{Im}\{(h_\nu h_\nu^\dagger)^2_{1i}\} \frac{M_1}{M_i}. \qquad (4.92)$$

Note that when N_k and N_j in the self-energy diagram (c) have near degenerate masses, there can be resonant enhancement in the contributions from the self-energy diagram to the asymmetry. Such resonant effect can allow M_1 to be much lower while still generating sufficient amount of the lepton number asymmetry. This will be discussed in Sec. 4.3.1.

To prevent the generated asymmetry given in Eq. 4.89 from being washed out by the inverse decay and scattering processes, the decay of the RH neutrinos has to be out-of-equilibrium. In other words, the condition

$$r \equiv \frac{\Gamma_1}{H|_{T=M_1}} = \frac{M_{pl}}{(1.7)(32\pi)\sqrt{g_*}} \frac{(h_\nu h_\nu^\dagger)_{11}}{M_1} < 1, \qquad (4.93)$$

has to be satisfied. This leads to the following constraint on the effective light neutrino mass

$$\widetilde{m}_1 \equiv (h_\nu h_\nu^\dagger)_{11} \frac{v^2}{M_1} \simeq 4\sqrt{g_*} \frac{v^2}{M_{pl}} \frac{\Gamma_{D_1}}{H}\bigg|_{T=M_1} < 10^{-3} \text{ eV}, \qquad (4.94)$$

where g_* is the number of relativistic degrees of freedom. For SM, $g_* \simeq 106.75$, while for MSSM, $g_* \simeq 228.75$. The wash-out effect is parameterized by the coefficient κ, and the final amount of lepton asymmetry is given by,

$$Y_L \equiv \frac{n_L - \bar{n}_L}{s} = \kappa \frac{\epsilon_1}{g_*}, \tag{4.95}$$

where κ parameterizes the amount of wash-out due to the inverse decays and scattering processes. The amount of wash-out depends on the size of the parameter r:

(1) If $r \ll 1$ for decay temperature $T_D \lesssim M_X$, the inverse decay and 2-2 scattering are impotent. In this case, the inverse decay width is given by,

$$\frac{\Gamma_{ID}}{H} \sim \left(\frac{M_X}{T}\right)^{3/2} e^{-M_X/T} \cdot r, \tag{4.96}$$

while the width for the scattering processes is,

$$\frac{\Gamma_S}{H} \sim \alpha \left(\frac{T}{M_X}\right)^5 \cdot r. \tag{4.97}$$

Thus the inverse decays and scattering processes can be safely ignored, and the asymmetry ΔB produced by decays is not destroyed by the asymmetry $-\Delta B$ produced in inverse decays and scatterings. At $T \simeq T_D$, the number density of the heavy particles X has thermal distribution, $n_X \simeq n_{\overline{X}} \simeq n_\gamma$. Thus the net baryon neumber density produced by out-of-equilibrium decays is

$$n_L = \epsilon_1 \cdot n_X \simeq \epsilon_1 \cdot n_\gamma. \tag{4.98}$$

(2) For $r \gg 1$, the abundance of X and $\overline{X}$ follows the equilibrium values, and there is no departure from thermal equilibrium. As a result, no lepton number may evolve, and the net lepton asymmetry vanishes,

$$\frac{n_\ell - n_{\bar{\ell}}}{dt} + 3H(n_\ell - n_{\bar{\ell}}) = \Delta \gamma^{eq} = 0. \tag{4.99}$$

In general, for $1 < r < 10$, there could still be sizable asymmetry. The wash out effects due to inverse decay and lepton number violating scattering processes together with the time evolution of the system is then accounted for by the factor κ, which is obtained by solving the Boltzmann equations

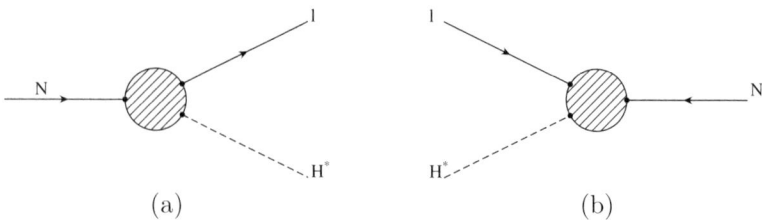

Fig. 4.8. Decay and inverse decay processes in the thermal bath.

for the system (see next section). An approximation is given by [19],

$$10^6 \lesssim r : \kappa = (0.1r)^{1/2} e^{-\frac{4}{3}(0.1)^{1/4}} \quad (< 10^{-7}) \quad (4.100)$$

$$10 \lesssim r \lesssim 10^6 : \kappa = \frac{0.3}{r(\ln r)^{0.8}} \quad (10^{-2} \sim 10^{-7}) \quad (4.101)$$

$$0 \lesssim r \lesssim 10 : \kappa = \frac{1}{2\sqrt{r^2+9}} \quad (10^{-1} \sim 10^{-2}) . \quad (4.102)$$

The EW sphaleron effects then convert Y_L into Y_B,

$$Y_B \equiv \frac{n_B - n_{\overline{B}}}{s} = cY_{B-L} = \frac{c}{c-1} Y_L , \quad (4.103)$$

where c is the conversion factor derived in Sec. 4.1.3.

4.2.1.2. Boltzmann Equations

As the decays of RH neutrinos are out-of-equilibrium processes, they are generally treated by Boltzmann equations. Main processes in the thermal bath that are relevant for leptogenesis include,

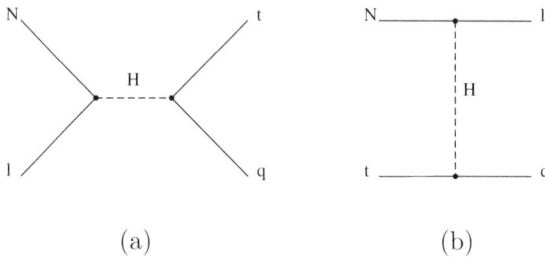

Fig. 4.9. The $\Delta L = 1$ scattering processes in the thermal bath.

(1) decay of N (Fig. 4.8 (a)):

$$N \to \ell + H, \quad N \to \overline{\ell} + \overline{H} \quad (4.104)$$

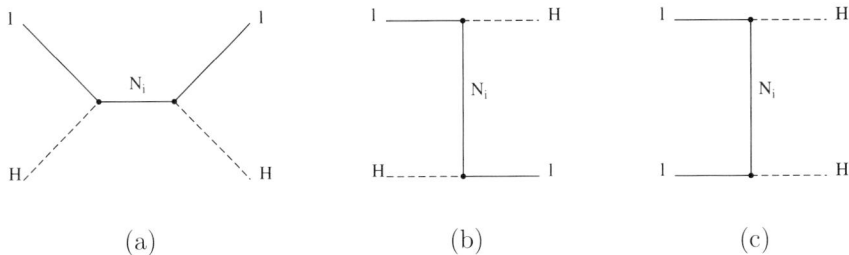

Fig. 4.10. The $\Delta L = 2$ scattering processes in the thermal bath.

(2) inverse decay of N (Fig. 4.8 (b)):

$$\ell + H \to N, \qquad \bar{\ell} + \overline{H} \to N \qquad (4.105)$$

(3) 2-2 scattering: These include the following $\Delta L = 1$ scattering processes (Fig. 4.9),

$$[\text{s-channel}]: \quad N_1 \ell \leftrightarrow t\bar{q} \;, \quad N_1 \bar{\ell} \leftrightarrow \overline{t}\,q \qquad (4.106)$$
$$[\text{t-channel}]: \quad N_1 t \leftrightarrow \bar{\ell} q \;, \quad N_1 \bar{t} \leftrightarrow \ell\bar{q} \qquad (4.107)$$

and $\Delta L = 2$ scattering processes (Fig. 4.10),

$$\ell H \leftrightarrow \bar{\ell}\,\overline{H} \;, \quad \ell\ell \leftrightarrow \overline{H}\,\overline{H}, \;\; \bar{\ell}\bar{\ell} \leftrightarrow HH \;. \qquad (4.108)$$

Basically, at temperatures $T \gtrsim M_1$, these $\Delta L = 1$ and $\Delta L = 2$ processes have to be strong enough to keep N_1 in equilibrium. Yet at temperature $T \lesssim M_1$, these processes have to be weak enough to allow N_1 to generate an asymmetry.

The Boltzmann equations that govern the evolutions of the RH neutrino number density and $B - L$ number density are given by [27],

$$\frac{dN_{N_1}}{dz} = -(D + S)(N_{N_1} - N_{N_1}^{eq}) \qquad (4.109)$$

$$\frac{dN_{B-L}}{dz} = -\epsilon_1 D(N_{N_1} - N_{N_1}^{eq}) - W N_{B-L} \;, \qquad (4.110)$$

where

$$(D, S, W) \equiv \frac{(\Gamma_D, \Gamma_S, \Gamma_W)}{Hz}, \quad z = \frac{M_1}{T} \;. \qquad (4.111)$$

Here Γ_D includes both decay and inverse decay, Γ_S includes $\Delta L = 1$ scattering processes and Γ_W includes inverse decay and $\Delta L = 1$, $\Delta L = 2$

scattering processes. The N_1 abundance is affected by the decay, inverse decay and the $\Delta L = 1$ scattering processes. It is manifest in Eq. 4.110 that the N_1 decay is the source for $(B - L)$, while the inverse decay and the $\Delta L = 1, 2$ scattering processes wash out the asymmetry. The generic behavior of the solutions to the Boltzmann equations is shown in Fig. 4.11.

4.2.1.3. Bounds on Neutrino Masses

In the case with strongly hierarchical right-handed neutrino masses, when the asymmetry ϵ_1 due to the decay of the lightest right-handed neutrino, N_1, contribute dominantly to the total asymmetry, leptogenesis becomes very predictive [1, 4, 27], provided that N_1 decays at temperature $T \gtrsim 10^{12}$ GeV. In particular, various bounds on the neutrino masses can be obtained.

For strongly hierarchyical masses, $M_1/M_2 \ll 1$, there is an upper bound

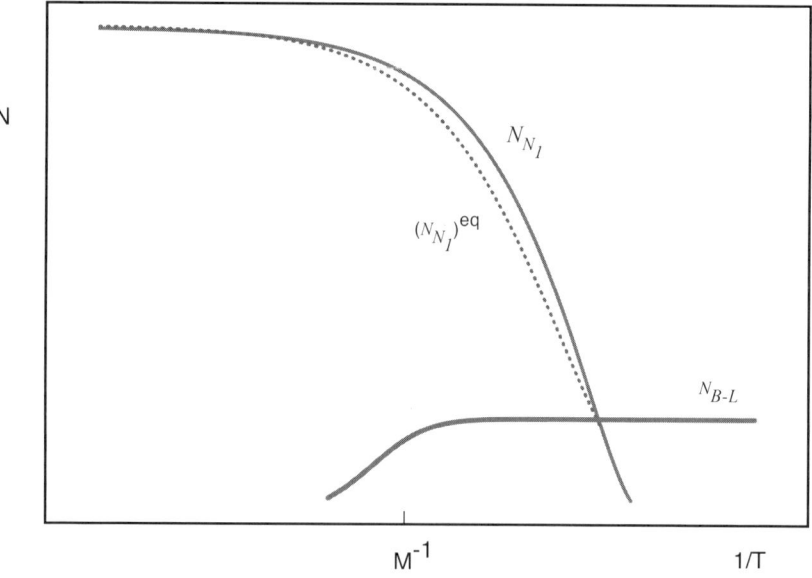

Fig. 4.11. Generic behavior of the solutions to Boltzmann equations. Here the functions N_{N_1} (red solid curve) and N_{B-L} (green solid curve) are solutions to Eq. 4.109 and 4.110. The function $(N_{N_1})^{eq}$ (blue dotted curve) is the equilibrium particle distribution.

on ϵ_1 [29], called the "Davidson-Ibarra" bound,

$$|\epsilon_1| \leq \frac{3}{16\pi} \frac{M_1(m_3 - m_2)}{v^2} \equiv \epsilon_1^{DI}, \qquad (4.112)$$

which is obtained by expanding ϵ_1 to leading order in M_1/M_2. Becuase $|m_3 - m_2| \leq \sqrt{\Delta m_{32}^2} \sim 0.05$ eV, a lower bound on M_1 then follows,

$$M_1 \geq 2 \times 10^9 \text{ GeV} . \qquad (4.113)$$

This bound in turn implies a lower bound on the reheating temperature, T_{RH}, and is in conflict with the upper bound from gravitino over production constraints if supersymmetry is incorporated. We will come back to this in Sec. 4.2.3. One should note that, in the presence of degenerate light neutrinos, the leading terms in an expansion of ϵ_1 in M_1/M_2 and M_1/M_3 vanish. However, the next to leading order terms do not vanish and in this case one has [30],

$$|\epsilon_1| \lesssim \text{Max}\left(\epsilon^{DI}, \frac{M_3^3}{M_1 M_2^2}\right). \qquad (4.114)$$

By requiring that there is no substantial washout effects, bounds on light neutrino masses can be derived. To have significant amount of baryon asymmetry, the effective mass $\widetilde{m}_1$ defined in Eq. 4.94 cannot be too large. Generally $\widetilde{m}_1 \lesssim 0.1 - 0.2$ is required. As the mass of the lightest active neutrino $m_1 \lesssim \widetilde{m}_1$, an upper bound on m_1 thus ensues. By further requiring the $\Delta L = 2$ washout effects be consistent with successful leptogenesis impose a bound on,

$$\sqrt{(m_1^2 + m_2^2 + m_3^2)} \lesssim (0.1 - 0.2) \text{ eV} , \qquad (4.115)$$

which is of the same order as the bound on $\widetilde{m}_1$. From these bounds, the absolute mass scale of neutrino masses is thus known up to a factor of ~ 3 to be in the range, $0.05 \lesssim m_3 \lesssim 0.15$ eV [4], if the observed baryonic asymmetry indeed originates from leptogenesis through the decay of N_1.

4.2.2. *Dirac Leptogenesis*

In the standard leptogenesis discussed in the previous section, neutrinos acquire their masses through the seesaw mechanism. The decays of the heavy right-handed neutrinos produce a non-zero lepton number asymmetry, $\Delta L \neq 0$. The electroweak sphaleron effects then convert ΔL partially into ΔB. This standard scenario relies crucially on the violation of lepton number, which is due to the presence of the heavy Majorana masses for the right-handed neutrinos.

It was pointed out [31] that leptogenesis can be implemented even in the case when neutrinos are Dirac fermions which acquire small masses through highly suppressed Yukawa couplings without violating lepton number. The realization of this depends critically on the following three characteristics of the sphaleron effects: (*i*) only the left-handed particles couple to the sphalerons; (*ii*) the sphalerons change $(B + L)$ but not $(B - L)$; (*iii*) the sphaleron effects are in equilibrium for $T \gtrsim T_{EW}$.

As the sphelarons couple only to the left-handed fermions, one may speculate that as long as the lepton number stored in the right-handed fermions can survive below the electroweak phase transition, a net lepton number may be generated even with $L = 0$ initially. The Yukawa couplings of the SM quarks and leptons to the Higgs boson lead to rapid left-right equilibration so that as the sphaleron effects deplete the left-handed $(B+L)$, the right-handed $(B + L)$ is converted to fill the void and therefore it is also depleted. So with $B = L = 0$ initially, no baryon asymmetry can be generated for the SM quarks and leptons. For the neutrinos, on the other hand, the left-right equilibration can occur at a much longer time scale compared to the electroweak epoch when the sphaleron washout is in effect. The left-right conversion for the neutrinos involves the Dirac Yukawa couplings, $\lambda \bar{\ell}_L H \nu_R$, where λ is the Yukawa coupling constant, and the rate for these conversion processes scales as,

$$\Gamma_{LR} \sim \lambda^2 T . \qquad (4.116)$$

For the left-right conversion not to be in equilibrium at temperatures above some critical temperature T_{eq}, requires that

$$\Gamma_{LR} \lesssim H , \quad \text{for} \quad T > T_{eq} , \qquad (4.117)$$

where the Hubble constant scales as,

$$H \sim \frac{T^2}{M_{\text{Pl}}} . \qquad (4.118)$$

Hence the left-right equilibration can occur at a much later time, $T \lesssim T_{eq} \ll T_{EW}$, provided,

$$\lambda^2 \lesssim \frac{T_{eq}}{M_{\text{Pl}}} \ll \frac{T_{EW}}{M_{\text{Pl}}} . \qquad (4.119)$$

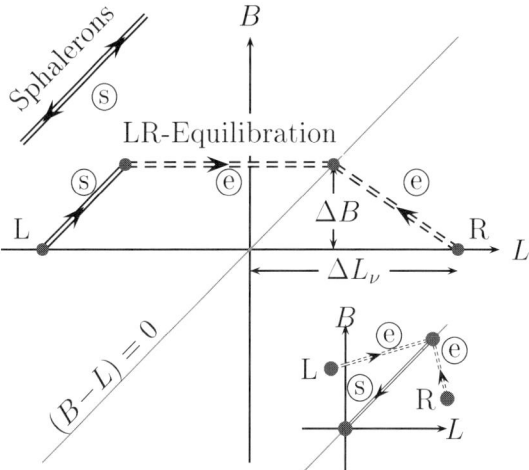

Fig. 4.12. With sufficiently small Yukawa couplings, the left-right equilibration occurs at a much later time, well below the electroweak phase transition temperature. It is therefore possible to generate a non-zero baryon number even if $B = L = 0$ initially. For the SM particles, as shown in the insert for comparison, the left-right equilibration takes place completely before or during the sphaleron processes. Thus no net baryon number can be generated if $B - L = 0$ initially. Figure taken from Ref [31].

With $M_{\rm Pl} \sim 10^{19}$ GeV and $T_{EW} \sim 10^2$ GeV, this condition then translates into

$$\lambda < 10^{-(8 \sim 9)} \ . \tag{4.120}$$

Thus for neutrino Dirac masses $m_D < 10$ keV, which is consistent with all experimental observations, the left-right equilibration does not occur until the temperature of the Universe drops to much below the temperature of the electroweak phase transition, and the lepton number stored in the right-handed neutrinos can then survive the wash-out due to the sphalerons [31].

Once we accept this, the Dirac leptogenesis then works as follows. Suppose that some processes initially produce a negative lepton number (ΔL_L), which is stored in the left-handed neutrinos, and a positive lepton number (ΔL_R), which is stored in the right-handed neutrinos. Because sphalerons only couple to the left-handed particles, part of the negative lepton number stored in left-handed neutrinos get converted into a positive baryon number by the electroweak anomaly. This negative lepton number ΔL_L with reduced magnitude eventually equilibrates with the positive lepton number, ΔL_R when the temperature of the Universe drops to $T \ll T_{EW}$. Because the equilibrating processes conserve both the baryon number B and the lepton number L separately, they result in a Universe with a total positive baryon number and a total positive lepton number. And hence a net baryon number can be generated even with $B = L = 0$ initially.

Such small neutrino Dirac Yukawa couplings required to implement Dirac leptogensis are realized in a SUSY model proposed in Ref. [32].

4.2.3. *Gravitino Problem*

For leptogenesis to be effective, as shown in Sec. 4.2.1.3, the mass of the lightest RH neutrino has to be $M_1 > 2 \times 10^9$ GeV. Figure 4.13 shows the lower bound on the lightest RH neutrino mass as a function of the low energy effective lightest neutrino mass, $\widetilde{m}_1$ [28, 33]. If RH neutrinos are produced thermally, the reheating temperature has to be greater than the right-handed neutrino mass, $T_{RH} > M_R$. This thus implies that $T_{RH} > 2 \times 10^9$ GeV, in order to generate sufficient baryon number asymmetry. Such a high reheating temperature is problematic as it could lead to overproduction of light states, such as gravitinos [34, 35]. If gravitinos are stable (i.e. LSP),

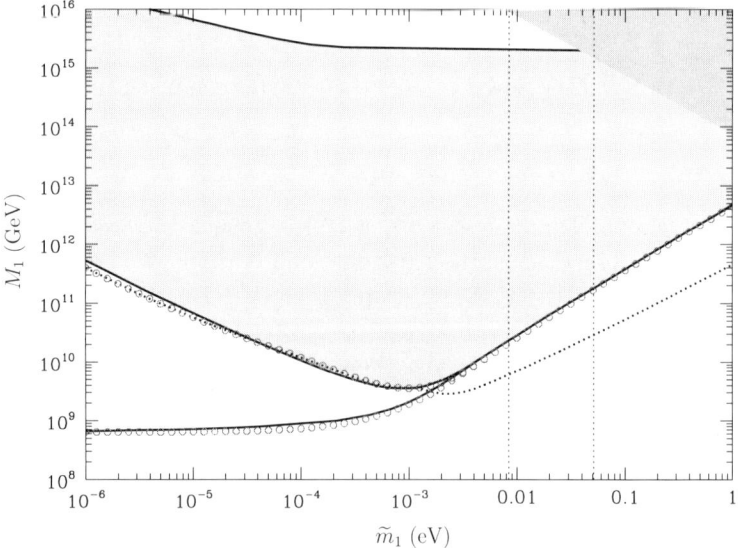

Fig. 4.13. Lower bound on the lightest RH neutrino mass, M_1 (circles) and the initial temperature, T_i (dotted line), for $m_1 = 0$ and $\eta_B^{CMB} = 6 \times 10^{-10}$. The red circles (solid lines) denote the analytical (numerical) results. The vertical dashed lines indicate the range ($\sqrt{\Delta m_{\text{sol}}^2}, \sqrt{\Delta m_{\text{atm}}^2}$). Figure taken from Ref. [28].

Table 4.3. Photo-dissociation reactions that the high energy photons can participate in. The light elements may be destroyed through these reactions and thus their abundance may be changed.

reaction	threshold (MeV)
$D + \gamma \to n + p$	2.225
$T + \gamma \to n + D$	6.257
$T + \gamma \to p + n + n$	8.482
$^3He + \gamma \to p + D$	5.494
$^4He + \gamma \to p + T$	19.815
$^4He + \gamma \to n + ^3He$	20.578
$^4He + \gamma \to p + n + D$	26.072

WMAP constraint on DM leads to stringent bound on gluino mass for any given gravitino mass $m_{3/2}$ and reheating temperature T_{RH}. (Bounds on other gaugino masses can also be obtained as discussed in [36].) On the other hand, if gravitinos are unstable, it has long lifetime and can decay during and after the BBN, and may have the following three effects on BBN [1]:

(1) These decays can speed up cosmic expansion, and increase the neutron to proton ratio and thus the ^{4}He abundance;
(2) Radiation decay of gravitinos, $\psi \to \gamma + \tilde{\gamma}$, increases the photon density and thus reduces the n_B/n_γ ratio;
(3) High energy photons emitted in gravitino decays can destroy light elements (D, T, ^{3}He, ^{4}He) through photo-dissociation reactions such as those given in Table 4.3;

The gravitino number density, $n_{3/2}$, during the thermalization stage after the inflation is governed by the following Boltzmann equation [35],

$$\frac{d}{dt}n_{3/2} + 3Hn_{3/2} \simeq \left\langle \sum_{\text{tot}} v \right\rangle \cdot n_{\text{light}}^2 \qquad (4.121)$$

where $\sum_{\text{tot}} \sim 1/M_{\text{Pl}}^2$ is the total cross section determining the production rate of gravitinos and $n_{\text{light}} \sim T^3$ is the number density of light particles in the thermal bath. As the thermalization is very fast, the friction term $3Hn_{3/2}$ in the above Boltzmann equation can be neglected. Using the fact that the Universe is radiation dominant, $H \sim t^{-1} \sim T^2/M_{\text{Pl}}$, it follows that,

$$n_{3/2} \sim \frac{T^4}{M_{\text{Pl}}}, \qquad (4.122)$$

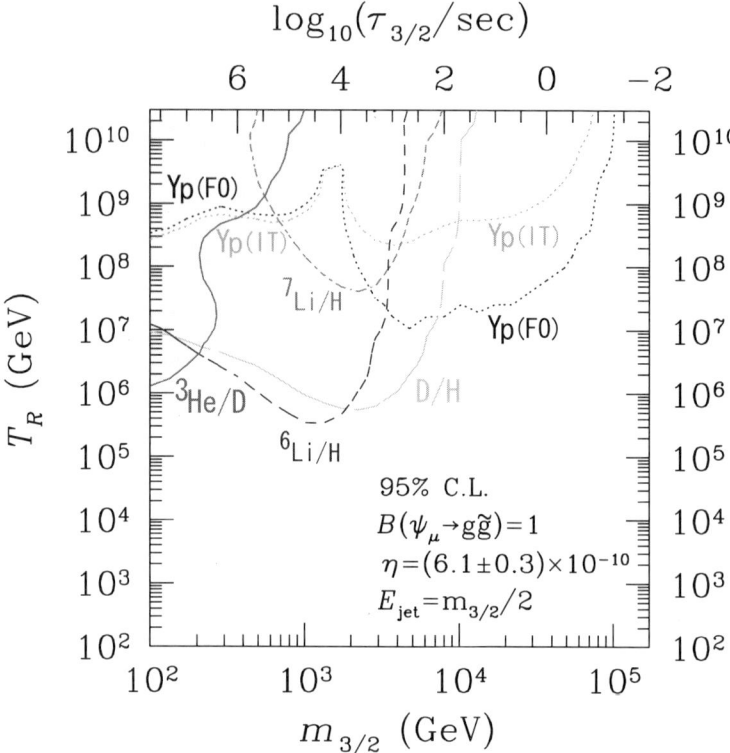

Fig. 4.14. Upper bound on reheating temperature as a function of the gravitino mass, for the case when gravitino dominant decays into a gluon-gluino pair. Figure taken from Ref. [37].

and the number density at thermalization in unit of entropy then reads,

$$\frac{n_{3/2}}{s} \simeq 10^{-2} \frac{T_{RH}}{M_{Pl}} \;. \tag{4.123}$$

The observed abundances for various light elements are,

$$0.22 < Y_p = (\rho_{^4He}/\rho_B)_p < 0.24 \;, \tag{4.124}$$

$$(n_D/n_H) > 1.8 \times 10^{-5} \;, \tag{4.125}$$

$$\left(\frac{n_D + n_{^3He}}{n_H}\right)_p < 10^{-4} \;. \tag{4.126}$$

Table 4.4. Upper bound on the reheating temperature for different values of gravitino mass.

Gravitino mass $m_{3/2}$	Upper bound on T_R
$\lesssim 100$ GeV	10^{6-7} GeV
100 GeV -1 TeV	10^{7-9} GeV
1 TeV -3 TeV	10^{9-12} GeV
3 TeV -10 TeV	10^{12} GeV

The most stringent constraint is from the abundance of (D + ^{3}He) which requires the gravitino number density to be

$$\frac{n_{3/2}}{s} \simeq 10^{-2}\frac{T_{RH}}{M_{\text{Pl}}} \lesssim 10^{-12}. \qquad (4.127)$$

The constraint $T_{RH} < 10^{8-9}$ GeV then follows. More recently, it has been shown that, for hadronic decay modes, $\psi \to g + \tilde{g}$, the bounds are even more stringent, $T_R < 10^{6-7}$ GeV, for gravitino mass $m_{3/2} \sim 100$ GeV [37]. Fig. 4.14 shows the upper bound on the reheating temperature, T_R, for different values of gravitino mass, $m_{3/2}$. Table 4.4 summarizes the numerical results for the upper bound on T_R for various values of $m_{3/2}$. It has also been pointed out [38] that including recent constraint from ^6Li, a new upper bound $T_R < 10^7$ GeV can be derived for the case of gravitino LSP in the constrained minimal supersymmetric Standard Model (CMSSM).

There is therefore a conflict between generation of sufficient amount of leptogenesis and not overly producing gravitinos. To avoid these conflicts, various non-standard scenarios for leptogenesis have been proposed. These are discussed in the next section.

4.3. Non-standard Scenarios

There are a few non-standard scenarios proposed to evade the gravitino over-production problem. In these new scenarios, the conflicts between leptogenesis and gravitino over-production problem are overcome by, (i) resonant enhancement in the self-energy diagrams due to near degenerate right-handed neutrino masses (resonant leptogenesis); (ii) relaxing the relation between the lepton number asymmetry and the right-handed neutrino mass (soft leptogenesis); (iii) relaxing the relation between the reheating temperature and the right-handed neutrino mass (non-thermal leptogenesis). These scenarios are discussed below.

4.3.1. Resonant Leptogenesis

Recall that in the standard leptogenesis discussed in Sec. 4.2, contributions to the CP asymmetry is due to the interference between the tree-level and the one-loop diagrams, that include the vertex correction and self-energy diagrams. It was pointed out in Ref. [39] that in the limit $M_{N_i} - M_{N_j} \ll M_{N_i}$, the self-energy diagrams dominate,

$$\epsilon_{N_i}^{\text{Self}} = \frac{Im[(h_\nu h_\nu^\dagger)_{ij}]^2}{(h_\nu h_\nu^\dagger)_{ii}(h_\nu h_\nu^\dagger)_{jj}} \left[\frac{(M_i^2 - M_j^2)M_i \Gamma_{N_j}}{(M_i^2 - M_j^2)^2 + M_i^2 \Gamma_{N_j}^2} \right]. \qquad (4.128)$$

When the lightest two RH neutrinos have near degenerate masses, $M_1^2 - M_2^2 \sim \Gamma_{N_2}^2$, the asymmetry can be enhanced. To be more specific, CP asymmetry of $\mathcal{O}(1)$ is possible, when

$$M_1 - M_2 \sim \frac{1}{2}\Gamma_{N_{1,2}}, \quad \text{assuming} \quad \frac{Im(h_\nu h_\nu^\dagger)_{12}^2}{(h_\nu h_\nu^\dagger)_{11}(h_\nu h_\nu^\dagger)_{22}} \sim 1. \qquad (4.129)$$

Due to this resonant effect, the bound on the RH neutrino mass scale from the requirement of generating sufficient lepton number asymmetry can be significantly lower. It has been shown that sufficient baryogenesis can be obtained even with $M_{1,2} \sim$ TeV [40].

4.3.2. Soft Leptogenesis

CP violation in leptogenesis can arise in two ways: it can arise in decays, which is the case in standard leptogenesis described in the previous section. It can also arise in mixing. An example of this is the soft leptogenesis. Recall that in the Kaon system, non-vanishing CP violation exists due to the mismatch between CP eigenstates and mass eigenstates (for a review, see for example, Ref. [41]). The CP eigenstates of the K^0 system are $\frac{1}{\sqrt{2}}(|K^0\rangle \pm |\overline{K}^0\rangle)$. The time evolution of the $(K^0, \overline{K}^0)$ system is described by the following Schrödinger equation,

$$\frac{d}{dt}\begin{pmatrix} K^0 \\ \overline{K}^0 \end{pmatrix} = \mathcal{H} \begin{pmatrix} K^0 \\ \overline{K}^0 \end{pmatrix} \qquad (4.130)$$

where the Hamiltonian $\mathcal{H}$ is given by $\mathcal{H} = \mathcal{M} - \frac{i}{2}\mathcal{A}$. Here, the off-diagonal matrix element $\mathcal{M}_{12}$ describes the dispersive part of the transition amplitude, while the element $\mathcal{A}_{12}$ gives the absorptive part of the amplitude. The physical (mass) eigenstates, $|K_{L,S}\rangle$, are given in terms of the flavor eigenstates, $|K^0\rangle$ and $|\overline{K}^0\rangle$, as

$$|K_L\rangle = p|K^0\rangle + q|\overline{K}^0\rangle \qquad (4.131)$$

$$|K_S\rangle = p|K^0\rangle - q|\overline{K}^0\rangle . \qquad (4.132)$$

To have non-vanishing CP violation requires that there exists a mismatch between the CP eigenstates and the physical eigenstates. This in turn implies,

$$\left|\frac{q}{p}\right| \neq 1, \quad \text{where} \quad \left(\frac{q}{p}\right)^2 = \left(\frac{2\mathcal{M}_{12}^* - i\mathcal{A}_{12}^*}{2\mathcal{M}_{12} - i\mathcal{A}_{12}}\right). \qquad (4.133)$$

For soft leptogenesis, the relevant soft SUSY Lagrangian that involves lightest RH sneutrinos $\widetilde{\nu}_{R_1}$ is the following,

$$-\mathcal{L}_{soft} = \left(\frac{1}{2}BM_1\widetilde{\nu}_{R_1}\widetilde{\nu}_{R_1} + A\mathcal{Y}_{1i}\widetilde{L}_i\widetilde{\nu}_{R_1}H_u + h.c.\right)$$
$$+\widetilde{m}^2\widetilde{\nu}_{R_1}^\dagger\widetilde{\nu}_{R_1} . \qquad (4.134)$$

This soft SUSY Lagrangian and the superpotential that involves the lightest RH neutrino, N_1,

$$W = M_1 N_1 N_1 + \mathcal{Y}_{1i} L_i N_1 H_u \qquad (4.135)$$

give rise to the following interactions

$$-\mathcal{L}_A = \widetilde{\nu}_{R_1}(M_1 Y_{1i}^* \widetilde{\ell}_i^* H_u^* + \mathcal{Y}_{1i}\widetilde{\overline{H}}_u \ell_L^i + A\mathcal{Y}_{1i}\widetilde{\ell}_i H_u) + h.c. \quad , \qquad (4.136)$$

and mass terms (to leading order in soft SUSY breaking terms),

$$-\mathcal{L}_\mathcal{M} = (M_1^2 \widetilde{\nu}_{R_1}^\dagger \widetilde{\nu}_{R_1} + \frac{1}{2}BM_1\widetilde{\nu}_{R_1}\widetilde{\nu}_{R_1}) + h.c. . \qquad (4.137)$$

Diagonalization of the mass matrix $\mathcal{M}$ for the two states $\widetilde{\nu}_{R_1}$ and $\widetilde{\nu}_{R_1}^\dagger$ leads to eigenstates $\widetilde{N}_+$ and $\widetilde{N}_-$ with masses,

$$M_\pm \simeq M_1\left(1 \pm \frac{|B|}{2M_1}\right), \qquad (4.138)$$

where the leading order term M_1 is the F-term contribution from the superpotential (RH neutrino mass term) and the mass difference between the two mass eigenstates $\widetilde{N}_+$ and $\widetilde{N}_-$ is induced by the SUSY breaking B term. The time evolution of the $\widetilde{\nu}_{R_1}$-$\widetilde{\nu}_{R_1}^\dagger$ system is governed by the Schrödinger equation,

$$\frac{d}{dt}\begin{pmatrix}\widetilde{\nu}_{R_1}\\ \widetilde{\nu}_{R_1}^\dagger\end{pmatrix} = \mathcal{H}\begin{pmatrix}\widetilde{\nu}_{R_1}\\ \widetilde{\nu}_{R_1}^\dagger\end{pmatrix}, \qquad (4.139)$$

where the Hamiltonian is $\mathcal{H} = \mathcal{M} - \frac{i}{2}\mathcal{A}$ with $\mathcal{M}$ and $\mathcal{A}$ being [42, 43],

$$\mathcal{M} = \begin{pmatrix} 1 & \frac{B^*}{2M_1} \\ \frac{B}{2M_1} & 1 \end{pmatrix} M_1 , \qquad (4.140)$$

$$\mathcal{A} = \begin{pmatrix} 1 & \frac{A^*}{M_1} \\ \frac{A}{M_1} & 1 \end{pmatrix} \Gamma_1 . \qquad (4.141)$$

For the decay of the lightest RH sneutrino, $\widetilde{\nu}_{R_1}$, the total decay width Γ_1 is given by, in the basis where both the charged lepton mass matrix and the RH neutrino mass matrix are diagonal,

$$\Gamma_1 = \frac{1}{4\pi}(\mathcal{Y}_\nu \mathcal{Y}_\nu^\dagger)_{11} M_1 . \qquad (4.142)$$

The eigenstates of the Hamiltonian $\mathcal{H}$ are $\widetilde{N}'_\pm = p\widetilde{N} \pm q\widetilde{N}^\dagger$, where $|p|^2 + |q|^2 = 1$. The ratio q/p is given in terms of $\mathcal{M}$ and Γ_1 as,

$$\left(\frac{q}{p}\right)^2 = \frac{2\mathcal{M}_{12}^* - i\mathcal{A}_{12}^*}{2\mathcal{M}_{12} - i\mathcal{A}_{12}} \simeq 1 + \text{Im}\left(\frac{2\Gamma_1 A}{BM_1}\right), \qquad (4.143)$$

in the limit $\mathcal{A}_{12} \ll \mathcal{M}_{12}$. Similar to the $K^0 - \overline{K}^0$ system, the source of CP violation in the lepton number asymmetry considered here is due to the CP violation in the mixing which occurs when the two neutral mass eigenstates $(\widetilde{N}_+, \widetilde{N}_-)$, are different from the interaction eigenstates, $(\widetilde{N}'_+, \widetilde{N}'_-)$. Therefore CP violation in mixing is present as long as the quantity $|q/p| \neq 1$, which requires

$$\text{Im}\left(\frac{A\Gamma_1}{M_1 B}\right) \neq 0 . \qquad (4.144)$$

For this to occur, SUSY breaking, *i.e.* non-vanishing A and B, is required. As the relative phase between the parameters A and B can be rotated away by an $U(1)_R$-rotation as discussed in Sec. 4.1.5, without loss of generality we assume from now on that the remaining physical phase is solely coming from the tri-linear coupling, A.

The total lepton number asymmetry integrated over time, ϵ, is defined as the ratio of the difference to the sum of the decay widths Γ for $\widetilde{\nu}_{R_1}$ and $\widetilde{\nu}_{R_1}^\dagger$ into final states of the slepton doublet $\widetilde{L}$ and the Higgs doublet H, or the lepton doublet L and the Higgsino $\widetilde{H}$ or their conjugates,

$$\epsilon = \frac{\sum_f \int_0^\infty [\Gamma(\widetilde{\nu}_{R_1}, \widetilde{\nu}_{R_1}^\dagger \to f) - \Gamma(\widetilde{\nu}_{R_1}, \widetilde{\nu}_{R_1}^\dagger \to \overline{f})]}{\sum_f \int_0^\infty [\Gamma(\widetilde{\nu}_{R_1}, \widetilde{\nu}_{R_1}^\dagger \to f) + \Gamma(\widetilde{\nu}_{R_1}, \widetilde{\nu}_{R_1}^\dagger \to \overline{f})]} . \qquad (4.145)$$

Here the final states $f = (\tilde{L}\, H)$, $(L\, \tilde{H})$ have lepton number $+1$, and $\bar{f}$ denotes their conjugate, $(\tilde{L}^\dagger\, H^\dagger)$, $(\overline{L}\, \overline{\tilde{H}})$, which have lepton number -1. After carrying out the time integration, the total CP asymmetry is [42, 43],

$$\epsilon = \left(\frac{4\Gamma_1 B}{\Gamma_1^2 + 4B^2}\right) \frac{\text{Im}(A)}{M_1} \delta_{B-F} \qquad (4.146)$$

where the additional factor δ_{B-F} takes into account the thermal effects due to the difference between the occupation numbers of bosons and fermions [44].

The final result for the baryon asymmetry is [42, 43],

$$\begin{aligned}\frac{n_B}{s} &\simeq -c_s\, d_{\tilde{\nu}_R}\, \epsilon\, \kappa\,, \\ &\simeq -1.48 \times 10^{-3} \epsilon\, \kappa\,, \\ &\simeq -(1.48 \times 10^{-3})\left(\frac{\text{Im}(A)}{M_1}\right) R\, \delta_{B-F}\, \kappa\,,\end{aligned} \qquad (4.147)$$

where $d_{\tilde{\nu}_R}$ in the first line is the density of the lightest sneutrino in equilibrium in units of entropy density, and is given by, $d_{\tilde{\nu}_R} = 45\zeta(3)/(\pi^4 g_*)$; the factor c_s, which characterizes the amount of $B - L$ asymmetry being converted into the baryon asymmetry Y_B, is defined in Eq. 4.57. The parameter κ is the efficiency factor given in Sec. 4.2.1.2. The resonance factor R is defined as the following ratio,

$$R \equiv \frac{4\Gamma_1 B}{\Gamma_1^2 + 4B^2}\,, \qquad (4.148)$$

which gives a value equal to one when the resonance condition, $\Gamma_1 = 2|B|$, is satisfied, leading to maximal CP asymmetry. As Γ_1 is of the order of $\mathcal{O}(0.1 - 1)$ GeV, to satisfy the resonance condition, a small value for $B \ll \tilde{m}$ is thus needed. Such a small value of B can be generated by some dynamical relaxation mechanisms [45] in which B vanishes in the leading order. A small value of $B \sim \tilde{m}^2/M_1$ is then generated by an operator $\int d^4\theta Z Z^\dagger N_1^2/M_{pl}^2$ in the Kähler potential, where Z is the SUSY breaking spurion field, $Z = \theta^2\, \tilde{m} M_{pl}$ [43]. In a specific SO(10) model constructed in Refs. [46, 47], it has been shown that with the parameter $B' \equiv \sqrt{BM_1}$ having the size of the natural SUSY breaking scale $\sqrt{\tilde{m}^2} \sim \mathcal{O}(1)$ TeV, a small value for B required by the resonance condition $B \sim \Gamma_1 \sim \mathcal{O}(0.1)$ GeV can be obtained.

4.3.3. *Non-thermal Leptogenesis*

The conflict between generating sufficient leptogenesis and not overly producing gravitinos in thermal leptogenesis arises due to strong dependence of the reheating temperature T_R on the lightest RH mass, M_{R_1}, in thermal leptogenesis. This problem may be avoided if the relation between the reheating temperature and the lightest RH neutrino mass is loosened. This is the case if the primordial RH neutrinos are produced non-thermally. One possible way to have non-thermal leptogenesis is to generate the primordial right-handed neutrinos through the inflaton decay [48].

Inflation solves the horizon and flatness problem, and it accounts for the origin of density fluctuations. Assume that the inflaton decays dominantly into a pair of lightest RH neutrinos, $\Phi \to N_1 + N_1$. For this decay to occur, the inflaton mass m_Φ has to be greater than $2M_1$. For simplicity, let us also assume that the decay modes into $N_{2,3}$ are energetically forbidden. The produced N_1 in inflaton decay then subsequently decays into $H + \ell_L$ and $H^\dagger + \ell_L^\dagger$. The out-of-equilibrium condition is automatically satisfied, if $T_R < M_1$. The CP asymmetry is generated by the interference of tree level and one-loop diagrams,

$$\epsilon = -\frac{3}{8\pi} \frac{M_1}{\langle H \rangle^2} m_3 \delta_{eff} , \qquad (4.149)$$

where δ_{eff} is given in terms of the neutrino Yukawa matrix elements and light neutrino masses as,

$$\delta_{eff} = \frac{Im\left\{ h_{13}^2 + \frac{m_2}{m_3} h_{12}^2 + \frac{m_1}{m_3} h_{11}^2 \right\}}{\left| h_{13} \right|^2 + \left| h_{12} \right|^2 + \left| h_{11} \right|^2} . \qquad (4.150)$$

Numerically, the asymmetry is given by [48],

$$\epsilon \simeq -2 \times 10^{-6} \left(\frac{M_1}{10^{10} \text{ GeV}} \right) \left(\frac{m_3}{0.05 \text{ eV}} \right) \delta_{eff} . \qquad (4.151)$$

The chain decays $\Phi \to N_1 + N_1$ and $N_1 \to H + \ell_L$ or $H^\dagger + \ell_L^\dagger$ reheat the Universe producing not only the lepton number asymmetry but also the entropy for the thermal bath. Taking such effects into account, the ratio of lepton number to entropy density after the reheating [48] is then,

$$\frac{n_L}{s} \simeq -\frac{3}{2} \epsilon \frac{T_R}{m_\Phi} \simeq 3 \times 10^{-10} \left(\frac{T_R}{10^6 \text{ GeV}} \right) \left(\frac{M_1}{m_\Phi} \right) \left(\frac{m_3}{0.05 \text{ eV}} \right) , \qquad (4.152)$$

assuming $\delta_{eff} = 1$. The ratio $n_B/s \sim 10^{-10}$ can thus be obtained with $M_1 \lesssim m_\Phi$, and $T_R \lesssim 10^6$ GeV.

4.4. Connection between leptogenesis and neutrino oscillation

As mentioned in Sec. 4.1.5, there is generally no connection between low energy CP violating processes, such as CP violation in neutrino oscillation and in neutrinoless double beta decay, and leptogenesis, which occurs at very high energy scale. This is due to the extra phases and mixing angles present in the heavy neutrino sector. One way to establish such connection is by reducing the inter-family couplings (equivalently, by imposing texture zero in the Yukawa matrix). This is the case for the 3×2 seesaw model. A more powerful way to obtain such connection is to have all CP violation, both low energy and high energy, come from the same origin. This ensues if CP violation occurs spontaneously. Below we described these two models in which such connection does exist.

4.4.1. *Models with Two Right-Handed Neutrinos*

One type of models where there exists connection between CP violating processes at high and low energies is models with only two RH neutrinos. In this case, the neutrino Dirac mass matrix is a 3×2 matrix. This 3×2 Yukawa matrix has six complex parameters, and hence six phases, out of which, three can be absorbed by the wave functions of the three charged leptons. Even though, the reduction in the number of right-handed neutrinos reduces the number of CP phases in high energy, it also reduces the number of CP phases at low energy to two. There is therefor still one high energy phase that cannot be determined by measuring the low energy phases. However, if one further assumes that the 3 Yukawa matrix has two zeros, there is then only one CP phase in the Yukawa matrix, making the existence of the connection possible.

The existence of two right-handed neutrinos is required by the cancellation of Witten anomaly, if a global leptonic $SU(2)$ family symmetry is imposed [49]. (For implications of non-anomalous gauge symmetry for neutrino masses, see Ref. [50]. This model provided the interesting possibility of probing the neutrino sector at the colliders through their couplings to the Z' gauge boson [51].) Along this line, Frampton, Glashow and Yanagida proposed a model, which has the following Lagrangian [52],

$$\mathcal{L} = \frac{1}{2}(N_1 N_2)\begin{pmatrix} M_1 & 0 \\ 0 & M_2 \end{pmatrix}\begin{pmatrix} N_1 \\ N_2 \end{pmatrix} + (N_1 N_2)\begin{pmatrix} a & a' & 0 \\ 0 & b & b' \end{pmatrix}\begin{pmatrix} \ell_1 \\ \ell_2 \\ \ell_3 \end{pmatrix} H + h.c.\,, \quad (4.153)$$

with the Yukawa matrix having two zeros in the $N_1 - \ell_3$ and $N_2 - \ell_1$ couplings. The effective neutrino mass matrix due to this Lagrangian is obtained, using the see-saw formula,

$$\begin{pmatrix} \frac{a^2}{M_1} & \frac{aa'}{M_1} & 0 \\ \frac{aa'}{M_1} & \frac{a'^2}{M_1} + \frac{b^2}{M_2} & \frac{bb'}{M_2} \\ 0 & \frac{bb'}{M_2} & \frac{b'^2}{M_2} \end{pmatrix}, \quad (4.154)$$

where a, b, b' are real and $a' = |a'|e^{i\delta}$. By takinging all of them to be real, with the choice $a' = \sqrt{2}a$ and $b = b'$, and assuming $a^2/M_1 \ll b^2/M_2$, the effective neutrino masses and mixing matrix are obtained

$$m_{\nu_1} = 0, \quad m_{\nu_2} = \frac{2a^2}{M_1}, \quad m_{\nu_3} = \frac{2b^2}{M_2} \quad (4.155)$$

$$U = \begin{pmatrix} 1/\sqrt{2} & 1/\sqrt{2} & 0 \\ -1/2 & 1/2 & 1/\sqrt{2} \\ 1/2 & -1/2 & 1/\sqrt{2} \end{pmatrix} \times \begin{pmatrix} 1 & 0 & 0 \\ 0 & \cos\theta & \sin\theta \\ 0 & -\sin\theta & \cos\theta \end{pmatrix}, \quad (4.156)$$

where $\theta \simeq m_{\nu_2}/\sqrt{2}m_{\nu_3}$, and the observed bi-large mixing angles and Δm^2_{atm} and $\Delta m^2_\odot$ can be accommodated. An interesting feature of this model is that the sign of the baryon number asymmetry ($B \propto \xi_B = Y^2 a^2 b^2 \sin 2\delta$) is related to the sign of the CP violation in neutrino oscillation (ξ_{osc}) in the following way

$$\xi_{osc} = -\frac{a^4 b^4}{M_1^3 M_2^3}(2 + Y^2)\xi_B \propto -B \quad (4.157)$$

assuming the baryon number asymmetry is resulting from leptogenesis due to the decay of the lighter one of the two heavy neutrinos, N_1. This idea can be realized in a $SO(10)$ with additional singlets [53].

4.4.2. Models with Spontaneous CP Violation (& Triplet Leptogenesis)

The second type of models in which relation between leptogenesis and low energy CP violation exists is the minimal left-right symmetric model with spontaneous CP violation (SCPV) [54]. The left-right (LR) model [55] is based on the gauge group, $SU(3)_c \times SU(2)_L \times SU(2)_R \times U(1)_{B-L} \times P$, where the parity P acts on the two $SU(2)$'s. (See also Kaladi Babu's lectures.) In this model, the electric charge Q can be understood as the sum of the two T^3 quantum numbers of the $SU(2)$ gauge groups,

$$Q = T_{3,L} + T_{3,R} + \frac{1}{2}(B-L). \tag{4.158}$$

The *minimal* LR model has the following particle content: In the fermion sector, the iso-singlet quarks form a doublet under $SU(2)_R$, and similarly for e_R and ν_R,

$$Q_{i,L} = \begin{pmatrix} u \\ d \end{pmatrix}_{i,L} \sim (1/2, 0, 1/3), \qquad Q_{i,R} = \begin{pmatrix} u \\ d \end{pmatrix}_{i,R} \sim (0, 1/2, 1/3)$$

$$L_{i,L} = \begin{pmatrix} e \\ \nu \end{pmatrix}_{i,L} \sim (1/2, 0, -1), \qquad L_{i,R} = \begin{pmatrix} e \\ \nu \end{pmatrix}_{i,R} \sim (0, 1/2, -1).$$

In the scalar sector, there is a bi-doublet and one triplet for each of the $SU(2)$'s,

$$\Phi = \begin{pmatrix} \phi_1^0 & \phi_2^+ \\ \phi_1^- & \phi_2^0 \end{pmatrix} \sim (1/2,\ 1/2,\ 0)$$

$$\Delta_L = \begin{pmatrix} \Delta_L^+/\sqrt{2} & \Delta_L^{++} \\ \Delta_L^0 & -\Delta_L^+/\sqrt{2} \end{pmatrix} \sim (1,\ 0,\ 2)$$

$$\Delta_R = \begin{pmatrix} \Delta_R^+/\sqrt{2} & \Delta_R^{++} \\ \Delta_R^0 & -\Delta_R^+/\sqrt{2} \end{pmatrix} \sim (0,\ 1,\ 2).$$

Under the parity P, these fields transform as,

$$\Psi_L \leftrightarrow \Psi_R, \quad \Delta_L \leftrightarrow \Delta_R, \quad \Phi \leftrightarrow \Phi^\dagger. \tag{4.159}$$

The VEV of the $SU(2)_R$ breaks the left-right symmetry down to the SM gauge group,

$$SU(3)_c \times SU(2)_L \times SU(2)_R \times U(1)_{B-L} \times P$$
$$\to SU(3)_c \times SU(2)_L \times U(1)_Y, \tag{4.160}$$

and the subsequent breaking of the electroweak symmetry is achieved by the bi-doublet VEV. In general,

$$\langle \Phi \rangle = \begin{pmatrix} \kappa e^{i\alpha_\kappa} & 0 \\ 0 & \kappa' e^{i\alpha_{\kappa'}} \end{pmatrix}, \tag{4.161}$$

$$\langle \Delta_L \rangle = \begin{pmatrix} 0 & 0 \\ v_L e^{i\alpha_L} & 0 \end{pmatrix}, \quad \langle \Delta_R \rangle = \begin{pmatrix} 0 & 0 \\ v_R e^{i\alpha_R} & 0 \end{pmatrix}.$$

To get realistic SM gauge boson masses, the VEV's of the bi-doublet Higgs must satisfy $v^2 \equiv |\kappa|^2 + |\kappa'|^2 \simeq 2M_w^2/g^2 \simeq (174 \text{ GeV})^2$. Generally, a non-vanishing VEV for the $SU(2)_L$ triplet Higgs is induced, and it is suppressed by the heavy $SU(2)_R$ breaking scale similar to the see-saw mechanism for the neutrinos,

$$<\Delta_L> = \begin{pmatrix} 0 & 0 \\ v_L e^{i\alpha_L} & 0 \end{pmatrix}, \qquad v_L v_R = \beta |\kappa|^2, \qquad (4.162)$$

where the parameter β is a function of the order $\mathcal{O}(1)$ coupling constants in the scalar potential and v_R, v_L, κ and κ' are positive real numbers in the above equations. (The presence of a triplet Higgs in warped extra dimensions can provide a natural way to generate small Majorana masses for the neutrinos [56].) Due to this see-saw suppression, for a $SU(2)_R$ breaking scale as high as 10^{15} GeV, which is required by the smallness of the neutrino masses, the induced $SU(2)_L$ triplet VEV is well below the upper bound set by the electroweak precision constraints [57]. The scalar potential that gives rise to the vacuum alignment described can be found in Ref. [58].

The Yukawa sector of the model is given by $\mathcal{L}_{Yuk} = \mathcal{L}_q + \mathcal{L}_\ell$, where $\mathcal{L}_q$ and $\mathcal{L}_\ell$ are the Yukawa interactions in the quark and lepton sectors, respectively. The Lagrangian for quark Yukawa interactions is given by,

$$-\mathcal{L}_q = \overline{Q}_{i,R}(F_{ij}\Phi + G_{ij}\tilde{\Phi})Q_{j,L} + h.c. \qquad (4.163)$$

where $\tilde{\Phi} \equiv \tau_2 \Phi^* \tau_2$. In general, F_{ij} and G_{ij} are Hermitian to preserve left-right symmetry. Because of our assumption of SCPV with complex vacuum expectation values, the matrices F_{ij} and G_{ij} are real. The Yukawa interactions responsible for generating the lepton masses are summarized in the following Lagrangian, $\mathcal{L}_\ell$,

$$\begin{aligned} -\mathcal{L}_\ell &= \overline{L}_{i,R}(P_{ij}\Phi + R_{ij}\tilde{\Phi})L_{j,L} \\ &+ if_{ij}(L_{i,L}^T C \tau_2 \Delta_L L_{j,L} + L_{i,R}^T C \tau_2 \Delta_R L_{j,R}) + h.c., \end{aligned} \qquad (4.164)$$

where $\mathcal{C}$ is the Dirac charge conjugation operator, and the matrices P_{ij}, R_{ij} and f_{ij} are real due to the assumption of SCPV. Note that the Majorana mass terms $L_{i,L}^T \Delta_L L_{j,L}$ and $L_{i,R}^T \Delta_R L_{j,R}$ have identical coupling because the Lagrangian must be invariant under interchanging $L \leftrightarrow R$. The complete Lagrangian of the model is invariant under the unitary transformation, under which the matter fields transform as

$$\psi_L \to U_L \psi_L, \qquad \psi_R \to U_R \psi_R \qquad (4.165)$$

where $\psi_{L,R}$ are left-handed (right-handed) fermions, and the scalar fields transform according to

$$\Phi \to U_R \Phi U_L^\dagger, \qquad \Delta_L \to U_L^* \Delta_L U_L^\dagger, \qquad \Delta_R \to U_R^* \Delta_R U_R^\dagger \qquad (4.166)$$

with the unitary transformations U_L and U_R being

$$U_L = \begin{pmatrix} e^{i\gamma_L} & 0 \\ 0 & e^{-i\gamma_L} \end{pmatrix}, \qquad U_R = \begin{pmatrix} e^{i\gamma_R} & 0 \\ 0 & e^{-i\gamma_R} \end{pmatrix}. \qquad (4.167)$$

Under these unitary transformations, the VEV's transform as

$$\kappa \to \kappa e^{-i(\gamma_L - \gamma_R)}, \qquad \kappa' \to \kappa' e^{i(\gamma_L - \gamma_R)}, \qquad (4.168)$$
$$v_L \to v_L e^{-2i\gamma_L}, \qquad v_R \to v_R e^{-2i\gamma_R}.$$

Thus by re-defining the phases of matter fields with the choice of $\gamma_R = \alpha_R/2$ and $\gamma_L = \alpha_\kappa + \alpha_R/2$ in the unitary matrices U_L and U_R, we can rotate away two of the complex phases in the VEV's of the scalar fields and are left with only two genuine CP violating phases, $\alpha_{\kappa'}$ and α_L,

$$<\Phi> = \begin{pmatrix} \kappa & 0 \\ 0 & \kappa' e^{i\alpha_{\kappa'}} \end{pmatrix}, \qquad (4.169)$$

$$<\Delta_L> = \begin{pmatrix} 0 & 0 \\ v_L e^{i\alpha_L} & 0 \end{pmatrix}, \qquad <\Delta_R> = \begin{pmatrix} 0 & 0 \\ v_R & 0 \end{pmatrix}.$$

The quark Yukawa interaction $\mathcal{L}_q$ gives rise to quark masses after the bi-doublet acquires VEV's

$$M_u = \kappa F_{ij} + \kappa' e^{-i\alpha_{\kappa'}} G_{ij}, \qquad M_d = \kappa' e^{i\alpha_{\kappa'}} F_{ij} + \kappa G_{ij}. \qquad (4.170)$$

Thus the relative phase in the two VEV's in the $SU(2)$ bi-doublet, $\alpha_{\kappa'}$, gives rise to the CP violating phase in the CKM matrix. To obtain realistic quark masses and CKM matrix elements, it has been shown that the VEV's of the bi-doublet have to satisfy $\kappa/\kappa' \simeq m_t/m_b \gg 1$ [59]. When the triplets and the bi-doublet acquire VEV's, we obtain the following mass terms for the leptons

$$M_e = \kappa' e^{i\alpha_{\kappa'}} P_{ij} + \kappa R_{ij}, \qquad M_\nu^{Dirac} = \kappa P_{ij} + \kappa' e^{-i\alpha_{\kappa'}} R_{ij} \qquad (4.171)$$
$$M_\nu^{RR} = v_R f_{ij}, \qquad M_\nu^{LL} = v_L e^{i\alpha_L} f_{ij}. \qquad (4.172)$$

The effective neutrino mass matrix, M_ν^{eff}, which arises from the Type-II seesaw mechanism, is thus given by

$$M_\nu^{eff} = M_\nu^{II} - M_\nu^{I} = \left(f e^{i\alpha_L} - \frac{1}{\beta} P^T f^{-1} P \right) v_L, \qquad (4.173)$$

$$\begin{aligned} M_\nu^I &= (M_\nu^{Dirac})^T (M_\nu^{RR})^{-1} (M_\nu^{Dirac}) & (4.174) \\ &= (\kappa P + \kappa' e^{-i\alpha_{\kappa'}} R)^T (v_R f)^{-1} (\kappa P + \kappa' e^{-i\alpha_{\kappa'}} R) \\ &\simeq \frac{v_L}{\beta} P^T f^{-1} P, \end{aligned}$$

$$M_\nu^I = v_L e^{i\alpha_L} f. \qquad (4.175)$$

Consequently, the connection between CP violation in the quark sector and that in the lepton sector, which is made through the phase $\alpha_{\kappa'}$, appears only at the sub-leading order, $\mathcal{O}(\kappa'/\kappa)$, thus making this connection rather weak. We will neglect these sub-leading order terms, and there is thus only one phase, α_L, that is responsible for all leptonic CP violation.

The three low energy phases δ, α_{21}, α_{31}, in the MNS matrix are therefore functions of the single fundamental phase, α_L. Neutrino oscillation probabilities depend on the Dirac phase through the leptonic Jarlskog invariant, which is proportional to $\sin\alpha_L$, $J_{CP}^\ell \propto \sin\alpha_L$. There are two ways to generate lepton number asymmetry. One is through the decay of the $SU(2)_L$ triplet Higgs, $\Delta^* \to \ell + \ell$, and the corresponding asymmetry is given by,

$$\epsilon = \frac{\Gamma(\Delta_L^* \to \ell + \ell) - \Gamma(\Delta_L \to \bar\ell + \bar\ell)}{\Gamma(\Delta_L^* \to \ell + \ell) + \Gamma(\Delta_L \to \bar\ell + \bar\ell)}. \qquad (4.176)$$

The asymmetry can also be generated through the decay of the lightest RH neutrinos, $N_1 \to \ell + H^\dagger$, and the asymmetry in this case is,

$$\epsilon = \frac{\Gamma(N_1 \to \ell + H^\dagger) - \Gamma(N_1 \to \bar\ell + H)}{\Gamma(N_1 \to \ell + H^\dagger) + \Gamma(N_1 \to \bar\ell + H)}. \qquad (4.177)$$

Whether N_1 decay dominates or Δ_L decay dominates depends upon if N_1 is heavier or lighter than Δ_L. As the mass of the triplet Higgs is typically at the scale of the LR breaking scale, it is naturally heavier than the lightest RH neutrino. As a result, N_1 decay dominates. With the particle content of this model, there are three diagrams at one loop that contribute to leptogeiesis, as shown in Fig. 4.15. The contribution from diagram (a) and (b) mediated by charged lepton and Higgs doublet, which appear also in standard leptogenesis with SM particle content, is given by [60],

$$\epsilon^{N_1} = \frac{3}{16\pi} \left(\frac{M_{R_1}}{v^2} \right) \cdot \frac{\mathrm{Im}\left(\mathcal{M}_D (M_\nu^I)^* \mathcal{M}_D^T \right)_{11}}{(\mathcal{M}_D \mathcal{M}_D^\dagger)_{11}}. \qquad (4.178)$$

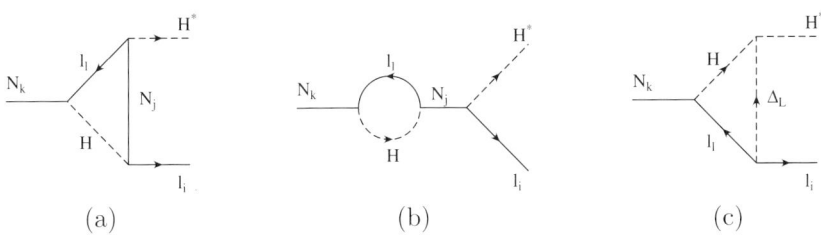

Fig. 4.15. Diagrams in the minimal left-right model that contribute to the lepton number asymmetry through the decay of the RH neutrinos.

Now, there is one additional one-loop diagram, Fig. 4.15 (c), mediated by the $SU(2)_L$ triplet Higgs. It contributes to the decay amplitude of the right-handed neutrino into a doublet Higgs and a charged lepton, which gives an additional contribution to the lepton number asymmetry [60],

$$\epsilon^{\Delta_L} = \frac{3}{16\pi} \left(\frac{M_{R_1}}{v^2}\right) \cdot \frac{\text{Im}(\mathcal{M}_D(M_\nu^{II})^* \mathcal{M}_D^T)_{11}}{(\mathcal{M}_D \mathcal{M}_D^\dagger)_{11}}, \quad (4.179)$$

where $\mathcal{M}_D$ is the neutrino Dirac mass term in the basis where the RH neutrino Majorana mass term is real and diagonal,

$$\mathcal{M}_D = O_R M_D, \quad f^{\text{diag}} = O_R f O_R^T. \quad (4.180)$$

Because there is no phase present in either $M_D = P\kappa$ or M_ν^I or O_R, the quantity $\mathcal{M}_D \left(M_\nu^I\right)^* \mathcal{M}_D^T$ is real, leading to a vanishing ϵ^{N_1}. This statement is true for *any* chosen unitary transformations U_L and U_R defined in Eq. (4.167). On the other hand, the contribution, ϵ^{Δ_L}, due to the diagram mediated by the $SU(2)_R$ triplet is proportional to $\sin\alpha_L$.

As all leptonic CP violation in this model come from one single origin, that is, the phase in the VEV of the LH triplet, $\langle \Delta_L \rangle$, strong correlation between leptogenesis and low energy CP violating processes can thus be established. In particular, both J_{CP}^ℓ and ϵ are proportional to $\sin\alpha_L$.

It has been found recently that, by lowering the left-right symmetry breaking scale with an additional $U(1)$ symmetry, the link between CP violation in the quark sector and that in the lepton sector can also be established [61].

4.5. Recent Progress and Concluding Remarks

Leptogenesis provides a very appealing way to generate the observed cosmological baryonic asymmetry. It has gained a significant amount of interests ever since the advent of the evidence of non-zero neutrino masses. In this scenario, the baryonic asymmetry is closely connected to the properties of the neutrinos, and the fact that the required neutrino mass scale for successful leptogenesis is similar to the scale observed in neutrino oscillations makes leptogenesis a very plausible source for the cosmological baryonic asymmetry. Even though there is so far no direct way to test leptogenesis, the search for leptonic CP violation in neutrino oscillations at very long baseline experiments [66] and to look for lepton number violation in neutrinoless double beta decay will inevitably further the credibility of leptogenesis as a source of the baryon asymmetry.

The recent developments in the subject of leptogenesis have been focused on the role of flavor. Recall that the total asymmetry given in Eq. 4.89 have summed over all three flavor indices,

$$\epsilon_1 = \sum_{\alpha=e,\mu,\tau} \epsilon^{\alpha\alpha} , \qquad (4.181)$$

where $\epsilon^{\alpha\alpha}$ is the CP asymmetry in the α-flavor. Correspondingly, previous solutions to the Boltzmann equations have summed over all the three flavors, e, μ, τ, and thus they did not include flavor dependence [62],

$$\frac{d(Y_{N_1} - Y_{N_1}^{eq})}{dz} = -\frac{z}{sH(M_1)}(\gamma_D + \gamma_{\Delta L=1})\left(\frac{Y_{N_1}}{Y_{N_1}^{eq}} - 1\right) \qquad (4.182)$$
$$- \frac{dY_{N_1}^{eq}}{dz},$$

$$\frac{dY_L}{dz} = \frac{z}{sH(M_1)}\left[\left(\frac{Y_{N_1}}{Y_{N_1}^{eq}} - 1\right)\epsilon_1 \gamma_D \qquad (4.183)\right.$$
$$\left. + - \frac{Y_L}{Y_L^{eq}}(\gamma_D \gamma_{\Delta L=1} + \gamma_{\Delta L=2})\right],$$

where Y_{N_1} and Y_L are the number density of the lightest right-handed neutrino N_1 and of the lepton number asymmetry, respectively, and γ's are the decay rates for the processes specified in the subscripts. It has recently been pointed out that flavor effects matter if heavy neutrino masses are hierarchical [62]. The Yukawa interactions of all three flavors, e, μ and τ, reach equilibrium at different temperatures. These temperatures are determined by the size of the Yukawa couplings, λ, as

$$\lambda^2 M_{Pl} = T_{eq} . \qquad (4.184)$$

Due to the relative large coupling constant, the τ Yukawa interactions reach equilibrium at $T \sim 10^{12}$ GeV, while the muon Yukawa interactions reach equilibrium at $T \sim 10^9$ GeV. If leptogenesis takes place at $T \sim M_1 > 10^{12}$ GeV, the Yukawa interactions of all three lepton flavors are out of equilibrium, and hence the three flavors are indistinguishable. In particular, the washout factor is universal for all three flavors. However, if leptogenesis takes place at temperature below 10^{12} GeV, which is generally the case for hierarchical right-handed neutrino masses, the three flavors are distinguishable and thus their effects should be included in the Boltzmann equations properly. Instead of a single evolution function for Y_L as given in Eq. 4.183, one should consider the evolution of the lepton number asymmetry, $Y^{\alpha\alpha}$, which is due to the decay of the lightest right-handed neutrino into charged lepton of flavor α with the corresponding asymmetry given by $\epsilon^{\alpha\alpha}$ and decay rate given by $\gamma_D^{\alpha\alpha}$ [62],

$$\frac{dY^{\alpha\alpha}}{dz} = \frac{z}{sH(M_1)} \left[\left(\frac{Y_{N_1}}{Y_{N_1}^{eq}} - 1\right) \epsilon^{\alpha\alpha} (\gamma_D^{\alpha\alpha} + \gamma_{\Delta L=1}) \right. \qquad (4.185)$$
$$\left. - \frac{Y^{\alpha\alpha}}{Y_L^{eq}} (\gamma_D^{\alpha\alpha} + \gamma_{\Delta L=1}) \right],$$

Note that in the above equation, there is no summation over the flavor index, α. By properly including the flavor effects, the amount of leptogenesis may be enhanced by a factor of 2 to 3 [62].

Except for the specific types of models [52, 54] discussed in Sec. 4.4, the general lack of connection between leptogenesis and low energy CP violation translates into the fact that the observation of the leptonic Dirac or Majorana phases at low energy does not imply non-vanishing leptogenesis. This statement is weakened in a framework when the right-handed neutrino sector is CP invariant and when the flavor effects are important [63]. This is elucidate by introducing the "orthogonal parametrization" for neutrino Dirac Yukawa matrix [64],

$$h = \frac{1}{v} M^{1/2} R m^{1/2} U^\dagger , \qquad (4.186)$$

where $m = \text{diag}(m_1, m_2, m_3)$ is the diagonal matrix of the light neutrino masses, M is the diagonal matrix of the right-handed neutrino masses and U is the MNS matrix. The orthogonal matrix R is defined by this equation as $R = vM^{-1/2}hUm^{-1/2}$. In the basis where the right-handed neutrino mass matrix and the charged lepton mass matrix are diagonal, the neutrino Dirac Yukawa matrix can be written as $h = V_R^{\nu\dagger}\text{diag}(h_1, h_2, h_3)V_L^\nu$. Therefore, the low energy CP violation in the lepton sector can arise from either the left-handed sector through V_L^ν, the right-handed sector through V_R^ν, or from both. From $hh^\dagger v^2 = V_R^{\nu\dagger}\text{diag}(h_1^2, h_2^2, h_3^2)V_R^\nu v^2 = M^{1/2}RmR^\dagger M^{1/2}$, it can be seen that the phases of R are related to those in the right-handed sector through V_R^ν. The asymmetry ϵ_1 given in Eq. 4.89, which is derived with one-flavor approximation, can be rewritten as follows [65],

$$\epsilon_1 = -\frac{3M_1}{16\pi v^2} \frac{\mathrm{Im}\left(\sum_\rho m_\rho^2 R_{1\rho}^2\right)}{\sum_\beta m_\beta |R_{1\beta}|^2} . \qquad (4.187)$$

Assuming the right-handed sector is CP invariant, low energy CP phases can then arise entirely from the left-handed sector and thus are irrelevant for ϵ_1, which vanishes because the orthogonal matrix R is real. If leptogenesis takes place at $T < 10^{12}$ GeV, the flavor effects must be taken into account. In this case the asymmetry in each flavor is given by [65],

$$\epsilon_\alpha = -\frac{3M_1}{16\pi v^2} \frac{\mathrm{Im}\left(\sum_{\beta\rho} m_\beta^{1/2} m_\rho^{3/2} U_{\alpha\beta}^* U_{\alpha\rho} R_{1\beta} R_{1\rho}\right)}{\sum_\beta m_\beta |R_{1\beta}|^2} . \qquad (4.188)$$

The contribution of each of these individual asymmetries to the total asymmetry is then weighted by the corresponding washout factor. Therefore, barring accidental cancellations, the presence of the MNS matrix elements in Eq. 4.188 signifies the need for low energy CP violation in order to have leptogenesis. Hence if leptonic CP violation in neutrino oscillations is observed at future very long baseline experiments [66] and if lepton number violation is established by observing neutrinoless double beta decay, it would even more strongly suggest than it has been that leptogenesis be the source for the origin of the cosmological baryon asymmetry.

Finally, a fundamental problem in the current treatment of leptogenesis is the fact that the Boltzmann equations utilized in the present calculations are purely classical treatment. However, the collision terms are zero-temperature S-matrix elements which involve quantum interference. In addition, the time evolution of the system should be treated quantum mechanically. These lead to the need of quantum Boltzmann equations which is based on Closed-Time-Path (**CTP**) formalism [67]. A more detailed discussion on this issue can be found in Refs. [6, 68].

Acknowledgments

I would like to thank the organizers, Sally Dawson, K. T. Mahanthappa and Rabi Mohapatra for inviting me to lecture and for organizing such an intellectually stimulating TASI summer school, which has been a very essential experience for graduate students in theoretical high energy physics.

I would also like to thank the student participants for their interesting questions and for their enthusiastic participation at the school.

References

[1] W. Buchmuller, R. D. Peccei and T. Yanagida, Ann. Rev. Nucl. Part. Sci. **55**, 311 (2005).
[2] A. Strumia, hep-ph/0608347.
[3] E. Nardi, hep-ph/0702033.
[4] Y. Nir, hep-ph/0702199.
[5] A. Riotto and M. Trodden, Ann. Rev. Nucl. Part. Sci. **49**, 35 (1999); M. Trodden and S. M. Carroll, "*TASI Lectures: Introduction to Cosmology*", published in *Boulder 2002, Particle Physics and Cosmology*, 703-793, the proceedings of the Theoretical Advanced Study Institute in Elementary Particle Physics – Particle Physics and Cosmology: The Quest for Physics Beyond the Standard Model(s), Boulder, Colorado, 2-28 Jun 2002, World Scientific, astro-ph/0401547.
[6] A. Riotto, hep-ph/9807454.
[7] M. Trodden, hep-ph/0411301.
[8] C. J. Copi, D. N. Schramm and M. S. Turner, Science **267**, 192 (1995).
[9] C. L. Bennett *et al.*, Astrophys. J. Suppl. **148**, 1 (2003).
[10] A. D. Sakharov, Pisma Zh. Eksp. Teor. Fiz. **5**, 32 (1967) [JETP Lett. **5**, 24 (1967); Sov. Phys. Usp. **34**, 392 (1991)].
[11] G. 't Hooft, Phys. Rev. Lett. **37**, 8 (1976); G. 't Hooft, Phys. Rev. **D14**, 3432 (1976) [Erratum-ibid. **D18**, 2199 (1978)].
[12] S. L. Adler, Phys. Rev. **177**, 2426 (1969); J. S. Bell and R. Jackiw, Nuovo Cimento **51**, 47 (1969).
[13] S. Dimopoulos and L. Susskind, Phys. Rev. **D18**, 4500 (1978); N. S. Manton, Phys. Rev. **D28**, 2019 (1983).
[14] T. P. Cheng and L. F. Li, "*Gauge Theory Of Elementary Particle Physics*", Oxford, UK: Clarendon (Oxford Science Publications), 536 pages, (1984).
[15] V. A. Kuzmin, V. A. Rubakov and M. E. Shaposhnikov, Phys. Lett. **B155**, 36 (1985).
[16] F. R. Klinkhamer and N. S. Manton, Phys. Rev. **D30**, 2212 (1984).
[17] P. Arnold and L. D. McLerran, Phys. Rev. **D36**, 581 (1987).
[18] P. Arnold, D. Son and L. G. Yaffe, Phys. Rev. **D55**, 6264 (1997); P. Arnold, D. T. Son and L. G. Yaffe, Phys. Rev. **D59**, 105020 (1999); D. Bodeker, Phys. Lett. **B426**, 351 (1998); D. Bodeker, Nucl. Phys. **B559**, 502 (1999); D. Bodeker, Nucl. Phys. **B559**, 502 (1999).
[19] E. W. Kolb and M. S. Turner, "The Early Universe," Redwood City, USA: Addison-Wesley (1988) 719 pp., (Frontier in Physics, 70).
[20] I. Affleck and M. Dine, Nucl. Phys. **B249**, 361 (1985).
[21] M. E. Shaposhnikov, JETP Lett. **44**, 465 (1986) [Pisma Zh. Eksp. Teor. Fiz. **44**, 364 (1986)]. see also, G. R. Farrar and M. E. Shaposhnikov, Phys. Rev. Lett. **70**, 2833 (1993) [Erratum-ibid. **71**, 210 (1993)].

[22] M.-C. Chen and K. T. Mahanthappa, Int. J. Mod. Phys. **A18**, 5819 (2003); M.-C. Chen and K. T. Mahanthappa, AIP Conf. Proc. **721**, 269 (2004) hep-ph/0311034.
[23] A. de Gouvea, in the proceedings of Theoretical Advance Study Institute in Elementary Particle Physics (TASI 2004): *Physics in D >= 4*, 197 - 258 pp, Boulder, Colorado, 6 Jun - 2 Jul 2004, hep-ph/0411274; R. N. Mohapatra *et al.*, hep-ph/0412099; R. N. Mohapatra *et al.*, hep-ph/0510213; C. H. Albright and M.-C. Chen, Phys. Rev. **D74**, 113006 (2006).
[24] P. Minkowski, Phys. Lett. **B67**, 421 (1977); M. Gell-Mann, P. Ramond and R. Slansky, in *Supergravity*, eds. D. Freedman and P. Van Niuwenhuizen (North Holland, Amsterdam, 1979), p. 315; T. Yanagida, in *Proceedings of the Workshop on Unified Theories and Baryon Number in the Universe*, eds. O. Sawada and A. Sugamoto (KEK, Tsukuba, Japan, 1979); S. L. Glashow in *1979 Cargèse Lectures in Physics – Quarks and Leptons*, eds. M. Lévy *et al.* (Plenum, New York, 1980), p. 707. See also R. N. Mohapatra and G. Senjanović, Phys. Rev. Lett. **44**, 912 (1980); J. Schechter and J.W.F. Valle, Phys. Rev. **D22**, 2227 (1980).
[25] M. Fukugita and T. Yanagida, Phys. Lett. **B174**, 45 (1986); M. A. Luty, Phys. Rev. **D45**, 455 (1992); M. Plumacher, Z. Phys. **C74**, 549 (1997); W. Buchmuller and M. Plumacher, Phys. Lett. **B389**, 73 (1996).
[26] S. Blanchet and P. Di Bari, JCAP **0606**, 023 (2006); G. Engelhard, Y. Grossman, E. Nardi and Y. Nir, hep-ph/0612187.
[27] W. Buchmuller, P. Di Bari and M. Plumacher, Nucl. Phys. **B643**, 367 (2002); R. Barbieri, P. Creminelli, A. Strumia and N. Tetradis, Nucl. Phys. **B575**, 61 (2000).
[28] W. Buchmuller, P. Di Bari and M. Plumacher, Annals Phys. **315**, 305 (2005).
[29] S. Davidson and A. Ibarra, Phys. Lett. **B535**, 25 (2002).
[30] T. Hambye, Y. Lin, A. Notari, M. Papucci and A. Strumia, Nucl. Phys. **B695**, 169 (2004).
[31] K. Dick, M. Lindner, M. Ratz and D. Wright, Phys. Rev. Lett. **84**, 4039 (2000).
[32] H. Murayama and A. Pierce, Phys. Rev. Lett. **89**, 271601 (2002).
[33] W. Buchmuller, P. Di Bari and M. Plumacher, Phys. Lett. **B547**, 128 (2002); S. Davidson and A. Ibarra, Phys. Lett. **B535**, 25 (2002); G. F. Giudice, A. Notari, M. Raidal, A. Riotto and A. Strumia, Nucl. Phys. **B685**, 89 (2004).
[34] M. Y. Khlopov and A. D. Linde, Phys. Lett. **B138**, 265 (1984); B. A. Campbell, S. Davidson and K. A. Olive, Nucl. Phys. **B399**, 111 (1993); S. Sarkar, hep-ph/9510369. G. G. Ross and S. Sarkar, Nucl. Phys. **B461**, 597 (1996); M. Y. Khlopov, "*Cosmoparticle Physics*", 577p, World Scientific, Singapore (1999).
[35] G. F. Giudice, A. Riotto and I. Tkachev, JHEP **9911**, 036 (1999).
[36] J. Pradler and F. D. Steffen, Phys. Rev. **D75**, 023509 (2007).
[37] M. Kawasaki, K. Kohri and T. Moroi, Phys. Rev. **D71**, 083502 (2005).
[38] J. Pradler and F. D. Steffen, to appear in Phys. Lett. **B**, hep-ph/0612291.
[39] A. Pilaftsis, Phys. Rev. **D56**, 5431 (1997).

[40] A. Pilaftsis and T. E. J. Underwood, Nucl. Phys. **B692**, 303 (2004).
[41] Y. Nir, hep-ph/0109090.
[42] Y. Grossman, T. Kashti, Y. Nir and E. Roulet, Phys. Rev. Lett. **91**, 251801 (2003); L. Boubekeur, hep-ph/0208003. L. Boubekeur, T. Hambye and G. Senjanovic, Phys. Rev. Lett. **93**, 111601 (2004); Y. Grossman, T. Kashti, Y. Nir and E. Roulet, JHEP **0411**, 080 (2004).
[43] G. D'Ambrosio, G. F. Giudice and M. Raidal, Phys. Lett. **B575**, 75 (2003).
[44] L. Covi, N. Rius, E. Roulet and F. Vissani, Phys. Rev. **D57**, 93 (1998).
[45] M. Yamaguchi and K. Yoshioka, Phys. Lett. **B543**, 189 (2002).
[46] M.-C. Chen and K. T. Mahanthappa, Phys. Rev. **D62**, 113007 (2000); *ibid.* **65**, 053010 (2002); *ibid.* **68**, 017301 (2003).
[47] M.-C. Chen and K. T. Mahanthappa, Phys. Rev. **D70**, 113013 (2004).
[48] M. Fujii, K. Hamaguchi and T. Yanagida, Phys. Rev. **D65**, 115012 (2002)
[49] R. Kuchimanchi and R. N. Mohapatra, Phys. Rev. **D66**, 051301 (2002); Phys. Lett. **B552**, 198 (2003).
[50] M.-C. Chen, A. de Gouvea and B. A. Dobrescu, Phys. Rev. **D75**, 055009 (2007).
[51] T. G. Rizzo, hep-ph/0610104.
[52] P. H. Frampton, S. L. Glashow and T. Yanagida, Phys. Lett. **B548**, 119 (2002).
[53] S. Raby, Phys. Lett. **B561**, 119 (2003).
[54] M.-C. Chen and K. T. Mahanthappa, Phys. Rev. **D71**, 035001 (2005).
[55] J. C. Pati and A. Salam, Phys. Rev. **D10**, 275 (1974); R. N. Mohapatra and J. C. Pati, Phys. Rev. **D11**, 566 (1975); R. N. Mohapatra and J. C. Pati, Phys. Rev. **D11**, 2558 (1975); G. Senjanovic and R. N. Mohapatra, Phys. Rev. **D12**, 1502 (1975).
[56] M.-C. Chen, Phys. Rev. **D71**, 113010 (2005).
[57] T. Blank and W. Hollik, Nucl. Phys. **B514**, 113 (1998); M. Czakon, M. Zralek and J. Gluza, Nucl. Phys. **B573**, 57 (2000); M. Czakon, J. Gluza, F. Jegerlehner and M. Zralek, Eur. Phys. J. **C13**, 275 (2000); M.-C. Chen and S. Dawson, Phys. Rev. **D70**, 015003 (2004); M.-C. Chen, S. Dawson and T. Krupovnickas, Int. J. Mod. Phys. **A21**, 4045 (2006); M.-C. Chen, Mod. Phys. Lett. **A21**, 621 (2006); M.-C. Chen, S. Dawson and T. Krupovnickas, Phys. Rev. **D74**, 035001 (2006).
[58] N. G. Deshpande, J. F. Gunion, B. Kayser and F. I. Olness, Phys. Rev. **D44**, 837 (1991); Y. Rodriguez and C. Quimbay, Nucl. Phys. **B637**, 219 (2002).
[59] P. Ball, J. M. Frere and J. Matias, Nucl. Phys. **B572**, 3 (2000).
[60] S. Antusch and S. F. King, Phys. Lett. **B597**, 199 (2004); A. S. Joshipura, E. A. Paschos and W. Rodejohann, Nucl. Phys. **B611**, 227 (2001); T. Hambye and G. Senjanovic, Phys. Lett. **B582**, 73 (2004); W. Rodejohann, Phys. Rev. **D70**, 073010 (2004)
[61] M.-C. Chen and K. T. Mahanthappa, Phys. Rev. **D75**, 015001 (2007).
[62] R. Barbieri, P. Creminelli, A. Strumia and N. Tetradis, Nucl. Phys. **B575**, 61 (2000); A. Abada, S. Davidson, A. Ibarra, F. X. Josse-Michaux, M. Losada and A. Riotto, JHEP **0609**, 010 (2006); A. Abada, S. Davidson, F. X. Josse-Michaux, M. Losada and A. Riotto, JCAP **0604**, 004 (2006).

[63] E. Nardi, Y. Nir, E. Roulet and J. Racker, JHEP **0601**, 164 (2006).
[64] J. A. Casas and A. Ibarra, Nucl. Phys. **B618**, 171 (2001).
[65] S. Pascoli, S. T. Petcov and A. Riotto, hep-ph/0609125; S. Pascoli, S. T. Petcov and A. Riotto, hep-ph/0611338.
[66] W. J. Marciano, hep-ph/0108181; M. V. Diwan et al., Phys. Rev. **D68**, 012002 (2003).
[67] K. T. Mahanthappa, Phys. Rev. **126**, 329 (1962); P. M. Bakshi and K. T. Mahanthappa, J. Math. Phys. **4**, 1 (1963); ibid. 12 (1963).
[68] A. De Simone and A. Riotto, hep-ph/0703175.

Chapter 5

Neutrino Experiments

J. M. Conrad

Department of Physics,
Columbia University,
New York, NY

This article is a summary of four introductory lectures on "Neutrino Experiments," given at the 2006 TASI summer school. The purpose was to sketch out the present questions in neutrino physics, and discuss the experiments that can address them. The ideas were then explored in depth by later lecturers.

This article begins with an overview of neutrinos in the Standard Model and what we know about these particles today. This is followed by a discussion of the direction of the field, divided into the three themes identified in the *APS Study on the Future of Neutrino Physics*.[1] This APS study represented the culmination of a year-long effort by the neutrino community to come to a consensus on future directions. The report is recommended reading for students, along with the accompanying working group white papers, especially the Theory Group Whitepaper.[2]

While these lectures used the APS Neutrino Study themes as the core, the emphasis here is different from the APS report. The point of a summer school is to teach specific ideas rather than provide a perfectly balanced overview of the field. The result is that, with apologies, some experiments were necessarily left out of the discussion. Students are referred to the Neutrino Oscillation Industry Website[3] for a complete list of all neutrino experiments, by category.

5.1. Neutrinos As We Knew Them

Neutrinos are different from the other fermions. Even before the recent evidence of neutrino mass, neutrinos were peculiar members of the Standard

Model. They are the only fermions

- to carry no electric charge.
- for which we have no evidence of a right-handed partner.
- that are defined as massless.

These ideas are connected by the fact that, unlike other spin 1/2 particles, neutrinos can only interact through the weak interaction.

Even though the Standard Model picture is now demonstrably wrong, this theoretical framework provides a good place to start the discussion. This section begins by expanding on the Standard Model picture of the neutrino sketched above. It then discusses how neutrinos interact. This is followed by an overview of neutrino sources and detectors.

5.1.1. Neutrinos in the Standard Model

Neutrinos are the only Standard Model fermions to interact strictly via the weak interaction. This proceeds through two types of boson exchange. Exchange of the Z^0 is called the neutral current (NC) interaction. Exchange of the $W^\pm$ is called the charged current (CC) interaction. When a W is emitted, charge conservation at the vertex requires that a charged lepton exits the interaction. We know the family of an incoming neutrino by the charged partner which exits the CC interaction. For example, a scattered electron tags a ν_e interaction, a μ tags a ν_μ interaction, and a τ tags a ν_τ interaction. The neutrino always emits the W^+ and the antineutrino always emits the W^- in the CC interaction. In order to conserve charge at the lower vertex, the CC interaction is flavor-changing for target quarks. For example, in a neutrino interaction, if a neutron, n, absorbs a W^+, a proton, p, will exit the interaction. The W has converted a d quark to a u quark. The first two diagrams shown on Fig. 5.1 illustrate a NC and a CC interaction, respectively.

In 1989, measurements of the Z^0 width at LEP[4] and SLD[5] determined that there are only three families of light-mass weakly-interacting neutrinos, although we will explore this question in more depth in section 3 of these lectures. These are the ν_e, the ν_μ, and the ν_τ. The interactions of the ν_e and ν_μ have been shown to be consistent with the Standard Model weak interaction. Until recently, there has only been indirect evidence for the ν_τ through the decay of the τ meson. In July 2000, however, the DoNuT Experiment (E872) at Fermilab presented direct evidence for ν_τ interactions.[6]

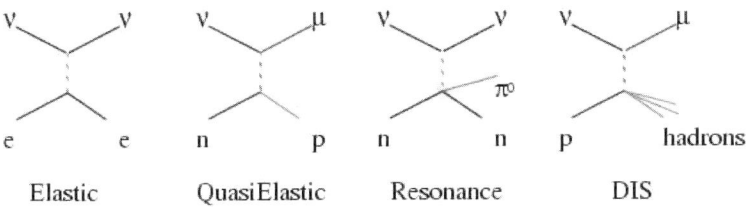

Fig. 5.1. Examples of the four types of neutrino interactions which appear throughout this discussion and are defined in Sec. 5.1.2. The first two diagrams show an NC and CC interaction, respectively.

Within the Standard Model, neutrinos are massless. This assumption is consistent with direct experimental observation. It is also an outcome of the feature of "handedness" associated with neutrinos. To understand handedness, it is simplest to begin by discussing "helicity," since for massless particles helicity and handedness are identical.

For a spin 1/2 Dirac particle, helicity is the projection of a particle's spin (Σ) along its direction of motion $\hat{p}$, with operator $\Sigma \cdot \hat{p}$. Helicity has two possible states: spin aligned opposite the direction of motion (negative, or "left helicity") and spin aligned along the direction of motion (positive or "right helicity"). If a particle is massive, then the sign of the helicity of the particle will be frame dependent. When one boosts to a frame where one is moving faster than the particle, the sign of the momentum will change but the spin will not, and therefore the helicity will flip. For massless particles, which must travel at the speed of light, one cannot boost to a frame where helicity changes sign.

Handedness (or chirality) is the Lorentz invariant (*i.e.*, frame-independent) analogue of helicity for both massless and massive Dirac particles. There are two states: "left handed" (LH) and "right handed" (RH). For the case of massless particles, including Standard Model neutrinos, helicity and handedness are identical. A massless fermion is either purely LH or RH, and, in principle, can appear in either state. Massive particles have both RH and LH components. A helicity eigenstate for a massive particle is a combination of handedness states. It is only in the high energy limit, where particles are effectively massless, that handedness and helicity coincide for massive fermions. Nevertheless, people tend to use the terms "helicity" and "handedness" interchangeably. Unlike the electromagnetic and strong interactions, the weak interaction involving neutrinos has a definite preferred handedness.

In 1956, it was shown that neutrinos are LH and outgoing antineutrinos are RH.[7] This effect is called "parity violation." If neutrinos respected parity, then an equal number of LH and RH neutrinos should have been produced in the 1956 experiment. The fact that all neutrinos are LH and all antineutrinos are RH means that, unlike all of the other fermions in the Standard Model, parity appears to be maximally violated for this particle. This is clearly very strange.

We need a method to enforce parity violation within the weak interaction theory. To this end, consider a fermion wavefunction, ψ, broken up into its LH and RH components:

$$\psi = \psi_L + \psi_R. \tag{5.1}$$

We can introduce a projection operator which selects out each component:

$$\gamma^5 \psi_{L,R} = \mp \psi_{L,R}. \tag{5.2}$$

To force the correct handedness in calculations involving the weak interaction, we can require a factor of $(1 - \gamma^5)/2$ at every weak vertex involving a neutrino. As a result of this factor, which corresponds to the LH projection operator, we often say the charged weak interaction (W exchange) is "left handed."

Note that by approaching the problem this way, RH neutrinos (and LH antineutrinos) could in principle exist but be undetected because they do not interact. They will not interact via the electromagnetic interactions because they are neutral, or via the strong interaction because they are leptons. RH Dirac neutrinos do not couple to the Standard Model W, because this interaction is "left handed," as discussed above. Because they are non-interacting, they are called "sterile neutrinos." By definition, the Standard Model has no RH neutrino.

With no RH partner, the neutrino can have no Dirac mass term in the Lagrangian. To see this, note that the free-particle Lagrangian for a massive, spin 1/2 particle is

$$\mathcal{L} = i\bar{\psi}\gamma_\mu \partial^\mu \psi - m\bar{\psi}\psi. \tag{5.3}$$

However, $\bar{\psi}\psi$ can be rewritten using

$$\psi_{L,R} = 1/2(1 \mp \gamma^5)\psi, \tag{5.4}$$

$$\bar{\psi}_{L,R} = 1/2\bar{\psi}(1 \pm \gamma^5), \tag{5.5}$$

giving

$$\bar{\psi}\psi = \bar{\psi}\left[\frac{1+\gamma^5}{2} + \frac{1-\gamma^5}{2}\right]\left[\frac{1+\gamma^5}{2} + \frac{1-\gamma^5}{2}\right]\psi = \bar{\psi}_L\psi_R + \bar{\psi}_R\psi_L. \tag{5.6}$$

In other words, an $m\bar{\psi}\psi$ ("mass") term in a Lagrangian mixes RH and LH states of the fermion. If the fermions have only one handedness (like νs), then the Dirac mass term will automatically vanish. In the Standard Model, there is no Dirac mass term for neutrinos.

5.1.2. Neutrino Interactions

Neutrino interactions in the Standard Model come in four basic types. Figure 5.1 shows examples of the four interactions. In *Elastic* scattering, "what goes is what comes out," just like two billiard balls colliding. An example is a NC interaction where the target is does not go into an excited state or break up, *e.g.*, $\nu_e + n \to \nu_e + n$. A more complicated example is electron-neutrino scattering from electrons, where the W exchange yields a final state which is indistinguishable from the Z exchange on an event-by-event basis, so this is categorized as an elastic scatter. *Quasi-elastic* scattering is, generally, the CC analogue to elastic scattering. Exchange of the W causes the incoming lepton and the target to change flavors, but the target does not go into an excited state or break apart. An example is $\nu_\mu + n \to \mu + p$. *Single pion* production may be caused by either NC or CC interactions. In resonant single pion production, the target becomes a Δ which decays to emit a pion. In coherent scattering, there is little momentum exchange with the nucleon and a single pion is produced diffractively in the forward direction. The case of NC single π^0 production is particularly important, because this forms a background in many neutrino oscillation searches. Finally, *DIS*, or Deep Inelastic Scattering, is the case where there is large 4-momentum exchange, breaking the nucleon apart. One can have NC or CC deep inelastic scattering.

Figure 5.2 summarizes the low energy behavior of σ/E for CC events (solid line), as predicted by the NUANCE neutrino event generator.[9] The quasi-elastic, single pion and deep inelastic contributions are indicated by the broken curves. The data indicate the state of the art for neutrino cross section measurements. One can see that if precision neutrino studies are to be pursued in the MeV to few GeV range, that more accurate measurements are essential. The MiniBooNE,[10] SciBooNE,[11] and MINERvA[12] experiments are expected to improve the situation in the near future.

Above a few GeV, the total neutrino cross section rises linearly with energy. The total cross section is the sum of many partial cross sections: quasi-elastic + single pion + two pions + three pions + etc. As the energy increases, each of these cross sections sequentially "turns on" and then

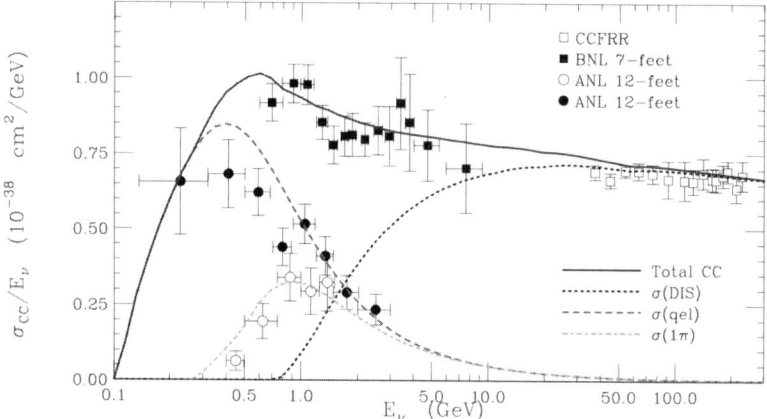

Fig. 5.2. Current status of ν_μ CC cross section measurements in the 1 to 100 GeV range. This plot shows σ/E, thus removing the linear energy dependence at high energies. Note the low energy cut-off due to the muon mass suppression. Components of the total cross section are indicated by the curves.[8]

becomes constant with energy. Thus the sum, which is the total cross section, increases continuously and linearly with E.

Nevertheless, even at high energies, this interaction is called "weak" for good reason. The total cross section for most neutrino scattering experiments is small. For 100 GeV ν_μ interactions with electrons, the cross section is $\sim 10^{-40}$ cm^2. For 100 GeV ν_μ interactions with nucleons, the cross section is $\sim 10^{-36}$ cm^2. This is many orders of magnitude less than the strong interaction. For example, for pp scattering, the cross section is $\sim 10^{-25}$ cm^2. The result is that a 100 GeV neutrino will have a mean free path in iron of 3×10^9 meters. Thus most neutrinos which hit the Earth travel through without interacting. It is only at ultra-high energies that the Earth becomes opaque to neutrinos, as discussed in Sec. 5.3.3.2.

In principle, the interactions of the ν_e, ν_μ, and ν_τ should be identical ("universal"). In practice, the mass differences of the outgoing leptons lead to considerable differences in the behavior of the cross sections. In the CC interaction, you must have enough CM energy to actually produce the outgoing charged lepton. Just above mass threshold, there is very little phase space for producing the lepton, and so production will be highly suppressed. The cross section increases in a non-linear manner until well above threshold. Consider, for example, a comparison of the ν_e and ν_μ CC quasielastic cross section on carbon, shown in Fig. 5.3. At very low energy

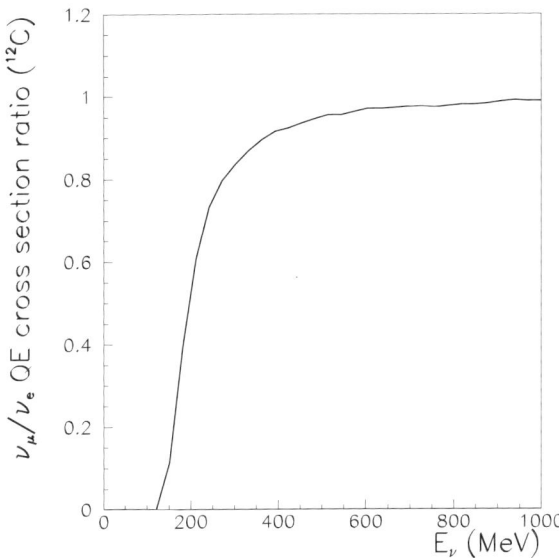

Fig. 5.3. The ratio of the ν_e to ν_μ CC cross sections as a function of neutrino energy, showing the suppression due to the lepton mass.

the CC ν_μ cross section is zero, while the ν_e cross section is non-zero, because the 105 MeV muon cannot be produced. The ratio approaches one at about 1 GeV. A similar effect occurs for the ν_τ CC interaction cross sections. The mass of the τ is 1.8 GeV, resulting in a cross section which is zero below 3.5 GeV and suppressed relative to the total ν_μ CC scattering cross section for ν_τ beam energies beyond 100 GeV. At 100 GeV, which corresponds to a center-of-mass energy of $\sqrt{2ME} \approx 14$ GeV, there is still a 25% reduction in the total CC ν_τ interaction rate compared to ν_μ due to leptonic mass suppression.

For low energy neutrino sources, the CC interaction may also be suppressed due to conversion of the nucleon at the lower vertex. For example, the CC interaction commonly called "inverse beta decay" (IBD), $\bar{\nu}_e p \to e^+ n$, which is crucial to reactor neutrino experiments, has a threshold of 1.084 MeV, driven by the mass difference between the proton and the neutron plus the mass of the positron. In the case of bound nuclei, the energy transferred in a CC interaction must overcome the binding energy difference between the incoming and outgoing nucleus as well as the mass suppression due to the charged lepton. This leads to nuclear-dependent

Table 5.1. Reactions from the Sun producing neutrinos.

Common Terminology	Reaction
"pp neutrinos"	$p + p \to\, ^2\text{H} + e^- + \nu_e$
"pep neutrinos"	$p + e^- + p \to\, ^2\text{H} + \nu_e$
"^{7}Be neutrinos"	$^7\text{Be} + e^- \to\, ^7\text{Li} + \nu_e$
"^{8}B neutrinos"	$^8\text{B} \to\, ^8\text{Be}^* + e^+ + \nu_e$
"hep neutrinos"	$^3\text{He} + p \to\, ^4\text{He} + e^+ + \nu_e$

thresholds for the CC interaction. For example:

$$^{35}\text{Cl}(75.8\%) \to\, ^{35}\text{Ar}: \quad 5.967 \text{ MeV};$$

$$^{37}\text{Cl}(24.2\%) \to\, ^{37}\text{Ar}: \quad 0.813 \text{ MeV};$$

$$^{69}\text{Ga}(60.1\%) \to\, ^{69}\text{Ge}: \quad 2.227 \text{ MeV};$$

$$^{71}\text{Ga}(39.9\%) \to\, ^{71}\text{Ge}: \quad 0.232 \text{ MeV}.$$

are the thresholds for isotopes which have been used as targets in past solar neutrino (ν_e) detectors.[14–16]

In discussing neutrino scattering at higher energies, several kinematic quantities are used to describe events. The squared center of mass energy is represented by the Mandelstam variable, s. The energy transferred by the boson is ν, and $y = \nu/E_\nu$ is the fractional energy transfer, or "inelasticity." The distribution of events as a function of y depends on the helicity. For neutrino scattering from quarks, the y-dependence is flat, but for antineutrinos, the differential cross section is peaked at low y. The variable Q^2 is the negative squared four-momentum transfer. Deep inelastic scattering begins to occur at $Q^2 \sim 1$ GeV^2. If x is the fractional momentum carried by a struck quark in a deep inelastic scatter, then $x = Q^2/2M\nu$, where M is the target mass. Elastic and quasielastic scattering occur at $x = 1$, hence $Q^2 = 2M\nu \approx sxy$, valid for large s.

5.1.3. *Sources of Neutrinos*

With such a small interaction probability, it is clear that intense neutrino sources are needed to have high statistics in a neutrino experiment. The primary sources of neutrino for interactions observed on Earth are the Sun, cosmic-ray interactions, reactors, and accelerator beams.

At present, there are two intense sources in the few MeV range that allow for low energy neutrino interaction studies. First, the interactions in the Sun produce a pure ν_e flux, as listed in Table 5.1. The energy distribution of neutrinos produced by these reactions is shown in Fig. 5.4. The sensitivity

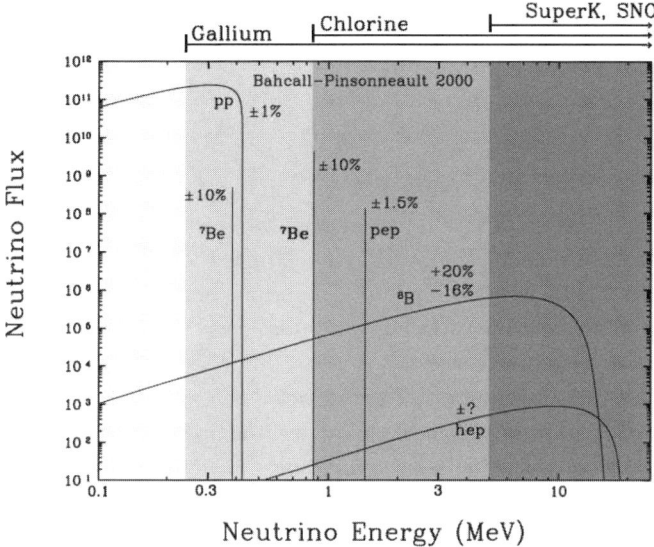

Fig. 5.4. The flux predicted by the Standard Solar Model.[13] The sensitivity of past solar neutrino detectors varies due to CC threshold in the target material.[14-16] The thresholds for various experiments is shown at the top of the plot.

of various solar neutrino experiments, due to the CC threshold, is shown at the top of the figure. There is no observable antineutrino content. The best limit on the solar neutrino $\bar{\nu}_e/\nu_e$ ratio for $E_\nu > 8.3$ MeV is 2.8×10^{-4} at 90% CL.[17] The second source is from reactors. In contrast to the Sun, reactors produce a nearly pure $\bar{\nu}_e$ flux. The energy peaks from ~ 3 to 7 MeV. Neutrinos from β decay of accelerated isotopes could, in principle, represent a third intense source of neutrinos in the MeV range (or higher), once the technical issues involved in designing such an accelerator are overcome. Such a "beta beam" would produce a very pure ν_e or $\bar{\nu}_e$ beam, depending on the accelerated isotope.[18]

At present, higher energy experiments use neutrinos produced at accelerators and in the atmosphere. In both cases, neutrinos are dominantly produced via meson decays. In the atmospheric case, cosmic rays hit atmospheric nuclei producing a shower of mesons which may decay to neutrinos along their path through the atmosphere to Earth. In a conventional neutrino beam, protons impinge on a target, usually beryllium or carbon, producing secondary mesons. In many experiments, the charged mesons are focussed (bent) toward the direction of the experiment with a magnetic

Table 5.2. Common sources of neutrinos in atmospheric and accelerator experiments.

2-body pion decay	$\pi^+ \to \mu^+ \nu_\mu$, $\pi^- \to \mu^- \bar{\nu}_\mu$
2-body kaon decay	$K^+ \to \mu^+ \nu_\mu$, $K^- \to \mu^- \bar{\nu}_\mu$
muon decay	$\mu^+ \to e^+ \bar{\nu}_\mu \nu_e$, $\mu^- \to e^- \nu_\mu \bar{\nu}_e$
K_{e3} decay	$K^+ \to \pi^0 e^+ \nu_e$, $K^- \to \pi^0 e^- \bar{\nu}_e$, $K^0 \to \pi^- e^+ \nu_e$, $K^0 \to \pi^+ e^- \bar{\nu}_e$

device called a horn. These devices are sign-selecting – they will focus one charge-sign and defocus the other – and so produce beams which are dominantly neutrinos or antineutrinos depending on the sign-selection. The beamline will have a long secondary meson decay region, which may be air or vaccuum. This is followed by a beam dump and an extended region of dirt or shielding to remove all particles except neutrinos. There are excellent reviews of methods of making accelerator-produced neutrino beams.[19] Table 5.2 summarizes the common sources of neutrino production in the atmosphere and conventional accelerator based beams.

Many atmospheric and accelerator-based neutrino experiments are designed to study 100 MeV to 10 GeV neutrinos. The atmospheric neutrino flux drops as a power-law with energy, and the 1 to 10 GeV range dominates the event rate. Accelerator beams can be tuned to a specific energy range and, using present facilities, can extend to as high as 500 GeV. From the viewpoint of sheer statistics, one should use the highest energy neutrino beam which is practical for the physics to be addressed, since the cross section rises linearly with energy. However, lower neutrino energy beams, from ∼ 1 to 10 GeV, are typically used for oscillation experiments. In these experiments, having a cleanly identified lepton in a low multiplicity event trumps sheer rate, and so ∼ 1 GeV beams are selected to assure that CCQE and single pion events dominate the interactions.

Both atmospheric and accelerator based neutrino sources are dominantly ν_μ-flavor. The main source of these neutrinos is pion decay. To understand why pions preferentially decay to produce ν_μ rather than ν_e, consider the case of pion decay to a lepton and an antineutrino: $\pi^- \to \ell^- \bar{\nu}_\ell$. The pion has spin zero and so the spins of the outgoing leptons from the decay must be opposite from angular momentum conservation. In the center of mass of the pion, this implies that both the antineutrino and the charged lepton have spin projected along the direction of motion ("right" or "positive" helicity). However, this is a weak decay, where the W only couples to the RH antineutrino and the LH component of the charged particle. The amplitude for the LH component to have right-helicity is proportional to m/E. Thus it is very small for an electron compared to the muon, producing a significant suppression for decays to electrons. Calculating the expected branching ratios:

$$R_{theory} = \frac{\Gamma(\pi^\pm \to e^\pm \nu_e)}{\Gamma(\pi^\pm \to \mu^\pm \nu_\mu)} \qquad (5.7)$$

$$= \left(\frac{m_e}{m_\mu}\right)^2 \left(\frac{m_\pi^2 - m_e^2}{m_\pi^2 - m_\mu^2}\right)^2 \qquad (5.8)$$

$$= 1.23 \times 10^{-4}; \qquad (5.9)$$

This compares well to the data:[20] $R_{exp} = (1.230 \pm 0.004) \times 10^{-4}$.

The above discussion assumed the neutrino was massless. If the neutrino is massive, then it too can be produced with wrong helicity with an amplitude proportional to m_ν/E and thus a probability proportional to $(m_\nu/E)^2$. As discussed in Sec. 5.2.2, below, neutrino mass is limited to be very small ($\sim$ eV) and thus the rate of wrong-helicity neutrino production is too low a level for any chance of observation in the near future.

Depending on the energy, there may also be significant neutrino production from kaon decays. The charged kaon preferentially decays to the ν_μ for the same reason as the charged pion. However, for equal energy mesons, the kinematic limit for a neutrino from K^+ decay is much higher than for π^+ decay: $E_\nu^{\max,K} = 0.98 E_K$ compared to $E_\nu^{\max,\pi} = 0.43 E_\pi$. Thus the neutrinos from kaon decays can be isolated by studying the highest energy component of a beam. Figure 5.5 shows the contributions of pion and kaon decays to the ν_μ flux in the MiniBooNE experiment, which uses an 8 GeV primary proton beam.

Electron neutrino flavors are produced in these beams through $K \to \pi \nu_e e$ (called "Ke3") decay and through the decay of the muons which were produced in the pion decay. These are three-body decays which avoid substantial helicity suppression. Helicity does, however, affect the energy spectrum of the outgoing decay products. In an accelerator-based experiment, the level of electron-flavor content can be regulated, at some level, by the choice of primary beam energy and the length of the decay region. A low primary beam energy will suppress kaon production because of the relatively high mass of this meson (494 MeV). A short decay pipe will suppress ν_e from μ decay, which tends to occur downstream, because it is produced in a multi-step decay chain ($\pi \to \mu \to \nu_e$). Both of these methods of suppressing ν_e production also lead to a reduction in the ν_μ production rate, so an experimenter must balance competing goals in the beam design. In the case of atmospheric neutrinos, the ratio is roughly 2:1 for ν_μ:ν_e, though the fraction ν_e's changes with energy (see Fig. 5.6). The atmospheric flux

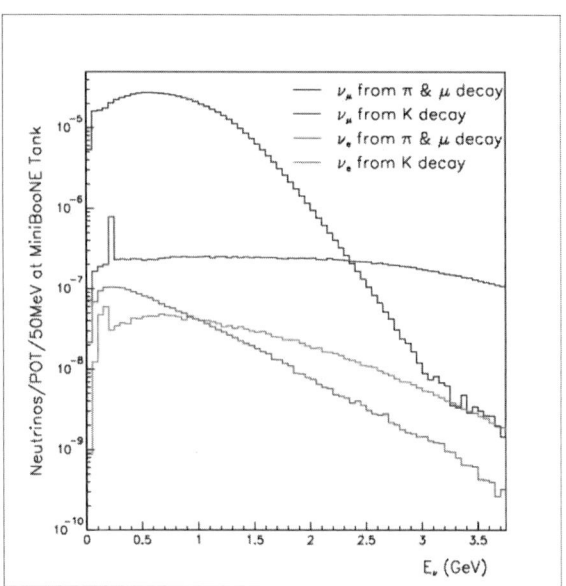

Fig. 5.5. The contributions from pion and kaon production to the total predicted ν_μ flux in the MiniBooNE experiment. The spikes at low energy in the K-produced fluxes are due to decays of stopped kaons in the beam dump.[10]

depends on the location of the detector because charged particle are bent by the Earth's magnetic field. The variation between fluxes at the Kamioka mine in Japan and Soudan mine in Minnesota are shown in Fig. 5.6.

As we move to a precision era in neutrino physics, precise "first-principles" predictions of the flux are becoming very important. For conventional accelerator-based neutrino beams and for the atmospheric flux, this requires well-measured cross-sections for production of secondary pions and kaons. This has motivated a range of secondary production experiments. The kinematic coverage is shown on Fig. 5.7.

The future of high intensity ν_μ and ν_e beams is likely to lie in beams produced from muon decay. Because of the potential for very high intensity, these beams are called "Neutrino Factories." The concept is very attractive because it produces beams which are very pure ν_μ and $\bar{\nu}_e$ from μ^- and vice versa from μ^+. Each flavor has no "wrong sign" (antineutrino-in-neutrino-beam or neutrino-in-antineutrino beam) background. However, neutrino factory designs[18] necessarily produce high energy neutrinos, since the muons must be accelerated to high energies in order to live long enough to be captured and circulated in an accelerator. The Neutrino Factory is

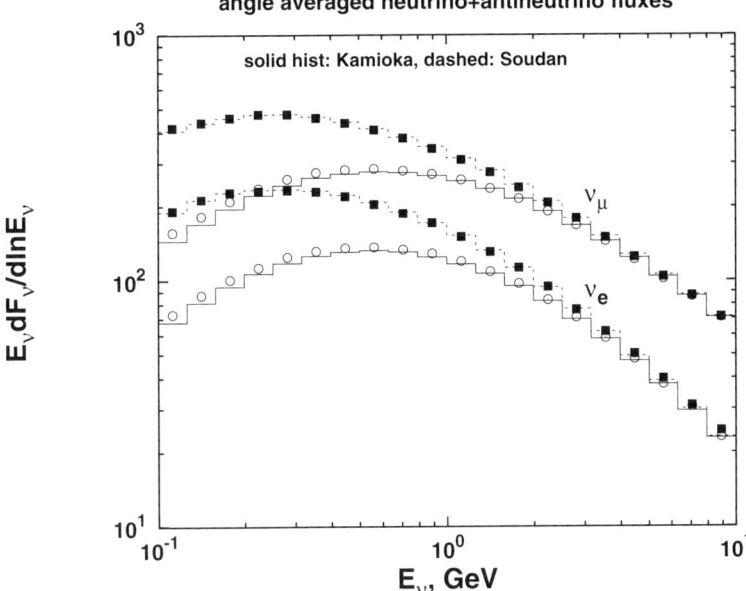

Fig. 5.6. The variation in the atmospheric neutrino flavor content as a function of energy for two locations, Japan (solid line, open circles) and Minnesota (dashed line, closed squares). The points are from a full 3-dimensional monte carlo of the flux, while the histograms are from a simpler model.[21]

seen as a promising first machine for testing ideas for a muon collider,[18] and thus has attracted interest beyond the neutrino community.

A beam enriched in ν_τ can be produced by impinging very high energy protons on a target to produce D_s-mesons which are sufficiently massive to decay to $\tau \, \nu_\tau$. The τ lepton is very massive, at 1.8 GeV, compared to the muon, at 106 MeV, and thus helicity considerations for the D_s decay strongly favor the $\tau \, \nu_\tau$ mode compared to $\mu \, \nu_\mu$, by a ratio of about 10:1. The τ then subsequently decays, also producing ν_τs.

Unfortunately, because of the short lifetime, it is not possible to separate D_s mesons from the other mesons prolifically produced by the primary interaction. As a result, the beam is dominated by the ν_μs produced by decays of other mesons. To reduce the production of ν_μ, experiments use a "beam dump" design where protons hit a very thick target where pions can be absorbed before decaying. The only enriched-ν_τ beam created to date was developed by DoNuT.[6] They used an 800 GeV proton on a beam dump, to produce a ratio of ν_e:ν_μ:ν_τ of about 6:9:1.

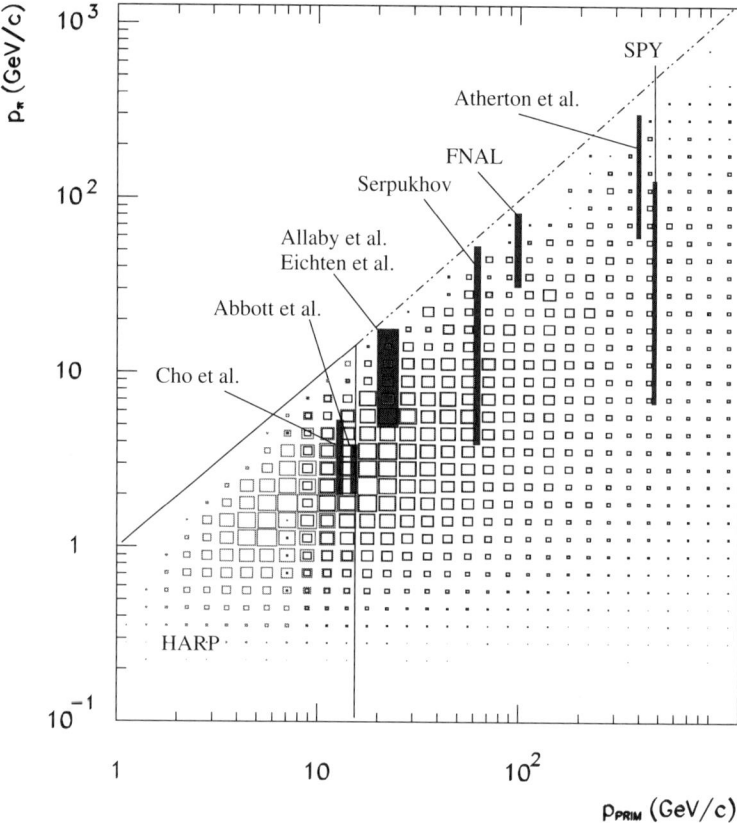

Fig. 5.7. The kinematic range covered by recent experiments measuring secondary pion and kaon production.[24]

5.1.4. *Typical Neutrino Detectors*

Because neutrinos interact so weakly, the options for detectors are limited to designs which can be constructed on a massive scale. There are several general styles in use today: unsegmented scintillator detectors, unsegmented Cerenkov detectors, segmented scintillator-and-iron calorimeters, and segmented scinitillator trackers. The most promising future technology is the noble-element based detector, which is effectively an electronic bubble chamber. Liquid argon detectors are likely to be the first large-scale working example of such technology. There are a few variations on these five themes, which are considered in later sections in the context of the measurement.

Unsegmented scintillator detectors are typically used for low energy antineutrino experiments. Recent examples include Chooz,[25] KamLAND[26] and LSND.[27] These consist of large tanks of liquid scintillator surrounded by phototubes. Usually the scintillator is oil based, hence the target material is CH_2 and its associated electrons. Often the tubes are in an pure oil buffer. This reduce backgrounds from radiation emitted from the glass which would excite scintillator. The free protons in the oil provide a target for the interaction, $\bar{\nu}_e p \rightarrow e^+ n$, which is the key for reactor experiments. The reaction threshold for this interaction is 1.806 MeV due to the mass differences between the proton and neutron and the mass of the positron. The scintillation light from the e^+, as well as light from the Compton scattering of the 0.511 MeV annihilation photons provide an initial ("prompt") signal. This is followed by n capture to produce deuterium and a 2.2 MeV. This sequence – positron followed by neutron capture – provides a clean signal for the interaction. Doping the liquid scintillator with gadolinium substantially increases the neutron capture cross section as well as the visible energy produced in the form of gammas upon neutron capture.

Unsegmented scintillator detectors are now being introduced for low energy solar neutrino measurements at Borexino,[28] KamLAND[29] and SNO+.[30] These provide energy information on an event-by-event basis, unlike most past solar neutrino experiments, such as Homestake,[14] SAGE[15] and GallEx,[16] which intergrated over time and energy. However, these are very difficult experiments to perform because a neutron is not produced and so the scattering does not produce a two-fold coincidence, but only a prompt flash of light.

Environmental backgrounds are by far the most important issue in low energy experiments. These fall into two categories: naturally occurring radioactivity and muon-induced backgrounds. To get a sense for what is expected, Fig. 5.8 shows the visible energy distribution of singles events from the KamLAND experiment with the sources of environmental background identified. The naturally occurring radioactive contaminants mainly populate the low energy range of Fig. 5.8, with isotopes from the U and Th chain extending to the highest energies. These isotopes must be kept under control by maintaining very high standards of cleanliness. The second source of environmental background, the β-decays of isotopes produced by cosmic ray muons. These dominate the background for $E_{vible} > 4$ MeV (see Fig. 5.8). These can only be eliminated by shielding the detector from cosmic rays. As a result, we must build deep underground laboratories with many thousands of meters-water-equivalent ("mwe") of rock shielding.

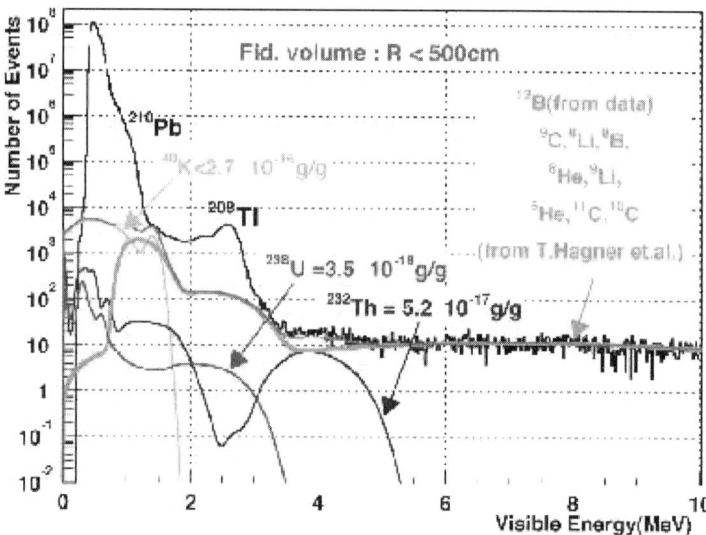

Fig. 5.8. Energy distribution and sources of singles events in KamLAND as a function of visible energy.[31]

In these scintillator detectors, the CC interaction with the carbon in the oil (which produces either nitrogen or boron depending on whether the scatterer is a neutrino or antineutrino) has a significantly higher energy threshold than scattering from free protons. $\nu_e + C \rightarrow e^- + N$ has a threshold energy of 13.369 MeV, which arises from the carbon-nitrogen mass difference (plus the mass of the electron). In the case of both reactor and solar neutrinos, the flux cuts off below this energy threshold.

Existing unsegmented Cerenkov detectors include MiniBooNE,[32] Super K,[33] and AMANDA.[34] These detectors make use of a target which is a large volume of a clear medium (undoped oil, water and ice, respectively) surrounded by or interspersed with phototubes. Undoped oil has the advantages of a larger refractive index, leading to larger Cerenkov opening angle, and of not requiring a purification system to remove living organisms. Water is the only affordable medium once a detector is larger than a few ktons. For ultra-high energy neutrino experiments, a vast natural target is needed. Sea water[35] and ice[34] have been used. Ice is, to date, more successful because it does not suffer from backgrounds from bioluminescence.

In most cases of these detectors, the tubes surround the medium and the projected image of the Cerenkov ring is used for particle identification.

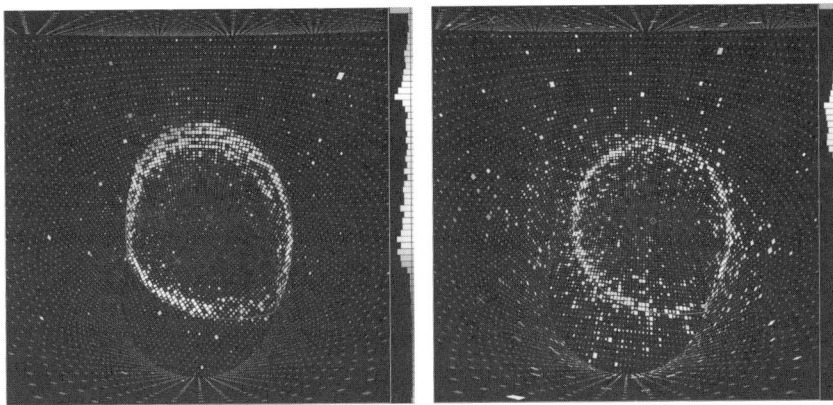

Fig. 5.9. An example of a muon ring (left) and electron ring (right) in the Super K Cerenkov detector.[33]

To understand how this works, first consider the case of a perfect, short track. This will project a ring with a sharp inner and outer edge onto the phototubes. Next consider an electron produced in a ν_e CC quasielastic interaction. Because the electron is low mass, it will multiple scatter and easily bremsstrahlung, smearing the light projected on the tubes and producing a "fuzzy" ring. A muon produced by a CC quasileastic ν_μ interaction is heavier and thus will produce a sharper outer edge to the ring. For the same visible energy, the track will also extend farther, filling the interior of the ring, and perhaps exit the tank. Fig. 5.9 compares an electron and muon ring observed in the Super K detector. If the muon stops within the tank and subsequently decays, the resulting "michel electron" provides an added tag for particle identification. In the case of the μ^-, 18% will capture in water, and thus have no michel electron tag, while only 8% will capture in oil.

Scintillator and iron calorimeters provide affordable detection for ν_μ interactions in the range of $\sim$1 GeV and higher. Recent examples include the MINOS[36] and NuTeV[37] experiments. In these detectors, the iron provides the target, while the scintillator provides information on energy deposition per unit length. This allows separation between the hadronic shower, which occurs in both NC and CC events, and the minimum ionizing track of an outgoing muon, which occurs in CC events. Transverse information can be obtained if segmented scintillator strips are used, or if drift chambers are interspersed. The light from scintillator strips is transported to tubes by mirrored wave-length-shifting fibers. Transverse information improves

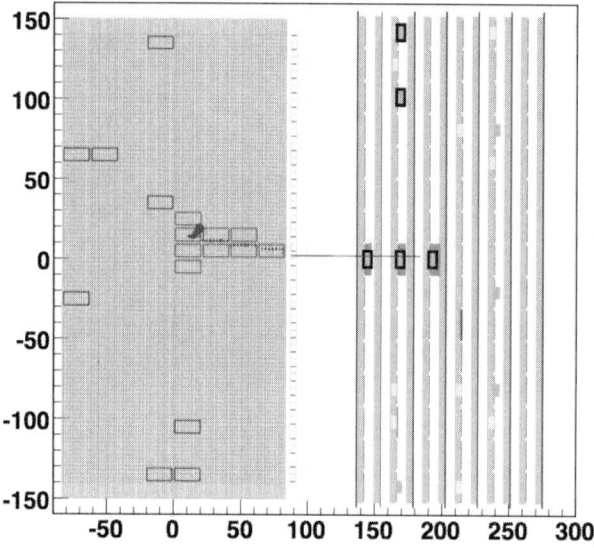

Fig. 5.10. A CCQE($\nu_\mu + n \to \mu + p$) event observed in the SciBooNE detector. The long, minimum-ionizing red track is identified as the muon, the short, heavily-ionizing red track is identifed as the proton.[11]

separation of electromagnetic and hadronic showers. The iron can be magnetized to allow separation of neutrino and antineutrino events based on the charge of the outgoing lepton.

In all three of the above detector designs, it is difficult to reconstruct multi-particle events. Tracking is not an option for an unsegmented scintillator detector. Cerenkov detectors can typically resolve two tracks per event. Segmented calorimeters reduce multiple hadrons to a shower, obscuring any track-by-track information other than from muons.

To address the problem of track reconstruction in low energy ($\lesssim 1$ GeV), low multiplicity events, there has been a move toward all-scintillator tracking detectors. This began with the SciBar detector in K2K.[38] This detector used scintillator strips, as in MINOS, but without interspersing iron. As a result, low energy (few MeV) tracks were clearly observable and quasielastic and single pion events could be fully reconstructed. SciBar has since been incorporated into the SciBooNE experiment at Fermilab.[11] The CCQE event in SciBooNE, shown in Fig. 5.10, makes clear the benefits of fine segmentation. The position of the vertex and the short track from the proton are well-resolved in the SciBar detector (green region). The technology has been taken further by the MINERvA experiment, which has attained 2 mm resolution with their prototype.[39] Scibar and MINERvA are relatively small (few ton) detectors. The first very large scale application of this technology will be NOvA, which is a future 15 kton detector.[40] This detector will use PVC tubes filled with liquid scintillator, which is more cost-effective than extruded scintillator strips for very large detectors. Their design also loops the wave-length shifting fiber, so that there are, effectively, two perfectly mirrored fibers are in each cell. This elegant solution increases the collected light by a factor of four, which is necessary for ~ 15 m strips.

The most promising new technology for high resolution track reconstruction in neutrino physics is the liquid argon TPC. A TPC, or time projection chamber, uses drift chambers to track in the x and y views and drift time to determine the z view. Liquid argon (LAr), which provides the massive target for the neutrino interaction, also scintillates, providing the start for the drift-time measurement. A key point for future neutrino experiments is the high efficiency for identifying electron showers (expected to be 80-90%) with a rejection factor of 70 for NC π^0 events. In particular, these detectors can differentiate between converted photons and electrons through the dE/dx in the first few centimeters of the track. Typical energy resolution for an electromagnetic shower is $3\%/\sqrt{E}$.

There is a great deal of activity on development of LAr detectors. Data have been taken successfully on a 50 liter LArTPC prototype in the NOMAD neutrino beam at CERN, resulting in reconstruction of ~ 100 CC quasielastic events.[41] Also, recently, a 600 ton Icarus module has been commissioned at Gran Sasso.[42] A 0.8 ton LAr test detector will begin taking data at Fermilab in January, 2008.[43] As discussed in Sec. 5.3.2, the microBooNE experiment is a proposed 100 ton detector which would take data in 2010.[44] In principle, these detectors can be scaled up to tens of ktons, as is discussed in the "Ash River Proposal".[45]

5.2. Neutrinos As We Know Them Now

The recent discovery of neutrino oscillations requires that we reconsider the Standard Model Lagrangian of Sec. 5.1.1. It must now incorporate, preferably in a motivated fashion, both neutrino mass and neutrino mixing. This represents both a challenge and an opportunity for the theory, which I will discuss in the following section. This section concentrates on the experimental discovery. It is interesting to note that while neither neutrino mass nor mixing were "needed" in the Standard Model theory, both are required for the discovery of neutrino oscillations. The probability for neutrino oscillations will be zero unless *both* effects are present.

The outcome of the observation of neutrino oscillations is typically summarized by the statement that "neutrinos have mass." To be clear: we still have no direct measurement of neutrino mass. At this point, we have clear evidence of mass differences between neutrinos from the observation of neutrino oscillations. A mass difference between two neutrinos necessarily implies that at least one of the neutrinos has non-zero mass. All experimental evidence indicates that the actual values of the neutrino masses are tiny in comparison to the masses of the charged fermions. At the end of this section, attempts at direct measurement of neutrino mass are described.

5.2.1. *Neutrino Oscillations*

Recent results on neutrino oscillations provide indisputable evidence that there is a spectrum of masses for neutrinos. In this section, I describe the formalism for neutrino oscillations, and then review the experimental results which have now been confirmed at the 5σ level. This is covered briefly because these results are well known and covered extensively elsewhere.[2]

5.2.1.1. *The Basic Formalism*

Neutrino oscillations requires that neutrinos have mass, that the difference between the masses be small, and that the mass eigenstates be different from the weak interaction eigenstates. In this case, the weak eigenstates can be written as mixtures of the mass eigenstates. For example, in a simple 2-neutrino model:

$$\nu_e = \cos\theta\, \nu_1 + \sin\theta\, \nu_2$$
$$\nu_\mu = -\sin\theta\, \nu_1 + \cos\theta\, \nu_2$$

where θ is the "mixing angle." In this case, a pure flavor (weak) eigenstate born through a weak decay can oscillate into another flavor as the state propagates in space. This oscillation is due to the fact that each of the mass eigenstate components propagates with different frequencies if the masses are different, $\Delta m^2 = |m_2^2 - m_1^2| > 0$. In such a two-component model, the oscillation probability for $\nu_\mu \to \nu_e$ oscillations is then given by:

$$\text{Prob}\,(\nu_\mu \to \nu_e) = \sin^2 2\theta\ \sin^2\left(\frac{1.27\,\Delta m^2\,(\text{eV}^2)\ L\,(\text{km})}{E\,(\text{GeV})}\right), \quad (5.10)$$

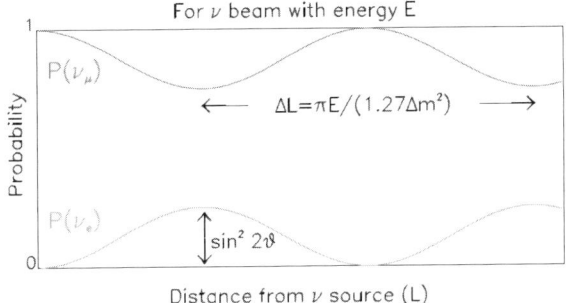

Fig. 5.11. Example of neutrino oscillations as a function of distance from the source, L. The wavelength depends upon the experimental parameters L and E (neutrino energy) and the fundamental parameter Δm^2. The amplitude of the oscillation is constrained by the mixing term, $\sin^2 2\theta$.

where L is the distance from the source, and E is the neutrino energy. As shown in Fig. 5.11, the oscillation wavelength will depend upon L, E, and Δm^2. The amplitude will depend upon $\sin^2 2\theta$.

Neutrino oscillations only occur if the two mass states involved have sufficiently small Δm^2 that the neutrino flavor is produced in a superposition of two mass states. If the mass splitting is sufficiently large, a given neutrino flavor would be produced in one or the other of the two mass eigenstates and interference (i.e., oscillations) would not occur.

Most neutrino oscillation analyses consider only two-generation mixing scenarios, but the more general case includes oscillations among all three neutrino species. This can be expressed as:

$$\begin{pmatrix} \nu_e \\ \nu_\mu \\ \nu_\tau \end{pmatrix} = \begin{pmatrix} U_{e1} & U_{e2} & U_{e3} \\ U_{\mu 1} & U_{\mu 2} & U_{\mu 3} \\ U_{\tau 1} & U_{\tau 2} & U_{\tau 3} \end{pmatrix} \begin{pmatrix} \nu_1 \\ \nu_2 \\ \nu_3 \end{pmatrix}.$$

This formalism is analogous to the quark sector, where strong and weak eigenstates are not identical and the resultant mixing is described conventionally by a unitary mixing matrix. The oscillation probability is then:

$$\text{Prob}(\nu_\alpha \to \nu_\beta) = \delta_{\alpha\beta} - 4 \sum_{j>i} U_{\alpha i} U_{\beta i}^* U_{\alpha j}^* U_{\beta j} \sin^2\left(\frac{1.27 \Delta m_{ij}^2 L}{E}\right), \quad (5.11)$$

where $\Delta m_{ij}^2 = m_j^2 - m_i^2$, α and β are flavor-state indices (e, μ, τ) and i and j are mass-state indices $(1, 2, 3)$.

For three neutrino mass states, there are three different Δm^2 parameters, although only two are independent since the two small Δm^2 parameters must sum to the largest. The neutrino mass states, ν_1, ν_2 and ν_3 are defined such that the difference between ν_1 and ν_2 always represents the smallest splitting. However, the mass of ν_3 relative to ν_1 and ν_2 is arbitrary and so the sign of the Δm^2 parameters which include the third mass state may be positive or negative. That is, if $\nu_3 > \nu_1, \nu_2$, then Δm^2_{23} will be positive, but if $\nu_1, \nu_2 > \nu_3$, then Δm^2_{23} will be negative. The former is called a "normal mass hierarchy" and the latter is the "inverted mass hierarchy." At this point, the sign is irrelevant because Δm^2 appears in a term which is squared. However, in Sec. 5.3.1, this point will become important.

The mixing matrix above can be described in terms of three mixing angles, θ_{12}, θ_{13} and θ_{23}:

$$U = \begin{pmatrix} c_{12}c_{13} & s_{12}c_{13} & s_{13} \\ -s_{12}c_{23} - c_{12}s_{23}s_{13} & c_{12}c_{23} - s_{12}s_{23}s_{13} & s_{23}c_{13} \\ s_{12}s_{23} - c_{12}c_{23}s_{13} & -c_{12}s_{23} - s_{12}c_{23}s_{13} & c_{23}c_{13} \end{pmatrix}, \quad (5.12)$$

where $c_{ij} \equiv \cos\theta_{ij}$ and $s_{ij} \equiv \sin\theta_{ij}$, with i and j referring to the mass states. In fits to the oscillation parameters, people variously quote the results in terms of the matrix element of U, sin-squared of the given angle, sin-squared of twice the angle and a variety of other forms, all of which are related. Using the 13 case as an example, the quoted parameters are related by:

$$U_{e3}^2 \approx \sin^2\theta_{13} \approx \frac{1}{4}\sin^2 2\theta_{13}. \quad (5.13)$$

Thus, in total, there are five free parameters in the simplest three-neutrino oscillation model, which can be taken to be Δm^2_{12}, Δm^2_{23}, θ_{12}, θ_{13} and θ_{23}.

Although in general there will be mixing among all three flavors of neutrinos, two-generation mixing is often assumed for simplicity. If the mass scales are quite different ($m_3 >> m_2 >> m_1$, for example), then the oscillation phenomena tend to decouple and the two-generation mixing model is a good approximation in limited regions. In this case, each transition can be described by a two-generation mixing equation. However, it is possible that experimental results interpreted within the two-generation mixing formalism may indicate very different Δm^2 scales with quite different apparent strengths for the same oscillation. This is because, as is evident from equation 5.11, multiple terms involving different mixing strengths and Δm^2 values contribute to the transition probability for $\nu_\alpha \to \nu_\beta$.

5.2.1.2. Matter Effects

The probability for neutrino oscillations is modified in the presence of matter. This is true in any material, however the idea was first explored for neutrino oscillations in the Sun, by Mikheyev, Smirnov and Wolfenstein. Therefore, matter effects are often called "MSW" effects.[46] In general, matter effects arise in neutrino-electron scattering. The electron neutrino flavor experiences both CC and NC elastic forward-scattering with electrons. However, the ν_μ and ν_τ experience only NC forward-scattering, because creation of the μ or τ is kinematically forbidden or suppressed (e.g. $\nu_\mu + e^- \to \nu_e + \mu^-$). This difference produces the matter effect.

For neutrinos propagating through a constant density of electrons, if V_e is the elastic forward scattering potential for the ν_e component, and V_{other} is the potential for the other neutrino flavors, then the additional scattering potential is

$$V = V_e - V_{other} = \sqrt{2} G_F n_e, \qquad (5.14)$$

where G_F is the Fermi constant and n_e is the electron density. This potential modifies the Hamiltonian, so that, if H_0 is the vacuum Hamiltonian, then in matter the Hamiltonian is $H_0 + V$. This means that the eigenstates are modified from those of a vacuum, ν_1 and ν_2, to become ν_{1m} and ν_{2m}. Effectively, the neutrino mass spectrum is not the same as in vacuum. The solutions to the Hamiltonian are also modified. From this, one can see that the presence of electrons may substantially change the oscillatory behavior of neutrinos.

The simplest outcome is that matter induces a shift in the mass state, which is a combination of flavor eigenstates, propagates through the material. This leads to a change in the oscillation probability:

$$\text{Prob}\,(\nu_e \to \nu_\mu) = \left(\sin^2 2\theta / W^2\right) \sin^2 \left(1.27 W \Delta m^2 L/E\right) \qquad (5.15)$$

where $W^2 = \sin^2 2\theta + (\sqrt{2} G_F n_e (2E/\Delta m^2) - \cos 2\theta)^2$ (Note that in a vacuum, where $n_e = 0$, this reduces to equation 5.10.) From this, one can see that if a neutrino, passing through matter, encounters an optimal density of electrons, a "resonance," or large enhancement of the oscillation probability, can occur. The Sun has a wide range of electron densities and thus is a prime candidate for causing matter effects. Also, neutrinos traveling through the Earth's core, which has a high electron density, might experience matter effects. This will produce a "day-night effect," or siderial variation, for neutrinos from the Sun.

For situations like the Sun, with very high electron densities which vary with the position of the neutrino (and hence the time which the neutrino has lived), the situation is complex. If the electron density is high and the density variation occurs slowly, or adiabatically, then transition (not oscillation!) between flavors in the mass state can occur as the neutrino propagates. Thus it is possible for neutrinos to be produced in the core of the Sun in a given mass and flavor state, and slowly evolve in flavor content until the neutrino exits the Sun, still in the same mass state. In other words, in the Sun, a ν_e produced in a mass eigenstate $\nu_{2m}(r)$, which depends on the local electron density at radius r, propagates as a ν_{2m} until it reaches the $r = R_{solar}$, where $\nu_{2m}(R_{solar}) = \nu_2$. This peculiar effect is called the Large Mixing Angle MSW solution.

5.2.1.3. Designing an Oscillation Experiment

From equation 5.10, one can see that three important issues confront the designer of the ideal neutrino experiment. First, if one is searching for oscillations in the very small Δm^2 region, then large L/E must be chosen in order to enhance the $\sin^2(1.27\Delta m^2 L/E)$ term. However if L/E is too large in comparison to Δm^2, then oscillations occur rapidly. Because experiments have finite resolution on L and E, and a spread in beam energies, the $\sin^2(1.27\Delta m^2 L/E)$ averages to 1/2 when $\Delta m^2 \gg L/E$ and one loses sensitivity to Δm^2. Finally, because the probability is directly proportional to $\sin^2 2\theta$, if the mixing angle is small, then high statistics are required to observe an oscillation signal.

There are two types of oscillation searches: "disappearance" and "appearance." To be simplistic, consider a pure source of neutrinos of type x. In a disappearance experiment, one looks for a deficit in the expected flux of ν_x. This requires accurate knowledge of the flux, which is often difficult to predict from first principles. Therefore, most modern disappearance experiments employ a near-far detector design. The near detector measures the flux prior to oscillation (the design goal is to effectively locate it at $L = 0$ in Fig. 5.11). This is then used to predict the unoscillated event rate in the far detector. A deficit compared to prediction indicates disappearance. Appearance experiments search for $\nu_\alpha \to \nu_\beta$ by directly observing interactions of neutrinos of type β. The case for oscillations is most persuasive if the deficit or excess has the (L/E) dependence predicted by the neutrino oscillation formula (equation 5.10).

The "sensitivity" of an experiment is defined as the average expected limit if the experiment were performed many times with no true signal (only background). Let us consider the sensitivity for a hypothetical perfect (no-systematic error) disappearance neutrino oscillation experiment with N events. A typical choice of confidence level is 90%, so in this case, the limiting probability, assuming there is no signal, is

$$P = \sigma\sqrt{N}/N. \tag{5.16}$$

There are two possible choices of σ associated with a 90% CL sensitivity, depending on the underlying philosophy. If one assumes there is no signal in the data, then one quotes the sensitivity based on 90% of a single-sided Gaussian, which is $\sigma = 1.28$. If the philosophy is that there is a signal which is too small to measure, then one quotes the sensitivity using $\sigma = 1.64$, which is appropriate for a double-sided Gaussian. Historically, $\sigma = 1.28$ was used in most publications. Physicists engage in arguments as to which is most correct, but what is most important from a practical point of view is for the reader to understand what was used. The reader can always scale between 1.28 and 1.64 depending on personal opinion.

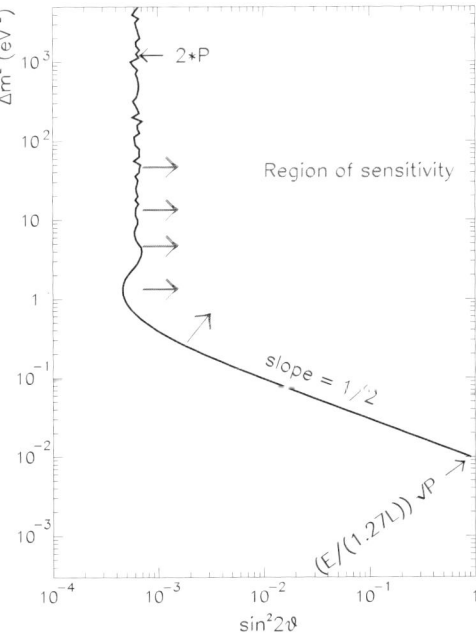

Fig. 5.12. An illustration of the sensitivity of an imaginary oscillation experiment. The region of sensitivity for an experiment depends on the oscillation probability, P, where one can set a limit at some confidence level. Most experiments use 90% CL. The boundaries depend on P, L and E.

There is only one measurement, P, and there are two unknowns, Δm^2 and $\sin^2 2\theta$; so this translates to a region of sensitivity within $\Delta m^2 - \sin^2 2\theta$ space. This is typically indicated by a solid line, with the allowed region on the right on a plot (see illustration in Fig 5.12). For the perfect (no-systematic error) experiment, the high Δm^2 limit on $\sin^2 2\theta$ is driven by the statistics. On the other hand, the L and E of the experiment drive the low Δm^2 limit, which depends on the fourth root of the statistics. If our perfect experiment had seen a signal, the indications of neutrino oscillations would appear as "allowed regions," or shaded areas on plots of Δm^2 vs. $\sin^2 2\theta$.

This rule of thumb – that statistics drives the $\sin^2 2\theta$-reach and L/E drives the Δm^2 reach – becomes more complicated when systematics are considered. The imperfections of a real experiment affect the limits which can be set. Systematic uncertainties in the efficiencies and backgrounds reduce the sensitivity of a given experiment. Background sources introduce multiple flavors of neutrinos in the beam. Misidentification of the interacting neutrino flavor in the detector can mimic oscillation signatures. In addition, systematic uncertainties in the relative acceptance versus distance and energy need to be understood and included in the analysis of the data.

For a real experiment, with both statistical and systematic errors, finding the sensitivity and final limit or allowed region requires a fit to the data. The data are compared to the expectation for oscillation across the range of oscillation parameters, and the set of parameters where the agreement is good to 90% CL are chosen. Historically, there are three main approaches which have been used in fits. The first method is the "single sided raster scan." In this case one chooses a Δm^2 value and scans through the $sin^2 2\theta$-space to find the 90% CL limit. The second method, the "global scan," explores Δm^2- and $sin^2 2\theta$-space simultaneously. Thus there are two parameters to fit and two degrees of freedom. The third method is the frequentist, or "Feldman-Cousins" approach,[47] in which one simulates "fake-experiments" for each Δm^2 and $sin^2 2\theta$ point, and determines the limit where, in 90% of the cases, no signal is observed. Each method has pros and cons and the choice is something of a matter of taste. As with the question of a single- or double-sided gaussian, what is important is to compare sensitivities, limits, and signals from like methods.

It is possible for an experiment which does not observe a signal to set a limit which is better than the sensitivity. This occurs if the experiment observed a downward fluctuation in the background. In this case, a limit is hard to interpret. The latest standard practice is to show the sensitivity and the limit on plots, and the readers can draw their own interpretation.[47]

5.2.1.4. *Experimental Evidence for Oscillations*

Two separate allowed regions in Δm^2-and-$\sin^2 2\theta$ -space for neutrino oscillations have been observed at the $> 5\sigma$ level. These are called the "Atmospheric Δm^2" and "Solar Δm^2" regions. The names are historical, as will be seen below. Many reviews have been written on these results (see, for example,,[2],[48] and[49]) and so here the results are briefly outlined.

The highest Δm^2 signal was first observed using neutrinos produced in the upper atmosphere. These atmospheric neutrinos are produced through collisions of cosmic rays with the atmosphere. The neutrinos are detected through their charged–current interactions in detectors on the Earth's surface.

The first evidence for atmospheric neutrino oscillations came from the Kamioka[50] and IMB[51] experiments. This was followed by the convincing case presented by the Super K experiment.[53] These were single detector experiments observing atmospheric neutrino interactions as a function of zenith angle (see Fig. 5.13). Several striking features were observed. The first was that the ν_μ flavor neutrinos showed clear evidence of disappearance while the ν_e flavor CC scatters were in good agreement with prediction. The second striking observation was that the apparent mixing was nearly maximal. In other words, the experiments were seeing a 50% reduction of the ν_μ event rate compared to expectation.

Complications in the analysis arise from the difficulty in understanding production of atmospheric neutrinos (affecting the understanding of E) and in the accurate reconstruction of events as a function of zenith angle (affecting knowledge of L). The Δm^2 extracted from the Kamoiokande data is an order of magnitude higher than that extracted from the Super K data, indicating a clear systematic effect. Thus, it was absolutely crucial for accelerator-based "long-baseline" neutrino experiments to confirm this result. In these experiments, the L is well defined by the distance from source to detector, and the E is well understood from a near detector measurement.

The challenge for long-baseline experiments is that the L/E required to access the atmospheric signal is on the order of 1000 km/GeV. If the beam is relatively low energy, so that the easy-to-reconstruct CCQE interaction dominates the events, then L is on the order of 1000 km. This leads to two major technical challenges. First, because the Earth is a sphere, if the source and detector are to be located on (or near) the surface, the beam must be directed downward, into the Earth. Engineering a beamline at a

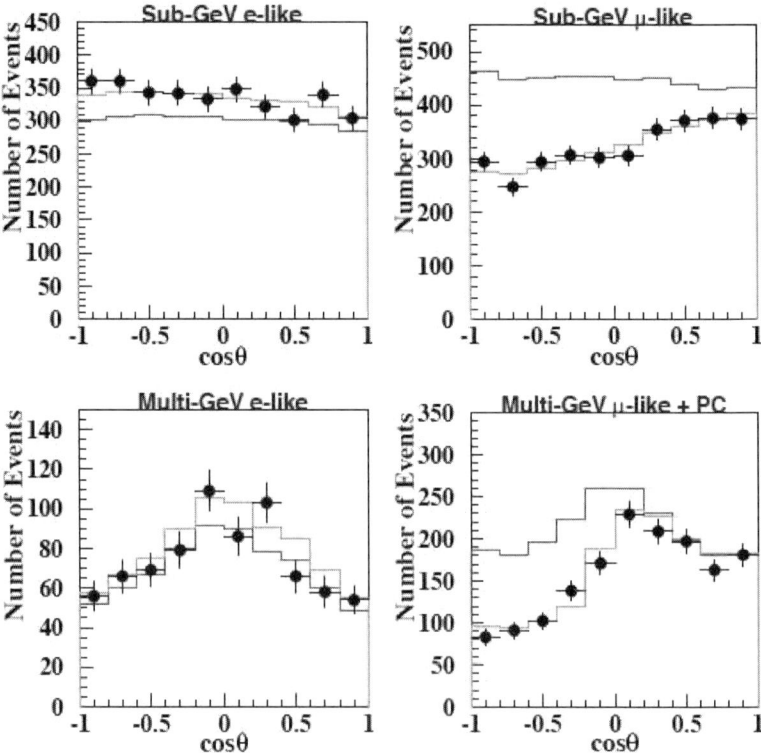

Fig. 5.13. Event rates observed in Super K as a function of zenith angle for two energy ranges. Candidate ν_e events are on the left, ν_μ are on the right. The red line indicates the predicted rate. The green line is the best fit including oscillations.[52]

steep angle requires overcoming substantial hurdles in tunneling. Second, the beam spreads as it travels outward from the source, resulting in low intensity at the detector. Therefore, very high rates are needed. However, these challenges have now been overcome at three accelerator complexes: KEK, FNAL and CERN, and a new long-baseline beam from the JPARC facility will be available soon. Making use of these lines, initial confirmation of the atmospheric neutrino deficit came from the KEK-to-Kamiokande (K2K) long baseline experiment.[54] This has since been followed up by the MINOS experiment to high precision.[55,56]

In the atmospheric data, the ν_e CC signal is in agreement with expectation, and in the long-baseline experiments, no ν_e excess has been observed. Therefore, one cannot interpret this oscillation signal as $\nu_\mu \to \nu_e$. This leaves only $\nu_\mu \to \nu_\tau$ as an explanation for the deficit in a three-neutrino model. Observation of ν_τ CC interactions is experimentally difficult in these experiments for a number of reasons. First, the L/E of the signal is such that for lengths available to present experiments, the energy of the beam must be low ($\lesssim 10$ GeV). As discussed in Sec. 5.1.2, because the τ mass is 1.8 GeV, there is substantial mass suppression for τ production at low energies, so the CC event rate is low. Second, the τ decays quickly, leaving behind a complicated event structure which can be easily confused with ν_μ and ν_e low multiplicity interactions in calorimeter or Cerenkov detectors. The difficulty of identifying ν_τ events even in a specialized emulsion-based detector with a high energy neutrino beam, was made clear by the DoNuT experiment,[6] which provided the first, and so far only, direct observation of CC ν_τ interactions. Thus, while some studies claim observation of ν_τ CC interactions in SuperK,[57] these results are not very convincing to this author. Fortunately, a specialized experiment called OPERA,[58] which is an emulsion-based long-baseline detector, is presently taking data. The average energy of the CNGS beam used by this experiment is 17 GeV, sufficiently high to produce ν_τ CC events. This experiment is expected to observe ~ 15 events in 5 years of running if the atmospheric neutrino deficit is due to $\nu_\mu \to \nu_\tau$ with $\Delta m^2 = 2.5 \times 10^{-3}$ eV^2.[59]

The lower Δm^2 signal is called the "Solar Neutrino Deficit," as it was first observed as a low rate of observed ν_e's from the Sun. The first observation of this effect was a ν_e deficit observed using a Cl target[14] by Ray Davis and collaborators at Homestake, using $\nu_e + \text{Cl} \to e + \text{Ar}$. Only about 1/3 of the total expected neutrino event rate was observed. By 1999, four additional experiments had confirmed these observations. The GALLEX[16] and SAGE[15] experiments confirmed a deficit for CC electron neutrino interactions in a Ga target producing Ge. The Super Kamiokande experiment observed a deficit for $\nu_e + e \to \nu_e + e$ reactions in water.[22] The deficit is shown on Fig. 5.14, indicated by the blue points. This plot shows the ratio to the Standard Solar Model prediction, which is indicated by the solid line at unity.

A few aspects of the initial solar neutrino deficit studies should be noted. First, the three types of experiments, chlorine-based, gallium-based, and water-based, measured different levels of deficit. Given that each type of nucleus has a different low energy threshold for observation of CC events, as previously discussed, one can interpret the varying levels of deficit as an energy dependent effect. Second, all of the above experiments rely upon the CC interaction. The energy of neutrinos from the Sun is so low, that

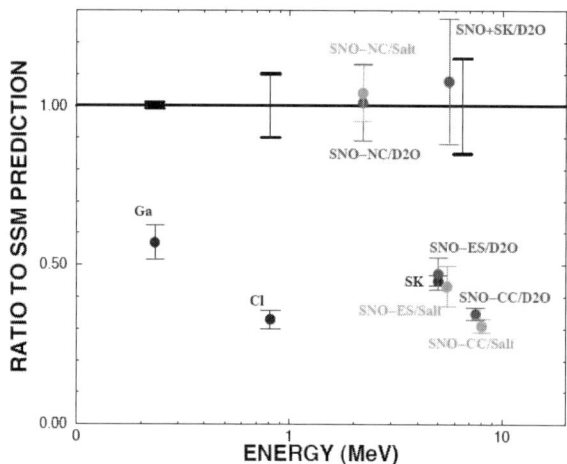

Fig. 5.14. Ratio of observed event rates in solar neutrino experiments compared to the Standard Solar Model. Experiments are plotted at the average energy of the detected signal, which varies due to detection threshold. Black error bars indicate Standard Solar Model error.[23]

should ν_μ or ν_τ be produced through oscillations, the CC interaction could not occur. This is because of the relatively high mass of the μ (106 MeV) and the τ (1.8 GeV). Thus all of these experiments can observe that ν_es disappeared, but they cannot observe if the neutrinos reappear as one of the other flavors. This makes a decisive statement that the effect is due to neutrino oscillations problematic.

For some time, people argued the apparent deficit was due to an incomplete picture of solar processes. The two important theoretical issues related to the solar neutrino fluxes were the fusion cross sections and the temperature of the solar interior. A comprehensive analysis of the available information on nuclear fusion cross sections important to solar processes has been compiled[60] and shows that the important cross sections are well-known. Results in helioseismology provided an important further test of the "Standard Solar Model".[61] The Sun is a resonant cavity, with oscillation frequencies dependent upon P/ρ, the ratio of pressure to density. Helioseismological data confirmed the SSM prediction of U to better than 0.1%.[62] With the results of these studies, most physicists were convinced that the Standard Solar Model was substantially correct. The error bars on the black line at unity in Fig. 5.14 shows the side of the estimated systematic error on the Standard Solar Model.

Interpreting the results as neutrino oscillations resulted in a complicated picture. The vacuum oscillation probability, calculated using equation 5.10, results in allowed regions of Δm^2 which are very low ($\Delta m^2 \sim 10^{-10} \text{eV}^2$). This is because the energy of the neutrinos is only a few MeV, and the Sun to the Earth pathlength is very long ($\sim 10^{11}$ m). On the other hand, the Sun has high electron content and density, so matter effects (Sec. 5.2.1.2) could interfere with the picture, allowing higher true values of Δm^2. The MSW effect yielded two solutions in fits to the data. One was at mixing angles of $\sim 10^{-3}$. Until very recently, this was regarded as the most likely solution based on analogy with mixing in the quark sector. The other solution gave a very large, although not maximal, mixing angle.

Two dramatic results of the early 2000's demonstrated that the solar neutrino deficit was due to oscillations with the MSW effect and with large mixing angle. The first result was from the SNO experiment.[63] SNO used a D_2O target which allowed for measurement of both CC ν_e interactions as well as $\nu + d \to \nu + n + p$. In the former measurement, SNO sees a deficit consistent with the other measurements within an oscillation interpretation, and which yields a ν_e flux of $(1.76 \pm 0.05(\text{stat}) \pm 0.09(\text{sys})) \times 10^6/\text{cm}^2\text{s}$.[64] The later measurement is an NC interaction, and thus is flavor-blind. It yields a total NC flux of $(5.09^{+0.44}_{-0.43}(\text{stat})^{+0.46}_{-0.43}(\text{sys})) \times 10^6/\text{cm}^2\text{s}$[64] which can be compared with the theoretical prediction of $(5.69 \pm 0.91) \times 10^6/\text{cm}^2\text{s}$.[65] In other words, SNO observed the expected total event rate, within errors. This implied that the ν_es are oscillating to neutrinos which participate in the NC interaction, ν_μs and/or ν_τs, with the total $\nu_\mu + \nu_\tau$ flux equal to $(3.41 \pm 0.45(\text{stat})^{+0.48}_{-0.45}(\text{sys})) \times 10^6/\text{cm}^2\text{s}$.[64] The results of two runs of the SNO experiment are shown by the red and green points on Fig. 5.14. The second result was from the KamLAND experiment. This was a reactor-based experiment located in Japan. Using many reactors which were hundreds of kilometers away, the KamLAND experiment was able to reach $L/E \sim 10^{-6}$ m/MeV. This covered the MSW allowed-Δm^2 solution. The statistics were on the order of hundreds of events, but this was enough to probe the large mixing-angle MSW solution. KamLAND expected 365 events and observed 258 events, and thus had clear evidence for oscillations with large mixing, $\tan^2\theta = 0.40^{+0.010}_{-0.07}$ and relatively high Δm^2, of $7.9^{+0.6}_{-0.5} \times 10^{-5}$ eV^2.[66] The energy distribution of the events observed in KamLAND is shown in Fig. 5.15.

Based on the atmospheric and solar studies, there are two squared mass differences: Δm^2_{solar} and Δm^2_{atmos}. The smaller is identified with the mass splitting between ν_1 and ν_2: $\Delta m^2_{12} = \Delta m^2_{solar}$. The atmospheric deficit

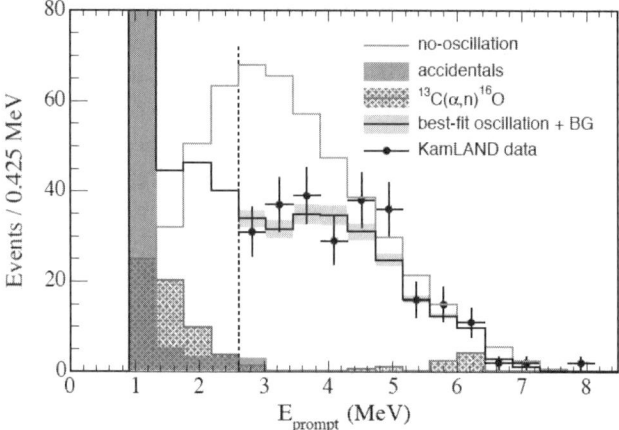

Fig. 5.15. Events in KamLAND as a function of energy. The grey line indicates the expectation for no oscillation.[66]

measures a combination of Δm_{23}^2 and Δm_{13}^2. However, since $\Delta m_{13}^2 = \Delta m_{12}^2 + \Delta m_{23}^2$ and Δm_{12}^2 is small, $\Delta m_{13}^2 \approx \Delta m_{23}^2 \approx \Delta m_{atmos}^2$.

A recent global analysis of the data[67] from the above experiments yields a consistent picture for three neutrino oscillations with five free parameters. The mass differences are: $\Delta m_{12}^2 = (7.9 \pm 0.3) \times 10^{-5} \text{eV}^2$ and $|\Delta m_{13}^2| = (2.5^{+0.20}_{-0.25}) \times 10^{-5} \text{eV}^2$, where the absolute value is indicated in the second case because the sign (i.e. the mass hierarchy) is unknown. The two well-measured mixing angles are determined to be: $\sin^2 \theta_{12} = 0.30^{+0.02}_{-0.03}$ and $\sin^2 \theta_{23} = 0.50^{+0.08}_{-0.07}$. One mixing angle, θ_{13} is yet to be measured, but a limit of $\sin^2 \theta_{13} < 0.025$ can be placed based on global fits.

Based on the measurements, the mixing matrix of Eq. 5.12, translates roughly into:

$$U = \begin{pmatrix} 0.8 & 0.5 & ? \\ 0.4 & 0.6 & 0.7 \\ 0.4 & 0.6 & 0.7 \end{pmatrix}. \tag{5.17}$$

This matrix, with its large off-diagonal components, looks very different from the quark-sector mixing matrix where the off-diagonal elements are all relatively small. In this matrix, the "odd element out" is U_{e3} which is clearly substantially smaller than the others. At this point, there is no consensus on what this matrix may be telling us about the larger theory, but there is a sense that the value of θ_{13} is an important clue. Theories which attempt to explain this matrix tend to fall into two classes – those where

Table 5.3. Selected predictions for $\sin^2 2\theta_{13}$.[104]

Model(s)	Refs.	approximate $\sin^2 2\theta_{13}$
Minimal SO(10)	68	0.13
Orbifold SO(10)	69	0.04
SO(10) + Flavor symmetry	70	$1.2 \cdot 10^{-6}$
	71	$7.8 \cdot 10^{-4}$
	72–74	0.01 .. 0.04
	75–77	0.09 .. 0.18
SO(10) + Texture	78	$4 \cdot 10^{-4}$ .. 0.01
	79	0.04
$SU(2)_L \times SU(2)_R \times SU(4)_c$	80	0.09
Flavor symmetries	81–83	0
	84,85,94	$\lesssim 0.004$
	87–89	10^{-4} .. 0.02
	90–94	0.04 .. 0.15
Textures	95	$4 \cdot 10^{-4}$ .. 0.01
	96–99	0.03 .. 0.15
3 × 2 see-saw	100	0.04
Anarchy	101	> 0.04
Renormalization group enhancement	102	0.03 .. 0.04
M-Theory model	103	10^{-4}

θ_{13} is just below the present limit and those with very small values. As an illustration of this point, Table 5.3 shows order of magnitude predictions for a variety of theories. Thus a measurement of $\sin^2 2\theta_{13}$ which is greater than about 1%, or a limit at this level, can point the way to the larger theory.

The best method for measuring θ_{13} is from reactor experiments which constrain this oscillation by searching for $\bar{\nu}_e$ disappearance. The oscillation probability is given by:

$$P_{reactor} \simeq \sin^2 2\theta_{13} \sin^2 \Delta + \alpha^2 \Delta^2 \cos^4 \theta_{13} \sin^2 2\theta_{12}, \quad (5.18)$$

with

$$\alpha \equiv \Delta m_{21}^2 / \Delta m_{23}^2 \quad (5.19)$$

$$\Delta \equiv \Delta m_{31}^2 L / (4 E_\nu). \quad (5.20)$$

Events are detected through the inverse beta decay (IBD) interaction. The CHOOZ experiment,[25] with a baseline of 1.1 km and typical neutrino event energies between 3 and 5 MeV ($\langle E \rangle = 3.5$ MeV) has set the best reactor-based limit to date, of $\sin^2 2\theta_{13} < 0.27$ at $\Delta m^2 = 2.5 \times 10^{-3}$ eV^2. This limit can be improved with a global fit, as quoted above.

Significant improvement is expected from the upcoming round of reactor experiments results due to introducing a near-far detector design. The near detector measures the unoscillated event rate, and the far detector is used to search for a deficit as a function of energy. The Double Chooz experiment, beginning in 2009, is expected to reach $\sin^2 2\theta_{13} \sim 0.03$.[105] This will be followed by the Daya Bay experiment which will reach ~ 0.01.[106]

5.2.2. *Direct Measurements of Neutrino Mass*

For neutrinos, there are no mass measurements, only mass limits. Observations of neutrino oscillations are sensitive to the mass differences between neutrinos, not the actual mass of the neutrino. Therefore, they do not fall into the category of a "direct measurement". One can, however, use these oscillation results to estimate the required sensitivity for a direct mass measurement. The upper limit comes from assuming that one of the neutrino masses is exactly zero. Given that the largest Δm^2 is $\sim 3 \times 10^{-3}$ eV^2, this implies there is a neutrino with mass $\sqrt{\Delta m^2} \sim 0.05$ eV. The mass of the neutrino can be directly measured from decay kinematics and from time of flight from supernovae. Neither method has reached the 0.05 eV range yet, although the next generation of decay-based experiments comes close.

We know from neutrino oscillations that there is a very poor correspondence between neutrino flavors and neutrino masses, *i.e.*, the mixings are large. However, it is easiest to conduct the discussion of these limits in terms of specific flavors. Thus, what is actually being studied is an average mass associated with each flavor. For example, for the ν_e mass measured from β decay, which will be expanded upon below, what is actually probed is:

$$m_\beta = \sqrt{\Sigma_i |U_{ei}|^2 m_i^2}. \tag{5.21}$$

The simplest method for measuring neutrino mass is applied to the ν_μ. The mass is obtained from the 2-body decay-at-rest kinematics of $\pi \to \mu \nu_\mu$. One begins in the center of mass with the 4-vector relationship: $p_\pi = p_\mu + p_\nu$. Squaring and solving for neutrino mass gives: $m_\nu^2 = m_\pi^2 + m_\mu^2 - \sqrt{4m_\pi^2(|\mathbf{p}_\mu|^2 + m_\mu^2)}$. From this, one can see that this technique requires accurate measurement of the muon momentum, $\mathbf{p}_\mu$, as well as the masses of the muon, m_μ and the pion, m_π. In fact, the uncertainty on the mass of the pion is what dominates the ν_μ mass measurement. As a result, a limit is set at $m_{\nu_\mu} < 170$ keV.[107,108]

The mass for the ν_τ is obtained from the kinematics of τ decays. The τ typically decays to many hadrons. However, the four vectors for each of the hadrons can be summed. Then the decay can be treated as a two-body

Table 5.4. Overview of ν_e squared mass measurements.

Experiment	measured m^2 (eV^2)	limit (eV), 95% C.L.	Year
Mainz[110]	-0.6± 2.2± 2.1	2.2	2004
Troitsk[111]	-1.0 ± 3.0± 2.1	2.5	2000
Mainz[112]	-3.7 ± 5.3 ± 2.1	2.8	2000
LLNL[113]	- 130 ± 20± 15	7.0	1995
CIAE[114]	- 31 ± 75± 48	12.4	1995
Zurich[115]	-24 ± 48± 61	11.7	1992
Tokyo INS[116]	- 65 ± 85± 65	13.1	1991
Los Alamos[117]	- 147 ± 68± 41	9.3	1991

problem with the neutrino as one 4-vector and the sum of the hadrons as the other vector. At this point, the same method described for the ν_μ can be applied. Measurements are again error-limited, so a limit on the mass is placed. The best limit, which is $m_{\nu_\tau} < 18.2$ MeV, comes from fits to $\tau^- \to 2\pi^-\pi^+\nu_\tau$ and $\tau^- \to 3\pi^-2\pi^+(\pi^0)\nu_\tau$ decays observed by the ALEPH experiment.[109]

The experimental situation for the ν_e mass measurement is more complicated. The endpoint of the electron energy spectrum from tritium β decay is used to determine the mass. Just as in the case of the ν_μ and ν_τ, the experiments measure a value of m^2. The problem is that the measurements have been systematically negative. A review of measurements, as a function of time, is given in Table 5.4. Recent measurements at Troitsk[111] and Mainz[112] are negative, but in agreement with zero. Following the Particle Data Group prescription for setting limit in the case of an unphysical results, $m^2 = 0$ is assumed, with the quoted errors. Based on these results, one can extract a limit of approximately < 2 eV for the mass of the ν_e.

The next big step in the measurement of neutrino mass from decay kinematics will come from the Katrin Experiment.[118] Katrin will use tritium beta decay to measure the mass of the neutrino to 0.2 eV. This does not reach the range of 0.05 eV, which our simplistic argument presented at the top of this section indicated. However, that argument assumed that the lightest neutrino had zero mass. A small offset from zero easily boosts the spectrum into the range observable by Katrin. On the other hand, Katrin is sensitive to only electron flavor. Thus, its sensitivity depends up on the amount of mixing of ν_e within the heaviest neutrino.

Another method for measuring neutrino mass from simple kinematics is to use time of flight for neutrinos from supernovae. Neutrinos carry away $\sim 99\%$ of the energy from a supernova. The mass limit is obtained from the spread in the propagation times of the neutrinos. The propagation time for a single neutrino is given by

$$t_{obs} - t_{emit} = t_0\left(1 + \frac{m^2}{2E^2}\right) \tag{5.22}$$

where t_0 is the time required for light to reach Earth from the supernova. Because the neutrinos escape from a supernova before the photons, we do not know t_{emit}. But we can obtain the time difference between 2 events:

$$\Delta t_{obs} - \Delta t_{emit} \approx \frac{t_0 m^2}{2}\left(\frac{1}{E_1^2} - \frac{1}{E_2^2}\right). \tag{5.23}$$

using the assumption that all neutrinos are emitted at the same time, one can obtain a mass limit of ~ 30 eV from the ~ 20 events observed from SN1987a at 2 sites.[119,120]

This is actually an oversimplified argument. The models for neutrino emission are actually quite complicated. The pulse of neutrinos has a prompt peak followed by a broader secondary peak with a long tail distributed over an interval which can be 4 s or more. The prompt peak is from "neutronization" and is mainly ν_e, while all three neutrino flavors populate the secondary peak. However, the rate of ν_e escape is slower compared to ν_μ and ν_τ produced at the same time, because the ν_es can experience CC interactions, while the kinematic suppression from the charged lepton mass prevents this for the other flavors. However, when all of the aspects of the modeling are put together, the bottom line remains the same: it will be possible to set stringent mass limits if we observe neutrinos from nearby supernovae.

Some argue that cosmology provides a "direct measurement." Cosmological fits have sensitivity to neutrino masses, but the results are dependent on the cosmological parameters[121] and the model for relic neutrino production. There are many examples of models with low relic neutrino densities which would significantly change the present interpretation of the cosmological data.[122] In the opinion of the author, until these issues are settled, cosmological measurements cannot convincingly compete with kinematic decays and supernova measurements, despite aggressive claims.

5.3. Neutrinos We Would Like to Meet

Now that we know that neutrinos have mass, and thus are outside of expectations, the obvious question is: "what other Beyond Standard Model properties do they possess?" *The APS Study on the Future of Neutrino Physics* focussed on this question. The plan for attack was divided into three fronts: (1) Neutrinos and the New Paradigm, (2) Neutrinos and the Unexpected and 3) Neutrinos and the Cosmos. The remainder of this paper follows this structure.

The consequences of the discovery of neutrino mass leads to a rich array of ideas. It is was beyond the scope of these lectures to cover the entire spectrum. So, in each of the three areas, two topics are chosen for extensive discussion. The reader is referred to the study[1] and the accompanying theory white paper[2] for further ideas.

5.3.1. *Neutrinos and the New Paradigm*

The first step in creating a "New Standard Model" is to incorporate neutrino mass. The simplest method is to introduce a Dirac mass, by analogy with the electron. This allows us to introduce a small neutrino mass, simply by arguing that the coupling to the Higgs is remarkably small. However, the unlikely smallness of the coupling has pushed theorists to look for other approaches. Among the oldest of these ideas is that neutrinos may be "Majorana particles," *i.e.*, they are their own antiparticle. This leads to a new type of mass term in the Lagrangian. Through the "see-saw" mechanism, which fits well with Grand Unified Theories, this can also give a motivation for the apparently small value of the neutrino masses.

A direct consequence of the Majorana See-Saw Model is a heavy neutrino, with mass near the GUT scale. Because the heavy neutrino gets its mass through the Majorana rather than Dirac term of the Lagrangian, this neutrino was massive during the earliest periods of the universe, before the electroweak phase transition. The decays of such a heavy lepton could be CP violating. This would provide a mechanism for producing the observed matter-antimatter imbalance seen today.

The tidiness of the the above theoretical ideas has caused this paradigm to emerge as the consensus favorite for the "New Standard Model." However, there is absolutely no experimental evidence for this theory at this time. We have no evidence for the Majorana nature of neutrinos. Nor do we have any evidence for CP violation in the neutrino system. The great challenge of the next few years, then, is to find any sign at all that this theory is correct.

This section reviews how one introduces mass into the Lagrangian. The search for evidence of the Majorana nature of neutrinos though neutrinoless double beta decay is considered. Then, the prospects for finding evidence for CP violation is considered.

5.3.1.1. How Neutrinos Might Get Their Mass

The simplest assumption is that the neutrino mass should appear in the Lagrangian in the same way as for the charged fermions – via a Dirac mass term. In general, the Dirac mass term in the Lagrangian will be of the form

$$m(\bar{\psi}_L \psi_R + \bar{\psi}_R \psi_L). \tag{5.24}$$

From the arguments presented in eqs. 5.1 through 5.6, we saw that the scalar "mass" term mixes the RH and LH states of the fermion. If the fermion has only one chirality, then the Dirac mass term will automatically vanish. For this reason, a standard Dirac mass term for the neutrino will require the RH neutrino and LH antineutrino states.

To motivate the mass term, the most straightforward approach is to use the Higgs mechanism, as was done for the electron in the Standard Model. In the case of the electron, when we introduce a spin-0 Higgs doublet ,(h^0, h^+), into the Lagrangian, we find terms like:

$$g_e \bar{\psi}_{e_R} (\psi_{\nu_L} (h^+)^\dagger + \psi_{e_L} (h^0)^\dagger) + h.c., \tag{5.25}$$

where g_e is the coupling constant and "h.c." is the Hermetian conjugate. The piece of this term proportional to $\bar{\psi}_{e_R} \psi_{e_L} (h^0)^\dagger$, combined with its Hermetian conjugate, can be identified with the Dirac mass term, $m_e \bar{\psi}_e \psi_e$. We set $\langle h^0 \rangle = v/\sqrt{2}$, so that we obtain $g \langle h^0 \rangle \bar{\psi}_e \psi_e$ and $m_e = g_e v/\sqrt{2}$. This is the Standard Model method for conveniently converting the *ad hoc* electron mass, m_e, into an *ad hoc* coupling to the Higgs, g_e and a vacuum expectation value (VEV) for the Higgs, v. Following the same procedure for neutrinos allows us to identify the Dirac mass term with $m_\nu = g_\nu v/\sqrt{2}$. The VEV, v, has to be the same as for all other leptons. Therefore, the small mass must come from a very small coupling, g_ν. This implies that $g_e > 5 \times 10^4 g_\nu$.

There are several troublesome features to this procedure. The first issue which is often raised is:

- Why would the Higgs coupling vary across eleven orders of magnitude (the approximate ratio of the neutrino mass to the top quark mass)?

In fact, this question is rather odd. Disregarding the neutrinos, the masses of the charged fermions varies across six orders of magnitude (from the electron mass to the top mass). If six orders of magnitude do not bother anyone, why should eleven? Turning this around, if the Higgs couplings alreay seemed stretched in the charged fermion case, the neutrinos stretch the argument much further. This leads to the second troublesome issue,

• Physically, what is occurring?

The Higgs mechanism really gives little physical insight. While it does introduce mass, it has simply shifted the arbitrariness of the magnitude of the mass into an arbitrary coupling to a new field.

These two questions have led theorists to look at other explanations for small neutrino mass. It has been noted that neutrinos have the unique feature of carrying no electric or strong charge. Thus, neutrinos, alone among the Standard Model fermions, may be their own antiparticle, *i.e.* they may be Majorana particles. The nice consequence of this is a somewhat more motivated theory of mass for neutrinos.

To understand this, first consider what is meant to be a Dirac versus a Majorana particle. If neutrinos are Dirac particles, then the ν and the $\bar{\nu}$ are distinct particles, just as the electron and positron are distinct. The particle, ν has lepton number $+1$ and the antiparticle, $\bar{\nu}$ has lepton number -1. Lepton number is conserved in an interaction. Thus, using the muon family as an example, νs ($L = +1$) must produce μ^- ($L = +1$) and $\bar{\nu}$s ($L = -1$) must produce μ^+ ($L = -1$). The alternative viewpoint is that the ν and $\bar{\nu}$ are two helicity states of the same "Majorana" particle, which we can call "ν^{maj}." The π^+ decay produces the left-handed ν^{maj} and the π^- decay produces the right-handed ν^{maj}. This model explains all of the data without invoking lepton number and has the nice feature of economy of total particles and quantum numbers, but it renders the neutrino different from all other Standard Model fermions.

Saying that the neutrino is its own antiparticle is equivalent to saying that the neutrino is its own charge conjugate, $\psi^c = \psi$. The operators which appear in the Lagrangian for the neutrino in this case are the set $(\psi_L, \psi_R, \psi_L^c, \psi_R^c)$ and $(\bar{\psi}_L, \bar{\psi}_R, \bar{\psi}^c{}_L, \bar{\psi}^c{}_R)$. Certain bilinear combinations of these in the Lagrangian can be identified as Dirac masses (*i.e.* $m(\bar{\psi}_L \psi_R +$ ...)). However, we also get a set of terms of the form:

$$(M_L/2)(\bar{\psi}_L{}^c \psi_L) + (M_R/2)(\bar{\psi}_R{}^c \psi_R) + \cdots \quad (5.26)$$

These are the "Majorana mass terms," which mix the pair of charge-conjugate states of the fermion. If the particle is not its own charge conjugate, then these terms automatically vanish and we are left with only the Dirac terms. Dirac particles have no Majorana mass terms, but Majorana particles will have Dirac mass terms.

The mass terms of the Lagrangian can be written in matrix form:

$$(1/2)(\bar{\psi}_L^c \ \bar{\psi}_R) \begin{pmatrix} M_L & m \\ m & M_R \end{pmatrix} \begin{pmatrix} \psi_L \\ \psi_R^c \end{pmatrix} + h.c., \qquad (5.27)$$

The Dirac mass, m, is on the off-diagonal elements, while the Majorana mass constants, M_L, M_R are on the diagonal. To obtain the physical masses, one diagonalizes the matrix.

One can now invoke "see-saw models" which motivate small observable neutrino masses. It turns out that GUT's motivate mass matrices that look like[124]:

$$\begin{pmatrix} 0 & m_\nu \\ m_\nu & M \end{pmatrix}, \qquad (5.28)$$

with $m_\nu \ll M$. When you diagonalize this matrix to obtain the physical masses, this results in two states which can be measured experimentally:

$$m_{light} \approx m_\nu^2/M, \qquad (5.29)$$

$$m_{heavy} \approx M \qquad (5.30)$$

Grand Unified Theories favor very large masses for the "heavy neutrino" (often called a "neutral heavy lepton"). It is argued that it is most "natural" to have M be at the GUT scale. If $M \sim 10^{25}$ eV, and $m_{light} < 1$ eV, as observed, then $m_\nu \sim 10^{12}$ eV, or is at the TeV scale. This is rather high compared to masses of other leptons, but not so far beyond the top quark mass to regard the connection as crazy. So while some arbitrariness remains in this model, nevertheless there is a general feeling in the theory community that this is an improvement.

In this theory neutrinos have only approximate handedness, where the light neutrino is mostly LH with a very small admixture of RH and the neutral heavy lepton is essentially RH. Thus we have a LH neutrino which is light, which matches observations, and a RH neutrino which is not yet observed because it is far too massive.

5.3.1.2. *Majorana vs. Dirac?*

How can we experimentally tell the difference between the Dirac ($\nu, \bar{\nu}$) and Majorana (ν^{maj}) scenarios? One can imagine a straight-forward thought experiment. First, produce left-handed neutrinos in π^+ decays. These may be νs or they may be ν_{LH}^{maj}s. Next, run the neutrino through a magic helicity-flipping device. If the neutrinos are Majorana, then what comes out of the flipping-device will be ν_{RH}^{maj}. These particles will behave like antineutrinos when they interact, showing the expected RH y-dependence for the cross section. But if the initial neutrino beam is Dirac, then what comes out of the flipping-device will be right-handed νs, which are sterile. They do not interact at all. Such a helicity-flipping experiment is presently essentially impossible to implement. If neutrinos do have mass, then they may have an extremely tiny magnetic moment and a very intense magnetic field could flip their helicity. But the design requirements of such an experiment are far beyond our capability at the moment. Therefore, at the moment, we do not know if neutrinos are Majorana or Dirac in nature.

Instead, experimentalists are pursing a different route. The Majorana nature of the neutrino can lead to an effect called neutrinoless double β decay: $(Z, A) \to (Z+2, A)+(e^- e^-)$. This is a beyond-the-Standard Model analogue to double β decay: $(Z, A) \to (Z + 2, A) + (e^- e^- \bar{\nu}_e \bar{\nu}_e)$. Double β decay is a standard nuclear decay process with a very low rate because there is a suppression proportional to $(G_F \cos\theta_C)^4$. Therefore, in most cases, if the weak decay is possible, single β decay $((Z, A) \to (Z + 1, A) + e^- + \bar{\nu}_e)$ will dominate. However, there are 13 nuclei, including ^{136}Xe $\to$ ^{136}Ba and ^{76}Ge $\to ^{76}$Se, for which single β decay is energetically disallowed. In these cases double β decay with two neutrinos has been observed.[123] If the neutrino were its own antiparticle, then the neutrinos produced in the double β decay process could annihilate, yielding neutrinoless double β decay.

If there are Majorana neutrinos, then the amplitude for $0\nu\beta\beta$ is proportional to the square of

$$m_{0\nu\beta\beta} = \sum U_{ei}^2 m_i. \tag{5.31}$$

This should be contrasted with Eq. 5.21. The $0\nu\beta\beta$ searches are probing different effective masses than the direct searches and the two results yield complementary information. Like the direct searches, the possibility of seeing $0\nu\beta\beta$ depends on the amount of electron-flavor mixed in the most massive neutrino state. If this is small, then the rate of decay will be very low. Thus the hierarchy of the neutrino states affects our ability to observe $0\nu\beta\beta$. To completely untangle the Dirac vs. Majorana question, three different experiments – direct mass measurement, hierarchy measurement and $0\nu\beta\beta$ measurement – may be required.[2]

Extracting $m_{0\nu\beta\beta}$ from a measured half-life leads to a theoretical error from the nuclear matrix element calculations. A favored style of calculation uses the "QRPA" (Quasiparticle Random Phase Approximation)[125-128] model. Using ^{100}Mo as an example, different matrix elements from QRPA

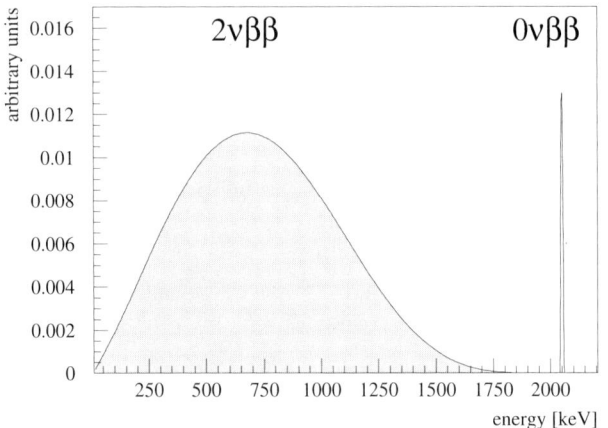

Fig. 5.16. Spectrum for two-neutrino double β decay and expected peak for neutrinoless double β decay.

calculations cause $m_{0\nu\beta\beta}$ to vary by up to 2 eV for a half-life of 4.5×10^{23} years.[129] So the error is significant.

The $0\nu\beta\beta$ events must be separated from the standard two-neutrino double β ($2\nu\beta\beta$) decay background. This can be done through simple kinematics cuts. The two-body nature of $0\nu\beta\beta$ decay will cause a peak at the endpoint of the $2\nu\beta\beta$ decay (4-body) spectrum, as shown in Fig. 5.16. An advantage of observing $2\nu\beta\beta$, however, is that measurement of its half-life allows direct measurement of the matrix element. At this point the $2\nu\beta\beta$ decay spectrum has been observed in 10 elements. In some cases, such as ^{100}Mo, the the $2\nu\beta\beta$ half-life is well measured and can be used to constrain nuclear matrix element calculation. For this case, NEMO-3 reports a half life of $(7.68 \pm 0.02(\text{stat}) \pm 0.54(\text{sys})) \times 10^{18}$ y.[130]

At present, no signal for $0\nu\beta\beta$ decay has been clearly observed. The present 90% CL limit on the lifetime from CUORICINO on ^{130}Te is 1.8×10^{24} years, corresponding to limit of $m_{0\nu\beta\beta} < 0.2 - 1.1$ eV.[131] The NEMO-3 experiment has set 90% CL limits of 4.6×10^{23} and 1.0×10^{23} on ^{100}Mo and ^{82}Se, respectively.[129] The corresponding limits on $m_{0\nu\beta\beta}$ are 0.7-2.8 eV and 1.7-4.9 eV.[129] There is a candidate signal observed at 4.2σ from a Germanium detector,[132] although the statistical significance is under debate.[2] The measured half-life was 1.19×10^{25} years. Until this result is confirmed by further experiments, it is best to reserve judgment.

Luckily, a range of future $0\nu\beta\beta$ decay experiments are on the horizon. These are expected to probe an order of magnitude further in lifetimes.

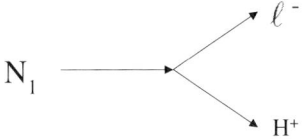

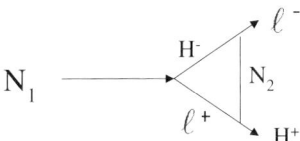

Fig. 5.17. Example of two diagrams for Neutral Heavy Lepton decay which can interfere to produce CP violation.

In particular, the germanium-based GERDA experiment,[133] will turn on soon and will address the existence of the possible signal. CUORE,[134] SuperNEMO,[135] EXO,[136] Majorana[137] and Moon[138] will extend the search even further using a wide range of elements. The reach of these near future $0\nu\beta\beta$ covers the prediction for the inverted mass hierarchy.

5.3.1.3. CP Violation in the Neutrino Sector

An intriguing aspect of the "new Standard Model" is the heavy GUT-scale neutrinos which gain mass through the Majorana terms in the Lagrangian. There could be more than one, and likely, given the trend in the Standard Model, there would be three, so we can label these N_1, N_2 and N_3. These heavy neutrinos have mass prior to the electroweak phase transition in which the Dirac terms appear. As a result, prior to the electroweak phase transition, decays shown in Fig. 5.17 are possible. Both decays produce the same final state, $N_1 \to \ell H$, where ℓ and H are oppositely charged. These diagrams interfere, and can lead to a different decay rate to ℓ^- and ℓ^+, which is CP violation.

This form of CP violation would lead to a lepton asymmetry in the early universe which could be transferred into a baryon asymmetry. A mechanism for this already appears in the Standard Model, in which but B, baryon number, and L, lepton number are not conserved, but the difference, $B-L$, is exactly conserved. B and L violation occurs in transitions between vacuum states at high energies, called the "sphaleron process." Variations on this mechanism, called "leptogenesis," may explain the matter-antimatter asymmetry we see today.

N_1, N_2, and N_3 are far too massive to be produced at accelerators in the near future. Thus observing CP violation in their decays is out of the question. However, observing CP violation in the light neutrino sector would be a plausible hint that the theory is correct.

To incorporate CP violation into the three-light-neutrino model, the leptonic mixing matrix is expanded and written as: $U^{with\ CP} = VK$. In this case, V is very similar to the U of Eq. 5.12, but with a CP violating phase, δ:

$$V = \begin{pmatrix} c_{12}c_{13} & s_{12}c_{13} & s_{13}e^{-i\delta} \\ -s_{12}c_{23} - c_{12}s_{23}s_{13}e^{i\delta} & c_{12}c_{23} - s_{12}s_{23}s_{13}e^{i\delta} & s_{23}c_{13} \\ s_{12}s_{23} - c_{12}c_{23}s_{13}e^{i\delta} & -c_{12}s_{23} - s_{12}c_{23}s_{13}e^{i\delta} & c_{23}c_{13} \end{pmatrix}. \tag{5.32}$$

This is analogous to the CKM matrix of the quark sector. The other term,

$$K = \text{diag}\left(1, e^{i\phi_1}, e^{i(\phi_2+\delta)}\right) \tag{5.33}$$

has two further CP violating phases, ϕ_1 and ϕ_2.

Now, we potentially have three non-zero CP violating parameters in the light neutrino sector, δ, ϕ_1 and ϕ_2, as well as one or more CP violating parameters in the heavy neutrino sector, where the number depends upon the total number of N. In the Lagrangian, these all come from a matrix of Yukawa coupling constants. In principle, all of these phases can take on the full range of values, including exactly zero. However, it is difficult to motivate a theory in which some are nonzero and some are exactly zero. It is expected that these parameters will either all have non-zero values or all be precisely zero. If the latter case, then the difference between the lepton sector, with no CP violation, and quark sector, with clear CP violation, must be motivated. As a result, observation of CP violation in the light neutrino sector is regarded as the "smoking gun" to CP violation in the heavy sector.

Returning to the light neutrino sector, how can the CP phases be measured? The ϕ phases arise as a direct consequence of the Majorana nature of neutrinos. Therefore, in principle, the the ϕ phase associated with the electron family is accessible in neutrinoless double beta decay. In practice, this will be extremely hard to measure because this term manifests itself as a change in the sum in Eq. 5.31, which is proportional to the $0\nu\beta\beta$ decay amplitude. Thus one seeks to measure a deviation of the (as-yet-unmeasured) $0\nu\beta\beta$ lifetime from the prediction which depends upon the mixing angles (with relatively large errors at present), the (unknown) neutrino masses, and the (poorly known) nuclear matrix element. Even if the effect is large, observation of the effect is clearly hopeless in the near future. On the other hand, δ, the "Dirac" CP violating term in V may be accessible though oscillation searches.

CP violation searches involve observing a difference in oscillation probability for neutrinos and antineutrinos. Only appearance experiments can observe CP violation. A difference between oscillations of neutrinos and antineutrinos in disappearance searches is CPT violating. In oscillation appearance searches, the K matrix does not affect the oscillation probability because this diagonal matrix is multiplied by its complex conjugate. On the other hand, non-zero δ can be observed. To test for non-zero δ, the oscillation probability must depend upon the U_{e3} component of Eq. 5.32. In other words, the search needs to involve transitions from or to electron flavor and involve the mass state ν_3. This combination of requirements – appearance signal, electron flavor involvement, and ν_3 mass state involvement – leads to one experimental option at present: comparison of $\nu_\mu \to \nu_e$ to $\bar{\nu}_\mu \to \bar{\nu}_e$ at the atmospheric Δm^2, which is Δm^2_{13}. The oscillation probability is given by:

$$\begin{aligned}P_{long-baseline} \simeq\ & \sin^2 2\theta_{13} \sin^2 \theta_{23} \sin^2 \Delta \\ & \mp \alpha\ \sin 2\theta_{13} \sin \delta_{CP} \cos \theta_{13} \sin 2\theta_{12} \sin 2\theta_{23} \sin^3 \Delta \\ & + \alpha\ \sin 2\theta_{13} \cos \delta_{CP} \cos \theta_{13} \sin 2\theta_{12} \sin 2\theta_{23} \cos \Delta \sin^2 \Delta \\ & + \alpha^2\ \cos^2 \theta_{23} \sin^2 2\theta_{12} \sin^2 \Delta, \end{aligned} \quad (5.34)$$

where α and Δ are defined in Eq. 5.20. The second term is negative for neutrino scattering and positive for antineutrino scattering.

Unfortunately, Eq. 5.34 convolutes two unknown parameters, the sign of Δm^2_{13} (the mass hierarchy) and the value of θ_{13}, with the parameter, δ, that we want to measure. The problem of the mass hierarchy can be mitigated by the experimental design. The sign of $\Delta m^2_{13} = m^2_3 - m^2_1$ affects the terms where Δm^2 is not squared. These terms arise from matter effects and can so be reduced if the pathlength in matter is relatively short. For long baseline experiments, which must shoot the beam through the Earth, this means that L must be relatively short. In order to retain the same L/E and, hence, the same sensitivity to Δm^2_{13}, E must be comparably reduced. On the other hand, the problem of θ_{13} cannot be mitigated. From Eq. 5.32, one sees that we are in the unfortunate situation of having the CP violating term multiplied by $\sin \theta_{13}$. The smaller this factor, the harder it will be to extract δ. If $\sin^2 2\theta_{13}$ is smaller than ~ 0.01 at 90% CL, then substantial improvements in beams and detectors will be required.

Equation 5.34 also depends on two other as-yet-poorly understood parameters, θ_{23} and the magnitude of Δm_{23}^2. Disappearance experiments measure $\sin^2 2\theta_{23} = 1.00^{+0.16}_{-0.14}$,[67] thus there is an ambiguity as to whether θ_{23}, is larger or smaller than 45°. Δm_{23}^2 is only known to about 10%.[56,67] This measurement is extracted from the location of the "dip" in the rate versus L/E distribution of disappearance experiments, and is already systematics-dominated. Improvement requires experiments with better energy resolution[56] and better understanding of the CCQE and background cross sections.[139] These errors lend a significant error to the analysis.

Lastly, it is difficult to measure a ν_e signal which is at the $\sim 1\%$ level. Most ν_μ beams have a substantial ν_e intrinsic contamination from μ and K decays. Given that θ_{13} is small, this contamination is a serious issue. One solution to this problem is to go to an off-axis beam design. This relies on the tight correlation between energy and off-axis angle, θ, in two-body decays. For pion decay, which dominates most beams,

$$E_\nu = \frac{0.43 E_\pi}{1 + \gamma^2 \theta^2}, \qquad (5.35)$$

where $\gamma = E_\pi/m_\pi$ is the Lorentz boost factor. The solution for two-body K decay replaces 0.43 with 0.96. Thus at $\theta = 0$, the relationship between E_ν and E_π is linear. However, for larger θ, above a moderate energy threshold, all values of E_π map to the same E_ν. This is illustrated in Fig. 5.18. The result is that an off axis ν_μ beam which comes largely from pion decay is tightly peaked in energy while the ν_e intrinsic background is spread across a range of energies. This also helps to reduce ν_μ events which are misreconstructed as ν_e scatters, such as NC π^0 production where a photon is lost. These "mis-ids" tend to be spread across a range of energies, since the true energy is misreconstructed. Thus a peaked signal, as one expects from an off-axis beam, is helpful in separating signal and background.

The major problem with this design is that off axis beams have substantially lower flux. The flux scales[140] as:

$$F = \left(\frac{2\gamma}{1 + \gamma^2 \theta^2} \right)^2 A/4\pi z^2. \qquad (5.36)$$

In this equation, A is the area of the detector and z is the distance to the detector. Two future long baseline experiments, NOvA[40] and T2K[141] are proposing off-axis beams for a $\nu_\mu \to \nu_e$ search. Because of the low flux, very large detectors are required.

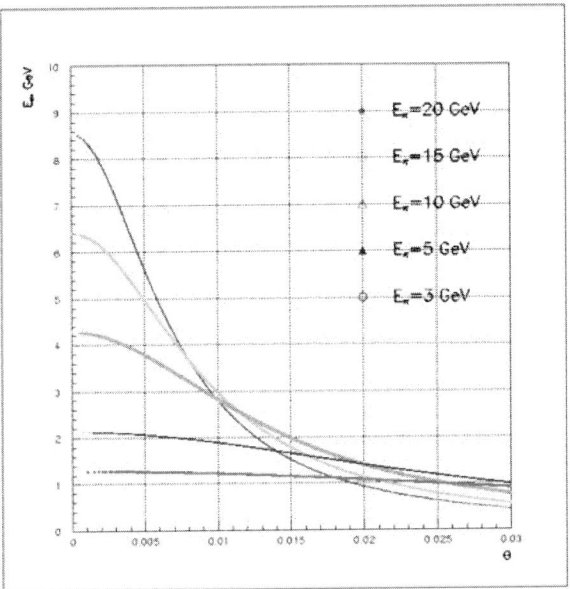

Fig. 5.18. Neutrino energy versus angle off-axis for various values of pion energy. In this example one can see that for moderate off-axis angles, between 15-30 mrad, all pion energies between 3 and 20 GeV map to approximately 1 GeV neutrino energy.

In summary, the path to a test for non-zero δ is clear but will take several steps and requires some luck. First, a clean measure θ_{13} from Double Chooz and Daya Bay is needed. If $\sin^2 2\theta_{13} < 0.005$ at 90% CL, then a significant measurement of CP violation is unlikely to be possible in the near future. At the same time, improvements in the θ_{23} and Δm_{23}^2 from disappearance $(\nu_\mu \to \nu_\mu)$ measurements at MINOS and T2K will improve the situation. T2K and NOvA[40] may be able to make a first exploration of CP parameter space, from ν_e appearance measurements, depending on statistics. NOvA may also be able to address the mass hierarchy question. This will open up the possibility of measuring CP violation to the next generation of very long baseline experiments.[142] The most sensitive of these use a beam originating at Fermilab and a LAr detector located at Ash River, Minnesota[45] or a Cerenkov or LAr detector located at the Deep Underground Science Laboratory at Homestake.[143]

At some point in the future, a beta beam or a neutrino factory beam could provide an intense source of ν_e and $\bar{\nu}_e$ fluxes, allowing comparison of $\nu_e \to \nu_\mu$ to $\bar\nu_e \to \bar\nu_\mu$. In this case one would search for events with wrong-sign muons in a calorimeter-style detector. This would be a striking signature with low background, especially in the case of a beta beam. This could allow a very precise measurement of δ.[144]

5.3.2. *Neutrinos and the Unexpected*

While it is nice to have a tidy, well-motivated theory of neutrino masses, it is disconcerting to have essentially no experimental evidence for this theory. Moreover, neutrino theories have a history of being incorrect. Only a decade ago, most theorists would have told you that neutrinos have no mass. Those who thought neutrinos might have mass believed it would be relatively large (> 5 eV), explaining dark matter. Most theorists also believed that if the solar neutrinos were experiencing oscillations, the correct solution would be the small mixing angle MSW solution, because the mixing matrix should look like the quark matrix. Using the same logic, the atmospheric neutrino deficit, which could only correspond to large mixing angle, was routinely dismissed as an experimental effect.

On the basis of this, it is wise not to constrain ourselves to the "New Paradigm." The reason the APS neutrino study chose to devote a chapter to "the Unexpected" was to emphasize the importance of being open to what nature is telling us about neutrinos. There are two ways to approach this idea: (1) the theory-driven approach: explore for properties which could, theoretically, exist and (2) the experiment-driven approach: follow up on anomalous results which have been observed.

For lack of time, I will only briefly consider two examples of the first case: searching for a neutrino magnetic moment and searching for CPT violation. In the Standard Model, the neutrino magnetic moment is expected to be $\sim 10^{-19} \mu_B$. Laboratory experiments and astrophysical limits are many orders of magnitude away from this level.[145] Nevertheless, if a new experiment could advance this measurement by an order of magnitude, that would be worth pursuing. A more startling discovery would be a difference in the oscillation disappearance probability of neutrinos versus antineutrinos. In a three-neutrino model, a difference in the rate of disappearance of neutrinos and antineutrinos would imply CPT violation. MINOS will be the next experiment to pursue such a search.[36] If CPT violation were discovered we would need to rethink the very basis of our theory. However, there are theorists exploring these ideas.

The remainder of this section will focus on the second approach, exploring "anomalies" which have appeared in various experiments. Physicists today are always cautious about pursuing deviations from the Standard Model. Most do not, in the end, point to new physics. The Standard Model has been very resilient Most anomalies are arguably more likely due to systematic effects or statistical fluctuations, than to new physics. However, those which do "pan out" completely change the way we think. The solar neutrino deficit is a perfect example. So, if a new, unexpected result withstands questions by the community on the systematics of the experiment, then the anomaly becomes worth pursuing further.

There are several examples of $> 3\sigma$ unexpected results in the neutrino sector which are worth pursuing and two cases are covered here. The first, the LSND anomaly, is being actively pursued. The second, the NuTeV anomaly, will require a new experiment. Unlike most of the topics in these lectures, the NuTeV anomaly is not directly related to neutrino oscillations and neutrino mass, and so expands the discussion, which has so far been rather narrowly focussed.

Along with the known discrepancies which have reached the level of full-fledged anomalies, there are also examples of "unexpected results to watch." These results which have not yet reached the 3σ level, but are showing interesting trends. For example, unconstrained fits to atmospheric oscillation data from a wide range of experiments consistently result in $\sin^2 2\theta_{23}$ best fit value greater than unity. While in each case, the best fit is $\sim 1\sigma$ from unity, it is the trend which is interesting, since the experiments involved are all very different. There is simply not enough space to cover this and other examples of "results to watch."

The take-away message of this section is: the neutrino sector is a rich place for new physics to appear, and physicists need to be alert and open-minded to what nature is saying.

5.3.2.1. *The LSND Anomaly*

The LSND experiment ran at the LAMPF accelerator at Los Alamos National Laboratory between 1993 and 1998. The decay-at-rest (DAR) beam was produced by impinging 800 MeV protons on a beam dump. These produced π^+s which stop and decay to produce μ^+s, which also stop and decay to produce $\bar{\nu}_\mu$ and ν_e. These were studied in the range of 20 to 55 MeV. The π^-s capture, so the beam has a $< 8 \times 10^{-4}$ contamination of $\bar{\nu}_e$. The neutrino events were observed in a detector located 30 m downstream of the beam dump. 1220 phototubes surrounded the periphery of a

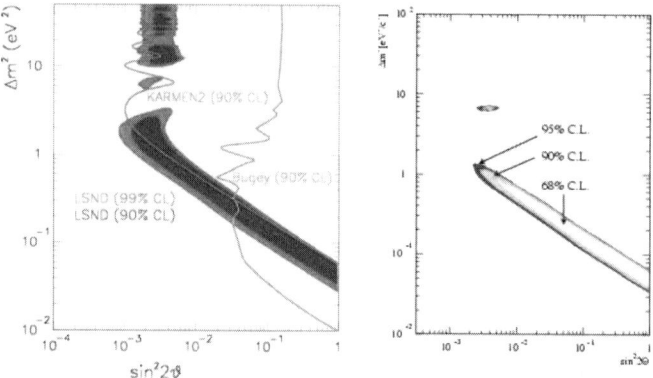

Fig. 5.19. Left: LSND allowed range compared to short baseline experiment limits. Right: Allowed range from the Karmen-LSND joint analysis.[149]

cylindrical detector filled with 167 tons of mineral oil, lightly doped with scintillator. The signature of a $\bar{\nu}_e$ appearance was $\bar{\nu}_e + p \to e^+ n$. This resulted in a two-component signature: the initial Cerenkov and scintillation light associated with the e^+, followed later by the scintillation light from the n capture on hydrogen, producing a 2.2 MeV γ. The experiment observed $87.9 \pm 22.4 \pm 6.0$ events, a 4σ excess[146] above expectation.

LSND is a short baseline experiment, with an $L/E \sim 1$ m/MeV. Thus, from the two-generation oscillation formula, Eq. 5.10, one can see that this experiment is sensitive to $\Delta m^2 \geq 0.1$ eV^2. Other experiments have searched for oscillations at high Δm^2, and the two most relevant to LSND are Karmen[147] and Bugey.[148] The KARMEN experiment, which also used a DAR muon beam and was located 17.7 m from the beam dump, had sensitivity to address only a portion of the LSND region, and did not see a signal there. Since the design of the experiments are very similar, one can think of the Karmen experiment as a "near detector," which measures the flux before oscillation. The results were combined in a joint analysis performed by collaborators from both experiments; the allowed range for oscillations is shown in Figure 5.19.[149] The Bugey experiment was a reactor-based $\bar{\nu}_e$ disappearance search which set a limit on oscillations. Because this is disappearance and not explicitly $\bar{\nu}_e \to \bar{\nu}_\mu$, its limit is applicable to LSND in many, but not all, oscillation models. This limit is shown in Figure 5.19.

Why can't we fit LSND into the three-neutrino theory? The LSND signal cannot be accommodated within the standard three-neutrino picture, given the solar and atmospheric oscillations. To see the incompatibility, first consider the case where the oscillation signals de-couple into, effectively, two-generation oscillations (Eq. 5.10). For three generations, then $\Delta m_{31}^2 = \Delta m_{32}^2 + \Delta m_{21}^2$, which is clearly not the the case for these three signals. The more general case allows the atmospheric result to be due to a mixture of high (LSND-range) and low (solar-range) Δm^2 values. In order for this model to succeed, the atmospheric Δm^2 from a shape analysis must shift up from its present value of $\sim 2 \times 10^{-3}$ eV^2 and the chlorine experiment must have overestimated the deficit of ^{7}Be solar neutrinos. However, the largest clash between data and this model arises from the Super-K ν_e events. This model requires that Super-K has missed a ν_e appearance signal of approximately the same size and shape as the ν_μ deficit before detector smearing and cuts. Neutrino measurements are experimentally difficult and parameters do sometimes shift with time as systematics are better understood, but it seems unlikely that all of the above results could change sufficiently to accommodate LSND.

Sterile Neutrinos as a Solution Additional neutrinos which do not interact via exchange of W or Z are called "sterile;" they may mix with active neutrinos, and thereby can be produced in neutrino oscillations. Experimental evidence of this would be the disappearance of the active flavor from the beam. In contrast to the GUT-scale sterile neutrinos we have already discussed, the sterile neutrinos which could explain LSND must be light (in the eV range), and this narrows the class of acceptable theories. Nevertheless, a number of possible explanations remain.[150]

Sterile neutrinos solve the LSND problem by adding extra mass splittings. The additional mass states must be mostly sterile, with only a small admixture of the active flavors in order to accommodate the limits on sterile neutrinos from the atmospheric and solar experiments. In principle, one might expect three sterile neutrinos. In practice, the data cannot constrain information on more than two sterile neutrinos. Therefore these are called "3+2" models. The method for fitting the data is described in reference.[151] One is fitting for two additional mass splittings, Δm_{14}^2 and Δm_{15}^2. In the fit, the three mostly active neutrinos are approximated as degenerate. The mixing matrix is also expanded by two rows and two columns.

The data that drive the fits are the "short baseline" experiments that provide information on high Δm^2 oscillations, summarized in Table 5.5. The combination of $\bar{\nu}_\mu \to \bar{\nu}_e$ (LSND,[146] Karmen II[147]), $\nu_\mu \to \nu_e$ (NOMAD[152]), ν_μ disappearance (CDHS,[153] CCFR84[154]), and ν_e disappearance

Table 5.5. Results used in 3 + 2 fit. $\sin^2 2\theta$ limit is 90% CL

Channel	Experiment	Lowest Δm^2	$\sin^2 2\theta$ at high Δm^2	Best reach in $\sin^2 2\theta$
$\nu_\mu \to \nu_e$	LSND	0.03 eV^2	$> 2.5 \times 10^{-3}$	$> 1.2 \times 10^{-3}$
	KARMEN	0.06 eV^2	$< 1.7 \times 10^{-3}$	$< 1.0 \times 10^{-3}$
	NOMAD	0.4 eV^2	$< 1.4 \times 10^{-3}$	$< 1.0 \times 10^{-3}$
ν_e disappearance	Bugey	0.01 eV^2	$< 1.4 \times 10^{-1}$	$< 1.3 \times 10^{-2}$
	Chooz	0.0001 eV^2	$< 1.0 \times 10^{-1}$	$< 5 \times 10^{-2}$
ν_μ disappearance	CCFR84	6 eV^2	NA	$< 2 \times 10^{-1}$
	CDHS	0.3 eV^2	NA	$< 5.3 \times 10^{-1}$

(Bugey,[148] CHOOZ[25]) must all be accommodated within the model. A constraint for Super K ν_μ disappearance is also included. None of the short baseline experiments except for LSND provide evidence for oscillations beyond 3σ. However, it should be noted that CDHS has a 2σ (statistical and systematic, combined) effect consistent with a high Δm^2 sterile neutrino when the data are fit for a shape dependence, and Bugey has a 1σ pull at $\Delta m^2 \sim 1\text{eV}^2$. As a result, these two experiments define the best fit combination of high and low Δm^2 for the 3+2 model. However, there are acceptable solutions with a combination of low Δm^2 values. The best fit,[151] has $\Delta m^2_{14} = 0.92$ eV^2, $\Delta m^2_{15} = 22$ eV^2, although there are combinations which work with two relatively low Δm^2 values. A wide range of mixing angles can be accommodated, and the best fit has $U_{e4} = 0.121$, $U_{\mu 4} = 0.204$, $U_{e5} = 0.036$ and $U_{\mu 4} = 0.224$. The other mixing angles involving the sterile states are not probed by the ν_μ disappearance, ν_e disappearance and $\nu_\mu \to \nu_e$ appearance experiments listed above. There is 30% compatibility for all other experiments and LSND.

Introducing extra neutrinos, including sterile ones, would have cosmological implications, compounded if the extra neutrinos have significant mass (>1 eV). However, there are several ways around the problem. The first is to note that while the best fit requires a high mass sterile neutrino, there are low-mass fits which work within the 3+2 model. The second is to observe that there are a variety of classes of theories where the neutrinos do not thermalize in the early universe.[122] In this case, there is no conflict with the cosmological data, since the cosmological neutrino abundance is substantially reduced.

If more than one Δm^2 contributes to an oscillation appearance signal, then the data can be sensitive to a CP-violating phase in the mixing matrix. Experimentally, for this to occur, the Δm^2 values must be within less than about two orders of magnitude of one another.

In 3+2 CP-violating models:[155]

$$P(\overset{(-)}{\nu_\mu} \to \overset{(-)}{\nu_e}) = 4|U_{e4}|^2|U_{\mu4}|^2 \sin^2 x_{41}$$
$$+ 4|U_{e5}|^2|U_{\mu5}|^2 \sin^2 x_{51}$$
$$+ 8|U_{e4}||U_{\mu4}||U_{e5}||U_{\mu5}|$$
$$\sin x_{41} \sin x_{51} \cos(x_{54} \mp \phi_{54}), \qquad (5.37)$$

where in the last line, the negative sign is for neutrino oscillations and the positive sign is for antineutrino oscillations, and defined:

$$x_{ji} \equiv 1.27 \Delta m_{ji}^2 L/E, \quad \phi_{54} \equiv arg(U_{e4}^* U_{\mu4} U_{e5} U_{\mu5}^*).$$

Thus the oscillation probability is affected by CP violation through the term ϕ_{54}. The CP conserving cases are $\phi_{54} = 0$ and 180 degrees.

MiniBooNE First Results The main purpose of the MiniBooNE experiment was to resolve the question of the LSND signal. First results of this experiment, presented in April, 2007, considered those explanations with a high expectation for $\nu_\mu \to \nu_e$ oscillations. This includes the CP conserving 3+2 model and many cases of CP violating 3+2 models described above. As will be described below, the first results are incompatible with $\nu_\mu \to \nu_e$ oscillations, but show an unexpected low energy excess, very much in keeping with the subsection title of "Neutrinos and the Unexpected."

The MiniBooNE experiment uses the Fermilab Booster Neutrino Beam, which is produced from 8 GeV protons incident on a beryllium target located within a magnetic focusing horn. The current of the horn can be reversed such that the beam is dominantly neutrinos or antineutrinos. The first results are from neutrino running. The MiniBooNE detector is located $L = 541$ m from the primary target, and the neutrino flux has average energy of ~ 0.75 GeV. The detector is located 541 m from the front of the beryllium target and consists of a spherical tank of radius 610 cm that is covered on the inside by 1520 8-inch photomultiplier tubes and filled with 800 tons of pure mineral oil (CH_2). Neutrino events in the detector produce both Cerenkov and scintillation light.

In order to test the LSND result, the MiniBooNE design maintains $L/E \sim 1$ m/MeV while substantially changing the systematic errors associated with the experiment. This is accomplished by increasing both L and E by an order of magnitude from the LSND design. This changes the source of the neutrinos (ν_μ from energetic pions rather than $\bar{\nu}_\mu$ from stopped muons), the signature for the signal, and the major backgrounds in the detector. In its first run, in neutrino mode, MiniBooNE collected over a million clean, neutrino events. About 99.5% of the MiniBooNE neutrino events are estimated to be ν_μ-induced, while 0.5% are estimated to be due to "intrinsic" ν_e background in the beam.

The initial MiniBooNE results were analyzed within an appearance-only, two neutrino oscillation context. While the LSND signal must be a result of a more complex oscillation model, in most cases a $\nu_\mu \rightarrow \nu_e$-like oscillation signal is predicted. After the complete ν_e event selection is applied, the total background was estimated to be 358 ± 35 events, while 163 ± 21 signal events were expected for the LSND central expectation of 0.26% $\nu_\mu \rightarrow \nu_e$ transmutation.

The top plot of Fig. 5.20 shows candidate ν_e events as a function of reconstructed neutrino energy (E_ν^{QE}). The vertical dashed line indicates the minimum E_ν^{QE} used in the two-neutrino oscillation analysis. There is no significant excess of events ($22 \pm 19 \pm 35$ events) for $475 < E_\nu^{QE} < 1250$ MeV; however, an excess of events ($96 \pm 17 \pm 20$ events) is observed below 475 MeV. In the top plot, the points show the statistical error, while the histogram is the expected background with systematic errors from all sources. The background subtracted excess as a function of E_ν^{QE} is shown in the bottom plot, where the points represent the data with total errors. Oscillation scenarios are indicated by the histograms.

The low-energy excess cannot be explained by a two-neutrino oscillation model, and its source is under investigation. The low energy events isolated by the cuts, including the excess events, are single-ring and electromagnetic-like, with no unusual detection issues. The low energy excess events are neither consistent with the spatial nor energy distributions of photons coming from interactions outside of the tank. Nor are they consistent with the energy distribution from single photons from radiative Δ decays ($\Delta \rightarrow N + \gamma$). Mis-identification of π^0 events is well constrained in MiniBooNE by the rate of reconstructed π^0 events, studied as a function of π^0 momentum.[156] This rate would need to be mis-measured by well over a factor of three to explain the excess, far outside of the error of the analysis.

With that said, as shown in Fig. 5.20, the excess is not in agreement with a simple two-neutrino $\nu_\mu \rightarrow \nu_e$ oscillation signal. This figure shows the predicted spectrum when the best-fit two-neutrino oscillation signal is added to the predicted background. The bottom panel of the figure shows background-subtracted data with the best-fit two-neutrino oscillation and two oscillation points from the favored LSND region.

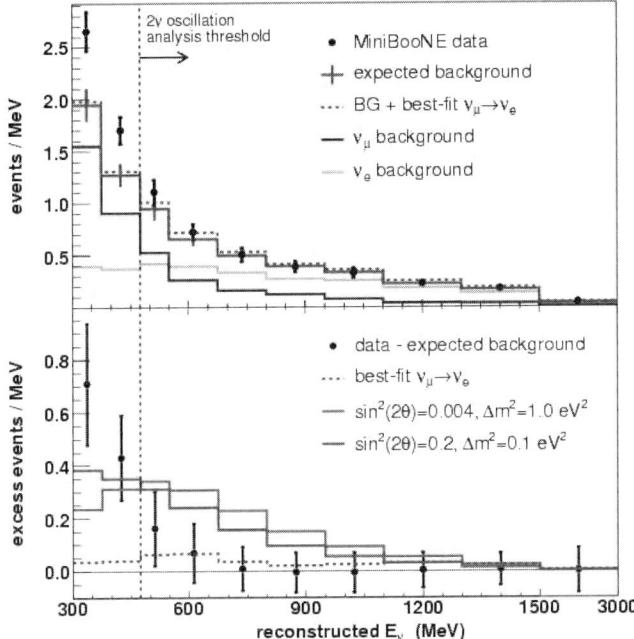

Fig. 5.20. The top plot shows the number of candidate ν_e events as a function of E_ν^{QE}. Also shown are the best-fit oscillation spectrum (dashed histogram) and the background contributions from ν_μ and ν_e events. The bottom plot shows the number of events with the predicted background subtracted as a function of E_ν^{QE}. The two histograms correspond to LSND solutions at high and low Δm^2.

A single-sided raster scan to a two neutrino appearance-only oscillation model is used in the energy range $475 < E_\nu^{QE} < 3000$ MeV to find the 90% CL limit corresponding to $\Delta\chi^2 = \chi^2_{limit} - \chi^2_{bestfit} = 1.64$. As shown by the top plot in Fig. 5.21, the LSND 90% CL allowed region is excluded at the 90% CL. A joint analysis as a function of Δm^2, using a combined χ^2 of the best fit values and errors for LSND and MiniBooNE, excludes at 98% CL two-neutrino appearance-only oscillations as an explanation of the LSND anomaly. The bottom plot of Fig. 5.21 shows limits from the KARMEN[147] and Bugey[148] experiments. This is plot represents an example of the problem of apples-to-apples comparisons raised in Sec. . The MiniBooNE and Bugey curves are 1-sided upper limits on $\sin^2 2\theta$ corresponding to $\Delta\chi^2 = 1.64$ – hence directly comparable – while the published KARMEN curve is a "Feldman Cousins" contour.

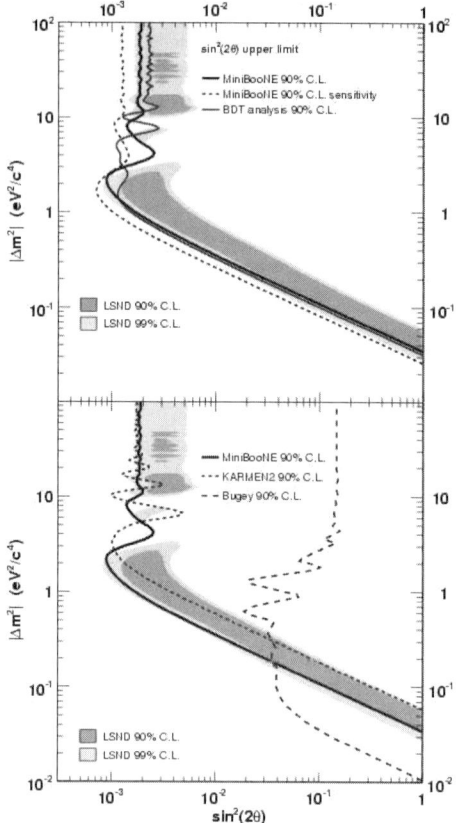

Fig. 5.21. The top plot shows the $\nu_\mu \to \nu_e$ MiniBooNE 90% CL limit (thick solid curve) and sensitivity (dashed curve) for events with $475 < E_\nu^{QE} < 3000$ MeV. Also shown is the limit from a second cross-check analysis (thin solid curve). The bottom plot shows the limits from the KARMEN[147] and Bugey[148] experiments. The shaded areas show the 90% and 99% CL allowed regions from the LSND experiment.

Next Steps From the initial MiniBooNE result, one can draw two conclusions: (1) there is excellent agreement between data and prediction in the analysis region originally defined for the two-neutrino oscillation search and (2) there is a presently unexplained discrepancy with data lying above background at low energy. This combination of information severely limits models seeking to explain the LSND anomaly.

Interpreting the MiniBooNE data as appearance-only and combining this result with other data in a 3+2 fit does not give a satisfactory result.[157] While MiniBooNE and LSND are compatible if CP violation is allowed in the 3+2 model, there is tension between these results and the ν_μ disappearance experiments. This might be addressed if a 3+2 interpretation of the MiniBooNE result were expanded to include the possibility of ν_μ disappearance and intrinsic ν_e disappearance. This analysis is underway by the MiniBooNE collaboration.[158] Most likely, if a good fit is obtained in a 3+2 scenario, it will require some level of CP violation. MiniBooNE is presently collecting data in antineutrino mode. However, this is a small data set ($\sim 2 \times 10^{20}$ protons on target producing the beam) and future running to reach roughly three times the statistics will be required to make a decisive statement. Other, alternative explanations are also being explored[159–161].

An upcoming result which will shed light on the question is the analysis of the MiniBooNE data from the NuMI beam. This beam is 110 mrad off-axis, with a π peak of average ν_μ energy of about 200 MeV and a K peak of about 2 GeV, and a length of 750 m. If an excess of events is observed in this analysis, this rules out mis-estimate of intrinsic ν_e in the Booster Neutrino Beam as the source of the MiniBooNE excess. Results from this study are expected in autumn, 2007.

If the unexplained excess persists after the above studies, then it will be valuable to introduce a detector which can differentiate between electrons and photons. That is the goal of MicroBooNE,[44] which uses a Liquid argon TPC (LArTPC) detector. This is particularly sensitive at low energies and nearly background-free. Specifically, this detector has a ν_e efficiency $> 80\%$ and rejects photons efficiently through dE/dx deposition in the first ~ 2 cm of the shower. With these qualities the detector can be an order of magnitude smaller in size than MiniBooNE, making quick construction feasible. A proposal for this experiment will be submitted to the Fermilab PAC in autumn 2007.

5.3.2.2. *The NuTeV Anomaly*

Neutrino scattering measurements offer a unique tool to probe the electroweak interactions of the Standard Model (SM). The NuTeV anomaly is a 3σ deviation of $\sin^2\theta_W$ from the Standard Model prediction.[162] $\sin^2\theta_W$ parameterizes the mixing between the weak interaction Z boson and the photon in electroweak theory. Deviations of measurements of this parameter, and its partner parameter, ρ, the relative coupling strength of the neutral-to-charged-current interactions, may indicate Beyond-Standard-Model physics. This section also highlights that fact that new neutrino

Table 5.6. left and right handed coupling constants.

f	ℓ_f	r_f
e^-	$-\frac{1}{2} + \sin^2\theta_W$	$\sin^2\theta_W$
u, c	$\frac{1}{2} - \frac{2}{3}\sin^2\theta_W$	$-\frac{2}{3}\sin^2\theta_W$
d, s	$-\frac{1}{2} + \frac{1}{3}\sin^2\theta_W$	$\frac{1}{3}\sin^2\theta_W$

properties may be revealed in TeV-scale interactions at LHC, which has not been addressed previously.

The NuTeV experiment represents a departure from the previous train of thought in several ways. NuTeV was a deep inelastic neutrino scattering experiment, and thus is performed at significantly higher energy than the experiments previously discussed. Also, while NuTeV did an oscillation search, it was mainly designed for another purpose: precision measurement of electroweak parameters. We will focus on that purpose here. As a result, this analysis allows new issues related to neutrino physics to be brought into the discussion.

$\sin^2\theta_W$ in Neutrino Scattering and Other Experiments

In neutrino scattering, the neutral current cross section depends upon $\sin^2\theta_W$. The dependence is a function of the neutrino flavor and the target. NuTeV was a muon-neutrino-flavor scattering experiment. In this case, the NC cross sections for scattering from a light fermion target are:

$$\frac{d\sigma(\nu_\mu f \to \nu_\mu f)}{dy} = \frac{G_F^2 s}{\pi} \left(\ell_f^2 + r_f^2(1-y)^2\right) \left(1 + \frac{sy}{M_Z^2}\right)^{-2}, \quad (5.38)$$

$$\frac{d\sigma(\bar{\nu}_\mu f \to \bar{\nu}_\mu f)}{dy} = \frac{G_F^2 s}{\pi} \left(\ell_f^2(1-y)^2 + r_f^2\right) \left(1 + \frac{sy}{M_Z^2}\right)^{-2}. \quad (5.39)$$

In this equation, f is the type of light fermion: $f = e^-, u, d, s, c$. Several constants appear: G_F is the Fermi constant, M_Z is the mass of the Z. The two kinematic variables are: s, the effective center of mass energy, which depends on the mass of f and y, the inelasticity (see definitions in Sec. 5.1.2). ℓ_f, r_f are left and right handed coupling constants which are given in Table 5.6.

While neutrino scattering has traditionally been a method for measuring $\sin^2\theta_W$, the "Standard Model Prediction" quoted in literature comes from the very precise measurements made by the LEP and SLD experiments, which have been summarized by the Electroweak Working Group.[163] $\sin^2\theta_W$ appears in various measurements from e^+e^- scattering at the Z pole. An example which leads to a highly precise measurement is the "left-right asymmetry" measured from polarized scattering at SLD:

$A_{LR} = (\sigma_L - \sigma_R)/(\sigma_L + \sigma_R)$ where σ_L and σ_R refer to the scattering cross sections for left- and right- polarized electrons, respectively. In this case the asymmetry is given by:

$$A_{LR}(Z^0) \equiv \frac{\left(\frac{1}{2} - \sin^2\theta_W^{((\text{eff}))}\right)^2 - \sin^4\theta_W^{(\text{eff})}}{\left(\frac{1}{2} - \sin^2\theta_W^{(\text{eff})}\right)^2 + \sin^4\theta_W^{(\text{eff})}}, \qquad (5.40)$$

When comparing quoted values of the weak mixing angle $\sin^2\theta_W$, care must be taken because this parameter is defined in various ways. The simplest definition is the "on shell" description:

$$1 - M_W^2/M_Z^2 \equiv \sin^2\theta_W^{(\text{on-shell})}. \qquad (5.41)$$

This is the definition commonly used in neutrino physics. In the discussion which follows, if not explicitly labeled, the on-shell definition for $\sin^2\theta_W$ is used. A variation on this definition uses the renormalized masses at some arbitrary scale μ which is usually taken to be M_Z:

$$1 - M_W(\mu)^2/M_Z(\mu)^2 \equiv \sin^2\theta_W^{(\overline{\text{MS}})}. \qquad (5.42)$$

However, the LEP experiments used the "effective" weak mixing angle which is related to the vector and axial vector couplings:

$$\frac{1}{4}(1 - g_V^l/g_A^l) \equiv \sin^2\theta_W^{(\text{eff})}. \qquad (5.43)$$

This is what appears in Eq. 5.40 above. One must convert between definitions, which have different radiative corrections and renormalization prescriptions, in order to make comparisons.

The parameter $\sin^2\theta_W$ evolves with Q^2, the squared 4-momentum transfer of the interaction. Fig. 5.22 illustrates this evolution. The highest Q^2 measurements are from LEP and SLD, with $Q^2 = M_Z^2$. There are several types of experiments, including neutrino experiments, which measure $\sin^2\theta_W$ with $Q^2 \ll m_Z^2$. NuTeV was performed at $Q^2 = 1$ to 140 GeV^2, $\langle Q_\nu^2 \rangle = 26$ GeV^2, $\langle Q_{\bar\nu}^2 \rangle = 15$ GeV^2. The lowest Q^2 measurements are from studies of atomic parity violation in the nucleus[164] (APV), which arises due to the electroweak interference of the photon and the Z in the boson exchange between the electrons and the nucleus. This samples $Q^2 \sim 0$. At higher Q^2, there is the result from SLAC E158, a Møller scattering experiment at average $Q^2 = 0.026$ GeV^2.[165] Using the measurements at the Z-pole with $Q^2 = M_Z^2$ to fix the value of $\sin^2\theta_W$, and evolving to low Q^2, Fig. 5.22[166] shows that APV and SLAC E158 are in agreement with the Standard Model.

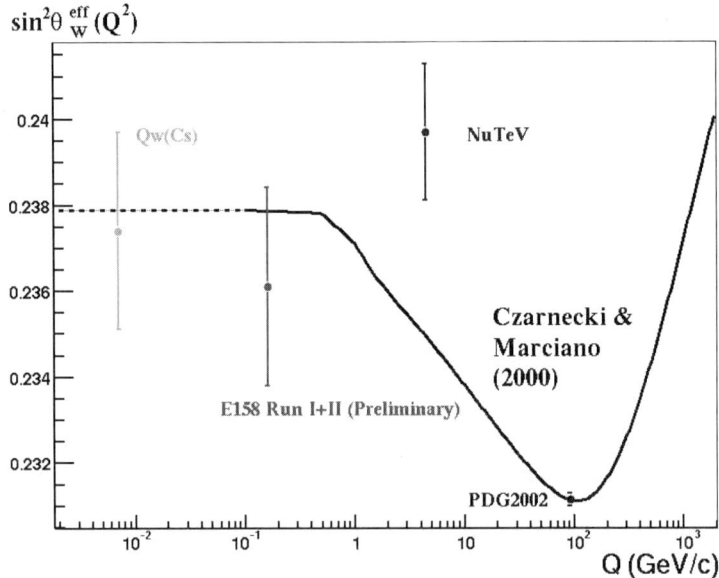

Fig. 5.22. Measurements of $\sin^2 \theta_W$ as a function of Q.[166] The curve shows the Standard Model expectation

NuTeV is strikingly off the prediction of Fig. 5.22. Neutrino scattering may measure a different result because new physics enters the neutrino process differently than the other experiments. Compared to the colliders, neutrino physics measures different combinations of couplings. Also neutrino scattering explores new physics through moderate space-like momentum transfer, as opposed to the time-like scattering at the colliders. With respect to the lower energy experiments, the radiative corrections to neutrino interactions allow sensitivity to high-mass particles which are complementary to the APV and Møller-scattering corrections.

The NuTeV Result The NuTeV experiment provides the most precise measurement of $\sin^2 \theta_W$ from neutrino experiments. The measurement relied upon deep inelastic scatter (DIS). It was performed using a "Paschos-Wolfenstein style"[167] analysis which is designed to minimize the systematic errors which come from our understanding of parton distributions and masses.

This method requires separated ν and $\bar{\nu}$ beams. In this case, the following ratios could be formed:

$$R^\nu = \frac{\sigma^\nu_{NC}}{\sigma^\nu_{CC}} \qquad (5.44)$$

$$R^{\bar\nu} = \frac{\sigma^{\bar\nu}_{NC}}{\sigma^{\bar\nu}_{CC}} \qquad (5.45)$$

$$(5.46)$$

Paschos and Wolfenstein[167] recast these as:

$$R^- = \frac{\sigma^\nu_{NC} - \sigma^{\bar\nu}_{NC}}{\sigma^\nu_{CC} - \sigma^{\bar\nu}_{CC}} = \frac{R^\nu - rR^{\bar\nu}}{1 - r}, \qquad (5.47)$$

where $r = \sigma^{\bar\nu}_{CC}/\sigma^\nu_{CC}$. In the case of R^-, many systematics cancel to first order. In particular, the quark and antiquark seas for $u, d, s,$ and c, which are less precisely known than the valence quark distributions, will cancel. Charm production only enters through $d_{valence}$ which is Cabbibo suppressed and at high x, thus the error from the charm mass is greatly reduced. One can also form R^+, but this will have much larger systematic errors, and so the strength of the NuTeV analysis lies in the measurement of R^-.

According to the "Paschos-Wolfenstein" method, an experiment should run in neutrino and antineutrino mode, categorize the events as CC or NC DIS, and then form R^- to extract $\sin^2\theta_W$. This requires identifying the CC or NC events properly in NuTeV's iron-scintillator/drift-chamber calorimeter. Most CC DIS events have an exiting muon, which causes a long string of hits in the scintillator and are therefore called "long." Most NC DIS events are relatively "short" hadronic showers. However, there are exceptions to these rules. A CC event caused by interaction of an intrinsic ν_e in the beam will appear short. An NC shower which contains a pion-decay-in-flight, producing a muon, may appear long. The connection between long vs. short and CC vs. NC must be made via Monte Carlo.

NuTeV measurement is in agreement with past neutrino scattering results, although these have much larger errors. However, the NuTeV result is in disagreement with the global fits to the electroweak data which give a Standard Model value of $\sin^2\theta_W = 0.2227$.[162]

Explanations In the case of any anomaly, it is best to start with the commonplace explanations. Three explanations for the NuTeV anomaly that are "within the Standard Model" have been proposed: the QCD-order of the analysis, isospin violation, and the strange sea asymmetry. The NuTeV analysis was not performed at a full NLO level. However, the effect of going to NLO on NuTeV can be estimated,[168] and the expected pull is away from the Standard Model. The NuTeV analysis assumed isospin symmetry, that is, $u(x)^p = d(x)^n$ and $d(x)^p = u(x)^n$. Various models for isospin violation have been studied and their pulls range from less than 1σ away from the Standard Model to $\sim 1\sigma$ toward the Standard Model.[169] Variations in the strange sea can either pull the result toward or away from the Standard Model expectation,[169] but not by more than one sigma.

With respect to Beyond-Standard-Model explanations, Chapter 14 of the APS Neutrino Study White Paper on Neutrino Theory[2] is dedicated to "The Physics of NuTeV" and provides an excellent summary. The discussion presented here is drawn from this source.

The NuTeV measurements of R^ν and $R^{\bar\nu}$, the NC-to-CC cross sections, are low compared to expectation. For this to be a Beyond-Standard-Model effect, it therefore requires introduction of new physics that suppresses the NC rate with respect to the CC rate. Two types of models produce this effect and remain consistent with the other electroweak measurements: (1) models which affect only the Z couplings, e.g., the introduction of a heavy Z' boson which interferes with the Standard Model Z; or (2) models which affect only the neutrino couplings, e.g., the introduction of moderate mass neutral heavy leptons which mix with the neutrino.

Any Z' model invoked to explain NuTeV must selectively suppress NC neutrino scattering, without significantly affecting the other electroweak measurements. This rules out most models, which tend to increase the NC scattering rate. Examples of successful models are those where the Z' couples to $B - 3L_\mu$[170] or to $L_\mu - L_\tau$.[171]

Moderate-mass neutral heavy leptons, a.k.a. "neutrissimos," can also produce the desired effect. Suppression of the coupling comes from intergenerational mixing of heavy states, so that the ν_μ is a mixture:

$$\nu_\mu = (\cos\alpha)\nu_{\text{light}} + (\sin\alpha)\nu_{\text{heavy}}. \quad (5.48)$$

The $Z\nu_\mu\nu_\mu$ coupling is modified by $\cos^2\alpha$ and the $W\mu\nu_\mu$ coupling is modified by $\cos\alpha$. Neutrissimos may have masses as light as ~ 100 GeV.[172] These new particles can play the role of the seesaw right-handed neutrinos, as long as one is willing to admit large tuning among the neutrino Yukawa couplings.[172] So this offers an alternative to the GUT-mass heavy neutrino model discussed in Sec. 5.3.1.

If neutrissimos exist, they would be expected to show up in other precision experiments. One must avoid the constraints on mixing from $0\nu\beta\beta$ (recall Eq. 5.31 to see why these experiments have sensitivity to the mixing). These experiment place a limit of $|U_{e4}|^2$ at less than a few $\times 10^{-5}$ for a 100 GeV right-handed neutrino. Rare pion and tau decays constrain $|U_{\mu 4}|^2$ to be less than 0.004 and $|U_{\tau 4}|^2$ to be less than 0.006, respectively.

Neutrissimos would be produced at LHC, thus neutrino physics can be done at the highest energy scales! However, they may be difficult to observe. One would naturally look for a signal of missing energy. However, neutrissiomos will not necessarily decay invisibly; for example one can have $N \to \ell + W$ and the W may decay to either two jets or a neutrino–charged-lepton pair. Only the latter case has missing energy. This may make them difficult to identify.

If the neutrissimo is a Majorana particle, then these could provide a clue to the mechanism for leptogenesis. The present models of leptogenesis require very high mass scales for the neutral lepton. However, theorists are identifying ways to modify the model to accommodate lower masses.[173] There also may be a wide mass spectrum for these particles, with one very heavy case that accommodates standard leptogenesis models, while the others have masses in the range observable at LHC.[174]

NuSOnG and Other Possibilities A new round of precision electroweak measurements can be motivated by the NuTeV anomaly as well as the imminent turn-on of LHC. These measurements are best done using neutrino-electron scattering, because this removes the quark-model related questions discussed in the previous section. Two possible methods for such a measurement are ν_μ scattering with higher statistics, using a NuTeV-style beam, or a $\bar{\nu}_e$s scattering measurement from a reactor. In either case, to provide a competitive measurement, the error from the best present neutrino-electron scattering measurement, from CHARM II, must be reduced by a factor of five.

The NuSOnG (Neutrino Scattering On Glass) Experiment[175] is proposed to run using a ν_μ beam produced by 800 GeV protons on target from the TeVatron. The plan is to use a design which is inspired by the CHARM II experiment: a target of SIO_2 in one quarter radiation length panels, with proportional tubes or scintillator to allow event reconstruction. The detector will have a 2.6 kton fiducial volume. The major technical challenge of such an experiment is in achieving the required rates from the TeVatron, as $\times 20$ the rate of NuTeV proton delivery is required.

Alternatively, a measurement of the weak mixing angle using antineutrinos from reactors may be possible.[176] The weak mixing angle can be extracted from the purely leptonic $\bar{\nu}_e e$ "elastic scatter" (ES) rate, which is normalized using the $\bar{\nu}_e p$ "inverse beta decay" (IBD) events, to reduce the error on the flux. Thus, a hydrocarbon (scintillator oil) based detector, which has free proton targets for the IBD events, is ideal. Gadolinium (Gd) doping is necessary for a high rate of neutron capture, which constitutes the signal for the IBD events. A window in visible energy of 3 to 5 MeV is selected to reduce backgrounds from contaminants in the oil and cosmic-muon-induced isotopes. In this energy range, the dominant contamination comes from the progeny of the uranium and thorium chain. This would clearly be an ambitious, state-of-the-art measurement, but could be done at a new reactor experiment where the detector is in close proximity to the source and had high shielding from cosmic rays.

5.3.3. *Neutrinos and the Cosmos*

Neutrinos are ubiquitous in the universe, and their presence and interactions must be incorporated into astrophysical and cosmological models. Nearly any new neutrino property will have direct consequences in these fields, which must be examined. As an illustrative example, the first discussion considers the impact of introducing of relatively light sterile neutrinos to the theory. Sterile neutrinos with keV-scale masses can explain dark matter as well other astrophysical questions.

As we improve our capability for detecting astrophysical neutrinos, these become a new source of neutrinos for study. The second example is a case in point: the search for ultra-high energy sources of neutrinos. The discovery of such sources would be of great interest to astrophysics, and the particle physics we can do with such a "beam" is remarkable.

While this section concentrates on the unknown, it is interesting to note that the known astrophysical sources of neutrinos are sufficiently intense that these neutrinos are already a possible background to other physics measurements. An example is the case of dark matter searches, which aim to measure cross sections as small as 10^{-46} cm^2. These experiments will have to contend with the background from coherent scattering ($\nu + N \rightarrow \nu + N$) of solar neutrinos, which has a cross section of 10^{-39} cm^2 and which produces a recoil nucleon that is very much like the expected dark matter signal.[177]

5.3.3.1. *Neutrinos as Dark Matter*

From the mid-1980's through mid-90's a 5 eV ν_τ was considered a likely candidate for dark matter. This was the motivation for the Chorus and NOMAD search for $\nu_\mu \to \nu_\tau$ oscillations in the > 10 eV^2 range[178,179] as well as the proposed COSMOS experiment.[180]

In the late 90's and early 2000's, two measurements led to a shift in opinion about neutrinos as candidates for dark matter. The first was the Super-K confirmation of ν_μ oscillations. As discussed in Sec. 5.2.1.4, the cleanest explanation, which fits within a three-neutrino model, is that this effect is $\nu_\mu \to \nu_\tau$. Combining this information with the direct limit on the ν_e implied that neutrinos were unlikely to have masses in the 5 to 10 eV range, as required for dark matter. Also, at the same time, studies of the large scale structure of the universe indicated that dark matter must be non-relativistic, or "cold." Relativistic, or "hot dark matter," like neutrinos, would smooth the large scale structure far beyond observations.[181]

The idea of neutrinos as dark matter candidates fell into disfavor. For some time, the more likely solution was assumed to be WIMPs, Weakly Interacting Massive Particles. The "weak" in this name is somewhat confusing, since it does not refer strictly to the weak interaction – other Beyond-Standard-Model interactions are involved. It is simply meant to say that the interaction rate is very low.

For some time, the lightest supersymmetric particle has been the most favored candidate for the WIMP. However, this is now starting to be questioned, as no evidence for supersymmetry has been observed at colliders.[182] This makes the formulation of the theory more awkward, and SUSY explanations for dark matter have been pushed from the "Minimal Super Symmetric Model" to the "Next-to-Minimal Super Symmetric Model," and even this is challenged.[183] If supersymmetry does not show up at LHC, then a new explanation for dark matter must be found.

As a result, neutrino models are being reconsidered.[184–193] The new models involve neutrinos which are mostly sterile, with a very tiny mixing with the light flavors and with keV masses ($0.5\,\text{keV} < m_\nu < 15\,\text{keV}$ and $\sin^2 2\theta \geq 10^{-12}$). Because of the high mass, they are not relativistic and are regarded as "warm" or "cold." At high mass, the large scale structure limits are much less stringent. A recent analysis from the Sloan Digital Sky Survey (SDSS) finds that a sterile neutrino mass above 9 keV can reproduce the power spectrum.[194] Only tiny mixing with the active neutrinos is required in order to produce the dark matter. With such small mixings, this model

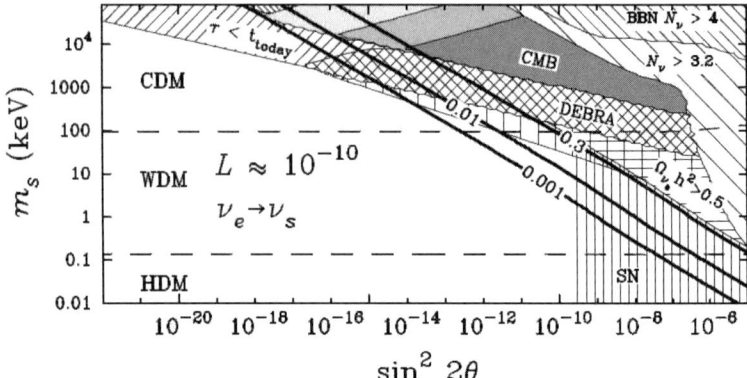

Fig. 5.23. Bounds for $\nu_e \to \nu_s$ oscillations from astrophysics and cosmology. Allowed regions for neutrino cold, warm and hot dark matter are shown.[187]

easily evades all accelerator-based bounds on $\nu_e \to \nu_s$ oscillations. As shown in Fig. 5.23, this model also escapes the cosmological limits on $\nu_e \to \nu_s$ from the CMB measurements, from Big Bang Nucleosynthesis (BBN), and supernova limits (SN), assuming negligible lepton number asymmetry, L, in the early universe.

Currently the only constraints on keV sterile neutrinos come from X-ray astronomy[193] which are searching for evidence of radiative decay of the massive neutrino into a lighter state, $\nu_2 \to \nu_1 + \gamma$. This proceeds through loop diagrams where the photon is coupling to a W or a charged lepton in the loop.[195] Because this is a 2-body decay, one is searching for a spectral line in the x-ray region. For a Dirac-type sterile neutrinos of mass m_s the decay rate is given by:

$$\Gamma_\gamma(m_s) = 1.36 \times 10^{-29} \text{s}^{-1} \left(\frac{\sin^2 2\theta}{10^{-7}} \right) \left(\frac{m_s}{1 \text{keV}} \right)^5, \quad (5.49)$$

which is clearly tiny, even for a keV scale neutrino. This is important as the dark matter neutrinos must be stable on the scale of the lifetime of the universe. No signal has been observed and the current mass limit, from the Chandra X-ray telescope observations, ranges from >3 to >6 keV depending on model assumptions.

Having motivated a keV-mass sterile state using dark matter, one can explore the consequences in other areas of cosmology and astrophysics. The small mixing allows these neutrinos to evade bounds from big bang nucleosynthesis.[196] Their presence may be beneficial to models of supernova explosions and pulsar kicks, as discussed below.

The existence of these neutrinos may improve the supernova models substantially. The problem faced by most models is that the supernova stalls and fails to explode. As modest increase in neutrino luminosity during the epoch when the stalled bounce shock is being reheated ($t_{\rm pb} < 1\,{\rm s}$) will incite the explosion.[197] As the supernova occurs, neutrinos will oscillate and even be affected by MSW resonances. If neutrinos oscillate to a sterile state, then their transport-mean-free-paths become larger. This increases the neutrino luminosity at the neutrino sphere and makes "the difference between a dud and an explosion." [197]

Also, sterile neutrinos in the 1 to 20 keV mass range can also be used to explain the origin of pulsar motion. Pulsars are known to have large velocities, from 100 to 1600 km/s. This is a much higher velocity than an ordinary star which typically has 30 km/s. Pulsars also have very high angular velocities. Apparently, there is some mechanism to give pulsars a substantial "kick" at birth, which sends them off with high translational and rotational velocities. One explanation for the kick is an asymmetric neutrino emission of sterile neutrinos during or moments after the explosion which forms the pulsar.[196] An asymmetry in neutrino emission occurs because the "urca reactions"

$$\nu_e + n \leftrightarrow p + e^-, \tag{5.50}$$

$$\bar{\nu}_e + p \leftrightarrow n + e^+, \tag{5.51}$$

are affected by magnetic fields which trap electrons and positrons. If the neutrinos oscillate to a sterile state, they stream out of the pulsar. If the sterile neutrinos have high mass, they can provide a significant kick. There are a number of solutions, either with standard neutrino oscillations, as described by Eq. 5.10 or with an MSW-type resonance. All require a sterile state in the 1 to 20 keV range with very small mixing, compatible with the dark matter scenario described above.

In summary, this is an example of how introducing a new neutrino property, *i.e.* sterile companions to the known neutrinos, can have a major impact on astrophysical models. This is an interesting case in point, because it is unlikely that these neutrinos will be observable in particle physics experiments in the near future. At present, the only detection method is through X-ray emission due to the radiative decay. Thus this is, at the moment, an example of a neutrino property which is entirely motivated and explored in the context of astrophysics.

5.3.3.2. Ultra High Energy Neutrinos

All of the experiments so-far discussed have used neutrino sources in the energy range of a few MeV to many GeV. We do not know how to produce neutrino beams at higher energies. However, nature clearly has high energy acceleration mechanisms, because cosmic rays with energies of 10^8 GeV have been measured. A new generation of neutrino experiments is now looking for neutrinos at these energies and beyond. These include AMANDA,[34] ICEcube,[198] Antares,[35] and Anita.[199]

These experiments make use of the fact that the Earth is opaque to ultra-high energy neutrinos. The apparent weakness of the weak interaction, which is due to the suppression by the mass of the W in the propagator term, is reduced as the neutrino energy increases. Amazingly, when you reach neutrino energies of 10^{17} eV, the Earth becomes opaque to neutrinos. To see this, recalling the kinematic variables defined in Sec. 5.1.2, consider the following back-of-the-envelope calculation. For a 10^8 GeV ν, $s = 2ME_\nu = 2 \times 10^8$ GeV^2. Most interactions occur at low x; and at these energies $x_{typical} \sim 0.001$. For neutrino interactions, the average y is 0.5. Therefore, using $Q^2 = sxy$, we find $Q^2_{typical} = (2 \times 10^8)(1 \times 10^{-3})(0.5) = 1 \times 10^5$ Gev^2. The propagator term goes as

$$\left(\frac{M_W^2}{M_W^2 + Q^2}\right)^2 = \left(\frac{1}{1+Q^2/M_W^2}\right)^2 \approx \frac{M_W^4}{Q^4}, \quad (5.52)$$

which is approximately 10^{-3} for our "typical" case. The typical cross section $\sigma_{typical}$ is:
$E \times (\sigma_{tot}/E) \times (\text{prop term}) = 10^8(\text{GeV})(0.6 \times 10^{-38}\text{cm}^2/\text{GeV})10^{-3}$
$= 0.6 \times 10^{-33}\text{cm}^2$. From this we can extract the interaction length, λ_0, on iron, by scaling from hadronic interactions, which tells us that at 30 mb ($= 0.3 \times 10^{-25}$ cm^2), $1\lambda_0 \sim 10$ cm. This implies that λ_0 for our very high energy νs is $\approx 5 \times 10^3$ km. However, the Earth is a few $\times 10^4$ km. Thus all of the neutrinos interact; the Earth is opaque to them.

This opens up the opportunity to instrument the Earth and use it as a neutrino target. One option is to choose a transparent region of the Earth, ice or water, and instrument it like a traditional neutrino detector. This has been the design chosen by AMANDA, ICECube and Antares, which use phototubes to sense the Cerenkov light produced when charged particles from neutrino interactions traverse the material. The largest of these detectors are on the order of (1 km)^3 of instrumented area. The second method exploits the Askaryan effect[200] in electromagnetic showers. Electron and positron scattering in matter have different cross sections. As the electromangetic shower develops, this difference leads to a negative charge asymmetry, inducing strong strong coherent Cerenkov radiation in the radio range. The pulse has unique and easy-to-distinguish broadband (0.2 - 1.1 GHz) spectral and polarization properties which can be received by detecting antennas launched above the target area. In ice, the radio attenuation length is 1 km. The Anita Experiment, which uses such a detector, can view 2×10^6 km^3 of volume. This makes it by far the world's largest tonnage experiment.

Neutrinos with energies above 10^4 GeV have yet to be observed. However, they are expected to accompany ultra high energy cosmic rays, which have been observed. Nearly all potential sources of ultra-high energy cosmic rays are predicted to produce protons, neutrinos, and gamma rays at roughly comparable levels. Ultra high energy protons, which have been observed by the HiRes[201] and Auger[202] experiments are guaranteed sources of neutrinos through the Gresein-Zatsepi-Kuzmin (GZK) interaction. In this effect, protons above $E_{GZK} = 6 \times 10^{10}$ GeV scatter from the cosmic microwave background: $p\gamma \to \Delta^+ \to n\pi^+$. This degrades the energy of protons above E_{GZK}, leading to an apparent cutoff in the flux called the "GZK cutoff." There are several Δ resonances and the CMB photons have an energy distribution, so the cutoff is not sharp. But it has been clearly observed by both HiRes and Auger.[203] As a result, ultra-high energy pions are produced and these must decay to ultra-high-energy neutrinos. There may be other, more exotic mechanisms for producing an ultra-high energy flux, possibly with energies beyond the GZK cutoff. Since, unlike the protons, these neutrinos do not interact with the cosmic microwave background, they can traverse long distances and can be messengers of distant point sources.

There many reviews of the exotic physics one can do with ultra high energy neutrino interactions. The opportunities include[204] gravitational lensing of neutrinos, the search for bumps or steps in the NC/CC ratio, the influence of new physics on neutrino oscillations at high energies, the search for neutrino decays, neutrino interaction with dark matter WIMPs, and the annihilation of the ultra high energy neutrinos by the cosmic neutrino background. This final example is interesting because significant limits have been set by Anita-lite, a small prototype for Anita that flew only 18.4 days. This illustrates the power of even a small experiment entering an unexplored frontier of particle physics.

The 1 eV mass neutrino implied by the LSND anomaly could be a candidate for the source of the ultra-high energy cosmic rays observed on Earth.[205] This neutrino, if produced at ultra-high energies, could annihilate on the cosmic neutrino background producing a "Z-burst" of ultra high energy hadrons. This was offered as an explanation for ultra-high energy cosmic which were observed by the AGASA experiment.[206] Scaling from the AGASA rate, a prediction for the flux of ultra-high neutrinos in the energy range of $10^{18.5} < E_\nu < 10^{23.5}$ eV for Z-burst models was made.[207,208] Based on this flux, Anita-lite was predicted to see between 5 and 50 events at >99% CL. During its short run, this prototype detector observed no events in the energy range and therefore could definitively rule out this model.[209] Shortly thereafter, the AGASA events were shown to be due to energy miscalibration.[210]

5.4. Conclusions

The goal of this review was to sketch out the present questions in neutrino physics, and discuss the experiments that can address them. Along the way, I have highlighted the experimental techniques and challenges. I have also tried to briefly touch on technological advances expected in the near future. This text followed the structure of the set of lectures entitled "Neutrino Experiments," given at the 2006 TASI Summer School.

Neutrino physics is an amalgam of astrophysics, cosmology, nuclear physics, and particle physics, making the field diverse and exciting, but hard to review comprehensively. In this paper, I have tried to touch on examples which are particularly instructive and have been forced to leave out a wide range of other interesting points. What should be clear, however, is that the recent discoveries by neutrino experiments have opened up a wide range of interesting questions and opportunities. This promises to be a rich field of research for both theorists and experimentalists for years to come.

Acknowledgments

I wish to thank A. Aguilar-Arevalo, G. Karagiorgi, B. Kayser, P. Nienaber, J. Spitz, and E. Zimmerman for their suggestions concerning this text.

References

1. https://www.interactions.org/cms/?pid=1009695

2. R.N. Mohapatra, et al., hep-ph/050213v2, 2005.
3. http://neutrinooscillation.org/
4. D. Decamp, et al., CERN-EP/89-169, Phys.Lett.B235:399, 1990.
5. H. Band, et al., SLAC-PUB-4990, published in the Proceedings of the Fourth Family of Quarks ,and Leptons, Santa Monica, CA, Feb 23-25, 1989.
6. http://fn872.fnal.gov/
7. C.S. Wu, et al. Phys. Rev. 105, 1413, 1957.
8. P. Lipari, Nucl. Phys. Proc. Suppl. **112**, 274 (2002) [arXiv:hep-ph/0207172].
9. D. Casper, Nucl. Phys. Proc. Suppl. 112:161, 2002.
10. http://www-boone.fnal.gov.
11. http://www-sciboone.fnal.gov/
12. http://minerva.fnal.gov/
13. http://www.sns.ias.edu/~jnb.
14. R. Davis, Prog. Part. Nucl. Phys. **32**, 13 (1994); B. T. Cleveland et al., Astrophys. J. **496**, 505 (1998).
15. J. N. Abdurashitov et al. [SAGE Collaboration], J. Exp. Theor. Phys. **95**, 181 (2002) [Zh. Eksp. Teor. Fiz. **122**, 211 (2002)] [arXiv:astro-ph/0204245].
16. W. Hampel et al. [GALLEX Collaboration], Phys. Lett. B **447**, 127 (1999).
17. K. Eguchi et al. [KamLAND Collaboration], Phys. Rev. Lett. **92**, 071301 (2004) [arXiv:hep-ex/0310047].
18. C. H. Albright et al. [Neutrino Factory/Muon Collider Collaboration], arXiv:physics/0411123.
19. See, for eaxmple, S. E. Kopp, Phys. Rept. **439**, 101 (2007) [arXiv:physics/0609129].
20. D.E. Groom et al, The European Physical Journal C15: 1, 2000.
21. G. D. Barr, T. K. Gaisser, P. Lipari, S. Robbins and T. Stanev, Phys. Rev. D **70**, 023006 (2004) [arXiv:astro-ph/0403630].
22. J. Hosaka et al. [Super-Kamkiokande Collaboration], Phys. Rev. D **73**, 112001 (2006) [arXiv:hep-ex/0508053].
23. H. Back et al., arXiv:hep-ex/0412016.
24. G. D. Barr, T. K. Gaisser, S. Robbins and T. Stanev, Phys. Rev. D **74**, 094009 (2006) [arXiv:astro-ph/0611266].
25. M. Apollonio et al., Eur. Phys. J. C **27**, 331 (2003) [arXiv:hep-ex/0301017].
26. T. Araki et al. [KamLAND Collaboration], [arXiv:hep-ex/0406035].
27. C. Athanassopoulos et al. [LSND Collaboration], Nucl. Instrum. Meth. A **388**, 149 (1997) [arXiv:nucl-ex/9605002].
28. http://pupgg.princeton.edu/ borexino/welcome.html
29. K. Nakamura [KamLAND Collaboration], AIP Conf. Proc. **721**, 12 (2004).
30. C. Kraus [SNO+ Collaboration], Prog. Part. Nucl. Phys. **57**, 150 (2006).
31. Karsten Heager, private communication.
32. E. Church et al. [BooNe Collaboration], "A proposal for an experiment to measure muon-neutrino → electron-neutrino oscillations and muon-neutrino disappearance at the Fermilab Booster: BooNE," FERMILAB-PROPOSAL-0898;
33. http://www-sk.icrr.u-tokyo.ac.jp/sk/index.html
34. http://amanda.berkeley.edu/

35. http://antares.in2p3.fr/
36. http://www-numi.fnal.gov/
37. http://www-e815.fnal.gov/
38. S. Yamamoto *et al.*, IEEE Trans. Nucl. Sci. **52**, 2992 (2005).
39. A. Pla-Dalmau, A. D. Bross, V. V. Rykalin and B. M. Wood [MINERvA Collaboration],
40. http://www-nova.fnal.gov/
41. A. Curioni, *et al.*, hep-ex/0603009.
42. S.Amerio, *et al.*, NIM A527: 329, 2004.
43. http://t962.fnal.gov/
44. http://www.fnal.gov/directorate/Longrange/Steering_Public/community_letters.html, see letter 15: "MicroBooNE - Fleming and Willis," June 12, 2007.
45. http://www.fnal.gov/directorate/Longrange/Steering_Public/community_letters.html, see letter 14: "Neutrino Expt with 5kton LAr TPC" Fleming and Rameika, June 12, 2007; D. Finley *et al.*,
46. L. Wolfenstein, Phys. Rev. D17, 2369 (1978); D20, 2634 (1979); S. P. Mikheyev and A. Yu. Smirnov, Yad. Fiz. 42, 1441 (1985) [Sov. J. Nucl. Phys. 42, 913 (1986)]; Nuovo Cimento 9C, 17 (1986).
47. G. J. Feldman and R. D. Cousins, Phys. Rev. D **57**, 3873 (1998) [arXiv:physics/9711021].
48. T. Schwetz, Acta Phys. Polon. B **36**, 3203 (2005) [arXiv:hep-ph/0510331].
49. B. Kayser, "Neutrino Mass, Mixing, and Flavor Change," available from the PDG website: http://pdg.lbl.gov/2007/reviews/contents_sports.html, W.-M. Yao, *et al.*, J. Phys. G **33**, 1 (2006).
50. Y. Totsuka, Nucl. Phys. A663, 218 (2000); Y.Fukuda, *et al.* Phys. Rev. Lett. 81, 1562 (1998)
51. D. Casper *et al.*, Phys. Rev. Lett. **66**, 2561 (1991); R. Becker-Szendy *et al.*, Phys. Rev. Lett. **69**, 1010 (1992).
52. E. Kearns, talk presented at Neutrino 2004.
53. Y. Fukuda *et al.* [Super-Kamiokande Collaboration], Phys. Lett. B **433**, 9 (1998) [arXiv:hep-ex/9803006]; Phys. Lett. B **436**, 33 (1998) [arXiv:hep-ex/9805006]; Phys. Rev. Lett. **81**, 1562 (1998) [arXiv:hep-ex/9807003]; Phys. Rev. Lett. **82**, 2644 (1999) [arXiv:hep-ex/9812014]; Phys. Lett. B **467**, 185 (1999) [arXiv:hep-ex/9908049]; Y. Ashie *et al.* [Super-Kamiokande Collaboration], Phys. Rev. D **71**, 112005 (2005) [arXiv:hep-ex/0501064].
54. S. H. Ahn *et al.* [K2K Collaboration], Phys. Lett. B **511**, 178 (2001) [arXiv:hep-ex/0103001]; Phys. Rev. Lett. **90**, 041801 (2003) [arXiv:hep-ex/0212007]; Phys. Rev. D **74**, 072003 (2006) [arXiv:hep-ex/0606032].
55. D. G. Michael *et al.* [MINOS Collaboration], [arXiv:hep-ex/0607088].
56. N. Saoulidou for the MINOS Collaboration, http://theory.fnal.gov/jetp/ July 19, 2007.
57. A. Habig [Super-Kamiokande Collaboration], arXiv:hep-ex/0106025.
58. http://www.cern.ch/opera.
59. J. Marteau [for the OPERA collaboration], arXiv:0706.1699 [hep-ex].

60. E. G. Adelberger et al. , "Solar Fusion Cross Sections," To be published in Rev. Mod. Phys., Oct. 1998, astro-ph/9805121.
61. For example, J. Christensen-Dalsgaard et al. , Science **272** 1286 (1996).
62. For example, Castellani et al. , Nucl. Phys. Proc. Suppl. **70** 301 (1998).
63. http://www.sno.phy.queensu.ca.
64. B. Aharmim et al. [SNO Collaboration], arXiv:nucl-ex/0610020.
65. J. N. Bahcall, M. H. Pinsonneault and S. Basu, Astrophys. J. **555**, 990 (2001) [arXiv:astro-ph/0010346].
66. T. Araki et al. [KamLAND Collaboration], Phys. Rev. Lett. **94**, 081801 (2005) [arXiv:hep-ex/0406035].
67. T. Schwetz, Phys. Scripta **T127**, 1 (2006) [arXiv:hep-ph/0606060].
68. H. S. Goh, R. N. Mohapatra and S. P. Ng, Phys. Rev. D **68**, 115008 (2003) [arXiv:hep-ph/0308197].
69. T. Asaka, W. Buchmuller and L. Covi, Phys. Lett. B **563**, 209 (2003) [arXiv:hep-ph/0304142].
70. K. S. Babu, J. C. Pati and F. Wilczek, Nucl. Phys. B **566**, 33 (2000) [arXiv:hep-ph/9812538].
71. C. H. Albright and S. M. Barr, Phys. Rev. D **64**, 073010 (2001) [arXiv:hep-ph/0104294].
72. T. Blazek, S. Raby and K. Tobe, Phys. Rev. D **62**, 055001 (2000) [arXiv:hep-ph/9912482].
73. G. G. Ross and L. Velasco-Sevilla, Nucl. Phys. B **653**, 3 (2003) [arXiv:hep-ph/0208218].
74. S. Raby, Phys. Lett. B **561**, 119 (2003) [arXiv:hep-ph/0302027].
75. R. Kitano and Y. Mimura, Phys. Rev. D **63**, 016008 (2001) [arXiv:hep-ph/0008269].
76. N. Maekawa, arXiv:astro-ph/0010559.
77. M. C. Chen and K. T. Mahanthappa, Phys. Rev. D **68**, 017301 (2003) [arXiv:hep-ph/0212375].
78. M. Bando and M. Obara, Prog. Theor. Phys. **109**, 995 (2003) [arXiv:hep-ph/0302034].
79. W. Buchmuller and D. Wyler, Phys. Lett. B **521**, 291 (2001) [arXiv:hep-ph/0108216].
80. P. H. Frampton and R. N. Mohapatra, JHEP **0501**, 025 (2005) [arXiv:hep-ph/0407139].
81. W. Grimus and L. Lavoura, JHEP **0107**, 045 (2001) [arXiv:hep-ph/0105212].
82. W. Grimus and L. Lavoura, Phys. Lett. B **572**, 189 (2003) [arXiv:hep-ph/0305046].
83. W. Grimus, A. S. Joshipura, S. Kaneko, L. Lavoura and M. Tanimoto, JHEP **0407**, 078 (2004) [arXiv:hep-ph/0407112].
84. M. C. Chen and K. T. Mahanthappa, arXiv:hep-ph/0409165.
85. I. Aizawa, M. Ishiguro, T. Kitabayashi and M. Yasue, Phys. Rev. D **70**, 015011 (2004) [arXiv:hep-ph/0405201].
86. R. N. Mohapatra, Pramana **63**, 1295 (2004).

87. S. Antusch and S. F. King, Nucl. Phys. B **705**, 239 (2005) [arXiv:hep-ph/0402121].
88. S. Antusch and S. F. King, Phys. Lett. B **591**, 104 (2004) [arXiv:hep-ph/0403053].
89. W. Rodejohann and Z. z. Xing, Phys. Lett. B **601**, 176 (2004) [arXiv:hep-ph/0408195].
90. K. S. Babu, E. Ma and J. W. F. Valle, Phys. Lett. B **552**, 207 (2003) [arXiv:hep-ph/0206292].
91. T. Ohlsson and G. Seidl, Nucl. Phys. B **643**, 247 (2002) [arXiv:hep-ph/0206087].
92. S. F. King and G. G. Ross, Phys. Lett. B **574**, 239 (2003) [arXiv:hep-ph/0307190].
93. Q. Shafi and Z. Tavartkiladze, Phys. Lett. B **594**, 177 (2004) [arXiv:hep-ph/0401235].
94. R. N. Mohapatra, JHEP **0410**, 027 (2004) [arXiv:hep-ph/0408187].
95. M. Bando, S. Kaneko, M. Obara and M. Tanimoto, Phys. Lett. B **580**, 229 (2004) [arXiv:hep-ph/0309310].
96. M. Honda, S. Kaneko and M. Tanimoto, JHEP **0309**, 028 (2003) [arXiv:hep-ph/0303227].
97. R. F. Lebed and D. R. Martin, Phys. Rev. D **70**, 013004 (2004) [arXiv:hep-ph/0312219].
98. A. Ibarra and G. G. Ross, Phys. Lett. B **575**, 279 (2003) [arXiv:hep-ph/0307051].
99. P. F. Harrison and W. G. Scott, Phys. Lett. B **594**, 324 (2004) [arXiv:hep-ph/0403278].
100. P. H. Frampton, S. L. Glashow and T. Yanagida, Phys. Lett. B **548**, 119 (2002) [arXiv:hep-ph/0208157].
101. A. de Gouvea and H. Murayama, Phys. Lett. B **573**, 94 (2003) [arXiv:hep-ph/0301050].
102. R. N. Mohapatra, M. K. Parida and G. Rajasekaran, Phys. Rev. D **69**, 053007 (2004) [arXiv:hep-ph/0301234].
103. R. Arnowitt, B. Dutta and B. Hu, Nucl. Phys. B **682**, 347 (2004) [arXiv:hep-th/0309033].
104. M. G. Albrow et al., arXiv:hep-ex/0509019.
105. http://doublechooz.in2p3.fr/
106. http://dayawane.ihep.ac.cn/
107. K. Assamagan, et al., Phys.Rev.D53:6065, 1996.
108. B. Jeckelman, et al., PL B3555 326, 1994.
109. Barate, et al, EPJ C2 395, 1998.
110. C. Kraus et al., Eur. Phys. J. C **40**, 447 (2005) [arXiv:hep-ex/0412056].
111. V. M. Lobashev, et al., Phys.Atom.Nucl.63: 962, 2000.
112. J. Bonn, et al., Phys. Atom. Nucl.63:969, 2000.
113. W. Stoeffl, et al., PRL 75: 3237, 1995.
114. C. Ching, et al., Int. Journ Mod. Phys. A10: 2841, 1995
115. E. Holzschuh, et al., Phys. Lett., B287: 381, 1992.
116. H. Kawakami, et al., Phys. Lett., B256: 105, 1991.

117. H. Robertson, *et al.*, Phys.Rev.Lett.67: 957, 1991; H. Robertson, PR D33: R6, 1991.
118. http://www-ik.fzk.de/ katrin/index.html
119. R. Bionta, *et al.*, PRL 58: 1494, 1987.
120. K. Hirata, *et al.*, PRL 58: 1490, 1987.
121. S. Hannestad, Prog. Part. Nucl. Phys. **57**, 309 (2006) [arXiv:astro-ph/0511595].
122. J.F. Beacom, N. F. Bell, S. Dodelson, astro-ph/0404585; Z. Chacko, L. J. Hall, S. J. Oliver, M. Perelstein, hep-ph/0405067; K. Abazajian, N. F. Bell, G. M. Fuller and Y. Y. Y. Wong, astro-ph/0410175.
123. A. Alessandrello, *et al.*, Phys. Lett. B486: 13, 2000; L. DeBraekelee, *et al.*, Phys. Atom Nucl 63:1214, 2000; R. Arnold, *et al.*, Nucl. Phys. A678:341, 2000; M. Alston-Garnjost, *et al.*, PR C55: 474, 1997; A. DeSilva, *et al.*, PR C56:2451, 1997; M. Gunther, *et al.*, PR D55:54, 1997; Arnold, *et al.*, Z. Phys. C72: 239, 1996; A. Balysh, *et al.*, PRL 77:5186, 1996.
124. A good discussion of GUT motivation for the sea-saw model appears in Kayser, Gibrat-Debu, and Perrier, *The Physics of Massive Neutrinos,* World Scientific Lecture Notes in Physics, 25, World Scientific Publishing, 1989.
125. V. A. Rodin, A. Faessler, F. Simkovic and P. Vogel, arXiv:nucl-th/0503063.
126. F. Simkovic, G. Pantis, J. D. Vergados and A. Faessler, Phys. Rev. C **60**, 055502 (1999) [arXiv:hep-ph/9905509].
127. M. Aunola and J. Suhonen, Czech. J. Phys. **48**, 145 (1998); M. Aunola and J. Suhonen, Czech. J. Phys. **48**, 145 (1998).
128. S. Stoica and H. V. Klapdor-Kleingrothaus, Nucl. Phys. A **694**, 269 (2001).
129. R. Arnold *et al.* [NEMO Collaboration], Phys. Rev. Lett. **95**, 182302 (2005) [arXiv:hep-ex/0507083].
130. R. Arnold *et al.* [the NEMO Collaboration], JETP Lett. **80**, 377 (2004) [Pisma Zh. Eksp. Teor. Fiz. **80**, 429 (2004)] [arXiv:hep-ex/0410021].
131. O. Cremonesi *et al.* [CUORICINO Collaboration], Phys. Atom. Nucl. **69** (2006) 2083.
132. H. V. Klapdor-Kleingrothaus, I. V. Krivosheina, A. Dietz and O. Chkvorets, Phys. Lett. B **586**, 198 (2004) [arXiv:hep-ph/0404088].
133. S. Schonert *et al.* [GERDA Collaboration], Phys. Atom. Nucl. **69**, 2101 (2006).
134. http://crio.mib.infn.it/wigmi/pages/cuore.php
135. http://nemo.in2p3.fr/supernemo/
136. http://0-www-project.slac.stanford.edu.ilsprod.lib.neu.edu/exo/
137. http://majorana.pnl.gov/
138. M. Nomachi *et al.*, Nucl. Phys. Proc. Suppl. **138**, 221 (2005).
139. C. Walter, Talk given at NuINT07, https://indico.fnal.gov/conferenceOtherViews.py?view=standard&confId=804, Session 1.
140. D. S. Ayres *et al.* [NOvA Collaboration], arXiv:hep-ex/0503053.
141. http://jnusrv01.kek.jp/public/t2k/
142. V. Barger *et al.*, arXiv:0705.4396 [hep-ph].
143. V. Barger, P. Huber, D. Marfatia and W. Winter, arXiv:hep-ph/0703029.
144. V. Barger, P. Huber, D. Marfatia and W. Winter, arXiv:hep-ph/0703029.

145. A. B. Balantekin, AIP Conf. Proc. **847**, 128 (2006) [arXiv:hep-ph/0601113];
146. A. Aguilar *et al.*, Phys. Rev. D **64**
147. B. Armbruster *et al.*, Phys. Rev. D **65** (2002) 112001.
148. B. Achkar *et al.*, Nucl. Phys. B **434** (1995) 503.
149. E. D. Church, K. Eitel, G. B. Mills,
150. B.H.J. McKellar, *et al.* hep-ph/0106121; R.N. Mohapatra, Phys. Rev. D **64** 091301, 2001; hep-ph/0107264; A. Ioannisian and J.W.F. Valle, Phys. Rev. D **63** 073002, 2001; E. Ma, G. Rajasekaran, and U. Sarkar, Phys. Lett. B **495** 363-368, 2000; hep-ph/0006340; Z. Berezhiani and R. Mohapatra, Phys. Rev. D **52** 6607, 1995; R. Fardon, A.E. Nelson, and N. Weiner, arXiv:astro-ph/0309800.
151. M. Sorel, J. Conrad, and M. Shaevitz, Phys. Rev. D70:073004,2004, hep-ph/0305255.
152. P. Astier, *et al.* Phys. Lett. B570:19, 2003; hep-ex/0306037.
153. F. Dydak *et al.*, Phys. Lett. B **134** 281, 1984..
154. I.E. Stockdale *et al.*, Phys. Rev. Lett. **52**, 1384 (1984); Z. Phys. C **27**, 53 (1985).
155. G. Karagiorgi, A. Aguilar-Arevalo, J. M. Conrad, M. H. Shaevitz, K. Whisnant, M. Sorel and V. Barger, Phys. Rev. D **75**, 013011 (2007) [arXiv:hep-ph/0609177].
156. A. Aguilar-Arevalo, *et al.*, paper in preparation. See also talk by J. Link, NuInt07.
157. M. Maltoni and T. Schwetz, arXiv:0705.0107 [hep-ph].
158. Paper in preparatuon, See talk at NuFact07 website: http://fphy.hep.okayama-u.ac.jp/nufact07/
159. T. Katori, A. Kostelecky and R. Tayloe, Phys. Rev. D **74**, 105009 (2006) [arXiv:hep-ph/0606154].
160. H. Pas, S. Pakvasa and T. J. Weiler, AIP Conf. Proc. **903**, 315 (2007) [arXiv:hep-ph/0611263].
161. X. Q. Li, Y. Liu and Z. T. Wei, arXiv:0707.2285 [hep-ph].
162. G. P. Zeller *et al.* Phys. Rev. Lett., **88** 091802, 2002.
163. http://lepewwg.web.cern.ch/LEPEWWG/
164. S. C. Bennett and Carl E. Wieman, Phys. Rev. Lett., **82** 2484–2487, 1999.
165. P. L. Anthony *et al.* [SLAC E158 Collaboration], Phys. Rev. Lett. **95**, 081601 (2005) [arXiv:hep-ex/0504049].
166. http://www.slac.stanford.edu/exp/e158/
167. E. A. Paschos and L. Wolfenstein, Phys. Rev. D **7**, 91 (1973).
168. K. S. McFarland and S. O. Moch, arXiv:hep-ph/0306052; S. Kretzer and M. H. Reno, Phys. Rev. D **69**, 034002 (2004) [arXiv:hep-ph/0307023]; B. A. Dobrescu and R. K. Ellis, Phys. Rev. D **69**, 114014 (2004) [arXiv:hep-ph/0310154].
169. M. Gluck, P. Jimenez-Delgado and E. Reya, arXiv:hep-ph/0501169; F. M. Steffens and K. Tsushima, Phys. Rev. D **70**, 094040 (2004) [arXiv:hep-ph/0408018]; J. T. Londergan and A. W. Thomas, arXiv:hep-ph/0407247; A. D. Martin, R. G. Roberts, W. J. Stirling and R. S. Thorne, Eur. Phys. J. C **39**, 155 (2005) [arXiv:hep-ph/0411040].

170. S. Davidson, J. Phys. G **29**, 2001 (2003) [arXiv:hep-ph/0209316].
171. E. Ma, D. P. Roy and S. Roy, Phys. Lett. B **525**, 101 (2002) [arXiv:hep-ph/0110146].
172. A. de Gouvea, arXiv:0706.1732 [hep-ph].
173. N. Sahu and U. A. Yajnik, Phys. Rev. D **71**, 023507 (2005) [arXiv:hep-ph/0410075].
174. A. de Gouvea, J. Jenkins and N. Vasudevan, Phys. Rev. D **75**, 013003 (2007) [arXiv:hep-ph/0608147].
175. http://www.fnal.gov/directorate/Longrange/Steering_Public/community_letters.html, see letter 3: "Precision Neutrino Scattering at the TeVatron - Conrad and Fisher," June 12, 2007. An expression of Interest is in preparation.
176. J. M. Conrad, J. M. Link and M. H. Shaevitz, Phys. Rev. D **71**, 073013 (2005) [arXiv:hep-ex/0403048].
177. J. Monroe and P. Fisher, arXiv:0706.3019 [astro-ph].
178. E. Eskut et al. [CHORUS Collaboration], Phys. Lett. B **497**, 8 (2001).
179. P. Astier et al. [NOMAD Collaboration], Nucl. Phys. B **611**, 3 (2001) [arXiv:hep-ex/0106102].
180. K. Kodama et al., 'Muon-neutrino to tau-neutrino oscillations: Proposal," FERMILAB-PROPOSAL-0803.
181. S. Bonometto, New Astron. Rev. **43**, 169 (1999).
182. M. Carena, D. Hooper and A. Vallinotto, Phys. Rev. D **75**, 055010 (2007) [arXiv:hep-ph/0611065]; D. Hooper and A. M. Taylor, JCAP **0703**, 017 (2007) [arXiv:hep-ph/0607086].
183. D. G. Cerdeno, E. Gabrielli, D. E. Lopez-Fogliani, C. Munoz and A. M. Teixeira, JCAP **0706**, 008 (2007) [arXiv:hep-ph/0701271].
184. S. Colombi, S. Dodelson and L. M. Widrow, Astrophys. J. **458**, 1 (1996) [arXiv:astro-ph/9505029]; S. Dodelson and L. M. Widrow, Phys. Rev. Lett. **72**, 17 (1994) [arXiv:hep-ph/9303287].
185. T. Asaka, M. Shaposhnikov and A. Kusenko, Phys. Lett. B **638**, 401 (2006) [arXiv:hep-ph/0602150].
186. X. D. Shi and G. M. Fuller, Phys. Rev. Lett. **83**, 3120 (1999) [arXiv:astro-ph/9904041].
187. K. Abazajian, G. M. Fuller and M. Patel, Phys. Rev. D **64**, 023501 (2001) [arXiv:astro-ph/0101524].
188. A. D. Dolgov and S. H. Hansen, arXiv:hep-ph/0103118.
189. A. D. Dolgov and S. H. Hansen, arXiv:hep-ph/0103118.
190. K. Abazajian, Phys. Rev. D **73**, 063506 (2006) [arXiv:astro-ph/0511630].
191. P. L. Biermann and A. Kusenko, Phys. Rev. Lett. **96**, 091301 (2006) [arXiv:astro-ph/0601004].
192. K. Abazajian and S. M. Koushiappas, Phys. Rev. D **74**, 023527 (2006) [arXiv:astro-ph/0605271].
193. K. N. Abazajian, M. Markevitch, S. M. Koushiappas and R. C. Hickox, Phys. Rev. D **75**, 063511 (2007) [arXiv:astro-ph/0611144].
194. P. McDonald, et al., Astrophys. J. Suppl. **163**, 80 (2006), astro-ph/0405013.
195. P. B. Pal and L. Wolfenstein, Phys. Rev. D **25**, 766 (1982).

196. G. M. Fuller, A. Kusenko, I. Mocioiu and S. Pascoli, Phys. Rev. D **68**, 103002 (2003) [arXiv:astro-ph/0307267].
197. J. Hidaka and G. M. Fuller, Phys. Rev. D **74**, 125015 (2006) [arXiv:astro-ph/0609425].
198. http://icecube.wisc.edu/
199. http://amanda.uci.edu/ anita/
200. G. A. Askaryan, JETP Lett. **50**, 478 (1989) [Pisma Zh. Eksp. Teor. Fiz. **50**, 446 (1989)].
201. S. Westerhoff [HiRes Collaboration], AIP Conf. Proc. **698**, 370 (2004).
202. http://www.auger.org/
203. R. Abbasi *et al.* [HiRes Collaboration], arXiv:astro-ph/0703099.
204. C. Quigg, arXiv:astro-ph/0603372.
205. T. J. Weiler, Astropart. Phys. 11:303 (1999).
206. http://www-akeno.icrr.u-tokyo.ac.jp/AGASA/
207. Z. Fodor, S. D. Katz and A. Ringwald, arXiv:hep-ph/0210123.
208. O. Kalashev, G. Gelmini and D. Semikoz, arXiv:0706.3847 [astro-ph].
209. S. W. Barwick *et al.* [ANITA Collaboration], Phys. Rev. Lett. **96**, 171101 (2006) [arXiv:astro-ph/0512265].
210. B. M. Connolly, S. Y. BenZvi, C. B. Finley, A. C. O'Neill and S. Westerhoff, Phys. Rev. D **74**, 043001 (2006) [arXiv:astro-ph/0606343].

Chapter 6

String Theory, String Model-Building, and String Phenomenology — A Practical Introduction

Keith R. Dienes

Department of Physics, University of Arizona, Tucson, AZ 85721, USA
dienes@physics.arizona.edu

> This is the written version of an introductory self-contained course on string model-building and string phenomenology given at the 2006 TASI summer school. No prior knowledge of string theory is assumed. The goal is to provide a practical, "how-to" manual on string theory, string model-building, and string phenomenology with a minimum of mathematics. These notes cover the construction of bosonic strings, superstrings, and heterotic strings prior to compactification. These notes also develop the ten-dimensional free-fermionic construction. A final lecture discusses general features of heterotic string models, Type I (open) string models, and recent trends of string phenomenology. and general features of low-energy string phenomenology.

6.0. Introduction

These lectures were delivered at the 2006 Theoretical Advanced Study Institute (TASI), to an audience of graduate students whose interests were primarily oriented towards high-energy phenomenology. Indeed, this school had a stated focus on neutrino physics, and consequently my goal was to present string theory in a way that ultimately might explain how a specific particle such as a neutrino might ultimately emerge from string theory. Of course, string theory contains a lot more than neutrino physics (and also, in some ways, a lot less!), and in the course of these lectures I will not really focus so much on neutrinos as on string theory as a whole. Nevertheless, I will continue to keep neutrinos as a running theme throughout these lectures as a way of reminding ourselves that our discussion of string theory is ultimately aimed at understanding something real and observable, such as an actual neutrino.

The title of these lectures indicates that these lectures are meant to serve as a practical introduction to string theory, string model-building, and string phenomenology. Let me explain, in a rough sense, what each of these words is meant to convey. We are all familiar with quantum field theory, which is a *language* through which we might construct particular *models* of physics (such as the Standard Model or the Minimal Supersymmetric Standard Model). Such models then have certain physical characteristics, certain *phenomenologies*. String theory, at least as I shall try to present it, can likewise be considered as a *language* for discussing physics: in this sense it replaces quantum field theory (a language based on point-particle physics) with a new language suitable for theories whose fundamental objects are the one-dimensional extended objects known as *strings*. However, from this perspective, string theory is still only a language: it is still necessary to take the next step and *use* this language to construct *models* that describe the everyday world. Therefore, although I will attempt to give a self-contained introduction to the language of string theory, these lectures will primarily focus on the model-building aspects of string theory and on the resulting phenomenologies that these models have. While there already exist many excellent reviews of string theory, there are relatively few that focus on its model-building and phenomenological aspects. These lecture notes will therefore hopefully help to fill the gap, especially for those readers who might care less for the formal aspects of string theory and more for their phenomenological implications.

Finally, I should explain the word "practical" which also appears in the title. The word "practical" refers to actual *practice* — the things that practitioners actually need to know in order to build *bona-fide* string models and/or comprehend their low-energy properties. Of course, string theory is a rich and beautiful subject, with many mathematical aspects that are compelling and ultimately essential for a deep understanding of the subject. However, the goal of these lectures is simply to present the basic features of string theory with a minimum of mathematics — as stated in the abstract, I am seeking to provide a "how-to" manual which cuts the subject to the bone and conveys only that information which will be important for phenomenology. Therefore, in many places the omissions will be substantial. Certainly they do not do justice to the subject. However, these lectures were designed for phenomenologically-oriented graduate students whose desire (I hope) was to learn something of string theory without being deluged by mathematical formalism. It is with them in mind that I designed these lectures to be as elementary as feasible, and to "get to the physics" as rapidly as possible. Therefore, I now issue the following

> **Warning:** These lectures are meant to cover a considerable amount of introductory material very rapidly and without mathematical sophistication. The purpose is to advance quickly to the model-building and phenomenological aspects of string theory, while still conveying an intuitive flavor of the essential issues. The target audience consists of people who have had no prior exposure to string theory, and who wish to understand the basic concepts from a purely phenomenological perspective.

Hopefully, the students came away with a sense that string theory is a real part of physics, one with direct relevance for the real world. Perhaps the reader will too. If so, then these lectures will have served their purpose.

6.1. Lecture #1: Why strings? — an overview

Why should we be interested in string theory? In this lecture, we shall review our present state of knowledge about the underlying constituents of matter, and discuss how string theory has the potential to extend that knowledge in a profoundly new direction. Since this lecture is meant only as an overview, we shall keep the discussion at an extremely superficial level and seek to present the intuitive *flavor* of string theory rather than its substance. We shall deal with the substance in subsequent lectures.

6.1.1. *From atoms to the Standard Model: A quick review*

Certainly we do not need to understand string theory in order to appreciate modern high-energy particle physics, or to understand or interpret the results of collider experiments. Why then should one study string theory, a subject whose connections to observable phenomena are usually considered rather tenuous at best?

The primary reason, of course, is that the goal of high-energy physics has always been to uncover the fundamental "elements" or building-blocks of the natural world. These consist of both the fundamental *particles* that make up the *matter*, and the fundamental *forces* that describe their *interactions*. In this way, we hope to expose the underlying laws of physics in their simplest forms.

But what is "fundamental"? Clearly, the answer depends on the energy scale, or equivalently the inverse length scale, at which these constituents are being probed. In order to establish our frame of reference, recall that 1 eV $\approx 1.6 \times 10^{-19}$ Joules $\approx (10^{-7}$ meters$)^{-1}$. At the eV scale, the fundamental objects are atoms, or nuclei plus electrons. But it turns out that there are many different types of atoms or nuclei — indeed, they fill out an entire periodic table, the complexity but regularity of which suggests a deeper substructure. And indeed such a deeper substructure exists: at the keV to MeV scale, the nuclei are no longer fundamental, but decompose into new fundamental objects — protons and neutrons. Thus, at this energy scale, the fundamental objects are protons, neutrons, and electrons. But once again, it is found that there are many different "types" of protons and neutrons — collectively they are called *hadrons*, and include not only the proton (p) and neutron (n), but also the pions (π), kaons (K), rho (ρ), omega (Ω), and so forth. Indeed, the "periodic table of the elements" at this energy scale is nothing but the Particle Properties Data Book! But once again, the complexity and regularity of these "elementary" particles suggests a deeper substructure, and indeed such a substructure is found, this time at the GeV scale: the proton and neutron are just made of two kinds of *quarks*, the so-called up and down quarks. Thus, at the GeV scale, the fundamental objects are up quarks, down quarks, and electrons. But once again complexity emerges: it turns out that there are many different "types" (flavors) of quarks: up, down, strange, charm, top, and bottom. Likewise, there are many different "types" of electrons (collectively called *leptons*): the electron, the muon, the tau, and their associated neutrinos. And indeed, once again there is a mysterious pattern, usually referred to as a family or generational structure. This once again suggests a deeper substructure.

Unfortunately, this is as far as we've come. Indeed, all of our present-day knowledge down to this energy scale is gathered together into the so-called *Standard Model* of particle physics. The primary features of the Standard Model are as follows. The fundamental *particles* are the quarks and leptons. They are all fermions, and are arranged into three generations of doublets:

$$\text{quarks}: \begin{pmatrix} u \\ d \end{pmatrix}, \begin{pmatrix} c \\ s \end{pmatrix}, \begin{pmatrix} t \\ b \end{pmatrix}$$

$$\text{leptons}: \begin{pmatrix} \nu_e \\ e \end{pmatrix}, \begin{pmatrix} \nu_\mu \\ \mu \end{pmatrix}, \begin{pmatrix} \nu_\tau \\ \tau \end{pmatrix}. \quad (6.1.1)$$

The fundamental *forces* also come in three varieties. First, there is the strong (or "color") force, associated with the non-abelian Lie group $SU(3)$.

Its fine-structure constant is $\alpha_3 \approx 1/8$ (as measured at energy scales of approximately 100 GeV), and it is responsible for binding quarks together to form hadrons and nuclei. As such, it is felt only by quarks. Its mediators or carriers are called *gluons*. Second, there is the electroweak force, associated with the non-abelian Lie group $SU(2)$. Its fine-structure constant is $\alpha_2 \approx 1/30$ (indeed, weaker than the strong force!), and it is responsible for β-decay. Unlike the strong force, it is felt by *all* of the fundamental particles. Finally, there is the "hypercharge" force, associated with the abelian Lie group $U(1)$, with fine-structure constant $\alpha_1 \approx 1/59$. Once again, this force is felt by essentially *all* particles, both quarks and leptons. The carriers of the latter two forces are the photon as well as the $W^{\pm}$ and Z particles. Indeed, ordinary electromagnetism is a combination of the electroweak and hypercharge forces, and is the survivor of electroweak symmetry breaking. This breaking is induced by the one remaining particle of the Standard Model, a boson called the *Higgs particle*. An excellent introduction to the physics of the Standard Model can be found in the TASI lectures of G. Altarelli (this volume).

6.1.2. *Beyond the Standard Model: Two popular ideas*

Is that all there is? Clearly, there are lots of reasons to believe in something deeper! First, the Standard Model contains many *arbitrary parameters*, such as the masses and "mixings" of fundamental particles. All of these must ultimately be *fit* to data rather than *explained*. Second, there are many *conceptual* questions. *Why* are there three generations? *Why* are there three kinds of forces? *Why* do these forces have different strengths and ranges? A fundamental theory should explain these features. Finally, there is also another force which we have not yet mentioned: the *gravitational* force. How do we incorporate the gravitational force into this framework? In other words, how do we "quantize" gravity?

There is only one conclusion we can draw from this state of affairs. Just as in each previous case, there must still be a deeper underlying principle. It is important to stress that this is not simply an issue of academic interest. Rather, it is one of practical importance, because the next generation of particle accelerators are being built right now! (Two of the most prominent that will be exploring physics beyond the Standard Model are Fermilab, where upgrades to the TeVatron are being implemented, and CERN, where construction of the Large Hadron Collider (LHC) is already underway.) The pressing question, therefore, is: What do we expect to see at these machines? What will high-energy physics be focusing on over the next ten to twenty years? It turns out that there are two very popular sets of ideas, both of which are thoroughly reviewed in the TASI lectures of N. Polonsky.

6.1.2.1. *Low-energy supersymmetry*

The first idea is *supersymmetry* (SUSY). This refers to a new kind of symmetry in physics, one which relates bosons (particles with integer spin) to fermions (particles with half-integer spin). Thus, for every known particle, there is a predicted new particle, its so-called superpartner:

$$\begin{aligned} \text{quarks} &\Longleftrightarrow \text{\textit{squarks}} \\ \text{leptons} &\Longleftrightarrow \text{\textit{sleptons}} \\ \text{gauge bosons} &\Longleftrightarrow \text{\textit{gauginos}} \,. \end{aligned} \qquad (6.1.2)$$

Clearly, this implies the existence of a *lot* of new particles and a *lot* of new interactions! Why then go through all this trouble?

Well, it turns out that supersymmetry can provide a number of striking benefits. First, through supersymmetry, we can explain the relative strengths of the forces ("gauge coupling unification"). Second, we can explain the origin of electroweak symmetry breaking. Third, supersymmetry has a number of favorable cosmological implications (for example, supersymmetry provides a natural set of dark-matter candidates). Finally, it turns out that supersymmetry is the only known answer to certain difficult theoretical puzzles in the Standard Model (chief among them the so-called "gauge hierarchy problem", *i.e.*, the difficulty of explaining the lightness of the Higgs particle, or equivalently to difficulty of explaining the stability of the scale of electroweak symmetry breaking against radiative corrections). In order to serve as an explanation of the gauge hierarchy problem, the energy scale associated with supersymmetry must not be too much higher than the scale of electroweak symmetry breaking. This is therefore called "low-energy supersymmetry", which refers to the common expectation that superparticles should exist at or near the TeV-scale.

Supersymmetry is a beautiful theory, both phenomenologically and mathematically. But it is *not* observed in nature. Therefore, supersymmetry must be broken. The problem, however, is that supersymmetry is very robust! It turns out to be quite hard to find mechanisms that can easily ("spontaneously") break supersymmetry at the expected energy scales. Therefore, we are faced with a major unsolved problem: *How do we break supersymmetry?* Indeed, we often have to resort to introducing

SUSY-breaking by hand, which requires the introduction of *many* additional unknown parameters. This is quite unpleasant, not only from an aesthetic point of view but also a phenomenological (predictive) point of view. However, it is often possible to consider only a minimal supersymmetric extension to the Standard Model (the so-called MSSM) where a minimal number of supersymmetry-breaking parameters are chosen.

6.1.2.2. *Grand unification*

The second popular idea for physics beyond the Standard Model concerns so-called *Grand Unified Theories* (GUTs). This refers to an attempt to realize the different forces and particles in nature as different "faces" or "aspects" of a single GUT force and a single GUT particle. An electromagnetic analogy here might be useful. Recall that the electric force is felt or caused by static charges, and that the magnetic force is felt or caused by moving charges. Are these therefore different forces? As we know, the answer is most definitely "no": we can Lorentz-boost from a rest frame to a moving frame, whereupon the distinction between the electric and magnetic forces melts away and these forces become intertwined. Thus, we conclude that the electric and magnetic forces are merely different aspects of *one* force, the "electromagnetic" force.

Is the same true for the strong, electroweak, and hypercharge forces? Is there a single "strong-weak-hypercharge" GUT force?

At first glance, this doesn't seem possible, because these different forces have different strengths. Recall their fine-structure constants: $\alpha_1 \approx 1/59$, $\alpha_2 \approx 1/30$, and $\alpha_3 \approx 1/8$. However, also recall that in quantum field theory, the strengths of forces ultimately depend on the energy scale through which they are measured. To see why this is so, let us think of placing a positive charge next to a dielectric. The positive charge draws some negative charge from within the dielectric towards it, so that the dielectric medium partially screens the positive charge. Therefore, in a rough sense, the less of the dielectric we see (*i.e.*, the more finely resolved our experimental apparatus to probe the original positive charge), the stronger our original positive charge seems to be. Thus, we see that at shorter distances (corresponding to higher energies), our electric charges (and therefore the corresponding electric forces) appear to be stronger. If this dielectric analogy serves as a good model for the results of a true quantum field-theoretic calculation (and in this case it does), we conclude that the electric force appears to grow stronger with increasing energy.

Of course, this is just a mechanical analogy. However, in the supersymmetric Standard Model, it turns out that the quantum field-theoretic vacuum itself indeed behaves like a dielectric for the hypercharge and weak forces. However, for the strong force, it behaves as an *anti*-dielectric. Thus, while the hypercharge and electroweak forces become *stronger* at higher energies, the strong force becomes *weaker* at higher energies. (This latter feature is the celebrated phenomenon of *asymptotic freedom*.) Together, these observations imply that these three forces have a chance of *unifying* at some energy scale if their strengths become equal, and indeed, carrying out the appropriate calculations, one finds the results shown in Fig. 6.1. We see from this figure that the forces appear to unify at the scale

$$M_{\rm GUT} \approx 2 \times 10^{16} \text{ GeV} . \qquad (6.1.3)$$

This would then be the natural energy scale for grand unification. Note that this unification also requires the existence of weak-scale supersymmetry in the form of weak-scale superpartners. Without such superpartners, the evolution of these fine-structure constants as a function of the energy scale is different, and they fail to unify at any scale. This then serves as another motivation for weak-scale supersymmetry.

GUTs would have numerous important effects on particle physics. First, by their very nature, they would imply new interactions that can mix the three fundamental forces. Second, this in turn implies that GUTs naturally lead to new, rare decays of particles. The most famous example of this is proton decay, the rate for which is experimentally known to be exceedingly small (since the proton lifetime is $\tau_p \gtrsim 10^{32}$ years). Third, GUTs would naturally explain the quantum numbers of all of the fundamental particles. Along the way, GUTs would also explain charge quantization. GUTs might also explain the origins of fermion mass. Finally, because they generally lead to baryon-number violation, GUTs even have the potential to explain the cosmological baryon/anti-baryon asymmetry. By combining GUTs with supersymmetry in the context of SUSY GUTs, it might then be possible to realize the attractive features of GUTs simultaneously with those of supersymmetry in a single theory.

Both the SUSY idea and the GUT idea are very compelling. Certainly, the SUSY idea (and indirectly the GUT idea, through measurements of proton decay and other rare decays) will be the focus of experimental high-energy physics over the next 20 years. But high-energy theorists also have plenty of work to do — we must build theories in order to interpret the

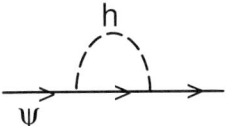

Fig. 6.1. One-loop evolution of the gauge couplings within the Minimal Supersymmetric Standard Model (MSSM), assuming supersymmetric thresholds at the Z scale. Here $\alpha_1 \equiv (5/3)\alpha_Y$, where α_Y is the hypercharge coupling in the conventional normalization. The relative width of each line reflects current experimental uncertainties.

data. But *how* do we build realistic SUSY theories? *How* do we build realistic GUT theories? *How* do we incorporate gravity?

Clearly, the possibilities seem endless. And even the SUSY or GUT ideas have not answered our most fundamental questions, such as why there are three gauge forces, or why there are three generations. Therefore, it is natural to hope that there is yet a deeper principle that can provide some theoretical guidance. And that's where string theory comes in.

6.1.3. *So what is string theory?*

The basic premise of string theory is very simple: all elementary particles are really closed vibrating loops of energy called *strings*. The length scale of these loops of energy is on the order of 10^{-35} meters (corresponding to 10^{19} GeV), so it is not possible to probe this stringy structure directly.

This idea has great power, because it provides a way to unify all of the particles and forces in nature. Specifically, each different elementary particle can be viewed as corresponding to a different vibrational mode of the string. A pictorial representation of this idea is given in Fig. 6.2, where we are schematically associating higher vibrational string modes with string loops containing more "wiggles". From the point of view of a low-energy observer who cannot make out this stringy structure, the different excitations each appear to be point particles. However, to such an observer, the states with more underlying "wiggles" appear to have higher spin. Thus, in this way we find that string theory predicts not only spin-1/2 and spin-1 states (which can be associated with the fermions and gauge bosons of the Standard Model respectively), but also a spin-2 state (which can naturally be associated with the graviton). Thus, through string theory, we see that the gauge interactions, particles, *and also gravity* are unified into a common quantized description as corresponding to different excitation modes of a single fundamental entity, the string itself.

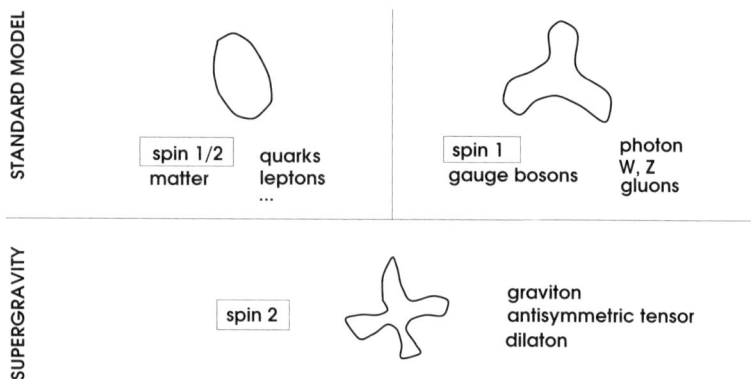

Fig. 6.2. The basic hypothesis of string theory is that the different elementary particles correspond to the different vibrational modes of a single fundamental entity, a closed loop of energy called a string. In this way one obtains not only spin-1/2 and spin-1 states which can be associated with the matter and gauge bosons of the Standard Model, but also a spin-2 state which can be identified with the graviton. Thus, string theory provides a way of unifying the Standard Model with gravity.

Of course, this is not the end of the story. Just as a violin string has an infinite number of harmonics, so too does a string give rise to an infinite tower of states corresponding to higher and higher vibrational modes. Since it takes more and more energy to excite these higher vibrational string modes, such states are increasingly massive. Indeed, because the fundamental string scale is on the order of $M_{\text{string}} \approx 10^{18}$ GeV, these string states are quantized in units of M_{string}. The states which we have illustrated in Fig. 6.2 are all *massless* with respect to M_{string}, and correspond, in some sense, to the ground states of the string. These are the so-called "observable states", and include not only the (supersymmetric) Standard Model and (super)gravity, but also may include various additional states (often called "hidden-sector states" which contain their own matter and gauge particles). However, there also exists an infinite tower of massive states with masses $M_n \approx \sqrt{n} M_{\text{string}}$, $n \in \mathbb{Z}^+$. In most discussions of the phenomenological properties of string theory, these massive states are ignored (since they are so heavy), and one concentrates on the phenomenology of the massless states. One then presumes that they accrue (relatively small) masses through other means, such as through radiative corrections.

Nevertheless, the passage from point particles to strings has tremendous consequences. Not only have we replaced the physics of zero-dimensional objects (elementary point particles) with the physics of one-dimensional objects (strings), but we have also replaced the physics of the one-dimensional worldlines that they sweep out with the physics of two-dimensional so-called *worldsheets*. Likewise, we have replaced the physics of Feynman diagrams with the physics of two-dimensional *manifolds*, so that a tree diagram corresponds to a genus-zero manifold (a sphere) and a one-loop diagram corresponds to a genus-one manifold (a torus). These comparisons are illustrated in Fig. 6.3. Note that the latter descriptions as spheres and tori correspond to shrinking the external strings to points, essentially "pinching off" the external legs. This is a valid description for reasons to be discussed in Lecture #2.

This is clearly a new language for doing physics. However, as we have seen, because string theory also includes gravity (which is exceedingly weak compared with the other forces), its fundamental mass scale is very high. Indeed, since the fundamental energy scale for gravity is the Planck mass

$$M_{\text{Planck}} \equiv \sqrt{\frac{\hbar c}{G_N}} \approx 10^{19} \text{ GeV} \approx (10^{-33} \text{ cm.})^{-1} , \qquad (6.1.4)$$

the string scale must also be very high. Indeed, to a first approximation, it turns out that

$$M_{\text{string}} \approx g_{\text{string}} M_{\text{Planck}} \qquad (6.1.5)$$

where g_{string} is the string coupling constant, typically assumed to be $\sim \mathcal{O}(1)$. Thus, we see that string theory is ultimately a theory of Planck-scale physics.

There are lots of "formal" reasons for being excited about string theory. First, it turns out that string theory requires the existence of extra space-time dimensions in order to be consistent, and consequently we now have to consider physics in different numbers of dimensions as well as all sorts of geometric questions pertaining to different possible "compactification" scenarios. Second, string theory gives us a new perspective on the structure of spacetime itself. For example, string theory gives rise to many novel Planck-scale effects. One of these is called T-duality: the physics of a closed string in a spacetime one of whose dimensions is compactified on a circle of radius R turns out to be *equivalent* to the physics of the same string in a spacetime in which the radius is M_{string}^2/R. Thus, T-duality interchanges large radii and small radii, and suggests that our naïve view of spacetime and its linear hierarchy of energy and length scales cannot ultimately be correct. Third, string theory also provides new types of strong/weak coupling dualities. These have proven useful for elucidating the strong-coupling

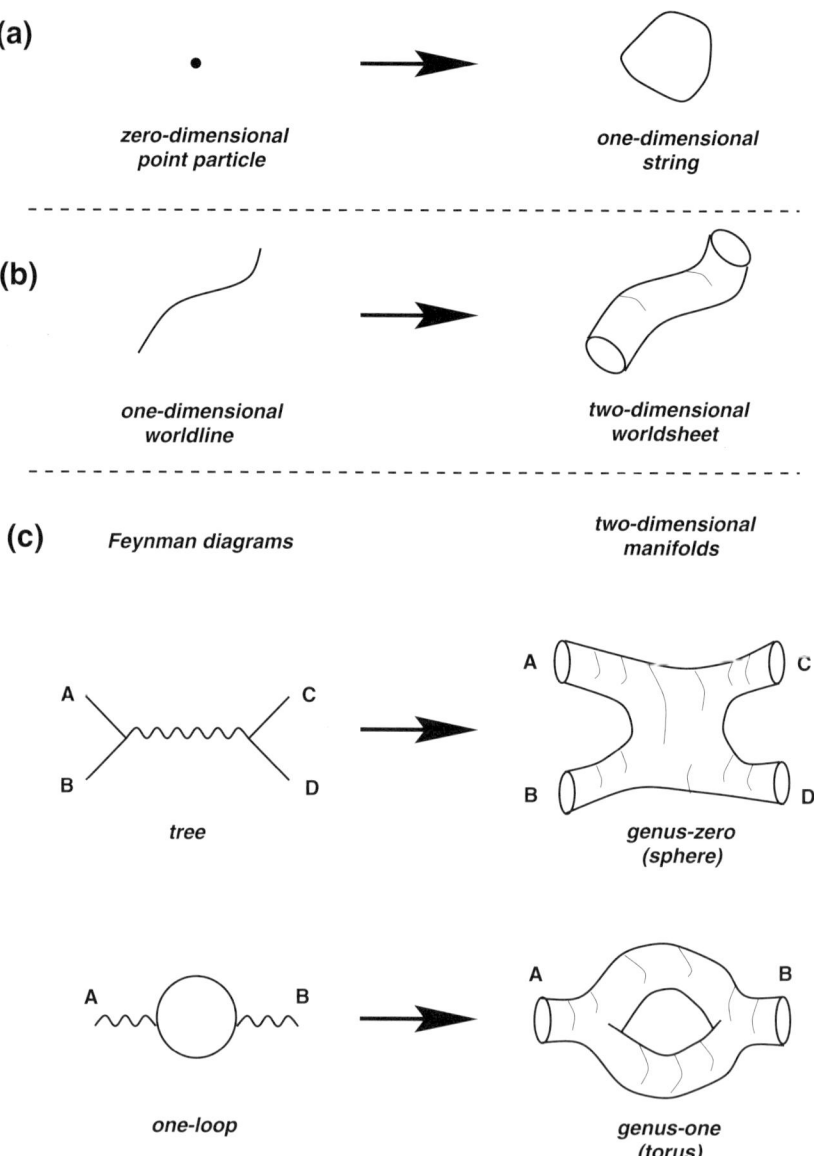

Fig. 6.3. In string theory, we replace (a) zero-dimensional elementary particles with one-dimensional strings; (b) one-dimensional worldlines with two-dimensional worldsheets; and (c) Feynman diagrams with two-dimensional manifolds. For example, tree diagrams correspond to genus-zero manifolds (spheres), and one-loop diagrams correspond to genus-one manifolds (tori).

dynamics of not only string theory, but also *field* theory. Finally, there have even been novel applications to black-hole physics. The most famous example of this is the fact that various non-perturbative string structures called D-branes have provided the first *statistical* (*i.e.*, microscopic) derivation of the Bekenstein-Hawking entropy formula $S = A/4$ that relates the entropy S of a black hole to its surface area A. Indeed, the above list only begins to scratch the surface of all of the many exciting recent formal developments in string theory.

But we are phenomenologists, so it is natural to ask about the rest of high-energy physics. How does string theory connect with the rest of particle physics?

Some of the answers to this question have already been given above. We have seen, in particular, that string theory is capable of reproducing the Standard Model as its low-energy limit. Moreover, as we have also seen, the Standard Model naturally emerges coupled with gravity. Furthermore, in many cases this entire structure is also joined with *supersymmetry*. Finally, this entire structure is also often joined with many properties of GUTs (such as gauge coupling unification). All of this comes out of the low-energy limit of string theory, in some sense automatically.

There are also many other benefits to considering the application of string theory to particle physics. First, string theory provides us with new kinds of symmetries (so-called "worldsheet symmetries") which lead to powerful new constraints on the resulting low-energy phenomenology. Second, in principle* string theory has *no free parameters*, which leads to a very predictive theory. Third, string theory has no divergences — in some sense, string theory is a completely *finite* theory in which many of the troublesome divergences associated with field theory are simply absent. Finally, it

---

*In this connection, we hasten to emphasize the phrase "in principle". Unfortunately, our relative inability to understand the non-perturbative structure of string theory often means that the pragmatic consequences of having no free parameters cannot be realized, and in practice one is often forced to introduce many parameters to reflect our ignorance of the underlying dynamics. This will be discussed in subsequent lectures. This situation is rather analogous to one that arises in the MSSM: we do not know how supersymmetry is broken, so we typically parametrize our ignorance through the introduction of various supersymmetry-breaking parameters. Likewise, in string theory, there are analogous questions which come under the heading of "vacuum selection": we do not know how the non-perturbative dynamics of string theory selects a particular vacuum state. Thus, in order to proceed to make phenomenological predictions, we are often forced to *assume* a certain vacuum state, or to parametrize the vacuum via the introduction of essentially unfixed parameters. The important point, however, is that string theory is a complete theory in that it should *in principle*, by virtue of its dynamics, uniquely fix the values of all of its fundamental parameters.

turns out that string theory can even give rise to a new perspective on the Standard Model itself, and often provides new and simpler ways to perform calculations.

These last three points (absence of free parameters, absence of divergences, and new ways to perform calculations) are truly remarkable. Therefore, let us pause to explain in an intuitive way why these features arise. First, let us explain why string theory has fewer free parameters. To do this, let us consider a Feynman diagram for a typical tree-level decay $A \to B+C$, as shown in Fig. 6.4(a). In field theory, such a process depends on many separate parameters ultimately associated with the separate propagators and vertices. Specifically, even though the propagators are determined once the masses and spins of the particles are specified, there still remains an *independent* choice as to the form of the vertex interaction. Thus, in a given field theory, there still remain many independent parameters to choose. In string theory, by contrast, there is no sharp distinction between propagators and vertices; they melt into each other, and are essentially the same. Thus, once the propagators are determined, the vertices are also intrinsically determined. This is one of the underlying reasons why string theory contains fewer free parameters than field theory.

Next, let us discuss why string theory is more finite than field theory. To do this, let us consider a typical one-loop Feynman diagram, as shown in Fig. 6.4(b). In field theory, the virtual interactions occur at sharp spacetime locations x and y. This is ultimately the origin of the ultraviolet (*i.e.*, short-distance) divergence as $x \to y$. In string theory, by contrast, we have seen that there are no such sharp interaction points — essentially the interaction is "smoothed out" by the presence of the string. Thus, there is no sense in which the dangerous $x \to y$ limit exists, for there are no precise means by which one can define such interaction locations x and y. It is in this manner that string theory automatically removes ultraviolet divergences: the string itself, through its extended geometry, acts as a (Planck-scale) ultraviolet regulator.

Finally, let us discuss why string theory can often give us simpler ways to perform calculations than in field theory. To do this, let us consider the total tree-level amplitude for a typical process $A + B \to C + D$, as illustrated in Fig. 6.4(c). As we know, in field theory there are two separate topologies of Feynman diagram that must be separately considered: the s-channel diagram and the t-channel diagram. In general, at any given order, there are many separate diagrams to evaluate, and one often finds that great simplifications and cancellations occur only when these individual

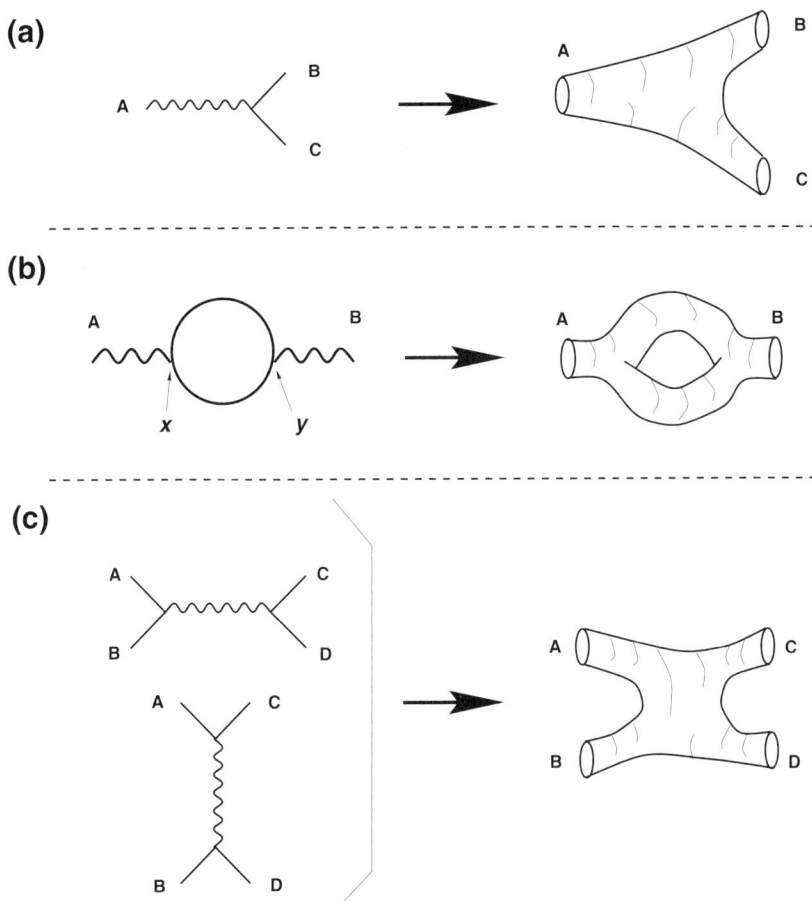

Fig 6.4. (a) Illustration of the fact that string propagators and string vertices are not independent. (b) Illustration of the fact that string theory lacks many of the ultraviolet divergences that arise in field theory from the short-distance limit $x \to y$. (c) Illustration of the fact that one string diagram often comprises many field-theoretic diagrams.

contributions are added together. In string theory, by contrast, there is only *one* corresponding diagram to evaluate at any given order. Thus, the sorts of simplifications or cancellations that might occur in field theory are automatically "built into" string theory from the very beginning. In some sense, string theory manages to find a way to reorganize the field-theory diagrams in a perturbative expansion in a useful and potentially profitable way. Indeed, this observation has even led to the development of many new techniques for evaluating complicated field-theoretic processes, particularly in QCD where the number of diagrams and the number of terms in each diagram can easily grow to otherwise unmanageable proportions.

We thus see that in a number of ways, string theory is a very useful language in which we might consider thinking about particle physics. Indeed, in various aspects (such as finiteness, fewer parameters, *etc.*) it is superior to field theory. But overall, the fundamental fact remains that if we are thinking about strings, we are abandoning our usual four-dimensional point of view of particle physics. Specifically, since each different particle in spacetime is now interpreted as a different quantum mode excitation of an underlying string, we see that four-dimensional (spacetime) physics is now ultimately the consequence of two-dimensional (worldsheet) physics. Thus, everything we ordinarily focus on in field theory (such as the four-dimensional particle spectrum, the gauge symmetries, the couplings, *etc.*) are now all ultimately determined or constrained by worldsheet symmetries.

And this brings us to string phenomenology.

6.1.4. *So what is string phenomenology?*

In order to understand what string phenomenology is, we can draw a useful analogy. Just as we are replacing the *language* of high-energy physics from field theory to string theory, we likewise replace field-theory phenomenology with string-theory phenomenology. The goals of string phenomenology are of course the same as those of ordinary field-theory phenomenology: both seek to reproduce, explain, and predict observable phenomena, and both seek to suggest or constrain new physics at even higher energy scales. Indeed, only the language in which we will carry out this procedure has changed. Thus, in some sense, string phenomenology is the "art" of using the new insights from string theory in order to understand, explain, and predict what physics at the next energy scale is going to look like. Or, recalling that string theory is ultimately a theory of Planck-scale physics, we can say that string phenomenology is the "interplay" or "meeting-ground" between Planck-scale physics and GeV-scale physics.

It is important to understand that we are not abandoning field theory completely. Nor would we want to. Field theory automatically incorporates many desirable features such as causality, spin-statistics relations, and CPT invariance (which in turn implies the existence of antiparticles). These are all generic predictions of field theory, and are the underlying reasons why field theory is the appropriate language for particle physics. However, since string theory ultimately reduces to field theory in its low-energy limit, all of these features will still be retained in string theory. Moreover, as we have seen, string theory *additionally* predicts or explains gravity, supersymmetry, and the absence of ultraviolet divergences. Furthermore, as we shall see, string theory also automatically predicts the existence of gauge symmetry, and even incorporates features such as gauge coupling unification. These are all generic predictions of string theory. It is for these reasons to believe that a change in language from field theory to string theory might be useful.

String theory will also provide us with new tools for model-building, new mechanisms and new guiding principles. Let us give some examples. In field theory, there are many well-known ideas that are part and parcel of the model-building game: one must enforce ABJ anomaly cancellation (to preserve gauge symmetries); one can employ the Higgs mechanism (to generate spontaneous symmetry breaking and give masses to particles); one has the GIM mechanism (to preserve flavor symmetries); and one has supersymmetry (to cancel quadratic divergences). Likewise, in string theory there are analogous sets of ideas, many of which are extensions of their field-theory counterparts. For example, one has the so-called "Green-Schwarz" mechanism for anomaly cancellation (to preserve gauge symmetries); one has string vacuum shifting via pseudo-anomalous $U(1)$ gauge symmetries (to generate spontaneous symmetry breaking and generate particle masses); one has spacetime compactification (to generate gauge symmetries); one has hidden string sectors (to break supersymmetry and impose selection rules); and one has massive towers of string states (to enforce finiteness). Thus, model-building proceeds, but with a different set of principles.

There is also a much more subtle effect of changing our language from field theory to string theory. Ultimately, since four dimensional physics is now derived from an underlying two-dimensional (worldsheet) theory, string phenomenology is ultimately much more constrained than field-theory phenomenology. One given worldsheet symmetry, which might serve as an "input", can have various seemingly unrelated effects in the resulting spacetime phenomenological "output". Thus, string theory not only leads to unexpected connections or correlations between seemingly disparate spacetime phenomena, but can also give rise to entirely new phenomenological scenarios that could not have been anticipated within field theory alone. We will see many examples of this in the coming lectures.

Thus, we see that string phenomenology does many things and has many goals:

- to provide a new framework for addressing and answering numerous phenomenological questions;
- to provide a rigorous test of string theory as a theory of physics;
- to explore the interplay between worldsheet physics and spacetime physics (*i.e.*, to ultimately determine which "patterns" of low-energy phenomenology are allowed or consistent with being realized as the low-energy limit of an underlying string theory); and
- to augment field theories of "low-energy" physics into the string framework so as to give them the full benefits of the language of string theory.

Because of these different roles, string phenomenology occupies a rather central position in high-energy physics: it allows the transmission of ideas from high-scale string theory to guide "low"-scale particle physics, and vice versa. This situation is illustrated in Fig. 6.5. At the lowest energies (lower left), string phenomenology has direct relevance for the Standard Model, where it can potentially explain features such as the choice of the gauge group, the number of generations, and numerous other parameters such as the masses and mixings of Standard-Model particles. At slightly higher energies (lower right), we see that string phenomenology can also suggest or constrain various extensions to the Standard Model, such as SUSY and SUSY-breaking, grand unification, and hidden-sector physics. At the highest energies (upper left), string phenomenology is also concerned with the more formal aspects of string theory: such important questions include string vacuum selection, non-perturbative string dynamics, string duality, and new mathematical structures and techniques. And string phenomenology even has relevance outside the strict confines of particle physics. For example, string theory should have a profound impact on cosmology (upper right), where important stringy issues include the role of the dilaton, the effects of many other light degrees of freedom (the so-called *moduli*), the possibility of extra spacetime dimensions, the cosmological constant problem, and even more exotic ideas such as topology change. As illustrated in Fig. 6.5, string phenomenology sits at the center of this web of ideas. Exploring the connections between the different corners of this figure is, therefore, the job of the string phenomenologist. Indeed, through string phenomenology, one "uses" string theory in order to open a window into the possibilities for physics beyond the Standard Model.

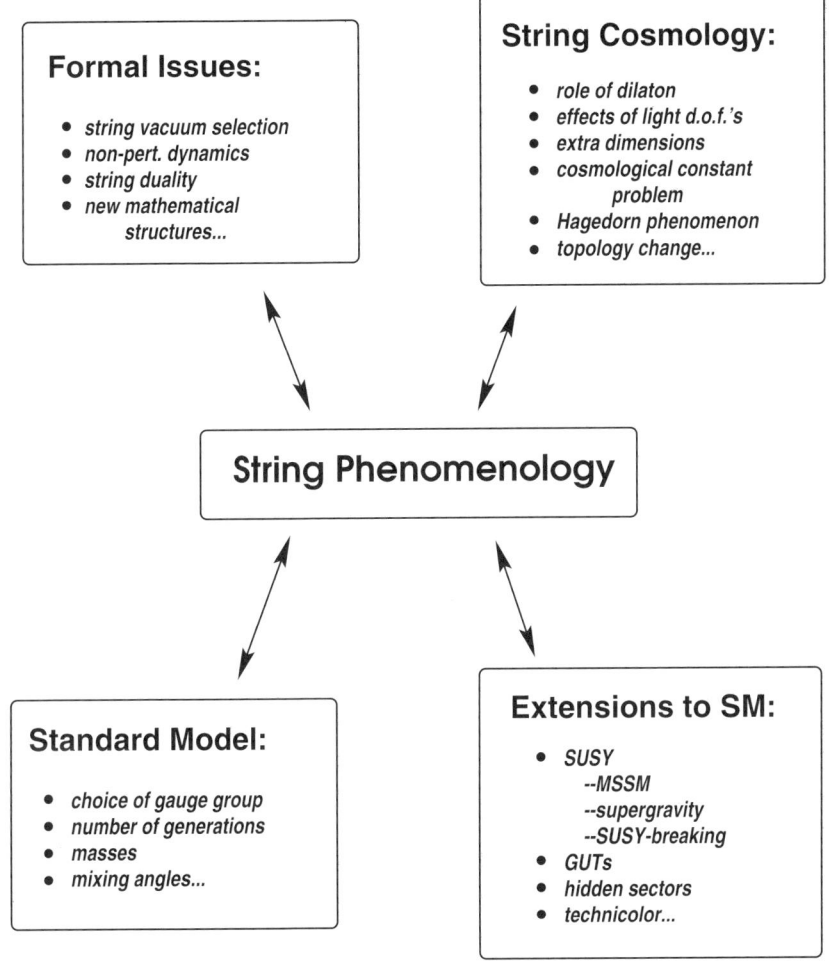

Fig. 6.5. String phenomenology is the central "meeting-ground" between Standard-Model physics, extensions to the Standard Model, formal string issues, and string cosmology.

6.1.5. *Plan of these lectures*

For much of the past decade, string phenomenology has been practiced assuming a particular type of underlying string theory, the so-called *perturbative heterotic string*. Therefore, this string will be the focal point of most of these lectures. However, it turns out that the heterotic string is built directly on the foundations of two other kinds of strings, the *bosonic string* and *Type II superstring*. Indeed, in a sense to be made more precise in Lecture #5, one can view the heterotic string as the "sum" of the bosonic string and the superstring string. Therefore, in these lectures, we will have to start at the beginning by studying first the bosonic string, then the Type II string, and finally the heterotic string. Indeed, this situation is analogous to the way in which one often studies quantum field theory: first one learns how to quantize the Klein-Gordon field, then the Dirac field, and finally the gauge field. In a certain sense, the bosonic string is the analogue of the Klein-Gordon field, while the Type II superstring is the analogue of the Dirac field and the heterotic string is the analogue of the gauge field. Of course, this analogy is only a pedagogical organizational one, since the heterotic string itself will ultimately contain *all* of the phenomenological properties (*e.g.*, scalars, fermions, and gauge symmetries) that we desire.

In Lecture #2, we will therefore give a brief introduction to the bosonic string, stopping only long enough to develop the ideas and techniques we will need for later applications. In Lectures #3 and #4, we will then proceed to develop the Type II superstring, once again focusing on only those aspects that will be useful for later applications. Finally, in Lecture #5, we will arrive at our destination: the heterotic string. In Lecture #6 we will construct some ten-dimensional heterotic string models, and in Lecture #7 we will develop a useful set of rules for heterotic string model-building.

It is important to note, however, that all of string phenomenology is not based on the heterotic string. Particularly over the past decade, there has been a profound shift in our understanding of both string theory and its phenomenological implications. One of the consequences of this so-called "second superstring revolution" has been a new emphasis on yet another class of strings, the *Type I (open) strings*. Within this class, so-called *intersecting D-brane models* have shown great promise in yielding chiral, Standard-Model-like spectra. Indeed, there has even emerged a new superstructure which promises to relate all of these strings to each other: this structure is called *M-theory*, and is deeply tied to many non-perturbative aspects of string theory which are still being understood. Needless to say, these recent developments have the potential to completely change the way we think about string theory and string phenomenology. We will therefore discuss some of these modern developments in the final Lecture #8. Nevertheless, the bulk of these lectures will primarily be focused on the more traditional aspects of string phenomenology that concern the weakly coupled heterotic string. Indeed, this affords the best introduction to string theory and string phenomenology, regardless of the future directions that string theory and string phenomenology might ultimately take.

We also remind the reader that our goal here is to provide an introduction to string theory that avoids mathematical complications wherever possible, and which "gets to the physics" as rapidly as possible. Therefore, in many places, we will simply assert a mathematical result to be true, leaving its derivation to be found in various textbooks on the subject. For this purpose, we recommend Volume I of the textbook *Superstring Theory*, by M.B. Green, J.H. Schwarz, and E. Witten (henceforth to be referred to as GSW†). In fact, our initial approach will be very similar to that of GSW, and we will continually refer back to this textbook as we proceed. Another recommended textbook with a more modern mathematical perspective is *Introduction to String Theory*, by J. Polchinski. Likewise, *A First Course in String Theory* by B. Zwiebach is particularly useful for students who may lack a full background in relativistic quantum field theory.

6.2. Lecture #2: Strings and their spectra: The bosonic string

6.2.1. *The action*

We begin by studying the simplest string of all: the bosonic string. As we discussed in Lecture #1, the physics of a string is ultimately described by the shape it takes (*e.g.*, its vibrational mode of oscillation) as it propagates through an external spacetime and thereby sweeps out a two-dimensional worldsheet. Therefore, we must first have a way of describing the shape of this worldsheet. To this end, we parametrize the worldsheet by two worldsheet coordinates (σ_1, σ_2) as illustrated in Fig. 6.6, and describe the embedding of this worldsheet into the external spacetime by giving the spacetime coordinates X^μ of any location (σ_1, σ_2) on the worldsheet. Thus, the physics of the string is ultimately encapsulated in the embedding functions $X^\mu(\sigma_1, \sigma_2)$, where $\mu = 0, 1, ..., D-1$. Here D is the total spacetime dimension, which we shall keep arbitrary for now.

Given these embedding functions, we can attempt to write down an appropriate action for the string. To do this, we first note that as we might

---

†Not to be confused with another great GSW trio, namely Glashow, Salam, and Weinberg. One can only hope that someday string theory will be as well-established, both theoretically and experimentally, as the GSW electroweak theory. This may sound a bit optimistic, but a possible new *experimental* direction for string theory and string phenomenology will be discussed in Lecture #8 in the context of the brane world.

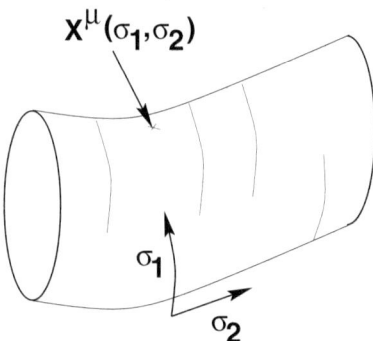

Fig. 6.6. The string worldsheet can be parametrized by two worldsheet coordinates (σ_1, σ_2). Thus, the location in the external spacetime of any point on the string worldsheet is described by a set of functions $X^\mu(\sigma_1, \sigma_2)$. It is convenient to think of σ_1 as a spacelike worldsheet coordinate, and σ_2 as a timelike worldsheet coordinate.

expect, strings have *tension* — i.e., strings generically have a non-zero energy per unit length. In other words, it takes energy to stretch a string and to give the worldsheet a larger area. Thus, as the string propagates along in spacetime, we expect on physical grounds that this string should choose a configuration that minimizes the area of the worldsheet. This leads us to identify the string action with the area of the corresponding worldsheet. Indeed, this results in the so-called *Nambu-Goto action*, which involves a non-trivial square root of the X^μ coordinates. For certain calculational purposes, however, this square root is often problematic. Fortunately, however, there exists an alternative action, the so-called *Polyakov action*, which is classically equivalent to the Nambu-Goto action but which does not involve fractional powers of the X coordinates. This action is given by

$$S = -\frac{1}{4\pi\alpha'} \int d^2\sigma \sqrt{h}\, h^{\alpha\beta}\, g_{\mu\nu}\, \partial_\alpha X^\mu \partial_\beta X^\nu \ . \tag{6.2.1}$$

Here $g_{\mu\nu}$ is the metric of the external spacetime, $h_{\alpha\beta}$ is the metric of the worldsheet, the worldsheet derivative is given by $\partial_\alpha \equiv \partial/\partial\sigma^\alpha$, and $h \equiv \det h_{\alpha\beta}$. In the prefactor, α' is a dimensionful constant (called the *Regge slope*) with units of (length)^2. Since these units are equivalent to length/energy, we see that α' is an inverse tension, and indeed the string tension T turns out to be related to α' via $T = (2\pi\alpha')^{-1}$. We shall discuss the numerical value of α' below. Note that the action (6.2.1) is manifestly spacetime Lorentz-invariant.

Before proceeding further, it may be useful to draw an analogy between this action and the analogous action for a *point* particle propagating through spacetime and sweeping out a *worldline* rather than a worldsheet. The worldline can be parametrized by a single coordinate σ, which functions as a proper time along the worldline. The point-particle action can then be written in the form

$$S_{\text{point particle}} = \tfrac{1}{2} \int d\sigma \, \left(e^{-1} g_{\mu\nu} \partial_\sigma X^\mu \partial_\sigma X^\nu - e \hat{m}^2 \right) \qquad (6.2.2)$$

where $\hat{m}$ is the mass of the point particle and where $e(\sigma)$ is an auxiliary field (a so-called *einbein*). Solving for $e(\sigma)$ through its equation of motion and substituting back into (6.2.2) yields an action proportional to the length of the worldline and involving a square root. Thus, we see that the string action (6.2.1) is nothing but the generalization of the point-particle action (6.2.2), where we have associated

$$e^{-1}(\sigma) \iff h^{\alpha\beta}(\sigma_1, \sigma_2) \,, \qquad \hat{m} = 0 \,. \qquad (6.2.3)$$

In other words, the string action (6.2.1) is the two-dimensional generalization of the action of a *massless* point particle, where the worldsheet metric functions as an auxiliary field (a "zweibein"). This masslessness property will be crucial shortly.

It is now possible to make some simplifications. Perhaps the most obvious is to restrict our attention to a flat spacetime and take $g_{\mu\nu} = \eta_{\mu\nu}$. We shall do this throughout these lectures. A much more subtle simplification, however, is to simplify the worldsheet metric. Let us therefore pause to discuss how this can be done.

One of the first things we realize is that the ultimate physics of the string should not depend on the particular choice of coordinate system (σ_1, σ_2) on the string worldsheet. After all, on purely physical grounds, we know that the particular choice of worldsheet coordinate system cannot have a physical effect, for the same worldsheet geometry can ultimately be described using an infinite variety of coordinate systems which differ from each other through relative reparametrizations or rescalings. (Indeed, in the point-particle case, we are likewise free to reparametrize our proper-time variable along the particle worldline.) Therefore, the string action should have a symmetry that makes it invariant under reparametrizations and rescalings of the worldsheet coordinates. Note, in particular, that the invariance under *rescalings* follows from the fact that we chose our string action (6.2.1) to generalize that of a *massless* point particle. In other words, we have taken $\hat{m} = 0$ in (6.2.3). While it is possible to add terms to the action of the bosonic string which mimic the effects of possible mass terms and which explicitly break the scale invariance of the bosonic string, we shall not need to consider such theories in these lectures.

The symmetry that comprises both reparametrizations and rescalings of the worldsheet coordinates is called *conformal symmetry*, and the bosonic string action (6.2.1) is thus said to be "conformally invariant". Clearly, this symmetry must hold not only at the classical level, but also at the quantum level, for we would not have a consistent theory if this symmetry were broken by quantum anomalies. Conformal invariance of the action is a very powerful physical tool which will play an important role throughout these lectures, and indeed the mathematical structure underlying conformal symmetry and its implications is a deep and beautiful subject which we will not have time or space to discuss here. A recommended starting point is *Applied Conformal Field Theory* (Proceedings of Les Houches, Session XLIX, 1988), by P. Ginsparg. Therefore, in order to proceed, we will have to make the first of many "great leaps", and take certain results on faith. Our first great leap will therefore be the following:

Great Leap #1: Conformal invariance of the string action allows us to replace the string metric $h_{\alpha\beta}$ with the two-dimensional Minkowski metric $\eta_{\alpha\beta}$ without loss of generality.

This then results in the simplified bosonic string action

$$S = -\frac{1}{4\pi\alpha'} \int d^2\sigma \, \partial_\alpha X^\mu \partial^\alpha X_\mu \, . \qquad (6.2.4)$$

Looking at the action (6.2.4), we see that it has two possible interpretations. The first interpretation is the one that we have already been following: minimizing this action is classically equivalent to minimizing the worldsheet area. This follows directly from the interpretation of $X^\mu(\sigma_1, \sigma_2)$ as the *spacetime coordinates* of a given worldsheet position (σ_1, σ_2). Note that this action is invariant under $SO(D-1,1)$ Lorentz transformations of the spacetime coordinates, with the index μ interpreted as a spacetime vector index relative to the Lorentz group. We shall refer to this as the *spacetime interpretation*.

There is, however, a completely different interpretation of (6.2.4): this is the action of a *two-dimensional quantum field theory* where the two dimensions refer to the worldsheet coordinates and where the "fields" are nothing but the functions $X^\mu(\sigma_1, \sigma_2)$, $\mu = 0, 1, ..., D-1$. Indeed, we see that these spacetime coordinate functions are simply a collection of D different massless bosonic Klein-Gordon fields which happen to exhibit an internal $SO(D-1,1)$ rotation symmetry (analogous to a gauge symmetry) between them. In such a case, the index μ is simply an internal symmetry index which tells us that the X^μ fields transform as vectors with respect to the internal $SO(D-1,1)$ symmetry. We shall refer to this as the *worldsheet interpretation*. Indeed, it is because this string action contains only bosonic worldsheet fields that we call this the *bosonic string*. In such a description, spacetime is not a fundamental concept but rather a "derived" concept: it results from the interpretation of various worldsheet fields as spacetime coordinates, and from the interpretation of an internal symmetry as a spacetime Lorentz symmetry. It is indeed remarkable that such different interpretations can be made of the same physics, and we shall often go back and forth between these different worldsheet and spacetime points of view.

Given these two descriptions of the action, we can also understand the origin of the Regge slope parameter α' on dimensional grounds. Let us first take the worldsheet point of view, so that our length dimensions are determined with respect to the coordinates (σ_1, σ_2). In such a case, we know that the ordinary Klein-Gordon action does not require any dimensionful prefactor, for $\int d^2\sigma (\partial_\alpha X^\mu)^2$ is indeed dimensionless when the Klein-Gordon field X^μ is itself dimensionless. However, from the spacetime point of view, we see that X^μ cannot be dimensionless, for we ultimately need to interpret this field as a spacetime coordinate with units of length. Thus, we are forced to compensate by inserting a dimensionful prefactor α' in front of the action. In other words, *the need for the dimensionful prefactor α' arises from the need to interpret our dimensionless (scale-free) worldsheet theory as a dimensionful (spacetime) theory.* Or, to put it slightly differently, the parameter α' is the dimensionful conversion factor that describes the overall scale of the embedding of the dimensionless worldsheet physics into the dimensionful spacetime. We shall see this phenomenon very often throughout these lectures: the worldsheet physics is by itself scale-invariant (since it generalizes the physics of a massless point particle with $\hat{m} = 0$), and it is only in the conversion to dimensionful *spacetime* quantities that the overall scale α' plays a role. Thus, α' sets the overall spacetime mass scale of string theory, often called the *string scale*:

$$M_{\text{string}} \equiv \frac{1}{\sqrt{\alpha'}} \, . \qquad (6.2.5)$$

A priori, this mass scale is unfixed, but we shall see shortly how this scale is ultimately determined.

Now that we have established the worldsheet picture and the spacetime picture, it is easy to see how they are related to each other: each quantum excitation of the Klein-Gordon worldsheet fields X^μ corresponds to a different particle in spacetime. Thus, the study of string theory can be reduced to the study of a two-dimensional quantum field theory! For example, particle scattering amplitudes in spacetime can be re-interpreted as the correlation functions of our two-dimensional worldsheet fields, evaluated on various two-dimensional manifolds. Of course, as we have stated above, this is not just *any* two-dimensional quantum field theory, for physical consistency also requires the presence of conformal symmetry. Thus, from this point of view, string theory is the study of two-dimensional *conformal* field theories. In two dimensions, it turns out conformal symmetry is extremely powerful, for it gives rise to an infinite number of conserved currents. Indeed, two-dimensional conformal symmetry is often sufficiently powerful to permit the *exact* evaluation for many scattering amplitudes.

In the case in question, the particular conformal field theory that concerns us is that of D free massless bosonic fields X^μ, $\mu = 0, 1, ..., D-1$. However, just as with any symmetry, there is always the danger of quantum anomalies. Nevertheless, it is straightforward to show that

> **Great Leap #2:** Conformal invariance of the string action is preserved at the quantum level (*i.e.*, all quantum anomalies are cancelled) if and only if $D = 26$.

This is clearly a big result, and we will not have space to provide a proper mathematical derivation of this fact. At the very least, however, we can give a guide as to the most useful way of thinking about this result. Note that our D bosonic fields are identical to each other and essentially decoupled from each other. Therefore, each contributes the same amount to any potential anomaly. This amount is called the *central charge*, and the central charge c of each bosonic field X will be denoted c_X. It turns out that $c_X = 1$, and therefore the total central charge from the D bosonic fields is $c_{\text{fields}} = D$. However, it can be shown that there also exists a "background" central charge (*i.e.*, a background quantum anomaly) of magnitude $c_{\text{background}} = -26$. Thus, the total anomaly is cancelled only if $D = 26$. Clearly, the most mysterious part of this discussion is the origin of this "background" central charge. In technical terms, it reflects the contributions of the conformal ghosts that arose when we used the conformal symmetry to set (or "gauge-fix") the worldsheet metric $h_{\alpha\beta} \to \eta_{\alpha\beta}$. However, all we will need to know for the future is that the value of the "background" anomaly $c_{\text{background}}$ depends on only the particular symmetry of the worldsheet action that we are dealing with. In the present case, this worldsheet symmetry is simply conformal invariance, and the corresponding background central charge corresponding to conformal invariance is $c_{\text{background}} = -26$. Therefore, we see that the total conformal anomaly is cancelled only if $D = 26$. This is typically called the *critical dimension* of the bosonic string.

We see, then, that string theory is able to determine the spacetime dimension as the result of an *anomaly cancellation argument*! It is worth reflecting on how this happened by considering an analogous situation in field theory, namely the cancellation of the triangle axial anomaly. We know that this anomaly is cancelled only for very particular combinations of particle representations (*e.g.*, we require complete generations of Standard-Model fields, with three colors of quark for every lepton). So we are used to the idea that anomalies are extremely sensitive to the field content of the theory. In string theory, however, we have seen that the analogous worldsheet field content is parametrized by the spacetime dimension. More worldsheeet fields correspond to more spacetime dimensions. Therefore, just as triangle anomaly cancellation requires three colors, conformal anomaly cancellation requires 26 dimensions.

Of course, our world does not consist of 26 flat spacetime dimensions, and we shall ultimately need to find a way of reducing this to a four-dimensional theory. For now, however, we can just think of the present bosonic string as a 26-dimensional toy model.

6.2.2. *Quantizing the bosonic string*

Let us now quantize this theory. Having already noted that the action (6.2.4) is nothing but the action of a set of 26 Klein-Gordon fields X^μ, we already know how to proceed: in the usual fashion, we introduce a Fourier-expansion of the fields X^μ, and interpret the coefficients of this expansion as creation and annihilation operators obeying canonical quantization relations.

Because we ultimately wish to interpret the fields X^μ as spacetime coordinates, we must first impose the constraint

$$X^\mu(\sigma_1 + \pi, \sigma_2) = X^\mu(\sigma_1, \sigma_2) \qquad (6.2.6)$$

where we have chosen to normalize the length of the closed string as π. In other words, the spacetime coordinates must be single-valued as we make one complete circuit around the closed string. This is the first place where we have essentially incorporated the requirement that we are dealing with closed strings whose topology is that of a circle. Moreover, because of this topology (and because of the linear nature of the wave equation resulting from the action (6.2.4)), we know that we can also decompose any possible quantum excitation of the wiggling string into a superposition of modes that travel clockwise around the string (in the direction of, say, decreasing σ_1) and those that travel counter-clockwise (in the direction of increasing σ_1). These are respectively called *left-movers* and *right-movers*. We can therefore decompose each of our Klein-Gordon fields into the form

$$X^\mu(\sigma_1,\sigma_2) = X^\mu_L(\sigma_1+\sigma_2) + X^\mu_R(\sigma_1-\sigma_2) \ . \qquad (6.2.7)$$

The most general mode-expansion consistent with the boundary condition (6.2.6) is then

$$X^\mu(\sigma_1,\sigma_2) = x^\mu + \ell^2 p^\mu \sigma_2 + \frac{i}{2}\ell \sum_{n\neq 0}\left[\frac{\alpha^\mu_n}{n}e^{-2in(\sigma_1+\sigma_2)} + \frac{\tilde{\alpha}^\mu_n}{n}e^{+2in(\sigma_1-\sigma_2)}\right], \qquad (6.2.8)$$

which decomposes into

$$X^\mu_L(\sigma_1+\sigma_2) = \tfrac{1}{2}x^\mu + \frac{\ell^2}{2}p^\mu(\sigma_1+\sigma_2) + \frac{i}{2}\ell\sum_{n\neq 0}\frac{\alpha^\mu_n}{n}e^{-2in(\sigma_1+\sigma_2)}$$

$$X^\mu_R(\sigma_1-\sigma_2) = \tfrac{1}{2}x^\mu - \frac{\ell^2}{2}p^\mu(\sigma_1-\sigma_2) + \frac{i}{2}\ell\sum_{n\neq 0}\frac{\tilde{\alpha}^\mu_n}{n}e^{+2in(\sigma_1-\sigma_2)} \qquad (6.2.9)$$

Here $\ell \equiv \sqrt{2\alpha'}$ is a fundamental length that has been inserted on dimensional grounds.

It is easy to interpret the different terms in (6.2.8) and (6.2.9). Clearly the final terms in each line represent the internal quantum vibrational oscillations of the string, where α^μ_n and $\tilde{\alpha}^\mu_n$ are the left-moving and right-moving creation/annihilation operators corresponding to vibrational modes of a given frequency n. We shall discuss these operators shortly. Note that the contribution from the "zero-mode" has been separated out and written explicitly in the form $x^\mu \pm \tfrac{1}{2}\ell^2(\sigma_1 \pm \sigma_2)$ for the left- and right-movers respectively. In the case when there are no quantum excitations (so that we can ignore the final exponential terms), these "zero-modes" are all that remain of the mode-expansion, whereupon we see from (6.2.8) that the total X^μ field takes the form $X^\mu = x^\mu + \ell^2 p^\mu \sigma_2$. Interpreting σ_2 as the timelike coordinate on the string worldsheet, we thus see that x^μ is nothing but the center-of-mass position of the string, and p^μ its center-of-mass momentum.

Let us now consider the quantization rules that we must impose. The first one (for the zero-modes) is easy: we simply impose the usual commutation relation $[x^\mu, p^\nu] = i\hbar \eta^{\mu\nu}$. We shall henceforth set $\hbar = 1$. The excited modes also have a similar commutation relation. First, note that because the X fields are interpreted as spacetime coordinates, they are necessarily *real*. This implies that we must identify $\alpha^\mu_{-n} = (\alpha^\mu_n)^\dagger$, with a similar result for the right-moving oscillator modes. In other words, the negative modes *create* excitations, while the positive modes *annihilate* the same excitations. Given this, we then can immediately write down the commutation relation for the creation/annihilation operators:

$$[\alpha^\mu_m, \alpha^\nu_n] = m\,\delta_{m+n}\,\eta^{\mu\nu}, \qquad [\tilde{\alpha}^\mu_m, \tilde{\alpha}^\nu_n] = m\,\delta_{m+n}\,\eta^{\mu\nu}. \qquad (6.2.10)$$

Here we have introduced the notation $\delta_x \equiv \delta_{x,0} \equiv 1$ if $x = 0$, and $\equiv 0$ if $x \neq 0$. Note that these are exactly the harmonic oscillator commutation relations, except that we have rescaled each mode α_n by its corresponding frequency n in (6.2.9). Thus, $a_n \equiv \alpha_n/\sqrt{n}$ obey the usual harmonic oscillator commutation relations. This rescaling has become conventional in string theory, and we shall retain it here. Likewise, it is often conventional to define the zero-mode $\alpha^\mu_0 \equiv \frac{1}{2}\sqrt{\alpha'}p^\mu$.

Given this mode-expansion, we can now construct the corresponding number operators

$$n > 0: \qquad N_n = \frac{1}{n}\alpha^\mu_{-n}\alpha_{n\mu}, \qquad \tilde{N}_n = \frac{1}{n}\tilde{\alpha}^\mu_{-n}\tilde{\alpha}_{n\mu} \qquad (6.2.11)$$

which count the number of excitations of the nth frequency modes of the string. Once again, this is completely analogous to the harmonic-oscillator creation/annihilation modes, after we take into account the rescaling $\alpha_n \equiv \sqrt{n}a_n$ and the hermiticity condition $\alpha_{-n} = \alpha^\dagger_n$.

Likewise, we can also write down the total *energy* of the system. To do this, let us consider the different contributions to the total energy. First, there is the energy associated with the internal quantum vibrational oscillations of the string. As we might expect, this is given by

$$L_0^{(\text{osc})} \equiv \sum_{n=1}^{\infty} n N_n = \sum_{n=1}^{\infty} \alpha^\mu_{-n}\alpha_{n\mu}$$

$$\bar{L}_0^{(\text{osc})} \equiv \sum_{n=1}^{\infty} n \tilde{N}_n = \sum_{n=1}^{\infty} \tilde{\alpha}^\mu_{-n}\tilde{\alpha}_{n\mu}. \qquad (6.2.12)$$

For convenience, we are defining these energy operators in such a way that they are dimensionless numbers (*i.e.*, they are *worldsheet* energies). These L_0 operators are often called *Virasoro generators*, which are more generally defined $L_m \equiv \sum_n \alpha^\mu_{m-n} \alpha_{n\mu}$. These generators are nothing but the different frequency modes of the total worldsheet stress-energy tensor, and together they satisfy the so-called *Virasoro algebra*. We shall only consider L_0 in these lectures.

Next, there is the energy of the zero-modes, which correspond to the net center-of-mass motion of the string. This is given by

$$L_0^{(\text{com})} \equiv \alpha_0^\mu \alpha_{0\mu} = \frac{\alpha'}{4} p^\mu p_\mu$$
$$\tilde{L}_0^{(\text{com})} \equiv \tilde{\alpha}_0^\mu \tilde{\alpha}_{0\mu} = \frac{\alpha'}{4} p^\mu p_\mu \ . \qquad (6.2.13)$$

Note that factors of α' must appear in order to counter-balance the fact that the center-of-mass momentum p^μ is a spacetime quantity, and hence dimensionful.

Finally, there is the possibility of an overall non-zero vacuum energy for both the left-movers and the right-movers. In other words, there is no reason to assume that the vacuum state (the state without any excitations) is exactly at zero energy. This is important, of course, since string theory is ultimately a theory which will contain gravity, and it is precisely in theories containing gravity that the overall zero of energy becomes important. Indeed, mathematically, one can imagine that due to the commutation relations (6.2.10), there can be an overall normal-ordering ambiguity in the definitions in (6.2.12), and this overall normal-ordering constant would be our "vacuum energy".

Thus, denoting the left- and right-moving vacuum energies as $a_{L,R}$, we have the total left- and right-moving energies

$$H \equiv L_0^{(\text{com})} + L_0^{(\text{osc})} + a_L \ , \qquad \tilde{H} \equiv \tilde{L}_0^{(\text{com})} + \tilde{L}_0^{(\text{osc})} + a_R \ . \qquad (6.2.14)$$

These are the total worldsheet Hamiltonians.

Clearly, the important thing to do at this stage is to determine the vacuum energies $a_{L,R}$. Of course, the symmetry between left-movers and right-movers requires $a_L = a_R$. Calculating this vacuum energy can be done in numerous ways, each of which would take too much space for our purposes. Once again, we refer the reader to Chapter 2 of GSW, where a full calculation is given. Therefore, it is time for another

Great Leap #3: Conformal invariance of the string action implies that $a_L = a_R = -1$.

Finally, in order to determine the total *spacetime mass* of a given string state, we must have a *mass-shell condition* for the string. Rather than provide a rigorous derivation (for which we again refer the curious reader to GSW), we can instead give an intuitive argument which suggests the proper answer. In a quantum field theory of point particles, the mass $\hat{m}$ is a parameter that appears in the Lagrangian through an explicit mass term that might be generated in some separate manner, *e.g.*, through the Higgs mechanism. Since a point particle has no internal degrees of freedom beyond those associated with its center-of-mass motion, such a mass parameter $\hat{m}$ would then be directly identified with M, the resulting physical mass of the particle. Such a physical mass M is the quantity satisfying the condition $p^\mu p_\mu = -M^2$, or equivalently the condition $L_0^{(\text{com})} = \bar{L}_0^{(\text{com})} = -\alpha' M^2/4$. In the special case of a massless particle (for which $\hat{m} = M = 0$), this mass-shell condition then takes the simple form $L_0^{(\text{com})} = \bar{L}_0^{(\text{com})} = 0$.

A similar condition emerges in string theory. We have already seen that our string action (6.2.1) generalizes that of a massless particle, which again suggests that our effective Lagrangian mass parameter $\hat{m}$ vanishes. Indeed, as we have discussed, this is the root of the scale invariance of the string action (6.2.1). However, unlike the point-particle case, a string *does* have additional, purely internal degrees of freedom — these are the oscillations of the string itself, whose additional energy contributions are represented by $L_0^{(\text{osc})}$, $\bar{L}_0^{(\text{osc})}$, and $a_{L,R}$. Thus, even though $\hat{m} = 0$, the resulting string state can still have a non-zero physical mass M in spacetime. Indeed, just as the mass-shell condition for massless point particles is given by $L_0^{(\text{com})} = \bar{L}_0^{(\text{com})} = 0$, the mass-shell condition for our scale-invariant string is generalized to $H = \bar{H} = 0$. This then becomes our scale-invariant mass-shell condition in string theory. Of course, spacetime Lorentz invariance still allows us to identify the physical spacetime mass M of a given string state via the relations $L_0^{(\text{com})} = \bar{L}_0^{(\text{com})} = -\alpha' M^2/4$. Thus, the string mass-shell conditions $H = \bar{H} = 0$ lead to the identifications

$$\frac{1}{4}\alpha' M^2 = L_0^{(\text{osc})} - 1 \,, \qquad \frac{1}{4}\alpha' M^2 = \bar{L}_0^{(\text{osc})} - 1 \,. \qquad (6.2.15)$$

Note that these two conditions can also be written in the form

$$\alpha' M^2 = 2\left(L_0^{(\text{osc})} + \bar{L}_0^{(\text{osc})} - 2\right) \qquad (6.2.16)$$

where we must obey the constraint

$$L_0^{(\text{osc})} = \bar{L}_0^{(\text{osc})} \,. \tag{6.2.17}$$

Interpreting the conditions (6.2.16) and (6.2.17) is easy. The condition (6.2.16) simply tells us that the physical spacetime mass M of a given string state (and thus the square of its center-of-mass momentum) is generated *solely* from its internal left- and right-moving vibrational excitations. The condition (6.2.17), by contrast, tells us that the mass of the string must come *equally* from left-moving and right-moving excitations. The latter condition (6.2.17) is often referred to as the *level-matching condition*, since it implies that a given string oscillator state is considered to be "on shell" (or "physical") only if the total excitation level of the left-movers matches the total excitation level of the right-movers. This condition implies that the string does not have an unbalanced "wobbling", for if such a wobbling existed, it could ultimately be used to determine a preferred coordinate system on the worldsheet (thereby breaking conformal invariance). Indeed, demanding invariance under shifts in the σ_1 variable leads directly to the condition (6.2.17). We remark, however, that states not satisfying (6.2.17) are nevertheless important for understanding the "off-shell" or "virtual" structure of string theory. Such "virtual" states contribute, for example, within loop amplitudes. In these lectures, however, we shall focus on only the so-called "tree-level" string spectrum for which the level-matching constraint (6.2.17) is imposed and the corresponding physical masses are given by (6.2.16).

6.2.3. *The spectrum of the bosonic string*

Having discussed the quantization of the bosonic string, we can now examine its spectrum. The procedure is simple: we simply consider all possible combinations of left- and right-moving mode excitations of the string worldsheet, subject to the level-matching constraint (6.2.17), and then we tensor these left- and right-moving states together to form the total resulting string state. The spacetime mass of this string state is then given by (6.2.16), and the properties of the state are deduced directly from the underlying vibrational configuration of the string.

The simplest state, of course, is the string vacuum state

$$|0\rangle_R \otimes |0\rangle_L \tag{6.2.18}$$

in which the right- and left-moving vacuum states are tensored together. This state trivially satisfies (6.2.17), which indicates that this state is indeed part of the physical string spectrum. Unfortunately, we see from (6.2.16) that this state has a negative squared mass — *i.e.*, the spacetime mass of this state is imaginary! This state is thus a *tachyon*. Making sense of this string state is problematic, and is one of the reasons that we shall not ultimately be interested in the bosonic string.

Let us continue, however. The first excited string state is

$$\tilde{\alpha}^{\mu}_{-1}|0\rangle_R \otimes \alpha^{\nu}_{-1}|0\rangle_L \ . \tag{6.2.19}$$

This state has $L_0^{(\text{osc})} = \bar{L}_0^{(\text{osc})} = 1$, and according to (6.2.16) is therefore massless. As evident from its Lorentz index structure, this state transforms under the spacetime Lorentz group as the tensor product of two spin-one Lorentz vectors. We can therefore decompose this tensor product into a spin-two state (the symmetric traceless component), a spin-one state (the antisymmetric component), and a spin-zero state (the trace). Mathematically, this is equivalent to the tensor-product rule for Lorentz transverse $SO(24)$ vector representations:

$$\mathbf{V}_{24} \otimes \mathbf{V}_{24} \ = \ \mathbf{1} \ \oplus \ \mathbf{276} \ \oplus \ \mathbf{299} \tag{6.2.20}$$

where $\mathbf{V}_8$ is the eight-dimensional vector representation, and where the $\mathbf{1}$ representation is the spin-zero state, the $\mathbf{276}$ representation is the spin-one state, and the $\mathbf{299}$ representation is the spin-two state.

How can we interpret these states? A massless spin-two state must, by Lorentz invariance, have equations of motion which are equivalent to the Einstein field equations of general relativity. Thus, we are forced to identify the spin-two (traceless symmetric) component of the state (6.2.19) as the *graviton* $g_{\mu\nu}$, which is the spin-two mediator of the gravitational interactions. The spin-one (antisymmetric) state within (6.2.19) is an antisymmetric tensor field, often denoted $B_{\mu\nu}$, and the spin-zero (trace) component is the so-called *dilaton*, denoted ϕ. Together, $(g_{\mu\nu}, B_{\mu\nu}, \phi)$ are called the *gravity multiplet*.

By identifying (6.2.19) with the gravity multiplet, we see that string theory becomes a theory that contains gravity! This in turn allows us to determine the value of our previously unfixed mass scale α'. We shall now sketch how this happens (with details available in GSW). It turns out that if one calculates loop amplitudes in string theory, one finds that $e^{-\phi}$ serves as a loop expansion parameter (*i.e.*, higher-loop amplitudes come multiplied by more powers of $e^{-\phi}$). Given this observation, it is natural to identify the string coupling constant as the vacuum expectation value of the dilaton:

$$g_{\text{string}} \ = \ e^{-\langle\phi\rangle} \ . \tag{6.2.21}$$

This string coupling constant describes the strength of string interactions. Given this definition, we then find that the graviton state couples to matter with the expected gravitational strength only if we choose

$$\alpha' = \frac{G_{\text{Newton}}}{g_{\text{string}}^2} \qquad (6.2.22)$$

where G_{Newton} is Newton's constant. Substituting this result into (6.2.5), we then find

$$M_{\text{string}} = g_{\text{string}} M_{\text{Planck}} , \qquad (6.2.23)$$

where $M_{\text{Planck}} \equiv 1/\sqrt{G_{\text{Newton}}}$. Thus, because it contains gravity, string theory becomes a theory whose fundamental mass scale is related to the Planck scale.

We can also construct more and more massive string states. Ultimately, these fill out an infinite tower of string states. It is clear that such additional states all have $\alpha' M^2 > 0$. Given the above value for α', this implies that these additional states all have Planck-scale masses. Such Planck-scale excited states are therefore not of direct relevance for string phenomenology. Let us note, however, one interesting fact about these states. For any given spacetime mass level M, the string state with maximum spin is achieved by exciting only the lowest vibrational modes α_{-1}^μ and $\tilde{\alpha}_{-1}^\mu$. We thus find that for a given spacetime mass M, the maximum spin J_{max} that can be realized is

$$\alpha' M^2 = 2J_{\text{max}} - 4 . \qquad (6.2.24)$$

For example, we see that the maximum spin that can be realized for a massless state is $J = 2$ (the graviton). The relation (6.2.24) was originally observed for hadron resonances, and historically gave rise to the so-called "dual resonance models" (which eventually became modern string theory). In such dual resonance models, the relation (6.2.24) describes a so-called "Regge trajectory", with α' serving as the so-called "Regge slope". It is for this reason that in modern string theory, we continue to refer to α' as the Regge slope.

Before concluding, let us briefly mention one further important issue. In ordinary four-dimensional quantum field theory, we know that a massless spin-one state (*e.g.*, a photon) naïvely has four distinct states (corresponding to the four components of a vector field A^μ). However, the underlying gauge invariance allows us to make a unitary gauge choice wherein only two of these states (the two helicity states) are truly physical. The timelike and longitudinal states decouple, leaving only the transverse components. In the above description of the string spectrum, however, we have taken a covariant approach analogous to the description of a photon as a four-component vector. One might then wonder which of these states are truly physical. This issue is an important one in string theory, and once again we cannot here provide a proper proof. We shall therefore make recourse to another

> **Great Leap #4:** The physical string states are those which are realized by exciting the oscillator modes of only the *transverse* coordinates X^i ($i = 1, ..., 24$).

Proving this statement requires showing that even after we have used conformal invariance to set the string worldsheet metric to $\eta_{\alpha\beta}$, there still remains sufficient freedom to make a further "gauge" choice wherein we set the oscillator modes of the timelike and longitudinal spacetime coordinates to zero. This gauge choice, which is called *light-cone gauge*, is thus the analogue of unitary gauge in quantum field theory, and essentially tells us that only the 24 transverse coordinates correspond to physical degrees of freedom in the string worldsheet action. An important by-product of this fact is that every remaining string state has a non-negative norm. This is non-trivial. For example, if our metric signature is chosen such that $\eta^{00} = -1$, then the state $\alpha_{-n}^{\mu=0}|0\rangle$ has a negative norm. However, one can demonstrate that in light-cone gauge all resulting states are physical and have non-negative norm.

6.2.4. *Summary*

Let us quickly review those features of the bosonic string that we shall need to bear in mind in subsequent lectures. We shall separate these features into worldsheet features and spacetime features.

Worldsheet: The worldsheet fields consist of D copies of the left- and right-moving spacetime coordinates X_L^μ and X_R^μ (the worldsheet bosons). The fact that these X coordinates are periodic as we traverse the closed string loop implies that they have integer modings α_n and $\tilde{\alpha}_n$, where $n \in \mathbb{Z}$. The relevant worldsheet symmetry is conformal invariance, which tells us that the number of these X^μ fields is $D = 26$ and also tells us that the vacuum energy corresponding to these fields is $a_L = a_R = -1$. As we have stated above, a useful way to think about these results is to imagine that there is a "background" conformal anomaly $c_{\text{background}} = -26$, and that each X^μ field makes a contribution $c_X = 1$. In general, the "background" conformal anomaly is only a function of the relevant worldsheet symmetry (in this case conformal invariance), and it will always remain true that $c_X = 1$. Thus, cancellation of the conformal anomaly requires $D = 26$. A similar interpretation can also be given to the vacuum energy. When calculating the vacuum energies, only the physical (*i.e.*, transverse) fields are relevant. It is a general result that each X field contributes $a_X = -1/24$ to the vacuum energy. Therefore, we find $a_L = a_R = 24 a_X = -1$.

Spacetime: The above worldsheet theory leads to the following features in spacetime. We find that the *spacetime dimension* (often called the *critical* spacetime dimension) is 26. The spectrum consists of a spinless tachyon, as well as a massless gravity multiplet consisting of the graviton $g_{\mu\nu}$, the antisymmetric tensor $B_{\mu\nu}$, and the dilaton ϕ. There is also an infinite tower of massive (Planck-scale) string states.

Comments: Two remarkable things have happened. First, we have a theory of quantized gravity! The graviton has emerged as the quantum excitation of a closed string. This alone is very exciting, but also somewhat mysterious. We started by assuming a closed string propagating through an external, fixed, flat spacetime. But this string itself includes a graviton mode, which implies a distortion in that background spacetime. This then acts back to change the worldsheet theory. Thus, in some sense, the string itself not only "creates" the spacetime in which it propagates, but is then affected by this change in the spacetime geometry. This coupling or interplay between the string and its spacetime is not fully understood, and is clearly at the heart of the many mysterious features of string theory as a theory of quantum gravity.

A second remarkable thing has also happened, although we have not demonstrated it explicitly. As indicated in (6.2.21), a coupling constant has been determined *not* as a free parameter, but rather *dynamically* as the vacuum expectation of a string field. It is in this sense that string theory contains no free parameters, and that all parameters such as coupling constants are determined dynamically.

There are, however, a number of drawbacks to this bosonic string theory. First, it contains a tachyonic state. We must somehow find a way to eliminate this. Second, all string excitations are spacetime *bosons* (*i.e.*, they have integer spin). We must find a way to obtain spacetime fermions. Third, there are no massless spin-one states (which we would wish to associate with gauge fields). Thus, there are no gauge symmetries. It is for these reasons that we shall go on to consider more complicated string theories.

And finally, there is another major drawback that we need to be aware of. Although it is compelling that the string coupling g_{string} is in principle determined dynamically, as the vacuum expectation value of the dilaton scalar field, in practice we do not understand how to calculate the potential of the dilaton field and thereby deduce its vacuum expectation value. In the bosonic string we are considering here, the dilaton potential $V(\phi)$ is actually divergent for all $\phi < \infty$, and so this question cannot be meaningfully addressed. However, even in the more realistic string theories to be discussed, this potential is either completely flat (as happens in a supersymmetric context), or generally takes a shape that sends $\langle\phi\rangle \to \infty$. This is the famous *dilaton runaway problem*. Solving this problem is perhaps one of the most important (unsolved) problems in string phenomenology.

How can we remedy these features? One possibility is prompted by the appearance of the tachyon. In ordinary quantum field theory, the existence of a tachyon (a state with a negative mass-squared) signals that the vacuum has been misidentified (as in the Higgs mechanism); the theory then "rolls" to a different vacuum configuration in which the tachyon is eliminated. So it is natural to speculate that perhaps the bosonic string theory also "rolls" to a new vacuum in such a way that the tachyon is no longer present and the dilaton is stabilized. Perhaps fermions and gauge fields might also appear in this new vacuum, as desired. However, as we have already indicated, it is not known how the bosonic string behaves in this context. We do not know if there exists a new ("stable") vacuum to roll to, and if so, what its properties might be. Of course, knowing the potential $V(\phi)$ would be extremely useful, yet as we indicated this potential is naïvely divergent and therefore requires some knowledge of the non-perturbative structure of string theory. So (at least for the time being) this option does not appear promising.

A second possibility, then, is simply to abandon the bosonic string and attempt to construct a new string theory altogether. And this is what we shall now do.

6.3. Lecture #3: Neutrinos are fermions: The superstring

As we saw in the last lecture, the bosonic string has two glaring failures: it contains a tachyon, and it does not give rise to spacetime fermions. Both of these features are troubling, especially since the announced goal of these lectures is to derive a neutrino from string theory, and we know that the neutrino is a fermionic object. We therefore seek to construct a new string theory which can give rise to excitations with half-integer spins.

6.3.1. *The action*

We have already seen that string theories are defined by their two-dimensional worldsheet actions. Thus, in order to construct a new string theory, we must construct a new worldsheet action. At the very least, this action should contain that of the bosonic string, since we still wish to retain the spacetime interpretatation that we had previously. Thus, our only option is to *introduce additional worldsheet fields* into the action:

$$S = -\frac{1}{4\pi\alpha'} \int d^2\sigma \; (\partial_\alpha X^\mu \partial^\alpha X_\mu + ...) \,. \tag{6.3.1}$$

What new fields can we add? If our goal is to produce spacetime fermions, a natural guess would be to add worldsheet fermions! These would complement the worldsheet bosonic fields X^μ that are already present. For the moment, let us denote such fermionic fields schematically as ψ. We would then attempt to consider an action of the form

$$S = -\frac{1}{4\pi\alpha'} \int d^2\sigma \; (\partial_\alpha X^\mu \partial^\alpha X_\mu + \bar\psi i \rho^\alpha \partial_\alpha \psi) \,. \tag{6.3.2}$$

Here $\psi(\sigma_1, \sigma_2)$ represents our two-dimensional fermionic fields, and ρ^α are an appropriate set of two-dimensional Dirac matrices (the analogues of the γ^μ matrices in four dimensions).

We then face a number of questions. First, how many ψ fields must we add? Second, what kinds of worldsheet fermions should these be? Should they be Dirac fermions, or Majorana fermions, or Majorana-Weyl fermions? Third, how should these two-dimensional spinors ψ transform under the (internal) $SO(D-1,1)$ spacetime Lorentz symmetry? We already know that the X^μ fields, for example, transform as vectors under this symmetry. Note that it is not obvious that the ψ fields should necessarily transform as spinors under $SO(D-1,1)$ and carry a spacetime spinor index. In particular, all we know thus far is that the ψ fields transform as spinors under worldsheet *two-dimensional* Lorentz transformations. This does not *a priori* give us any information about their *spacetime* transformation properties.

There is also another potential worry that appears if we try to add new worldsheet fields. We have already seen in the bosonic string that worldsheet conformal invariance was sufficiently powerful a symmetry to allow us to choose a light-cone gauge and thereby eliminate all negative-norm states. However, the presence of new worldsheet fields implies the existence of new quantum excitation modes in the resulting string spectrum, and some of these new states may also have negative norm. Thus, conformal symmetry may no longer be sufficient (and indeed would not be sufficient) to allow us to eliminate these states as well.

It turns out that all of these questions have a common answer: we can impose an extra symmetry beyond simple worldsheet conformal invariance. Indeed, the extra symmetry that we shall impose is nothing but worldsheet (*i.e.*, two-dimensional) *supersymmetry*. Specifically, we shall require that the ψ fields be the two-dimensional superpartners of the X fields, so that the resulting action has a manifest worldsheet (two-dimensional) supersymmetry.* This new theory will be called the *superstring*.

It is important to stress that this supersymmetry that we will be discussing is *not* the spacetime supersymmetry that might be seen in the next round of accelerator experiments. Instead, this is a *worldsheet* supersymmetry which stems directly from the worldsheet interpretation of the original Polyakov action (6.2.4), and which relates the worldsheet bosons X to worldsheet fermions ψ via a worldsheet supercurrent J.

Imposing this worldsheet supersymmetry then answers all of the questions we previously raised. How many ψ fields? The answer is D, one for each boson X^μ. What kind of ψ spinor? The answer is a Majorana (two-component) spinor. How does the ψ field transform under the $SO(D-1,1)$ spacetime Lorentz symmetry? The answer is that the ψ field must transform as a *vector* under the Lorentz symmetry, since the X^μ field (for which it is the worldsheet superpartner) also transforms as a vector. In other words, the *worldsheet* supersymmetry commutes with the *spacetime* Lorentz symmetry, and thus does not change the Lorentz index structure.

---

*We remark that this is only one possible choice, and will ultimately lead us to the so-called Ramond/Neveu-Schwarz (RNS) formalism. Another possible choice would be to demand *spacetime* supersymmetry, and to imagine that the ψ fields are the Grassmann coordinates θ of a super-spacetime. This possibility would then lead to the so-called Green-Schwarz (GS) formalism. It turns out that these two formalisms are ultimately equivalent, however, and both provide suitable descriptions of the resulting superstring theory. This equivalence is possible because the RNS superstring ultimately also has spacetime supersymmetry (as we shall discover below). In these lectures, however, we shall restrict our attention to the RNS formulation in which the ψ fields are *worldsheet* (rather than spacetime) superpartners of the X^μ fields. Aside from being more useful for string phenomenology, the RNS formalism has the philosophical advantage that it treats the *string* as the fundamental object, with the spacetime structure emerging as a *derived consequence*. The RNS formalism thus reinforces one of the central themes of these lectures, namely that we define a string theory by its worldsheet properties alone, and then deduce the spacetime effects of these properties as consequences. The GS formalism, on the other hand, has the benefit of being manifestly spacetime supersymmetric from the very beginning.

Thus, the ψ fields transform as spacetime vectors, and carry a spacetime vector index: $\psi^\mu(\sigma_1, \sigma_2)$.

This last point may initially seem confusing, so we reiterate: the ψ fields are worldsheet fermions, but spacetime bosons! They transform as spinors under worldsheet Lorentz transformations, but as vectors under the spacetime Lorentz transformations.

Given this, we can now explicitly write down the superstring action:

$$S = -\frac{1}{4\pi\alpha'} \int d^2\sigma \left(\partial_\alpha X^\mu \partial^\alpha X_\mu - i\bar{\psi}_\mu \rho^\alpha \partial_\alpha \psi^\mu \right) . \quad (6.3.3)$$

Our worldsheet fields are $X^\mu(\sigma_1, \sigma_2)$ and $\psi^\mu(\sigma_1, \sigma_2)$, and the μ index (with $\mu = 0, 1, 2, ..., D-1$) is a vector index with respect to the internal symmetry $SO(D-1, 1)$. From the worldsheet perspective, each X^μ is a scalar field (containing one component), while each ψ^μ is a two-component spinor. The ρ^α are two-dimensional Dirac matrices satisfying the two-dimensional Clifford algebra $\{\rho^\alpha, \rho^\beta\} = -2\eta^{\alpha\beta}$, and $\bar{\psi} \equiv \psi^\dagger \rho^0$. One can then show that the action (6.3.3) is invariant under the worldsheet supersymmetry transformations $\delta X^\mu = \bar{\epsilon}\psi^\mu$, $\delta \psi^\mu = -i\rho^\alpha \epsilon \partial_\alpha X^\mu$, where ϵ is a constant anticommuting spinor that parametrizes the "magnitude" of the supersymmetry transformation. The corresponding generator of this worldsheet supersymmetry transformation is the worldsheet supercurrent $J_\alpha = \frac{1}{2}\rho^\beta \rho_\alpha \psi^\mu \partial_\beta X_\mu$.

It is convenient to choose a particular Weyl (chiral) representation for the two-dimensional ρ^α matrices:

$$\rho^0 = \begin{pmatrix} 0 & -i \\ i & 0 \end{pmatrix}, \quad \rho^1 = \begin{pmatrix} 0 & i \\ i & 0 \end{pmatrix} \implies \rho^0 \rho^1 = \begin{pmatrix} 1 & 0 \\ 0 & -1 \end{pmatrix} . \quad (6.3.4)$$

Here the product $\rho^0 \rho^1$ plays the role of the chirality operator (the analogue of γ_5 in four dimensions), and thus in this basis we can identify the upper and lower components of the two-component Majorana spinor ψ as being left-moving and right-moving respectively. Our worldsheet action (6.3.3) then decomposes into the form

$$S = -\frac{1}{4\pi\alpha'} \int d^2\sigma \left(\partial_\alpha X^\mu \partial^\alpha X_\mu - \psi_{\mu R} \partial_- \psi_R^\mu - \psi_{\mu L} \partial_+ \psi_L^\mu \right) \quad (6.3.5)$$

where $\partial_\pm$ are derivatives with respect to the left- and right-moving worldsheet coordinates $\sigma_1 \pm \sigma_2$. The worldsheet content of this theory therefore consists of D left-moving worldsheet bosons X_L, D right-moving worldsheet bosons X_R, D left-moving worldsheet Majorana-Weyl (one-component) fermions ψ_L, and D right-moving worldsheet Majorana-Weyl (one-component) fermions ψ_R. There are two worldsheet supercurrents in this theory:

$$J_L = \psi_{\mu L}\, \partial_+ X_L^\mu\,, \qquad J_R = \psi_{\mu R}\, \partial_- X_R^\mu\,. \qquad (6.3.6)$$

Note that our original goal in constructing the superstring had been to obtain spacetime fermions. However, it may seem from the above that we have failed in this regard, since we have only introduced new fields ψ which themselves are spacetime vectors. How then are we to obtain spacetime fermions? It turns out that this will happen in a surprising way.

Let us proceed to analyze this string following the same steps as we used for the bosonic string. First, we see that our worldsheet symmetry has been enlarged: rather than simply have conformal invariance, we now have conformal invariance *plus* worldsheet supersymmetry. Together, this is called *superconformal invariance*, which is a much larger symmetry than conformal invariance alone.

This enlargement of the worldsheet symmetry changes many of the features of the resulting string. The most profound is the value of the spacetime dimension D. Recall from our discussion of the bosonic string that associated with each worldsheet symmetry there is a particular "background" conformal (central charge) anomaly, and that it is necessary to choose a sufficient number of worldsheet fields so as to cancel this anomaly and ensure that conformal invariance is maintained even at the quantum level. The same argument applies here as well, except that

> **Great Leap #5:** The "background" conformal anomaly associated with *superconformal* invariance is not $c = -26$ but rather $c = -15$. Likewise, the conformal anomaly contribution from each worldsheet Majorana fermion is $c = 1/2$.

We can understand the origin of the "background" conformal anomaly $c = -15$ as follows. Just as in the bosonic string, a certain contribution $c = -26$ is attributable to the conformal ghosts resulting from conformal gauge fixing. The new feature here is that we now have an additional contribution $+11$ which is attributable to the worldsheet *superpartners* of these ghosts. Together, this produces a background anomaly $c = -15$. What this means is that we must choose the number D of worldsheet bosons and fermions such that this "background" anomaly is cancelled. We have already seen that the anomaly contribution from each worldsheet boson X^μ is $c_X = 1$. Since the anomaly contribution from each Majorana fermion is $c_\psi = 1/2$, we must satisfy

$$D\left(1 + \tfrac{1}{2}\right) - 15 = 0 \quad \Longrightarrow \quad D = 10 \,. \tag{6.3.7}$$

Thus, we see that the critical dimension of the superstring is $D=10$ rather than $D=26$. Moreover, just as for the bosonic string, the *super*conformal symmetry of the superstring worldsheet action again allows us to choose a light-cone gauge in which only *eight* transverse bosons and *eight* transverse fermions represent the truly physical propagating worldsheet fields.

6.3.2. *Quantizing the superstring*

Let us now quantize the superstring, just as we did for the bosonic string. The boundary conditions (6.2.6) for the X^μ fields remain valid even for the superstring, since the X^μ continue to have the interpretation of spacetime coordinates. Therefore the mode-expansions (6.2.9) continue to apply.

The only new feature, then, is the mode-expansion for the fermionic fields ψ^μ. However, unlike the bosonic fields X^μ which must be periodic because of their interpretation as spacetime coordinates, these fermionic fields ψ^μ do not have any immediate interpretation in spacetime. Therefore, the only boundary conditions that might be imposed on these fields are those that are required directly from the symmetries of the action. In particular, we must choose boundary conditions for the ψ^μ fields so as to maintain the single-valuedness of the action as we traverse the closed string (*i.e.*, as $\sigma_1 \to \sigma_1 + \pi$), and so as to maintain the worldsheet supersymmetry of the action (whose algebra includes a requirement that the supercurrent square to the Hamiltonian, *i.e.*, $J \cdot J \sim H$). It turns out that are only two choices of boundary conditions that satisfy these requirements. One possibility is that the ψ^μ fields are periodic under $\sigma_1 \to \sigma_1 + \pi$:

$$\text{Ramond:} \quad \psi^\mu(\sigma_1 + \pi, \sigma_2) \;=\; +\, \psi^\mu(\sigma_1, \sigma_2) \,. \tag{6.3.8}$$

Such periodic boundary conditions are typically called "Ramond" (R) boundary conditions, after P. Ramond (who introduced these fermionic boundary conditions in 1971). The second possibility is that the ψ^μ fields are *anti*-periodic under $\sigma_1 \to \sigma_1 + \pi$:

$$\text{Neveu-Schwarz:} \quad \psi^\mu(\sigma_1 + \pi, \sigma_2) \;=\; -\, \psi^\mu(\sigma_1, \sigma_2) \,. \tag{6.3.9}$$

Such periodic boundary conditions are typically called "Neveu-Schwarz" (NS) boundary conditions, after A. Neveu and J. Schwarz (who introduced these fermionic boundary conditions in 1971). As we shall see in Lecture

#4, both of these boundary conditions are ultimately required for the self-consistency of the superstring.

In the case of periodic (Ramond) boundary conditions, the mode-expansion of the ψ^μ field resembles that of the X^μ field:

$$\text{Ramond:} \quad \psi_L^\mu(\sigma_1 + \sigma_2) = \sum_{n \in \mathbb{Z}} b_n^\mu \, e^{-2in(\sigma_1+\sigma_2)}$$

$$\psi_R^\mu(\sigma_1 - \sigma_2) = \sum_{n \in \mathbb{Z}} \tilde{b}_n^\mu \, e^{+2in(\sigma_1-\sigma_2)} \ . \quad (6.3.10)$$

Here $b_n^\mu, \tilde{b}_n^\mu$ are the (fermionic) creation and annihilation operators, satisfying the *anti*-commutation relations

$$\{b_m^\mu, b_n^\nu\} = \eta^{\mu\nu} \delta_{m+n} \quad (6.3.11)$$

where we recall the hermiticity condition $b_{-n}^\mu = (b_n^\mu)^\dagger$. The same relations hold for the right-moving modes as well. This hermiticity condition follows from the fact that the ψ fields are Majorana (*i.e.*, real) fields. Note that unlike the bosonic mode-expansion (6.2.9), we have joined the zero-modes together with the excited modes in (6.3.10).† There is also no "center-of-mass" term in the mode-expansion (a fermionic analogue of x^μ) because the ψ fields are Grassmann variables and thus lack a classical limit. Finally, also note that unlike the bosonic α_n^μ modes, which are rescaled relative to the usual harmonic oscillator modes by powers of the mode frequency n, the fermionic b_n^μ modes are defined without this rescaling and hence satisfy the usual harmonic-oscillator commutation relations (6.3.11) directly. This too is traditional in string theory.

In the case of anti-periodic (Neveu-Schwarz) boundary conditions, the mode-expansion of the ψ^μ field involves *half-integer* rather than integer modes:

$$\text{Neveu-Schwarz:} \quad \psi_L^\mu(\sigma_1 + \sigma_2) = \sum_{r \in \mathbb{Z}+1/2} b_r^\mu \, e^{-2ir(\sigma_1+\sigma_2)}$$

$$\psi_R^\mu(\sigma_1 - \sigma_2) = \sum_{r \in \mathbb{Z}+1/2} \tilde{b}_r^\mu \, e^{+2ir(\sigma_1-\sigma_2)} \ . \quad (6.3.12)$$

---

†We are cheating slightly here, since the treatment of Ramond zero-modes for Majorana worldsheet fermions is actually quite subtle. In some sense, each Majorana fermion has only "half" a zero-mode. We will provide a rigorous discussion of this fact in Lecture #5. In the meantime, it will suffice to ignore this subtlety.

Once again, $b_r^\mu, \tilde{b}_r^\mu$ are the (fermionic) creation and annihilation operators, satisfying the *anti*-commutation relations

$$\{b_r^\mu, b_s^\nu\} = \eta^{\mu\nu} \delta_{r+s} \tag{6.3.13}$$

where we have the hermiticity condition $b_{-r}^\mu = (b_r^\mu)^\dagger$.

The expressions for the total energy of a given string configuration now receive contributions from not only the bosonic oscillator modes, as in (6.2.12), but also the fermionic oscillator modes. These new contributions are given by

$$\text{R:} \qquad L_0^{(\text{osc})} = \sum_{n=0}^\infty n\, b_{-n}^\mu b_{n\mu}$$

$$\text{NS:} \qquad L_0^{(\text{osc})} = \sum_{r=1/2}^\infty r\, b_{-r}^\mu b_{r\mu} \,, \tag{6.3.14}$$

with similar expressions for the right-movers.

Finally, we must consider the vacuum energies a_L and a_R for the superstring. Recall that for the bosonic string, each of the 24 transverse X^μ fields contributed $a_X = -1/24$, yielding a total of $a_L = a_R = -1$. This contribution from each bosonic field remains the same for the superstring, so we continue to have $a_X = -1/24$. It therefore only remains to determine the vacuum-energy contributions from the worldsheet Majorana fermions, and it is found that

> **Great Leap #6:** Each Ramond fermion contributes vacuum energy $a_\psi = +1/24$, whereas each Neveu-Schwarz fermion contributes vacuum energy $a_\psi = -1/48$.

We thus see that like the bosons, the Neveu-Schwarz fermions contribute negative vacuum energies, while Ramond fermions contribute positive vacuum energies.

Given these mode-expansions and commutation relations, it is instructive to consider the Fock space of an individual Ramond (R) or Neveu-Schwarz (NS) fermion. It turns out to be simplest to consider the Fock space of an individual (left- or right-moving) NS fermion first. The two lowest-lying states are

$$\begin{array}{rll} \text{vacuum:} & |0\rangle & L_0^{(\text{osc})} = 0 \\ \text{first-excited state:} & b_{-1/2}|0\rangle & L_0^{(\text{osc})} = 1/2 \,. \end{array} \tag{6.3.15}$$

Note that relative to the vacuum, all further excited states are reached through only half-integer excitations. Also note that the vacuum of the NS Fock space is unique, just like that of the bosons X^μ. What this means is that from the spacetime perspective, the vacuum is spinless (and hence a spacetime bosonic state), and that all subsequent excitations of the vacuum are also spacetime bosons. Recall, in this connection, that the fermion mode operators b are only fermionic from the worldsheet perspective; they are still bosonic operators (just like the fields ψ^μ themselves) relative to *spacetime* Lorentz symmetries.

Let us now consider the corresponding Fock space for the Ramond fermions with periodic boundary conditions. Once again, we have a tower of states

$$\text{vacuum:} \quad |0\rangle \quad L_0^{(\text{osc})} = 0$$
$$\text{first-excited state:} \quad b_{-1}|0\rangle \quad L_0^{(\text{osc})} = 1 \quad (6.3.16)$$

which now continues upwards through integer, rather than half-integer, steps. However, in this case it is important to observe that we also have a *zero-mode* in the theory. The existence of this zero-mode means that it is possible to excite this zero-mode without increasing the overall energy of the state. We therefore have the additional tower of states

$$\text{vacuum:} \quad b_0^\dagger|0\rangle \quad L_0^{(\text{osc})} = 0$$
$$\text{first-excited state:} \quad b_{-1} b_0^\dagger|0\rangle \quad L_0^{(\text{osc})} = 1 \,. \quad (6.3.17)$$

(Note that b_0 and $b_0^\dagger$ are equivalant.) In other words, combining (6.3.16) and (6.3.17), we see that the Ramond vacuum consists of *two degenerate states*,

$$|0\rangle \quad \text{and} \quad b_0^\dagger|0\rangle \,, \quad (6.3.18)$$

and that all further excitations maintain this two-fold degeneracy.

How can we interpret this two-fold degeneracy of the Ramond vacuum? It may seem, at first, that both of the states in (6.3.18) cannot be considered as the true vacuum, because the second state in (6.3.18) appears to be realized as a zero-mode excitation of the first. However, let us define the first state in (6.3.18) as $|V_0\rangle$ and let us also define $|V_1\rangle \equiv \sqrt{2} b_0^\dagger|0\rangle$, which is a rescaling of the second state in (6.3.18). Then using (6.3.11), it is easy to show that

$$|V_1\rangle = \sqrt{2} b_0^\dagger|V_0\rangle \,, \quad |V_0\rangle = \sqrt{2} b_0^\dagger|V_1\rangle \,. \quad (6.3.19)$$

Thus, we see that neither state in (6.3.18) is more fundamental than the other, and there exists an unbroken symmetry between them — they are realized as zero-mode excitations of each other. The interpretation of this fact is that the true Ramond vacuum state is a two-component object, a spacetime spinor! It then follows that all of the excited states in the Ramond spectrum are also spacetime spinors, since they are realized as non-zero-mode excitations of a spinorial ground state.

Of course, the above discussion is only suggestive, since we have not proven that these two vacuum states actually form a Lorentz spinor representation with respect to the spacetime Lorentz algebra. However, it is easy to see that this is indeed the case. Observe from (6.3.11) that the zero-modes satisfy the algebra $\{b_0^\mu, b_0^\nu\} = \eta^{\mu\nu}$. Thus, if we define $\Gamma^\mu \equiv \sqrt{2}ib_0^\mu$, then we see that $\{\Gamma^\mu, \Gamma^\nu\} = -2\eta^{\mu\nu}$, which is nothing but the spacetime Clifford algebra. In other words, the zero-modes act as spinorial gamma-matrices. This implies that all states built upon such a vacuum state will transform in spinor representations of the spacetime Lorentz symmetry group $SO(D-1,1)$, and hence will be spacetime fermions.

This is a remarkable result. Even though we have introduced worldsheet ψ^μ fields which are spacetime bosons and which carry a spacetime Lorentz *vector* index, the algebra of zero-modes in the case of Ramond boundary conditions has managed to change these vector indices into spinor indices and thereby produce spacetime fermions. Of course, this is completely analogous to what happens in the usual four-dimensional Dirac equation, where the γ^μ matrices are matrices in a spinor space but nevertheless carry vector indices. Thus, we see that by choosing Ramond boundary conditions for worldsheet fermions, string theory affords us with the same possibility. We therefore now see that string theory can indeed give rise to spacetime fermions: while excitations of worldsheet Neveu-Schwarz fermions give rise to spacetime bosons, excitations of worldsheet Ramond fermions give rise to spacetime fermions.

6.4. Lecture #4: Some famous superstrings

The next step is to determine the spectrum of the full superstring, just as we did for the bosonic string. However, the presence of two possibilities (Neveu-Schwarz and Ramond) for the modings of the fermions introduces several new complications relative to the bosonic string, and enables us to make different choices for what kind of superstring we wish to construct. These different choices are typically called different "string models", and so we are finally in a position to begin to discuss string model-building. That is the subject of the present lecture.

6.4.1. *String sectors*

Recall from the previous lecture that in light-cone gauge, the worldsheet field content of the ten-dimensional superstring consists of eight right-moving bosons X_R, eight right-moving Majorana-Weyl (one-component) fermions ψ_R, and a similar set of left-moving fields X_L and ψ_L. The bosons X_L and X_R must have periodic (integer) modings because of their interpretation as spacetime coordinates, but their worldsheet fermionic superpartners ψ_L and ψ_R can have either Ramond (periodic, integer) or Neveu-Schwarz (anti-periodic, half-integer) modings. The question then immediately arises: What rules govern the possible self-consistent choices of fermion modings? *A priori*, the appearance of 16 distinct fermions would seem to lead to 2^{16} different choices.

It is easy to see that not all possibilities are allowed, however. One quick way to see this is to realize that if some of the right-moving fermions had different periodicities than other right-moving fermions, then these different periodicities would necessarily break spacetime Lorentz invariance because these fermions carry a spacetime vector index μ. A similar situation would also hold for the left-moving fermions. This would then imply that all of the right-moving fermions should have the same periodicity as each other, and that all of the left-moving fermions should have the same periodicity as each other (though not necessarily the same as that of the right-moving fermions). However, this argument is not really satisfactory because we do not necessarily wish to preserve the full *ten*-dimensional Lorentz invariance (or even its eight-dimensional transverse subgroup); after all, our sole phenomenological requirement is that *four*-dimensional Lorentz invariance must be maintained. Moreover, it goes against the spirit of string theory (as we have been presenting it) that we should demand a certain phenomenological property of the resulting *spacetime* physics when formulating our worldsheet theory. In string theory the spacetime physics is a *consequence* of the worldsheet physics, and we would ultimately like to base our worldsheet choices directly on worldsheet symmetries.

Fortunately, it is easy to find a worldsheet argument that leads to the same constraint. Recall that the worldsheet symmetry that we must maintain is superconformal invariance. The worldsheet supersymmetry that makes up superconformal invariance is generated by the two worldsheet supercurrents given in (6.3.6). Because these two supercurrents are also worldsheet fermionic, they may also be either periodic or anti-periodic as we traverse the closed string. Indeed, each individual term $\psi^\mu \partial X_\mu$ in these supercurrents will have the periodicity property of the fermion ψ^μ. However, in order for each of these supercurrents J_R and J_L to have a unique, well-defined periodicity as we traverse the closed string, we see that it is necessary that all right-moving fermions have the same periodicity as each other, and that all left-moving fermions have the same periodicity as each other. This is required in order to preserve worldsheet supersymmetry. Thus, we have our first constraints on fermion modings:

- All right-moving fermions ψ_R^μ must have the same periodicity as each other, either Ramond or Neveu-Schwarz.
- All left-moving fermions ψ_L^μ must have the same periodicity as each other, either Ramond or Neveu-Schwarz.

Note that there is no requirement that the right- and left-moving periodicities be the same.

Table 6.1. The four possible sectors of the ten-dimensional superstring, numbered 1 through 4. Here 'NS' and 'R' respectively indicate Neveu-Schwarz (anti-periodic) and Ramond (periodic) boundary conditions for worldsheet fermions, and a_R and a_L respectively denote the corresponding right- and left-moving vacuum energies.

#	$\psi_R^{i=1,\ldots,8}$	$\psi_L^{i=1,\ldots,8}$	a_R	a_L
1	NS	NS	$-1/2$	$-1/2$
2	R	R	0	0
3	R	NS	0	$-1/2$
4	NS	R	$-1/2$	0

Given these constraints, we see that we are left with four distinct periodicity choices for our sixteen Majorana-Weyl worldsheet fermions, as shown in Table 6.1. Each individual choice is called a *sector* or *spin structure* of the superstring, so we see that the ten-dimensional superstring has four possible sectors. For future convenience, these sectors have been numbered in Table 6.1. We have also indicated the corresponding right- and left-moving vacuum energies of these sectors. Recall from the previous lecture (in particular, Great Leap #6) that the vacuum-energy contribution of each Ramond fermion is $+1/24$, while that of each Neveu-Schwarz fermion is $-1/48$ and that of each worldsheet boson is $-1/24$. Therefore, generally assuming $n_{\rm NS}$ Neveu-Schwarz fermions and $n_{\rm R}$ Ramond fermions, we can add these individual contributions to find

$$a = -\frac{1}{24}\left(8 - n_{\rm R} + \tfrac{1}{2}n_{\rm NS}\right) = -\frac{n_{\rm NS}}{16} \ . \qquad (6.4.1)$$

The second equality results from setting $n_{\rm R} = 8 - n_{\rm NS}$. Of course, as discussed above, in the ten-dimensional superstring we are restricted to the cases $n_{\rm NS} = 0, 8$ for both the right- and left-moving fermions.

6.4.2. *Modular invariance and GSO projections*

The next question that arises is whether we are free to pick any one of these sectors to construct our superstring theory, or whether we must consider all of them together, superposing the spectrum from each sector separately in order to construct the full superstring spectrum. What rules govern the choices of sectors?

Ultimately, it turns out that a special form of conformal invariance known as *modular invariance* will give us the answer. In keeping with the spirit of these lectures, we will not be able to provide a proper mathematical discussion of modular invariance. (Indeed, doing so would require a preliminary discussion of string partition functions and the modular group.) However, we can discuss the relevance and implications of modular invariance at a conceptual level.

Recall from Lecture #2 that our string actions always have a certain symmetry known as conformal invariance, which reflects the fact that the action should be invariant under *local* reparametrizations and rescalings of the coordinates (σ_1, σ_2) that parametrize the string worldsheet. For *tree*-level string interactions, demanding this *local* symmetry is sufficient to ensure that the resulting physics is indeed invariant under arbitrary coordinate reparametrizations. This is because any tree-level string interaction has the topology of a sphere (a genus-zero surface, with no handles), and on a sphere it can be shown that any possible net coordinate reparametrization can be generated or "built up" in small steps as the cumulative effect of small, local coordinate reparametrizations. Geometrically, this is equivalent to saying that any closed loop on the surface of a sphere can be continuously shrunk to a point, as illustrated in Fig. 6.7(a), by sliding the loop along the surface of the sphere towards one side. Thus, demanding invariance under *local* coordinate reparametrizations (*i.e.*, conformal invariance) by itself is sufficient to guarantee consistency for tree-level string amplitudes.

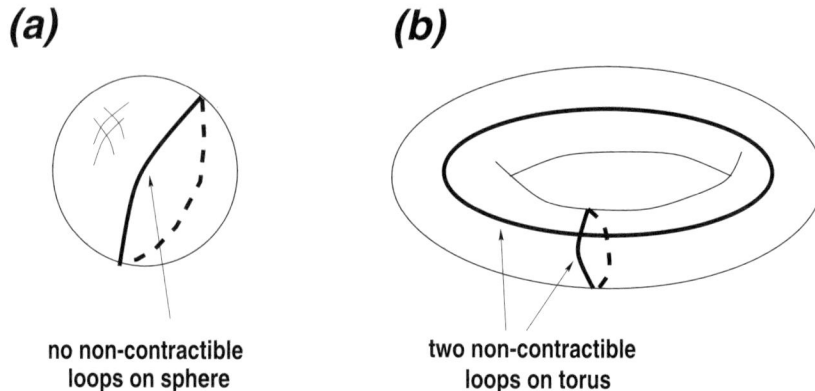

Fig. 6.7. (a) On a sphere, all closed loops can be continuously shrunk to a point. (b) On a torus, there exist two topologically distinct non-contractible loops.

However, this situation changes drastically if we now consider one-loop amplitudes. As discussed in Lecture #1, these amplitudes have the worldsheet topology of a *torus* (a genus-*one* surface), and we see from Fig. 6.7(b) that on a torus there exist *two* types of closed loops that cannot be continuously shrunk to a point. Such loops are said to be *non-contractible*, which is indeed the defining property of such higher-genus surfaces. The presence of these non-contractible loops means that for torus diagrams, there exist possible coordinate reparametrizations that *cannot* be built up from local coordinate reparametrizations alone. Indeed, these reparametrizations nontrivially involve "large", discrete mappings around these non-contractible loops. Thus, we see that demanding conformal invariance alone is not sufficient to ensure that one-loop string amplitudes are truly invariant under worldsheet coordinate reparametrizations: we must also demand an invariance under these "large" discrete mappings around these non-contractible loops. This additional global invariance is called "modular invariance", and just like conformal invariance, it too stems from our need to maintain the overall invariance of the string under reparametrizations and rescalings of the worldsheet coordinates.

One might wonder, at this stage, why we are suddenly worrying about modular invariance, whereas we did not need to consider modular invariance in Lecture #2 when we discussed the bosonic string. The truth of the matter is that we must *always* consider modular invariance in addition to conformal invariance, regardless of the type of (closed) string we are discussing. However, in the simple case of the 26-dimensional bosonic string, it turns out that all amplitudes are trivially modular-invariant, so we did not need to make recourse to modular invariance in order to distinguish between different possibilities. However, for the superstring (and particularly for the heterotic string to be discussed later), the possible sector choices become quite numerous, and it turns out that modular invariance is the powerful tool by which we are able to narrow down the self-consistent possibilities.

What, then, are the effects of modular invariance? It turns out that at the level of string model-building, modular invariance has two primary effects:

- it forces us to consider only certain selected *sets* or *combinations* of underlying sectors, and
- it produces new *constraints* (beyond the level-matching constraint $L_0 = \bar{L}_0$) that govern which Fock-space excitations are allowed in each sector.

These new constraints are called *GSO constraints*, after F. Gliozzi, J. Scherk, and D. Olive who first imposed some of these constraints in 1977. The important point is that these conditions stem directly from modular invariance, and thus they follow from the worldsheet physics of the string and do not represent any additional arbitrary input. We will provide many explicit examples of such combinations and constraints shortly.

In order to construct a fully consistent string model, therefore, our procedure is as follows. First, we must determine which are the allowed sectors that need to be considered as part of our set. For each of these sectors in our allowed set, we then determine the corresponding Fock space of physical states by applying not only the usual level-matching constraint, but also the GSO constraints appropriate for that sector. In this way each underlying sector then gives rise to a different Fock space of states, and the full Hilbert space of states for the full string theory (*i.e.*, for the resulting string "model") is nothing but the direct sum of these different Fock spaces corresponding to each of the underlying sectors in the specified set. This then yields a fully self-consistent (and in particular, modular-invariant) theory.

This is an important point, so it is worth repeating: the full Hilbert space of string states is given by the direct sum of the different Fock spaces corresponding to different underlying boundary conditions for worldsheet fields. In order to better understand this fact, an analogy with QCD may be useful. Recall that Yang-Mills quantum field theory contains non-perturbative instanton solutions, and therefore one can imagine doing quantum field theory in an n-instanton background $|n\rangle$. Of course, as we know, the full vacuum state of QCD is not composed of any one of these $|n\rangle$ vacua by itself, but rather by an appropriately weighted *combination* of these vacua:

$$|\theta\rangle = \sum_n e^{in\theta} |n\rangle \,. \qquad (6.4.2)$$

This is the famous θ-vacuum of QCD. The situation that we now face in string theory is somewhat analogous. The fact that the string worldsheet fermions can have different boundary conditions (thereby giving rise to different sectors) is in some sense analogous to the fact that QCD can have different instanton backgrounds. Indeed, each underlying string sector is analogous to a different n-instanton vacuum state $|n\rangle$, and the different "combinations of sectors" that we are now being forced to consider are analogous to the different QCD θ-vacua. In this sense, then, each different "string model" that we will be constructing can be viewed as a different θ-vacuum of string theory! Of course, this analogy with the QCD θ-vacuum can take us only so far. One important difference is that whereas the θ-vacuum necessarily involves *all* of the $|n\rangle$ states regardless of the value of θ, in string theory our "vacuum" may consist of more complicated combinations of sectors which may or may not include all possible sectors. In fact, the more sectors that are included in our "combination of sectors", the more GSO constraints there are for each sector. But the important lesson that emerges from all of this is that no single sector by itself forms a consistent string vacuum; rather, we must select an appropriate combination of sectors and add together their corresponding Fock spaces in order to produce the fully self-consistent string model.

In Lecture #7, we shall provide an explicit set of rules which will enable us to quickly determine the appropriate sector combinations and GSO constraints that can be chosen in order to yield self-consistent theories. For the time being, however, we shall defer a discussion of these rules and proceed directly with the construction of actual string models in order to deduce their physical properties. Therefore, even though we shall simply assert certain sector combinations and GSO projections to be required by modular invariance, we stress that all of these features can (and ultimately will) be derived using the rules to be presented in Lecture #7.

6.4.3. Ten-dimensional superstring models

In the case of the ten-dimensional superstring, we have already seen that the four possible sectors are listed in Table 6.1. It then only remains to determine the particular sector combinations and GSO constraints that are required by modular invariance. In this case, it turns out that there are only two possible combinations or sets of sectors that can be considered:

- we consider the contributions from only Sectors #1 and #2, or
- we consider the contributions from *all* Sectors #1 through #4.

Moreover, for each of the above cases, it turns out that there are two possible choices of GSO projections that may be imposed in each sector. Thus, combining all of these possibilities, we see that there are four distinct possible superstring "models" that can be constructed in ten dimensions. We shall therefore now turn to a construction of these models.

6.4.3.1. *The Type 0 strings*

Let us begin by considering the first option, taking our set of sectors to consist only of Sectors #1 and #2. For each of these sectors, we need to determine the appropriate GSO constraints that must be applied in addition to the usual level-matching constraint. In order to write down these GSO constraints, let us first recall that for a given left-moving worldsheet fermion (with either Ramond or Neveu-Schwarz boundary conditions), the corresponding number operator is defined by

$$\text{R}: \quad N^{(i)} = \sum_{n=0}^{\infty} b^i_{-n} b_{ni}$$

$$\text{NS}: \quad N^{(i)} = \sum_{r=1/2}^{\infty} b^i_{-r} b_{ri} \ . \quad (6.4.3)$$

Here the index $i = 1, ..., 8$ labels the individual fermion. For right-moving fermions, the analogous number operators $\bar{N}^{(i)}$ are constructed using the right-moving mode operators $\tilde{b}_n, \tilde{b}_r$. Let us also define N_L and N_R respectively as the total left- and right-moving number operators, *i.e.*,

$$N_L \equiv \sum_{i=1}^{8} N^{(i)} \ , \qquad N_R \equiv \sum_{i=1}^{8} \bar{N}^{(i)} \ . \quad (6.4.4)$$

Note that these number operators are defined to include only the contributions of the worldsheet *fermions*, and in particular do not include the contributions of the worldsheet bosons. It then turns out (and we shall see in Lecture #7) that if we choose our set of sectors to consist only of Sectors #1 and #2, then the appropriate GSO constraints in each sector are as follows:

$$\text{Sector \#1:} \quad N_L - N_R = \text{even}$$
$$\text{Sector \#2:} \quad N_L - N_R = \left\{ \begin{array}{c} \text{odd} \\ \text{even} \end{array} \right\} . \quad (6.4.5)$$

In the second line, we have used a brace notation to indicate a further choice: we can choose to impose *either* the 'odd' constraint, or the 'even' constraint. As we shall see, this is a residual choice that is not fixed by modular invariance (or by any other worldsheet symmetry), leading to two equally valid possibilities. Thus, we see that if we choose our set of sectors to consist of only Sectors #1 and #2, then this leads to *two different string models* depending on our subsequent choice of which GSO constraint we choose to impose in (6.4.5).

Let us now determine the spectra of these two models, beginning with the states that arise from Sector #1. Note that in this sector, both models have the same states (because both models have the same GSO constraint for Sector #1). As with the bosonic string, our procedure is to consider all possible excitations of the worldsheet fields (in this case, the worldsheet fermions as well as the worldsheet bosons). These excitations are subject to the level-matching constraint $L_0 = \bar{L}_0$ (which ensures that the total bosonic and fermionic worldsheet *energy* is distributed equally between left- and right-moving excitations) and the GSO constraint $N_L - N_R = \text{even}$ (which is a constraint on the worldsheet *number* operators of the worldsheet fermions only). In general, the mass-shell condition for the superstring is

$$\alpha' M^2 = 2(L_0 + \bar{L}_0 + a_L + a_R) \quad (6.4.6)$$

where a_L and a_R are the individual left- and right-moving vacuum energies, and where L_0 and $\bar{L}_0$ include the contributions from not only the worldsheet bosons, but also the worldsheet fermions. Note from Table 6.1 that the left- and right-moving vacuum energies in Sector #1 are $a_L = a_R = -1/2$.

We see that the tachyonic vacuum state $|0\rangle_R \otimes |0\rangle_L$ satisfies both constraints, and thus it remains in the spectrum. However, unlike the tachyon in the bosonic string (which has spacetime mass $\alpha' M^2 = -4$), we see from (6.4.6) that the tachyonic state in the superstring has spacetime mass $\alpha' M^2 = -2$. This is the result of the smaller (less negative) vacuum energy of the superstring compared to that of the bosonic string.

Because the vacuum energies in Sector #1 are $a_L = a_R = -1/2$, we see that massless states cannot be obtained by exciting the quantum modes of worldsheet bosons, for each of these excitations would add a full unit of energy. Instead, massless states can be obtained only by adding a half-unit of energy. Fortunately, this is possible in Sector #1 because in this sector, all worldsheet fermions have Neveu-Schwarz boundary conditions and therefore have half-integer modings. The first excited states in Sector #1 are therefore

$$\tilde{b}^\mu_{-1/2}|0\rangle_R \otimes b^\nu_{-1/2}|0\rangle_L \; . \tag{6.4.7}$$

Note that these states satisfy both the level-matching constraint (since $L_0 = \bar{L}_0 = 1/2$) as well as the GSO constraint (since $N_L = N_R = 1$). The interpretation of these states is precisely the same as in the bosonic string: these states give us the gravity multiplet, consisting of the graviton $g_{\mu\nu}$, dilaton ϕ, and anti-symmetric tensor $B_{\mu\nu}$. Mathematically, this is equivalent to the tensor-product rule for Lorentz transverse $SO(8)$ vector representations:

$$\mathbf{V_8} \otimes \mathbf{V_8} \;=\; \mathbf{1} \;\oplus\; \mathbf{28} \;\oplus\; \mathbf{35} \tag{6.4.8}$$

where $\mathbf{V_8}$ is the eight-dimensional vector representation, and where the $\mathbf{1}$ representation is the spin-zero state, the $\mathbf{28}$ representation is the spin-one state, and the $\mathbf{35}$ representation is the spin-two state. It is indeed a general principle that *all* weakly coupled closed strings contain at least these massless states, and this is a useful cross-check of the GSO constraints.

Let us now turn to the states from Sector #2. Before concerning ourselves with the implication of the GSO constraints in (6.4.5), let us first understand the general structure of the states from this sector. In this sector, the vacuum energy (according to Table 6.1) is $(a_R, a_L) = (0,0)$, so we see immediately that this sector contains no tachyons. Indeed, the ground state is already massless, so all that will concern us here is the nature of this ground state. As we discussed at the end of Lecture #3, the left- and right-moving ground states in this sector are each spacetime *spinors* since all worldsheet fermions in this sector have Ramond boundary conditions. Because the nature of these spinors will be important to us, let us pause to review some properties of these spinors.

Since we are considering these ten-dimensional strings in light-cone gauge, the Lorentz group that concerns us here is the transverse ("little") Lorentz group $SO(8)$. In general, the groups $SO(2n)$ share a number of properties. Their smallest representations, of course, are simply the identity representations. These are singlets, which will be denoted **1**. The next representations are the vector representations, which are $(2n)$-dimensional, and which will be denoted $\mathbf{V}_{2n}$. Along with these are the spinor representations, which are (2^{n-1})-dimensional. In general, there are two types of spinor representations, **S** and **C**, the so-called "spinor" and "conjugate spinor" representations. In the special case of $SO(8)$, the vector, spinor, and conjugate spinor representations are all eight-dimensional, and will be denoted $\mathbf{V}_8$, $\mathbf{S}_8$, and $\mathbf{C}_8$ respectively. The distinction between $\mathbf{S}_8$ and $\mathbf{C}_8$ is one of spacetime *chirality*, but the choice of which is to be associated with a given physical chirality is a matter of convention.

The ground state of Sector #2 has the structure

$$\{\tilde{b}_0^\mu\} \, |0\rangle_R \otimes \{b_0^\nu\} \, |0\rangle_L \qquad (6.4.9)$$

where the notation $\{b_0^\mu\}$ (and similarly for the right-movers) indicates that each of the individual Ramond zero-modes can be either excited or not excited.

How can we interpret (6.4.9) physically? This issue is actually quite subtle, and we shall not have the space to give a proper discussion. Moreover, as we have already indicated, we are not giving a fully rigorous treatment of Ramond zero-modes in these lectures, since our aim is to focus more on the physics than the formalism. However, it is possible to understand the appropriate physical interpretation intuitively. First, let us *count* the number of states in (6.4.9). *A priori*, it would seem that we have 2^{16} individual states, since each Ramond fermion zero-mode can either be excited or not excited. However, this is not correct because (as we shall discuss more completely in Lecture #5, and as we have already hinted in the footnote in Sec. 3.2), one should really count only one zero-mode per *pair* of Ramond Majorana-Weyl fermions. Thus, we can imagine that there are only four independent zero-modes for the right-movers, and four for the left-movers. Therefore, (6.4.9) consists of only $2^8 = 128$ states.

All combinations of these zero-mode excitations already satisfy the level-matching constraint (since $L_0 = \bar{L}_0 = 0$). Imposing either of the GSO constraints for Sector #2 in (6.4.5) then reduces the number of allowed states by a factor of two. Specifically, if we impose the constraint $N_L - N_R =$ odd, then we can choose only an even number of right-moving zero-mode excitations together with an odd number of left-moving zero-mode excitations, or an odd number of right-moving excitations together with an even number of left-moving excitations. Choosing the constraint $N_L - N_R =$

even has the opposite effect, pairing even numbers of excitations for left- and right-movers with each other, and likewise pairing odd numbers with each other.

Interpreting these results is therefore quite simple. As we discussed at the end of Lecture #3, the left-moving states and right-moving states are spacetime spinors, and we have already seen that there are two possible spinors, $\mathbf{S}_8$ and $\mathbf{C}_8$. At this stage, the names assigned to each are arbitrary, so we shall now establish the following convention: spinors realized by an even number of zero-mode excitations will be identified with $\mathbf{C}_8$, and those realized by an odd number of zero-mode excitations will be identified with $\mathbf{S}_8$. Of course, only the relative difference between these two spinors is physically significant (having the interpretation of spacetime chirality).

Given these definitions, we see that if we choose the first GSO constraint $N_L - N_R = \text{odd}$, the 128 states in (6.4.9) decompose into

$$(\bar{\mathbf{C}}_8 \otimes \mathbf{S}_8) \;\oplus\; (\bar{\mathbf{S}}_8 \otimes \mathbf{C}_8) \;, \tag{6.4.10}$$

whereas if we choose the second GSO constraint $N_L - N_R = \text{even}$, these states instead decompose into

$$(\bar{\mathbf{C}}_8 \otimes \mathbf{C}_8) \;\oplus\; (\bar{\mathbf{S}}_8 \otimes \mathbf{S}_8) \;. \tag{6.4.11}$$

If we wish to further decompose these states into representations of the Lorentz group, we can use the $SO(8)$ tensor-product relations

$$\mathbf{S}_8 \otimes \mathbf{S}_8 = \mathbf{1} \oplus \mathbf{28} \oplus \mathbf{35'}$$
$$\mathbf{C}_8 \otimes \mathbf{C}_8 = \mathbf{1} \oplus \mathbf{28} \oplus \mathbf{35''}$$
$$\mathbf{S}_8 \otimes \mathbf{C}_8 = \mathbf{V}_8 \oplus \mathbf{56} \;. \tag{6.4.12}$$

Here the **28** representation is the anti-symmetric component of the spinor tensor product (spin-one), while the **35'** and **35''** representations are the symmetric components of the spinor tensor product (also spin-one). (These latter representations are not to be confused with the spin-two **35** graviton representation in (6.4.8).) Likewise, the **56** is a certain vectorial (spin-one) higher-dimensional representation.* However, for our present purposes it

---

*For the mathematically inclined reader, we can succinctly describe all of these states as follows. Recall that a given representation is called a p-form if it can be realized as the totally anti-symmetric combination within the tensor product of p different vector indices of $SO(8)$, with resulting dimension $8 \times 7 \times 6 \times ... \times (9-p)/p!$. Using this language, we see that singlet states are zero-forms, the **28** representations are two-forms, and the **35'** and **35''** representations are "self-dual" four-forms. (The self-duality condition eliminates exactly half of the degrees of freedom in the four-form.) Likewise, the $\mathbf{V}_8$ state is a one-form, and the **56** representation is a three-form. These different forms (and the

will be sufficient to think of these states in the tensor-product forms (6.4.10) and (6.4.11). Note that in each case, the tensor product of two spacetime fermionic (spinor) states produces a spacetime bosonic state. Thus, just as in Sector #1, the states emerging in Sector #2 are spacetime bosons.

Thus, summarizing, we see that the spectra of our two resulting superstring models are as follows. First, from Sector #1, we have the tachyonic state $|0\rangle_R \otimes |0\rangle_L$. In the notation of $SO(8)$ Lorentz representations, this state may be denoted $\bar{\mathbf{1}} \otimes \mathbf{1}$; this tachyon is a Lorentz singlet. Next, we have the massless gravity multiplet. In the notation of $SO(8)$ Lorentz representations, this state takes the form $\bar{\mathbf{V}}_8 \otimes \mathbf{V}_8$. Finally, from Sector #2, we have massless states whose form depends on the particular choice of the GSO projection. In the first case, we have the states given in (6.4.10), while in the second case, we have the states given in (6.4.11). There are then, as usual, an infinite tower of massive (Planck-scale) states above these.

The string model produced by the first GSO projection is called the *Type 0A* string model, and the second is called the *Type 0B* string model. Collectively, these are sometimes simply called the Type 0 strings. As we see, both of these strings are tachyonic, and moreover they contain only bosonic states. Furthermore, as is evident from (6.4.10) and (6.4.11), both of these strings are non-chiral. In other words, they are invariant under the transposition $\mathbf{S}_8 \leftrightarrow \mathbf{C}_8$ for the left- and right-movers. These string theories were first constructed by N. Seiberg and E. Witten in 1985. Although not relevant for phenomenology, they are currently proving to have an important role in understanding certain non-perturbative aspects of non-supersymmetric string theory.

6.4.3.2. *The Type II strings*

Let us now turn to the second choice outlined at the beginning of Sec. 4.3, namely the case in which we consider the contributions from *all* of the sectors in Table 6.1. This will result in the so-called *Type II strings*. As we discussed at the end of Sec. 4.2, it is a general property that the larger the set of sectors that we consider, the more GSO constraints there are that must be imposed in each sector. Thus, the introduction of new sectors generally leads to new GSO constraints in each of the sectors (old and new), and likewise the introduction of new GSO constraints in a given sector requires the introduction of entire new sectors to compensate.

---

so-called *D-branes* whose existence they imply) are important when considering the non-perturbative structure of these string theories.

It turns out (and we shall see explicitly in Lecture #7) that if we consider the full set of sectors in Table 6.1, then the appropriate GSO constraints in each sector are given as follows:

$$\begin{aligned}
\text{Sector \#1:} \quad & N_L - N_R = \text{odd}, \quad N_R = \text{odd} \\
\text{Sector \#2:} \quad & N_L - N_R = \left\{ \begin{array}{c} \text{odd} \\ \text{even} \end{array} \right\}, \quad N_R = \text{odd} \\
\text{Sector \#3:} \quad & N_L - N_R = \text{even}, \quad N_R = \text{odd} \\
\text{Sector \#4:} \quad & N_L - N_R = \left\{ \begin{array}{c} \text{odd} \\ \text{even} \end{array} \right\}, \quad N_R = \text{odd}.
\end{aligned} \quad (6.4.13)$$

Note that in each case where a choice is possible, these choices are correlated: we simultaneously choose either the top lines within all braces, or the bottom lines. Thus, once again there are two sets of GSO conditions that can be imposed, resulting in two distinct string models.

Before proceeding further, it is useful to note the *pattern* of these GSO projections. In the case of the Type 0 strings, we considered only Sectors #1 and #2; as shown in Table 6.1, these were the sectors for which the right-moving fermions were always identical to the left-moving fermions and shared the same boundary conditions. The corresponding GSO projections in (6.4.5) likewise did not distinguish between right- and left-moving fermions. (In this context, note that the GSO projections in (6.4.5) can equivalently be written with minus signs replaced by plus signs.) Thus, in some sense, the Type 0 strings are symmetric under exchange of left- and right-movers. However, for the Type II strings, we have now introduced two additional sectors (Sectors #3 and #4) whose structure explicitly breaks this symmetry between left- and right-movers. No longer does each sector individually exhibit this left/right symmetry. As we see from (6.4.13), the effect of this breaking is to introduce additional GSO conditions which mirror this broken symmetry by becoming sensitive to right- or left-moving number operators *by themselves*. The technical word for this breaking of symmetry is "twisting" or "orbifolding", for by including Sectors #3 and #4, we see that we have twisted the left-movers relative to the right-movers by allowing them to have oppositely moded boundary conditions. Thus, the Type II strings that will result can be viewed as twisted (or orbifolded) versions of the Type 0 strings. This twisting procedure ultimately serves as the means by which more and more complicated (and more and more phenomenologically realistic) string models may be constructed, and will be discussed more fully in Lecture #7.

Given the GSO constraints in (6.4.13), we can proceed to determine the resulting spectrum just as we did for the Type 0 strings. Let us begin with Sector #1 (this is often called the "NS-NS sector"). Because the boundary conditions of the worldsheet fermions are the same in this sector as they were for the Type 0 strings, the possible states that arise are the same as they were for the Type 0 strings, and consist of the tachyon $|0\rangle_R \otimes |0\rangle_L$ as well as the gravity multiplet (6.4.7). The only difference is that we must now impose the additional GSO constraint $N_R =$ odd. It is immediately clear that the effect of this new GSO constraint is that *the tachyon is projected out of the spectrum*, while the gravity multiplet is retained. Thus, by "twisting" the Type 0 strings in just this way, we have succeeded in curing one of the major problems of the bosonic and Type 0 strings, namely the appearance of tachyons. Moreover, we have done this *without* eliminating the desirable gravity multiplet.

Let us now consider the states from Sector #2 (this is often called the "Ramond-Ramond" sector). Once again, if we impose only the first GSO constraint in (6.4.13), we obtain the states in either (6.4.10) or (6.4.11). Imposing the additional GSO constraint in (6.4.13) then enables us to project out half of these states, so that we retain only the states

$$\bar{\mathbf{S}}_8 \otimes \left\{ \begin{matrix} \mathbf{C}_8 \\ \mathbf{S}_8 \end{matrix} \right\} . \tag{6.4.14}$$

These states are spacetime bosons.

Finally, let us consider the states that arise in the new Sectors #3 and #4. In Sector #4, the vacuum energy is $(a_R, a_L) = (-1/2, 0)$. Therefore, in order to have level-matching ($L_0 = \bar{L}_0$), we see that we are immediately forced to excite a half-unit of energy for the right-movers while not increasing the energy of the left-movers. This is the only way to produce a massless state. This also ensures that this sector does not give rise to tachyons. Fortunately, since the right-moving fermions have Neveu-Schwarz boundary conditions in this sector, these fermions have half-integer modings, and thus by exciting their lowest modes we can indeed introduce a half-unit of energy. The left-moving fermions have Ramond boundary conditions in this sector, and hence their ground state is the Ramond zero-mode state. The massless states in Sector #4 therefore take the form

$$\tilde{b}^{\mu}_{-1/2} |0\rangle_R \otimes \{b^{\nu}_0\} |0\rangle_L . \tag{6.4.15}$$

At this stage, of course, these states satisfy only the level-matching constraint. Imposing the GSO constraints then leaves us with the state in which we excite only an even (or odd) number of left-moving Ramond zero modes.

How can we interpret this state? First, we notice that this state is a spacetime *fermion* because it results from tensoring a right-moving Neveu-Schwarz state with a left-moving Ramond state. Thus, we now have a string theory that contains spacetime fermions! This is yet another benefit of performing the "twist" that takes us from the Type 0 strings to the Type II strings. However, let us examine this state a bit more closely. Clearly, it has the Lorentz structure

$$\overline{\mathbf{V}_8} \otimes \begin{Bmatrix} \mathbf{C}_8 \\ \mathbf{S}_8 \end{Bmatrix} \qquad (6.4.16)$$

where we have retained the spinor-labelling conventions that we employed for the Type 0 strings. The relevant tensor-product decompositions in this case are given by

$$\overline{\mathbf{V}_8} \otimes \mathbf{C}_8 = \mathbf{S}_8 \oplus \mathbf{56}'$$
$$\overline{\mathbf{V}_8} \otimes \mathbf{S}_8 = \mathbf{C}_8 \oplus \mathbf{56}'' \qquad (6.4.17)$$

where the $\mathbf{S}_8$ and $\mathbf{C}_8$ representations are spin-1/2 and where the $\mathbf{56}'$ and $\mathbf{56}''$ representations are spin-3/2. Thus, we see that the Type II strings contain a massless, spin-3/2 object! Just as a massless spin-two object satisfies the Einstein field equations and must be interpreted as the graviton, a massless spin-3/2 object must be interpreted as a *gravitino* — i.e., a superpartner of the graviton. This implies that this string not only gives rise to spacetime bosons *and* fermions, but actually gives rise a spectrum which exhibits *spacetime supersymmetry*! This is yet another phenomenologically compelling feature.

Finally, let us now consider Sector #3. This sector has vacuum energies $(a_R, a_L) = (0, -1/2)$, so now we must excite right-moving zero-modes and left-moving $b^\mu_{-1/2}$ modes. This then leads to states of the form

$$\{\tilde{b}^\mu_0\} |0\rangle_R \otimes b^\nu_{-1/2} |0\rangle_L , \qquad (6.4.18)$$

and imposing the GSO projections results in states with the Lorentz structure $\overline{\mathbf{S}}_8 \otimes \mathbf{V}_8$. Once again, this also contains a gravitino!

So what do we have in the end? The first choice of GSO projections results in the so-called *Type IIA string*, while the second choice results in the *Type IIB string*. Both of these strings are tachyon-free, and their spectra contain both bosons and fermions. Moreover, these strings exhibit *spacetime* supersymmetry. This is most easily seen in the following suggestive way. Let us collect together the states from all four sectors, retaining our Lorentz-structure tensor-product notation:

$$\bar{\mathbf{V}}_8 \otimes \mathbf{V}_8\,,\quad \bar{\mathbf{S}}_8 \otimes \left\{ \begin{matrix} \mathbf{C}_8 \\ \mathbf{S}_8 \end{matrix} \right\}\,,\quad \bar{\mathbf{V}}_8 \otimes \left\{ \begin{matrix} \mathbf{C}_8 \\ \mathbf{S}_8 \end{matrix} \right\}\,,\quad \bar{\mathbf{S}}_8 \otimes \mathbf{V}_8\,. \qquad (6.4.19)$$

Together, this collection of states can be written in the factorized form

$$(\bar{\mathbf{V}}_8 \oplus \bar{\mathbf{S}}_8) \,\otimes\, \left(\mathbf{V}_8 \oplus \left\{ \begin{matrix} \mathbf{C}_8 \\ \mathbf{S}_8 \end{matrix} \right\} \right)\,. \qquad (6.4.20)$$

We thus see that there are *two* spacetime supersymmetries exhibited in this massless spectrum: the first exchanges $\bar{\mathbf{V}}_8 \leftrightarrow \bar{\mathbf{S}}_8$ amongst the right-movers, while the second exchanges

$$\mathbf{V}_8 \leftrightarrow \left\{ \begin{matrix} \mathbf{C}_8 \\ \mathbf{S}_8 \end{matrix} \right\} \qquad (6.4.21)$$

amongst the left-movers. Thus, the massless spectrum exhibits $N = 2$ supersymmetry. This is, of course, consistent with the appearance of two gravitinos in the massless spectrum (one from Sector #3 and one from Sector #4). Another way to understand this $N = 2$ supersymmetry is to realize that the first supersymmetry relates the bosonic states in Sector #1 to the fermionic states in Sector #3 (and the bosons in Sector #2 to the fermions in Sector #4), while the second supersymmetry relates the bosons in Sector #1 to the fermions in Sector #4 (and the bosons in Sector #2 to the fermions in Sector #3). In either case, we thus see that we have two independent spacetime supersymmetries.

It is important to note that we did not demand spacetime supersymmetry when constructing the superstring. We merely introduced *worldsheet* supersymmetry, and found that spacetime supersymmetry emerged naturally as the result of certain GSO projections. This further illustrates the fact that in string theory, spacetime properties such as supersymmetry emerge only as the consequences of deeper, more fundamental *worldsheet* symmetries. Another important point is that the same "twist" which eliminated the tachyon has introduced spacetime supersymmetry. While this is certainly an interesting phenomenon that arises for ten-dimensional superstrings, it is certainly *not* a general property that the elimination of the tachyon requires spacetime supersymmetry. In particular, we shall see in Lecture #6 that it is possible to construct string theories whose tree-level spectra lack spacetime supersymmetry but nevertheless are tachyon-free.

One might question whether we have really demonstrated the existence of $N = 2$ supersymmetry, since we have examined only the massless spectrum. However, it can be shown that any unitary theory which contains a massless spin-$3/2$ state necessarily exhibits supersymmetry, and hence must be supersymmetric at all mass levels (*i.e.*, for all massive, excited states as well). Of course, this is still not a proof, since we do not *a priori* know (and would therefore need to verify) that string theory is a consistent theory in this sense. However, it is possible to construct (two) explicit spacetime supercurrent operators and to demonstrate that they commute with the full (massless and massive) spectrum of the string. Another approach (as indicated in the footnote in Sec. 3.1) is to develop an alternative formulation of the superstring in which *spacetime* (rather than worldsheet) supersymmetry is manifest at the level of the string action, and to demonstrate the equivalence of the two formulations. Indeed, both approaches have been successfully carried out, thereby demonstrating that the Type II spectrum is indeed $N = 2$ supersymmetric. It is for this reason that these strings are referred to as Type II strings.

One important distinction between these two strings is their chirality. The Type IIA string, as we see, contains two supersymmetries of opposite chiralities, interchanging $\bar{\mathbf{V}}_8 \leftrightarrow \bar{\mathbf{S}}_8$ for the right-movers and $\mathbf{V}_8 \leftrightarrow \mathbf{C}_8$ for the left-movers. Equivalently, the two gravitinos associated with these supersymmetries are of opposite chiralities (because the **56'** and **56''** representations in (6.4.17) are of opposite chiralities). Because it contains supersymmetries of both chiralities, this string is ultimately non-chiral, and its low-energy (field-theoretic) limit consists of so-called *Type IIA supergravity* (whose discovery predates that of the Type IIA string). It is for this reason that this string is called the Type IIA string. The Type IIB string, by contrast, contains two supersymmetries (or two gravitinos) of the *same* chirality, exchanging $\mathbf{V}_8 \leftrightarrow \bar{\mathbf{S}}_8$ and $\mathbf{V}_8 \leftrightarrow \mathbf{S}_8$ respectively. Thus, this string theory is *chiral*, and has a low-energy field-theoretic limit consisting of Type IIB supergravity.

We conclude, then, that by introducing a twist relative to the Type 0 strings, we have constructed a set of strings (the Type IIA and Type IIB strings) that exhibit a number of compelling features: they are tachyon-free, they contain both bosons and fermions in their spacetime spectra, they contain gravity, and they are spacetime $N = 2$ supersymmetric. Despite this success, however, there is still something that we lack: we do not, as yet, have gauge symmetries. Specifically, there are no gauge bosons (such as photons, gluons, or W and Z particles). Likewise, there are no states which carry gauge charges. Therefore, once again, we shall need to construct a new kind of string.

6.5. Lecture #5: Neutrinos have gauge charges: The heterotic string

6.5.1. *Motivation and alternative approaches*

Thus far in these lectures, we have shown how string theory can give rise to quantized gravity, spacetime bosons and fermions, spacetime supersymmetry, and tachyon-free spectra. There is, however, one important phenomenological feature that is still missing: *gauge symmetry*. In other words, we wish to have massless gauge bosons, *i.e.*, spacetime vectors that transform in the adjoint representation of some internal symmetry group. As a side issue, we would also like to find a way of breaking $N = 2$ supersymmetry to $N = 1$ supersymmetry (if our goal is to reproduce the MSSM) or even to $N = 0$ supersymmetry (if our goal is to reproduce the Standard Model).

It is worth considering why such gauge-boson states fail to appear for the ten-dimensional Type II strings discussed in the previous lecture. The problem is the following. In order to produce worldsheet bosons, we are restricted to considering only the NS-NS or Ramond-Ramond sectors (Sectors #1 and #2 in Table 6.1). In the NS-NS sector (Sector #1), the vacuum energy is $(a_R, a_L) = (-1/2, -1/2)$, so we must excite the half-energy fermionic mode oscillators $\tilde{b}^\mu_{-1/2}, b^\mu_{-1/2}$ for the both the left- and right-movers. This produces a state with *two* vector indices rather than one, and as we see from the vector-vector tensor-product decomposition in (6.4.8), this does not contain a vectorial state. In the Ramond-Ramond sector (Sector #2), by contrast, the vacuum energy is $(a_R, a_L) = (0, 0)$, which implies that our massless states comprise the tensor product of two Ramond spinors as in (6.4.10) for the Type IIA string, or as in (6.4.11) for the Type IIB string. In the case of the Type IIB string, we see from (6.4.12) that the tensor product $\bar{\mathbf{S}}_8 \otimes \mathbf{S}_8$ does not contain a vector state $\mathbf{V}_8$. Thus, the Type IIB string contains no massless vectors. In the case of the Type IIA string, we observe from (6.4.17) that indeed $\bar{\mathbf{S}}_8 \otimes \mathbf{C}_8 \supset \mathbf{V}_8$, and thus the Type IIA string does contain a massless vector. (This state is often called a "Ramond-Ramond gauge boson".) However, the $U(1)$ "gauge" symmetry associated with this state is too small to contain the Standard-Model gauge group, and moreover it can be shown that no states in the perturbative spectrum of the Type IIA string spectrum can carry this Ramond-Ramond charge.*

In each case, the fundamental obstruction that we face is that we need to generate representations of a *gauge group* (*i.e.*, an internal symmetry group) that is *different* from the Lorentz group. Until now, all of our worldsheet fields (such as $X^{\mu}_{L,R}$ and $\psi^{\mu}_{L,R}$) have carried Lorentz indices associated with the $SO(D-1,1)$ Lorentz symmetry. In order to produce a separate gauge symmetry, we therefore need fields which do *not* carry a Lorentz index but which carry a purely internal index. (Note that these fields cannot carry a Lorentz index because we ultimately want our gauge symmetries to commute with the Lorentz symmetries.)

How can we do this? One idea is to *compactify* the Type II strings that we constructed in the previous lecture. Although this approach ultimately fails for phenomenological reasons, it will be instructive to briefly explain this idea. Recall that for the superstring, the critical dimension $D = 10$ emerges as the result of an anomaly cancellation argument: each worldsheet boson X contributes $c_X = 1$, each Majorana fermion ψ^{μ} contributes $c_{\psi} = 1/2$, and thus ten copies of each are necessary in order to cancel the "background" central charge associated with the worldsheet superconformal symmetry. But, even though we require ten bosons and ten fermions, there is no reason why we must endow *all* of them with Lorentz vector indices μ. Since we are ultimately interested in four-dimensional string theories, one natural idea is to consider these ten bosons and ten fermions in two groups, four with indices $\mu = 0, 1, 2, 3$, and the remaining six with purely internal indices $i = 1, ..., 6$. This internal symmetry could then be interpreted as a gauge symmetry.

This idea is in fact reminiscent of the original Kaluza-Klein idea whereby gauge symmetries are realized from higher-dimensional gravitational theories upon compactification. Moreover, this idea does succeed in producing gauge bosons (and gauge symmetries) in dimensions $D < 10$. However, the problem is that this idea fails to produce *enough* gauge symmetry. Specifically, although we obtain gauge symmetries that are large enough to contain the Standard Model gauge symmetry $SU(3) \times SU(2) \times U(1)$, we cannot obtain massless representations that simultaneously transform as

---

*Despite this fact, Ramond-Ramond charge plays a crucial role in recent developments concerning string duality. While none of the states in the *perturbative* Type IIA string spectrum carry Ramond-Ramond charge, these strings also contain non-trivial *solitonic* states (so-called *D-branes*) which do carry Ramond-Ramond charge. We shall briefly discuss D-branes in Lecture #8.

triplets of $SU(3)$ and doublets of $SU(2)$. Such "quark" representations are required phenomenologically. Thus, even though this compactification idea is interesting as a way of generating certain amounts of gauge symmetry, it cannot be used in order to save the superstring.

What we require, then, is a different way of introducing worldsheet fields without Lorentz vector indices. Since we will (temporarily) abandon the idea of removing Lorentz indices from our ten worldsheet bosons and fermions, what this means is that we require a way of obtaining *even more worldsheet fields* in ten dimensions. In other words, if we want bigger gauge symmetries in $D = 4$, then we require more than six extra fields with internal indices i, which in turn means that we already want extra fields even in the original ten-dimensional interpretation.

But how can we introduce extra worldsheet fields without violating our previous conformal anomaly cancellation arguments? Just adding extra fields will reintroduce the conformal anomaly at the quantum level.

6.5.2. The heterotic string: Constructing the action

The idea, of course, is to abandon the Type II string and proceed to construct a new kind of string that can accomplish the goal. This string is called the *heterotic* string, and it is this string that will be our focus for the remainder of these lectures. This string was first introduced by D. Gross, J. Harvey, E. Martinec, and R. Rohm in 1985, and for more than a decade dominated (and still continues to play a pivotal role in) discussions of string phenomenology.

Let us begin by recalling the action of the bosonic string:

$$S_{\text{bosonic}} = -\frac{1}{4\pi\alpha'} \int d^2\sigma \left\{ (\partial_- X_R^\mu)^2 + (\partial_+ X_L^\mu)^2 \right\} . \tag{6.5.1}$$

Here the worldsheet symmetry is simply conformal invariance, which requires that we take $\mu = 0, 1, ..., 25$ in order to cancel the conformal anomaly. Clearly, this action contains lots of worldsheet fields. However, we saw in Lecture #2 that this string does not give rise to spacetime fermions.

Next, we considered the superstring, whose action is given by:

$$S_{\text{super}} = -\frac{1}{4\pi\alpha'} \int d^2\sigma \left\{ (\partial_- X_R^\mu)^2 - \psi_R^\mu \partial_- \psi_{R\mu} + (\partial_+ X_L^\mu)^2 - \psi_L^\mu \partial_+ \psi_{L\mu} \right\} . \tag{6.5.2}$$

Here the worldsheet symmetry is *superconformal* invariance, which requires that we take $\mu = 0, 1, ..., 9$ in order to cancel the superconformal anomaly. Unlike the bosonic string, this string gives rise to spacetime fermions. But as we have just explained, this string does not contain enough worldsheet fields to give rise to appropriate gauge symmetries.

Clearly, each of these strings has an advantage lacked by the other. The natural solution, then, is to attempt to "weld" them together, to "crossbreed" them in such a way as to retain the desirable attributes of each. But how can this be done?

The fundamental observation is that we are always dealing with *closed* strings, and for closed strings, we have seen that the left- and right-moving modes are essentially independent of each other and form separate theories. Indeed, only the level-matching constraint $L_0 = \bar{L}_0$ serves to relate these two halves to each other, but even this constraint applies at the level of the physical Fock space rather than the level of the action. Therefore, since these two halves are essentially independent, a natural idea is to construct a new hybrid string whose left-moving half is the left-moving half of the bosonic string, but whose right-moving half is the right-moving half of the superstring. As we shall see, this fundamental idea is just what we need. The resulting string is therefore called a *heterotic* string, where the prefex *hetero-* indicates the joining of two different things.

Given this idea, let us now see how the action for the heterotic string can be constructed. We shall do this in three successive attempts. Our first attempt would be to write an action of the form

$$S = -\frac{1}{4\pi\alpha'} \int d^2\sigma \left\{ (\partial_- X_R^\mu)^2 - \psi_R^\mu \partial_- \psi_{R\mu} + (\partial_+ X_L^\mu)^2 \right\} . \quad (6.5.3)$$

In this case, the worldsheet symmetry would be conformal invariance for the left-movers, but superconformal invariance for the right-movers.

But what is the spacetime dimension of such a string? If we consider the right-moving sector, then just as in the superstring we would require $D = 10$, so that $\mu = 0, 1, ..., 9$. But given this, how do we interpret the left-moving side of the heterotic string? On the left-moving side, cancellation of the *conformal* (rather than superconformal) anomaly requires that we still retain 26 X_L fields! But if only ten of these fields are spacetime coordinates, then the remaining sixteen must be mere internal scalar fields. In other words, rather than carry the μ index (which would imply that these X fields would transform as vectors under the spacetime Lorentz group $SO(9,1)$), these sixteen extra fields must instead carry a purely internal index $i = 1, ..., 16$. So our second attempt at writing a heterotic string action would result in an action of the form

$$S = -\frac{1}{4\pi\alpha'} \int d^2\sigma \,\{(\partial_- X_R^\mu)^2 - \psi_R^\mu \partial_- \psi_{R\mu} + (\partial_+ X_L^\mu)^2 + (\partial_+ X_L^i)^2\} \tag{6.5.4}$$

where we have explicitly separated the left-moving bosons into two groups, with $\mu = 0, 1, ..., 9$ and $i = 1, ..., 16$.

But there still remains a subtlety. We cannot simply *decide* to remove the μ index from the X fields and make no other changes, because these X^i fields would continue to have a mode-expansion of the form (6.2.9) with the μ index replaced by an internal index i. While the interpretation of the oscillation exponential terms in (6.2.9) is not problematic, how would we interpret the "zero-mode" terms $x^i + \ell^2 p^i (\sigma_1 + \sigma_2)$? In the case of the spacetime coordinate fields X^μ, recall that these "zero-mode" quantities x^i and p^i are interpreted as the center-of-mass position and momentum of the string. But for purely internal fields X^i, this interpretation is problematic. To clarify this difficulty, let us consider the worldsheet energy $L_0^{(\text{com})}$ associated with these degrees of freedom, as in (6.2.13). Just as in the case of the spacetime coordinates X^μ, these worldsheet energies for the X^i fields would *a priori* take *continuous* values, thereby leading to a continuous spectrum even in $D = 10$. A continuous spectrum, of course, indicates nothing but the appearance of extra spacetime dimensions, so even though we may have replaced the index μ with the index i, we have not really solved the fundamental problem that there are too many uncompactified degrees of freedom amongst the left-movers.

Therefore, we still must find a way to replace this continuous spectrum with a discrete one. Because the following discussion is slightly technical and outside the main line of the development of the heterotic string action, we shall separate it from the main flow of the text. The reader uninterested in the following details can skip them completely and proceed directly to the resumption of the main text.

In order to eliminate this continuous spectrum, we must compactify these extra sixteen dimensions. This is analogous to discretizing the continuous spectrum of a free particle (plane wave) by localizing it in a box. In the present case, we can choose to compactify each of these extra spacetime "coordinates" X^i on a circle of radius R_i. What this means, operationally, is that we make the following topological identification in *spacetime*:

$$X^i \iff X^i + 2\pi R_i . \tag{6.5.5}$$

For simplicity (and as we shall see, without loss of generality), we shall take $R_i = R$ for all i. Thus, rather than demand simple periodicity of the X^i "coordinates" as in (6.2.6) as we traverse the closed string worldsheet, we must allow for the more general possibility

$$X^i(\sigma_1 + \pi, \sigma_2) = X^i(\sigma_1, \sigma_2) + 2\pi n_i R , \quad n_i \in \mathbb{Z} \tag{6.5.6}$$

where the integer n_i is called the "winding number". The interpretation of this condition is that as we traverse the closed string once on the *worldsheet* (i.e., as $\sigma_1 \to \sigma_1 + \pi$), the spacetime "coordinate" field X^i traverses the compactified spacetime circle n_i times. In other words, the closed string "winds" around the ith compactified *spacetime* circle n_i times. Because of this compactification, we see that the momentum p^i is now quantized (as we would expect for any particle in a periodic box of length R), and is restricted to take the values $p^i = m_i/R$, $m_i \in \mathbb{Z}$. Indeed, working out the most general mode-expansion consistent with (6.5.6), we find that a given such coordinate X^i takes the form

$$X(\sigma_1, \sigma_2) = x + 2nR\sigma_1 + \ell^2 \frac{m}{R}\sigma_2 + \text{oscillators} , \tag{6.5.7}$$

where $\ell \equiv \sqrt{2\alpha'}$ is our fundamental length scale and where 'oscillators' generically denotes the higher frequency modes. This decomposes into left- and right-moving components

$$X_{L,R}(\sigma_1 \pm \sigma_2) = \tfrac{1}{2}x + \left(\frac{\alpha' m}{R} \pm nR\right)(\sigma_2 \pm \sigma_1) + \text{oscillators} . \tag{6.5.8}$$

Comparing (6.5.8) with (6.2.9) enables us to identify the left- and right-moving compactified momenta

$$p_{L,R} \equiv \frac{m}{R} \pm \frac{nR}{\alpha'} . \tag{6.5.9}$$

We would then simply keep X_L in our heterotic theory.

Let us pause here to note an interesting phenomenon: this mode-expansion is invariant under the simultaneous exchange $R \leftrightarrow \alpha'/R$, $m \leftrightarrow n$. This is a so-called *T-duality*. What this means is that unlike point particles, strings cannot distinguish between extremely large spacetime compactification radii and extremely small spacetime compactification radii. Indeed, although the usual momentum m/R is extremely small in the first case and extremely large in the second, we see from the above mode-expansions that there is another contribution to the momentum, a "winding-mode momentum" nR/α', which compensates by growing large in the first case and small in the second. Since there is no physical way of distinguishing between these two types of momenta, the string spectrum is ultimately invariant under this T-duality symmetry. This duality underlies many of the unexpected physical properties of strings relative to point particles, and has important (and still not well-understood) implications for string cosmology. More importantly, however, this duality dramatically illustrates the breakdown of the traditional (field-theoretic) view of the linearly ordered progression of length scales and energy scales as we approach the string scale.

Having succeeded in avoiding the consequences of a continuous momentum p^i, our final question is the size of the radius R. It would certainly be aesthetically undesirable if we were forced to incorporate a new, fundamental, unfixed parameter R into our string theory. Fortunately, it turns out that in $D = 10$, there are only a very restricted set of possibilities that lead to consistent theories, and these restrictions imply that we can restrict our attention to the simple case $R = \ell = \sqrt{2\alpha'}$ *without loss of generality*. Thus, we see that R can be taken to be at the string scale, and hence essentially unobservable to "low-energy" measurements.

In order to see what is special about this radius, recall that the conformal anomaly contribution for each worldsheet boson is $c_X = 1$, while the conformal anomaly contribution for each worldsheet Majorana (real) fermion is $c_\psi = 1/2$. This suggests that the spectrum of a single compactified boson X might somehow be related to the spectrum of two Majorana fermions ψ_1, ψ_2, and this is indeed the case. Such a relation is typically referred to as a "boson-fermion equivalence" (which is possible in two dimensions because the usual spin-statistics distinction between bosons and fermions does not apply in two dimensions). In general, the spectrum of a compactified boson is identical to the spectrum of two Majorana fermions which are *coupled* to each other in a radius-dependent manner, and $R = \sqrt{2\alpha'}$ is the only value of the radius for which this coupling vanishes. Thus, if X is compactified on a circle of radius $R = \sqrt{2\alpha'}$, then the spectrum of quantum excitations of X is identical to the spectrum of quantum excitations of two *free* Majorana fermions ψ_1, ψ_2 (or equivalently those of one *complex* fermion $\Psi \equiv \psi_1 + i\psi_2$).[†] In fact, at a mathematical level, it turns out that this equivalence takes the form of an actual *equality* between the product $\psi_1 \psi_2$ and the partial derivative ∂X. Note, however, that while this specific radius is special from the point of view of boson/fermion equivalence, this is *not* the self-dual radius with respect to the T-duality transformation $R \leftrightarrow \alpha'/R$.

The upshot, then, is that in the action (6.5.4), we are free to replace the worldsheet bosons X^i ($i = 1, ..., 16$) with *complex* worldsheet fermions Ψ^i ($i = 1, ..., 16$). For ten-dimensional heterotic strings, we shall see that this replacement can be made *without loss of generality*. This replacement suffices to make the center-of-mass "momenta" associated with the X^i fields discrete rather than continuous, as we require. Given this, the final action for the heterotic string takes the form:

$$S_{\text{heterotic}} = -\frac{1}{4\pi\alpha'} \int d^2\sigma \left\{ (\partial_+ X_L^\mu)^2 - \bar{\Psi}_L^i \partial_+ \Psi_L^i + (\partial_- X_R^\mu)^2 - \psi_R^\mu \partial_- \psi_{R\mu} \right\} \quad (6.5.10)$$

[†]We are again cheating slightly here. The rigorous statement is that we must compactify the X boson on a so-called $\mathbb{Z}_2$ orbifold with this radius in order for the spectrum of X to be identical to that of two free Majorana fermions. The equivalence between these bosonic and fermionic systems can be demonstrated explicitly at the level of their full underlying left/right two-dimensional conformal field theories. By contrast, compactifying X on a *circle* of this radius yields the spectrum of a single *complex* fermion, and the full left/right conformal field theory corresponding to a single complex fermion actually differs from that corresponding to two real fermions. These distinctions between circles and orbifolds, and likewise between a single complex fermion and two real fermions, will not be relevant for what follows.

where ψ_R are Majorana-Weyl (real) right-moving worldsheet fermions, where Ψ_L are complex Weyl left-moving fermions, and where $\mu = 0, 1, ..., 9$ and $i = 1, ..., 16$.

6.5.3. Quantizing the heterotic string

The next step, then, is to quantize the worldsheet fields of the heterotic string. The quantization of the bosonic fields X^μ and worldsheet Majorana fermions ψ_R^μ was discussed in previous lectures, and does not change in this new setting. The only new feature, then, are the mode-expansion and quantization rules for the *complex* fermions Ψ_L^i.

Once again, there are two possible mode expansions for the left-moving complex fermions Ψ, depending on whether we choose Neveu-Schwarz (anti-periodic) or Ramond (periodic) boundary conditions.[‡] In the case of anti-periodic boundary conditions, recall that our mode-expansion (6.3.12) for left-moving *real* (Majorana) fermions can be written in the form

$$\psi(\sigma_1 + \sigma_2) = \sum_{r=1/2}^{\infty} \left[b_r e^{-ir(\sigma_1+\sigma_2)} + b_r^\dagger e^{+ir(\sigma_1+\sigma_2)} \right] \quad (6.5.11)$$

where we recall the hermiticity condition $b_{-r} = b_r^\dagger$. Thus, for a left-moving *complex* fermion, our analogous mode-expansion takes the form

$$\Psi(\sigma_1 + \sigma_2) = \sum_{r=1/2}^{\infty} \left[b_r e^{-ir(\sigma_1+\sigma_2)} + d_r^\dagger e^{+ir(\sigma_1+\sigma_2)} \right] \quad (6.5.12)$$

which of course implies

$$\Psi^\dagger(\sigma_1 + \sigma_2) = \sum_{r=1/2}^{\infty} \left[b_r^\dagger e^{+ir(\sigma_1+\sigma_2)} + d_r e^{-ir(\sigma_1+\sigma_2)} \right] . \quad (6.5.13)$$

For $r > 0$, b_r destroys fermionic excitations and $b_r^\dagger$ creates them, while d_r destroys *anti-fermionic* excitations and $d_r^\dagger$ creates them. Thus, as expected, the only new feature is the presence of twice as many mode degrees of freedom, one set associated with fermionic excitations and the other

---

[‡]Because there is no worldsheet supersymmetry that relates these left-moving fermions to corresponding left-moving bosons X^μ, more general boundary conditions may actually be imposed in this case. However, for heterotic strings in ten dimensions, it turns out that we can restrict our attention to periodic or anti-periodic boundary conditions without loss of generality. Fermions with generalized worldsheet boundary conditions will be discussed further in Lecture #7.

with their anti-fermionic counterparts. These modes satisfy the usual anti-commutation relations

$$\{b_r^\dagger, b_s\} = \{d_r^\dagger, d_s\} = \delta_{rs} \ . \tag{6.5.14}$$

The corresponding number operator and worldsheet energy contributions are then given by

$$N = \sum_{r=1/2}^{\infty} \left(b_r^\dagger b_r - d_r^\dagger d_r\right)$$

$$L_0 = \sum_{r=1/2}^{\infty} r \left(b_r^\dagger b_r + d_r^\dagger d_r\right) \ . \tag{6.5.15}$$

Note that the anti-particle excitations *subtract* from the number operator yet *add* to the total energy. Finally, as expected, the vacuum energy contribution from each complex Neveu-Schwarz fermion is twice that for each real Neveu-Schwarz fermion: $a_\Psi = 2a_\psi = -1/24$.

The Ramond case, of course, is more subtle because of the zero-mode. It turns out that the complex-fermion mode-expansion is given by

$$\Psi(\sigma_1 + \sigma_2) = \sum_{n=1}^{\infty} [b_n e^{-in(\sigma_1+\sigma_2)} + d_n^\dagger e^{+in(\sigma_1+\sigma_2)}] + b_0$$

$$\Psi^\dagger(\sigma_1 + \sigma_2) = \sum_{n=1}^{\infty} [b_n^\dagger e^{+in(\sigma_1+\sigma_2)} + d_n e^{-in(\sigma_1+\sigma_2)}] + b_0^\dagger, \tag{6.5.16}$$

with the anti-commutation relations

$$\{b_m^\dagger, b_n\} = \{d_m^\dagger, d_n\} = \delta_{mn} \ . \tag{6.5.17}$$

In (6.5.16), we have explicitly separated out the zero-mode from the higher-frequency modes. The number operator and worldsheet energy conributions are given by

$$N = \sum_{r=1/2}^{\infty} \left(b_r^\dagger b_r - d_r^\dagger d_r\right) + b_0^\dagger b_0$$

$$L_0 = \sum_{r=1/2}^{\infty} r \left(b_r^\dagger b_r + d_r^\dagger d_r\right) \ . \tag{6.5.18}$$

Note that there is no worldsheet energy contribution from the zero-modes. Finally, the vacuum energy contribution from each complex Ramond fermion is twice that for each real Ramond fermion: $a_\Psi = 2a_\psi = +1/12$.

One might wonder, at first, why there is no *anti-particle* zero-mode d_0. However, such an anti-particle zero-mode d_0 would be *equivalent* to the *particle* zero-mode b_0. The easiest way to see this is to realize that ultimately (6.5.16) represents a Fourier-decomposition of the $\Psi(\sigma_1+\sigma_2)$ into different harmonic frequencies (exponentials). By its very nature, the zero-mode is the constant term in such a decomposition (since it corresponds to zero frequency), and this constant term is nothing but b_0. However, there can only be *one* degree of freedom associated with a given constant term. Having an additional zero-mode d_0 would thus represent a redundant (non-independent) degree of freedom. Of course, whether we associate b_0 or d_0 with the constant term is purely a matter of convention.

Given this observation, we are finally in a position to explain our counting of zero-mode states in Lectures #3 and #4. Since there is only one zero-mode degree of freedom for each *complex* worldsheet fermion, there can really be only "half" a zero-mode for each *real* worldsheet fermion. This explains the footnote in Sec. 3.2, and also explains why (in the paragraph following (6.4.9)) we counted only one zero-mode excitation per *pair* of Majorana fermions. This also explains why, ultimately, the treatment of the Ramond zero-mode for a *real* worldsheet fermion is rather subtle: essentially we must take a "square root" of the complex Ramond zero-mode b_0. There does exist a consistent method for taking this square root, but this is beyond the scope of these lectures. For our purposes, it will simply be sufficient to recall that there is only one zero-mode state for each complex worldsheet fermion, or for each pair of real worldsheet fermions.

6.6. Lecture #6: Some famous heterotic strings

Our next step is to construct actual heterotic string *models*, just as we did for the superstring. This will be the subject of the present lecture.

6.6.1. *General overview*

Before plunging into details, it is worthwhile to consider the general features that will govern the construction of our heterotic string models. Recall from the previous lecture that the worldsheet fields of the heterotic string in light-cone gauge consist of eight right-moving worldsheet bosons X_R^μ, eight left-moving worldsheet bosons X_L^μ, eight right-moving Majorana (real) worldsheet fermions ψ_R^μ, and sixteen left-moving complex worldsheet fermions Ψ_L^i ($i = 1, ..., 16$).

The role of the right-moving fermions ψ_R^μ is the same as in the superstring: if they have Neveu-Schwarz modings, the corresponding states are spacetime bosons, and if they Ramond modings, the corresponding states are spacetime fermions. Indeed, by properly stitching these sectors together, it may also be possible to obtain spacetime supersymmetry (as in the superstring). Note that unlike the superstring, however, these boson/fermion identifications hold *regardless* of the modings of the left-moving complex fermions Ψ_L^i. This is because only the right-moving fermions carry spacetime Lorentz indices μ, and hence only these fermions determine the representations of the spacetime Lorentz algebra.

The role of the left-moving complex fermions Ψ_L^i is analogous. Because they carry internal indices rather than spacetime Lorentz indices, the symmetries they carry are also internal, and as we shall see, they can be interpreted as gauge symmetries. Indeed, these Ψ_L^i fields are precisely the internal fields we were hoping to obtain in Sec. 5.1. When they have Neveu-Schwarz modings, these fermions provide "vectorial" (scalar, vector, tensor) representations of the internal gauge symmetry. When they have Ramond modings, by contrast, they provide "spinorial" representations of the internal gauge symmetry. Thus, we expect a rich gauge representation structure in these models as well.

As with the superstring, different models can be constructed depending on how the different modings are joined together to form our set of underlying sectors, and how the corresponding GSO constraints are implemented. We shall construct explicit models below. But it is already apparent that the heterotic string contains all the ingredients we require for successful phenomenology. By choosing certain combinations of right-moving fermionic modings with left-moving fermionic modings, we can control which gauge-group representations are bosonic and which are fermionic. Moreover, by choosing the relative modings *amongst* the left-moving complex fermions, we can even control the gauge group that is ultimately produced.

6.6.2. *Sectors and GSO constraints*

Just as in the superstring, we begin the process of model-building by choosing an appropriate set of underlying sectors and corresponding GSO constraints. Moreover, just as in the superstring, we know that preservation of the right-moving worldsheet supersymmetry (or equivalently spacetime Lorentz invariance) requires that we choose our eight right-moving fermions ψ_R^μ to all have the same boundary condition in each sector. This implies that we can, if we wish, combine these right-moving fermions to form four complex right-moving fermions which we can denote Ψ_R^μ. (We retain the index μ to remind ourselves that these fields carry indices with respect to the spacetime Lorentz algebra, even though strictly speaking only the real fields ψ_R^μ carry such vectorial indices.) However, unlike the superstring, there is no longer any such restriction on the boundary conditions of the left-moving fermions Ψ_L^μ. Thus, there remains substantial freedom in choosing the boundary conditions of these left-moving fermions. Ultimately this choice becomes the choice of the gauge group for the particular model in question.

In the next lecture, we shall provide a detailed discussion of the rules by which one can choose these boundary conditions and determine their associated GSO constraints. Therefore, for the time being, we shall simply restrict our attention to the sectors listed in Table 6.2. Note that the corresponding vacuum energies are also listed in Table 6.2. In order to compute these energies, we can continue to use the middle expression in (6.4.1) where we recall that n_R and n_{NS} count the number of *real* worldsheet fermions. Thus, for complex fermions, these numbers are doubled.

Table 6.2. Eight possible sectors for ten-dimensional heterotic strings, numbered 1 through 8. Here 'NS' and 'R' respectively indicate Neveu-Schwarz (anti-periodic) and Ramond (periodic) boundary conditions for worldsheet fermions, and a_R and a_L respectively denote the corresponding right- and left-moving vacuum energies.

#	$\psi_R^{i=1,\ldots,8}$	$\Psi_L^{i=1,\ldots,8}$	$\Psi_L^{i=9,\ldots,16}$	a_R	a_L
1	NS	NS		$-1/2$	-1
2	R	R		0	$+1$
3	NS	R		$-1/2$	$+1$
4	R	NS		0	-1
5	NS	NS	R	$-1/2$	0
6	NS	R	NS	$-1/2$	0
7	R	NS	R	0	0
8	R	R	NS	0	0

Before proceeding further, we can immediately deduce some physical properties of the string states that would emerge in each sector. First, we see that Sector #1 is the only sector from which tachyons can possibly emerge. This is because the level-matching constraints prevent tachyons in any other sector (*i.e.*, there is no other sector which for which both a_L and a_R are negative). Second, we observe that Sectors #2 and #3 cannot give rise to massless states. This again follows from the level-matching constraints, and implies that (for phenomenological purposes) we will not need to consider the states arising in these sectors. Finally, we observe that Sectors #1,3,5,6 give rise to spacetime bosons, while Sectors #2,4,7,8 give rise to spacetime fermions.

In some sense, Sectors #1–4 are the direct analogues of the four possible sectors in Table 6.1 for the superstring. Thus, the heterotic models that result from these sectors will be the analogues of the Type 0 and Type II superstring models. However, the additional Sectors #5–8 represent new sectors that arise only for heterotic strings. We hasten to add that these sectors are not unique, and others could equally well have been chosen. We will discuss these possibilities in the next lecture.

The next issue we face is to determine which *combinations* of sectors form self-consistent sets. It turns out (following the rules to be discussed in Lecture #7) that there are three different possibilities:

- Case A: we consider Sectors #1 and #2 by themselves;
- Case B: we consider Sectors #1 through #4 by themselves; or
- Case C: we consider *all* Sectors #1 through #8.

For each of these cases, there is then a different set of GSO constraints for each sector. As we have seen in our discussion of the superstring, the more sectors we have in our model, the more GSO constraints there are in each sector. In particular, each time the number of sectors doubles, the number of GSO constraints in each sector increases by one. For completeness, Table 6.3 lists the GSO constraints that apply in each sector for each of these three cases.

Once again, observe the *pattern* of the GSO constraints. In Case A, we have only Sectors #1 and #2, for which all right-moving and left-moving boundary conditions are identical. Thus, the GSO constraints that apply in Case A combine N_L and N_R together. (Recall that since $N_{L,R} \in \mathbb{Z}$, we can just as easily write the GSO constraint for Case A as $N_L + N_R =$ odd.) When we move from Case A to Case B, we introduce two new sectors (Sectors #3 and #4 in Table 6.2) which "twist" the boundary conditions of the right-movers relative to those of the left-movers. This has the effect of introducing a new GSO constraint in each sector, one which distinguishes separately between N_L and N_R. Finally, when we move from Case B to Case C, we introduce four new sectors (Sectors #5 through #8) which introduce an additional "twist" that distinguishes between the first eight left-moving fermions $\Psi_L^{i=1,\ldots,8}$ and the second eight left-moving fermions

$\Psi_L^{i=9,\ldots,16}$. The corresponding new GSO constraint in each sector is then one which is sensitive only to $^{(8)}N_L \equiv \sum_{i=1}^{8} N^{(i)}$. This suggests (and we shall see explicitly in Lecture #7) that the set of sectors is deeply correlated with the set of GSO constraints that are applied in each sector: each new "twist" introduces both a new set of sectors and a new GSO constraint in each sector. The fact that we are considering only Ramond or Neveu-Schwarz boundary conditions for our left-moving complex fermions Ψ_L^i means that each successive twist doubles the number of sectors and introduces one new GSO constraint in each sector. These are called $\mathbb{Z}_2$ twists. If we were to consider more general "multi-periodic" boundary conditions for the left-moving fermions (which is possible because they are not related to the left-moving worldsheet bosons by worldsheet supersymmetry), then we could introduce so-called "higher-order" twists that would result in more complicated GSO constraints. However, it turns out that in ten dimensions, we lose no generality by restricting our attention to such $\mathbb{Z}_2$ twists.

6.6.3. *Four ten-dimensional heterotic string models*

It is apparent from Table 6.3 that Case A and Case B each correspond to one heterotic string model, while Case C corresponds to two separate heterotic string models. Thus, the GSO constraints in Table 6.3 together give rise to four distinct heterotic string models. In the remainder of this lecture, we shall work out the physical properties of these four models.

6.6.3.1. *The non-supersymmetric SO(32) string*

Let us begin by considering Case A, which consists of only Sectors #1 and #2. Only Sector #1 (the so-called "NS-NS sector") can contain massless states. As indicated in Table 6.1, the vacuum energy in this sector is $(a_R, a_L) = (-1/2, -1)$. Thus, at the bare minimum, the level-matching constraint $L_0 = \bar{L}_0$ forces us to excite at least a half-unit of energy on the left-moving side. This can be accomplished by exciting any of the left-moving half-unit fermionic modes, since in this sector the left-moving fermions all have Neveu-Schwarz boundary conditions and thus contain half-integer modings. This produces the 32 possible states

$$|0\rangle_R \otimes b^i_{-1/2}|0\rangle_L \quad \text{and} \quad |0\rangle_R \otimes d^i_{-1/2}|0\rangle_L \,. \tag{6.6.1}$$

Note that these states also satisfy the single applicable GSO constraint $N_L - N_R = \text{odd}$, so they remain in the spectrum. From (6.4.6), we see that these states are tachyonic with $\alpha' M^2 = -2$.

Table 6.3. GSO constraints for each of the eight heterotic string sectors in Table 6.2. Here the notation $^{(8)}N_L \equiv \sum_{i=1}^{8} N^{(i)}$ indicates the total left-moving number operator for only the first *eight* left-moving complex fermions. As before, the braces indicate different *correlated* choices of GSO projections, so that we simultaneously choose either the upper choice or the lower choice for all sets.

Sector #	Case A	Case B	Case C
1	$N_L - N_R =$ odd	$N_L - N_R =$ odd $N_L =$ even	$N_L - N_R =$ odd $N_L =$ even $^{(8)}N_L =$ even
2	$N_L - N_R =$ odd	$N_L - N_R =$ odd $N_L =$ even	$N_L - N_R =$ odd $N_L =$ even $^{(8)}N_L =$ even
3	—	$N_L - N_R =$ odd $N_L =$ even	$N_L - N_R =$ odd $N_L =$ even $^{(8)}N_L = \left\{ \begin{array}{c} \text{odd} \\ \text{even} \end{array} \right\}$
4	—	$N_L - N_R =$ odd $N_L =$ even	$N_L - N_R =$ odd $N_L =$ even $^{(8)}N_L = \left\{ \begin{array}{c} \text{odd} \\ \text{even} \end{array} \right\}$
5	—	—	$N_L - N_R =$ odd $N_L = \left\{ \begin{array}{c} \text{odd} \\ \text{even} \end{array} \right\}$ $^{(8)}N_L = \left\{ \begin{array}{c} \text{odd} \\ \text{even} \end{array} \right\}$
6	—	—	$N_L - N_R =$ odd $N_L = \left\{ \begin{array}{c} \text{odd} \\ \text{even} \end{array} \right\}$ $^{(8)}N_L =$ even
7	—	—	$N_L - N_R =$ odd $N_L = \left\{ \begin{array}{c} \text{odd} \\ \text{even} \end{array} \right\}$ $^{(8)}N_L =$ even
8	—	—	$N_L - N_R =$ odd $N_L = \left\{ \begin{array}{c} \text{odd} \\ \text{even} \end{array} \right\}$ $^{(8)}N_L = \left\{ \begin{array}{c} \text{odd} \\ \text{even} \end{array} \right\}$

Further states are realized by exciting higher worldsheet modes. Because our worldsheet modes are quantized in minimum half-integer steps, we see that the next excited states in this model are massless. These states come in two varieties:

$$\tilde{b}^{\mu}_{-1/2}|0\rangle_R \otimes \alpha^{\nu}_{-1}|0\rangle_L \qquad (6.6.2)$$

and

$$\tilde{b}^{\mu}_{-1/2}|0\rangle_R \otimes \begin{cases} b^i_{-1/2} b^j_{-1/2} |0\rangle_L \\ b^i_{-1/2} d^j_{-1/2} |0\rangle_L \\ d^i_{-1/2} b^j_{-1/2} |0\rangle_L \\ d^i_{-1/2} d^j_{-1/2} |0\rangle_L \end{cases} . \quad (6.6.3)$$

In (6.6.2), we have excited the lowest mode of the left-moving worldsheet boson X^{μ}_L, whereas in (6.6.3) we have excited two of the lowest modes of the left-moving fermions $\Psi^{i,j}_L$. Note that it is possible to excite both the particle and anti-particle modes from the same fermion Ψ^i, and thus there is no restriction that $i \neq j$. Also note that all of these states in (6.6.2) and (6.6.3) satisfy the GSO constraint $N_L - N_R =$ odd. While $N_R = 1$ in all cases, we have $N_L = 0$ in (6.6.2) (since the number operators are defined not to include the contributions from worldsheet bosons), and $N_L = 2$ in (6.6.3).

How do we interpret these states? Once again, the states (6.6.2) are easily recognized as our gravity multiplet, consisting of the spin-two graviton $g_{\mu\nu}$, the spin-one anti-symmetric tensor $B_{\mu\nu}$, and the spin-zero dilaton ϕ. It is interesting to note that this state (6.6.2) is realized as a hybrid of the gravity multiplet state in the bosonic string (6.2.19) and in the superstring (6.4.7). This reflects the underlying construction of the heterotic string, and ensures that the heterotic string, like its predecessors, is also a theory of quantized gravity. Once again, the appearance of the gravity multiplet is a useful cross-check of the GSO constraints.

The states in (6.6.3) have a different interpretation, however. Clearly, their Lorentz structure indicates that they are massless Lorentz vectors. Thus, they are to be interpreted as spacetime *gauge bosons*. Thus, we see that the heterotic string has succeeded in providing us with spacetime gauge symmetry, just as we had originally hoped.

But what is the gauge group? Of course, the gauge group is ultimately determined from the i, j indices, and since (in Cases A and B) we have not destroyed the rotational symmetry in the space of the 16 complex left-moving fermions Ψ^i_L (or the 32 real left-moving fermions into which they can be decomposed), we immediately suspect that the gauge symmetry should be $SO(32)$. There are number of ways to deduce that this is correct. Perhaps the easiest way is simply to *count* the gauge boson states in (6.6.3). If we restrict our attention to the cases $i \neq j$, then there are $(2 \cdot 16)(2 \cdot 15)/2$ states. The first factor $(2 \cdot 16)$ reflects the fact that for each of the 16 possible choices of Ψ_L^i, we can excite either the fermion or anti-fermion mode. The second factor $(2 \cdot 15)$ reflects the same set of options for the second fermion Ψ_L^j, and we divide by two as the interchange symmetry factor. There are also the cases with $i = j$: from such cases we obtain 16 possible states, reflecting the 16 different fermions Ψ_L^i whose fermion and anti-fermion modes are jointly excited. The total number of states is then

$$\frac{(2 \cdot 16)(2 \cdot 15)}{2} + 16 = 496 = \dim SO(32) \,. \qquad (6.6.4)$$

Of course, the above counting method for determining the gauge group is hardly precise, for there are a number of gauge groups with the same overall dimension (and we shall come across another such gauge group very soon). We therefore require a more sophisticated method which also generalizes to more complicated cases. By definition, of course, the gauge group can be determined by explicitly examining the charges of the gauge boson states and determining which Lie algebra (*i.e.*, which root system) they fill out. We therefore need a way of determining the charges of the gauge boson states. Since our gauge symmetry is ultimately associated with the left-moving worldsheet fermions Ψ_L^i, the relevant current in this case is simply the worldsheet current $J^i \equiv \bar{\Psi}_L^i \Psi_L^i$. From this, we can deduce the associated charge Q_i. It turns out that

> **Great Leap #7:** The charge associated with each worldsheet fermion Ψ_L^i for a given string state with fermionic excitation number $N^{(i)}$ is given by $Q_i \equiv N^{(i)} + q_i$. Here q_i is a "background" charge which is 0 if Ψ_L^i is a Neveu-Schwarz fermion and $-1/2$ if Ψ_L^i is a Ramond fermion.

Given this result, we can easily deduce the gauge group for the case in question. For simplicity, let us first imagine that there are only *two* left-moving fermions $\Psi_L^{i=1,2}$. In this case, (6.6.3) reduces to six states:

$$b_{-1/2}^1 b_{-1/2}^2 |0\rangle_L \,, \quad b_{-1/2}^1 d_{-1/2}^2 |0\rangle_L \,, \quad d_{-1/2}^1 b_{-1/2}^2 |0\rangle_L \,,$$
$$d_{-1/2}^1 d_{-1/2}^2 |0\rangle_L \,, \quad b_{-1/2}^1 d_{-1/2}^1 |0\rangle_L \,, \quad b_{-1/2}^2 d_{-1/2}^2 |0\rangle_L \,. \qquad (6.6.5)$$

For each of these states, there are two charges, Q_1 and Q_2, associated with each of the two complex fermions. If we denote these states as A through F respectively, we can plot the charges of these six states as in Fig. 6.8. The resulting diagram is easily recognized as the root system (or equivalently the weight system of the adjoint representation) of the Lie group $SO(4)$. Generalizing from two complex fermions to n complex fermions analogously

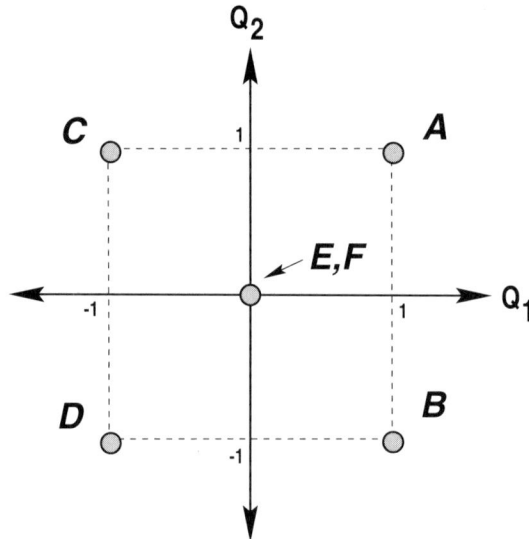

Fig. 6.8. The two-dimensional "charge lattice" associated with the six string states A through F in (6.6.5). Note that the two states E and F fill out the Cartan subalgebra of the root system. For a ten-dimensional heterotic string, the charge lattice is always sixteen-dimensional (generally implying a gauge group of rank 16), with a Cartan subalgebra consisting of sixteen gauge boson states.

yields the gauge group $SO(2n)$, provided that all n complex fermions have the same modings. Thus, in the case of 16 complex fermions, we find the gauge group $SO(32)$.

Note that this argument suffices to show that the gauge bosons fill out the adjoint representation of $SO(32)$. However, it does not demonstrate that all other string states in the model fall into representations of this gauge group. Of course, this is required for the consistency of the string. However, such a result can indeed be proven mathematically by constructing the current operators associated with the gauge group in question (as discussed above), and demonstrating that all states surviving the appropriate GSO constraints transform appropriately under these currents. For example, the 32 tachyonic states in (6.6.1) transform in the vector representation of $SO(32)$, and the gravity multiplet (6.6.2) transforms as a singlet of $SO(32)$ (as it must). However, a proof that this holds for all states in both the massless and massive string spectrum is beyond the scope of these lectures.

We should also point out that what emerges in such closed string theories is not simply the algebra associated the gauge symmetry in question, but rather an infinite-dimensional extension (or "affinization") of it. Such affine Lie algebras are discussed in Ginsparg (reference given at the end of Lecture #1), and play an important role in the consistency and phenomenology of such heterotic string theories.

To summarize, then, we see that Case A results in a tachyonic string model with quantum gravity and $SO(32)$ gauge symmetry. In addition to 32 scalar tachyons transforming in the vector representation of $SO(32)$, this model contains massless gauge bosons transforming in the adjoint representation of $SO(32)$ as well as the usual gravity multiplet. This non-supersymmetric $SO(32)$ heterotic string model is the heterotic analogue of the Type 0 string models in Lecture #4.

6.6.3.2. The supersymmetric $SO(32)$ string

Let us now proceed to Case B. In this case there are four sectors (#1 through #4 in Table 6.2), and we must impose the GSO constraints listed in the second column of Table 6.3.

Let us begin by considering the states from Sector #1. These are the same as those considered in Case A, except that we must now impose the additional GSO constraint $N_L = $ even. This projects out the tachyonic states (6.6.1), but preserves the gravity multiplet as well as the gauge bosons.

As we discussed previously, Sectors #2 and #3 contain no massless states. Therefore, all that remains is to consider the states from Sector #4. Here the vacuum energy is $(a_R, a_L) = (0, -1)$. The right-moving ground state in this sector is the Ramond zero-mode ground state, which we have previously denoted $\{\tilde{b}_0^\mu\}|0\rangle_R$, and thus massless states are realized only through non-zero excitations of the left-movers. The possible states are

$$\{\tilde{b}_0^\mu\}|0\rangle_R \otimes \begin{cases} \alpha^\nu_{-1}|0\rangle_L \\ b^i_{-1/2} b^j_{-1/2}|0\rangle_L \\ b^i_{-1/2} d^j_{-1/2}|0\rangle_L \\ d^i_{-1/2} b^j_{-1/2}|0\rangle_L \\ d^i_{-1/2} d^j_{-1/2}|0\rangle_L \end{cases} \quad (6.6.6)$$

In each case, the GSO constraints imply that we can excite only an odd number of right-moving zero-modes. According to our previous conventions, this indicates that the right-moving ground state corresponds to the spacetime Lorentz spinor $\bar{\mathbf{S}}_8$ (rather than the conjugate spinor $\bar{\mathbf{C}}_8$).

It is, by now, easy to interpret the states in (6.6.6). The first state provides the superpartner states to the gravity multiplet, and contains a gravitino. This implies that the model has spacetime supersymmetry. Likewise, the remaining states correspond to the superpartners of the $SO(32)$ gauge bosons, and contain the $SO(32)$ gauginos. The chirality of these spinor states is fixed by the GSO constraint and the right-moving ground state $\bar{\mathbf{S}}_8$.

Summarizing, we see that this model therefore consists of the following states. We shall describe these states using the notation $\bar{R}_1 \otimes (R_2; R_3)$ where R_1, R_2 are representations of the spacetime Lorentz group, and where R_3 is a representation of the $SO(32)$ gauge group. These states consist of

$$\bar{\mathbf{V}}_8 \otimes (\mathbf{V}_8; \mathbf{1}) \ , \quad \bar{\mathbf{V}}_8 \otimes (\mathbf{1}; \mathbf{adj}) \ , \quad \bar{\mathbf{S}}_8 \otimes (\mathbf{V}_8; \mathbf{1}) \ , \quad \bar{\mathbf{S}}_8 \otimes (\mathbf{1}; \mathbf{adj}) \ , \quad (6.6.7)$$

where the first and third states form the $N = 1$ supergravity multiplet and the second and fourth states form the $SO(32)$ gauge boson supermultiplet. Together these states can be written in the factorized form

$$(\bar{\mathbf{V}}_8 \oplus \bar{\mathbf{S}}_8) \ \otimes \ \{(\mathbf{V}_8; \mathbf{1}) \oplus (\mathbf{1}; \mathbf{adj})\} \ , \quad (6.6.8)$$

thereby explicitly exhibiting the supersymmetry $\bar{\mathbf{V}}_8 \leftrightarrow \bar{\mathbf{S}}_8$.

This string is the famous supersymmetric $SO(32)$ heterotic string. Although not directly relevant for string phenomenology, this string plays a vital role in recent developments in string duality (to be discussed briefly in Lecture #8).

6.6.3.3. The $SO(16) \times SO(16)$ and $E_8 \times E_8$ strings

Let us now proceed to Case C. As discussed in Sec. 6.2, this case differs from Case B because we have now "twisted" the second group of eight left-moving complex worldsheet fermions relative to the first set. *A priori*, it is easy to imagine that this twist will break the gauge symmetry $SO(32) \to SO(16) \times SO(16)$. However, there a few surprises still in store for us.

We begin in Sector #1, which previously gave rise to the states given in (6.6.2) and (6.6.3). Introducing the third GSO constraint [8] $N_L \equiv \sum_{i=1}^{8} N^{(i)} =$ even does not affect the gravity multiplet (6.6.2), but has a drastic effect on the remaining gauge boson states. We now see that we cannot excite arbitrary combinations of (i, j) fermions; instead we must choose either $(i, j) = 1, ..., 8$ or $(i, j) = 9, ..., 16$. In string-theory parlance, all of the other states have been "projected out of the spectrum". It is in this manner that we remove gauge boson states and break gauge symmetries in string theory. (There are other methods for doing this in string theory, but this is the only method at tree-level.) It is easy to see (following the arguments given above) that the remaining gauge boson states fill out the adjoint representation of two copies of $SO(16)$, and thus the gauge group is *a priori* $SO(16) \times SO(16)$. Therefore, we shall henceforth denote our string states in the notation $\bar{R}_1 \otimes (R_2; R_3, R_4)$ where $\bar{R}_1, R_2$ are the representations of the Lorentz group from the right- and left-movers, and where R_3, R_4 are the representations with respect to the two gauge group factors of $SO(16)$ respectively. Thus, we see that Sector #1 gives rise to the states

$$\bar{\mathbf{V}}_8 \otimes (\mathbf{V}_8; \mathbf{1}, \mathbf{1}) \,, \quad \bar{\mathbf{V}}_8 \otimes (\mathbf{1}; \mathbf{adj}, \mathbf{1}) \,, \quad \bar{\mathbf{V}}_8 \otimes (\mathbf{1}; \mathbf{1}, \mathbf{adj}) \,, \qquad (6.6.9)$$

where the first states form the gravity multiplet and the second and third states are the $SO(16) \times SO(16)$ gauge bosons.

As before, Sectors #2 and #3 do not give rise to massless states. Let us now consider what happens in Sector #4. The states that previously emerged in Sector #4 are given in (6.6.6). We now must impose the remaining GSO constraint $^{(8)}N_L = \begin{Bmatrix} \text{odd} \\ \text{even} \end{Bmatrix}$. Let us consider each case separately. If we impose the odd choice, then the gravitino state in (6.6.6) is projected out of the spectrum, indicating that *supersymmetry is broken*. Likewise, we find that the gaugino states are also affected: we can now excite only those states for which $i = 1, ..., 8$ and $j = 9, ..., 16$. This spinor state transforms in the $(\mathbf{16}, \mathbf{16})$ representation of $SO(16) \times SO(16)$ (*i.e.*, as the vector-vector bifundamental). By contrast, if we impose the even choice, then the gravitino state in (6.6.6) remains in the spectrum, indicating that *supersymmetry is preserved*. Likewise, the gaugino states are affected only by the new requirement that either $i, j = 1, ..., 8$ or $i, j = 9, ..., 16$. Thus, the new GSO projection projects our $SO(32)$ gauginos down to $SO(16) \times SO(16)$ gauginos, as expected. Summarizing, we find that in the "even" case, the states from Sector #4 are

$$\bar{\mathbf{S}}_8 \otimes (\mathbf{V}_8; \mathbf{1}, \mathbf{1}) \,, \quad \bar{\mathbf{S}}_8 \otimes (\mathbf{1}; \mathbf{adj}, \mathbf{1}) \,, \quad \bar{\mathbf{S}}_8 \otimes (\mathbf{1}; \mathbf{1}, \mathbf{adj}) \,. \qquad (6.6.10)$$

Let us now consider Sector #5. As indicated in Table 6.2, in this sector the vacuum energy is $(a_R, a_L) = (-1/2, 0)$ and the first eight left-moving complex fermions are Neveu-Schwarz while the second eight are Ramond. Choosing the "odd" GSO constraints projects all possible massless states out of the spectrum (because there is no simultaneous solution to all three GSO constraints in the "odd" case). By contrast, choosing the "even" GSO constraints yields the states

$$\tilde{b}^{\mu}_{-1/2}|0\rangle_R \otimes \{b_0^i\}|0\rangle_L \qquad (i=9,...,16) \qquad (6.6.11)$$

where we must choose an even number of zero-mode excitations on the left-moving side. This produces a massless vector state which transforms in a (128-dimensional) *spinorial* representation of the second $SO(16)$ gauge group factor. Following our previous conventions, we shall refer to this spinor as $\mathbf{C}_{128}$ rather than its conjugate $\mathbf{S}_{128}$. This state can therefore be denoted as

$$\bar{\mathbf{V}}_8 \otimes (\mathbf{1}; \mathbf{1}, \mathbf{C}_{128}) . \qquad (6.6.12)$$

We shall discuss the physical interpretation of this state shortly.

Sector #6 is similar to Sector #5, except that now the first eight left-moving complex fermions are Ramond and the second eight are Neveu-Schwarz. In a similar way we then find that there are no states in the "odd" case, while in the "even" case we find the states

$$\bar{\mathbf{V}}_8 \otimes (\mathbf{1}; \mathbf{C}_{128}, \mathbf{1}) . \qquad (6.6.13)$$

We now turn to Sector #7. Here the vacuum energy is $(a_R, a_L) = (0,0)$, which implies that if we restrict our attention to massless states, we can tolerate only zero-mode excitations amongst both the left- and right-movers. In the "odd" case, we find the states

$$\{\tilde{b}_0^{\mu}\}|0\rangle_R \otimes \{b_0^i\}|0\rangle_L \qquad (i=9,...,16) \qquad (6.6.14)$$

where the GSO projections restrict us to an even number of zero-mode excitations on the right-moving side and an odd number on the left-moving side. According to our conventions, this produces the state $\bar{\mathbf{C}}_8 \otimes (\mathbf{1}; \mathbf{1}, \mathbf{S}_{128})$. In the "even" case, by contrast, we are restricted to (6.6.14) where now we must have an even number of zero-mode excitations on the right-moving side and an odd number of the left-moving side. This produces the state

$$\bar{\mathbf{S}}_8 \otimes (\mathbf{1}; \mathbf{1}, \mathbf{C}_{128}) . \qquad (6.6.15)$$

Finally, in Sector #8, we similiarly find the states $\bar{\mathbf{C}}_8 \otimes (\mathbf{1}; \mathbf{S}_{128}, \mathbf{1})$ in the "odd" case and

$$\bar{\mathbf{S}}_8 \otimes (\mathbf{1}; \mathbf{C}_{128}, \mathbf{1}) \qquad (6.6.16)$$

in the "even" case.

What are we to make of these results? Collecting our states for the "odd" case, we find a string model with the following massless spectrum:

$$\bar{\mathbf{V}}_8 \otimes (\mathbf{V}_8; \mathbf{1}, \mathbf{1}) \;, \quad \bar{\mathbf{V}}_8 \otimes (\mathbf{1}; \mathbf{adj}, \mathbf{1}) \;, \quad \bar{\mathbf{V}}_8 \otimes (\mathbf{1}; \mathbf{1}, \mathbf{adj})$$
$$\bar{\mathbf{S}}_8 \otimes (\mathbf{1}; \mathbf{V}_{16}, \mathbf{V}_{16}) \;, \quad \bar{\mathbf{C}}_8 \otimes (\mathbf{1}; \mathbf{S}_{128}, \mathbf{1}) \;, \quad \bar{\mathbf{C}}_8 \otimes (\mathbf{1}; \mathbf{1}, \mathbf{S}_{128}) \;. \quad (6.6.17)$$

This is clearly a non-supersymmetric spectrum consisting of a gravity multiplet, vector bosons transforming of the adjoint of $SO(16) \times SO(16)$, one spinor transforming as a vector-vector bifundamental with respect to the gauge group, and two additional spinors of opposite chirality transforming in the spinor representations of the gauge group. This is the non-supersymmetric $SO(16) \times SO(16)$ heterotic string model, first constructed in 1986. Note that this spectrum configuration is anomaly-free, as required for a self-consistent string theory. Also note that this string is tachyon-free even though it is non-supersymmetric. This example thus proves that *not all non-supersymmetric strings have tachyons* (although it is certainly true that all supersymmetric strings lack tachyons). While this is the only non-supersymmetric tachyon-free heterotic string in ten dimensions, there exist a plethora of such strings in lower dimensions. We shall discuss some of the properties of such strings in Lecture #8, but this raises an interesting issue: Does string theory *predict* spacetime supersymmetry? As this example makes clear, string theory certainly does not predict spacetime supersymmetry on the basis of tachyon-avoidance. However, the general answer to this question is unknown.

Even more interesting is the model that results in the "even" case. Collecting our states from (6.6.9), (6.6.10), (6.6.12), (6.6.13), (6.6.15), and (6.6.16), we find that the total massless spectrum of this string can be written in the factorized form

$$(\bar{\mathbf{V}}_8 \oplus \bar{\mathbf{S}}_8) \;\otimes\; \{(\mathbf{V}_8; \mathbf{1}, \mathbf{1}) \oplus (\mathbf{1}; \{\mathbf{adj} \oplus \mathbf{C}_{128}\}, \mathbf{1}) \oplus (\mathbf{1}; \mathbf{1}, \{\mathbf{adj} \oplus \mathbf{C}_{128}\})\} \;. \quad (6.6.18)$$

The appearance of the right-moving factor $\bar{\mathbf{V}}_8 \oplus \bar{\mathbf{S}}_8$ indicates that this model has $N = 1$ supersymmetry, as expected from the appearance of a single gravitino in the massless spectrum. The left-moving factor, by contrast, contains three terms. The first term combines with the right-moving factor to produce the supergravity multiplet. The second two terms formerly gave rise to the $SO(16) \times SO(16)$ gauge supermultiplet. However, we now see that for each $SO(16)$ gauge group factor, the massless vector states transform in the **adj** $\oplus$ $\mathbf{C}_{128}$ representation rather than simply in the **adj** representation. While the **adj** contribution is easy to interpret (giving rise to the usual gauge bosons of $SO(16)$), the extra massless vector states transforming in the $\mathbf{C}_{128}$ representation of each gauge group factor appear to cause an inconsistency, for we know that all massless vector states must be interpreted as gauge bosons, and hence such states can only transform in the adjoint representation. Thus, the only possible way that this string can be consistent is if the massless vector states in this model somehow combine to fill out the adjoint representation of some *other* group G:

$$\mathbf{adj}_{SO(16)} \oplus \mathbf{C}_{128} \stackrel{?}{=} \mathbf{adj}_G \;. \qquad (6.6.19)$$

Remarkably, this is precisely what occurs: the group G is nothing but the exceptional Lie group E_8! Indeed, the 120 states of the adjoint representation of $SO(16)$ together with the 128 states of the spinor representation of $SO(16)$ combine to produce the 248 states of the adjoint representation of E_8! In string parlance, we thus say that the presence of the "twisted" states (6.6.12), (6.6.13), (6.6.15), and (6.6.16) has *enhanced* the total gauge group from $SO(16) \times SO(16)$ to $E_8 \times E_8$. This, then, is the famous supersymmetric $E_8 \times E_8$ heterotic string.

Unlike the supersymmetric $SO(32)$ string, this string is generally considered to have excellent phenomenological prospects. It has $N=1$ spacetime supersymmetry, quantum gravity, and an $E_8 \times E_8$ gauge symmetry. E_8 is a compelling gauge group for phenomenology because it contains E_6 as a subgroup, and E_6 is a group that contains chiral representations which can be associated with grand unification and which thereby contain all of the particle content of the Standard Model. (Of course, it is still necessary to obtain actual *matter* representations from this string, but these can arise upon compactification.) Moreover, while we can imagine the Standard Model to reside entirely within one of the E_8 gauge group factors, the other factor may be interpreted as a "hidden" sector which can also have important phenomenological uses (such as triggering supersymmetry breaking, providing dark-matter candidates, and enforcing string selection rules). Thus, historically, much of the original work in string phenomenology began with a study of the compactification of this model down to four dimensions. However, it is possible to construct heterotic string models directly in four dimensions, and to obtain models which do not necessarily have an interpretation as arising via the compactification of any particular string model in ten dimensions. Thus, as we shall see, the prospects for phenomenological heterotic string model-building are broader than merely studying the compactifications of the $E_8 \times E_8$ heterotic string.

6.6.4. More ten-dimensional heterotic strings

So far, we have constructed four heterotic string models in ten dimensions. Of these, two have spacetime supersymmetry, and two do not. However, it is readily apparent that further models can be constructed by introducing further "twists" which further enlarge the set of sectors in Table 6.2 and which further break the gauge group into smaller factors (or which break the original $SO(32)$ gauge group in entirely different ways). The question that arises, then, is whether there exist other ten-dimensional heterotic strings with spacetime supersymmetry, or whether there exist other non-supersymmetric strings in ten dimensions that are tachyon-free. The answer to both questions turns out to be "no". A complete list of ten-dimensional heterotic strings is given in Table 6.4.

Table 6.4. The complete set of ten-dimensional heterotic string models. Two have spacetime supersymmetry, one is non-supersymmetric but tachyon-free, and the remaining six are non-supersymmetric and tachyonic.

gauge group	spacetime SUSY?	tachyon-free?
$SO(32)$	yes	yes
$E_8 \times E_8$	yes	yes
$SO(16) \times SO(16)$	no	yes
$SO(32)$	no	no
$SO(16) \times E_8$	no	no
$SO(8) \times SO(24)$	no	no
$(E_7)^2 \times [SU(2)]^2$	no	no
$U(16)$	no	no
E_8	no	no

The presence of the last string in Table 6.4 might seem surprising. After all, the rank of the gauge group for this string is only eight rather than sixteen, which implies that its construction must differ substantially from that of the previous strings. It turns out that this is indeed the case.[*] We briefly indicate in Lecture #7 how such strings may be constructed.

---

[*]Unlike the other ten-dimensional heterotic strings, this string involves splitting each complex worldsheet fermion into a pair of two real worldsheet fermions and then introducing relative "twists" within each pair. In technical language, this results in a gauge group whose rank is reduced but whose so-called *affine level* is increased relative to those of the other strings. This increase in the affine level is important for string GUT model-building, and will be discussed in subsequent lectures.

6.7. Lecture #7: Rules for string model-building

In the last several lectures, we constructed many different string models. Amongst the superstring models, we constructed the Type 0A, Type 0B, Type IIA, and Type IIB models, while amongst the heterotic string models, we constructed the non-supersymmetric $SO(32)$ model, the non-supersymmetric $SO(16) \times SO(16)$ model, and the supersymmetric $SO(32)$ and $E_8 \times E_8$ models. In each case, we simply *asserted* a set of sectors (combinations of Neveu-Schwarz and Ramond modings) and a set of GSO constraints in each sector. Of course, each of these sets of sectors and GSO constraints conspires to yield a self-consistent string model, and occasionally it is even possible to see intuitively which choices can lead to self-consistent string models. However, we ultimately wish to construct semi-realistic string models where the groups are broken down to much smaller pieces than we have been dealing with thus far (*e.g.*, $SU(3) \times SU(2) \times U(1)$, or even $SU(5)$ or $SO(10)$), and this is going to require more complicated twists than we have thus far been using. Furthermore, all of our string models thus far have been in ten dimensions, yet we are ultimately going to wish to compactify our string models to four dimensions. It turns out that this will introduce even further choices for modings, twists, and their associated GSO projections. (In geometric language, these further choices amount the choice of compactification manifold.)

The question that arises, then, is to determine the minimal set of parameters that govern these choices. What we require is a way to *systematize* the whole process of string model-construction, so that we will know precisely which choices govern the construction of a string model and guarantee its internal self-consistency. In other words, we require *rules for string model-building*. This is the subject of the present lecture.

Once we learn the rules for the construction of ten-dimensional string models, it will be relatively straightforward to generalize these rules for the construction of models in four dimensions. We will then have the tools whereby we may finally construct semi-realistic four-dimensional string models.

6.7.1. *Generating the sector combinations: The 20-dimensional lattice*

The first issue we face is that of choosing the appropriate sector combinations. For example, let us recall the possible heterotic string sectors in

Table 6.2. As we discussed in Sec. 6.2, this set of sectors permits only three distinct sector combinations: either we choose Sectors #1 and #2 only, or we choose Sectors #1 through #4 only, or we choose Sectors #1 through #8. How can we know which combinations are allowed, and which sectors are required in each grouping? In Sec. 6.2, we discussed how modular invariance ultimately governs these choices. Here, however, we shall develop a rule which we can use in order to deduce these sector combinations rather quickly and which can easily be generalized to more complicated situtions.

First, let us introduce some notation. Since it is rather awkward to consider left-moving complex fermions Ψ_L^i at the same time as right-moving *real* (Majorana) fermions ψ_R^μ, let us "complexify" our right-moving Majorana fermions so that *all* of our worldsheet fermions are complex. This means that instead of having eight left-moving real fermions ψ_R^μ in light-cone gauge, we have instead four complex ones Ψ_R^μ formed by pairing the left-moving real fermions in groups of two. (We retain the index μ to remind ourselves that these fields carry indices with respect to the spacetime Lorentz algebra, even though strictly speaking it is only their real component fields ψ_R^μ that carry such vectorial indices.)

We also need a more general notation for discussing the possible boundary conditions and modings that any such complex worldsheet fermion can take. In general, we can parametrize any possible worldsheet boundary condition in the form

$$\Psi(\sigma_1 + \pi, \sigma_2) = -e^{-2\pi i v} \Psi(\sigma_1, \sigma_2) \qquad (6.7.1)$$

where $-\frac{1}{2} \leq v < \frac{1}{2}$. Thus the quantity v parametrizes the boundary condition of the individual fermion, with

$$v = 0: \quad \text{anti-periodic} \quad \text{(Neveu-Schwarz)}$$
$$v = -1/2: \quad \text{periodic} \quad \text{(Ramond)} \,. \qquad (6.7.2)$$

General values of v correspond to so-called "multi-periodic fermions". For example, the general moding of a multi-periodic left-moving complex fermion is given by

$$\Psi_L(\sigma_1 + \sigma_2) = \sum_{n=1}^{\infty} \left[b_{n+v-1/2} e^{-i(n+v-1/2)(\sigma_1+\sigma_2)} \right.$$
$$\left. + d^\dagger_{n-v-1/2} e^{+i(n-v-1/2)(\sigma_1+\sigma_2)} \right], \qquad (6.7.3)$$

and the corresponding number operator and worldsheet energy are defined accordingly. Note that these modings generalize those given in Sec. 5.3.

Likewise, the vacuum energy contribution from such a fermion is given by

$$a_\Psi = \frac{1}{2}\left(v^2 - \frac{1}{12}\right). \qquad (6.7.4)$$

This too generalizes our previous results.

In ten dimensions, it turns out that we lose no generality by considering only the specific cases $v = 0, -\frac{1}{2}$ for all worldsheet fermions. What this means is that all self-consistent ten-dimensional string models can ultimately be realized using worldsheet fermions with only Neveu-Schwarz or Ramond boundary conditions. In lower dimensions, by contrast, other choices are possible. Therefore, even though we shall primarily focus our attention on the cases $v \in \{0, -\frac{1}{2}\}$, we shall develop our formalism in such a way that it holds for arbitrary values of v.

Given this parametrization, we can describe the boundary conditions within any sector rather succinctly by specifying twenty v-values, four for the complex right-movers Ψ_R^μ and sixteen for the complex left-movers Ψ_L^i. We can group these twenty v-values to form a "boundary-condition" vector

$$\mathbf{V} = [\bar{v}_1, \bar{v}_2, \bar{v}_3, \bar{v}_4 \,|\, v_1, ..., v_{16}], \qquad (6.7.5)$$

and thus we may associate a vector with each underyling string sector. For example, the sectors in Table 6.2 now correspond to the vectors shown in Table 6.5. Note that in Table 6.5, we have used a shorthand notation in which superscripts indicate repeated components. We have also dropped the minus signs from the Ramond entries $v = -\frac{1}{2}$. We stress, however, that even though we shall no longer explicitly indicate the Ramond minus sign, it should continue to be implicitly understood for all Ramond boundary conditions. (This minus sign can play an important role for string models in lower dimensions.)

What, then, are the self-consistent combinations of sectors? Recall from the previous lecture that the first self-consistent combination of sectors comprises Sectors #1 and #2 only. Let us therefore study this simplest combination. Sector #1 (the so-called NS-NS sector) corresponds to the *zero-vector* $\mathbf{0}$, the vector whose entries all vanish. Thus, in this sense, we might associate the NS-NS sector with the *origin* in a twenty-dimensional vector space. Sector #2 (the so-called Ramond-Ramond sector) then corresponds to some other point in the vector space which is some distance away from the origin. Let us call this other location $\mathbf{V}_0 \equiv [(\frac{1}{2})^4 | (\frac{1}{2})^{16}]$.

If we were to consider $\mathbf{V}_0$ to be a lattice basis vector, a natural question would be to determine the lattice that is generated by this basis vector.

Table 6.5. The eight possible sectors for ten-dimensional heterotic strings from Table 6.2, written in the boundary-condition vector notation of (6.7.5). Here the superscripts indicate repeated components, and we have dropped the minus sign for Ramond boundary conditions.

Sector #	V
1	$[(0)^4 \mid (0)^{16}]$
2	$[(\frac{1}{2})^4 \mid (\frac{1}{2})^{16}]$
3	$[(0)^4 \mid (\frac{1}{2})^{16}]$
4	$[(\frac{1}{2})^4 \mid (0)^{16}]$
5	$[(0)^4 \mid (0)^8 (\frac{1}{2})^8]$
6	$[(0)^4 \mid (\frac{1}{2})^8 (0)^8]$
7	$[(\frac{1}{2})^4 \mid (0)^8 (\frac{1}{2})^8]$
8	$[(\frac{1}{2})^4 \mid (\frac{1}{2})^8 (0)^8]$

Because there is only one such non-zero vector, this would clearly be a one-dimensional "lattice". Since $\mathbf{V}_0 \equiv [(\frac{1}{2})^4 \mid (\frac{1}{2})^{16}]$, the next point in the lattice would be $2\mathbf{V}_0 \equiv [(1)^4 \mid (1)^{16}]$. How can we interpret this point? Recall from (6.7.1) that the components of such vectors (*i.e.*, the values of v) are defined only modulo 1 (*i.e.*, they are restricted to the unit interval $-\frac{1}{2} \leq v < \frac{1}{2}$). Thus, we see that $v = 1$ is physically the same as $v = 0$, once again implying a Neveu-Schwarz boundary condition. In other words, we should only add our vectors *modulo 1*. Given this, we find that $2\mathbf{V}_0 \stackrel{1}{=} \mathbf{0}$, where we have introduced the notation $\stackrel{1}{=}$ to indicate equality modulo 1. Likewise, $3\mathbf{V}_0 \stackrel{1}{=} \mathbf{V}_0$, and so forth. Thus, we see that $\mathbf{V}_0$ generates a "lattice" consisting of only two physically distinct "points":

$$\{\mathbf{0}, \mathbf{V}_0\} . \tag{6.7.6}$$

However, these are precisely the two "points" that comprised our first self-consistent set of sectors (Case A in Lecture #6), and which led to our first string model!

It turns out that this is a general property: *All self-consistent choices of string sectors are those that correspond to the "points" in a twenty-dimensional lattice generated by a set of basis vectors.* To illustrate this principle, let us consider the next case (Case B in Lecture #6). In this case, we included only Sectors #1 through #4. This indicates that we need a larger lattice, which in turn implies the existence of not just the single lattice-generating basis vector $\mathbf{V}_0$, but also an additional basis vector $\mathbf{V}_1$.

One choice is:
$$\mathbf{V}_0 = [(\tfrac{1}{2})^4 \,|\, (\tfrac{1}{2})^{16}]$$
$$\mathbf{V}_1 = [(0)^4 \,|\, (\tfrac{1}{2})^{16}] \,. \tag{6.7.7}$$

Using these choices, we can see that indeed all four of these sectors can be generated as the different "points" in the resulting lattice: Sector #1 corresponds to the origin $\mathbf{0}$, Sector #2 corresponds to $\mathbf{V}_0$ itself, Sector #3 corresponds to $\mathbf{V}_1$ itself, and Sector #4 corresponds to the remaining lattice point $\mathbf{V}_0 + \mathbf{V}_1$. Note that no other points exist in this lattice, since $2\mathbf{V}_0 \stackrel{1}{=} 2\mathbf{V}_1 \stackrel{1}{=} \mathbf{0}$. Thus, we see that the introduction of the additional basis vector $\mathbf{V}_1$ is physically equivalent to the "twist" that shifts the boundary conditions of the left-moving fermions relative to those of the right-moving fermions in Sectors #3 and #4.

Finally, let us consider the full set (Case C) consisting of Sectors #1 through #8. It is easy to see that this set is generated by the *three* basis vectors:
$$\mathbf{V}_0 = [(\tfrac{1}{2})^4 \,|\, (\tfrac{1}{2})^{16}]$$
$$\mathbf{V}_1 = [(0)^4 \,|\, (\tfrac{1}{2})^{16}]$$
$$\mathbf{V}_2 = [(0)^4 \,|\, (\tfrac{1}{2})^8 (0)^8] \,. \tag{6.7.8}$$

Once again, the introduction of the new basis vector $\mathbf{V}_2$ implements the "twist" that separates the boundary conditions of the first set of eight left-moving fermions from those of the second set.

This procedure can be continued. Each additional basis vector introduces a new twist, increases the size of the resulting lattice, and leads to the introduction of new physical string sectors (so-called "twisted sectors"). For example, one further basis vector that might be introduced is $\mathbf{V}_3 \equiv [(0)^4 | (\tfrac{1}{2})^4 (0)^4 (\tfrac{1}{2})^4 (0)^4]$. This vector would have the effect of introducing a further twist amongst the left-moving fermions within each group of eight.

Clearly, given a set of N basis vectors $\mathbf{V}_i$ ($i = 0, ..., N-1$), the procedure for generating the full set of resulting string sectors is to consider all possible lattice vectors $\sum_{i=0}^{N-1} \alpha_i \mathbf{V}_i$ where $\alpha_i \in \{0, 1\}$. Note that this restriction on the values of α_i assumes that we are considering only Neveu-Schwarz or Ramond boundary conditions for the worldsheet fermions; generalizations to multi-periodic fermions will be discussed shortly. We shall henceforth denote a given string sector as $\alpha \mathbf{V} \equiv \sum_i \alpha_i \mathbf{V}_i$. For example, the NS-NS sector (*i.e.*, Sector #1) always corresponds to $\alpha = (0, 0, ...)$ and the Ramond-Ramond sector (*i.e.*, Sector #2) corresponds to $\alpha = (1, 0, ...)$.

At this stage, we now know how to generate the full set of underlying string sectors once we are given a "primordial" set of basis vectors $\mathbf{V}_i$. The next issue that arises is to determine the rules that govern the allowed choices of these basis vectors. Of course, we have already derived one such rule: each basis vector $\mathbf{V}_i$ must take the form

$$\mathbf{V}_i = [(\bar{v})^4 \,|\, v_1, ..., v_{16}] \tag{6.7.9}$$

where the right-moving fermions all have *same* moding $\bar{v} \in \{0, -\frac{1}{2}\}$. Indeed, as we saw in Lectures #5 and #6, this requirement is necessary for the preservation of the right-moving worldsheet supersymmetry (so that the right-moving worldsheet supercurrent has a unique moding in each sector). This is also necessary for the preservation of spacetime Lorentz invariance, since the right-moving worldsheet fermions carry Lorentz spacetime indices.

As we might expect, there are still several additional conditions that our basis vectors $\mathbf{V}_i$ must satisfy. But before we can discuss these conditions, we must turn to the generation of the GSO constraints in each sector.

6.7.2. *Generating the GSO constraints*

We have already seen in previous lectures that the appearance of new string sectors is correlated with the appearance of new GSO constraints in each sector. We are now in a position to formulate this correlation more precisely: *in each string sector, there is one GSO projection for each basis vector*. Our task, then, is to find a simple way to generate the exact forms of these GSO projections.

Let us return to Case A, and consider the model consisting of only Sectors #1 and #2. As we have seen above, this model is generated by the single basis vector $\mathbf{V}_0 \equiv [(\frac{1}{2})^4|(\frac{1}{2})^{16}]$, resulting in the two sectors $\mathbf{0}$ (Sector #1) and $\mathbf{V}_0$ (Sector #2). In each of these sectors, recall from Table 6.3 that we then had the single GSO constraint $N_L - N_R = $ odd, or equivalently

$$\sum_{i=1}^{16} N^{(i)} - \sum_{j=1}^{4} \bar{N}^{(j)} = \text{odd} \,. \tag{6.7.10}$$

(Here we have used the j-index to span our four complex right-moving fermions, while the i-index spans our sixteen complex left-moving fermions.) It is this GSO constraint that we now wish to write in a more transparent manner.

Given our success in using the lattice idea and modular arithmetic in order to generate the complete set of string sectors, let us attempt to write (6.7.10) in a form that makes use of both ideas. Let us first concentrate on the modular arithmetic idea. Since all of our basis vectors are defined only modulo one, let us cast (6.7.10) into the form of a modulo-one relation. Since (6.7.10) is already a modulo-two relation, this can be achieved by dividing by two:

$$\tfrac{1}{2}\sum_{i=1}^{16} N^{(i)} - \tfrac{1}{2}\sum_{j=1}^{4} N^{(j)} \stackrel{1}{=} \tfrac{1}{2} \qquad (6.7.11)$$

where we have used the notation $\stackrel{1}{=}$ to indicate equality modulo 1.

Let us now try to incorporate the lattice idea. To do this, let us make a *vector* out of our twenty number operators:

$$\mathbf{N} \equiv [\bar{N}^{(1)}, \bar{N}^{(2)}, \bar{N}^{(3)}, \bar{N}^{(4)} \,|\, N^{(1)}, ..., N^{(16)}] \,. \qquad (6.7.12)$$

Clearly, each different possible string state in a given sector corresponds to a different $\mathbf{N}$-vector, and the physical (surviving) string states are those satisfying (6.7.11). Let us now attempt to write (6.7.11) in a vector notation. Neglecting the minus sign in (6.7.11) for the moment, we see that (6.7.11) involves a sum of vector components, which reminds us of a vector dot product. Thus, if we define the "signature" of our twenty-dimensional lattice to be $[(-)^4 \,|\, (+)^{16}]$, we can write (6.7.11) in the form of a vector dot product:

$$[(\tfrac{1}{2})^4 \,|\, (\tfrac{1}{2})^{16}] \cdot \mathbf{N} \stackrel{1}{=} \tfrac{1}{2} \qquad (6.7.13)$$

where we have introduced a vector each of whose components is equal to $\tfrac{1}{2}$. However, this vector is nothing but $\mathbf{V}_0$, the basis vector that generates the lattice for this model! Thus, we see that if our model is generated by the basis vector $\mathbf{V}_0$, then in each of the resulting sectors $\{\mathbf{0}, \mathbf{V}_0\}$ the GSO projections take the form

$$\mathbf{V}_0 \cdot \mathbf{N} \stackrel{1}{=} \tfrac{1}{2} \,. \qquad (6.7.14)$$

This produces the non-supersymmetric $SO(32)$ string model from Lecture #6!

Let us now consider Case B, consisting of Sectors #1 through #4. As we saw in Lecture #6, this produces the *supersymmetric* $SO(32)$ heterotic string model, and is generated by the set of two basis vectors given in (6.7.7). In each of the four resulting sectors $\{\mathbf{0}, \mathbf{V}_0, \mathbf{V}_1, \mathbf{V}_0 + \mathbf{V}_1\}$, the *two*

GSO projections were $N_L - N_R = $ odd and $N_L = $ even. (Recall Table 6.3.) These now take the form

$$\mathbf{V}_0 \cdot \mathbf{N} \stackrel{1}{=} \tfrac{1}{2}\,, \qquad \mathbf{V}_1 \cdot \mathbf{N} \stackrel{1}{=} \tfrac{1}{2}\,. \qquad (6.7.15)$$

Similarly, Case C is generated by the *three* basis vectors in (6.7.8), and the three GSO constraints in each sector take the general form

$$N_L - N_R = \ldots\,, \qquad N_L = \ldots,\qquad {}^{(8)}N_L = \ldots\,. \qquad (6.7.16)$$

Here ${}^{(8)}N_L \equiv \sum_{i=1}^{8} N^{(i)}$, and we shall momentarily defer a discussion of the values of the right sides of these constraint equations. We then find that these three GSO constraints take the general forms

$$\mathbf{V}_0 \cdot \mathbf{N} \stackrel{1}{=} \ldots\,, \qquad \mathbf{V}_1 \cdot \mathbf{N} \stackrel{1}{=} \ldots\,, \qquad \mathbf{V}_2 \cdot \mathbf{N} \stackrel{1}{=} \ldots\,. \qquad (6.7.17)$$

Depending on the right sides of these equations, this generates either the supersymmetric $E_8 \times E_8$ string or the non-supersymmetric $SO(16) \times SO(16)$ string.

The final question, then, is to determine what appears on the right sides of these GSO constraint equations. In general, this will be some value x which satisfies $-\tfrac{1}{2} \leq x < \tfrac{1}{2}$. This x-value is called a *GSO projection phase*, and is generally different for each sector. Thus, we know that x must itself depend on α, where (as discussed in Sec. 7.1) α parametrizes the particular sector in question. We also know from our prior experience (in particular, from Table 6.3) that x must also contain some additional *free* parameters because we occasionally still had the freedom to make choices such as $\{\text{evenodd}\}$ when constructing our GSO constraints.

It turns out the final result is the following. Within any given string sector $\alpha \mathbf{V} \equiv \sum_{i=0}^{N-1} \alpha_i \mathbf{V}_i$, the states that survive are those whose number operator vectors $\mathbf{N}$ satisfy the equations

$$\mathbf{V}_i \cdot \mathbf{N} \stackrel{1}{=} \sum_{j=0}^{N-1} k_{ij}\alpha_j \,+\, s_i \,-\, \mathbf{V}_i \cdot (\alpha \mathbf{V})\,, \qquad 0 \leq i \leq N-1\,. \qquad (6.7.18)$$

This is therefore the full set of GSO constraint equations for the sector $\alpha \mathbf{V}$. In (6.7.18), the notation is as follows. There are N different equations here, depending on the value of i. In the last term, the dot product $\mathbf{V}_i \cdot (\alpha \mathbf{V})$ is the dot product between $\mathbf{V}_i$ and the sector $\alpha \mathbf{V}$ for which the GSO constraint is being applied. In the second-to-last term, s_i is defined as the first component (*i.e.*, the first of the right-moving components) of the vector $\mathbf{V}_i$:

$$s_i \equiv \mathbf{V}_i^{(1)}\,. \qquad (6.7.19)$$

Thus s_i parametrizes the *spacetime statistics* of the sector $\mathbf{V}_i$, with $s_i = 0$ indicating spacetime bosons and $s_i = -\frac{1}{2}$ indicating spacetime fermions. Likewise, the sum $\sum \alpha_i s_i$ (mod 1) indicates the statistics of the sector $\alpha \mathbf{V}$. In the remaining term, k_{ij} denotes a certain $N \times N$ matrix of numbers (so-called *GSO projection phases*) satisfying $-\frac{1}{2} \leq k_{ij} < \frac{1}{2}$. These are therefore the remaining degrees of freedom that enter into our GSO constraints. In the case of $\mathbb{Z}_2$ twists (for which all fermionic boundary conditions have either Neveu-Schwarz or Ramond boundary conditions), one has $k_{ij} \in \{0, -\frac{1}{2}\}$ only. The case of multi-periodic fermions will be discussed shortly.

Thus, if we are given a set of parameters $\{\mathbf{V}_i, k_{ij}\}$, we can now generate the resulting string model and the entire corresponding spectrum! These parameters are ultimately the parameters that physically describe a given string model.

6.7.3. *Self-consistency constraints*

We finally turn to the remaining question: what determines how the parameters $\{\mathbf{V}_i, k_{ij}\}$ are to be chosen? What are the rules that guarantee a self-consistent choice?

Clearly, as we have discussed earlier, modular invariance is one of many symmetries that govern these choices. Other requirements for self-consistency include proper spacetime spin-statistics relations (so that all Ramond states are indeed anti-commuting spacetime fermions, and all Neveu-Schwarz states are commuting spacetime bosons) and physically sensible GSO projections (so that unitarity is not violated, among other things). It is important to stress that these are not *additional* constraints that need to be imposed in order to guarantee the consistency of the string in spacetime; rather these constraints are intrinsic to string theory itself at the worldsheet level, emerging as string self-consistency constraints, and together imply these features in spacetime.

We have already discussed the first contraint that governs the choices of the basis vectors: they must all have the form (6.7.9), with all right-moving fermions sharing the same boundary condition. Second, these vectors must all be linearly independent with respect to addition (modulo 1); otherwise, at least one of these vectors is redundant. The third constraint also turns out to be quite simple: among our set of basis vectors, we must always start with the vector

$$\mathbf{V}_0 \equiv [(\tfrac{1}{2})^4 \,|\, (\tfrac{1}{2})^{16}]\,. \tag{6.7.20}$$

The presence of this vector ensures that the resulting string model contains at least a Ramond-Ramond sector in addition to a NS-NS sector.

The remaining constraints serve to correlate the $\mathbf{V}_i$ vectors with the GSO projection phases k_{ij}, and take the form:

$$k_{ij} + k_{ji} \stackrel{1}{=} \mathbf{V}_i \cdot \mathbf{V}_j$$
$$k_{ii} + k_{i0} \stackrel{1}{=} \tfrac{1}{2}\mathbf{V}_i \cdot \mathbf{V}_i - s_i \ . \quad (6.7.21)$$

Note that given a set of boundary condition vectors $\mathbf{V}_i$, the constraints (6.7.21) imply that only the elements k_{ij} with $i > j$ are independent parameters. The first equation in (6.7.21) then enables us to uniquely determine k_{ij} with $i < j$, and the second equation in (6.7.21) enables us to uniquely determine the diagonal elements k_{ii}.

6.7.4. *Summary, examples, and generalizations*

Let us now summarize the rules for heterotic string model-building in $D = 10$. We begin by choosing a set of linearly independent basis vectors $\mathbf{V}_i$ ($i = 0, ..., N-1$) and a corresponding matrix of GSO projection phases k_{ij} ($i, j = 0, ..., N-1$). Our set of basis vectors may be as large as we desire; since each vector corresponds to an additional twist, larger sets of vectors lead to more complicated string models. Among our choice of basis vectors must always appear the vector $\mathbf{V}_0$ defined in (6.7.20), and every basis vector is required to have the form (6.7.9). We must also ensure that our choices of basis vectors $\mathbf{V}_i$ and GSO projection phases k_{ij} are properly *correlated* according to (6.7.21). If there does not exist a solution for k_{ij}, then our original choice of $\mathbf{V}_i$ must be discarded or repaired. These are the only constraints that govern the choices of the parameters $\{\mathbf{V}_i, k_{ij}\}$.

Given such a self-consistent choice of parameters $\{\mathbf{V}_i, k_{ij}\}$, we are then guaranteed to have a self-consistent string model. The different sectors of this model are generated as all combinations $\sum_i \alpha_i \mathbf{V}_i$ that fill out the twenty-dimensional lattice, where $\alpha_i \in \{0, 1\}$. In each sector $\alpha \mathbf{V} \equiv \sum_i \alpha_i \mathbf{V}_i$, the allowed states are then those whose number operator vectors $\mathbf{N}$ simultaneously satisfy the constraints (6.7.18) for $i = 0, ..., N-1$. This is often called the *spectrum-generating formula*.

It is straightforward to see how this formalism can be applied in practice. We shall leave it as an exercise to verify that the choice

$$\mathbf{V}_0 \equiv [(\tfrac{1}{2})^4 \,|\, (\tfrac{1}{2})^{16}] \ , \qquad k_{00} = (0) \quad (6.7.22)$$

generates the non-supersymmetric $SO(32)$ heterotic string model; that the choice

$$\begin{cases} \mathbf{V}_0 \equiv [(\tfrac{1}{2})^4 \,|\, (\tfrac{1}{2})^{16}] \\ \mathbf{V}_1 \equiv [(0)^4 \,|\, (\tfrac{1}{2})^{16}] \end{cases} \qquad k_{ij} = \begin{pmatrix} 0 & 0 \\ 0 & 0 \end{pmatrix} \qquad (6.7.23)$$

generates the *supersymmetric* $SO(32)$ heterotic string model; and that the choices

$$\begin{cases} \mathbf{V}_0 \equiv [(\tfrac{1}{2})^4 \,|\, (\tfrac{1}{2})^{16}] \\ \mathbf{V}_1 \equiv [(0)^4 \,|\, (\tfrac{1}{2})^{16}] \\ \mathbf{V}_2 \equiv [(0)^4 \,|\, (\tfrac{1}{2})^{8}(0)^{8}] \end{cases} \qquad k_{ij} = \begin{pmatrix} 0 & 0 & 0 \\ 0 & 0 & k \\ 0 & k & 0 \end{pmatrix} \qquad (6.7.24)$$

generate the supersymmetric $E_8 \times E_8$ string model if we choose $k = 0$, and the non-supersymmetric $SO(16) \times SO(16)$ string model if we choose $k = 1/2$. Indeed, it is a general property that if we choose our vector $\mathbf{V}_1$ as above, then spacetime supersymmetry is preserved if $k_{i0} = k_{i1}$ for all $i = 0, 1, ..., N-1$, and broken otherwise. Thus, we see that we now have a very compact notation and procedure for generating and analyzing ten-dimensional heterotic string models! We should also stress that these are not the only parameter choices of $\{\mathbf{V}_i, k_{ij}\}$ that will lead to these models. In fact, there is often a great redundancy in this procedure, so that a given physical string model can have many different representations in terms of the worldsheet parameters $\{\mathbf{V}_i, k_{ij}\}$. However, a given set of parameters always corresponds to a single, unique, self-consistent string model in spacetime.

The formalism that we have presented in this lecture is called the "free-fermionic construction", and was developed in 1986 by H. Kawai, D.C. Lewellen, and S.-H.H. Tye and by I. Antoniadis, C. Bachas, and C. Kounnas. The name stems from the fact that the fundamental degrees of freedom on the string worldsheet (in addition to the spacetime coordinate fields X^μ) are taken to be the free fermionic fields Ψ. Even though we have presented this formalism for the case of ten-dimensional heterotic strings, there also exists a straightforward generalization of this formalism to *four-dimensional* heterotic string models.

As we have indicated, this formalism also carries over directly to the case of multi-periodic complex fermions for which the boundary condition parameter v in (6.7.1) can be an arbitrary rational number in the range $-\tfrac{1}{2} \le v < \tfrac{1}{2}$. For each resulting boundary-condition vector $\mathbf{V}_i$, let us define

m_i to be the smallest integer such that if we multiply each element in $\mathbf{V}_i$ by m_i, we obtain a vector of integer entries. In general, m_i is called the "order" of the vector $\mathbf{V}_i$, and is also the order of the corresponding physical twist introduced by that vector. For example, in the case of only Neveu-Schwarz or Ramond fermions, we have $m_i = 2$ for all i, implying only $\mathbb{Z}_2$ twists. Nevertheless, even for general multi-periodic boundary conditions, the above constraints continue to apply exactly as written. Indeed, the only small change is that we now must take $\alpha_i \in \{0, 1, ..., m_i - 1\}$ when generating our lattice of corresponding string sectors. Likewise, each GSO projection phase k_{ij} must now also be chosen such that $m_j k_{ij} \in \mathbb{Z}$.

In this regard, it is important to note that the only fermions which can possibly have such generalized boundary conditions are those which are *not* the worldsheet superpartners of worldsheet bosons. This restriction arises because the structure of the worldsheet supersymmetry algebra itself restricts the corresponding fermions to have only Neveu-Schwarz or Ramond boundary conditions. For example, in the case of the ten-dimensional heterotic string, only the left-moving worldsheet fermions are *a priori* permitted to have generalized boundary conditions. By contrast, the right-moving fermions are restricted by the right-moving worldsheet supersymmetry algebra to have either Neveu-Schwarz or Ramond boundary conditions. This in turn implies that $s_i \in \{0, -\frac{1}{2}\}$, so that a given string sector continues to give rise to only spacetime bosons or spacetime fermions. Also note that although we are capable *in principle* of utilizing multi-periodic fermions while constructing ten-dimensional heterotic string models, in practice it turns out that this does not lead to new models which are physically distinct from those using only Ramond or Neveu-Schwarz fermions. It is for this reason that we can ultimately restrict ourselves to these simpler boundary conditions in ten dimensions without loss of generality. In lower dimensions, by contrast, this is no longer true, and the number of possible models grows dramatically.

This formalism can also be carried over to the case of ten-dimensional *superstrings* (rather than heterotic strings). For superstrings, the boundary-condition vectors take the simpler form

$$\mathbf{V}_i = [(\bar{v})^4 \,|\, (v)^4] \tag{6.7.25}$$

where $v, \bar{v} \in \{0, \frac{1}{2}\}$. Our mandatory vector $\mathbf{V}_0$ then takes the form $[(\frac{1}{2})^4|(\frac{1}{2})^4]$, and we define $s_i \equiv v + \bar{v} \pmod{1}$ as our new spacetime statistics parameter, replacing (6.7.19). The results (6.7.21) and (6.7.18) then continue to apply directly. Of course, this formalism is fairly trivial in the case of *ten-dimensional* superstrings, for the maximal set of linearly independent basis vectors of the form (6.7.25) consists of only $\mathbf{V}_0$ and $\mathbf{V}_1 \equiv [(0)^4|(\frac{1}{2})^4]$. As we have seen in Lecture #4, this results in only four distinct superstring models in ten dimensions: omitting $\mathbf{V}_1$ from our basis set generates the Type 0 models, while including $\mathbf{V}_1$ in our basis set generates the Type II models. However, just as for the heterotic strings, this formalism can also be generalized to the case of four-dimensional superstring models where the possibilities become much richer.

It turns out that the free-fermionic formalism can be extended still further. For example, one can also extend this formalism to compactifications of the *bosonic* string. Moreover, one can even extend this formalism to special types of superstring and heterotic string models whose worldsheet actions must be represented in terms of real rather than complex fermions. Likewise, there even exist generalizations to string models involving non-free worldsheet fermions (*i.e.*, models whose worldsheet actions involve additional Thirring-type interactions between the worldsheet fermions). In fact, even though there exist alternative model-construction formalisms that do not involve free worldsheet fermions at all, the free-fermionic construction can often yield models that are physically equivalent to those that are constructed through these other means.

How general, then, is the free-fermionic construction? It turns out that for *ten-dimensional* string models, this construction is completely general. What this means is that all known physically consistent superstring and heterotic string models in ten dimensions can be realized via this construction (*i.e.*, as stemming from an underlying set of free-fermionic parameters $\{\mathbf{V}_i, k_{ij}\}$). In lower dimensions, by contrast, this construction is *not* completely general — there exist self-consistent lower-dimensional string models which cannot be written or constructed in this manner. However, the free-fermionic construction does comprise a *vast set* of semi-realistic string models. Moreover, the free-fermionic construction has the great advantage that the rules for construction are relatively simple, and that they enable one to *systematically* construct many string models and examine their phenomenological properties. Indeed, many computer programs have been written that use this formalism in order to scan the space of string models and analyze their low-energy phenomenologies. Thus, for these reasons, the free-fermionic construction has played a very useful role as the underlying method through which the majority of string model-building has historically been pursued.

6.7.5. *Assessment:*

At this point, it is perhaps useful to assess the position in which we now find ourselves. Clearly, through these constructions, we are able to produce *many* string models. In fact, as we shall see, the number of self-consistent string models in $D < 10$ is virtually infinite, and there exists a whole space of such models. This space of models is called a *moduli space*, where the so-called moduli are various continuous parameters which can be adjusted in order to yield different models. (Of course, we have seen that we have only discrete parameter choices in ten dimensions, but these parameters can become continuous in lower dimensions.) Moreover, each of these models has a completely different spacetime phenomenology. What, then, is the use of string theory as an "ultimate" theory, if it does not lead to a single, unique model with a unique low-energy phenomenology?

To answer this question, we should recall our discussion at the beginning of these lectures. Just as field theory is a language for building certain models (one of which, say, is the Standard Model), string theory is a new and deeper language by which we might also build models. The advantages of using this new language, as discussed in Lecture #1, include the fact that our resulting models incorporate quantum gravity and Planck-scale physics. Of course, in field theory, many parameters enter into the choice of model-building. These parameters include the choice of fields (for example, the choice of the gauge group, and whether or not to have spacetime supersymmetry), the number of fields (for example, the number of generations), the masses of particles, their mixing angles, and so forth. These are all *spacetime* parameters. In string theory, by contrast, we do not choose these spacetime parameters; we instead choose a set of *worldsheet* parameters. For example, in the free-fermionic construction, we choose the parameters $\{\mathbf{V}_i, k_{ij}\}$. All of the phenomenological properties in spacetime are then derived as consequences of these more fundamental choices. But still, just as in field theory, we are faced with the difficult task of model-building.

Is this progress, then? While opinions on this question may differ, one can argue that the answer is still definitely "yes". Recall that quantum gravity is automatically included in these string models. This is one of the benefits of model-building on the worldsheet rather than in spacetime. Also recall that string theory is a finite theory, and does not contain the sorts of ultraviolet divergences that plague us in field theory. This is another benefit of worldsheet, rather than spacetime, model-building. Moreover, worldsheet model-building ultimately involves choosing *fewer* parameters than we would have to choose in field theory — for example, we have seen that an entire infinite tower of string states, their gauge groups and charges and spins, are all ultimately encoded in a few underlying worldsheet parameters such as $\{\mathbf{V}_i, k_{ij}\}$. Furthermore, because of this drastic reduction in the number of free parameters, string phenomenology is in many ways more tightly constrained than ordinary field-theoretic phenomenology. Thus, it is in this way that string theory can guide our choices and expectations for physics beyond the Standard Model. Indeed, from a string perspective, we see that we should favor only those patterns of spacetime physics that can ultimately be derived from an underlying set of worldsheet parameters such as $\{\mathbf{V}_i, k_{ij}\}$. These would then serve as a "minimal set" of parameters which would govern all of spacetime physics!

Of course, at a theoretical or philosophical level, this state of affairs is still somewhat unsatisfactory. After all, we still do not know *which* self-consistent choice of string parameters ultimately corresponds to reality. However, *in principle*, string theory should be able to predict this dynamically. Indeed, even though there exists a whole moduli space of self-consistent string models, there should exist an energy or potential function in this space (*i.e.*, some function $V(\{\phi\})$ of all the moduli $\{\phi\}$) which should dynamically select a particular point in moduli space (*e.g.*, as a local or global minimum of V). This would then fix all of the moduli to specific values, or equivalently (in the language of the free-fermionic construction) tell us which choices of parameters $\{\mathbf{V}_i, k_{ij}\}$ are preferred dynamically.

Unfortunately, we do not understand the dynamics of string theory well enough to carry out such an ambitious undertaking. Certainly, at the level of perturbative (weakly coupled) string theory, we have no way to distinguish amongst the possible low-energy models by calculating such a function $V(\{\phi\})$. This is particularly true for string models exhibiting spacetime supersymmetry, for which $V = 0$ exactly to all orders in perturbation theory. Even if the spacetime supersymmetry is broken, the resulting potential $V(\{\phi\})$ often turns out not to have a stable minimum. This is the so-called "runaway problem", to be discussed further in Lecture #8. Of course, one might hope that recent advances in understanding the *non-perturbative* structure of string theory will ultimately be able to provide guidance in this direction. However, as we shall discuss briefly in Lecture #8, although these non-perturbative insights (particularly those concerning string duality) have thus far changed our understanding of the size and shape of this moduli space, they have not yet succeeded in leading us to an explanation of which points in this moduli space are dynamically selected.

So where do we stand? As string phenomenologists, we can do two things. First, we can pursue *model-building*: we can search through the moduli space of self-consistent string models in order to determine how close to realistic spacetime physics we can come. This is, in some sense, a direct test of string theory as a phenomenological theory of physics. Of course, this approach to string phenomenology is ultimately limited by many factors: we have no assurance that our model-construction techniques are sufficiently powerful or general to include the "correct" string model (assuming that one exists); we have no assurance that our model-construction techniques will not lead to physically distinct models which nevertheless "agree" as far as their testable low-energy predictions are concerned; and we have no assurance that the most important phenomenological features that describe our low-energy world (such as the pattern of supersymmetry-breaking) are to be found in perturbative string theory rather than in non-perturbative string theory. For example, it may well be (and it has indeed been argued) that the true underlying string theory that describes nature is one which is intrinsically non-perturbative, and which would therefore be beyond the reach of the sorts of approaches typically followed in studies of string phenomenology.

Another option, then, is to temporarily abandon string model-building somewhat, and to seek to extract general phenomenological theorems or correlations about spacetime physics that follow directly from the general structure of string theory itself. Clearly, we would wish such information to be *model-independent*, *i.e.*, independent of our particular location in moduli space or the values of particular string parameters such as $\{\mathbf{V}_i, k_{ij}\}$. For example, if some particular configuration of spacetime physics (some pattern of low-energy phenomenology) can be shown to be inconsistent with being realized from an underlying set of $\{\mathbf{V}_i, k_{ij}\}$ parameters, and if such a demonstration can be made to transcend the particular free-fermionic construction so that it relies on only the primordial string symmetries themselves, then such patterns of phenomenology can be ruled out. In this way, one can still use string theory in order to narrow the list of possibilities for physics at higher energies, and to correlate various seemingly disconnected phenomenological features with each other. Such correlations would then be viewed as "predictions" from string theory, and we shall see many examples of this phenomenon in subsequent lectures.

In summary, then, we have seen that there exist powerful ways of constructing string models and surveying their low-energy phenomenologies, but that this leads to the problem of selecting the true model (*i.e.*, the true "ground state" or "vacuum") of string theory. Despite recent advances in understanding various non-perturbative aspects of string theory, our inability to answer the fundamental question of vacuum selection persists. Until this challenge is overcome, string phenomenology therefore must content itself with answering questions of a *relative* nature (such as questions concerning relative *patterns* of phenomenology) rather than the sorts of absolute questions (such as calculating the mass of the electron) that one would also ideally like to ask. Nevertheless, as we shall see, string theory can still provide us with considerable guidance for physics beyond the Standard Model.

6.8. Lecture #8: A final lecture

Up to this point, we have primarily discussed string *model-building* — i.e., the art of building string models. Hopefully, we have given the reader some sense of the complexity of the many constraints that are involved. In this final lecture, however, we shall depart from the somewhat "linear" development we have followed thus far in order to discuss *string phenomenology* — the study of the low-energy physical attributes of these models.

In the first part of this final lecture, we shall outline some general properties of four-dimensional heterotic string models. Then, we shall contrast these with the phenomenological properties of open-string D-brane models. Finally, we shall provide general comments concerning string phenomenology as a whole, and conclude with a brief discussion of some new, recent directions in string phenomenology.

6.8.1. *General properties of perturbative $D = 4$ heterotic string models*

In previous lectures, we have discussed the construction of perturbative heterotic string models. Here, we shall now turn the general low-energy properties that emerge from these constructions.

First, such models all have big gauge groups. For perturbative heterotic strings in four dimensions, we find that

$$\text{rank}(G) \leq 22 \ . \tag{6.8.1}$$

This is the four-dimensional analogue of the observation that the maximum rank in 10 dimensions is 16, such as for the $SO(32)$ and $E_8 \times E_8$ heterotic strings. The additional six units of rank emerge from the Kaluza-Klein reduction from $D = 10$ to $D = 4$.

If the string model in question is "realistic", then typically we can write

$$G = G_1 \times G_2 \qquad (6.8.2)$$

where G_1 contains $SU(3) \times SU(2) \times U(1)_{\text{hypercharge}}$. Here G_1 is called the "observable-sector" gauge group: e.g., G_1 could be $SU(3) \times SU(2) \times U(1)$, $SO(6) \times SO(4)$, $SU(5)$, $SO(10)$, E_6, etc. By contrast, G_2 is called the "hidden-sector" gauge group.

Second, there are typically lots of massless ("observable") states! These can be classified into several categories:

- Typical representations will carry charges under both G_1 and G_2 (*i.e.*, transform as non-singlet representations of these groups). In general, we will only have spinors, vectors (*i.e.*, fundamentals), and adjoints at the massless string level. (This is indeed a theorem: the allowed representations are closely tied to something called the "affine level" of the gauge group.) These states will typically fall into two subsets. First, there may be states that can be identified as (MS)SM quarks and leptons. In such cases, all gauge symmetry groups under which the SM gauge particles transform as singlets are considered to be part of G_2, *i.e.*, the hidden sector. Second, there can be extra states *beyond* the (MS)SM. There will typically be a lot of such states as well. They may be identified as exotic quarks and leptons. They will typically have fractional electric charge. This could cause problems (see below).
- Many gauge-*singlet* states (*i.e.*, states carrying no gauge charges) will also exist in the string model. For example, such states include the graviton, antisymmetric tensor, and dilaton ϕ. Recall that $g_{\text{string}} \sim \exp(-\langle\phi\rangle)$. Thus, the dilaton must be stabilized to yield a fixed value for the string coupling, and to avoid the so-called "dilaton runaway problem" (wherein $\langle\phi\rangle \to \infty$, or $g_{\text{string}} \to 0$).

The dilaton is just one example of a generic class of Lorentz-singlet particles called string "moduli". The effective potential for such models is *flat* to all orders in perturbation theory. Thus, non-perturbative string effects must somehow introduce a potential for these fields, *i.e.*, lift the degeneracy of string "ground states" and select a string vacuum. But how does this happen? This is a major unsolved problem, with lots of ideas in the literature. This is critically important for string phenomenology, since the vevs of the moduli set the values for gauge couplings, particle masses, and so forth. Without knowing the values of these couplings, the best we can look for is string-constrained *patterns* (textures) in these parameters.

Third, there will be infinite towers of Planck-scale string states! These states come in increasingly larger representations of gauge groups, and likewise have higher and higher Lorentz spins. These states are the means by which string theory maintains finiteness. They propagate in all string loop diagrams, and their contributions cancel the divergences of the massless states. They are the result of conformal invariance (really its one-loop extension, called "modular invariance").

One interesting fact about these states is that the number of such massive states with spacetime mass M grows *exponentially*:

$$g_M \sim \exp(cM\sqrt{\alpha'}) \qquad (6.8.3)$$

where c is a fixed positive constant. One of the implications of an exponentially growing degeneracy is as follows. Let us consider the thermodynamic partition function:

$$Z \equiv \sum_M g_M \exp(-M/kT) \qquad (6.8.4)$$

where T is the temperature and k is Boltzmann's constant. This gives:

$$Z = \sum_M \exp[M(c\sqrt{\alpha'} - 1/kT)] \ . \qquad (6.8.5)$$

Thus, if T is bigger than a critical value

$$T_c \equiv (kc\sqrt{\alpha'})^{-1} \ , \qquad (6.8.6)$$

then the thermodynamic partition function *diverges*! This is the so-called "Hagedorn" phenomenon.

Does this signal a phase transition? Or is there instead a limiting ("Hagedorn") temperature for string theory beyond which one cannot go? What happens to a box of strings (*i.e.*, the "universe") if we pump in lots of energy and try to raise the temperature?

The answers to these questions are really not known. The current belief is that we have a phase transition in which all extra energy gets dumped into long string modes. But the nature of this phase transition is generally unclear. Indeed, this Hagedorn phenomenon is one of the central hallmarks of the the subject of *string thermodynamics*. As might be imagined, this subject is of critical importance for string cosmology and for string-based studies of the early universe.

Fourth, such heterotic string models will typically give rise to a single "pseudo-anomalous" $U(1)$ gauge group! Recall that in field theory, given a set of states with $U(1)$ charges Q_i, we must have $\sum_i Q_i = 0$ in order to cancel axial (triangle) anomalies. In particular, certainly the hypercharge $U(1)$ must be anomaly-free.

However, in (many/most) heterotic string models, gauge groups are big and there can be *extra* $U(1)$ gauge groups. One finds, upon summing over massless spectrum, that *one* of these gauge groups, typically denoted $U(1)_X$, has corresponding states with charges Q_X such that $\sum Q_X \neq 0$. Thus, from the field-theory point of view, this $U(1)_X$ appears to be anomalous!

In fact, however, this gauge group is not anomalous (since string theory is always anomaly-free); there are extra contributions to the apparent anomaly which come from anomalous transformations of the string "axion field" (related to the antisymmetric tensor $B_{\mu\nu}$) which cancel this anomaly. This is an intrinsically "stringy" mechanism (called the Green-Schwarz mechanism) for cancelling an anomaly.

Fifth, such models typically give rise to automatic gauge coupling unification (regardless of existence of any GUT symmetry in the string model). In fact, the gauge couplings are even unified with the gravitational coupling!

It is easy to understand why this is the case. Recall that our original "untwisted" four-dimensional heterotic string model has a "unified" gauge group $SO(44)$, with one gauge gauge coupling whose value is set by the dilaton vev. (This is the analogue of $SO(32)$ in ten dimensions.) When we break subsequently break the gauge symmetry by introducing twists ("orbifolding"), this does not affect the gauge couplings. They are still all set by the same dilaton vev (ultimately because there is only *one* dilaton to which all gauge groups can couple). Thus, gauge coupling unification is automatic in heterotic string theory.

One important question is the *scale* of the unification. Clearly, by dimensional analysis, this can be nothing but the string scale. At tree-level, we have already seen in Lecture #1 that the string scale is given by $M_{\text{string}} = g_{\text{string}} M_{\text{Planck}}$. However, work by Kaplunovsky has shown that at one-loop order, and with the usual GUT assumption of $g_{\text{string}} \approx 0.7$, this result is shifted down to an approximate value, $M_{\text{string}} \approx 5.27 \times 10^{17}$ GeV. This is generally a problem, since the expected GUT value for the unification scale is $M_{\text{GUT}} = 2 \times 10^{16}$ GeV. How then do we explain this factor-of-20 discrepancy between M_{string} and M_{GUT}? This is currently an open question, with many potential solutions. A comprehensive review of this subject can be found in K.R. Dienes, Phys Reports 287 (1997) 447 = hep-th/9602045.

Sixth, such models typically give rise to states with fractional electric charge. We already referred to this above. Indeed, extra states beyond the MSSM will typically have $SU(2) \times U(1)$ quantum numbers which imply non-integer values for the electric charge. In fact, one can prove (see theorem by A.N. Schellekens) that if the model has a gauge symmetry $SU(3) \times SU(2) \times U(1)$ rather than a GUT, then the string will *necessarily* give rise to such fractionally charged states. This is a result of conformal invariance and modular invariance.

One possible resolution to this problem is that such fractionally charged states might be able to *confine* to form integer-charged states under the influence of non-abelian gauge symmetries beyond the SM. However, if this is not possible in a given string model, then that model is generally considered to be phenomenologically inconsistent.

General theorems exist which enable one to classify the different types of fractional charges one can expect to find in a given string model and which can be "confined" away. (We refer the reader to papers by Schellekens; also by Dienes, Faraggi, March-Russell.)

Seventh, it turns out that such string models *cannot* contain any exact global symmetries! For example, in heterotic string theory, baryon- and lepton-number conservation, as well as other discrete symmetries, must all be parts of *local* symmetries (gauge symmetries) or be only approximate symmetries (*i.e.*, accidental).

Eighth, such heterotic string models will either exhibit spacetime supersymmetry, or they will be non-supersymmetric. If non-supersymmetric, however, they nevertheless have a hidden symmetry called a "misaligned supersymmetry" which governs how the bosons and fermions are arranged at all mass levels so that finiteness is preserved, even without SUSY. Even the supertraces, when evaluated over the entire Fock space of string states, continue to vanish. [References include K.R. Dienes, Nucl. Phys. B429 (1994) 533; K.R. Dienes, M. Moshe, and R.C. Myers, Phys. Rev. Lett. 74 (1995) 4767.]

However, it is not known whether such non-supersymmetric strings can ever be stable beyond tree level. This is an important open question in string theory. However, if such stable non-SUSY strings exist, then this could provide a whole new framework for thinking about the gauge hierarchy problem, SUSY-breaking, questions of finiteness, the role of effective field theories and in particular the massive Planck-scale states, and gauge coupling unification. This may even provide an alternative, "stringy" approach towards the hierarchy problem which does not involve either supersymmetry or extra spacetime dimensions. [For some speculative ideas along this direction, see K.R. Dienes, hep-th/0104274.]

Ninth, the spacetime string spectrum can exhibit certain dualities. Indeed, there are several kinds of duality which, taken together, form an interconnected web of relations between different kinds of string theories.

- One kind of duality is called "T-duality". Consider string #1, compactified on a circle of radius R, and string #2, compactified on a circle of radius $\sqrt{\alpha'}/R$. It turns out that these strings are indistinguishable, in the sense that they have exactly the same spacetime spectrum! What would be considered a momentum state in string #1 would be considered a winding-mode state in string #2, and vice versa. This is clearly a very "stringy" symmetry! In fact, this symmetry transcends the mere tree-level spectrum, and holds to all orders. It also applies for all correlation functions, scattering amplitudes, both perturbatively and even non-perturbatively. This is an exact symmetry of closed string theories.

  One important implication of T-duality is that closed string theory (unlike point-particle field theory) cannot distinguish between large and small compactification radii! An interesting question is what this might imply about string cosmology. Likewise, what are the implications about our ultimate ability to derive *effective field theories* from the string?

- There are also other kinds of dualities which exist amongst the different string theories. For example, there is a duality called "S-duality" which flips the sign of the dilaton and thus relates theories at weak coupling to theories at strong coupling! Under such a mapping, perturbative string states (such as the ones we have been considering all along) are exchanged with non-perturbative string states (which have not considered at all, but which are "solitons" = D-branes in the theory). Under this mapping, for example, the $SO(32)$ heterotic (closed) string theory is mapped into the $SO(32)$ Type I (open) string theory. Combined with T-duality, one finds that all the different kinds of ten-dimensional strings are ultimately related to each other, becoming part of a larger superstructure called "M-theory".

The study of string dualities is a vast subject which easily deserves its own lectures, and which comprises the so-called "second" superstring revolution, dating from 1995. Indeed, insights have enabled us, in many cases, to "solve" for the strong-behavior of string theory!

I cannot give a proper introduction to M-theory here, but I will simply give some general comments. M-theory is a conjectured eleven-dimensional theory (of strings? of membranes? – we don't know) which can be *defined* through its three fundamental properties:

- The low-energy limit of M-theory is eleven-dimensional SUGRA (recall that $D = 11$ is the maximum dimension for SUGRA).
- Compactifying M-theory on a circle of radius R yields the Type IIA string with a coupling that is a growing function of R. So, at strong coupling, the Type IIA string begins to "see" an extra dimension and become eleven-dimensional.
- Compactifying M-theory on a line segment of length L yields the $E_8 \times E_8$ heterotic string with a coupling that grows with L. This is why one does not see this 11th dimension in studies of the perturbative heterotic string: the very act of taking the string coupling to be small reduces the 11th dimension to zero size!

Studying M-theory and its compactifications (and its phenomenological properties, such as how SUSY-breaking may be realized in this framework) has been a hot topic in the string literature. In particular, one may ask whether it is possible to compactify M-theory to four dimensions in ways that do *not* pass through an intermediate realization in terms of a $D = 10$ heterotic string, thereby constructing new classes of four-dimensional string models? The answer is to this question is 'yes'. Thus, even without knowing the precise nature of M-theory, it has already been possible to use insights gleaned from the mere existence of such a theory in order to generate new classes of string models.

Taken together, these developments have led to the realization that many of our cherished "fundamental" string symmetries (such as conformal invariance, modular invariance, etc.) are only *effective* weak-coupling symmetries, applicable only for closed strings. Thus, as the string coupling grows in closed string theories, we expect to see *deviations* from the constraints that come from these symmetries. This could be very useful in "freeing up" certain undesirable predictions of string phenomenology, even within closed strings.

6.8.2. General properties of $D = 4$ open-string models

Many of the above phenomenological features of heterotic strings change when one deals with Type I theories (*i.e.*, theories which include *open* strings). Some of the most viable models in this class that have chiral spectra include so-called "intersecting D-brane Models" as well as models with D-branes at singularities. Unfortunately, we do not have the space here to discuss such constructions. However, there are excellent reviews available, In particular, we refer the reader to R. Blumenhagen, M. Cvetic, P. Langacker, and G. Shiu, hep-th/0502005 and to M. Grana, hep-th/0509003.

There are many reasons to examine such Type I theories. Of course, they are interesting in their own right since they are among the possible allowed string constructions. However, as a result of the various string dualities discussed above, such strings often represent the strong-coupling limits of the heterotic models (this is "heterotic/Type I duality", a component of S-duality). Thus, by studying Type I string models, one is often really analyzing the strong-coupling limit of a closed heterotic string model.

We cannot provide a complete discussion of such Type I models here. However, the basic ideas are simple. Unlike heterotic string models, which realize their gauge symmetries along the closed strings through Kaluza-Klein reductions from 10 or 26 dimensions (as discussed above), gauge symmetries are realized in open strings through so-called "Chan-Paton factors" which reside at the endpoints of the open strings. These are the analogues of "quarks" at the ends of the open strings, and they carry the gauge charges associated with the string states.

Nowadays these Chan-Paton factors are reinterpreted as the labels associated with D-branes, so that open strings are considered to have endpoints which are restricted to lie on D-branes. Indeed, one definition of a D-brane is that it represents a solitonic membrane-like object on which an open strings can end. A single D-brane corresponds to a $U(1)$ gauge symmetry (the corresponding photon being represented by an open string which starts and ends on the brane), while non-abelian $U(N)$ gauge symmetries are realized through stacks of N coincident D-branes. In such configurations, the non-abelian gauge bosons are realized as strings which start and end on different D-branes within the stack. The Higgs mechanism (by which certain gauge symmetries can be broken and certain corresponding gauge bosons get heavy) can be realized in this framework by separating branes within the stack; those strings which start and end on different D-branes get stretched as a result of this separation, and thus become massive as a result of the tension involved in that stretching.

In general, within such constructions, one might realize the Standard Model through an $SU(3)$ stack of D-branes and an $SU(2)$ stack of D-branes. In such a scenario, quarks (which carrying non-trivial $SU(3)$ and $U(2)$ gauge charges) would be represented by strings stretching from the $SU(3)$ stack to the $SU(2)$ stack; such states can indeed be light (or massless) if these stacks of branes intersect, and the strings lie near that intersection. Of course, gravitational physics continues to be represented by closed strings which, having no endpoints, are not tied to particular branes and can therefore propagate freely in the "bulk". In general, only those states which are neutral with respect to all gauge symmetries (such as gravitons) are permitted to wander freely in the entire volume both within and transverse to the branes. In certain constructions, other possible closed-string states might include right-handed neutrinos (which are also completely neutral with respect to all Standard-Model gauge symmetries).

In general, the requirements of spacetime supersymmetry imply that the theory contain combinations of D-branes of only certain dimensionalities; likewise, the relative positions and/or geometric intersections of these D-branes are highly constrained. There are also generally other extended objects in these theories (beyond D-branes): these include anti-Dbranes, orientifold planes, and other types of branes (such as NS branes). Anomaly cancellation considerations end up playing a huge role in determining which configurations of all of these objects are required to form self-consistent string models.

Other than these constraints, however, one has tremendous freedom in designing D-brane configurations, compactifying the theory, wrapping the D-branes around the compactification manifolds, and so forth. This is then the art of Type I model building. Because of the tremendous range of allowed D-brane configurations and dimensionalities, and because the closed-string and open-string sectors have very different properties, Type I string phenomenology turns out to be *very rich* and *unconstrained* compared to heterotic string phenomenology.

In particular, even without providing details concerning such constructions, it is possible to summarize some of the major phenomenological differences between these string models and the closed (heterotic) strings discussed above.

First, the rank of the gauge group no longer restricted to 22! Indeed, non-perturbative effects can give rise to new gauge interactions that can increase the total rank beyond 22, and there is no bound to how large these gauge groups can become! (You can decide for yourself whether you consider this to be a good thing...)

Second, the fundamental scale of the theory (M_{string}) is no longer tied to M_{Planck}. The usual heterotic relation $M_{\text{string}} = g_{\text{string}} M_{\text{Planck}}$ no longer applies to open strings. The reason is that for closed strings, both gauge forces and the gravitational force emerge together. However, for Type I strings, the gravitational force emerges from the closed-string sector, while the gauge forces typically emerge from the open-string sectors. This difference introduces an undetermined "rescaling" factor between the different sectors, and therefore allows one to "dial" M_{string} as we wish in such theories. [For more details, see Chapter 10 of K.R. Dienes, Phys Reports 287 (1997) 447 = hep-th/9602045, which summarizes the original proposal of Witten: E. Witten, Nucl Phys B 471, 135 (1996)]. One could conceivably dial the Type I string scale all the way down to the TeV range – see, e.g., J. Lykken, PRD 54, 3693 (1996); K.R. Dienes, E. Dudas, T. Gherghetta, Nucl. Phys. B537, 47 (1999); G. Shiu, S.-H.H. Tye, Nucl. Phys. B548, 180 (1999).

This freedom to adjust the string scale and realize the Standard Model as an open string living on a brane while gravitational fields correspond to closed strings living in the bulk is the primary reason why Type I strings provide the natural realization (and inspiration) for extra-dimensional "brane-world" scenarios.

Third, for weakly coupled heterotic strings, there is only one dilaton-like field which couples to all gauge groups and matter fields in a universal way. However, in Type I theories there can generally be *multiple* dilaton-like fields.

Fourth, for Type I theories, gauge coupling unification is no longer automatic. This is a consequence of the existence of multiple dilaton-like fields. Each gauge coupling can be determined by the vev of a different dilaton field, and likewise the gauge theories living on different D-branes can experience different transverse volumes which also affect the values of their respective gauge couplings.

Fifth, in heterotic strings, there was only one anomalous $U(1)$ because there was only one dilaton to cancel this anomaly through the Green-Scwharz mechanism. However, in Type I theories there can be multiple anomalous $U(1)$'s because the presence of multiple dilatons in Type I theories implies that there can be a generalized Green-Schwarz mechanism which cancels multiple $U(1)$ anomalies.

Sixth, it turns out that whole new types of spacetime compactifications are possible. In heterotic strings, one must compactify on a so-called "Calabi-Yau" manifold if one wishes to preserve N=1 spacetime supersymmetry. (See Polchinski's textbook for a complete discussion: technically CY manifolds are six-dimensional complex manifolds with $SU(3)$ holonomy or equivalently vanishing first Chern class.) While the simple cases of tori (and orbifolds thereof) are well understood, the general full class of CY manifolds is not well understood (not even classified by mathematicians) and it is hard to perform detailed calculations of the resulting low-energy phenomenologies that emerge when heterotic strings are compactified on such spaces.

Type I string models are different. Because the matter arises locally (on branes) rather than globally (in the bulk), the compactification geometry is less constrained. For example, chirality no longer requires compactification on an orbifold, since chirality can instead emerge directly from D-brane intersections even when the compactification space is a smooth manifold.

For further discussions of these differences between the phenomenologies of open and closed strings, good references are: L.E. Ibanez, hep-th/9804236; F. Quevedo, Trieste String School Lectures, March 2002.

6.8.3. *String model-building and string phenomenology: General practice and goals*

Having discussed the different types of phenomenological features of these different types of string models, we now outline the basic way in which the string model-building game is played. Of course, the following steps are merely caricatures, with many details omitted. Nevertheless, they do indicate the rough methodology that a string phenomenologist must follow in order to claim to have a realistic string model.

The first step, as always, is to build the candidate string model itself. We have discussed how to do this in great detail in previous lectures. This is the string "model-building" aspect of string phenomenology. How one goes about doing this will depend on the particular string framework one has in mind, whether closed or open strings are involved, whether one is dealing with perturbative or non-perturbative constructions, and so forth. Each construction will carry with it its own constraints, its own techniques, and its own unique advantages and difficulties.

Once one has a particular string model in hand, one then extracts the gauge symmetry, the particle content (massless spectrum only, if one cares only about questions pertaining to observable low-energy states), and all associated charges and couplings.

The next step, if necessary (such as in heterotic strings), is to do a so-called "string vacuum shift". This is a technical step. Recall that there often exists a pseudo-anomalous $U(1)_X$ gauge symmetry. Although this is not really anomalous, it leads to an effective Fayet-Iliopoulos D-term which can break spacetime SUSY and destabilize the string vacuum. So, in order to "fix" this problem, one shifts the ground state slightly: one assigns a vev to certain moduli in the theory in order to break the $U(1)_X$ gauge symmetry and cancel the D-term. This makes the model stable again. This vev may often also break other gauge symmetries in addition to $U(1)_X$. It also can generate intermediate mass scales for various light states in the string model.

The third step is to write down an effective Lagrangian of these light fields that are derived from the string. Typically we will write something of the form

$$\mathcal{L} = \mathcal{L}_{\text{SUGRA}} + \mathcal{L}_{\text{matter}} + \mathcal{L}_{\text{couplings}} . \qquad (6.8.7)$$

These different pieces are as follows.

- $\mathcal{L}_{\text{SUGRA}}$: This must be appropriate for the given model in question, e.g., $N = 0$ (non-susy), or $N = 1$, or Type IIA or IIB SUGRA, etc.
- $\mathcal{L}_{\text{matter}}$: This will consist of the kinetic terms for all light fields (including an appropriate dilaton dependence).
- $\mathcal{L}_{\text{couplings}}$: Here, we must include all couplings allowed by string symmetries, given the charges that these states have under both observable and hidden-sector gauge symmetries (i.e., selection rules). These will include renormalizable *and* non-renormalizable couplings, where the non-renormalizable ones are suppressed by powers of the string scale M_{string}. In principle, one should calculate the coefficients that pre-multiply these terms by explicitly evaluating the appropriate corresponding string diagrams.

All together, this is the "effective Lagrangian from the string model". One must make sure that it is consistent with all string symmetries (e.g., T-duality, S-duality, others) if you are going to ask physics questions for which those symmetries are likely to be important.

The final step is to proceed to analyze the physics of the string model by analyzing the effective field theory of the effective Lagrangian derived from the string. We treat this Lagrangian as describing the physics at the string scale, and use RGE's to pass to lower energy scales (as we would in ordinary field theory). Along the way (*i.e.*, at intermediate scales), various new features can arise. For example, although Standard Model gauge groups will hopefully stay perturbative, the *hidden-sector* gauge couplings *may*, depending on the particle content, become strong and non-perturbative at some intermediate scale. This can trigger the corresponding gauginos to condense ("gaugino condensation"). which in turn can trigger SUSY-breaking. This is indeed an elegant string-inspired but field-theoretic means of breaking SUSY at intermediate energy scales. Likewise, extra matter beyond the MSSM (with masses determined by vacuum shifting, as discussed) can decouple. Clearly, the analysis for this sep is generally very model-dependent!

Ultimately, we seek to reproduce the low-energy world at the TeV-scale — *i.e.*, we wish to reproduce the Standard Model, and then study the phenomenological implications of the extra string-inspired particles or interactions that are predicted at higher scales. For example, one might construct string GUT models (realizing standard field-theory GUT scenarios from string theory), or realize the Standard Model directly at the string scale without an intervening GUT, or...

Given this procedure as outlined, one might wonder what the goals of string phenomenology ultimately are. Is it sufficient to try to construct semi-realistic string models, or are there are other goals as well? While a conversation on this topic can easily yield as many opinions as there are people in the conversation, the following represent the personal opinions of your humble lecturer. Therefore, the reader is forewarned about the potential bias of the lecturer.

Clearly, one important and undeniable goal must be to try to construct realistic string models, *i.e.*, to see how far one can really "push" the embedding of the low-energy world into string theory, to test the extent to which one can really make string theory consistent with the real world.

Unfortunately, this is very model-dependent. Also, given the large (infinite) "moduli" space of all possible string models, it is hard (impossible?) to believe that we would really be lucky enough to stumble across the right string model (assuming one exists).

Also, although we discussed one particular method of model-construction in these lectures (the so-called "free-fermionic construction" for closed strings), its applications and scope are limited (it essentially only hits discrete points in moduli space). They are points of enhanced symmetry, so they may indeed be special, but we don't know the structure of moduli space well enough to have a feeling for whether other, more compelling points might exist. And for open strings, we have seen the possibilities are even more varied!

Therefore, an alternative goal might be to try to uncover *model-independent* phenomenological truths from string theory. For example, one might ask questions such as

- What "patterns" of low-energy phenomenology are consistent with coming from or being realized from an underlying string theory?
- What "patterns" of low-energy phenomenology can be excluded?
- What sorts of "correlations" does string theory predict between phenomenological features that would otherwise appear to be completely independent from a field theory point of view? As an example, string theory predicts correlations between gauge groups and fractional charges, *etc.* These correlations are ultimately the reflections of the deeper string symmetries (*i.e.*, worldsheet symmetries) from which all spacetime physics is ultimately derived. For further editorializing along these lines, see the comments about the string landscape at the end of this lecture. In this way, we can then ask the question:
- What *guidance* does string theory provide for answering or addressing questions of physics beyond the Standard Model?

Of course, string theory also has the potential to provide insights of a completely different nature. For example, just as field theory provides certain mechanisms for addressing long-standing questions of particle physics, string theory (viewed as a general theory of extended objects) has the potential to provide new, additional, intrinsically geometric mechanisms for solving some of these same problems. Moreover, these mechanisms may also be able to generate new approaches to solving long-statnding problems that ordinary field theories based on point particles cannot reach. Thus, string phenomenology may be able to enlarge the domain of problems that a particle physicist might hope to address, and provide new tools for this endeavor.

But finally, perhaps the most important unresolved problem within string phenomenology is to understand what selects the string vacuum. Clearly, in order to make full progress in our understanding of string theory and its low-energy phenomenological predictions, we must eventually uncover the dynamics (presumably non-perturbative or semi-perturbative, or involving a mix of perturbative and non-perturbative physics) which ultimately pushes the universe towards the true ground state of string theory, the one in which we live.

Progress along these lines will be very hard, but is very important. Perhaps insight will come (or even may be coming) from recent developments in string duality. This problem seems tied up with the whole issue of how SUSY is broken, and the cosmological constant problem, so it is likely to take some time.

6.8.4. *New/current directions in string phenomenology*

We close this final lecture with a brief discussion of three new/current directions in string phenomenology. Once again, the following list is hardly complete. However, it does capture several of the main thrusts of string phenomenology research over the past few years, and the directions which are likely to hold the attention of string phenonenologists for the forseeable future.

Large-radius compactifications / TeV-scale strings

Many of you probably consider higher-dimensional "brane world" scenarios as something separate from string theory. But in truth, much of this work is really a branch of string phenomenology: one is studying the properties of string theories in a corner of the parameter space where the compactification radii are large, or where the Standard Model is restricted to a brane (stack) as in Type I models! Indeed, the whole setup of much of this work (SM restricted to a brane, gravity propagating in the bulk, and so forth) really emerges from Type I string theories where the SM is realized through open strings (whose endpoints therefore must lie on D-branes) and the graviton is realized through closed strings (which have no endpoints and which are therefore free to wander throughout the full higher-dimensional spacetime).

Thus, when one studies issues of flavor physics or develops new higher-dimensional mechanisms for understanding hierarchies, supersymmetry breaking, proton stability, *etc*, one is really developing an understanding of the phenomenology of open strings in a particular corner of compactification parameter space! In other words, the *brane world* is nothing but a branch of string phenomenology, studied through an effective field theory approach which might ultimately emerge from an underlying (Type I) string.

In addition to Kaustubh Agashe's excellent lectures on this subject at this year's TASI school, I recommend (of course) my own TASI 2002 lectures on the brane world: K.R. Dienes, 2002 TASI Lectures: New Directions for New Dimensions: An Introduction to Kaluza-Klein Theory, Large Extra Dimensions, and the Brane World", available at http://scipp.ucsc.edu/haber/tasi_proceedings/dienes.ps.

Flux compactifications

Another line of intense research in recent years concerns the possibility of so-called *flux compactifications*. There are compactifications in which various background fluxes associated with different p-form gauge fields in the theory are actually turned on. (Previous work had always assumed that such fluxes were zero.) It turns out that turning on such fluxes has a number of important effects. For example, the constraints on the allowed compactification geometries are modified, and the extra flux contributions allow us to go beyond the simple class of Calabi-Yau compactifications.

However, the most important phenomenological aspect of such flux compactifications is that they provide a framework leading to new methods of moduli stabilization. Indeed, within the framework of flux compactifications, it has been been possible to build semi-realistic string models in which the vast majority of complex and Kähler moduli are completely frozen!

Flux compactifications thus provide a new arena in which to address the all-important issues of moduli stabilization and vacuum selection. Indeed, work of Kachru, Kallosh, Linde, and Trivedi (KKLT) has even provided a framework in which it might be possible to realize meta-stable string vacua with deSitter (dS) geometries. This is of critical importance if string theory is to make contact with cosmological evolution.

The string theory "landscape"

Finally, as we have seen repeatedly throughout these lectures, one of the most serious problems faced by practitioners of string phenomenology is the multitude of possible, self-consistent string vacua. That there exist large numbers of potential string solutions has been known since the earliest days of string theory; these result from the large numbers of possible ways in which one may choose an appropriate compactification manifold (or orbifold), an appropriate set of background fields and fluxes, and appropriate expectation values for the plethora of additional moduli to which string theories generically give rise. Although historically these string solutions were not completely stabilized, it was tacitly anticipated for many years that some unknown vacuum stabilization mechanism would ultimately lead to a unique vacuum state. Unfortunately, recent developments suggest that there continue to exist huge numbers of self-consistent string solutions (*i.e.*, string "models" or "vacua") even after stabilization. Thus, a picture emerges in which there exist huge numbers of possible string vacua, all potentially stable (or sufficiently metastable), with apparently no dynamical principle to select amongst them. Indeed, each of these potential vacua can be viewed as sitting at the local minimum of a complex terrain of possible string solutions dominated by hills and valleys. This terrain has come to be known as the "string-theory landscape".

The existence of such a landscape has tremendous practical significance because, as we have seen, the specific low-energy phenomenology that can be expected to emerge from string theory depends critically on the particular choice of vacuum state. Detailed quantities such as particle masses and mixings, and even more general quantities and structures such as the choice of gauge group, number of chiral particle generations, magnitude of the supersymmetry-breaking scale, and even the cosmological constant can be expected to vary significantly from one vacuum solution to the next. Thus, in the absence of some sort of vacuum selection principle, it is natural to tackle a secondary but perhaps more tractable question concerning whether there might exist generic string-derived correlations between different phenomenological features. In this way, one can still hope to extract phenomenological predictions from string theory.

Over the past two years, this idea has triggered a surge of activity concerning the *statistical* properties of the landscape. Investigations along these lines have focused on diverse phenomenological issues including the value of the supersymmetry-breaking scale, the value of the cosmological constant, and the preferred rank of the corresponding gauge groups, the prevalence of the Standard-Model gauge group, and possible numbers of chiral generations. Discussions of the landscape have also led to various theoretical paradigm shifts, ranging from alternative landscape-based notions of naturalness and novel cosmological inflationary scenarios to the use of anthropic arguments to constrain the set of viable string vacua. There have even been proposals for field-theoretic analogues of the string-theory landscape, as well as discussions concerning whether a landscape of sufficiently stable string vacua actually exists.

The implications of a landscape (if it exists) have been hotly debated in the string community. Undoubtedly, if the string landscape exists, it is a very rich place, full of unanticipated properties and characteristics. Nevertheless, at the very least, the possible existence of such a landscape has focused the attention of the string community on the fundamental question which has plagued string theory over the past twenty years, namely the issue of vacuum selection.

One might argue that the landscape is simply too large to permit any reasonable analysis. Indeed, one might even argue that if such a landscape exists, string theory is doomed as a predictive theory of physics, and that the answers to some of the most fundamental questions in physics might find their answers in random environmental selection (or as the result of cosmological chance).

However, it is also true that the direct examination of actual string models uncovers features and behaviors that might not otherwise be expected. Moreover, through direct enumeration, we gain valuable experience in the construction and analysis of phenomenologically viable string vacua. Finally, as string phenomenologists, we must ultimately come to terms with the landscape (if it exists). Just as in other fields ranging from astrophysics and botany all the way to zoology, the first step in the analysis of a large data set is enumeration and classification. Indeed, this is how science begins. Thus, properly interpreted, statistical landscape studies might be useful and relevant in this overall endeavor of connecting string theory to the real world.

Acknowledgments

I would like to thank the organizers of the 2006 TASI school, especially Sally Dawson, Rabi Mohapatra, and K.T. Mahanthappa, for their invitation to visit Boulder, Colorado, and deliver these lectures in such a pleasant and stimulating environment. I would also like to thank Sally and Rabi for their extreme patience waiting for the written version of these lectures to appear. But most importantly, I also wish to thank the TASI students themselves for their questions and sustained interest in these lectures. There is nothing more pleasant for a lecturer than an enthusiastic and inquisitive audience. This work was supported in part by the U.S. National Science Foundation under Grant PHY/0301998, by the U.S. Department of Energy under Grant DE-FG02-04ER-41298, and by a Research Innovation Award from Research Corporation.

Chapter 7

Theoretical Aspects of Neutrino Masses and Mixings

R. N. Mohapatra

Department of Physics, University of Maryland, College Park, MD, 20742, USA

Neutrino oscillation experiments have yielded valuable information on the nature of neutrino masses and mixings and have provided the first glimpse of new physics beyond the standard model. Even though we are far from a complete understanding of the new physics implied by them, some tell-tale hints are emerging which have narrowed the direction of the new physics and have provided some insight into the flavor problem. In these lectures, I provide a panoramic overview of the current thinkings in neutrino model building.

7.1. Introduction

For a long time, it was believed that neutrinos are massless, spin half particles, making them drastically different from their other standard model spin half cousins such as the charged leptons (e, μ, τ) and the quarks (u, d, s, c, t, b), which are known to have mass. This myth has however been shattered by the accumulating evidence for neutrino mass from the solar and atmospheric neutrino observations compiled in the nineties as well as several terrestrial experiments in the new century. One must therefore now be free to look beyond the massless neutrino idea to explore new physics as we proceed to understand the neutrino mass.

The possibility of a nonzero neutrino mass at phenomenological level goes back almost 50 years. In the context of gauge theories, they were discussed extensively in the 70's and 80's long before there was any firm evidence for it. For instance the left-right symmetric theories of weak interactions introduced in 1974 and discussed in those days in connection with the structure of neutral current weak interactions, predicted nonzero neutrino mass as a necessary consequence of parity invariance and quark lepton symmetry.

The existence of a nonzero neutrino mass makes neutrinos more like the quarks, and allows for mixing between the different neutrino species leading to the phenomenon of neutrino oscillation, an idea first discussed by Pontecorvo[1] and Maki et al.[1] in the 1960's, unleasing a whole new realm of particle physics phenomena to explore beyond the standard model. At the present time, we are of course far from a complete picture of the masses and mixings of the various neutrinos and cannot therefore have a full outline of the theory of neutrino masses. However there exist enough information that we can surmise some viable possibilities for the theories beyond the standard model. Combined with other ideas outside the neutrino arena such as supersymmetry and unification, the possibility narrows even further. Many clever experiments now under way will soon clarify or rule out many of the allowed models. In these lectures, I give a panoramic view of what we may have learned about physics beyond the standard model and new symmetries of flavor as we attempt to understand neutrino masses.[2]

For simplicity, I will focus on a widely discussed framework for understanding of the small neutrino masses, the seesaw mechanism, which employs a minimal extension of the standard model by adding two or three right handed neutrinos, which are super-heavy and Majorana type. I will touch briefly on some specific models that are based on the above general framework but attempt to provide an understanding of the detailed mass and mixing patterns using family symmetries which must supplement the seesaw mechanism. We will also present a class of SO(10) grand unified theories where there is no need for family symmetries to understand large mixings. These works are instructive for several reasons: first they provide an existence proof that this is a sensible way to proceed in tackling the hard problem of understanding large lepton mixings; second they often illustrate the kind of assumptions needed and through that provide a unique insight into which directions the next step should be; finally of course nature may be generous in picking one of those models as the final message bearer.

The fact that the neutrino has no electric charge endows it with certain properties not shared by other fermions of the standard model. One can write two kinds of Lorentz invariant mass terms for the neutrino, the Dirac and Majorana masses, whereas for the charged fermions, conservation of electric charge allows only Dirac type mass terms. In the four component notation for the fermions, the Dirac mass has the form $\bar{\psi}\psi$, whereas the Majorana mass is of the form $\psi^T C^{-1} \psi$, where ψ is the four component spinor and C is the charge conjugation matrix. One can also discuss the two different kinds of mass terms using the two component notation for the spinors, which provides a very useful way to discuss neutrino masses. We therefore present some of the salient concepts behind the two component description of the neutrino.

7.1.1. *Two component notation for neutrinos*

Before we start the discussion of the 2-component neutrino, let us write down the Dirac equation for an electron:[3]

$$i\gamma^\lambda \partial_\lambda \psi - m\psi = 0 \qquad (7.1)$$

This equation follows from a free Lagrangian

$$\mathcal{L} = i\bar{\psi}\gamma^\lambda \partial_\lambda \psi - m\bar{\psi}\psi \qquad (7.2)$$

and leads to the relativistic energy momentum relation $p^\lambda p_\lambda = m^2$ for the spin-half particle only if the four γ_λ's anticommute. If we take γ_λ's to be $n \times n$ matrices, the smallest value of n for which four anticommuting matrices exist is four. Therefore ψ must be a four component spinor. The physical meaning of the four components is as follows: two components for particle spin up and down and same for the antiparticle.

A spin-half particle is said to be a Majorana particle if the spinor field ψ satisfies the condition of being self charge conjugate, i.e.

$$\psi = \psi^c \equiv C\bar{\psi}^T, \qquad (7.3)$$

where C is the charge conjugation matrix and has the property $C\gamma_\lambda C^{-1} = -\gamma^{\lambda T}$. This constraint reduces the number of independent components of the spinor by a factor of two, since the particle and the antiparticle are now the same particle. Using this condition, the mass term in the Lagrangian in Eq. (2) can be written as $\psi^T C^{-1} \psi$, where we have used the fact that C is a unitary matrix. Writing the mass term in this way makes it clear that if a field carries a $U(1)$ charge and the theory is invariant under those $U(1)$ transformations, then the mass term breaks this symmetry. This means that one cannot impose the Majorana condition on a particle that has a gauge charge. Since the neutrinos do not have electric charge, they can be Majorana particles unlike the quarks, electron or the muon. It is of course well known that the gauge boson interactions in a gauge theory Lagrangian conserve a global $U(1)$ symmetry known as lepton number with the neutrino and electron carrying the same lepton number. If lepton number were to be established as an exact symmetry of nature, the Majorana mass for the neutrino would be forbidden and the neutrino, like the electron, would be a Dirac particle.

The properties of a Majorana fermion can be seen in its free field expansion in terms of creation and annihilation operators:

$$\psi(x) = \int \frac{d^3p}{\sqrt{(2\pi)^3 2E_p}} \Sigma_s \left(a_s(\mathbf{p}) u_s(\mathbf{p}) e^{-ip\cdot x} + a_s^\dagger v_s(\mathbf{p}) e^{ip\cdot x} \right). \tag{7.4}$$

In the gamma matrix convention where $\gamma_i = \begin{pmatrix} 0 & \sigma_i \\ \sigma_i & 0 \end{pmatrix}$ and $\gamma_0 = \begin{pmatrix} 0 & I \\ I & 0 \end{pmatrix}$, the u_s and v_s are given by

$$u_s(\mathbf{p}) = \frac{m}{\sqrt{E}} \begin{pmatrix} \alpha_s \\ \frac{E - \sigma \cdot \mathbf{p}}{m} \alpha_s \end{pmatrix} \tag{7.5}$$

and

$$v_s(\mathbf{p}) = \frac{m}{\sqrt{E}} \begin{pmatrix} -\frac{E + \sigma \cdot \mathbf{p}}{m} \alpha_s' \\ \alpha_s' \end{pmatrix}. \tag{7.6}$$

α_s and α_s' are two component spinors.

If we choose $\alpha_s' = \sigma_2 \alpha_s$, we get the relation among the spinors $u_s(\mathbf{p})$ and $v_s(\mathbf{p})$ $C\gamma_0 u_s^*(\mathbf{p}) = v_s(\mathbf{p})$ and the Majorana condition follows. Note that if ψ were to describe a Dirac spinor, then we would have had a different creation operator $b^\dagger$ in the second term in the free field expansion above.

The origin of the two component neutrino is rooted in the isomorphism between the Lorentz group and the SL(2, C) group. The latter is defined as the set of 2×2 complex matrices with unit determinant, whose generators satisfy the same Lie algebra as that of the Lorentz group. Its basic representations are 2 and 2* dimensional. These are the spinor representations and can be used to describe spin half particles.

We can therefore write the familiar 4-component Dirac spinor used in the text books to describe an electron can be written as $\psi = \begin{pmatrix} \phi \\ i\sigma_2 \chi^* \end{pmatrix}$, where χ and ϕ two two component spinors. A Dirac mass is the given by $\chi^T \sigma_2 \phi$ whereas a Majorana mass is given by $\chi^T \sigma_2 \chi$, where σ_a are the Pauli matrices. To make correspondence with the four component notation, we point out that ϕ and $i\sigma_2 \chi^*$ are nothing but the ψ_L and ψ_R respectively. It is then clear that χ and ϕ have opposite electric charges; therefore the Dirac mass $\chi^T \sigma_2 \phi$ maintains electric charge conservation (as well as any other kind of charge like lepton number etc.).

2-component neutrino is described by the following Lagrangian:

$$\mathcal{L} = \nu^\dagger i\sigma^\lambda \partial_\lambda \nu - \frac{im}{2} e^{i\delta} \nu^T \sigma_2 \nu + \frac{im}{2} e^{-i\delta} \nu^\dagger \sigma_2 \nu^*. \tag{7.7}$$

This leads to the following equation of motion for the field χ

$$i\sigma^\lambda \partial_\lambda \chi - im\sigma_2 \chi^* = 0. \tag{7.8}$$

As is conventionally done in field theories, we can now give a free field expansion of the two component Majorana field in terms of the creation and annihilation operators:

$$\chi(x,t) = \sum_{p,s}[a_{\mathbf{p},s}\alpha_{p,s}e^{-ip.x} + a^\dagger_{\mathbf{p},s}\beta_{p,s}e^{ip.x}], \tag{7.9}$$

where the sum on s goes over the spin up and down states.

Exercise 1: Using the field equations for a free massive two component Majorana spinor, show that its expansion in terms of the creation and annihilation operators and two component spinors $\alpha = \begin{pmatrix} 1 \\ 0 \end{pmatrix}$ and $\beta = \begin{pmatrix} 0 \\ 1 \end{pmatrix}$ is given by the following expression:

$$\chi(x,t) = \sum_p [a_{\mathbf{p},+}e^{-ip.x} - a^\dagger_{\mathbf{p},-}e^{ip.x}\alpha]\sqrt{E+p}$$
$$+ \sum_p [a_{\mathbf{p},-}e^{-p.x} + a^\dagger_{\mathbf{p},+}e^{ip.x}\alpha]\sqrt{E-p}. \tag{7.10}$$

Note that in a beta decay process, where a neutron is annihilated and proton is created, the leptonic weak current that is involved is $\bar{e}\nu$ (dropping gamma matrices); therefore, along with the electron, what is created predominantly is a right handed particle (with a wave function α), the amplitude being of order $\sqrt{E+p} \approx \sqrt{2E}$. This is the right handed anti-neutrino. The left handed neutrino is produced with a much smaller amplitude $\sqrt{E-p} \approx m_\nu/E$. Similarly, in the fusion reaction in the core of the Sun, what is produced is a left handed state of the neutrino with a very tiny i.e. $O(m_\nu/E)$ admixture of the right handed helicity (or the "anti-neutrino" component).

7.1.2. *Neutrinoless double beta decay and neutrino Majorana mass*

As already noted a Majorana neutrino breaks lepton number by two units. This has the experimentally testable prediction that it leads to the process of neutrino-less double beta decay, that involves the decay of an even-even nucleus i.e. $(Z,N) \to (Z+2, N-2) + 2e^-$. We will now show by using the above property of the Majorana neutrino that if light neutrino exchange is responsible for this process, then the amplitude is proportional to the neutrino mass.

Double beta decay involves the change of two neutrons to two protons and therefore has to be a second order weak interaction process. Since each weak interaction process emits an antineutrino, in second order weak interaction, the final state will involve two anti-neutrinos. But in neutrino-less double beta decay, there are no neutrinos in the final state; therefore the two neutrinos must go into the vacuum state. Vacuum state by definition has no spin whereas the antineutrino emitted in a beta decay has spin. Consider the antineutrino from one of the decays: it must be predominantly right handed. But to disappear into vacuum, it must combine with a lefthanded antineutrino so that the left and right handed spin projections add up to zero. In the previous paragraph, we showed that the fraction of left handed spin projection in a neutrino emitted in beta decay is m_ν/E. Therefore, $\bar{\nu}_e\bar{\nu}_e \to |0>$ must be proportional to the neutrino mass. Thus neutrino-less double beta decay is therefore a very sensitive measure of neutrino mass.

7.1.3. Neutrino mass in two component notation

Let us now discuss the general neutrino mass for Majorana neutrinos. We saw earlier that for Majorana neutrinos, there are two different ways to write a mass term consistent with relativistic invariance. This richness in the possibility for neutrino masses also has a down side in the sense that in general, there are more parameters describing the masses of the neutrinos than those for the quarks and leptons. For instance for the electron and quarks, dynamics (electric charge conservation) reduces the number of parameters in their mass matrix. As an example, using the two component notation for all fermions, for the case of two two component spinors, a charged fermion mass will be described only by one parameters whereas for a neutrino, there will be three parameters. This difference increases rapidly e.g. for 2N spinors, to describe charged fermion masses, we need N^2 parameters (ignoring CP violation) whereas for neutrinos, we need $\frac{2N(2N+1)}{2}$ parameters. What is more interesting is that for a neutrino like particle, one can have both even and odd number of two component objects and have a consistent theory.

In this article, we will use two component notation for neutrinos. Thus when we say that there are N neutrinos, we will mean N two-component neutrinos.

In the two component language, all massive neutrinos are Majorana particles and what is conventionally called a Dirac neutrino is really a very specific choice of mass parameters for the Majorana neutrino. Let us give some examples: If there is only one two component neutrino (we will drop the prefix two component henceforth), it can have a mass $m\nu^T\sigma_2\nu$ (to be called $\equiv m\nu\nu$ in shorthanded notation). The neutrino is now a self conjugate object which can be seen if we write an equivalent 4-component spinor ψ:

$$\psi = \begin{pmatrix} \nu \\ i\sigma_2\nu^* \end{pmatrix} \tag{7.11}$$

Note that this 4-component spinor satisfies the condition

$$\psi = \psi^c \equiv C\bar{\psi}^T \tag{7.12}$$

This condition implies that the neutrino is its own anti-particle, a fact more transparent in the four- rather than the two-component notation. The above exercise illustrates an important point i.e. given any two component spinor, one can always write a self conjugate (or Majorana) 4-component spinor. Whether a particle is really its own antiparticle or not is therefore determined by its interactions. To see this for the electrons, one may solve the following exercise i.e. if we wrote two Majorana spinors using the two two-component spinors that describe the electron, then until we turn on the electromagnetic interactions and the mass term, we will not know whether the electron is its own antiparticle or not. Once we turn on the electromagnetism, this ambiguity is resolved since electric charge conservation will allow a mass term that connects the two 2-component spinors and no mass term connecting either of the two component spinors with themselves.

Let us now go one step further and consider two 2-component neutrinos (ν_1, ν_2). The general mass matrix for this case is given by:

$$\mathcal{M}_{2\times 2} = \begin{pmatrix} m_1 & m_3 \\ m_3 & m_2 \end{pmatrix} \tag{7.13}$$

Note first that this is a symmetric matrix and can be diagonalised by orthogonal transformations. The eigen-states which will be certain admixtures of the original neutrinos now describe self conjugate particles. One can look at some special cases:

Case i:

If we have $m_{1,2} = 0$ and $m_3 \neq 0$, then one can assign a charge +1 to ν_1 and -1 to ν_2 under some $U(1)$ symmetry other than electromagnetism and the theory is invariant under this extra $U(1)$ symmetry which can be identified as the lepton number and the particle is then called a Dirac neutrino. The point to be noted is that the Dirac neutrino is a special case of for two Majorana neutrinos. In fact instead of calling this a Dirac neutrino, we could call this a case with two Majorana neutrinos with equal and opposite (in sign) mass. Since a complex mass term in general refers to its C transformation property (i.e. $\psi^c = e^{i\delta_m}\psi$, (where δ_m is the phase of the complex mass term), the two two-component fields of a Dirac neutrino having opposite sign mass would be equivalent to having opposite charge conjugation properties.

Case ii:

If we have $m_{1,2} \ll m_3$, this case is called pseudo-Dirac neutrino since this is a slight departure from case (i). In reality, in this case also the neutrinos are Majorana neutrinos with their masses $\pm m_0 + \delta$ with $\delta \ll m_0$. The two component neutrinos will be maximally mixed. Thus this case is of great current physical interest in view of the atmospheric (and perhaps solar) neutrino data.

Case iii:

There is third case where one may have $m_1 = 0$ and $m_3 \ll m_2$. In this case the eigenvalues of the neutrino mass matrix are given respectively by: $m_\nu \simeq -\frac{m_3^2}{m_2}$ and $M \simeq m_2$. One may wonder under what conditions such a situation may arise in a realistic gauge model. It turns out that if ν_1 transforms as an $SU(2)_L$ doublet and ν_2 is an $SU(2)_L$ singlet, then the value of m_3 is limited by the weak scale whereas m_2 has no such limit and $m_1 = 0$ if the theory has no $SU(2)_L$ triplet field (as for instance is the case in the standard model). Choosing $m_2 \gg m_3$ then provides a natural way to understand the smallness of the neutrino masses. This is known as the seesaw mechanism.[4] Since this case is very different from the case (i) and (ii), it is generally said that in grand unified theories, one expects the neutrinos to be Majorana particles. The reason is that in most grand unified theories there is a higher scale which under appropriate situations provides a natural home for the large mass m_2.

While we have so far used only two neutrinos to exemplify the various cases including the seesaw mechanism, these discussions generalize when $m_{1,2,3}$ are each $N \times N$ matrices (which we denote by $m_3 \equiv m_D$ and $m_2 \equiv M_R$). For example, the seesaw formula for this general situation can be written as

$$\mathcal{M}_\nu \simeq m_1 - m_D^T M_R^{-1} m_D \qquad (7.14)$$

where the subscripts D and R are used in anticipation of their origin in gauge theories where M_D turns out to be the Dirac matrix and M_R is the mass matrix of the right handed neutrinos and all eigenvalues of M_R are much larger than the elements of M_D. It is also worth pointing out that Eq. 7.14 can be written in a more general form where the Dirac matrices are not necessarily square matrices but $N \times M$ matrices with $N \neq M$. We give such examples below.

Although there is no experimental proof that the neutrino is a Majorana particle, the general belief is that since the seesaw mechanism provides such a simple way to understand the glaring differences between the masses of the neutrinos and the charged fermions, neutrino is indeed most likely to be a Majorana particles as implied by it.

Even though in many situations, the difference between the Dirac and Majorana neutrinos is not manifest, there are some physical processes where differences becomes explicit: one such process is when the two neutrinos annihilate. For Dirac neutrinos, the particle and the antiparticle are distinct and therefore their annihilation is not restricted by Pauli principle in any manner. However, for the case of Majorana neutrinos, the identity of neutrinos and antineutrinos plays an important role and one finds that the annihilation to the Z-bosons occurs only via the P-waves. Similarly in the decay of the neutrino to any final state, the decay rate for the Majorana neutrino is a factor of two higher than for the Dirac neutrino.

7.1.4. *Experimental indications for neutrino masses*

There have been other lectures at this school on the experimental evidences for neutrino masses and their analyses to determine the current favorite values for the various mass differences as well as mixing angles. I will therefore only summarize the main results: (For detailed discussion and references, see[5] and lectures by J. Conrad[6]).

The evidence for neutrino masses and mixings have come from neutrino oscillation experiments involving neutrinos from the Sun, the cosmic rays as well as from accelerators. Neutrino oscillation is a phenomenon where neutrinos of one flavor transmute to neutrinos of another flavor. Since such transmutation can occur only if the neutrinos have masses and mixings, these experiments provide evidence for neutrino mass. To see this, note the expression for vacuum oscillation probability for neutrinos of a given energy E that have travelled a distance L is given by:

$$P_{\alpha\beta} = \sum_{i,j} |U_{\alpha i} U^*_{\beta i} U^*_{\alpha j} U_{\beta j}| \cos\left(\frac{\Delta_{ij} L}{2E} - \phi_{\alpha\beta,ij}\right) \qquad (7.15)$$

This can be derived on the basis of simple quantum mechanical superposition principle and the equations of time evolution of free particles. The observed neutrino oscillation probabilities therefore yield information about the mass difference squares of the neutrinos ($\Delta_{ij} = m_i^2 - m_j^2$) and mixing angles $U_{\alpha i}$. If the neutrino propagates in dense matter, the oscillation probability is changed by the so-called Mikheyev-Smirnov-Wolfenstein effect however the new probability depends on the same parameters i.e. mass difference square and the $U_{\alpha i}$ in addition to depending on the density of matter through which neutrinos travel. The analysis of the data for the atmospheric neutrinos where the neutrino propagates in vacuum and that for solar neutrinos where the effect of dense matter in the Sun is included lead to the following picture values for mass differences and mixings:

7.1.4.1. Atmospheric neutrinos:

The Super-Kamiokande experiment observed the oscillations of the atmospheric muon neutrinos to tau neutrinos (although the tau neutrinos from the oscillations have not been confirmed yet). This oscillation of ν_μ to ν_τ has now been confirmed by the accelerator observations in the K2K and the MINOS experiments. From the existing data several important conclusions can be drawn: (i) the data cannot be fit assuming oscillation between ν_μ and ν_e nor $\nu_\mu - \nu_s$, where ν_s is a sterile neutrino which does not any direct weak interaction; (ii) the oscillation scenario that fit the data best is $\nu_\mu - \nu_\tau$ for the mass and mixing parameters

$$\Delta m^2_{\nu_\mu - \nu_\tau} \simeq (2 - 8) \times 10^{-3} \text{ eV}^2; \qquad (7.16)$$

$$in^2 2\theta_{\mu-\tau} \geq 0.92$$

7.1.4.2. Solar neutrinos

The second evidence for neutrino oscillation comes from the several experiments that have observed a deficit in the flux of neutrinos from the Sun as compared to the predictions of the standard solar model championed by Bahcall and his collaborators[5] and more recently studied by many groups. The experiments responsible for this discovery are the Chlorine experiment of Ray Davis, Kamiokande, Gallex, SAGE, Super-Kamiokande, SNO, GNO experiments conducted at the Homestake mine, Kamioka in Japan, Gran Sasso in Italy and Baksan in Russia and Sudbery in Canada respectively.

The different experiments see different parts of the solar neutrino spectrum. The details of these considerations are discussed in other lectures.

As far as the final state goes, it can either be one of the two remaining active neutrinos, ν_μ and ν_τ or it can be the sterile neutrino ν_s. SNO neutral current data announced recently has very strongly constrained the second possibility (i.e. the sterile neutrino in the final state). The global analyses of all solar neutrino data seem to favor the so called large mixing angle MSW solution with parameters: $\Delta m^2 \simeq 1.2 \times 10^{-5} - 3.1 \times 10^{-4} \text{eV}^2$; $\sin^2 2\theta \simeq 0.58 - 0.95$.

This result has been confirmed by the terrestrial KamLand experiment which looked for oscillation for reactor neutrinos from several reactors with a detector at an average distance of about 100 Km from the various sources. It eliminated solutions to the solar neutrino puzzle based e.g. on spin flavor precession as well as the so called low solution and confirmed the large angle MSW resolution as the most plausible one.

7.1.4.3. Search for the mixing angle θ_{13}

The remaining mixing angle θ_{13} has been probed by the reactor experiments that used the French reactor CHOOZ and the US reactor in Palo-Verde by looking for the oscillation of reactor electron anti-neutrinos with a detector at a distance of about a kilo-meter. In the simple three neutrino picture, the dominant oscillation in this case would be to the tau neutrinos since $\Delta m_{31}^2 \gg \Delta m_{21}^2$. The absence of a signal put an upper limit on the mixing angle of about $\theta_{13} \leq 0.17$. There are several reactor as well as long baseline experiments now being prepared which will conduct a higher precision search for θ_{13} during the next decade.[7]

7.1.4.4. LSND and MiniBooNe

Finally, we come to the last indication of neutrino oscillation from the Los Alamos Liquid Scintillation Detector (LSND)[6] experiment, where neutrino oscillations both from a stopped muon (DAR) as well as the one accompanying the muon in pion decay (known as the DIF) have been observed. The evidence from the DAR is statistically more significant and is an oscillation from $\bar{\nu}_\mu$ to $\bar{\nu}_e$. The mass and mixing parameter range that fits data is[6]:

$$LSND: \Delta m^2 \simeq 0.2 - 2eV^2; \sin^2 2\theta \simeq 0.003 - 0.03 \qquad (7.17)$$

There are also points at higher masses specifically at 6 eV^2 which are also allowed by the present LSND data for small mixings. KARMEN experiment at the Rutherford laboratory has very strongly constrained the allowed parameter range of the LSND data. Recently the Miniboone experiment at Fermilab has announced the results of its search for $\nu_\mu - \nu_e$ oscillation. They have not found any evidence for oscillation with characteristic mass diffrerence square in the eV range.[8,9]

7.1.4.5. *Neutrino-less double beta decay and Tritium decay experiment*

Oscillation experiments only depend on the difference of mass squares of the different neutrinos and the mixing angles. Therefore, in order to have a complete picture of neutrino masses, we need other experiments. Two such experiments are the neutrino-less double beta decay searches and the search for neutrino mass from the analysis of the end point of the electron energy spectrum in tritium beta decay. They have been discussed at this institute by P. Vogel,[11] which is referred to for more details and references.

Neutrino-less double beta decay measures the following combination of masses and mixing angles:

$$<m>_{\beta\beta} = \sum_i U_{ei}^2 m_i \qquad (7.18)$$

Therefore naively speaking it is sensitive to the overall neutrino mass scale. But in practice, as we will see below, for the case of both normal and inverted hierarchies, it is unlikely to settle the question of the overall mass scale at the presently contemplated level of sensitivity in double beta decay searches. Only if the neutrino mass patterns are inverse hierarchical does one expect a visible signal in $\beta\beta_{0\nu}$ decay. We do not get into great details into this issue except to mention that in drawing any conclusions about neutrino mass from this process, one has to first have a good calculation of nuclear matrix elements of the various nuclei involves such as ^{76}Ge, ^{136}Xe, ^{100}Mo etc.; secondly, another confusing issue has to do with alternative physics contributions to $\beta\beta_{0\nu}$ which are unrelated to neutrino mass. Nevertheless, neutrino-less double beta decay is a fundamental experiment and a nonzero signal will establish the important result that neutrino is a Majorana particle and that lepton number symmetry is violated. Regardless of whether it tells us anything about the neutrino masses, it would provide an indication in favor of the seesaw mechanism. Presently two experiments

Heidelberg-Moscow and IGEX that use enriched ^{76}Ge have published limits of ≤ 0.3 eV. More recently, evidence for a double beta signal in the Heidelberg-Moscow data has been claimed.[12] Several experiments are now under planning e.g. EXO, Majorana and Cuore etc. which are expected to improve the sensitivity to the Majorana mass of the neutrino to the level of 20 milli-eV. This can for example test the hypothesis that neutrino mass ordering may be inverted if they are Majorana fermions.

Another important result in further understanding of neutrino mass physics could come from the tritium end point searches for neutrino masses. This experiment will measure the parameter $m_\nu = \sqrt{\sum_i |U_{ei}|^2 m_i^2}$. This involves a different combination of masses and mixing angles than $<m>_{\beta\beta}$. Presently, the KATRIN proposal for a high sensitive search for for m_ν has been made and it is expected that it can reach a sensitivity of 0.2 eV.

A third source of information on neutrino mass will come from cosmology, where more detailed study of structure in the universe is expected to provide an upper limit on $\sum_i m_i$ of less than an eV. Present WMAP data appears to provide a limit of $\sum_i m_{\nu_i} \leq 0.6 - 1$ eV.[13]

Our goal now is to study the theoretical implications of these discoveries. We will proceed towards this goal in the following manner: we will isolate the mass patterns that fit the above data and search for patterns and symmetries that lead to observed mixing angles. We then look for plausible models that can first lead to the general feature that neutrinos have tiny masses; then we would try to understand in simple manner some of the features indicated by data in the hope that these general ideas will be part of our final understanding of the neutrino masses. As mentioned earlier on, to understand the neutrino masses one has to go beyond the standard model. First we will sharpen what we mean by this statement. Then we will present some ideas which may form the basic framework for constructing the detailed models.

7.2. Physics of neutrino mass

There are two distinct aspects to neutrino mass physics: first is the absolute overall magnitude and second is the flavor structure. Understanding the first will reveal the gross features of the new physics such as the presence of a new symmetry and its scale responsible for the smallness of neutrino mass compared to masses of other fundamental fermions, whereas understanding the flavor pattern is likely to throw light on possible new family symmetries of matter which may in turn be relevant to unravelling the mystery of quark flavor. It could be (probably likely) that both are related to each other.

To begin this discussion, we first list the puzzles of neutrino mass physics; we then discuss the neutrino mass matrix which is the starting point of many attempts to understand the neutrino mixings and then discuss ideas that have been proposed to understand these patterns before going to a discussion of the overall scale.

7.2.1. *Puzzles of neutrino mass physics*

The present neutrino discoveries have posed a list of puzzles for physics beyond the standard model, whose resolution will provide an unmistakable path beyond it. Below we give a list of these puzzles.

- *Ultra-light-ness of neutrinos:* Why are the neutrino masses so much lighter than the quark and charged lepton masses?
- *Bi-large mixing:* How to understand simultaneously two large mixing angles one for the $\mu - \tau$ and another for $e - \mu$?
- *Smallness of $\Delta m_\odot^2 / \Delta m_A^2$:* Experimentally, $\Delta m_\odot^2 \simeq 10^{-2} \Delta m_A^2$. How does one understand this in a natural manner?
- *Smallness of U_{e3}:* The reactor results also seem to indicate that the angle $\theta_{13} \equiv U_{e3}$ is a very small number. One must also understand this in a framework that simultaneously explains all other puzzles.

Possible other puzzles include a proper understanding of neutrino mass degeneracy if there is a large positive signal for the neutrinoless double beta decay and of course, when we have evidence for CP violating phases in the mass matrix, we must understand their magnitude.

In order to take the first step towards understanding these puzzles, we discuss the flavor pattern of leptons that may be at the root of large mixings and defer the discussion of the origin of mass scale to the next section and finally focus on specific unification models which address both the issues as examples how one may proceed to unravel the grand picture of physics beyond standard model inspired by neutrino physics. It could of course be that large neutrino mixings result from a joint effect of both the charged lepton and the neutrino matrix since $U_{PMNS} = U_l^\dagger U_\nu$ and we do not know apriori whether the large neutrino mixings come from the charged lepton sector or the neutrino sector or both. However, we first follow the line of thinking that in the fundamental theory, charged lepton mass matrix is diagonal or near diagonal and all mixings result from the neutrino mass matrix. This point of view is not so unreasonable since charged lepton mass matrix is likely to be similar to the quark sector and the small observed CKM mixings pretty much guarantees that quark mass matrices are near diagonal. We will also present grand unified models where this conjecture is borne out.

7.2.2. *Notation*

We will assume two component neutrinos and therefore their masses will in general be Majorana type. Let us also give our notation to facilitate further discussion: the neutrinos emitted in weak processes such as the beta decay or muon decay are weak eigenstates and are not mass eigenstates. The mass eigenstates determine how a neutrino state evolves in time. Similarly, in the detection process, it is the weak eigenstate that is picked out. This is of course the key idea behind neutrino oscillation and the formula presented in the last section. To set the notation, let us express the weak eigenstates in terms of the mass eigenstates. We will denote the weak eigenstate by the symbol α, β or simply e, μ, τ etc whereas the mass eigenstate will be denoted by the symbols i, j, k etc. To relate the weak eigenstates to the mass eigenstates, let us start with the mass terms in the Lagrangian for the neutrino and the charged leptons:

$$\mathcal{L}_m = \nu_L^T \mathcal{M}_\nu \nu_L + \bar{E}_L M_\ell E_R + h.c. \tag{7.19}$$

Here the ν and E which denote the column vectors for neutrinos and charged leptons are in the weak basis. To go to the mass basis, we diagonalize these matrices as follows:

$$U_L^T \mathcal{M}_\nu U_L = d_\nu \tag{7.20}$$
$$V_L M_\ell V_R^\dagger = d_\ell$$

The physical neutrino mixing matrix is then given by:

$$\mathbf{U} = V_L U_L \tag{7.21}$$

$\mathbf{U}_{\alpha i}$ and relate the two sets of eigenstates (weak and mass) as follows:

$$\begin{pmatrix} \nu_e \\ \nu_\mu \\ \nu_\tau \end{pmatrix} = U \begin{pmatrix} \nu_1 \\ \nu_2 \\ \nu_3 \end{pmatrix} \tag{7.22}$$

Using this equation, one can derive the well known oscillation formulae for the survival probability of a particular weak eigenstate α discussed in the previous section.

To see the general structure of the mixing matrix **U**, let us recall that the matrix $\mathcal{M}_\nu$ is complex and symmetric and therefore has six complex parameters describing it for the case of three generations. But since the neutrino is described by a complex field, we can redefine the phases of three fields to remove three parameters. That leaves nine parameters. In terms of observables, there are three mass eigenvalues (m_1, m_2, m_3) and three mixing angles and phases in the mixing matrix **U**. The three phases can be split into one Dirac phase, which is analogous to the phase in the quark mixing matrix and two Majorana phases. We can then write the matrix **U** as

$$\mathbf{U} = U^{(0)} \begin{pmatrix} 1 & & \\ & e^{i\phi_1} & \\ & & e^{i\phi_2} \end{pmatrix} \quad (7.23)$$

The matrix $\mathbf{U}^{(0)}$ has three real angles $\theta_{12}, \theta_{23}, \theta_{13}$ and a phase. The goal of experiments is to determine all nine of these parameters. The knowledge of the nine observables allows one to construct the mass matrix for the neutrinos and from there one can go in search of the new physics beyond the standard model that leads to such a mass matrix.

The neutrino mass observables given above can be separated into two classes: (i) oscillation observables and (ii) non-oscillation observables. The first class of observables are those accessible to neutrino oscillation experiments and are the two mass differences $\Delta m_\odot^2$ and Δm_A^2; three mixing angles θ_{12} (or $\theta_\odot$); θ_{23} (or θ_A) and θ_{13} (the reactor angle, also called U_{e3}) and the CP phase δ in $U^{(0)}$. The remaining three observables i.e. the lightest mass of the three neutrinos and the two Majorana phases $\phi_{1,2}$ can only be probed by nonoscillation experiments such as $\beta\beta_{0\nu}$ decay, beta decay spectrum at the endpoint and cosmological observations etc.

7.3. Neutrino mixing matrix and mass patterns

A good starting point for the exploration of new physics such as new symmetries, new scaleshidden in neutrino observations is to construct the neutrino mass matrix in the basis where the charged leptons are mass eigenstates (or near mass eigen states.) One can then look for symmetries which could be responsible for this form of the neutrino mass matrix in extensions of the standard model.

In order to construct the neutrino mass matrix, we will use the following experimental numbers: $\Delta m_A^2 \simeq 0.0021$ eV^2; for solar neutrinos, it gives $\Delta m_\odot^2 \simeq (7.21-8.63) \times 10^{-5}$ eV^2. It also provides information on the angles in **U** which can be summarized by the following mixing matrix (neglecting all CP phases):

$$\mathbf{U} = \begin{pmatrix} c & s & \epsilon \\ -\frac{s+c\epsilon}{\sqrt{2}} & \frac{c-s\epsilon}{\sqrt{2}} & \frac{1}{\sqrt{2}} \\ -\frac{s-c\epsilon}{\sqrt{2}} & \frac{-c-s\epsilon}{\sqrt{2}} & \frac{1}{\sqrt{2}} \end{pmatrix} \tag{7.24}$$

where from the discussion above $\epsilon \leq 0.17$; $s = \sin\theta_{12}$ is in the range $0.267 \leq \sin^2\theta_{12} \leq 0.371$ with the central value being near 0.314. We have chosen the atmospheric mixing angle θ_{23} to be maximal.

A particularly interesting form of the mixing matrix which seems to be in accord with data is the so-called tri-bi-maximal form[15] where:

$$U_{PMNS} = \begin{pmatrix} \sqrt{\frac{2}{3}} & \frac{1}{\sqrt{3}} & 0 \\ -\frac{1}{\sqrt{6}} & \frac{1}{\sqrt{3}} & \frac{1}{\sqrt{2}} \\ -\frac{1}{\sqrt{6}} & \frac{1}{\sqrt{3}} & -\frac{1}{\sqrt{2}} \end{pmatrix} \tag{7.25}$$

As far as the mass pattern goes however, there are three possibilities:

- (i) normal hierarchy: $m_1 \ll m_2 \ll m_3$;
- (ii) inverted hierarchy: $m_1 \simeq -m_2 \gg m_3$ and
- (iii) approximately degenerate pattern[14] $m_1 \simeq m_2 \simeq m_3$,

where m_i are the eigenvalues of the neutrino mass matrix. In the first case, the atmospheric and the solar neutrino data give direct information on m_3 and m_2 respectively. On the other hand, in the last case, the mass differences between the first and the second eigenvalues will be chosen to fit the solar neutrino data and the second and the third to fit the atmospheric neutrino data.

7.3.1. *Neutrino mass textures*

From the mixing matrix in Eq. 7.24, we can write down the allowed neutrino mass matrix for any arbitrary mass pattern assuming the neutrino is a Majorana fermion. Denoting the matrix elements of $\mathcal{M}_\nu$ as $\mu_{\alpha\beta}$ for $\alpha, \beta =$

1, 2, 3, we have (Recall that $\mu_{\alpha\beta} = \mu_{\beta\alpha}$):

$$\mu_{11} = [c^2 m_1 + s^2 m_2 + \epsilon^2 m_3]$$
$$\mu_{12} = \frac{1}{\sqrt{2}}[-c(s+c\epsilon)m_1 + s(c-s\epsilon)m_2 + \epsilon m_3]$$
$$\mu_{13} = \frac{1}{\sqrt{2}}[-c(s-c\epsilon)m_1 - s(c+s\epsilon)m_2 + \epsilon m_3]$$
$$\mu_{22} = \frac{1}{2}[(s+c\epsilon)^2 m_1 + (c-s\epsilon)^2 m_2 + m_3] \quad (7.26)$$
$$\mu_{23} = \frac{1}{2}[-(s^2 - c^2\epsilon^2)m_1 - (c^2 - s^2\epsilon^2)m_2 + m_3]$$
$$\mu_{33} = \frac{1}{2}[(s-c\epsilon)^2 m_1 + (c+s\epsilon)^2 m_2 + m_3].$$

In certain limits for the mixing angles in the above mass matrix, symmetries leptons can manifest. Below I give some examples:

7.3.2. $\mu - \tau$ exchange symmetry

The current observations that θ_{23} is very close to maximal (with $\theta_{23} = \frac{\pi}{4}$ being the best fit solution in many analyses) and the fact that θ_{13} could be vanishing can be understood if the neutrino Majorana mass matix has a Z_2 symmetry that interchanges $\mu - \tau$.[16] The corresponding mass matrix is the special case of the matrix below with $c = 1$, and $a = b$.

$$\mathcal{M}_\nu = \sqrt{\Delta m_A^2} \begin{pmatrix} d\epsilon^n & b\epsilon & a\epsilon \\ b\epsilon & 1+\epsilon & 1 \\ a\epsilon & 1 & 1+c\epsilon \end{pmatrix} \quad (7.27)$$

where a, b, c, d are parameters of order one and $\epsilon \sim \sqrt{\frac{\Delta m_\odot^2}{\Delta m_A^2}} \sim 0.2$. Since at the moment it is not certain whether $\mu - \tau$ symmetry is exact or approximate, the mass matrix in Eq. 7.27 includes small breaking terms characterized by $a \neq b$ and $c \neq 1$. These small departures lead to non-zero values for θ_{13} and $\theta_{23} - \frac{\pi}{4}$ which are correlated with each other.[17] We have ignored leptonic CP violation in this discussion. One can have other ways of introducing $\mu - \tau$ symmetry breaking using CP phases.[18]

Overall, this symmetry has a good chance be part of the final theory of neutrino mixing since there seems to be good experimental support for it. This has therefore led to a great deal of model building activity[19] most of whom predict departures from the exact symmetry limit and can provide insight into which way to proceed in assimilating this symmetry as part of the quark-lepton world.

7.3.3. Mass matrix for tri-bi-maximal mixing and associated approximate flavor symmetries

In the $\mu - \tau$ symmetric models, the value of the solar mixing angle θ_{12} remains large but arbitrary. The indication that the value of $\sin\theta_{12} \simeq \frac{1}{\sqrt{3}}$ may be an indication of higher symmetries of the lepton world. This is called tri-bi-maximal mixing pattern.[15] Clearly these must have $\mu - \tau$ symmetry as a subgroup. A typical neutrino mass matrix that leads to the tri-bi-maximal mixing pattern is:

$$\mathcal{M}_\nu = \begin{pmatrix} a & b & b \\ b & a+c & b-c \\ b & b-c & a+c \end{pmatrix} \quad (7.28)$$

where a, b, c are arbitrary parameters. The charged lepton mass matrix is chosen to be diagonal. Diagonalizing this matrix leads to the U_{PMNS} in the tri-bi-maximal form and the neutrino masses: $m_1 = a - b; m_2 = a + 2b$ and $m_3 = a - b + 2c$. Clearly if $a \simeq b \ll c$, we get a normal hierarchy for masses. A symmetry that has been found to lead to this mass matrix is S_3.[20]

Another way to get the tri-bi-maximal form is to use specific forms for the charged lepton and neutrino mass matrices in such a way that the combined diagonalization leads to the desired lepton mixing matrix. An example is to have a charged lepton mass matrix of the form:

$$\mathcal{M}_\nu = \begin{pmatrix} a & 0 & 0 \\ 0 & a & b \\ 0 & b & a \end{pmatrix}$$

$$M_\ell = \frac{1}{\sqrt{3}} \begin{pmatrix} 1 & 1 & 1 \\ 1 & \omega^2 & \omega \\ 1 & \omega & \omega^2 \end{pmatrix} \begin{pmatrix} m_e & 0 & 0 \\ 0 & m_\mu & 0 \\ 0 & 0 & m_\tau \end{pmatrix} \quad (7.29)$$

where $\omega = e^{\frac{2\pi i}{3}}$. This for may appear much too contrived to arise from some symmetries, but remarkably enough it has been shown that this can emerge under certain assumptions if the assumed symmetry is A_4,[21] which is group of even permutations of four elements.

Detailed theory for all these cases involves many Higgs multiplets and probably should not be taken literally but the important message is the presence of the hidden symmetry and its implications. Generally such symmetries are hard to grand unify and further research along this direction is going to be important.

7.3.4. *Inverted hierarchy and $L_e - L_\mu - L_\tau$ symmetry*

In this sub-section, we consider another interesting clue to model building present in neutrino data if the mass arrangement is inverted. It starts with the observation that if the neutrino mass matrix has the form

$$\mathcal{M}_\nu = \begin{pmatrix} 0 & A & A \\ A & 0 & 0 \\ A & 0b & 0 \end{pmatrix} \qquad (7.30)$$

this leads to two degenerate neutrinos with mass $\pm\sqrt{2}A$ and one massless neutrino. The atmospheric mass difference is given by $\Delta m_A^2 = 2A^2$ and mixing angle $\theta_A = \pi/4$. As far as the solar ν_e oscillation is concerned, the $\sin^2 2\theta_\odot = 1$ but $\Delta m_\odot^2 = 0$. While this is unphysical, this raises the hope that as corrections to this mass matrix are taken into account, it may be possible understand the smallness of $\Delta m_\odot^2/\Delta m_A^2$ naturally.

In fact this hope is fortified by the observation that this mass matrix has the leptonic symmetry $L_e - L_\mu - L_\tau$; therefore one might hope that as this symmetry is broken by small terms, one will end up with a situation that fits data well.

This question was studied in two papers.[25,26] To proceed with the discussion, let us consider the following mass matrix for neutrinos where small $L_e - L_\mu - L_\tau$ violating terms have been added.

$$\mathcal{M}_\nu = m \begin{pmatrix} z & c & s \\ c & y & d \\ s & d & x \end{pmatrix}. \qquad (7.31)$$

The charged lepton mass matrix is chosen to have a diagonal form in this basis and $L_e - L_\mu - L_\tau$ symmetric. In the perturbative approximation, there are sum rules involving the neutrino observables and the elements of the neutrino mass matrix,[26] which first of all imply (i) a close connection between the measured value of the solar mixing angle and the neutrino mass measured in neutrino-less double beta decay; (ii) the present values for the solar mixing angle can be used to predict the $m_{\beta\beta}$ for a value of the $\Delta m_\odot^2$. For instance, for $\sin^2 2\theta_\odot = 0.9$, we would predict $(\frac{\Delta m_\odot^2}{4\Delta m_A^2} - z) = 0.3$. For small $\Delta m_\odot^2$, this implies $m_{\beta\beta} \simeq 0.01$ eV. This is expected to be within the reach of new double beta decay experiments contemplated.[27] In fact now there exist more thorough numerical analyses[23] of general $L_e - L_\mu - L_\tau$ broken models which imply that the inverted hierarchy for neutrinos can be tested in neutrinoless double beta experiments in the next decade or so. This way of breaking $L_e - L_\mu - L_\tau$ symmetry also implies a value for $\sin^2 2\theta_{12} \geq 0.9$, which is now almost ruled out. Of course there could be other ways of breaking this symmetry using charged lepton sector etc. which can still lead to lower solar angle.

If the value of $\sin^2 2\theta_\odot$ is ultimately determined to be less than 0.9, the question one may ask is whether the idea of $L_e - L_\mu - L_\tau$ symmetry is dead. The answer is in the negative since so far we have explored the breaking of $L_e - L_\mu - L_\tau$ symmetry only in the neutrino mass matrix. It was shown in[25] that if the symmetry is broken in the charged lepton mass, one can lower the $\sin^2 2\theta_\odot$ as long as the value of U_{e3} is sizable. However given the present upper limit on U_{e3}, the smallest value is somewhere around $\sin^2 2\theta_\odot \simeq 0.8$.

7.3.5. *Scale invariant mass matrix*

Most of the above forms for the mass matrices are scale dependent in the sense that once radiative corrections are taken into account, their forms can change. This is specially relevant because in many theories neutrino mass matrix is predicted at a high scale due to physics at this scale. They have to be extrapolated to the weak scale to compare with experiments. This process relies on the nature of physics between the neutrino mass generation scale and the weak scale. Thus connection between fundamental physics responsible for neutrino masses and observations gets interrupted. Luckily the radiative correction effects are not significant if neutrino mass hierarchy is normal (see discussion later). However both for inverted and degenerate spectra, they are. It is therefore interesting to search for neutrino flavor structure that is not affected by such effects. One such form arises when the neutrino mass matrices satisfy certain scaling properties.[28] An example of such a mass matrix is:

$$\mathcal{M}_\nu = m_0 \begin{pmatrix} A & B & B/c \\ B & D & D/c \\ B/c & D/c & D/c^2 \end{pmatrix}. \quad (7.32)$$

It is called $\mu - \tau$ scaling whose most important phenomenological is that it leads to an inverted hierarchy with $m_3 = 0$ and $U_{e3} = 0$. Atmospheric neutrino mixing is governed by the "scaling factor" c via $\tan^2 \theta_{23} = 1/c^2$, i.e., is in general non-maximal because c is naturally of order, but not equal to, one. These results are scale independent predictions and do not depend on extraneous physics between the neutrino mass generation scale and the weak scale. It is interesting to note that current data analyzes (though at the present stage statistically not very significant) yield non-maximal $\tan^2\theta_{23} = 0.89$ as the best-fit point.[5]

7.3.6. *CP violation*

A not very well explored aspect of neutrino physics at the moment is CP violation in lepton physics. Unlike the quark sector, CP violation for Majorana neutrinos allows for more phases than in the quark sector. Since the Majorana neutrino mass matrix is symmetric, for N generations of neutrinos, there are in general $\frac{N(N+1)}{2}$ phases in it. When the mass matrix is diagonalized, these phases will appear in the unitary matrix U_L that does the diagonalization (i.e. $U^T \mathcal{M}_\nu U_L = d_\nu$). If we are working in a basis where the charged lepton mass matrix is diagonal, then U_L is the leptonic weak mixing matrix. As we saw this has $N(N+1)/2$ phases. Out of them, redefinition of the charged lepton fields in the weak current allows the removal of N phases; so there are $N(N-1)/2$ phases in the neutrino masses. In the quark sector, both up and down fields could be redefined allowing for the number of physical phases that appear in the end to be smaller. However for Majorana neutrinos, redefinition of the fields does not remove the phases entirely from the theory but rather shifts them to other places where they can manifest themselves physically.[29] The detailed discussion of CP violation is in other lectures at this school.

7.4. **Neutrino mass scale and physics beyond the standard model**

In the standard model (SM), the neutrino mass vanishes to all orders in perturbation theory as well as nonperturbatively implying that observation of neutrino masses is the first laboratory evidence for physics beyond the standard model. To clarify this point, note that the SM is based on the gauge group $SU(3)_c \times SU(2)_L \times U(1)_Y$ under which the quarks and leptons, Higgs bosons and gauge bosons transform as described in the Table I.

Table I

Field	gauge transformation
Quarks Q_L	$(3, 2, \frac{1}{3})$
Righthanded up quarks u_R	$(3, 1, \frac{4}{3})$
Righthanded down quarks d_R	$(3, 1, -\frac{2}{3})$
Lefthanded Leptons L	$(1, 2 - 1)$
Righthanded leptons e_R	$(1, 1, -2)$
Higgs Boson **H**	$(1, 2, +1)$
Color Gauge Fields G_a	$(8, 1, 0)$
Weak Gauge Fields $W^\pm$, Z, γ	$(1, 3+1, 0)$

Table caption: The assignment of particles to the standard model gauge group $SU(3)_c \times SU(2)_L \times U(1)_Y$.

The electro-weak symmetry $SU(2)_L \times U(1)_Y$ is broken by the vacuum expectation of the Higgs doublet $<H^0> = \frac{v_{wk}}{\sqrt{2}} \simeq 186$ GeV, which gives mass to the gauge bosons. The fermion masses arise from the Yukawa couplings:

$$\mathcal{L}_Y = h_u \bar{Q}_L H u_R + h_d \bar{Q}_L \tilde{H} d_R + h_e \bar{L} \tilde{H} e_R + h.c. \qquad (7.33)$$

when H^0 acquires a vev. Note that since there are no right handed neutrinos in the theory, there is no term in Eq. 7.33 that can give mass to the neutrinos. Thus they remain massless at the tree level.

There are several questions that arise at this stage. What happens when one goes beyond the above simple tree level approximation? Secondly, do non-perturbative effects change this tree level result? Finally, how to judge whether this result will be modified when the quantum gravity effects are included?

The first and second questions are easily answered by using the B-L symmetry of the standard model. The point is that since the standard model has no $SU(2)_L$ singlet neutrino-like field, the only possible mass terms that are allowed by Lorentz invariance are of the form $\nu_{iL}^T C^{-1} \nu_{jL}$, where i, j stand for the generation index and C is the Lorentz charge conjugation matrix. Since the ν_{iL} is part of the $SU(2)_L$ doublet field and has lepton number $+1$, the above neutrino mass term transforms as an $SU(2)_L$ triplet and furthermore, it violates total lepton number (defined as $L \equiv L_e + L_\mu + L_\tau$) by two units. However, a quick look at the standard model Lagrangian convinces one that the model has exact lepton number symmetry after symmetry breaking; therefore such terms can never arise in perturbation theory. Thus to all orders in perturbation theory, the neutrinos are massless. As far as the nonperturbative effects go, the only known source is the weak instanton effects. Such effects could change the result if they broke the lepton number symmetry. One way to see if such breaking occurs is to look for anomalies in lepton number current conservation from triangle diagrams. Indeed it is easy to convince oneself that $\partial_\mu j_\ell^\mu = cW\tilde{W} + c'B\tilde{B}$ due to the contribution of the leptons to the triangle involving the lepton number current and W's or B's. Luckily, it turns out that the anomaly contribution to the baryon number current nonconservation has also an identical form, so that the $B-L$ current j_{B-L}^μ is conserved to all orders in the gauge couplings. As a consequence, nonperturbative effects from the gauge sector cannot induce $B-L$ violation. Since the neutrino mass operator described above violates also $B-L$, this proves that neutrino masses remain zero even in the presence of nonperturbative effects.

Let us now turn to the effect of gravity. Clearly as long as we treat gravity in perturbation theory, the above symmetry arguments hold since all gravity coupling respect $B-L$ symmetry. However, once nonperturbative gravitational effects e.g black holes and worm holes are included, there is no guarantee that global symmetries will be respected in the low energy theory. The intuitive way to appreciate the argument is to note that throwing baryons into a black hole does not lead to any detectable consequence except through a net change in the baryon number of the universe. Since one can throw in an arbitrary number of baryons into the black hole, an arbitrary information loss about the net number of missing baryons would prevent us from defining a baryon number of the visible universe- thus baryon number in the presence of a black hole can not be an exact symmetry. Similar arguments can be made for any global charge such as lepton number in the standard model. A field theoretic parameterization of this statement is that the effective low energy Lagrangian for the standard model in the presence of black holes and worm holes etc must contain baryon and lepton number violating terms. In the context of the standard model, the only such terms that one can construct are nonrenormalizable terms of the form $LHLH/M_{P\ell}$. After gauge symmetry breaking, they lead to neutrino masses; however these masses are at most of order $v_{wk}^2/M_{P\ell} \simeq 10^{-5}$ eV.[30] But as we discussed in the previous section, in order to solve the atmospheric neutrino problem, one needs masses at least three orders of magnitude higher.

Thus one must seek physics beyond the standard model to explain observed evidences for neutrino masses. While there are many possibilities that lead to small neutrino masses of both Majorana as well as Dirac kind, here we focus on the possibility that there is a heavy right handed neutrino (or neutrinos) that lead to a small neutrino mass. The resulting mechanism is known as the seesaw mechanism[4] and leads to neutrino being a Majorana particle.

The nature and origin of the seesaw mechanism can also be tested in other experiments and we will discuss them below. This will be dependent on the kind of operators that play a role in generating neutrino masses. If the leading order operator is of dimension 5, then the scale necessarily is very high (of order 10^{12} GeV or greater). On the other hand, in theories with extra space dimensions, this operator may be forbidden and one may be forced to go to higher dimensional operators, in which case the scale could be lower or it could be that neutrino Dirac Yukawa couplings are of order 10^{-6} (similar to the electron Yukawa coupling in the standard model) in which case even the seesaw scale could be in the TeV range.

The seesaw mechanism raises a very important question: since we require the mass of the right handed neutrino to be much less than the Planck scale, a key question is "what symmetry keeps the right handed neutrino mass lighter?" We will give two examples of symmetries that can do this.

7.4.1. Seesaw and the right handed neutrino

The simplest possibility extension of the standard model that leads to nonzero mass for the neutrino is one where only a right handed neutrino is added to the standard model. In this case ν_L and ν_R can form a mass term; but apriori, this mass term is like the mass terms for charged leptons or quark masses and will therefore involve the weak scale. If we call the corresponding Yukawa coupling to be Y_ν, then the neutrino mass is $m_D = Y_\nu v/\sqrt{2}$. For a neutrino mass in the eV range requires that $Y_\nu \simeq 10^{-11}$ or less which is far below even the small electron Yukawa coupling of SM. Introduction of such small coupling constants into a theory is generally considered unnatural and a sound theory must find a symmetry reason for such smallness. As already already alluded to before, seesaw mechanism,[4] where we introduce a singlet Majorana mass term for the right handed neutrino is one way to achieve this goal. The effective mass terms for the (ν_L, ν_R) system is given by (suppressing the generation index):

$$\mathcal{L}_m = m_D^* \bar{\nu}_L \nu_R + M_R^* \nu_R^T C^{-1} \nu_R + h.c. \tag{7.34}$$

where $m_D = Y_\nu v_{wk}$, Y_ν is the lepton doublet coupling to the right handed neutrinos (defined as $\bar{\nu}_R Y_\nu L$). Suppose we write ν spinor in terms of its two component spinors as $\nu = \begin{pmatrix} \nu \\ i\sigma_2 N^* \end{pmatrix}$, then $\nu_L = \begin{pmatrix} \nu \\ 0 \end{pmatrix}$ and $\nu_R = \begin{pmatrix} 0 \\ i\sigma_2 N^* \end{pmatrix}$. This gives remembering that $\gamma_0 = \begin{pmatrix} 0 & I \\ I & 0 \end{pmatrix}$ and $C^{-1} = \begin{pmatrix} i\sigma_2 & 0 \\ 0 & -i\sigma_2 \end{pmatrix}$

$$\mathcal{L}_m = im_D^* \nu^\dagger \sigma_2 N^* + iM_R^* N^\dagger \sigma_2 N^* + h.c. \tag{7.35}$$

. We can write the neutrino mass matrix as:

$$\mathcal{L}_m = i\begin{pmatrix} \nu^T & N^T \end{pmatrix} \sigma_2 \mathbf{M} \begin{pmatrix} \nu \\ N \end{pmatrix} + h.c. \tag{7.36}$$

where $\mathbf{M}$ has the form:

$$\mathbf{M} = \begin{pmatrix} 0 & m_D \\ m_D^T & M_R \end{pmatrix} \tag{7.37}$$

Since M_R is not constrained by the standard model symmetries, it is natural to choose it to be at a scale much higher than the weak scale. Now diagonalizing this mass matrix, we get a set of heavy eigenstate N_R and a set of light eigenstates with mass matrix given by:

$$\mathcal{M}_\nu \simeq -m_D^T M_R^{-1} m_D \tag{7.38}$$

This provides a natural way to understand a small neutrino mass without any unnatural adjustment of parameters of a theory. In a subsequent section, we will discuss a theory which connects the scale M_R to a new symmetry of nature beyond the standard model. This formula for neutrino masses is called type I seesaw formula.

7.4.1.1. Why is $M_R \ll M_{P\ell}$?

The question "why $M_R \ll M_{P\ell}$?" is in many ways similar to the question in the standard model i.e. "why is $M_{Higgs} \ll M_{P\ell}$?" It is well known that searches for answer to this latter question has led to many interesting possibilities for physics beyond the standard model including supersymmetry, extra dimensions and technicolor. It is hoped that answering this question for ν_R can also lead us to new insight into new symmetries beyond the standard model. There are two interesting answers to our question that I will elaborate later on.

$B-L$:

If one adds three right handed neutrinos to implement the seesaw mechanism, the model admits an anomaly free new symmetry i.e. $B-L$. One can therefore extend the standard model symmetry to either $SU(2)_L \times U(1)_{I_{3R}} \times U(1)_{B-L}$ or its left-right symmetric extension $SU(2)_L \times SU(2)_R \times U(1)_{B-L}$. In either case the right handed neutrino carries the B-L quantum number and its Majorana mass breaks this symmetry. Therefore, the mass of the ν_R can at most be the scale of $B-L$ symmetry breaking, hence answering the question "why $M_{\nu_R} \ll M_{P\ell}$?".

$SU(2)_H$:

While local $B-L$ is perhaps the most straight forward and natural symmetry that keeps ν_R lighter than the Planck scale, another possibility has recently been suggested in Ref. 24. The main observation here is that is the standard model is extended by including a local $SU(2)_H$ symmetry acting on the first two lepton generations including the right handed charged leptons, then global Witten anomaly freedom dictates that there must be at least two right handed neutrinos which transform as a doublet under the $SU(2)_H$ local symmetry. In this class of models, in the limit of exact $SU(2)_H$ symmetry, the ν_R's are massless and as soon as the $SU(2)_H$ symmetry is broken, they pick up mass. Therefore "lightness" of the ν_R's compared to the Planck scale in these models is related to an $SU(2)_H$ symmetry. These comments are elaborated with explicit examples later on in this review.

7.4.2. Double seesaw mechanism with ν_R and B-L singlet neutral fermions

As we saw from the previous discussion, the conventional seesaw mechanism requires rather high mass for the right handed neutrino and therefore a correspondingly high scale for B-L symmetry breaking. The right handed neutrinos are B-L non-singlet fields. There is however no way at present to know what the scale of B-L symmetry breaking is. For lower B-L scale models, one must either find a mechanism to suppress the Dirac mass in the conventional seesaw formula or extend the theory some other way. A particularly simple way is to introduce, B-L singlet heavy neutrinos S and use a double seesaw mechanism suggested in Ref. 31 where one writes a three by three neutrino mass matrix in the basis (ν, N, S) of the form:

$$M = \begin{pmatrix} 0 & m_D & 0 \\ m_D & 0 & M \\ 0 & M & \mu \end{pmatrix} \qquad (7.39)$$

This comes from an effective mass Lagrangian of the form:

$$\mathcal{L}'_m = m_D^\dagger \bar{\nu}_L N_R + M^\dagger \bar{N}_R S + \mu S_L^T C^{-1} S_L + h.c. \qquad (7.40)$$

It is possible to have extra symmetries that guarantees the above form for the Lagrangian. (It will be a good exercise to discover these symmetries.) For the case $\mu \ll M \approx M_{B-L}$, (where M_{B-L} is the $B-L$ breaking scale) this matrix has one light and two heavy neutrinos per generation and the latter two form a pseudo-Dirac pair with mass of order M_{B-L}. The important thing for us is that the light mass eigenvalue is given by $m_D^2 \mu/M^2$; for $m_D \approx \mu \simeq$ GeV, a 10 TeV $B-L$ scale is enough to give neutrino masses in the eV range. For the case of three generations, the formula for the light neutrino mass matrix is given by:

$$\mathcal{M}_\nu = m_D^T M^{-1} \mu M^{-1} m_D. \qquad (7.41)$$

7.4.3. *High mass Higgs triplet induced neutrino masses*

As already noted, one way to generate nonzero neutrino masses without using righthanded neutrinos is to extend the standard model by the addition of an $SU(2)_L$ triplet Higgs field with $Y=2$ so that the electric charge profile of the members of the multiplet is given as follows: $(\Delta^{++}, \Delta^+, \Delta^0)$. This allows an additional Yukawa coupling of the form $f_L L^T \tau_2 \tau L.\Delta$, where the Δ^0 couples to the neutrinos. Clearly Δ field has $L=2$. When Δ^0 field has a nonzero vev, it breaks lepton number by two units and leads to Majorana mass for the neutrinos. There are two questions that arise now: one, how does the vev arise in a model and how does one understand the smallness of the neutrino masses in this scheme. There are two answers to the first question: One can maintain exact lepton number symmetry in the model and generate the vev of the triplet field via the usual "mexican hat" potential. There are two problems with this case. This leads to the triplet Majoron which has been ruled out by LEP data on Z-width. In any case, in this model smallness of the neutrino mass is not naturally understood.

Another way to generate the induced vev is to keep a large but positive mass (M_Δ) for the triplet Higgs boson and allowing for a lepton number violating coupling $M\Delta^* HH$.² In this case, minimization of the potential induces a vev for the Δ^0 field when the doublet field acquires a vev[36]:

$$v_T \equiv <\Delta^0> = \frac{Mv_{wk}^2}{M_\Delta^2} \qquad (7.42)$$

Since the mass of the Δ field is invariant under $SU(2)_L \times U(1)_Y$, it can be very large connected perhaps with some new scale of physics. If we assume that $M_\Delta \sim M \sim 10^{13}$ GeV or so, we get $v_T \sim$ eV. Now in the Yukawa coupling $f_L L^T \tau_2 \tau L.\Delta$, since the Δ^0 couples to the neutrinos, its vev leads to a neutrino mass We will see later when we discuss the seesaw models that unlike those models, the neutrino mass in this case is not hierarchically dependent on the charged fermion masses. Note further the high mass suppression in Eq. 7.42 leading to a new kind of seesaw suppression. This is called type II seesaw.

7.5. Left right symmetric unification: a natural realization of the seesaw

Let us now explore the implications of including the righthanded neutrinos into the extensions of the standard model to understand the small neutrino mass by the seesaw mechanism. As already emphasized, if we assume that there are no new symmetries beyond the standard model, the right handed neutrino will have a natural mass of order of the Planck scale making the light neutrino masses too small to be of interest in understanding the observed oscillations. We must therefore search for new symmetries that can keep the RH neutrinos at a lower scale than the Planck scale. A new symmetry always helps in making this natural.

To study this question, let us note that the inclusion of the right handed neutrinos transforms the dynamics of the gauge models in a profound way. To clarify what we mean, note that in the standard model (that does not contain a ν_R) the $B-L$ symmetry is only linearly anomaly free i.e. $\text{Tr}[(B-L)Q_a^2] = 0$ where Q_a are the gauge generators of the standard model but $\text{Tr}(B-L)^3 \neq 0$. This means that $B-L$ is only a global symmetry and cannot be gauged. However as soon as the ν_R is added to the standard model, one gets $\text{Tr}[(B-L)^3] = 0$ implying that the B-L symmetry is now gaugeable and one could choose the gauge group of nature to be either $SU(2)_L \times U(1)_{I_{3R}} \times U(1)_{B-L}$ or $SU(2)_L \times SU(2)_R \times U(1)_{B-L}$, the latter being the gauge group of the left-right symmetric models.[37] Furthermore the presence of the ν_R makes the model quark lepton symmetric and leads to a Gell-Mann-Nishijima like formula for the electric charges[39] i.e.

$$Q = I_{3L} + I_{3R} + \frac{B-L}{2} \qquad (7.43)$$

The advantage of this formula over the charge formula in the standard model charge formula is that in this case all entries have a physical meaning. Furthermore, it leads naturally to Majorana nature of neutrinos as can be seen by looking at the distance scale where the $SU(2)_L \times U(1)_Y$ symmetry is valid but the left-right gauge group is broken. In that case, one gets

$$\Delta Q = 0 = \Delta I_{3L} : \qquad (7.44)$$
$$\Delta I_{3R} = -\Delta \frac{B-L}{2}$$

We see that if the Higgs fields that break the left-right gauge group carry righthanded isospin of one, one must have $|\Delta L| = 2$ which means that the neutrino mass must be Majorana type and the theory will break lepton number by two units.

Let us now proceed to give a few details of the left-right symmetric model and demonstrate how the seesaw mechanism emerges in this model.

The gauge group of the theory is $SU(2)_L \times SU(2)_R \times U(1)_{B-L}$ with quarks and leptons transforming as doublets under $SU(2)_{L,R}$. In Table II, we denote the quark, lepton and Higgs fields in the theory along with their transformation properties under the gauge group.

Table II

Fields	$SU(2)_L \times SU(2)_R \times U(1)_{B-L}$ representation
Q_L	$(2,1,+\frac{1}{3})$
Q_R	$(1,2,\frac{1}{3})$
L_L	$(2,1,-1)$
L_R	$(1,2,-1)$
ϕ	$(2,2,0)$
Δ_L	$(3,1,+2)$
Δ_R	$(1,3,+2)$

Table caption Assignment of the fermion and Higgs fields to the representation of the left-right symmetry group.

The first task is to specify how the left-right symmetry group breaks to the standard model i.e. how one breaks the $SU(2)_R \times U(1)_{B-L}$ symmetry so that the successes of the standard model including the observed predominant V-A structure of weak interactions at low energies is reproduced. Another question of naturalness that also arises simultaneously is that since the charged fermions and the neutrinos are treated completely symmetrically (quark-lepton symmetry) in this model, how does one understand the smallness of the neutrino masses compared to the other fermion masses.

It turns out that both the above problems of the LR model have a common solution. The process of spontaneous breaking of the $SU(2)_R$ symmetry that suppresses the V+A currents at low energies also solves the problem of ultralight neutrino masses. To see this let us write the Higgs fields explicitly:

$$\Delta = \begin{pmatrix} \Delta^+/\sqrt{2} & \Delta^{++} \\ \Delta^0 & -\Delta^+/\sqrt{2} \end{pmatrix}; \quad \phi = \begin{pmatrix} \phi_1^0 & \phi_2^+ \\ \phi_1^- & \phi_2^0 \end{pmatrix} \quad (7.45)$$

All these Higgs fields have Yukawa couplings to the fermions given symbolically as below.

$$\begin{aligned} \mathcal{L}_Y &= h_1 \bar{L}_L \phi L_R + h_2 \bar{L}_L \tilde{\phi} L_R \\ &+ h_1' \bar{Q}_L \phi Q_R + h_2' \bar{Q}_L \tilde{\phi} Q_R \\ &+ f(L_L L_L \Delta_L + L_R L_R \Delta_R) + h.c. \end{aligned} \quad (7.46)$$

The $SU(2)_R \times U(1)_{B-L}$ is broken down to the standard model hypercharge $U(1)_Y$ by choosing $<\Delta_R^0> = v_R \neq 0$ since this carries both $SU(2)_R$ and $U(1)_{B-L}$ quantum numbers. It gives mass to the charged and neutral righthanded gauge bosons i.e. $M_{W_R} = gv_R$ and $M_{Z'} = \sqrt{2}gv_R \cos\theta_W/\sqrt{\cos 2\theta_W}$. Thus by adjusting the value of v_R one can suppress the right handed current effects in both neutral and charged current interactions arbitrarily leading to an effective near maximal left-handed form for the charged current weak interactions.

The fact that at the same time the neutrino masses also become small can be seen by looking at the form of the Yukawa couplings. Note that the f-term leads to a mass for the right handed neutrinos only at the scale v_R. Next as we break the standard model symmetry by turning on the vev's for the ϕ fields as $Diag <\phi> = (\kappa, \kappa')$, we not only give masses to the W_L and the Z bosons but also to the quarks and the leptons. In the neutrino sector the above Yukawa couplings after $SU(2)_L$ breaking by $<\phi> \neq 0$ lead to the so called Dirac masses for the neutrino connecting the left and right handed neutrinos. In the two component neutrino language, this leads to the following mass matrix for the ν, N using the notation earlier with the four component $\nu = \begin{pmatrix} \nu \\ i\sigma_2 N^* \end{pmatrix}$.

$$M = \begin{pmatrix} 0 & h\kappa \\ h\kappa & fv_R \end{pmatrix} \quad (7.47)$$

Note that m_D in previous discussions of the seesaw formula (see Eq. ()) is given by $m_D = h\kappa$, which links it to the weak scale and the mass of the RH neutrinos is given by $M_R = fv_R$, which is linked to the local B-L symmetry. This justifies keeping RH neutrino mass at a scale lower than the Planck mass. It is therefore fair to assume that seesaw mechanism coupled with observations of neutrino oscillations are a strong indication of the existence of a local B-L symmetry far below the Planck scale.

By diagonalizing this 2×2 matrix, we get the light neutrino eigenvalue to be $m_\nu \simeq \frac{(h\kappa)^2}{fv_R}$ and the heavy one to be fv_R. Note that typical charged fermion masses are given by $h'\kappa$ etc. So since $v_R \gg \kappa, \kappa'$, the light neutrino mass is automatically suppressed. This way of suppressing the neutrino masses is called the seesaw mechanism.[4] Thus in one stroke, one explains the smallness of the neutrino mass as well as the suppression of the V+A currents.

In deriving the above seesaw formula for neutrino masses, it has been assumed that the vev of the lefthanded triplet is zero so that the $\nu_L \nu_L$ entry of the neutrino mass matrix is zero. However, in most explicit models such as the left-right model which provide an explicit derivation of this formula, there is an induced vev for the Δ_L^0 of order $<\Delta_L^0> = v_T \simeq \frac{v_{wk}^2}{v_R}$. In the left-right models, this this arises from the presence of a coupling in the Higgs potential of the form $\Delta_L \phi \Delta_R^\dagger \phi^\dagger$. In the presence of the Δ_L vev, the seesaw formula undergoes a fundamental change and takes the form

$$M_\nu = fv_L - h_\nu^T f_R^{-1} h_\nu \left(\frac{v_{wk}^2}{v_R} \right) \quad (7.48)$$

which includes both the type I and the type II seesaw contributions. In Fig. 1, the two mdiagrams responsible for type I and type II seesaw are given:

This left-right symmetric seesaw formula has recently been shown to exhibit some interesting duality properties[38] which can perhaps be used to restrict some of the arbitrariness in its applications.

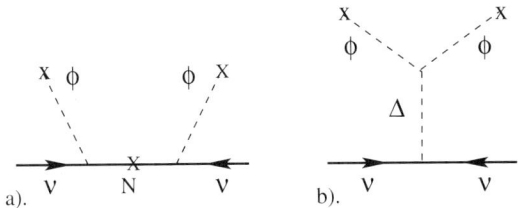

Fig. 7.1. The type I and type II contribution to seesaw formula for neutrino masses.

Note that in the type I seesaw formula, what appears is the square of the Dirac neutrino mass matrix which in general expected to have the same hierarchical structure as the corresponding charged fermion mass matrix. In fact in some specific GUT models such as SO(10), $M_D = M_u$. This is the origin of the common statement that neutrino masses given by the seesaw formula are hierarchical i.e. $m_{\nu_e} \ll m_{\nu_\mu} \ll m_{\nu_\tau}$ and even a more model dependent statement that $m_{\nu_e} : m_{\nu_\mu} : m_{\nu_\tau} = m_u^2 : m_c^2 : m_t^2$.

On the other hand if one uses the type II seesaw formula, there is no reason to expect a hierarchy and in fact if the neutrino masses turn out to be degenerate as discussed before as one possibility, one possible way to understand this may be to use the type II seesaw formula.

Secondly, the type II seesaw formula is a reflection of the parity invariance of the theory at high energies. Evidence for it would point more strongly towards left-right symmetry at high energies.

7.5.1. *Understanding detailed mixing pattern for neutrinos using the seesaw formula*

Let us now address the question: to what extent one can understand the details of the neutrino masses and mixings using the seesaw formulae. The answer to this question is quite model dependent. While there exist many models which fit the observations, none (except a few) are completely predictive and almost always they need to invoke new symmetries or new assumptions. The problem in general is that the seesaw formula of type I, has 12 parameters in the absence of CP violation (six parameters for a symmetric Dirac mass matrix and six for the M_R) which is why its predictive power is so limited. In the presence of CP violation, the number of parameters double making the situation worse. Specific predictions can be made only under additional assumptions.

For instance, in a class of seesaw models based on the SO(10) group that embodies the left-right symmetric unification model or the $SU(4)$-color, the mass the tau neutrino mass can be estimated provided one assumes the normal mass hierarchy for neutrinos and a certain parameter accompanying a higher dimensional operator to be of order one. To see this, let us assume that in the SO(10) theory, the B-L symmetry is broken by a **16**-dim. Higgs boson. The RH neutrino mass in such a model arises from the nonrenormalizable operator $\lambda(\mathbf{16_F \overline{16}_H})^2/\mathbf{M_{P\ell}}$. In a supersymmetric theory, if **16**-Higgs is also responsible for GUT symmetry breaking, then after symmetry breaking, one obtains the RH neutrino mass $M_R \simeq \lambda(2 \times 10^{16})^2/M_{P\ell} \simeq 4\lambda 10^{14}$ GeV. In models with $SU(4)_c$ symmetry, $m_{\nu_\tau,D} \simeq m_t(M_U) \sim 100$ GeV. Using the seesaw formula then, one obtains for $\lambda = 1$, tau neutrino mass $m_{\nu_\tau} \simeq 0.025$ eV, which is close to the presently preferred value of 0.05 eV. The situation with respect to other neutrino masses is however less certain and here one has to make assumptions.

The situation with respect to mixing angles is much more complicated. In generic seesaw models, one needs additional family symmetries to understand the largeness of both solar and atmospheric mixing angles, as has been commented before. It could of course very well be that the Dirac coupling in the seesaw formula is similar to the quark Yukawas but large neutrino mixings owe their origin to the flavor structure of right handed neutrino mass matrix. Or it could be that it is the type II seesaw term (the triplet Higgs contribution) dominates the neutrino mass decoupling neutrino masses completely from the charged lepton and quark mixings.

Essentially, one has to arrive at matrices similar to the above examples. There are however some exceptional situations such as in a class of minimal SO(10) models described below where the overall unification constraints on Yukawa textures is enough to explain desired large mixings, without the need for any family symmetry.

7.5.2. *General consequences of the seesaw formula for neutrino masses*

In this section, we will consider some implications of the seesaw mechanism for understanding neutrino masses. We will discuss two main points. One is the nature of the right handed neutrino spectrum as dictated by the seesaw mechanism and secondly, ways to get an approximate $L_e - L_\mu - L_\tau$ symmetric neutrino mass matrix using the seesaw mechanism and its possible implications for physics beyond the standard model.[33]

For this purpose, we use the type I seesaw formula along with the assumption of a diagonal Dirac neutrino mass matrix to obtain the right handed neutrino mass matrix M_R:

$$\mathcal{M}_{R,ij} = m_{D,i}\mu_{ij}^{-1}m_{D,j} \qquad (7.49)$$

with

$$\begin{aligned}
\mu_{11}^{-1} &= \frac{c^2}{m_1} + \frac{s^2}{m_2} + \frac{\epsilon^2}{m_3} \\
\mu_{12}^{-1} &= -\frac{c(s+c\epsilon)}{\sqrt{2}m_1} + \frac{s(c-s\epsilon)}{\sqrt{2}m_2} + \frac{\epsilon}{\sqrt{2}m_3} \\
\mu_{13}^{-1} &= \frac{c(s-c\epsilon)}{\sqrt{2}m_1} - \frac{s(c+s\epsilon)}{\sqrt{2}m_2} + \frac{\epsilon}{\sqrt{2}m_3} \\
\mu_{22}^{-1} &= \frac{(s+c\epsilon)^2}{2m_1} + \frac{(c_s\epsilon)^2}{2m_2} + \frac{1}{2m_3} \\
\mu_{23}^{-1} &= -\frac{(s^2-c^2\epsilon^2)}{2m_1} - \frac{(c^2-s^2\epsilon^2)}{2m_2} + \frac{1}{2m_3} \\
\mu_{33}^{-1} &= \frac{(s-c\epsilon)^2}{2m_1} + \frac{(c+s\epsilon)^2}{2m_2} + \frac{1}{2m_3}.
\end{aligned} \qquad (7.50)$$

Since for the cases of normal and inverted hierarchy, we have no information on the mass of the lightest neutrino m_1, we could assume it in principle to be quite small. In that case, the above equation enables us to conclude that quite likely one of the three right handed neutrinos is much heavier than the other two, leading to the so-called two right handed neutrino dominance model.[32] The situation is of course completely different for the degenerate case. This kind of separation of the RH neutrino spectrum is very suggestive of a symmetry. In fact we have recently argued that,[24] this indicates the possible existence of an $SU(2)_H$ horizontal symmetry, that leads in the simplest case to an inverted mass pattern for light neutrinos. A scenario which realizes this is given below.

7.5.2.1. Approximate $L_e - L_\mu - L_\tau$ symmetric mass matrix from seesaw

In this section, we discuss how an approximate $L_e - L_\mu - L_\tau$ symmetric neutrino mass matrix may arise within a seesaw framework. Consider a simple extension of the standard model by adding two additional singlet right handed neutrinos,[34] N_1, N_2 assigning them $L_e - L_\mu - L_\tau$ quantum numbers of +1 and −1 respectively. Denoting the standard model lepton doublets by $\psi_{e,\mu,\tau}$, the $L_e - L_\mu - L_\tau$ symmetry allows the following new couplings to the Lagrangian of the standard model:

$$\mathcal{L}' = (h_3\bar{\psi}_\tau + h_2\bar{\psi}_\mu)HN_2 + h_1\bar{\psi}_e HN_1 + MN_1^T C^{-1}N_2 + h.c. \quad (7.51)$$

where H is the Higgs doublet of the standard model; C^{-1} is the Dirac charge conjugation matrix. We add to it the symmetry breaking mass terms for the right handed neutrinos, which are soft terms, i.e.

$$\mathcal{L}_B = \epsilon(M_1 N_1^T C^{-1} N_1 + M_2 N_2^T C^{-1} N_2) + h.c. \quad (7.52)$$

with $\epsilon \ll 1$. These terms break $L_e - L_\mu - L_\tau$ by two units but since they are dimension 3 terms, they are soft and do not induce any

with $\epsilon \ll 1$. These terms break $L_e - L_\mu - L_\tau$ by two units but since they are dimension 3 terms, they are soft and do not induce any new terms into the theory.

It is clear from the resulting mass matrix for the ν_L, N system that the linear combination $h_2\nu_\tau - h_3\nu_\mu$) is massless and the atmospheric oscillation angle is given by $tan\theta_A = h_2/h_3$; for $h_3 \sim h_2$, the θ_A is maximal. The seesaw mass matrix then takes the following form (in the basis $(\nu_e, \tilde{\nu}_\mu, N_1, N_2)$ with $\tilde{\nu}_\mu \equiv h_2\nu_\mu + h_3\nu_\tau$):

$$M = \begin{pmatrix} 0 & 0 & m_1 & 0 \\ 0 & 0 & 0 & m_2 \\ m_1 & 0 & \epsilon M_1 & M \\ 0 & m_2 & M & \epsilon M_2 \end{pmatrix} \quad (7.53)$$

The diagonalization of this mass matrix leads to the mass matrix of the form discussed before.

7.5.2.2. Tri-bi-maximal mixing from seesaw

In this section, we present an example of a seesaw model for the mass matrix for neutrinos (Eq. 7.28) that leads to the tri-bi-maximal mixing.[35] It was shown that the Majorana neutrino mass matrix in 7.28 can be realized in a combined type I type II seesaw model with soft-broken S_3 family symmetry for leptons. The type II contribution comes from an S_3 invariant coupling of lepton doublets to the triplet field Δ i.e. $f_{\alpha\beta}L_\alpha L_\beta \Delta$. The most general S_3 invariant form for f is:

$$f = \begin{pmatrix} f_a & f_b & f_b \\ f_b & f_a & f_b \\ f_b & f_b & f_a \end{pmatrix} \quad (7.54)$$

After the triplet Higgs field Δ gets vev and decouples, its contribution to the light neutrino mass can written as

$$M_{II} = \begin{pmatrix} a' & b' & b' \\ b' & a' & b' \\ b' & b' & a' \end{pmatrix} \quad (7.55)$$

where $a' = \frac{v^2 \sin^2 \beta \lambda}{M_T} f_a$ and $b' = \frac{v^2 \sin^2 \beta \lambda}{M_T} f_b$. We denote M_T as the mass of the triplet Higgs and λ as the coupling constant between the triplet and doublets in the superpotential.

Coming to the type I contribution, the Dirac mass matrix for neutrinos comes from an S_3 invariant Yukawa coupling of the form:

$$\mathcal{L}_D = h_\nu [\overline{\nu_{R1}} H(L_e - L_\mu) + \overline{\nu_{R2}} H(L_\mu - L_\tau) \\ + \overline{\nu_{R3}} H(L_\tau - L_e)] + h.c. \quad (7.56)$$

leading to

$$Y_\nu = \begin{pmatrix} h & -h & 0 \\ 0 & h & -h \\ -h & 0 & h \end{pmatrix}. \quad (7.57)$$

In the limit of $|M_{R1,R3}| \gg |M_{R2}|$, where a single right-handed neutrino dominates the type I contribution, the mixed type I+II seesaw formula

$$\mathcal{M}_\nu = M_{II} - M_D^T M_{\nu R}^{-1} M_D, \quad (7.58)$$

then leads to Eq. 7.28 which gives the tri-bi-maximal mixing matrix. It turns out that,[35] the charged lepton mass matrix in this case can be made diagonal if one of the two lepton Yukawa couplings is set to zero.

7.6. Neutrino mass and grand unification

One of the interesting features of the seesaw mechanism is that if one assumes the Dirac masses to be roughly of order of the up-quark masses, then the atmospheric neutrino mass difference would directly measure the mass of m_3 and can be used to get a rough idea of how high the seesaw scale is. In order to do this one can use the rough relation $m_{\nu_3} \sim \frac{m_t^2}{M_R}$ which then yields $M_R \sim 10^{14}$ GeV. This is of course not a rigorous argument at all and can therefore only be used as a suggestive one. If however one takes this seriously, then it suggests that the seesaw scale could be related to the scale of grand unification which from arguments of coupling constant unification is also of order 10^{16} GeV. In view of other theoretical arguments in favor of GUTs, one may try to understand the neutrino masses within a grand unified theory framework.

The minimal GUT group that appears to have many desirable properties is the SO(10) group,[41] whose spinor representation is **16** dimensional and is just right for all then SM fermions of one generation plus the right handed neutrino needed for implementing the seesaw mechanism. This has therefore been extensively studied as a way to understand neutrino properties.

7.6.1. *SO(10) Grand Unification of seesaw mechanism and predictions for neutrino masses*

In addition to the fermion unification by the **16** dimensional spinor representation, SO(10) contains the B-L as a subgroup and seesaw mechanism requires that the process of symmetry breaking down to the standard model must break the B-L at a high scale. One implication of this is a natural understanding of the seesaw scale as being connected to the GUT scale. Secondly in the context of supersymmetric SO(10) models, the way B-L breaks has profound consequences for low energy physics. For instance, if B-L is broken by a Higgs field belonging to the **16** dimensional Higgs field, then the field that acquires a nonzero vev has the quantum numbers of the ν_R field i.e. B-L breaks by one unit. In this case higher dimensional operators of the form $\Psi\Psi\Psi\Psi_H$ will lead to R-parity violating operators in the effective low energy MSSM theory such as $QLd^c, u^c d^c d^c$ etc which can lead to large breaking of lepton and baryon number symmetry and hence unacceptable rates for proton decay. This theory also has no dark matter candidate without making additional assumptions. Furthermore, since non-renormalizable operators are an essential part of this approach, there are many more parameters, making it non-predictive in the absence of additional assumptions.[42]

On the other hand, one may break B-L by a **126** dimensional Higgs field.[43,44] The member of this multiplet that acquires vev has $B-L=2$ and leaves R-parity as an automatic symmetry of the low energy Lagrangian. This then gives a naturally stable dark matter. Furthermore, in this approach, since one considers only renormalizable couplings, the number of Yukawa parameters are quite limited so that the model is quite predictive.[43] The predictivity clearly arises from one irreducible **16** dimensional spinor multiplet containing all fermions of each family or complete fermion unification.

In order to study the predictions of the model, we first note that since the SO(10) model contains the left-right subgroup, the seesaw formula takes the modified form as in Eq.7.48 that we repeat below.[36]

$$M_\nu = fv_L - h_\nu^T f_R^{-1} h_\nu \left(\frac{v_{wk}^2}{v_R}\right) \qquad (7.59)$$

It turns out that if the B-L symmetry is broken by **16** Higgs fields, the first term in the type II seesaw (effective triplet vev induced term) becomes very small compared to the type I term. On the other hand, if B-L is broken by a **126** field, then the first term in the type II seesaw formula is not necessarily small and can in principle dominate in the seesaw formula. As we discuss below, this leads to predictions for neutrino masses and mixings that are in excellent agreement with experiments.

The basic ingredients of this model[43] are that one considers only two Higgs multiplets that contribute to fermion masses i.e. one **10** and one **126**. A unique property of the **126** multiplet is that it not only breaks the B-L symmetry and therefore contributes to right handed neutrino masses, but it also contributes to charged fermion masses by virtue of the fact that it contains MSSM doublets which mix with those from the **10** dimensional multiplets and survive down to the MSSM scale. This leads to a tremendous reduction of the number of arbitrary parameters.

There are only two Yukawa coupling matrices in this model: (i) h for the **10** Higgs and (ii) f for the **126** Higgs. SO(10) has the property that the Yukawa couplings involving the **10** and **126** Higgs representations are symmetric. Therefore if we assume that CP violation arises from other sectors of the theory (e.g. squark masses) and work in a basis where one of these two sets of Yukawa coupling matrices is diagonal, then it will have only nine parameters. Noting the fact that the (2,2,15) submultiplet of **126** has a pair of standard model doublets that contributes to charged fermion masses, one can write the quark and lepton mass matrices as follows[43]:

$$\begin{aligned} M_u &= h\kappa_u + fv_u \\ M_d &= h\kappa_d + fv_d \\ M_\ell &= h\kappa_d - 3fv_d \\ M_{\nu_D} &= h\kappa_u - 3fv_u \end{aligned} \qquad (7.60)$$

where $\kappa_{u,d}$ are the vev's of the up and down standard model type Higgs fields in the **10** multiplet and $v_{u,d}$ are the corresponding vevs for the same doublets in **126**. Note that there are 13 parameters in the above equations and there are 13 inputs (six quark masses, three lepton masses and three quark mixing angles and weak scale). Thus all parameters of the model that go into fermion masses are determined.

To determine the light neutrino masses, we use the seesaw formula in Eq. 7.48, where the **f** is nothing but the **126** Yukawa coupling. Thus all parameters that give neutrino mixings except an overall scale are determined. A simple way to see how large mixings arise in this model is to note that when the triplet term dominates the seesaw formula, we have the neutrino mass matrix $M_\nu \propto f$, where f matrix is the **126** coupling to fermions discussed earlier.

$$M_\nu = c(M_d - M_\ell) \tag{7.61}$$

All the quark mixing effects are then in the up quark mass matrix i.e. $M_u = U_{CKM}^T M_u^d U_{CKM}$. Note further that the minimality of the Higgs content leads to the following sum-rule among the mass matrices:

$$k\tilde{M}_\ell = r\tilde{M}_d + \tilde{M}_u \tag{7.62}$$

where the tilde denotes the fact that we have made the mass matrices dimensionless by dividing them by the heaviest mass of the species. We then find that we have

$$M_{d,\ell} \approx m_{b,\tau} \begin{pmatrix} \lambda^3 & \lambda^3 & \lambda^3 \\ \lambda^3 & \lambda^2 & \lambda^2 \\ \lambda^3 & \lambda^2 & 1 \end{pmatrix} \tag{7.63}$$

where $\lambda \sim 0.22$ and the matrix elements are supposed to give only the approximate order of magnitude. An important consequence of the relation between the charged lepton and the quark mass matrices in Eq. 7.62 is that the charged lepton contribution to the neutrino mixing matrix i.e. $U_\ell \simeq 1 + O(\lambda)$ or close to identity matrix. As a result the neutrino mixing matrix is given by $U_{PMNS} = U_\ell^\dagger U_\nu \simeq U_\nu$, since in U_ℓ, all mixing angles are small. Thus the dominant contribution to large mixings will come from U_ν, which in turn will be dictated by the sum rule in Eq. 7.61.

As we extrapolate the quark masses to the GUT scale, due to the fact that $m_b - m_\tau \approx m_\tau \lambda^2$ for a wide range of values of $\tan\beta$, the neutrino mass matrix $M_\nu = c(M_d - M_\ell)$ takes roughly the form

$$M_\nu = c(M_d - M_\ell) \approx m_0 \begin{pmatrix} \lambda^3 & \lambda^3 & \lambda^3 \\ \lambda^3 & \lambda^2 & \lambda^2 \\ \lambda^3 & \lambda^2 & \lambda^2 \end{pmatrix} \tag{7.64}$$

It is easy to see that both the θ_{12} (solar angle) and θ_{23} (the atmospheric angle) are now large. The detailed magnitudes of these angles of course depend on the details of the quark masses at the GUT scale. Using the extrapolated values of the quark masses and mixing angles to the GUT scale, the predictions of this model for various oscillation parameters are given in.[26] Some of the salient features are: (i) the atmospheric mixing angle θ_{23} is not maximal and the maximum value for it is around 38^0; (ii) the prediction for $sin\theta_{13} \equiv U_{e3}$ is near 0.18, a value within the reach of MINOS as well as other planned Long Base Line neutrino experiments such as Numi-Off-Axis, JPARC etc.

7.6.2. *CP violation in the minimal SO(10) model*

In the discussion given above, it was assumed that CP violation is non-CKM type and resides in the soft SUSY breaking terms of the Lagrangian. The overwhelming evidence from experiments seem to be that CP violation is perhaps of CKM type with CKM phase of about 60^0. Success of the above approach in understanding neutrino mixings suggests that we should consider extending the above simple model to accommodate CKM CP violation. Several such attempts have been made in recent literature.[19,47]

One approach discussed by us[47] employs a slight extension of the **10+126** model by adding a **120** Higgs field. A further Z_2 symmetry is imposed in such a way that the **10** and **126** couplings are real whereas the **120** couplings turn out to be imaginary. This will add a new piece to all fermion masses but in such a way that the $b-\tau$ mass convergence still leads to large atmospheric mixing as in the 10+126 case.

The new model is still predictive in the neutrino sector. Of the three new parameters, one is determined by the CP violating quark phase. the two others are determined by the solar mixing angle and the solar mass difference squared. Therefore we lose the prediction for these parameters. However, we can predict in addition to θ_A which is now close to maximal, $\theta_{13} \geq 0.1$ (see figure below) and the Dirac phase for the neutrinos, also close to 90°.

In the above discussion, we assumed type II seesaw terms to dominate.

This model has been reanalyzed using type I seesaw term to dominant in a recent paper[19] and a fit to the fermion masses as well as neutrino mixings exists for this case.

Finally, a few comments on what really will constitute a true test of the grand unification theories: a key prediction of simple grand unified theories such as $SU(5)$ and SO(10) is the existence of proton decay. In supersymmetric theories, proton decay turns from an exciting prediction to somewhat of a challenge since the presence of super-partners at the TeV scale generates "dangerous" operators such as $\tilde{Q}\tilde{Q}QL/M_U$ which could lead to very rapid proton decay. In fact it is the appearance of these kind of operators that has ruled out minimal SUSY $SU(5)$ model. For this reason, in the last two papers by us,[47] a Yukawa texture was chosen that is in accord with current experimental bounds on proton lifetime resulting from dimension five operators as the one given above and a fit to neutrino masses as well as charged fermions etc was found with type II seesaw. A proton decay check is therefore needed for the type I fit to fermions carried out in Ref. 48 and proton decay predictions for the model of[47] need to be worked out.

7.6.3. *Type II seesaw and Quasi-degenerate neutrinos*

In this subsection we like to discuss some issues related to the degenerate neutrino hypothesis, which will be necessary if there is evidence for neutrinoless double beta decay at a significant level(see for example the recent results from the Heidelberg-Moscow group[12]) and assuming that no other physics such as R-parity breaking or doubly charged Higgs etc are not the source of this effect). Thus it is appropriate to discuss how such models can arise in theoretical schemes and how stable they are under radiative corrections.

There are two aspects to this question: one is whether the degeneracy arises within a gauge theory framework without arbitrary adjustment of parameters and the second aspect being that given such a degeneracy arises at some scale naturally in a field theory, is this mass degeneracy stable under renormali9zation group extrapolation to the weak scale where we need the degeneracy to be present. In this section we comment on the first aspect.

It was pointed out long ago[14] that degenerate neutrinos arise naturally in models that employ the type II seesaw since the first term in the mass formula is not connected to the charged fermion masses. One way that has been discussed is to consider schemes where one uses symmetries such as SO(3) or $SU(2)$ or permutation symmetry S_4 so that the Majorana Yukawa couplings f_i are all equal. This then leads to the dominant contribution to all neutrinos being equal. This symmetry however must be broken in the charged fermion sector in order to explain the observed quark and lepton masses. Such models consistent with known data have been constructed based on SO(10) as well as other groups. The interesting point about the SO(10) realization is that the dominant contributions to the Δm^2's in this model comes from the second term in the type II seesaw formula which in simple models is hierarchical. It is of course known that if the MSW solution to the solar neutrino puzzle is the right solution (or an energy independent solution), then we have $\Delta m^2_{solar} \ll \Delta m^2_{ATMOS}$. In fact if we use the fact true in SO(10) models that $M_u = M_D$, then we have $\Delta m^2_{ATMOS} \simeq m_0 \frac{m_t^2}{fv_R}$ and $\Delta m^2_{SOLAR} \simeq m_0 \frac{m_c^2}{fv_R}$ where m_0 is the common mass for the three neutrinos. It is interesting that for $m_0 \sim$ few eV and $fv_R \approx 10^{15}$ GeV, both the Δm^2's are of right order to the required values.

Outside the seesaw framework, there could also be electroweak symmetries that guarantee the mass degeneracy.

The second question of stability under RGE of such a pattern is discussed in a subsequent section.

7.7. Some other consequences of seesaw paradigm

Generic seesaw models have several other important implications that we go into now. For simplicity, we first consider the type I seesaw formula 7.38. The first question one can ask is that given low energy information, to what extent we can discover the high scale physics associated with the seesaw mechanism such as the spectrum of right handed neutrinos, the structure of the Dirac mass matrix m_D (or equivalently Y_ν). A simple parameter counting shows that the neutrino masses and mixings (including CP phases) are characterized by nine observables whereas seesaw formula involves eighteen parameters (in the basis where RH neutrinos are mass eigenstates, there are three masses and 15 parameters characterizing m_D. Thus we need nine more pieces of low energy inputs to completely determine the seesaw physics (granted that all observables in the neutrino mass matrix are determined). Radiative leptonic decays such as $\mu \to e + \gamma, \tau \to \mu, e + \gamma$ including both CP violating and conserving channels could provide six pieces of information; three electric dipole moments of the charged leptons could provide the remaining three. Thus in principle, all the seesaw parameters could be determined from low energy observations.

In discussing the connection between high scale and low scale physics for neutrinos, it is often convenient to use a parameterization suggested by Casas and Ibarra.[49]

$$Y_\nu v_{wk} = iM_R^{1/2} O (\mathcal{M}_\nu^d)^{1/2} U^\dagger \tag{7.65}$$

where O is a complex matrix with the property that $OO^T = 1^{49}$ and U is the neutrino mixing matrix; M_ν^d is the diagonal neutrino mass matrix. The set of matrices O in fact form a group analogous to the complex extension of the Lorentz group. Note that six parameters (or three complex angles) characterize O, three needed each for M_R and $\mathcal{M}_\nu$ and six for U giving a total of 18 as we counted above. In special cases where there are symmetries e.g. $\mu - \tau$ symmetry, the number of complex angles in reduces to only one making the direct connection between high and low energy phases somewhat closer.

For the case of type II seesaw, the corresponding relation is:

$$Y_\nu v_{wk} = iM_R^{1/2} O[U^* \mathcal{M}_\nu^d U^\dagger - M_R \zeta]^{1/2} \qquad (7.66)$$

where $\zeta = \frac{v_L}{v_R}$.

It is clear from Eq.7.65 that in general the neutrino mixing matrix is only indirectly related to the details of Y_ν due to the unknown matrix O. In a given model however, when Y_ν is given, O and U get related.

7.7.1. SUSY seesaw and lepton flavor violation

In the standard model, the masslessness of the neutrino implies that that there is no lepton flavor changing effects unlike in the quark sector. Thus the leptons are completely "flavor sterile" and do not throw any light on the flavor puzzle. Once one includes the right handed neutrinos N_R one for each family, there is lepton mixing and this activates the lepton flavor. A simple phenomenological consequence of this "flavor activation" is that there appear lepton flavor changing effects such as $\mu \to e + \gamma$, $\tau \to e, \mu + \gamma$ etc. However, a simple estimate of the one loop contribution to such effects shows that the amplitude is of order

$$A(\ell_j \to \ell_i + \gamma) \simeq \frac{eG_F m_{\ell_j} m_e m_\nu^2}{\pi^2 m_W^2} \mu_B \qquad (7.67)$$

This leads to an unobservable branching ratio (of order $\sim 10^{-40}$) for the rare radiative decay modes for the leptons given above.

The situation however changes drastically as soon as the seesaw mechanism for neutrino masses is combined with supersymmetry. It has been noted in many papers already that in supersymmetric theories, the lepton flavor changing effects get significantly enhanced. They arise from the the mixings among sleptons (superpartners of leptons) of different flavor caused by the renormalization group extrapolations which via loop diagrams lead to lepton flavor violating (LFV) effects at low energies.[50]

The way this happens is as follows. In the simplest N=1 supergravity models,[51] the supersymmetry breaking terms at the Planck scale are taken to have only few parameters: a universal scalar mass m_0, universal A terms, one gaugino mass $m_{1/2}$ for all three types of gauginos. Clearly, a universal scalar mass implies that at Planck scale, there is no flavor violation anywhere except in the Yukawa couplings (or when the Yukawa terms are diagonalized, in the CKM angles). However as we extrapolate this theory to the weak scale, the flavor mixings in the Yukawa interactions induce non universal flavor violating scalar mass terms (i.e. flavor violating slepton and squark mass terms). In the absence of neutrino masses, the Yukawa matrices for leptons can be diagonalized so that there is no flavor violation in the lepton sector even after extrapolation down to the weak scale. On the other hand, when neutrino mixings are present or when the quarks and leptons are unified in such a way that this diagonalization becomes impossible, there is no basis where all leptonic flavor mixings can be made to disappear. In fact, in the most general case, of the three matrices Y_ℓ, the charged lepton coupling matrix, Y_ν, RH neutrino Yukawa coupling and M_{N_R}, the matrix characterizing the heavy RH neutrino mixing, only one can be diagonalizd by an appropriate choice of basis and the flavor mixing in the other two remain. In a somewhat restricted case where the right handed neutrinos do not have any interaction other than the Yukawa interaction and an interaction that generates the Majorana mass for the right handed neutrino, one can only diagonalize two out of the three matrices (i.e. Y_ν, Y_ℓ and M_R). Thus there will always be lepton flavor violating terms in the basic Lagrangian, no matter what basis one chooses. These LFV terms can then induce mixings between the sleptons of different flavor and lead to LFV processes. If we keep the M_ℓ diagonal by choice of basis, searches for LFV processes such as $\tau \to \mu + \gamma$ and/or $\mu \to e + \gamma$ can throw light on the RH neutrino mixings/or family mixings in M_D, as has already been observed.

Since in the absence of CP violation, there are at least six mixing angles (nine if M_D is not symmetric) in the seesaw formula and only three are observable in neutrino oscillation, to get useful information on the fundamental high scale theory from LFV processes, it is assumed that M_{N_R} is diagonal so that one has a direct correlation between the observed neutrino mixings and the fundamental high scale paramters of the theory. The important point is that the flavor mixings in Y_ν then reflect themselves in the slepton mixings that lead to the LFV processes via the RGEs.

From the point of view of the LFV analysis, there are essentially two classes of neutrino mass models that need to be considered: (i) the first class is where it is assumed that the RH neutrino mass M_{N_R} is either a mass term in the basic Lagrangian or arises from nonrenormalizable terms such as $\nu^c \chi^{c2}/M_{P\ell}$, as in a class of SO(10) models; (we will such models Dirac type) and (ii) a second class where the Majorana mass of the right handed neutrino itself arises from a renormalizable Yukawa coupling e.g. $f\nu^c\nu^c\Delta$ (we will call them Majorana type models). In Dirac type models, in principle, one could decide to have all the flavor mixing effects in the right handed neutrino mass matrix and keep the Y_ν diagonal. In that case, RGEs would not induce any LFV effects. However we will bar this possibility and consider the case where all flavor mixings are in the Y_ν so that RGEs can induce LFV effects. In Majorana type models on the other hand, there will always be an LFV effect, although its magnitude will depend on the choice of the seesaw scale (v_{BL}).

Examples of class two models are models for neutrino mixings such as SO(10) with a **126** Higgs field[43] or models with a triplet Higgs, whose vev is the seesaw scale.

In both these examples, the equations that determine the extent of lepton flavor violation in leading order, for the case $A = 0$ are:

$$\frac{dm_L^2}{dt} \simeq \frac{1}{16\pi^2}[3m_0^2(Y_\nu^\dagger Y_\nu)] \qquad (7.68)$$

In the Majorana case, this equation will have contributions from the renormalizable f couplings that give Majorana masses to the right handed neutrinos[52] in the sense that generation mixing elements in Y_ν will be generated by f's even if they were absent in the beginning. Using these equations, one can obtain the branching ratios for the radiative lepton flavor violating processes using the formula below:

$$\mathcal{L} = iem_j \left(\bar{\ell}_{jL}\sigma_{\mu\nu}\ell_{iR}C_L + \bar{\ell}_{jR}\sigma_{\mu\nu}\ell_{iL}C_R \right) F^{\mu\nu} + h.c. \qquad (7.69)$$

then the Branching ratio for the decay $\ell_j \to \ell_i + \gamma$ is given by the formula

$$B(\ell_j \to \ell_i + \gamma) = \frac{48\pi^3 \alpha_{em}}{G_F^2}(|C_L|^2 + |C_R|^2)B(\ell_j \to \ell_i + 2\nu). \qquad (7.70)$$

7.7.2. *Renormalization group evolution of the neutrino mass matrix*

In the seesaw models for neutrino masses, the neutrino mass arises from the effective operator

$$\mathcal{O}_\nu = -\frac{1}{4}\kappa_{\alpha\beta}\frac{L_\alpha H L_\beta H}{M} \qquad (7.71)$$

after symmetry breaking $< H^0 > \neq 0$; here L and H are the leptonic and weak doublets respectively. α and β denote the weak flavor index. The matrix κ becomes the neutrino mass matrix after symmetry breaking i.e. $< H^0 > \neq 0$. This operator is defined at the scale M since it arises after the heavy field N_R is integrated out. On the other hand, in conventional oscillation experiments, the neutrino masses and mixings being probed are at the weak scale. One must therefore extrapolate the operator down from the seesaw scale M to the weak scale M_Z.[53] The form of the renormalization group extrapolation of course depends on the details of the theory. For simplicity we will consider only the supersymmetric theories, where the only contributions come from the wave function renormalization and is therefore easy to calculate. The equation governing the extrapolation of the $\kappa_{\alpha\beta}$ matrix is given in the case of MSSM by:

$$\frac{d\kappa}{dt} = [-3g_2^2 + 6Tr(Y_u^\dagger Y_u)]\kappa + \frac{1}{2}[\kappa(Y_e^\dagger Y_e) + (Y_e^\dagger Y_e)\kappa] \quad (7.72)$$

We note two kinds of effects on the neutrino mass matrix from the above formula: (i) one that is flavor independent and (ii) a part that is flavor specific. If we work in a basis where the charged leptons are diagonal, then the resulting correction to the neutrino mass matrox is given by:

$$\mathcal{M}_\nu(M_Z) = (1+\delta)\mathcal{M}(M_{B-L})(1+\delta) \quad (7.73)$$

where δ is a diagonal matrix with matrix elements $\delta_{\alpha\alpha} \simeq -\frac{m_\alpha^2 \tan^2\beta}{16\pi^2 v^2}$ In more complicated theories, the corrections will be different. Let us now study some implications of this corrections. For this first note that in the MSSM, this effect can be sizable if $\tan\beta$ is large (of order 10 or bigger).

7.7.3. Radiative magnification of neutrino mixing angles

A major puzzle of quark lepton physics is the diverse nature of the mixing angles. Whereas in the quark sector the mixing angles are small, for the neutrinos they are large. One possible suggestion in this connection is that perhaps the mixing angles in both quark and lepton sectors at similar at some high scale; but due to renormalization effects, they may become magnified at low scales. It was shown in Ref. 54 that this indeed happens if the neutrino spectrum os degenerate. This can be seen in a simple way for the $\nu_\mu - \nu_\tau$ sector.[54]

Let us start with the mass matrix in the flavor basis:

$$\mathcal{M}_\mathcal{F} = U^* \mathcal{M}_\mathcal{D} U^\dagger$$
$$= \begin{pmatrix} C_\theta & S_\theta \\ -S_\theta & C_\theta \end{pmatrix} \begin{pmatrix} m_1 & 0 \\ 0 & m_2 e^{-i\phi} \end{pmatrix} \begin{pmatrix} C_\theta & -S_\theta \\ S_\theta & C_\theta \end{pmatrix}. \quad (7.74)$$

Let us examine the situation when $\phi = 0$ (i.e. CP is conserved), which corresponds to the case when the neutrinos ν_1 and ν_2 are in the same CP eigenstate. Due to the presence of radiative corrections to m_1 and m_2, the matrix $\mathcal{M}_\mathcal{F}$ gets modified to

$$\mathcal{M}_\mathcal{F} \to \begin{pmatrix} 1+\delta_\alpha & 0 \\ 0 & 1+\delta_\beta \end{pmatrix} \mathcal{M}_\mathcal{F} \begin{pmatrix} 1+\delta_\alpha & 0 \\ 0 & 1+\delta_\beta \end{pmatrix}. \quad (7.75)$$

The mixing angle $\bar{\theta}$ that now diagonalizes the matrix $\mathcal{M}_\mathcal{F}$ at the low scale μ (after radiative corrections) can be related to the old mixing angle θ through the following expression:

$$\tan 2\bar{\theta} = \tan 2\theta \, (1 + \delta_\alpha + \delta_\beta) \, \frac{1}{\lambda}, \quad (7.76)$$

where

$$\lambda \equiv \frac{(m_2 - m_1)C_{2\theta} + 2\delta_\beta(m_1 S_\theta^2 + m_2 C_\theta^2) - 2\delta_\alpha(m_1 C_\theta^2 + m_2 S_\theta^2)}{(m_2 - m_1)C_{2\theta}}. \quad (7.77)$$

If

$$(m_1 - m_2)\, C_{2\theta} = 2\delta_\beta(m_1 S_\theta^2 + m_2 C_\theta^2) - 2\delta_\alpha(m_1 C_\theta^2 + m_2 S_\theta^2), \quad (7.78)$$

then $\lambda = 0$ or equivalently $\bar{\theta} = \pi/4$; i.e. maximal mixing. Given the mass heirarchy of the charged leptons: $m_{l_\alpha} \ll m_{l_\beta}$, we expect $|\delta_\alpha| \ll |\delta_\beta|$, which reduces (7.78) to a simpler form:

$$\epsilon = \frac{\delta m C_{2\theta}}{(m_1 S_\theta^2 + m_2 C_\theta^2)} \quad (7.79)$$

In the case of MSSM, the radiative magnification condition can be satisfied provided provided

$$h_\tau(MSSM) \approx \sqrt{\frac{8\pi^2 |\Delta m^2(\Lambda)| C_{2\theta}}{\ln(\frac{\Lambda}{\mu}) m^2}}. \quad (7.80)$$

For $\Delta m^2 | simeq \Delta m_A^2$, this condition can be satisfied for a very wide range of $\tan\beta$.

It is important to emphasize that this magnification occurs only if at the seesaw scale the neutrino masses are nearly degenerate. A similar mechanism using the right handed neutrino Yukawa couplings instead of the charged lepton ones has been carried out recently.[55] Here two conditions must be satisfied: (i) the neutrino spectrum must be nearly degenerate (i.e. $m_1 \simeq m_2$ as in Ref. 54) and (ii) there must be a hierarchy between the right handed neutrinos.

7.7.3.1. *An explicit example of a neutrino mass matrix unstable under RGE*

In this section, we give an explicit example of a neutrino mass matrix unstable under RGE effects. Consider the following mass matrix with degenerate neutrino masses and a bimaximal mixing.[56]

$$\mathcal{M}_\nu = \begin{pmatrix} 0 & \frac{1}{\sqrt{2}} & \frac{1}{\sqrt{2}} \\ \frac{1}{\sqrt{2}} & \frac{1}{2} & -\frac{1}{2} \\ \frac{1}{\sqrt{2}} & \frac{1}{2} & \frac{1}{2} \end{pmatrix} \tag{7.81}$$

The eigenvalues of this mass matrix are $(1, -1, 1)$ and the eigenvectors:

$$V_1 = \begin{pmatrix} 0 \\ \frac{1}{\sqrt{2}} \\ \frac{1}{\sqrt{2}} \end{pmatrix} ; V_2 = \begin{pmatrix} \frac{1}{\sqrt{2}} \\ -\frac{1}{2} \\ -\frac{1}{2} \end{pmatrix} ; V_3 = \begin{pmatrix} \frac{1}{\sqrt{2}} \\ \frac{1}{2} \\ \frac{1}{2} \end{pmatrix} \tag{7.82}$$

After RGE to the weak scale, the mass matrix becomes

$$\mathcal{M}_\nu = \begin{pmatrix} 0 & \frac{1}{\sqrt{2}} & \frac{1}{\sqrt{2}}(1+\delta) \\ \frac{1}{\sqrt{2}} & \frac{1}{2} & -\frac{1}{2}(1+\delta) \\ \frac{1}{\sqrt{2}}(1+\delta) & \frac{1}{2}(1+\delta) & \frac{1}{2}(1+2\delta) \end{pmatrix} \tag{7.83}$$

It turns out that the eigenvectors of this matrix become totally different and are given by:

$$V_1 = \begin{pmatrix} \frac{1}{\sqrt{3}} \\ \frac{2}{\sqrt{3}} \\ 0 \end{pmatrix} ; V_2 = \begin{pmatrix} \frac{1}{\sqrt{2}} \\ -\frac{1}{2} \\ -\frac{1}{2} \end{pmatrix} ; V_3 = \begin{pmatrix} \frac{1}{\sqrt{6}} \\ -\frac{1}{2\sqrt{3}} \\ \frac{\sqrt{3}}{2} \end{pmatrix} \tag{7.84}$$

We thus see that the neutrino mixing pattern has become totally altered, although the eigenvalues are only slightly perturbed from their unperturbed value.

7.7.4. *Seesaw paradigm and leptogenesis*

Finally let us comment that in models where the light neutrino mass is understood via the seesaw mechanism using heavy right handed neutrinos, there is a very simple mechanism for the generation of baryon asymmetry of the universe. Since the righthanded neutrino has a high mass, it decays at a high temperature which in combination with CP violation in Y_ν generates a lepton asymmetry.[57] This lepton asymmetry is converted to baryon asymmetry via the sphaleron effects[58] above the electroweak phase transition temperature since sphalerons break B+L conservation. It also turns out that one of the necessary conditions for sufficient leptogenesis is that the right handed neutrinos must be heavy as is required by the seesaw mechanism. To see this note that one of Sakharov conditions for leptogenesis is that the right handed neutrino decay must be slower than the expansion rate of the universe at the temperature $T \sim M_{N_R}$. The corresponding condition is:

$$\frac{h_\ell^2 M_{N_R}}{16\pi} \leq \sqrt{g^*}\frac{M_{N_R}^2}{M_{P\ell}} \tag{7.85}$$

This implies that $M_{N_R} \geq \frac{h_\ell^2 M_{P\ell}}{16\pi\sqrt{g^*}}$. Translating this into a reliable bound on the masses of the right handed neutrinos is quite model dependent since the Yukawa texture i.e. h_ℓ values in the above equation depends on the particular way to understand large mixings as well as the neutrino mass hierarchy.

To proceed further, we start with the expression for lepton asymmetry in these scenarios:[59]

$$\varepsilon_i^I = -\frac{1}{8\pi}\frac{1}{[Y_\nu Y_\nu^\dagger]_{ii}}\sum_j \mathrm{Im}[Y_\nu Y_\nu^\dagger]_{ij}^2 F\left(\frac{M_j^2}{M_i^2}\right), \tag{7.86}$$

One can draw several conclusions from this expression: first using Eq. 7.65 and 7.86, we see that, ε_i^I is independent of the low energy CP phases. This implies that in principle observation of CP violation in neutrino oscillation may not throw any light on the origin of matter.

Second point to notice is that for hierarchical masses for RH neutrinos i.e. $M_1 \ll M_{2,3}$, one can write 7.86 as

$$\varepsilon_i^I = -\frac{1}{8\pi}\frac{1}{[Y_\nu Y_\nu^\dagger]_{ii}}\sum_j \mathrm{Im}[Y_\nu \mathcal{M}_\nu^* Y_\nu^T]_{ii}. \tag{7.87}$$

From this it follows that if the neutrino masses are strictly degenerate, then using 7.65, it is easy to see that $\varepsilon_i^I = 0$. This is an interesting result although this cannot strictly be used to rule out the possibility of degenerate neutrino masses since, it is more natural (as emphasized earlier) for a degenerate neutrino spectrum to arise from a type II seesaw rather than type I seesaw which has been used in drawing this conclusion.

Another consequence of 7.86 is that combining 7.85 and 7.86, it is possible to obtain a reliable lower bound on the lightest RH neutrino mass[60] and it turns out to be: $M_{N_1} \geq 10^9$ GeV. This is bound is somewhat strengthened if one further demands that there is a $\mu - \tau$ exchange symmetry in the left as well as the right handed neutrino sector for which there is some observational indication (since θ_{13} appears to be very small). The bound then becomes $M_{N_1} \geq 6 \times 10^9$ GeV.[61]

In the presence of triplet contributions to seesaw (type II), there are new contributions to the lepton asymmetry given by: The type II contribution has been calculated and is given in Refs. 45 and 62 to be

$$\varepsilon_i^{II} = \frac{3}{8\pi} \frac{\text{Im}[Y_\nu f^* Y_\nu^T \mu]_{ii}}{[Y_\nu Y_\nu^\dagger]_{ii} M_i} \ln\left(1 + \frac{M_i^2}{M_T^2}\right), \qquad (7.88)$$

where $\mu \equiv \lambda M_T$ and λ is the coupling between triplet and two doublets in the superpotential and f is the triplet coupling to leptons. Thus baryogenesis via leptogenesis is a very sensitive way to probe the neutrino mass mechanisms. For more details on this see the lectures by M. C. Chen.[63]

7.8. Conclusions and outlook

In summary, the neutrino oscillation experiments have provided the first evidence for new physics beyond the standard model. The field of neutrino physics, along with the search for the origin of mass, dark matter has therefore become central to the study of new physics at the TeV scale and beyond. Another area which is foremost in the minds of many theorists is supersymmetry which stabilizes the Higgs mass, provides a way to understand the electro-weak symmetry breaking and possibly a dark matter candidate. In discussing consequences of seesaw mechanism as well as in seeking theories of neutrino mass, we have assumed supersymmetry. An exception is the last section, where we consider low scale extra dimensional models for understanding Higgs mass and its possibility as an alternative to seesaw mechanism.

What have we learned so far? One thing that seems very clear is that there is probably a set of three right handed neutrinos which restore quark lepton symmetry to physics; secondly there must be a local $B-L$ symmetry at some high scale beyond the standard model that keeps the RH neutrinos so far below the Planck scale. While there are very appealing arguments that the scale of $B-L$ symmetry is close to 10^{14}-10^{16} GeV's, in models with extra dimensions, one cannot rule out the possibility that it is around a few TeVs, although the present TeV scale models generally require many near TeV particles with sometimes undesirable consequences for flavor violation. Third thing that one may suspect is that the right handed neutrino spectrum may be split into a heavier one and two others which are nearby. If this suspicion is confirmed, that would point towards an $SU(2)_H$ horizontal symmetry or perhaps even an $SU(3)_H$ symmetry which breaks into an $SU(2)_H$ symmetry (although simple anomaly considerations prefer the first alternative).

The correct theory should explain:

(i) Why both the solar mixing angle is large and atmospheric mixing angle maximal?

(ii) Why the $\Delta m^2_\odot \ll \Delta m^2_A$ and what is responsible for the smallness of U_{e3}? While in the inverted hierarchy models and models with $\mu - \tau$ symmetry, the smallness of U_{e3} is natural, in general it is not. In fact high precision search for U_{e3} may hold the clue to possible family symmetries vrs simple grad unification.

(iii) What is the nature of CP phases in the lepton sector and what is their relation to the CP phases possibly responsible for baryogenesis via leptogenesis?

(iv) What is the complete mass spectrum for neutrinos?

These and other questions are likely to prove to be very exciting challenges to both theory and experiment in neutrino physics for the next two decades. These lectures are meant to be a very cursory overview of what seems to be the simplest way to understand neutrino masses and mixings i.e. seesaw mechanism and some related physics. Even at that, only a few selected topics are covered and for more details and references, we refer the reader to other excellent reviews in the literature.

Acknowledgment

This work is supported by the National Science Foundation grant number PHY-0354401. The author is grateful to K. T. Mahanthappa and the organizing committee of TASI 2006 for the invitation to co-direct the school and providing the unique opportunity to engage in many lively interactions with the students.

References

1. B. Pontecorvo, Sov. Phys. JETP **6**, 429 (1958); Z. Maki, M. Nakagawa, S. Sakata, Prog. Theor. Phys. **28**, 870 (1962).
2. For a recent review of new physics from neutrinos, see R. N. Mohapatra and A. Yu Smirnov, Ann. Rev. Nucl. and Part. Sc. **56**, 569 (2006); G. Altarelli, arXiv:0711.0161 [hep-ph].
3. We follow the discussion in B. Kayser and R. N. Mohapatra, in *Current Aspects of Neutrino Physics*, ed. D. Caldwell (Springer, 2001); pp.17.
4. P. Minkowski, Phys. Lett. **B 67**, 421(1977); M. Gell-Mann, P. Ramond and R. Slansky, in *Supergravity*, eds. P. van Niewenhuizen and D.Z. Freedman (North Holland 1979); T. Yanagida, in Proceedings of *Workshop on Unified Theory and Baryon number in the Universe*, eds. O. Sawada and A. Sugamoto (KEK 1979); S. L. Glashow, Erice lectures (1979); R. N. Mohapatra and G. Senjanović, Phys. Rev. Lett. **44**, 912 (1980).
5. M. C. Gonzalez-Garcia and M. Maltoni, arXiv:0704.1800 [hep-ph].
6. J. Conrad, lectures at this institute.
7. E. Abouzaid *et al.*, LBNL-56599.
8. A. A. Aguilar-Arevalo *et al.* [The MiniBooNE Collaboration], Phys. Rev. Lett. **98**, 231801 (2007)
9. For a theoretical analysis of MiniBooNe experiment in terms of sterile neutrinos, see M. Maltoni and T. Schwetz, arXiv:0705.0107 [hep-ph].
10. For a history, see H. Klapdor-Kleingrothaus, *Sixty years of double beta decay*, World Scientific (1999).
11. P. Vogel, TASI 2006 lectures, this volume.
12. H. Klapdor-Kleingrothaus *et al.*, Mod. Phys. Lett. **A16**, 2409 (2001).
13. S. Hannestad, Phys. Rev. Lett. **95**, 221301 (2005); U. Seljak, A. Slosar and P. McDonald, JCAP **0610**, 014 (2006); For a review and more references, see O. Elgaroy, hep-ph/0612097.
14. D. Caldwell and R. N. Mohapatra, Phys. Rev. **D 48**, 3259 (1993); A. Joshipura, Phys. Rev. **D51**, 1321 (1995).
15. L. Wolfenstein, Phys. Rev. **D 18**, 958 (1978); P. F. Harrison, D. Perkins and W. G. Scott, Phys. Lett. **B 530**, 167 (2002); P. F. Harrison and W. G. Scott, Phys. lett. **B 535**, 163 (2002); Z. z. Xing, Phys. Lett. **B 533**, 85 (2002); X. G. He and A. Zee, Phys. Lett. B **560**, 87 (2003).
16. T. Fukuyama and H. Nishiura, hep-ph/9702253; R. N. Mohapatra and S. Nussinov, Phys. Rev. **D 60**, 013002 (1999); E. Ma and M. Raidal, Phys. Rev. Lett. **87**, 011802 (2001); C. S. Lam, hep-ph/0104116.
17. W. Grimus, A. S.Joshipura, S. Kaneko, L. Lavoura, H. Sawanaka, M. Tanimoto, hep-ph/0408123; R. N. Mohapatra, SLAC Summer Inst. lecture; http://www-conf.slac.stanford.edu/ssi/2004; hep-ph/0408187; JHEP, **0410**, 027 (2004); A. de Gouvea, Phys. Rev. **D69**, 093007 (2004).
18. R. N. Mohapatra and W. Rodejohann, Phys. Rev. **D 72**, 053001 (2005); T. Kitabayashi and M. Yasue, Phys. Lett,. **B 621**, 133 (2005).
19. A. Ghosal, hep-ph/0304090; W. Grimus and L. Lavoura, hep-ph/0305046; 0309050; K. Matsuda and H. Nishiura, hep-ph/0511338; A. Joshipura, hep-

ph/0512252; R. N. Mohapatra, S. Nasri and H. Yu, Phys. Lett. **B 636**, 114 (2006).
20. F. Caravaglios and S. Morisi, hep-ph/0503234; R. N. Mohapatra, S. Nasri and H. Yu, Phys. Lett. **B 639**, 318 (2006); O. Felix, A. Mondragon, M. Mondragon and E. Peinado, hep-ph/0610061. .
21. E. Ma, Phys. Rev. **D70**, 031901R (2004); **D 72**, 037301 (2005); K. S. Babu and X. G. He, hep-ph/0507217; G. Altarelli and F. Feruglio, Nucl. Phys. **B720**, 64 (2005); A. Zee, Phys. Lett. **B630**, 58 (2005).
22. S. Petcov, Phys. Lett. **B110**, 245 (1982); R. Barbieri, L. J. Hall, D. R. Smith, A. Strumia and N. Weiner, JHEP **9812**, 017 (1998); A. Joshipura and S. Rindani, Eur.Phys.J. **C14**, 85 (2000); R. N. Mohapatra, A. Perez-Lorenzana, C. A. de S. Pires, Phys. Lett. **B474**, 355 (2000); Q. Shafi and Z. Tavartkiladze, Phys. Lett. **B 482**, 1451 (2000);T. Kitabayashi and M. Yasue, Phys. Rev. **D 63**, 095002 (2001); Phys. Lett. **B 508**, 85 (2001); hep-ph/0110303; L. Lavoura, Phys. Rev. D 62, 093011 (2000); W. Grimus and L. Lavoura, Phys. Rev. D 62, 093012 (2000); J. High Energy Phys. 09, 007 (2000); J. High Energy Phys. 07, 045 (2001); R. N. Mohapatra, Phys. Rev. **D 64**, 091301 (2001). K. S. Babu and R. N. Mohapatra, Phys. Lett. **B 532**, 77 (2002); H. S. Goh, R. N. Mohapatra and S.-P. Ng, Phys. Lett. **B542**, 116 (2002)); Duane A. Dicus, Hong-Jian He, John N. Ng, Phys. Lett. **B 536**, 83 (2002); G. Altarelli and R. Franceschini, JHEP **0603**, 047 (2006).
23. F. Feruglio, A. Strumia and F. Vissani, Nucl. Phys. B **637**, 345 (2002); S. Pascoli, S.T. Petcov and L. Wolfenstein, Phys. Lett. **B524** (2002) 319; S. Pascoli, S. T. Petcov and W. Rodejohann, Phys. Lett. B **558**, 141 (2003).
24. R. Kuchimanchi and R. N. Mohapatra, hep-ph/0107110 Phys. Rev. **D 66**, 051301 (2002); and hep-ph/0107373.
25. K. S. Babu and R. N. Mohapatra, Phys. Lett. **B 532**, 77 (2002).
26. H. S. Goh, R. N. Mohapatra and S.-P. Ng, ref..[22]
27. GENIUS collaboration, H. Klapdor-Kleingrothaus *et al.*, J. Phys. G. **G 24**, 483 (1998); CUORE collaboration, E. Fiorini, Phys. Rep. **307**, 309 (1998); MAJORANA collaboration, C. E. Aalseth *et al.* hep-ex/0201021; EXO collaboration, G. Gratta *et al.* Invited talk at the *Neutrino miniworkshop at INT, Seattle*, April, 2002; MOON collaboration, H. Ejiri *et al.* Phys. Rev. lett. **85**, 2917 (2000); For a review, see O. Cremonesi, Proceedings of *Neutrino 2002*, Munchen.
28. R. N. Mohapatra and W. Rodejohann, Phys. Lett. B **644**, 59 (2007); A. Blum, R. N. Mohapatra and W. Rodejohann, arXiv:0706.3801 [hep-ph].
29. B. Kayser, in *CP violation*, ed. C. Jarlskog (World Scientific, 1988), p. 334; A. Barroso and J. Maalampi, Phys. Lett. **132B**, 355 (1983).
30. R. Barbieri, J. Ellis and M. K. Gaillard, Phys. Lett. **90 B**, 249 (1980); E. Akhmedov, Z. Berezhiani and G. Senjanović, Phys. Rev. Lett. **69**, 3013 (1992).
31. R. N. Mohapatra, Phys. Rev. lett. **56**, 561 (1986); R. N. Mohapatra and J. W. F. Valle, Phys. Rev. **D 34**, 1642 (1986).
32. S. F. King, JHEP **0209**, 011 (2002); S. F. King and N. N. Singh, Nucl. Phys. B **591**, 3 (2000).

33. For discussions of the seesaw formula and attempts to understand neutrino mixings see E. Akhmedov, G. Branco and M. Rebelo, Phys. Rev. lett. **84**, 3535 (2000); G. Altarelli, F. Feruglio and I. Masina, Phys. Lett. B **472**, 382 (2000); D. Falcone, hep-ph/0204335; M. Jezabek and P. Urban, hep-ph/0206080.
34. W. Grimus and L. Lavoura, Phys. Rev. **D 62**, 093012 (2000); JHEP **07**, 045 (2001).
35. R. N. Mohapatra, S. Nasri and H. Yu, Phys. Lett. **B 639**, 318 (2006).
36. G. Lazarides, Q. Shafi and C. Wetterich, Nucl.Phys.**B181**, 287 (1981); R. N. Mohapatra and G. Senjanović, Phys. Rev. **D 23**, 165 (1981).
37. J. C. Pati and A. Salam, Phys. Rev. **D10**, 275 (1974); R. N. Mohapatra and J. C. Pati, Phys. Rev. **D 11**, 566, 2558 (1975); G. Senjanović and R. N. Mohapatra, Phys. Rev. **D 12**, 1502 (1975).
38. E. K. Akhmedov and M. Frigerio, Phys. Rev. Lett. **96**, 061802 (2006);P. Hosteins, S. Lavignac and C. A. Savoy, Nucl. Phys. B **755**, 137 (2006); E. K. Akhmedov, M. Blennow, T. Hallgren, T. Konstandin and T. Ohlsson, arXiv:hep-ph/0612194.
39. R. N. Mohapatra and R. E. Marshak, Phys. Lett. **B 91**, 222 (1980); A. Davidson, Phys. Rev. **D20**, 776 (1979).
40. See for instance, R. N. Mohapatra, *Unification and Supersymmetry*, 3rd edition, Springer-Verlag, 2002.
41. H. Georgi, in *Particles and Fields*, (ed. C. Carlson), A. I. P. (1975); H. Fritzsch and P. Minkowski, Ann. Phys. **93**, 193 (1975).
42. C. Albright, K. S. Babu and S. Barr, Phys. Rev. lett. **81**, 1167 (1998); C. Albright and S. Barr, Phys. Rev. **D 58**, 013002 (1998); K. S. Babu, J. C. Pati and F. Wilczek, hep-ph/9812538; X. Ji, Y. Li and R. N. Mohapatra, Phys. Lett. **B 633**, 755 (2006).
43. K. S. Babu and R. N. Mohapatra, Phys. Rev. Lett. **70**, 2845 (1993).
44. C. S. Aulakh and R. N. Mohapatra, Phys. Rev. **D 28**, 217 (1983); D. G. Lee and R. N. Mohapatra, Phys. Rev. **D 51**, 1353 (1995); Y. Achiman, hep-ph/9812389; arXiv:hep-ph/0612138; M. Bando, T. Kugo and K. Yoshioka, Phys. Rev. Lett. **80**, 3004 (1998); K. Oda, E. Takasugi, M. Tanaka and M. Yoshimura, Phys. Rev. **D59**, 055001 (1999); M. C. Chen and K. T. Mahanthappa, Phys. Rev. **D 62** , 113007 (2000); T. Fukuyama and N. Okada, hep-ph/0206118; T. Fukuyama, Y. Koide, K. Matsuda and H. Nishiura, Phys. Rev. **D 65**, 033008 (2002); B. Bajc, A. Melfo, G. Senjanovic and F. Vissani, Phys. Lett. **B 634**, 272 (2006); C. S. Aulakh and S. K. garg, hep-ph/0512224; K. S. Babu and C. macesanu, Phys. Rev. **D 72**, 115003 (2005); S. Bertolini, M. Malinsky and T. Schwetz, Phys. Rev. **D 73**, 115012 (2006).
45. B. Bajc, G. Senjanović and F. Vissani, Phys. Rev. Lett. **90**, 051802 (2003) hep-ph/0210207.
46. H. S. Goh, R. N. Mohapatra and S. P. Ng, Phys. Lett. B **570**, 215 (2003) [arXiv:hep-ph/0303055].
47. B. Dutta, Y. Mimura and R. N. Mohapatra, hep-ph/0406262, Phys. Lett. **B 603** , 35 (2004) ; Phys. Rev. Lett. **94**, 091804 (2005); Phys. Rev. **D 72**, 075009 (2005).

48. W. Grimus and H. Kuhbock, hep-ph/0612132.
49. J. A. Casas and A. Ibarra, Nucl. Phys. B **618**, 171 (2001).
50. F. Borzumati and A. Masiero, Phys. Rev. Lett. **57**, 961 (1986); A. Masiero, S. Vempati and O. Vives, hep-ph/0209303.
51. A. H. Chamseddine, R. Arnowitt and P. Nath, Phys. Rev. Lett. **49**, 970 (1982); R. Barbieri, S. Ferrara and C. A. Savoy, Phys. Lett. B **119**, 343 (1982).
52. K. S. Babu, B. Dutta and R. N. Mohapatra, Phys. Rev. **D67**, 076006 (2003).
53. K.S. Babu, C.N. Leung and J. Pantaleone, Phys. Lett. **B319**, 191 (1993); P.H. Chankowski and Z. Pluciennik, Phys. Lett. **B316**, 312 (1993); S. Antusch, M. Drees, J. Kersten, M. Lindner and M. Ratz Phys. Lett. **B519**, 238 (2001).
54. K. R. S. Balaji, A. Dighe, R. N. Mohapatra and M. K. Parida, Phys. Rev. Lett. **84**, 5034 (2000); Phys. Lett. B **481**, 33 (2000); R. N. Mohapatra, M. K. Parida and G. Rajasekaran, Phys. Rev. **D 69**, 053007 (2004); S. Agarwalla et al. hep-ph/0611225.
55. S. Antusch, J. Kersten, M. Lindner and M. Ratz, hep-ph/0206078.
56. H. Georgi and S. L. Glashow, hep-ph/9808293.
57. M. Fukugita and T. Yanagida, Phys. Lett. **74 B**, 45 (1986).
58. V. Kuzmin, V. Rubakov and M. Shaposnikov. Phys. Lett. **185B**, 36 (1985).
59. W. Buchmuller, P. Di Bari and M. Plumacher, Nucl. Phys. B **643**, 367 (2002).
60. S. Davidson and A. Ibarra, Nucl. Phys. **B 648**, 345 (2003).
61. R. N. Mohapatra, S. Nasri and H. Yu, Phys. Lett. **B615**, 231 (2005).
62. T. Hambye and G. Senjanovic, Phys. Lett. B **582**, 73 (2004); S. Antusch and S. F. King, Phys. Lett. B **597**, 199 (2004).
63. M. C. Chen, this volume.

Chapter 8

Searching for the Higgs Boson

D. Rainwater

*Dept. of Physics and Astronomy, University of Rochester,
Rochester, NY, USA*

These lectures on Higgs boson collider searches were presented at TASI 2006. I first review the Standard Model searches: what LEP did, prospects for Tevatron searches, the program planned for LHC, and some of the possibilities at a future ILC. I then cover in-depth what comes after a candidate discovery at LHC: the various measurements one has to make to determine exactly what the Higgs sector is. Finally, I discuss the MSSM extension to the Higgs sector.

8.1. Introduction

Despite all the remarkable progress made early in the 21st century formulating possible explanations for the weakness of gravity relative to the other forces, the nature of dark matter (and dark energy), what drove cosmological inflation, why neutrino masses are so small, and what might unify the gauge forces, we still have not yet answered the supposedly more readily accessible problem of electroweak symmetry breaking. Just what, exactly, gives mass to the weak gauge bosons and the known fermions? Is it weakly-coupled and spontaneous, involving fundamental scalars, or strongly-coupled, involving composite scalars? Is the flavor problem linked? Do we discover the physics behind dark matter (and *its* mass), gauge unification and flavor at the same time? Or are those disconnected problems?

Our starting point is unitarity, the conservation of probability: the weak interaction of the Standard Model (SM) of particle physics violates it at about 1 TeV [1]. The theory demands at least one new propagating scalar state with gauge coupling to weak bosons to keep this under control. The same problem holds for fermion–boson interactions [2–5], only at much higher energy, so is generally less often discussed.[a] While the variety of explanations for electroweak symmetry breaking (EWSB) is vast, what we call the Standard Model (SM) assumes the existence of a single fundamental scalar field which spontaneously acquires a vacuum expectation value to generate all fermion and boson masses. It is a remarkably compact and elegant explanation, simple in the extreme. Yet while it tidies up the immediate necessities of the SM, it suffers from glaring theoretical pathologies that drive much of the model-building behind more ambitious explanations.

Numerous lectures and review articles already exist, covering the SM Higgs sector and the minimal supersymmetric (MSSM) extension [6–9], which are useful both for learning nitty-gritty theoretical details and serving as formulae references. These lectures are instead a crash-course tour of theory in practical application: previous, present and planned Higgs searches, what happens after a candidate Higgs discovery, and an overview of MSSM Higgs phenomenology as a perturbation of that for SM Higgs. They are not comprehensive, but do provide a solid grounding in the basics of Higgs hunting. They should be read only after one has become intimate with the SM Higgs sector and its underlying theoretical issues. Within TASI 2006, this means you should already have studied Sally Dawson's lectures. After both of these you should also be able to explain to your friends how we look for a Higgs boson at colliders (if they care), how to confirm it's a Higgs and figure out what variety it is (since we care), and describe how some basic extensions to the SM Higgs sector behave as a function of their parameter space (nature might not care for the SM).

Herein I'll assume that nature prefers fundamental scalars and spontaneous symmetry breaking. This is a strong bias, but one that provides a solid framework for phenomenology. The ambitious student who wants to really learn all the varieties of EWSB should also study strong dynamics [16], dimensional deconstruction [17], extra-dimensional Higgsless constructions [18] and the Little Higgs [19] and Twin Higgs mechanisms [20]. In many of these classes of theories the Higgs sector appears to be very SM-like, but in some no Higgs appears and one instead would pay great attention to weak boson scattering around a TeV.

---

[a]The original study [2] was clearly incorrect, but the correct line of reasoning is a work in progress [3–5].

8.2. Collider searches for the Standard Model Higgs

Even though the SM Higgs sector doesn't explain flavor (why all the fermion masses are scattered about over 12 orders of magnitude in energy) and has a disconcerting radiative stability problem that surely must involve new physics beyond the SM, it's a suitable jumping-off point for formulating Higgs phenomenology. That is, the study of physical phenomena associated with a theory, exploring the connection between theory and experiment. Without this connection, experiments would not make sense and theory would flail about, untested. To survey SM Higgs collider physics we need to recall a few fundamentals about the SM Higgs boson.

1. The Higgs boson unitarizes weak boson scattering, $VV \to VV$, so its interaction with weak bosons is very strictly defined to be the electroweak gauge coupling times the vacuum expectation value (vev); i.e., proportional to the weak boson masses.
2. The Higgs also unitarizes $VV \to f\bar{f}$ scattering, so its fermion couplings (except ν_i) are proportional to the fermion mass, with a strictly defined universal coefficient.
3. Because of the coupling strengths, the Higgs is dominantly produced by or in association with massive particles (including loop-induced processes, as we'll see in Sec. 8.2.1.1), and prefers to decay to the most massive particles kinematically allowed.
4. The Higgs boson mass itself is a free parameter[b], but influences EW observables, so we can fit EW precision data to make a prediction for its mass.

We may thus define the SM Higgs sector by its vacuum expectation value, v, measured via M_W, G_F, etc., and the known electroweak gauge couplings; 9 Yukawa couplings (fermion mass parameters, ignoring neutrinos and CKM mixing angles); and one free parameter, M_H.

Prior to the Large Electron Positron (LEP) collider era starting around 1990, Higgs searches involved looking for resonances amongst the low energy hadronic spectra in e^+e^- collisions. These were in fact non-trivial searches, mostly involving decays of hadrons to Higgs plus a photon, but are generally regarded as comprehensive and set a lower mass bound of $M_H \gtrsim 3$ GeV.

Higgs hunting in the 1990s was owned by LEP, an e^+e^- collider at CERN which steadily marched up in energy over the decade. It found no

---

[b]We know it is not massless, due to the absence of additional long-range forces.

Higgs bosons[c]. Attention then turned to the long-delayed Tevatron Run II program, proton–antiproton collisions at 2 TeV, which got off to a shaky start but is now performing splendidly. It so far sees nothing Higgs-like, either, but has not yet gathered enough data to be able to say much. The proton–proton Large Hadron Collider (LHC) at CERN is also many years behind schedule, but its construction is now nearing completion and we may expect physics data within a few years.

Our survey begins with LEP from a historical perspective and some general statements about Higgs boson behavior as a function of its mass. Next we turn our attention to the ongoing Tev2 search, for which the prospects hinge critically on machine performance. Then we delve into the intricacies of LHC Higgs pheno, which is far more complicated than either LEP or Tevatron, yet essentially guarantees an answer to our burning questions.

8.2.1. *The LEP Higgs search*

An obvious question to ask is, can we produce the Higgs directly in e^+e^- collisions? We could then probe Higgs masses up to our machine energy, which for LEP-II eventually reached 209 GeV. Recalling that the Higgs–electron coupling is proportional to the electron mass, which is quite a bit smaller than the electroweak vev of 246 GeV, the coupling strength is about 1.5×10^{-6}, or teeny-tiny in technical parlance. A quick calculation reveals that it would take about 4 years running full-tilt to produce just one Higgs boson. This one event would have to be distinguished from the general scattering cross section to fermion pairs in the SM, which is beyond hopeless.

Instead, we think of what process involves something massive, with vastly larger Higgs coupling, so that the interaction rate is large enough to produce a statistically useful number of Higgs bosons. The two obvious possibilities are $e^+e^- \to W^+W^-H$ (two W's required for charge conservation) and $e^+e^- \to ZH$. The first process will obviously have less reach in M_H as the two W bosons require far more energy than a single Z boson to produce. LEP Higgs searches therefore focused on the latter process, shown as a Feynman diagram in Fig. 8.1: the electron and positron annihilate to form a virtual Z, far above its mass shell, which returns on-shell by spitting off a Higgs boson. This process is generically known as Higgsstrahlung, analogous to bremsstrahlung radiation. Both the Higgs and

---

[c]This may be a somewhat controversial statement, depending on what lunch table you're sitting at. See Sec. 8.2.1.3.

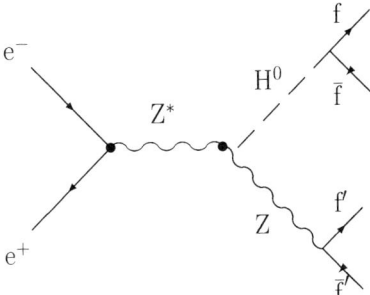

Fig. 8.1. Feynman diagram for the process $e^+e^- \to ZH$ with subsequent Higgs and Z boson decays to fermion pairs. All LEP Higgs searches were based primarily on this process, with various fermion combinations in the final state composing the different search channels.

Z immediately decay to an asymptotic final state of SM particles. For the Higgs this is preferentially to the most massive kinematically-allowed pair, while Z decays are governed by the fermion gauge couplings.[d] In brief, the Z decays 70% of the time to jets, 20% of the time invisibly (to neutrinos, which the detectors can't see), and about 10% to charged leptons, which are the most distinctive, "clean" objects in a detector.

8.2.1.1. *Momentary diversion: Higgs decays*

What, precisely, are the Higgs branching ratios (BRs)? To find these, we first need the Higgs partial widths; that is, the inverse decay rates to each final state kinematically allowed. Everyone should calculate these once as an exercise.

Let's start with the easiest case: Higgs decay to fermion pairs, which is a very simple matrix element. The general result at tree-level is:

$$\Gamma_{f\bar{f}} = \frac{N_c G_F m_f^2 M_H}{4\sqrt{2}\pi} \beta^3 \quad \text{where } \beta = \sqrt{1 - \frac{4m_f^2}{M_H^2}} \quad (8.1)$$

One factor of the fermion velocity β comes from the matrix element and two factors come from the phase space. I emphasize that this is at tree-level because there are significant QCD corrections to colored fermions. The bulk of these corrections are absorbed into a running mass (see Ref. [8]). For

---
[d]See the PDG [21] for Z boson branching ratios, which you should memorize.

calculations we should always use $m_q(M_H)$, the quark mass renormalized to the Higgs mass scale, rather than the quark pole mass. Programs such as HDECAY [22] will calculate these automatically given SM parameter inputs, greatly simplifying practical phenomenology.

Note that the partial width to fermions is linear in M_H, modulo the cubic fermion velocity dependence, which steepens the ascent with M_H near threshold. Partial widths for various Higgs decays are shown in Fig. 8.2. While the total Higgs width above fermion thresholds grows with Higgs mass, Higgs total widths below W pair threshold are on the order of tens of MeV – quite narrow. The only complicated partial width to fermions is that for top quarks, for which we must treat the fermions as virtual (at least near threshold) and use the matrix elements to the full six-fermion final state, integrated over phase space. This is slightly more complicated, but easily performed numerically.

Before the decay to top quarks is kinematically allowed, however, the decays to weak bosons turn on. A few W/Z widths above threshold the W and Z may be treated as on-shell asymptotic final states, making

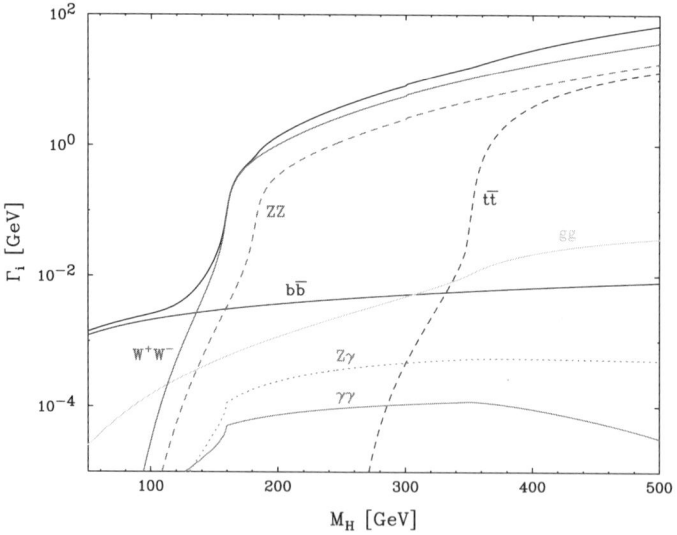

Fig. 8.2. Select Standard Model Higgs boson partial widths, as a function of mass, M_H. Individual partial widths are labeled, while the total width (sum of all partial widths, some minor ones not shown) is the black curve. Widths calculated with HDECAY [22].

the partial width calculation easier. We find:

$$\Gamma_{VV} = \frac{G_F M_H^3}{16\sqrt{2}\pi} \delta_V \beta \left(1 - x_V + \frac{3}{4}x_V^2\right) \quad \text{where} \quad \begin{cases} \delta_{W,Z} = 2, 1 \\ \beta = \sqrt{1-x_V} \\ x_V = \frac{4M_V^2}{M_H^2} \end{cases} \quad (8.2)$$

The factor of β comes from phase space, while the matrix elements give the more complicated function of x_V. The partial width is dominantly cubic in M_H, although the factors of beta and x_V enhance this somewhat near threshold, as in the fermion case. We can see this in Fig. 8.2: the partial widths to VV gradually flatten out to cubic behavior above threshold. The reason for this stronger M_H dependence compared to fermions is that a longitudinal massive boson wavefunction is proportional to its energy in the high-energy limit, which enhances the coupling by a factor E/M_V. (Recall that it is this property of massive gauge bosons that requires the Higgs, lest their scattering amplitude rise as E^2/M_V^2, violating unitarity. The Higgs in fact generates the longitudinal modes.) This much stronger dependence on M_H leads to a very rapid total width growth with M_H, which reaches 1 GeV around $M_H = 190$ GeV. We'll return to this when discussing Higgs couplings measurements in Sec. 8.3.3. The bottom line is that bosons "win" compared to fermions. Thus, even though the top quark has a larger mass than W or Z, it cannot compete for partial width and thus BR. Note that the partial widths to VV are non-trivial below threshold: the W and Z are unstable and therefore have finite widths; they may be produced off-shell. The Higgs can decay to these virtual states because its coupling is proportional to the daughter pole masses (or, in the case of quarks, the running masses), not the virtual q^2, which can be much smaller. Below threshold the analytical expressions are known [23] (see Ref. [7] for a summary), but are not particularly insightful to derive as an exercise.

The astute reader will have noticed by now that Fig. 8.2 contains curves for Higgs partial widths to *massless* final states! (Have another look if you didn't notice.) We know the Higgs couples to particles proportional to their masses, so this requires some explanation. Recall that loop-induced transitions can occur at higher orders in perturbation theory. Such interactions typically are important to calculate only when a tree-level interaction doesn't exist. They are responsible for rare decays of various mesons, for instance, and are in some cases sensitive to new physics which may appear in the loop. Here, we consider only SM particles in the loop. Which ones

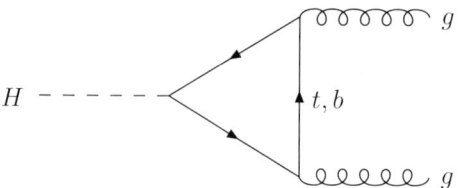

Fig. 8.3. Feynman diagram for the loop-induced process $H \to gg$ in the SM. All quarks enter the loop, but contribute according to their Yukawa coupling squared (mass squared). In the SM, only the top quark is important.

are important? Recall also once again that the Higgs boson couples proportional to particle mass. Thus, the top quark and EW gauge bosons are most important. For $H \to gg$, then, that means only the top quark, while for $H \to \gamma\gamma$ it is both the top quark and W loops (there is no $ZZ\gamma$ vertex). The $H \to gg$ expression (for the Feynman diagram of Fig. 8.3) is [24]:

$$\Gamma_{gg} = \frac{\alpha_s^2 G_F M_H^3}{16\sqrt{2}\,\pi^3} \left| \sum_i \tau_i [1 + (1-\tau_i) f(\tau_i)] \right|^2 \quad (8.3)$$

with $\tau_i = \dfrac{4 m_f^2}{M_H^2}$ and $f(\tau) = \begin{cases} \left[\sin^{-1}\sqrt{1/\tau}\right]^2 & \tau \geq 1 \\ -\frac{1}{4}\left[\ln\frac{1+\sqrt{1-\tau}}{1-\sqrt{1-\tau}} - i\pi\right]^2 & \tau < 1 \end{cases}$ (8.4)

which is for a general quark in the loop with SM Yukawa coupling. It's easy to see that in the SM the b quark contribution, which is second in size to that of the top quark, is inconsequential. Remember to use the running mass $m_f(M_H)$ to take into account the largest QCD effects. When you derive this expression yourself as an exercise, take care to solve the loop integral in $d > 4$ dimensions, otherwise you miss a finite piece. The $H \to \gamma\gamma, Z\gamma$ expressions have a similar form [25], but with two loop functions, since it can also be mediated a W boson loop (which interferes destructively with the top quark loop!):

$$\Gamma_{\gamma\gamma} = \frac{\alpha^2 G_F M_H^3}{128\sqrt{2}\,\pi^3} \left| \sum_i N_{c,i} Q_i^2 F_i \right|^2 \quad (8.5)$$

$F_1 = 2 + 3\tau[1 + (2-\tau)f(\tau)], \quad F_{1/2} = -2\tau[1 + (1-\tau)f(\tau)], \quad F_0 = \tau[1 - \tau f(\tau)]$ (8.6)

where $N_{c,i}$ is the number of colors, Q_i the charge, and F_j the particle's spin.

Now look again more closely at Fig. 8.2. The important feature to notice is that these loop-induced partial widths are ostensibly proportional to M_H^3, like the decays to gauge bosons. However, the contents of the brackets, specifically the $f(\tau_i)$ function, can alter this in non-obvious ways. For $H \to gg$, Fig. 8.2 shows a slightly more than cubic dependence at low masses, leveling of to approximately M_H^3, and flattening out to approximately quadratic a bit above the top quark pair threshold. We see from Eq. 8.3 that the functional form changes at that threshold, albeit fairly smoothly, by picking up a constant imaginary piece when the top quarks in the loop can be on-shell.

The partial widths to $\gamma\gamma$ and $Z\gamma$ behave very differently than gg. For M_H below W pair threshold, the interference between top quark and W loops produces an extremely sharp rise with M_H, which transitions to something slightly more than linear in M_H at W pair threshold where the W bosons in the loop go on-shell. There is is a smoother transition at the top quark pair threshold, where they can similarly go on-shell. The $\gamma\gamma$ and $Z\gamma$ partial widths behave differently because of the different $t\bar{t}\gamma$ and $t\bar{t}Z$ couplings: the partial width to $Z\gamma$ at large M_H is almost a constant, but falls off for $\gamma\gamma$ almost inverse cubic in M_H.

Once we've calculated all the various possible partial widths, we sum them up to find the Higgs total width. Each BR is then simply the ratio Γ_i/Γ_{tot}. These are shown in Fig. 8.4; note the log scale. If it wasn't obvious from the partial width discussion, it should be now: near thresholds, properly including finite width effects can be very important to get the BRs correct. Observe how the BR to WW^* (at least one W is necessarily off-shell) is 50% at $M_H = 140$ GeV, 20 GeV below W pair threshold. $\text{BR}(H \to b\bar{b}) \sim \text{BR}(H \to W^+W^-)$ at $M_H = 136$ GeV.

8.2.1.2. *A brief word on statistics – the simple view*

Now that we understand the basics of Higgs decay, and production in electron-positron collisions, we should take a moment to consider statistics. The reason we must resort to statistics is that particle detectors are imperfect instruments. It is impossible to precisely measure the energy of all outgoing particles in every collision. The calorimeters are sampling devices, which means they don't capture all the energy; rather they're calibrated to give an accurate central value at large statistics, with some Gaussian uncertainty about the mean for any single event. Excess energy can also appear, due to cosmic rays, beam–gas or beam secondary interactions. Quark final

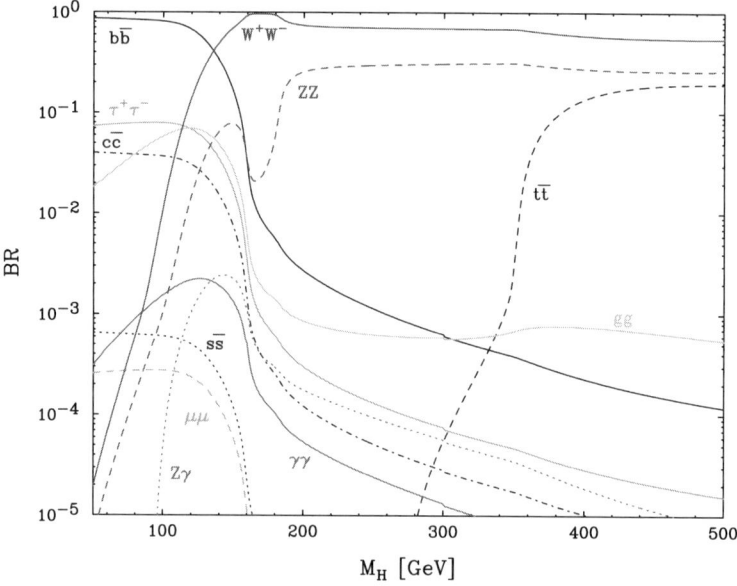

Fig. 8.4. Select Standard Model Higgs boson branching ratios as a function of mass, M_H [22]. The Higgs prefers to decay to the most massive possible final state. The ratio of fermionic branching ratios are proportional to fermion masses squared, modulo color factors and radiative corrections.

states hadronize, resulting in the true final state in the detector (a jet) being far more complicated and difficult even to identify uniquely. The electronics can suffer hiccups, and software *always* has bugs, leading to imperfect analysis. Thus, we would never see two or three events at precisely the Higgs mass of, say, 122.6288... GeV, and pop the champagne. Rather, we'll get a distribution of masses and have to identify the central value and its associated uncertainty.

In any experiment, event counts are quantum rolls of the dice. For a sufficient number of events, they also follow a Gaussian distribution about the true mean:

$$f(x; \mu, \sigma) = \frac{1}{\sigma\sqrt{2\pi}} \exp\left(-\frac{(x-\mu)^2}{2\sigma^2}\right) \quad (8.7)$$

The statistical uncertainty in the rate then goes as $1/\sqrt{N}$, where N is the number of events. This is "one sigma" of uncertainty: 68.2% of identically-conducted experiments would obtain N within $\sigma \approx \pm\sqrt{N}$ about $\mu = N_{\text{true}}$, representing the true cross section. Figure 8.5 shows the fractional prob-

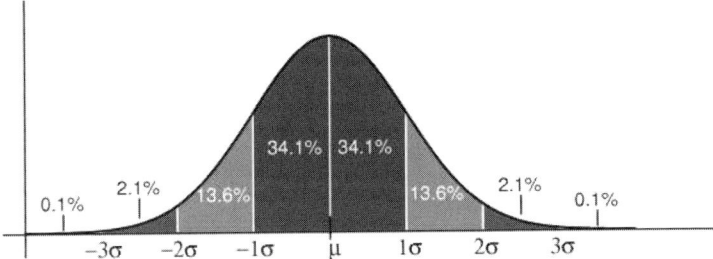

Fig. 8.5. Gaussian distribution about a mean μ, showing the fractional probability of events within one, two and three standard deviations of the mean.

abilities for various "sigma", or number of standard deviations from the true mean. To claim observation of a signal deviating from our expected background, we generally use a 5σ criteria for discovery. This means, if systematic errors have been properly accounted for, that there is only a 0.00006% chance that the signal is due to a statistical fluctuation. However, this threshold is subjective, and you will often hear colleagues take 4σ or even 3σ deviations seriously. Since particle physics has seen dozens of three sigma deviations come and go over the decades, I would encourage you to regard 3σ as "getting interesting", and 4σ as "pay close attention and ask lots of questions about systematics".

Because SM processes can produce the same final state as any ZH combined BR, we must know accurately what the background rate is for each signal channel (final state) and how it is distributed in invariant mass, then look for a statistically significant fluctuation from the expected background over a fixed window region. The size of the window is determined by detector resolution: the better the detector, the narrower the window, so the smaller the background, yielding a better signal-to-background rate. Generally, the window is adjusted to accept one or two standard deviations of the hypothesized signal (68–95%).

Analyses are then defined by two different Gaussians: that governing how many signal (and background) events were produced, and that parameterizing the detector's measurement abilities. The event count N in our above expression is the actual number of events observed, in an experiment. But in performing calculations ahead of time for expected signal and background, it is variously taken as just B, the number of background events expected, or $S + B$, expected signal included, depending on the relative sizes of S and B. For doing phenomenology, trying to decide which signals to study and calculate more precisely, the distinction is often ignored.

The statistical picture I've outlined here is quite simplified. Not all experiments have sufficient numbers of events to describe their data by Gaussians – Poisson statistics may be more appropriate. (An excellent text on statistics for HEP is Ref. [26].) Not all detector effects are Gaussian-distributed. Nevertheless, it gets across the main point: multiple sources of randomness introduce a level of uncertainty that must be parameterized by statistics. Only when the probability of a random background fluctuation up or down to the observed number of events is small enough, perhaps in some distribution, can signal observation be claimed. Exactly where this line lies is admittedly a little hazy, but there's certainly a point of several sigmas at which everybody would agree.

8.2.1.3. *LEP Higgs data and results*

Now to the actual LEP search. Electrons and positrons have only electroweak interactions, so backgrounds and a potential Higgs signal are qualitatively of the same size. (We'll see shortly in Sec. 8.2.2 how this is not so at a hadron collider, which has colored initial states.) LEP thus had the ability to examine almost all Z and H decay combinations: $b\bar{b}jj$, $b\bar{b}\ell^+\ell^-$, $b\bar{b}\nu\bar{\nu}$, $\tau^+\tau^-jj$, $jjjj$, etc. The largest of these is $b\bar{b}jj$, as it combines the largest BRs of both the Z and H. It's closely followed by $b\bar{b}\nu\bar{\nu}$, since a Z will go to neutrinos 20% of the time. Neutrinos are missing energy, however, so not precisely measured, making it possible that any observed missing energy didn't in fact come from a Z. Jets are much less well-measured than leptons, so a narrower mass window can be used for the Z in $b\bar{b}\ell^+\ell^-$ events than $b\bar{b}jj$; the smaller backgrounds in the narrower window might beat the smaller statistics of the leptonic final state.

The exact details of each LEP search channel are not so important, as lack of observation means we're more interested in channels' signal and background attributes at hadron colliders. For these lectures I just present the final LEP result combining all four experiments. The interested student should read Eilam Gross' "Higgs Statistics for Pedestrians", which goes into much more depth, and with wonderful clarity [27].

The money plot is shown in Fig. 8.6. It shows the expected confidence level (CL) for the signal+background hypothesis as a function of Higgs mass. The thin solid horizontal line at CL=0.05 signifies a 5% probability that a true signal together with the background would have fluctuated down

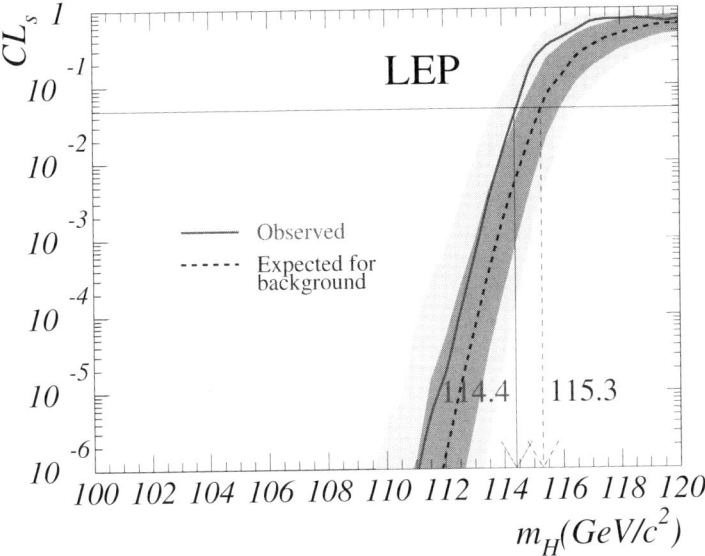

Fig. 8.6. Four-experiment combined result of the LEP Standard Model Higgs search. No signal was observed, establishing a lower limit of 114.4 GeV. See text of Ref. [27] for explanation.

in number of events to not be discriminated from the expected background. The green and yellow regions are the 1σ and 2σ *expected* uncertainty bands as a function of M_H, taking into account all sources of uncertainty, calculational as well as detector effects. Where the central value (dashed curve) crosses 0.05 defines the 95% CL expected exclusion (lower mass limit). This is essentially the available collision energy minus the Z mass minus a few extra GeV to account for the Z finite width – it may be produced slightly off-shell with some usable rate. The solid red curve is the actual experimental result, which is slightly above the experimental result everywhere, meaning that the experiments gathered a couple more events than expected in the 115-116 GeV mass bin.

The end of LEP running involved a certain amount of histrionics. At first, the number of excess event at the kinematic machine limit was a few, but more careful analyses removed most of these. For example, one particularly notorious event originally included in one experiment's analysis had more energy than the beam delivered. Another experiment removed a candidate event because some of the outgoing particles traveled down a poorly-instrumented region of the detector which was not normally used in

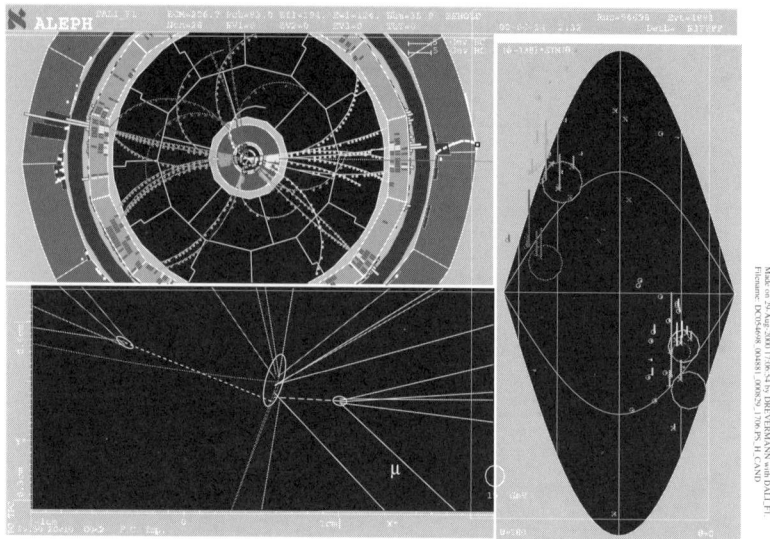

Fig. 8.7. Event display of an interesting candidate $Z \to j\bar{j}$, $H \to b\bar{b}$ event in the Aleph detector at the end of LEP-II running and at the machine's kinematic limit [28].

analysis. The final, most credible enumeration was one candidate event in one experiment, show in Fig. 8.7.

8.2.2. Prospects at Tevatron

With the end of the LEP era, all eyes turned to Run II of the upgraded Fermilab Tevatron. Its energy increased from 1.8 to 1.96 GeV, and is expected to gather many tens of times the amount of data in Run I. Higgs-hunting hopes were high [29], although it was clear that the machine and both detectors have to perform exceptionally well to have a chance, as Tevatron's Higgs mass reach will not be all that great, and will have significant observability gaps in the mass region expected from precision EW data.

To understand the details and issues, we first need to identify how a Higgs boson may be produced in proton-antiproton collisions. Like the electron, the light quarks have too small a mass (Yukawa coupling) to produce a Higgs directly with any useful rate, discernible against the large QCD backgrounds produced in hadron collisions.[e] Quarks may annihilate, however, to EW gauge bosons, which have large coupling to the Higgs; and

---

[e]For example, $H \to b\bar{b}$ is the dominant BR of a light Higgs, but QCD b jet pair production in hadron collisions is many orders of magnitude larger. Cf. Fig. 8.10.

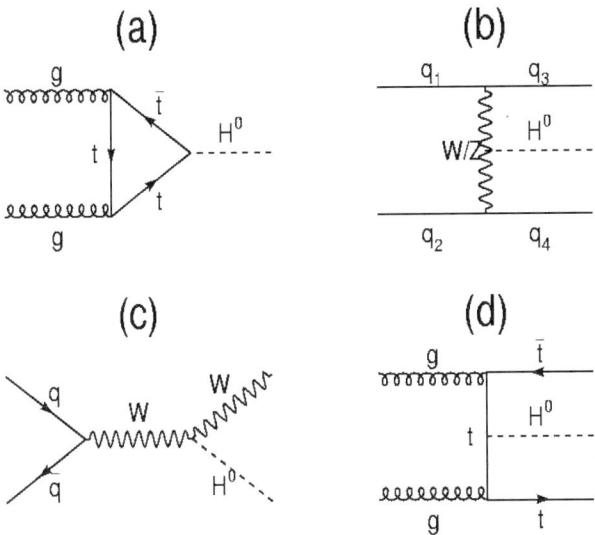

Fig. 8.8. Feynman diagrams for the four dominant Higgs production processes at a hadron collider.

likewise to a top quark pair. Incoming quarks may also emit a pair of gauge bosons which fuse to form a Higgs, a process known as weak boson fusion (WBF). But high energy protons also possess a large gluon content; recall that gluons have a loop-induced coupling to the Higgs. Figure 8.8 displays Feynman diagrams for all four of these processes at hadron colliders. The questions are, what are their relative sizes, and what are their backgrounds? Because of the partonic nature of hadron collisions, the Higgs couplings are not enough to tell us the relative sizes; we also need to take into account incoming parton fluxes and final state phase space – single Higgs production is much less greedy than $t\bar{t}H$ associated production, for instance. In addition, the internal propagator structure of the processes is important: WH, ZH bremsstrahlung are s-channel suppressed, but no other process is.

The various rates, updated in 2006 with the latest theoretical calculations [30, 31], are shown in Fig. 8.9 for a light SM Higgs boson. Students not already familiar with hadron collider Higgs physics will probably be surprised to learn that $gg \to H$, gluon fusion Higgs production, dominates at Tevatron energy. This is partly because the coupling is actually not all that small, partly because high-energy protons contain a plethora of gluons, and partly because there is no propagator suppression, and much less phase space suppression, compared to other processes. Higgsstrahlung

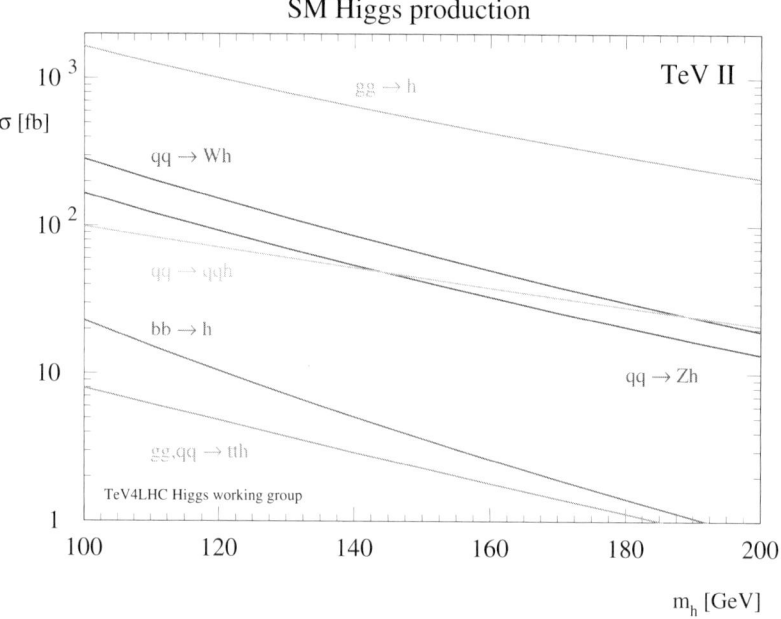

Fig. 8.9. Cross sections for Higgs production in various channels at Tevatron Run II ($\sqrt{s} = 2$ TeV). Note the log scale. Figure from the Tev4LHC Higgs working group [30].

(Fig. 8.8(c)) is still important at Tevatron, analogous to LEP. Note that the smaller cross sections have more complicated final states, therefore potentially less background, and possibly distinctive kinematic distributions that could assist in separating a signal from the background. It's not obvious that the largest rate is the most useful channel! Considering that the Higgs decays predominantly to different final states as a function of its mass, it's also not obvious that the optimal channel at one mass is optimal for all masses. In fact, that's definitely not the case.

Not knowing the answer, we naturally start by considering the largest cross section times branching ratio, $gg \to H \to b\bar{b}$. Just how large is the background, QCD $p\bar{p} \to b\bar{b}$ production? Figure 8.10 shows a variety of SM cross section for hadron collisions of various energy, and marks off in particular Tevatron and LHC. (The discontinuity in some curves is because Tevatron is $p\bar{p}$ and LHC is pp.) We immediately notice that the $b\bar{b}$ inclusive rate is almost nine orders of magnitude larger than inclusive $H \to b\bar{b}$. Of course the background will be smaller in a finite window about the Higgs mass. But jets are not so well-measured, necessitating a fairly large window,

proton - (anti)proton cross sections

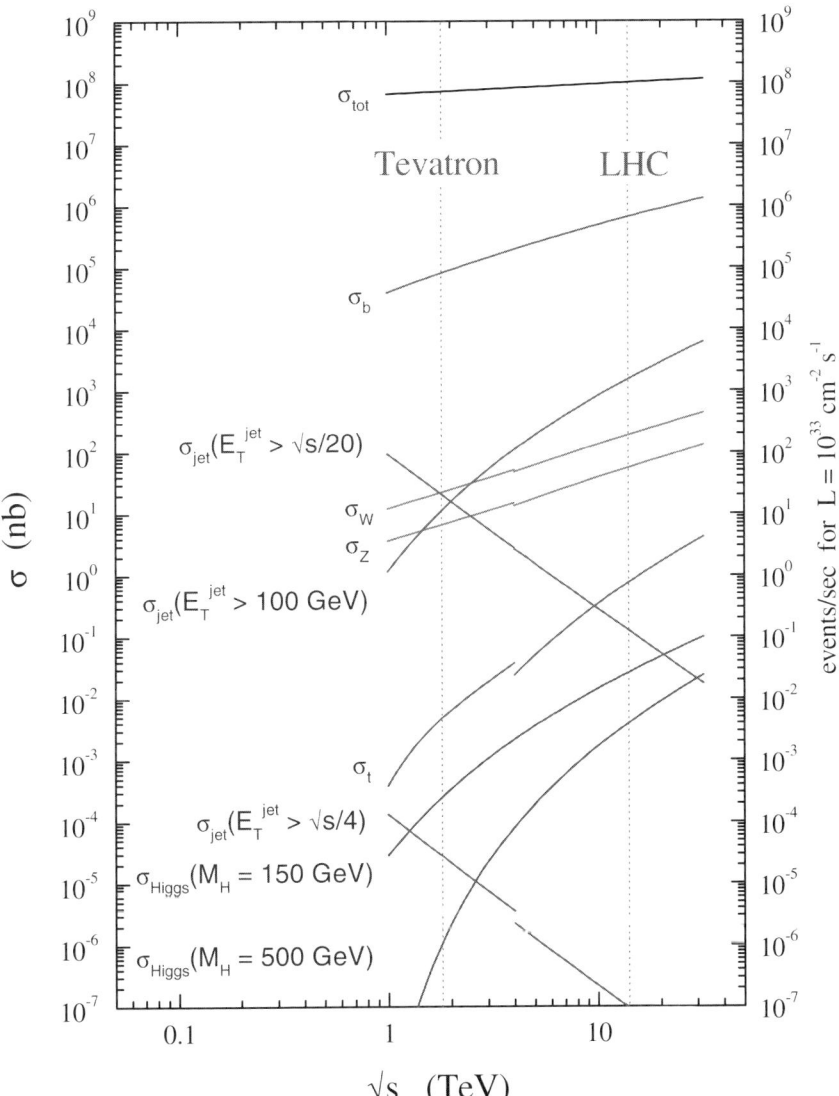

Fig. 8.10. Various SM "standard candle" cross sections at hadron colliders of varying energy, with Tevatron and LHC marked in particular. Note the log scale. Discontinuities are due to the difference between $p\bar{p}$ for Tevatron and pp for LHC. Figure from Ref. [32].

~15–20 GeV either side of the central value. We lose only a few orders of magnitude of the background, taking us from "laughable" to just terminally hopeless.

The general rule of thumb at hadron collider experiments is to require a final state with at least one high-energy lepton. This means lower backgrounds because the event had at least some EW component, such as a W or Z, or came from a massive object, such as the top quark, which is not produced in such great abundance due to phase space suppression.

Tevatron's Higgs search is rate-limited. We can see this by multiplying the 150 GeV Higgs cross section from Fig. 8.9 by the expected integrated luminosity of 4–8 fb^{-1} during Run II. Because of this, and the very low efficiency of identifying final-state taus in a hadron collider environment (unlike at LEP), Tevatron's experiments CDF and DØ focus on the $H \to b\bar{b}$ final state where that decay dominates the BR, and Higgsstrahlung to obtain the lepton tag. For larger Higgs masses, where $H \to W^+W^-$ dominates, gluon fusion Higgs production is the largest rate, but Higgsstrahlung has some analyzing power. To summarize [29]:

$M_H \lesssim 140$ GeV: $H \to b\bar{b}$ dominates, so we use:

- $WH \to \ell^\pm \nu b\bar{b}$
- $ZH \to \ell^+\ell^- b\bar{b}$
- $WH, ZH \to jjb\bar{b}$
- $ZH \to \nu\bar{\nu}b\bar{b}$

$M_H \gtrsim 140$ GeV: $H \to W^+W^-$ dominates, so we use:

- $gg \to H \to W^+W^-$ (dileptons)
- $WH \to W^\pm W^+W^-$ (2ℓ and 3ℓ channels)

8.2.2.1. $VH, H \to b\bar{b}$ at Tevatron

While a lepton tag gets rid of most QCD backgrounds, it doesn't automatically eliminate top quarks: they decay to Wb, thus the event often contains one lepton and two jets, or two leptons and missing energy, in addition to the b jet pair. This is the same final state as our Higgs signal, with either extra jets or transverse energy imbalance. Kinematic cuts help, but because the detectors are imperfect some top quark events will leak through. Jet mismeasurement gives fake missing energy, for example (and is one of the most difficult uncertainties to quantify in a hadron collider experiment). In addition, QCD initial-state radiation from the incoming partons can give extra jets. Thus top quark and Higgs signal events qualitatively become very similar. To control this further the experiments have to look at other observables, such as angular distributions of the b jets and leptons. Other backgrounds to consider are QCD $Wb\bar{b}$ production, weak bosons pairs where one decays to $b\bar{b}$ (and thus has invariant mass close to the Higgs signal window).

Figure 8.11 shows the results of a CDF simulation study of WH and ZH Higgsstrahlung events at Run II for M_H = 115 GeV (right at the LEP Higgs limit) [33]. First note how the top quark pair and diboson backgrounds peak very close to the Higgs mass. Eyeballing the plots and simplistically applying our knowledge of Gaussian statistics, we could easily believe that this could yield a four or five sigma signal, perhaps combined with DØ results. However, carefully observe that the shape of the invariant mass distribution for background alone and with signal are extremely similar: they are both steeply falling; the Higgs signal is not a stand-out peak above a fairly flat background. Therein lies a hidden systematic! This means that we must understand the kinematic-differential shape of the QCD backgrounds to a very high degree of confidence. This is not just knowing the SM background at higher orders in QCD, differentially, but also the detector response. This criticality is not appreciated in most discussions of a potential discovery at Tevatron. It should be obvious that an excess in one of these channels would cause a scramble of cross-checking and probably further theoretical work to ensure confidence, in spite of the statistics alone. We'll run into this feature again with one of the LHC channels in Sec. 8.2.3.1, but quantified.

CDF has in fact already observed an interesting candidate Higgs event in Run II, in the first few hundred pb^{-1}. It is in the $ZH \to \nu\bar{\nu}b\bar{b}$ channel (a b jet pair plus missing transverse energy). The event display and key kinematic information are shown in Fig. 8.12. Given the very low b jet pair invariant mass, it's much more likely that the event came from EW ZZ or QCD $Zb\bar{b}$ production (cf. Fig. 8.11). It therefore doesn't generate the kind of excitement that the handful of events at LEP did. Nevertheless, finding this event was a milestone, showing that CDF could perform such an analysis and find Higgs-like events with good efficiency.

Table 8.1 summarizes the 2000 Tevatron Higgs Working Group Report predictions for Higgsstrahlung reach in Run II [29]. The results are quoted for one detector and per fb^{-1}, hence the rather small significances. CDF and DØ will eventually combine results, giving a factor of two in statistics. However, it's not known how much data they'll eventually collect by 2009 or 2010, when LHC is expected to have first physics results and CDF & DØ

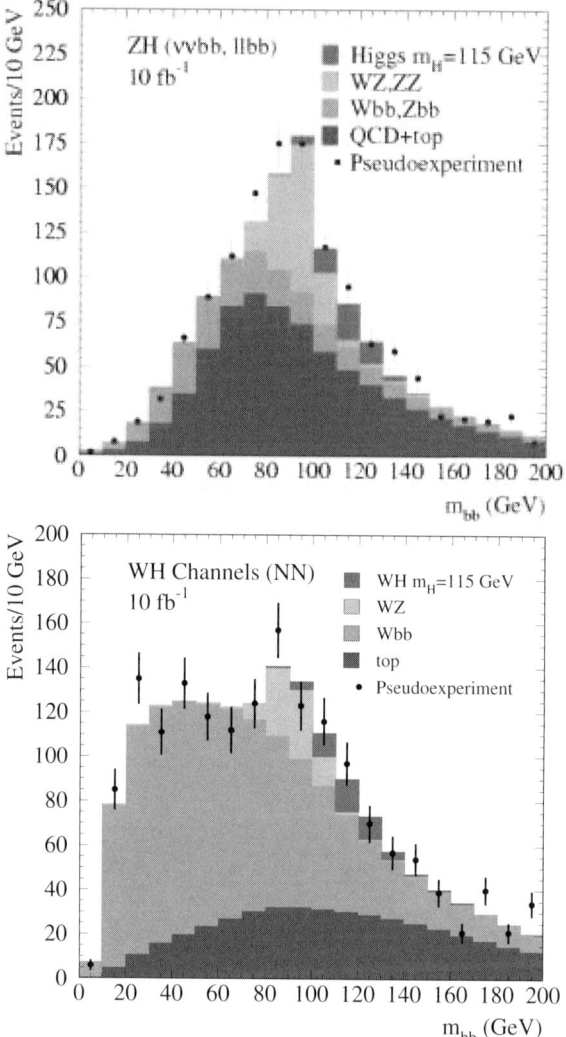

Fig. 8.11. CDF simulations of a 115 GeV signal at Tevatron Run II in ZH (left) and WH (right) Higgsstrahlung production with Higgs decays $H \to b\bar{b}$ and assuming 10 fb^{-1} is collected [33].

detector degradation becomes an issue. Fairly low Higgs masses are shown, because when the report was written nobody expected LEP to perform as well as it did, greatly exceeding its anticipated search reach. It should be obvious that a clear discovery would require a large amount of data,

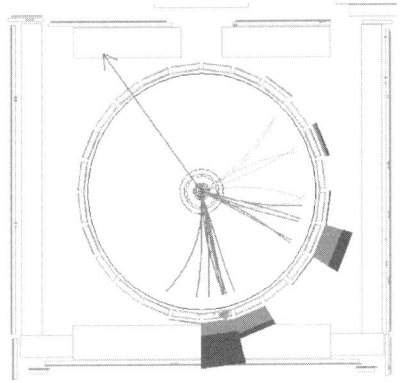
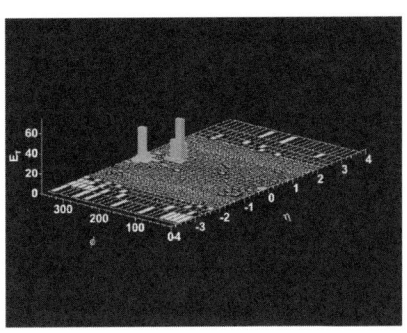

Two b-tagged jets

Jet_1 E_T= 100.3 GeV
Jet_2 E_T= 54.7 GeV

m_{jj} = 82 GeV

Missing E_T=145 GeV

Could be ZZ

Fig. 8.12. Interesting $bb\not{p}_T$ event at CDF in Tevatron Run II [34].

Table 8.1. Predicted signal significances at Tevatron Run II, for one detector and 1 fb^{-1}, for various VH, $H \to b\bar{b}$ searches, taken from Ref. [29].

Channel	Rate	Higgs Mass (GeV/c^2)				
		90	100	110	120	130
$\ell^{\pm}\nu b\bar{b}$	S	8.7	9.0	4.8	4.4	3.7
	B	28	39	19	26	46
	$S/\sqrt{B}$	1.6	1.4	1.1	0.9	0.5
$\nu\nu b\bar{b}$	S	12	8	6.3	4.7	3.9
	B	123	70	55	45	47
	$S/\sqrt{B}$	1.1	1.0	0.8	0.7	0.6
$\ell^+\ell^- b\bar{b}$	S	1.2	0.9	0.8	0.8	0.6
	B	2.9	1.9	2.3	2.8	1.9
	$S/\sqrt{B}$	0.7	0.7	0.5	0.5	0.4
$q\bar{q}b\bar{b}$	S	8.1	5.6	3.5	2.5	1.3
	B	6800	3600	2800	2300	2000
	$S/\sqrt{B}$	0.10	0.09	0.07	0.05	0.03

combining multiple channels, and the Higgs boson happening to be fairly light; not to mention the QCD shape systematic concern I described earlier (but is not quantified). In spite of this apparent pessimism, however, CDF and DØ seem to be performing modestly better than expected – higher efficiencies for b tagging and phase space coverage, better jet resolution, etc. There is as yet no detailed updated report with tables such as this, but there are some newer graphically-presented expectations I'll show as a summary.

8.2.2.2. $gg \to H \to W^+W^-$ at Tevatron

For $M_H \gtrsim 140$ GeV, a SM Higgs will decay mostly to W pairs (cf. Fig. 8.4), which has a decent rate to dileptons and has very little SM background – essentially just EW W pair production, with some background from top quark pairs where both b jets are lost. This channel has some special characteristics due to how the Higgs decay proceeds. There is a marked angular correlation between the outgoing leptons which differs from the SM backgrounds: they prefer to be emitted together, that is close to the same flight direction in the center-of-mass frame [35].

To understand this correlation, consider what happens if the Higgs decays to a pair of transversely-polarized W bosons. For W decays, the lepton angle with respect to the $W^\pm$ spin follows a $(1 \pm \cos\theta_{\ell\pm})^2$ distribution. That is, the positively-charged lepton prefers to be emitted with the W spin, while the negatively-charged lepton prefers to be emitted opposite the W spin. Since the Higgs is a scalar (spin-0), the W spins are anti-correlated, thus the leptons are preferentially emitted in the same direction. For longitudinal W bosons, the lepton follows a $\sin^2\theta_\ell$ distribution. The W spins are still correlated, however, and the matrix element squared (an excellent exercise for the student) is proportional to $(p_{\ell^-} \cdot p_\nu)(p_{\ell^+} \cdot p_{\bar\nu})$. Since a charged lepton and neutrino are emitted back-to-back in the W rest frame, this is again maximized for the charged leptons emitted together. This correlation is shown visually by the schematic of Fig. 8.13. Projected onto the azimuthal plane (transverse to the beam), its efficacy is shown in Fig. 8.14 by comparison to various backgrounds [29, 36].

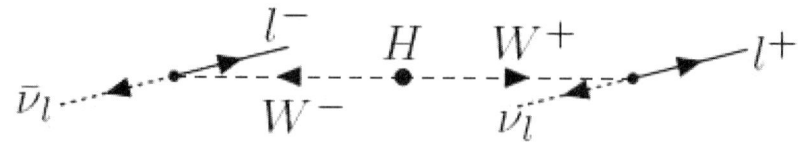

Fig. 8.13. Diagram showing the preferred flight direction of charged leptons in $H \to \ell^+\nu\ell^-\bar\nu$.

In addition to this angular correlation, we may also construct a transverse mass (M_T) for the system, despite the fact that two neutrinos go missing [37]. We first write down the transverse energy (p_T) of the dilepton and missing transverse energy ($\not{E}_T$) systems,

$$E_{T_{\ell^+\ell^-}} = \sqrt{\vec{p}_{T_{\ell^+\ell^-}}^2 + m_{\ell^+\ell^-}^2}, \qquad \not{E}_T = \sqrt{\vec{\not{p}}_T^2 + m_{\ell^+\ell^-}^2} \qquad (8.8)$$

where I've substituted the dilepton invariant mass $m_{\ell^+\ell^-}^2$ for $m_{\nu\bar{\nu}}^2$. This is exact at $H \to WW$ threshold, and is a very good approximation for Higgs masses below about 200 GeV and where this decay mode is open. The W pair transverse mass is now straightforward:

$$M_{T_{WW}} = \sqrt{(\not{E}_T + E_{T_{\ell^+\ell^-}})^2 - (\vec{p}_{T_{\ell^+\ell^-}} + \vec{\not{p}}_T)^2} \qquad (8.9)$$

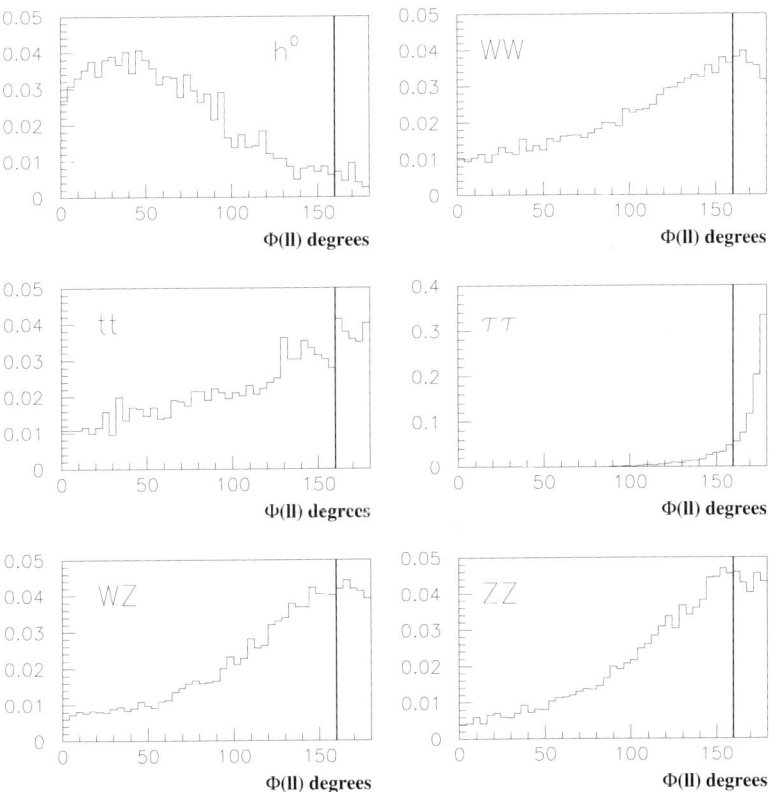

Fig. 8.14. Dilepton azimuthal angular correlation for a $H \to W^+W^- \to \ell^+\nu\ell^-\bar{\nu}$ signal and its backgrounds. The efficacy of the cut (vertical line) can easily be estimated visually. From the Tevatron Run II Higgs Working Group Report [29].

This gives a nice Jacobian peak for the Higgs signal, modulo detector missing-transverse-energy resolution, whereas the SM backgrounds tend to be comparatively flat.

Utilizing these techniques gives Tevatron some reach for a heavier Higgs boson, mostly in the mass range $150 \lesssim M_H \lesssim 180$ GeV, where the BR to WW is significant and the Higgs production rate is not too small.

8.2.2.3. *Tevatron Higgs summary expectations*

Tevatron Higgs physics expectations have changed since the 2000 Report, as DØ and CDF have better understood their detectors and made analysis improvements. As yet, the only progress summary is from 2003, shown in Fig. 8.15. It compares the original Report's findings, shown by the thick curves, with improved findings for the low-mass region, shown by the thinner lines. However, the new results do not yet include systematic uncertainties, which may be considerable. We should expect some form of a new summary expectation sometime in 2007. A final note on the undiscussed WBF production mode: some study has been done (see Sec. II.C.4 of Ref. [29]), but DØ and CDF both lack sufficient coverage of the forward region to use this mode. This is not the case at LHC.

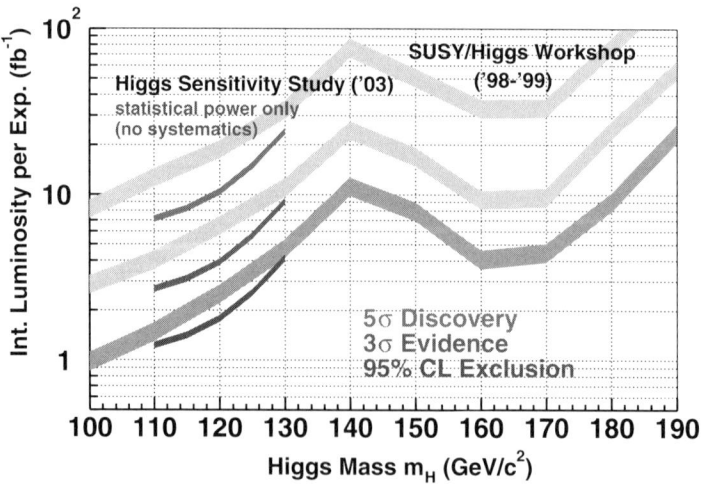

Fig. 8.15. Expected required integrated luminosity per experiment required in Run II to observe a SM Higgs as a function of M_H [33].

Run II now has about 1 fb^{-1} of analyzed data, and a Higgs search summary progress report is available in Ref. [38], which updates each channel's expectations.

8.2.3. *Higgs at LHC*

Higgs physics at LHC will be similar to that at Tevatron. There is the slight difference that LHC will be pp collisions rather than $p\bar{p}$. The biggest difference, however, is the increased energy, from 2 to 14 TeV. Particle production in the 100 GeV mass range will be at far lower Feynman x, where the gluon density is much larger than the quark density. In fact, it's useful (for Higgs physics) to think of the LHC as a gluon collider to first order. The ratio between gluon fusion Higgs production and Higgsstrahlung is thus larger than at Tevatron. Figure 8.16 displays the various SM Higgs cross sections, only over a much larger range of M_H – at LHC, large-M_H cross sections are not trivially small, compared to at the Tevatron. There are huge QCD corrections to the $gg \to H$ rate (also at Tevatron), but these

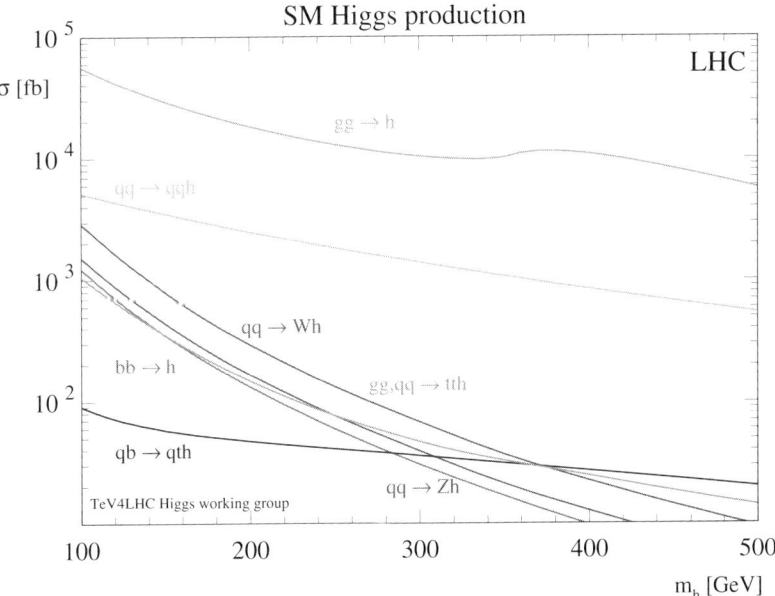

Fig. 8.16. Cross sections for Higgs production in various channels at LHC ($\sqrt{s} = 14$ TeV) [30].

are now known at NNLO and under control [39] (and included in Fig. 8.16). They don't affect the basic phenomenology, however. Knowing that LHC is plans to collect several hundred fb^{-1} of data, a quick calculation reveals that the LHC will truly be a Higgs factory, producing hundreds of thousands of light Higgs bosons, or tens of thousands if it's heavy.

Looking back at Fig. 8.10, we see that while the Higgs cross section rises quite steeply with collision energy ($gg \to H$ is basically a QCD process), so do important backgrounds like top quark production. The inclusive b cross section is still too large to access to $gg \to H \to b\bar{b}$, but note that the EW gauge boson cross sections do not rise as swiftly with energy. Immediately we realize that channels like $gg \to H \to W^+W^-$ should have a much better signal-to-background (S/B) ratio. (In fact it suffers from non-trivial single-top quark [40] and $gg \to W^+W^-$ [41] backgrounds, but is still an excellent channel for $M_H \gtrsim 150$ GeV.) The figure does not show cross sections like $Wb\bar{b}$ or $Zb\bar{b}$, which grow QCD-like and thus become a terminal problem for WH and ZH channels.

Obviously there are a few significant differences between Tevatron and LHC with implications for Higgs physics. We'll lose access to WH and ZH at low mass, at least for Higgs decay to b jets. What about rare decays, since the production rate is large? The $t\bar{t}H$ cross section is large and would yield a healthy event rate. It's complexity is distinctive, so one might speculate that perhaps it could be useful. WBF production is also accessible due to better detectors, and likewise its more complex signature is worthy of a look. It will in fact turn out to be perhaps the best production mode at LHC.

As with Tevatron, we need to understand both the signal and background for each Higgs channel we wish to examine. As a prelude to Chapter 8.3, Higgs measurements, at LHC we won't want to just find the Higgs in one mode. Rather, we'll want to observe it in as many production and decay modes as possible, to study all its properties, such as couplings.

8.2.3.1. $t\bar{t}H, H \to b\bar{b}$

Let's begin by discussing a very complex channel, top quark associated production at low mass, $t\bar{t}H, H \to b\bar{b}$. This was studied early on in the ATLAS TDR [42] and in various obscure CMS notes, and found to be a sure-fire way to find a light Higgs. Figure 8.17 shows a schematic of such an event, with multiple b jets from both top quarks and the Higgs, at least one lepton from a W for triggering, and possibly extra soft jets from QCD

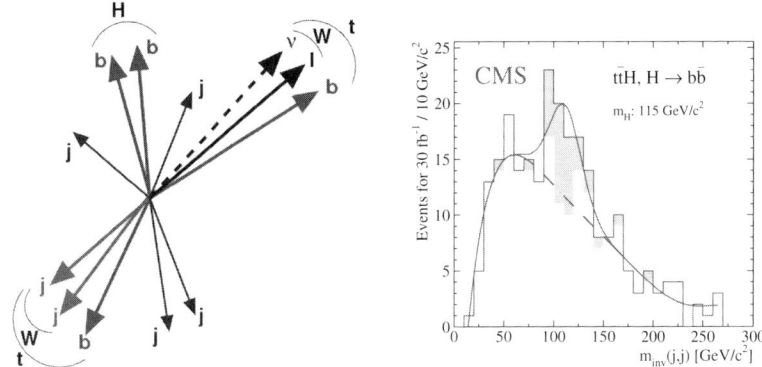

Fig. 8.17. Left: schematic of the outgoing particles in a typical $t\bar{t}H, H \to b\bar{b}$ event at LHC [47]. Right: early CMS study expectations for a $b\bar{b}$ mass peak in such events, for $M_H = 115$ GeV [44, 45].

radiation. The schematic is a bit fanciful in the neatness of separation of the decay products, but is useful to get an idea of what's going on.

These early studies [42–45] were too ambitious, however. The backgrounds to this signal are $t\bar{t}b\bar{b}$ and $t\bar{t}jj$[f] production, pure QCD processes. The extra (b) jets must be fairly energetic, or hard, because the signal is a 100+ GeV-mass object which decays to essentially massless objects. Despite this being a known problem [46], these backgrounds were calculated using the soft/collinear approximation for extra jet emission implemented in standard Monte Carlo tools such as PYTHIA or HERWIG. This greatly underestimated the backgrounds.

The left panel of Fig. 8.18 shows the results of a repeated study by ATLAS using a proper background calculation [47]. (Recent CMS studies found similar results, and the new CMS TDR [48] does not even bother to discuss this channel.) There is no longer any clearly-visible mass peak, and S/B is now about 1/6, much poorer. While the figure reflects only 1/10 of the expected total integrated luminosity at LHC, statistics is not the problem. Rather, it is systematic: uncertainty on the exact shape of the QCD backgrounds.

Therein lies the sleeping dragon. Now is a good time to explain how systematic errors may enter our estimate of signal significance. Our simple

---

[f]Non-b jets can fake b jets with a probability of about 1% or a little less.

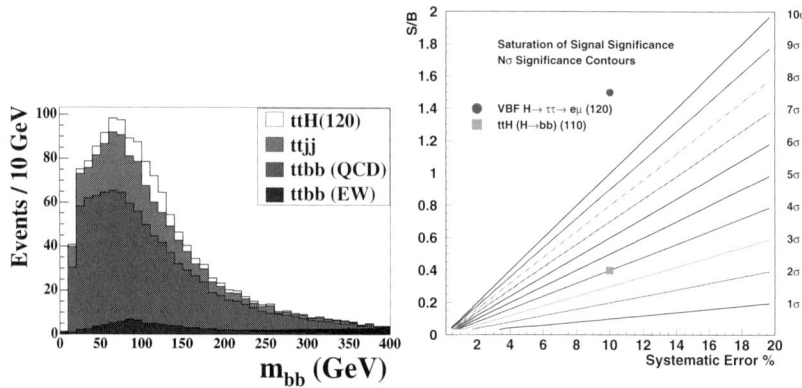

Fig. 8.18. Left: results of a more up-to-date ATLAS studying of $t\bar{t}H$, $H \to b\bar{b}$ production at LHC, for 30 fb^{-1} of data and a Higgs mass of $M_H = 120$ GeV [47]. The QCD backgrounds were calculated with exact matrix elements rather than in the soft/collinear approximation. Right: maximum achievable signal significance for two LHC Higgs channels as a function of S/B and shape systematic uncertainty $\triangle$ [49], as discussed in the text.

formula is modified:

$$\frac{S}{\sqrt{B}} \to \frac{S}{\sqrt{B(1+B\triangle^2)}} \xrightarrow{\mathcal{L}\to\infty} \frac{S/B}{\triangle} \qquad (8.10)$$

where $\triangle$ is the shape uncertainty in the background, a kind of normalization uncertainty. In the limit of infinite data, if S/B is fixed (which it is), signal significance saturates. The only way around this is to perform higher-order calculations of the background to reduce $\triangle$ (and hope you understand the residual theoretical uncertainties). The right panel of Fig. 8.18 shows the spectrum of possibilities [49]. For the known 10% QCD shape systematic for $t\bar{t}H$, even an infinite amount of data would never be able to grant us more than about a 3σ significance. This could still potentially be useful for a coupling measurement, albeit poorly, but will not be a discovery channel unless higher-order QCD calculations can improve the situation. Calculating even just $t\bar{t}b\bar{b}$ at NLO is currently beyond the state of the art, but is likely to become feasible within a few years.

While I don't discuss it here, top quark associated Higgs production does show some promise for the rare Higgs decays to photons. Photons are very clean, well-measured, and the detectors have good rejection against QCD jet fakes. The final word probably hasn't been written on this, but the CMS TDR [48] does have updated simulation results which the interested student may read up on.

8.2.3.2. $gg \to H \to \gamma\gamma$

We've just seen that QCD can be a really annoying problem for Higgs hunting at LHC. A logical alternative for a low-mass Higgs is to look for its rare decays to EW objects, e.g. photons. The BR is at about the two per-mille level for a light Higgs, $110 \lesssim M_H \lesssim 140$ GeV. The LHC will certainly produce enough Higgses, but what are the backgrounds like?

It turns out that the loop-induced QCD process $gg \to \gamma\gamma$ is a non-trivial contribution, but we also have to worry about single and double jet fakes from QCD $j\gamma$ and jj production. This occurs when a leading π^0 from jet fragmentation goes to photons, depositing most of the energy in the EM calorimeter, thereby looking like a real photon. Fortunately, because photons and jets are massless, the invariant mass distribution obeys a very linear $1/m_{\gamma\gamma}$ falloff in our region of interest. The experiments can in that case normalize the background very precisely from the sidebands, where we know there is no Higgs signal. Shape systematics are not much of a concern, thus avoiding the pitfalls of the $t\bar{t}H, H \to b\bar{b}$ case.

Figure 8.19 shows the results of an ATLAS study for this channel using 30 fb^{-1} of data [42], 1/10 of the LHC run program or 3 years at low-luminosity running. The exact expectations are still uncertain, mostly due to an ongoing factor of two uncertainty in the fake jet rejection efficiency. A conservative estimate shows that this channel isn't likely to be the first

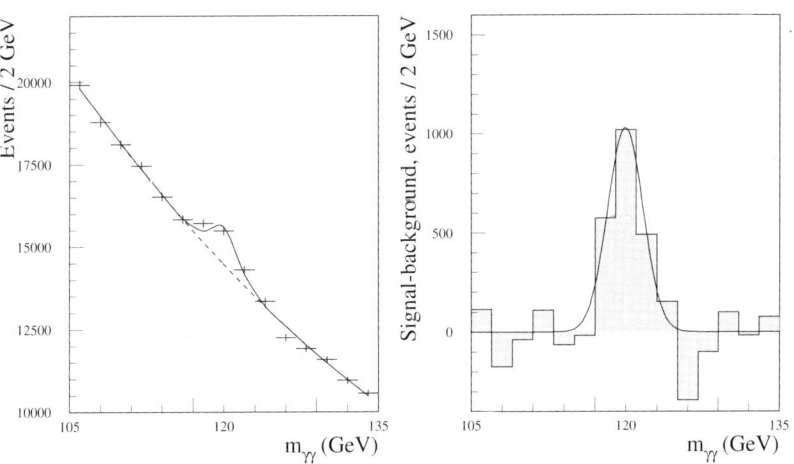

Fig. 8.19. ATLAS simulation of $gg \to H \to \gamma\gamma$ at LHC for $M_H = 120$ GeV and 30 fb^{-1} of data [42]. The right panel is the mass distribution after background subtraction, normalized from sidebands.

discovery mode, but would be crucial for measuring the Higgs mass precisely at low M_H, to about 1% [42, 48]. Photon energy calibration nonlinearity in the detector may be an issue for the ultimate precision, but is generally regarded as minor. We'll come back to this point in Chapter 8.3 on Higgs property measurements.

While I focus here on the SM, keep in mind that because $H \to \gamma\gamma$ is a rare decay, it can be very sensitive to new physics. Recall that the coupling is induced via both top quark and W loops which mostly cancel. Depending on how the new physics alters couplings, or what new particles appear in the loop, the partial width could be greatly suppressed or enhanced. (Anticipating Chapter 8.4, the interested student could peruse Ref. [51] and references therein to see how this can happen in supersymmetry.)

8.2.3.3. Weak boson fusion Higgs production

Let us explore this other production mechanism I said isn't accessible at Tevatron, weak boson fusion (WBF). It was long ignored for LHC light Higgs phenomenology because its rate is about an order of magnitude smaller than $gg \to H$ there. However, it has quite distinctive kinematics and QCD properties that make it easy to suppress backgrounds, for all Higgs decay channels. The process itself is described by an incoming pair of quark partons which brem a pair of weak gauge bosons, which fuse to produce a Higgs; see Fig. 8.20.

The first distinctive characteristic of WBF[g] is that the quarks scatter with significant transverse momentum, and will show up as far forward and backward jets in the hadronic calorimeters of CMS and ATLAS. The Higgs boson is produced centrally, however, so its decay products, regardless of decay mode, typically show up in the central detector region. This is shown in the lego plot schematic in the right panel of Fig. 8.20[h].

The reason for this scattering behavior comes from the W (or Z) propagator, $1/(Q^2 - M^2)$. For t-channel processes, Q^2 is necessarily always negative. Thus the propagator suppresses the amplitude least when Q^2 is small. For small Q^2, we have $Q^2 = (p_f - p_i)^2 \approx E_q^2(1-x)\theta^2$, where x is

---

[g]Some experimentalists refer to this as vector boson fusion (VBF), even though the vector QCD boson (gluon) process of Fig. 8.22 is not included. This will cause increasing confusion as time goes by.

[h]The angle ϕ is the azimuthal angle perpendicular to the beam axis. Pseudorapidity η is a boost-invariant description the polar scattering angle, $\eta = -\log(\tan\frac{\theta}{2})$. The lego plot is a Cartesian map of the finite-resolution detector in these coordinates, as if the detector had been sliced lengthwise and unrolled.

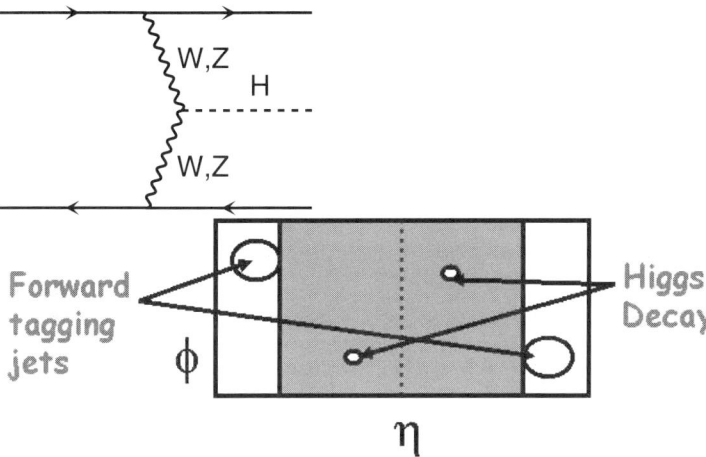

Fig. 8.20. WBF Higgs production Feynman diagram and lego plot schematic of a typical event.

the fraction of incoming quark energy the weak boson takes with it, and is small. Thus θ prefers to be small, translating into large pseudorapidity. One quark will be scattered in the far forward detector, the other far backward, and the pseudorapidity separation between them will tend to be large. We call these "tagging" jets. QCD processes with an extra EW object(s) which mimics a Higgs decay, on the other hand, have a fundamentally different propagator structure and prefer larger scattering angles [52, 53], including at NLO [54]. The differences between the two are shown in Fig. 8.21 [55].

The second distinctive characteristic is QCD radiation [56]. Additional jet activity in WBF prefers to be forward of the scattered quarks. This is because it occurs via bremsstrahlung off color charge, which is scattered at small angles, with no connection between them. In contrast, QCD production always involves color charge being exchanged between the incoming partons: acceleration through 180 degrees. QCD bremsstrahlung thus takes place over large angles, covering the central region. Central jet activity can be vetoed, giving large background suppression [57]. We won't discuss it further, due to theoretical uncertainties; the interested student may learn more from Ref. [58].

We'll see in the next few subsections that WBF Higgs channels are extremely powerful even without a central jet (minijet) veto[i]. Eventually

---

[i]A technical topic outside our present scope: see Refs. [53, 57–59] and the literature they reference.

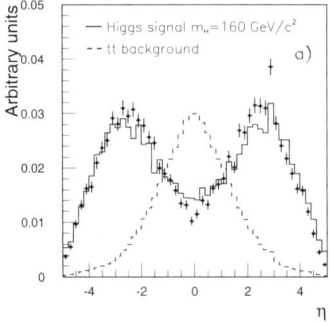 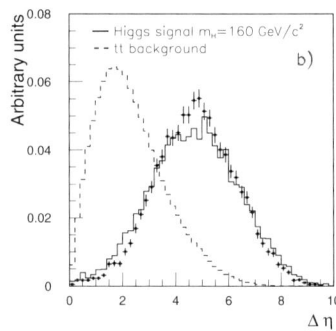

Fig. 8.21. Tagging jet rapidity (left) and separation (right) for WBF Higgs production v. QCD $t\bar{t}$ production [55].

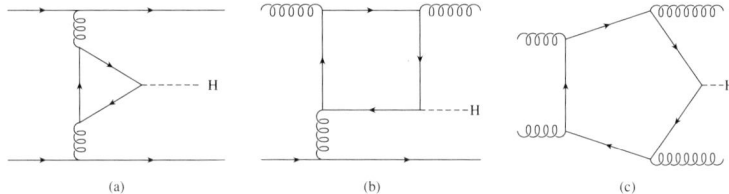

Fig. 8.22. Representative Feynman diagrams for gluon fusion Higgs plus two jets production [60].

a veto will be used, after calibration from observing EW v. QCD Zjj production in the early running of LHC [53]. There is however another lingering theoretical uncertainty, coming from Higgs production itself!

QCD Higgs production via loop-induced couplings may itself give rise to two forward tagging jets, which would then fall into the WBF Higgs sample [60]. Some representative Feynman diagrams for this process are shown in Fig. 8.22. After imposing WBF-type kinematic cuts (far forward/backward, well-separated jets, central Higgs decay products), this contribution to the WBF sample adds about another third for a light Higgs, or doubles it for a very heavy Higgs, $M_H \gtrsim 350$ GeV, as shown in the left panel of Fig. 8.23. The residual QCD theoretical cross section uncertainty is about a factor of two, however, and being QCD it will produce far more central jets, which will be vetoed to reject QCD backgrounds. Naïvely, then, gluon fusion Hjj is an $\sim 10\%$ contribution to WBF, but with a huge uncertainty.

This contribution is a mixed blessing. It's part of the signal, so would hasten discovery. Yet it creates confusion, since at some point we want

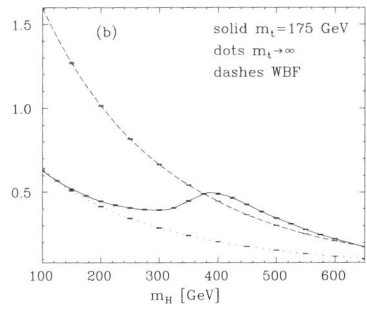 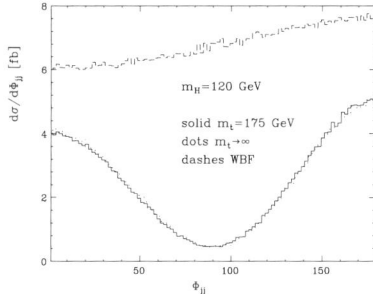

Fig. 8.23. Left: WBF and gluon fusion contributions to the forward-tagged Hjj sample at LHC. Right: azimuthal angular distributions for the same two processes, showing distinctive differences. Figures taken from Ref. [60].

to measure couplings, and the WBF and gluon fusion components arise from different couplings. Fortunately, there is a difference! WBF produces an almost-flat distribution in ϕ_{jj}, the azimuthal tagging jet separation, but gluon fusion has a suppression at 90 degrees [60]; cf. right panel of Fig. 8.23.

8.2.3.4. Weak boson fusion $H \to \tau^+\tau^-$

Now we know that the WBF signature can strongly suppress QCD backgrounds because of its unique kinematic characteristics. We expect that $H \to \gamma\gamma$ is visible in WBF [48, 61, 62], but being a rare decay in a smaller-rate channel, it's not expected to lead to discovery. Rather, it would be a useful additional channel for couplings measurements. Let's now instead discuss a decay mode we haven't yet considered, $H \to \tau^+\tau^-$. This is subdominant to $H \to b\bar{b}$ in the light Higgs region, $M_H \lesssim 150$ GeV, but the backgrounds are more EW than QCD. We thus have some hope to see it, whereas $H \to b\bar{b}$ remains frustratingly hopeless.

We first have to realize that taus decay to a variety of final states:

- 35% $\tau \to \ell\nu_\ell\nu_\tau$, ID efficiency $\epsilon_\ell \sim 90\%$
- 50% $\tau \to h_1\nu_\tau$ "1-prong" hadronic (one charged track), ID efficiency $\epsilon_h \sim 25\%$
- 15% $\tau \to h_3\nu_\tau$ "3-prong" hadronic (three charged tracks), which are thrown away

The obvious problem is that with at least two neutrinos escaping, the Higgs cannot be reconstructed from its decay products. Or can it?

Let's assume the taus decay collinearly. This is an excellent approximation: since 50+ GeV energy taus have far more energy than their mass, so their decay products are highly collimated. We then have two unknowns, x_+ and x_-, the fractions of tau energy that the charged particles take with them. What experiment measures is missing transverse energy in the x and y directions. Two unknowns with two measurements is exactly solvable. For our system this gives [63]:

$$m_{\tau^+\tau^-}^2 = \frac{m_{\ell^+\ell^-}^2}{x_+ x_-} + 2m_\tau^2 \qquad (8.11)$$

(an excellent exercise for all students to get a grip on kinematics and useful tricks at hadron colliders). An important note is that this doesn't work for back-to-back taus (the derivation will reveal why), but WBF Higgses are typically kicked out with about 100 GeV of p_T, so this almost never happens in WBF. This trick can't be used in the bulk of $gg \to H$ events because there it is produced mostly at rest with nearly all taus back-to-back.

We need a lepton trigger, so consider two channels: $\tau^+\tau^- \to \ell^\pm h$ and $\tau^+\tau^- \to \ell^+\ell'^-$ ($\ell = e, \mu$). The main backgrounds are EW and QCD Zjj production (really Z/γ^*), top quark pairs, EW & QCD $WWjj$ and QCD $b\bar{b}jj$ production. But after reconstruction, the non-Z backgrounds look very different than the signal in x_+–x_- space, as shown in Fig. 8.24.

ATLAS and CMS have both studied these channels with full detector simulation and WBF kinematic cuts, but no minijet veto, and found extremely promising results [55]. Figure 8.25 shows invariant mass distributions for a reconstructed Higgs in the two different decay channels, assuming only 30 fb^{-1} of data. The Higgs peak is easily seen above the backgrounds and away from the Z pole. Mass resolution is expected to be a few GeV.

But this joint study by CMS and ATLAS [55] is not the best we can do. The joint study ignored the minijet veto, for instance. While that will assuredly improve the situation further, we're just not sure precisely how much. Putting this aside for the moment, there are yet further tricks to play to improve the situation.

The leading idea zeroes in on the fact that missing transverse momentum ($\not{p}_T$) has some uncertainty due to jet energy mismeasurement (those imperfect detectors). Using a χ^2 test, one determines which is more likely: $Z \to \tau^+\tau^-$ or $H \to \tau^+\tau^-$, *using a fixed Higgs mass constraint* [65]. Examining the schematics in Fig. 8.26, we see this is tantamount to deciding which fit is closer to the center of the $\not{p}_T$ uncertainty region. Early indications are that this technique would improve S/B by about a factor *four*, in addition to recovering some signal lost using more traditional strict kinematic cuts on x_+ and x_- (recall Fig. 8.24). This would approximately halve

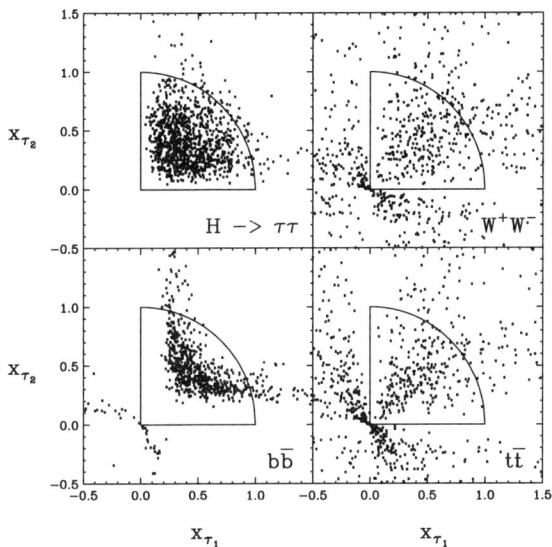

Fig. 8.24. Reconstructed x_+ v. x_- (x_1, x_2) for a WBF $H \to \tau^+\tau^-$ signal v. non-Z backgrounds [64].

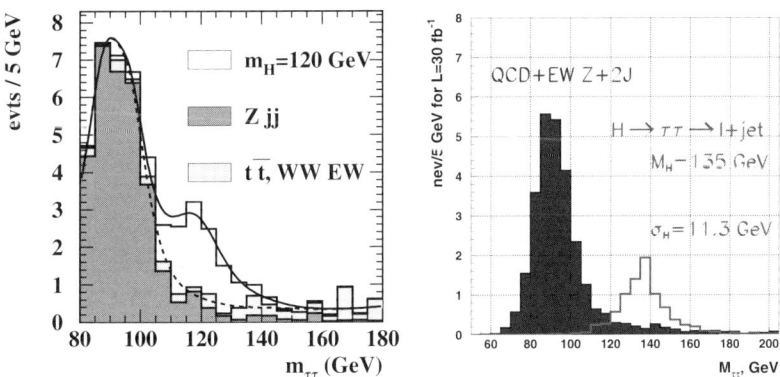

Fig. 8.25. ATLAS (left) and CMS (right) simulations of WBF $H \to \tau^+\tau^-$ events after 30 fb^{-1} of data at LHC. The Higgs resonance clearly stands out from the background. Figures from Ref. [55].

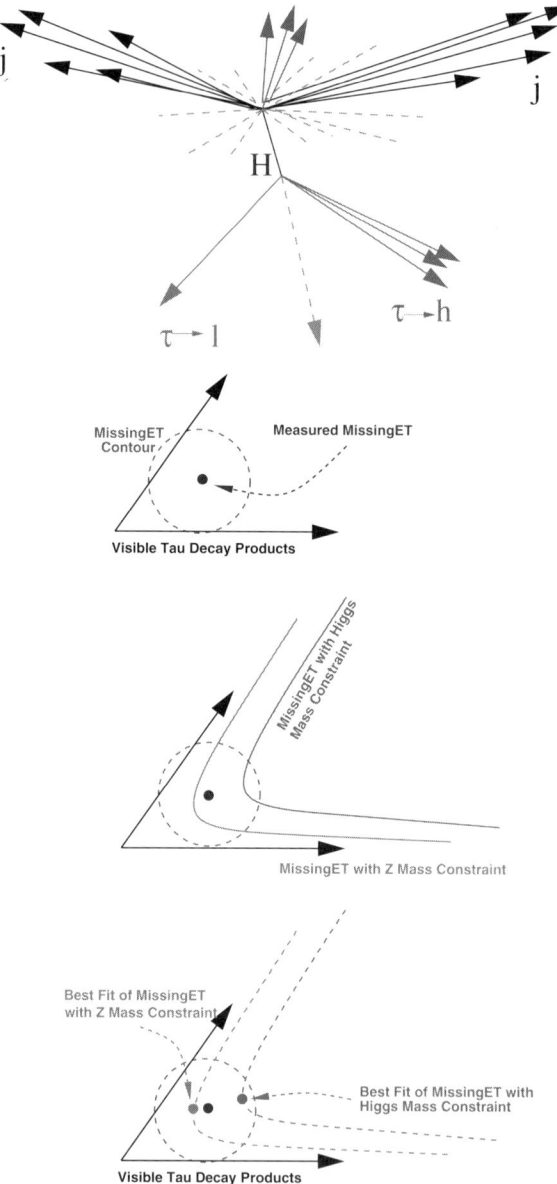

Fig. 8.26. Top: schematic azimuthal projection of WBF $H \to \tau^+\tau^-$ events at LHC. Bottom: diagram illustrating the 1σ uncertainty region (due to jet mismeasurement) of missing p_T, and how a Z mass or Higgs mass hypothesis can be best fit using a χ^2 test. Figures from Ref. [50].

the data required to discover a light SM Higgs boson using this channel. Keep it in mind when we see the current official discovery expectations in Sec. 8.2.3.7. Further improvements might also be expected from neural-net type analyses, which are coming to the fore now that Tevatron has demonstrated their viability.

A final word on systematic uncertainties. Unlike the tortuous case of $t\bar{t}H, H \to b\bar{b}$, we don't have to worry about shape systematics here. The dominant background is Zjj production. We can separately examine $Z \to ee, \mu\mu$, which produces an extremely sharp, clean peak, precisely calibrating Zjj production in Monte Carlo. The only uncertainty then is tau decay modeling, which is very well understood from the LEP era.

8.2.3.5. Weak boson fusion $H \to W^+W^-$

A natural question to ask is, how well does WBF Higgs hunting work for $M_H \gtrsim 140$ GeV, where $H \to W^+W^-$ dominates? We should expect fairly well, since it's the production process characteristics that supply most of the background suppression, leaving us only to look for separated reconstructed mass peaks.

For $H \to W^+W^-$ we'll consider only the dilepton channel, as it has relatively low backgrounds, while QCD gives a large rate for the other possible channel, one central lepton plus two central jets (and the minijet veto will likely not work). We'll therefore rely on exactly the same angular correlations and transverse mass variable we encountered in the Tevatron case [37] (cf. Eqs. 8.8,8.9). The only critical distinction is then $e\mu$ v. ee, $\mu\mu$ samples, as the latter have a continuum background (Z^*/γ^*). These are not too much of a concern, however.

Without going too much into detail, I'll simply say that top quarks are a major background, and they have the largest uncertainty. The largest component comes from $t\bar{t}j$ production, where the extra hard parton is far forward and ID'd as one tagging jet; a b jet from top decay gives the other tagging jet, and the other b jet is unobserved. This background requires care to simulate, because the soft/collinear approximation in standard codes is no good. There is also a significant contribution from single-top production, and off-shell effects are crucial to simulate, which is not normally an issue for backgrounds at LHC [66]. Work is still needed in this area to be fully prepared for this particular search channel. Fortunately, we may expect an NLO calculation of $t\bar{t}j$ before LHC start [67].

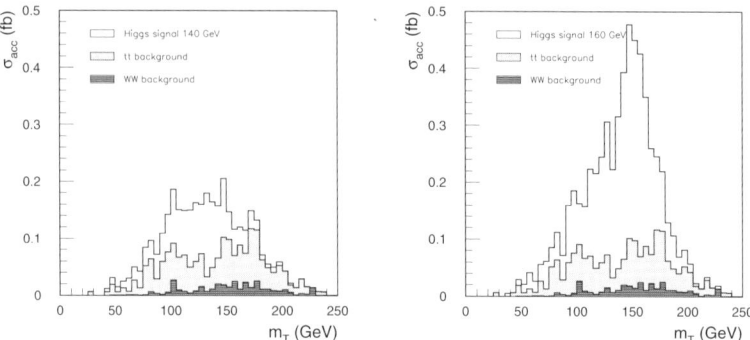

Fig. 8.27. ATLAS simulations of WBF $H \to W^+W^-$ events after 30 fb^{-1} of data at LHC for $M_H = 140$ GeV (left) and 160 GeV (right). The Higgs signal clearly stands out from the background in both cases, although the Jacobian peak is easier to identify closer to threshold. Figures taken from Ref. [55].

Figure 8.27 shows the results of the same ATLAS/CMS joint WBF Higgs study for this channel [55]. The results are extremely positive, with $S/B > 1/1$ without a minijet veto over a large mass range; even for $M_H = 120$ GeV, $S/B \sim 1/2$, allowing for Higgs observation even down to the LEP limit in this channel. The transverse mass variable works extremely well for Higgs masses near WW threshold, and reasonably well for lower masses, where the W bosons are off-shell.

8.2.3.6. $t\bar{t}H, H \to W^+W^-$ at higher mass

A late entry to the Higgs game at LHC is top quark associated production, but with Higgs decaying to W bosons. Representative Feynman diagrams are shown in Fig. 8.28. Obviously this is intended to apply to larger Higgs masses, but turns out to work fairly well even below W pair threshold [68, 69]. The key is to use same-sign dilepton and trilepton subsamples. The backgrounds then don't come from pure QCD production, rather from mixed QCD-EW top quark pairs plus W, Z/γ^*, W^+W^-, etc. We would be especially eager to observe this channel because, if the HWW coupling is measured elsewhere, it provides the only viable direct measurement of the top quark Yukawa coupling. More on this in Chapter 8.3.

A noteworthy features of this channel is that while the $t\bar{t}H$ cross section falls with increasing M_H, BR($H \to W^+W^-$) rises with increasing M_H in our mass region of interest, and the two trends coincidentally approximately balance each other. From a final-state rate perspective, this channel is

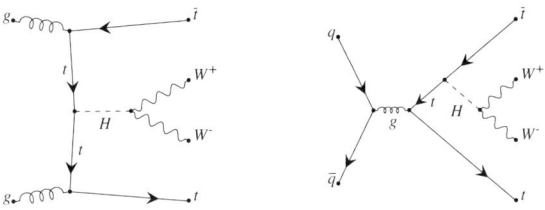

Fig. 8.28. Representative Feynman diagrams for $t\bar{t}H$, $H \to W^+W^-$ production at LHC.

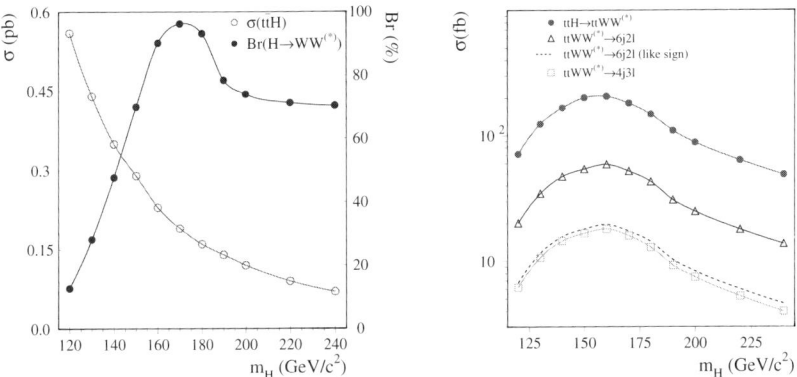

Fig. 8.29. Left: $t\bar{t}H$ cross section and BR($H \to W^+W^-$) as a function of M_H. Right: the cross section to three different final states after top quark and Higgs decays. Figures from Ref. [69].

approximately constant over a wide mass range, up to about 200 GeV. Figure 8.29 shows this numerically. Figure 8.30 shows ATLAS's expected statistical uncertainty on the top quark Yukawa coupling. It ranges from about 20% over a broad mass range for 30 fb^{-1} of data, to about 10% from the full LHC run. Systematic uncertainties are currently unexplored.

8.2.3.7. *LHC Higgs in a nutshell*

LHC Higgs phenomenology has come a long way in the decade since the first comprehensive studies were reported (e.g. the ATLAS TDR [42]). The old studies give a seriously misleading picture of LHC capabilities. Students should refer to newer ATLAS Notes and the new CMS TDR [48]. Solid

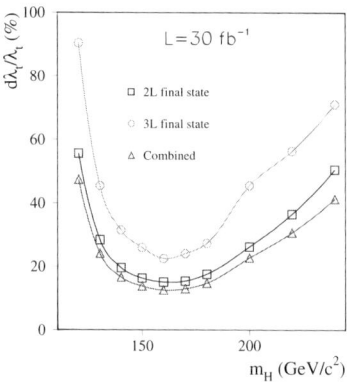 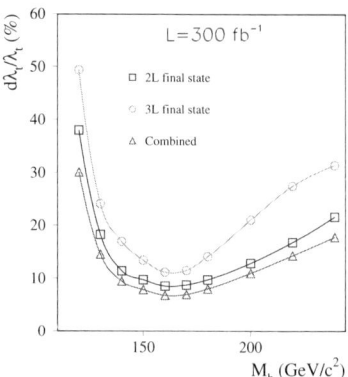

Fig. 8.30. ATLAS prediction [69] for the top quark Yukawa coupling measurement uncertainty (statistical only) from $t\bar{t}H, H \to W^+W^-$, for separate leptonic final-state channels and combined.

grounds exist for expecting even more improvements. Fig. 8.31 summarizes ATLAS's projections for multiple Higgs channels as a function of Higgs mass. Note especially the new dominance of WBF channels and degradation of $t\bar{t}H$.

8.3. Is it the Standard Model Higgs?

Imagine yourself in 2010 (hey, we're optimists!), squished shoulder-to-shoulder in the CERN auditorium, waiting for the speaker to get to the punchline. Rumors have been circulating for months about excess events showing up in some light Higgs channels, but not all that would be expected. LHC has 40 fb^{-1}, after all. Your experimental friends tell you that both collaborations have been scrambling madly, independent groups cross-checking the original first analyses. Then the null result slides start passing by. No diphoton peaks anywhere. Nothing in the WW or ZZ channels. Even CMS's invisible Higgs search (WBF – tagging jets with no central objects at all) doesn't show anything. Numerous standard MSSM Higgs results fly by, invariant mass spectra fitting the SM predictions perfectly. The audience becomes restless, irritated. People around you mutter that there must not be a Higgs after all. But you realize that the speaker skipped mention of the WBF $H \to \tau^+\tau^-$ channel. Then suddenly it appears, and there's a peak above the Z pole, centered around 125 GeV, broader than

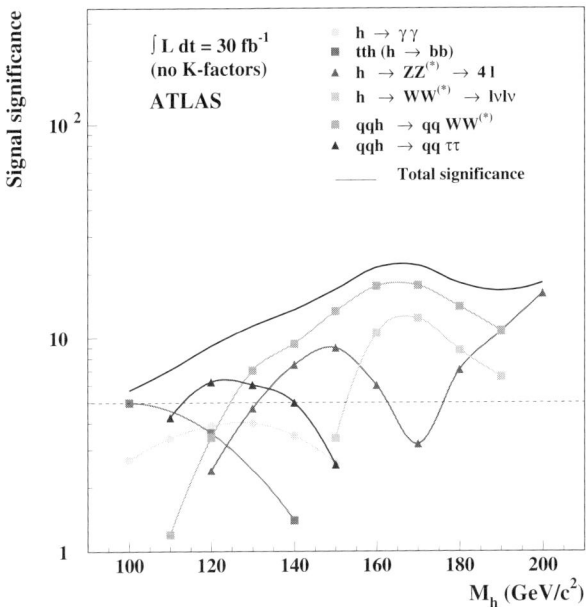

Fig. 8.31. ATLAS significance projections in multiple Higgs channels w/ 30 fb^{-1} of LHC data [55].

you'd expect but the speaker says something about resolution will improve with further refinement of the tau reconstruction algorithms. It's also a too-small rate, less than half what's expected.

So what is this beast? The bump showed up in a Higgs search channel, but at that mass it should have shown up in several others as well. If it's Standard Model, that is. At 125 GeV there should be $H \to W^+W^-$ in WBF, and $H \to \gamma\gamma$ both inclusively and in WBF, although maybe they're still marginal. Photons turned out to be hard at first, and QCD predictions weren't quite on the mark. Quite a few people are on their cell phones already. You hear a dozen different exclamations, ranging from "We found the Higgs!" to "The Standard Model is dead!". Quite obviously this is a new physics discovery, but what exactly is going on?

By now you should get the point of this imaginary scenario: finding a new bump is merely the start of real physics. For numerous reasons you've heard at this summer school, some better than others, finding a SM Higgs really isn't very likely. But as we'll see in Chapter 8.4, SM Higgs phenomenology is a superb base for beyond-the-SM (BSM) Higgs sectors. They're variations on a theme in some sense, with the occasional special channel thrown in, like the invisible Higgs search alluded to above. Our job will be to figure out what any new resonance is. But how do we go about doing that in a systematic way that's useful to theorists for constructing the New Standard Model?

For starters, we want to know the complete set of quantum numbers for any Higgs candidate we find. Standard Model expectations will probably prejudice us as to what they are (roughly, at least) based on which search channel a bump shows up in. But for the scenario above, I can envision at least three very reasonable yet completely different models that would give that kind of a result in early LHC running. We should keep in mind that further data may reveal more resonances – not everything is easy to see against backgrounds, or is produced with enough rate to emerge with only 1/10 of the planned LHC data. In some cases we would have to wait much longer, using data from the planned LHC luminosity upgrade (SLHC) [70]. New physics could also mean new quantum numbers that we don't yet know about, so we should be prepared to expand our list of measurements needed to sort out the theory, and spend time *now* thinking about what kinds of observables are even possible at the LHC. Some measurements will almost certainly require the clean environment of a future high-energy electron-positron machine like an ILC [71, 72]. The most complete picture would emerge only after combining results [73], which could take than a decade. In the meantime we might get a good picture of the new physics, but not its details.

Let's prepare a preliminary list of quantum numbers we need to measure for a candidate Higgs resonance, which I'll generically call ϕ. In brackets is the SM expectation. I'll order them in increasing level of difficulty. (See also the review article of Ref. [74].)

- electric charge [neutral]
- color charge [neutral]
- mass [free parameter]
- spin [0]
- CP [even]
- gauge coupling (g_{WWH}) [$SU(2)_L$ with tensor structure $g^{\mu\nu}$]
- Yukawa couplings [m_f/v]
- spontaneous symmetry breaking potential (self-couplings) [fixed by the mass]

Of course, the first two of those, electric and color charge, are known immediately from the decay products. (A non-color-singlet scalar is a radically different beast than the SM Higgs and would have dramatically different couplings and signatures.) Mass is also almost immediate, with some level of uncertainty that depends almost purely on detector effects. Spin and CP are related to some degree, and not entirely straightforward if the Higgs sector is non-minimal and contains CP violation. Gauge and Yukawa couplings are generally regarded as the most crucial observables, and in some sense I would agree. However, I would argue that the linchpin of spontaneous symmetry breaking (SSB) is the existence of a Higgs potential, which requires Higgs self-couplings. Measuring these and finding they match to some gauge theory with a SSB Higgs sector would to me be the most definitive proof of SSB, and strongly suggest that the Higgs is a fundamental scalar, not composite. It is also the most difficult task – perhaps not even possible.

A cautionary note: the results I show in this section are in general applicable only to the Standard Model Higgs! This point is often lost in many presentations highlighting the capabilities of various experiments, but it is very easy to understand. For example, if for some reason the Higgs sector has suppressed couplings to colored fermions, then any measurement of, say, the b Yukawa coupling, will be less precise, simply because the signal rate is lower, yet the background remains fixed. It's statistics!

8.3.1. Mass measurement

As already noted, our Higgs hunt pretty much gets us this quantum number immediately, but with some slop driven by detector performance. We want to measure it as accurately as possible, but in practice a GeV or so is good enough, because theoretical uncertainties in parameter fits tend to dominate for most BSM physics. (This is a long-standing problem in SUSY scenarios, for example. It may be that we need to know the Higgs mass theoretical prediction to four loops [75]; at present only a partial three-loop calculation is known [76], and only two-loop results exist in usable code [77].) Figure 8.32 shows the CMS and ILC expected Higgs mass precision as a function of M_H [78]. It varies, of course, because different decay modes are accessible at different M_H, and detector resolution depends on the final state. In general, photon pairs ($H \to \gamma\gamma$) and four leptons coming from Z pairs ($H \to ZZ \to \ell^+\ell^-\ell'^+\ell'^-$) will give the most precise measurement. As a rule of thumb, we may expect per-mille precision over a broad mass range, translating typically to a few hundred MeV.

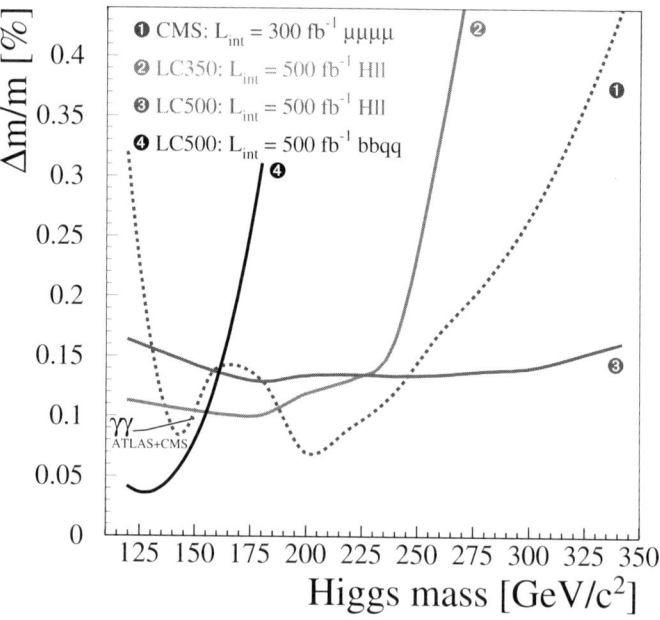

Fig. 8.32. Expected SM Higgs mass precision at LHC (for CMS; ATLAS will be slightly different but comparable) and a future ILC, as a function of Higgs mass [78].

8.3.2. Spin & CP measurement

Spin and CP (J^{PC}) experimental measurements are linked, because both require angular distributions to obtain. Numerous techniques have been proposed to address this, with significant overlap but also some unique features with each method. I'll highlight the leading proposals which garner the most attention from LHC experimentalists today.

From the observed final state we can tell that the Higgs candidate is a boson. We'll start by assuming that it may be spin 0, 1 or 2, but no higher[j]. Then we recall that the Yang-Landau Theorem [80] forbids a coupling between three $S = 1$ bosons if two of them are identical. Thus, if we observe $\phi \to \gamma\gamma$, then our new object cannot be spin-1, and $C = 1$. For the very curious student who wants to delve deeper, there is a recent report on CP Higgs studies at colliders [81].

---

[j]$S \geq 3$ fundamental particles are believed to have deep problems in renormalizable field theory [79].

8.3.2.1. Nelson technique

The first method is the oldest, developed by Nelson [82]. It assumes the object is a scalar or pseudoscalar[k] and relies on the decay angular distributions to a pair of EW gauge bosons, which decay further. The most practical aspect relevant for LHC Higgs physics is in essence a measurement of the relative azimuthal angle between the decay planes of two Z bosons in turn coming from the scalar decay, in the scalar particle's rest frame. See Fig. 8.33 for clarity. One bins the data in this distribution and fits to the equation:

$$F(\phi) = 1 + \alpha \cos(\phi) + \beta \cos(2\phi) \tag{8.12}$$

For a scalar, such as the SM Higgs, the coefficients α and β are functions of the scalar mass, and further we have the constraint that $\alpha(M_\phi) > \frac{1}{4}$. In contrast, for a pseudoscalar, $\alpha = 0$ and $\beta = -0.25$, independent of the mass.

Reference [83] was the first to apply this to the LHC Higgs physics program using detector simulation. Assuming 100 fb^{-1} of data, the study found that LHC could readily distinguish a SM Higgs from a pseudoscalar for $M_H > 200$ GeV, and from a spin-1 boson of either CP state from a little above that, but not right at 200 GeV; see Fig. 8.34. Applying this technique to $M_H < 200$ but above ZZ threshold was not examined.

As a practical matter, $H \to ZZ^{(*)}$ observation is assured only for both Z bosons decaying to leptons (e or μ), where there is essentially zero background. Unfortunately, this is an extremely tiny branching ratio, only 0.05% of all $H \to ZZ$ events. Some studies consider $jj\ell^+\ell^-$ channels, which is a ten-times larger sample, in an attempt to increase statistics, but this suffers from non-trivial QCD backgrounds.

8.3.2.2. CMMZ technique

Reference [84] provides an extension to the Nelson technique below ZZ threshold. Its full analysis is far more in-depth, discussing the angular behavior of the matrix elements for arbitrary boson spin and parity. It first demonstrates how objects of odd normality (spin times parity) can be discriminated via angular distributions, but for even normality require a further discriminant. That is, a $J^P = 2^+$ boson could mimic a SM Higgs in angular distribution below ZZ threshold. (Exotic higher spin states can

---

[k]A pseudoscalar doesn't couple at tree-level to W or Z, but can have a (large) loop-induced coupling.

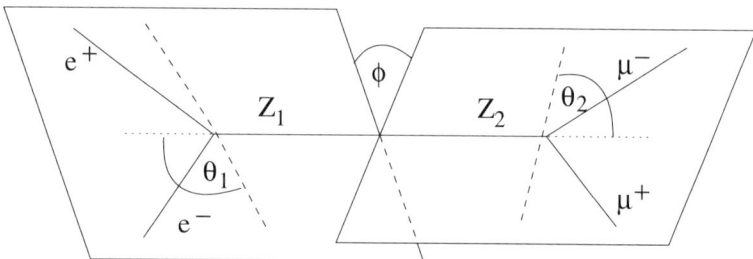

Fig. 8.33. Schematic of the azimuthal angle between the decay planes of Z bosons arising from massive scalar decay. All angles are in the scalar rest frame. Figure from Ref. [83].

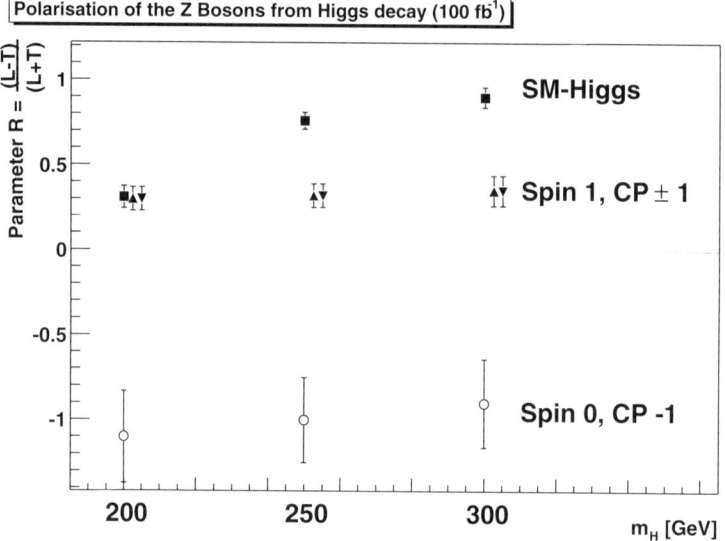

Fig. 8.34. Results of the LHC expectations spin/CP study of Ref. [83], showing how a SM Higgs could be distinguished from a pseudoscalar or spin-1 boson as a function of M_H.

be trivially ruled out via the lack of angular correlation between the beam and the object's flight direction.)

The key discriminant is the differential partial decay rate for the off-shell Z boson[1] It depends on the invariant mass of the final-state lepton

---

[1]Typically only one Z boson is off-shell for $M_H < 2M_Z$, but this ceases to be a good approximation at much lower (but observable) masses.

pair and is linear in Z^* velocity:

$$\frac{d\Gamma_H}{dM_*^2} \sim \beta \sim \sqrt{(M_H - M_Z)^2 - M_*^2} \qquad (8.13)$$

Figure 8.35 shows the predicted distributions for 150 GeV spin-0,1,2 even-normality objects as a function of M_*, the off-shellness of the $Z^*\ell^+\ell^-$. The histogram represents about 200 events that a SM Higgs would give in this channel after 300 fb^{-1} of data at LHC. Unfortunately there are no error bars, although one can estimate the statistical uncertainty for each bin as $\sqrt{N}$ and observe that the measurement is likely not spectacular. We can expect that CMS and ATLAS will eventually get around to quantifying the discriminating power, but it would not be surprising to learn that this measurement requires far more data, e.g. at the upgraded SLHC [70].

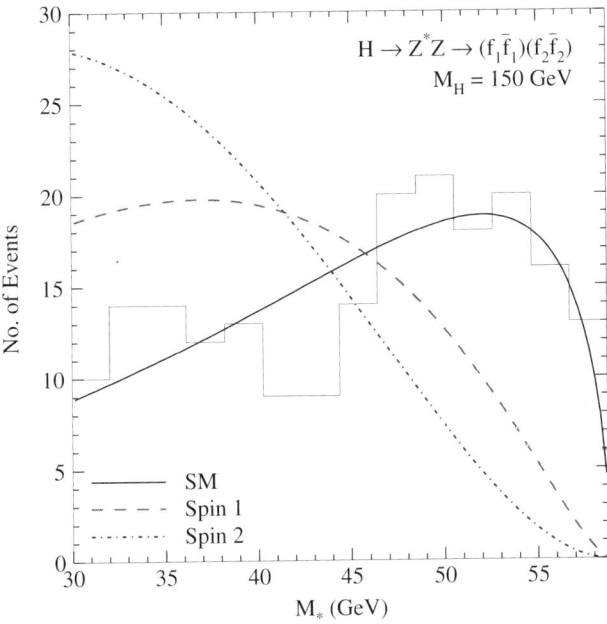

Fig. 8.35. Differential decay rate as a function of dilepton invariant mass of the off-shell Z^* in ZZ events, for a 150 GeV SM Higgs v. spin-1 and spin-2 objects of even normality and the same mass. The histogram is the SM Higgs case for 300 fb^{-1} of data at LHC. Figure from Ref. [84].

8.3.2.3. *CP and gauge vertex structure via WBF*

A third technique [85] takes a different approach, but addressing spin and CP in a slightly different way. Rather than examine Higgs decays, it notes that WBF Higgs production is observable for *any* Higgs mass, regardless of decay mode. Furthermore, the same HVV vertex appears on the production side for all masses, also independent of decay. More precisely, this vertex has the structure $g^{\mu\nu}HV_\mu V_\nu$ ($V = W, Z$). This tensor structure is not gauge invariant by itself. It must come from a gauge-invariant kinetic term $(D_\mu \Phi)^\dagger (D^\mu \Phi)$. Identifying it in experiment would go a long way to establishing that the scalar field is a remnant of spontaneous symmetry breaking.

For a scalar field which couples via higher-dimensional operators to two gauge bosons, however, we may write down the CP-even and CP-odd gauge-invariant D6 operators [86]:

$$\mathcal{L}_6 = \frac{g^2}{2\Lambda_{6,e}}(\Phi^\dagger\Phi)W^+_{\mu\nu}W^{-\mu\nu} + \frac{g^2}{2\Lambda_{6,o}}(\Phi^\dagger\Phi)\widetilde{W}^+_{\mu\nu}W^{-\mu\nu} \qquad (8.14)$$

where Λ_6 is the scale of new physics that is integrated out, $W^{\mu\nu}$ is the W boson field strength tensor, and $\widetilde{W} = \epsilon_{\alpha\beta\mu\nu}W^{\alpha\beta}$ is its dual. After expanding Φ with a vev and radial excitation, we obtain two D5 operators:

$$\mathcal{L}_5 = \frac{1}{\Lambda_{5,e}}HW^+_{\mu\nu}W^{-\mu\nu} + \frac{1}{\Lambda_{5,o}}H\widetilde{W}^+_{\mu\nu}W^{-\mu\nu} \qquad (8.15)$$

where Λ_5 are dimensionful but now parameterize both the D6 coefficients and the Φ vev.

These two D5 operators produce very distinctive matrix element behavior. Recalling that the external gauge bosons in WBF are actually virtual and connect to external fermion currents, the initial-state scattered quarks, we derive the following approximate relations for the CP-even operator, using $J_{1,2}$ for the incoming fermion currents:

$$\mathcal{M}_{e,5} \propto \frac{1}{\Lambda_{e,5}} J_1^\mu J_2^\nu \left[g_{\mu\nu}(q_1 \cdot q_2) - q_{1,\nu}q_{2,\mu} \right] \sim \frac{1}{\Lambda_{e,5}}[J_1^0 J_2^0 - J_1^3 J_2^3]\, \vec{p}_T^{\,j1} \cdot \vec{p}_T^{\,j2} \qquad (8.16)$$

That is, the amplitude is proportional to the tagging jets' transverse momentum dot product. This is easy to measure experimentally – we just plot the azimuthal angular distribution, i.e. angular separation in the plane perpendicular to the beam. It will be minimal, nearly zero, for $\phi_{jj} = \pi/4$. In contrast, the $g^{\mu\nu}$ tensor structure of the SM Higgs mechanism does not correlate the tagging jets. The CP-odd D5 operator is different and more

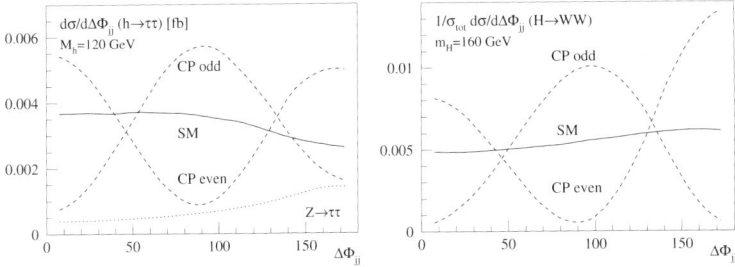

Fig. 8.36. Azimuthal angular distributions of the tagging jets in WBF production of a SM Higgs v. scalar field coupled to weak bosons via CP-even/odd D6 operators. The dotted line in the left panel is the SM background, which is added to the signal curves. Figures from Ref. [85].

complex, but may be understood by noting that it contains a Levi-Civita tensor $\epsilon^{\mu\nu\rho\delta}$ connecting the external fermion momenta. This is non-zero only when the four external momenta are independent, i.e. not coplanar. Thus this distribution will be zero for $\phi_{jj} = 0, \pi$.

Figure 8.36 shows the results of a parton-level simulation for scalars in both the mass range where decays to taus would be used, and where $\phi \to W^+W^-$ dominates. The SM signal curve is not entirely flat due to kinematic cuts imposed on the final state to ID all objects. The D5 operators produce behavior qualitatively distinct from spontaneous symmetry breaking, with minima for the distributions exactly where expected, and orthogonal from each other. It would be essentially trivial to distinguish the cases from each other shortly after discovery, regardless of M_H and the particular channel used to discover the Higgs candidate. A key requirement for this, of course, is that the discovery searches don't use this distribution to separate signal from background.

Now, what happens if the Higgs indeed arises from SSB, but new physics generates sizable D6 operators? Since $H_{\rm SM}$ is CP-even, a CP-even D5 operator would interfere with the SM amplitude, while a CP-odd contribution would remain independent. This is illustrated in the left panel of Fig. 8.37. The obvious thing to do is create an asymmetry observable sensitive to this interference:

$$A_\phi = \frac{\sigma(\Delta\phi_{jj} < \pi/2) - \sigma(\Delta\phi_{jj} > \pi/2)}{\sigma(\Delta\phi_{jj} < \pi/2) + \sigma(\Delta\phi_{jj} > \pi/2)} \quad (8.17)$$

With only 100 fb^{-1} of data at LHC (one experiment), this asymmetry would have access to $\Lambda_6 \sim 1$ TeV, which is itself within the reach of LHC,

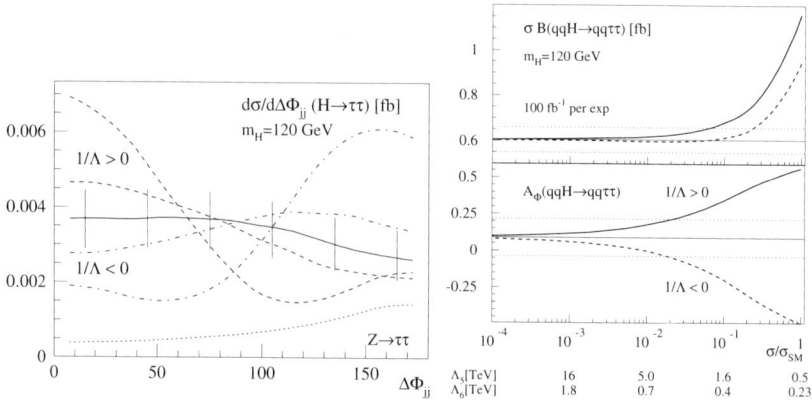

Fig. 8.37. Left: As in Fig. 8.36, but with interference between the SM Higgs and a CP-even D5 operator. Right: the effective reach in $\Lambda_{5,e}$ for 100 fb^{-1} at LHC, using only the rate information (top) or the asymmetry (bottom). Figures from Ref. [85].

likely resulting in new physics observation directly. One caveat: the study Ref. [85] was done before the $gg \to Hgg$ contamination [60] was known, which will complicate this measurement.

8.3.2.4. *Spin and CP at an ILC*

The much cleaner, low-background environment of e^+e^- collisions would be an excellent environment to study a new resonance's spin and CP properties. J^{PC} can in fact be determined completely model-independently. Recalling the LEP search, the canonical production mechanism is $e^+e^- \to ZH$. We would identify the Z via its decay to leptons, and sum over all Higgs decays (this is possible using the recoil mass technique, coming up in Sec. 8.3.4). J and P are completely determined by a combination of the cross section rise at threshold and the polar angle of the Z flight direction in the lab, shown in the left panel of Fig. 8.38. The differential cross section is [71]:

$$\frac{d\sigma}{d\cos\theta_Z} \propto \beta\left[1 + a\beta^2\sin^2\theta_Z + b\eta\beta\cos\theta_Z + \eta^2\beta^2(1+\cos^2\theta_Z)\right] \quad (8.18)$$

where a and b depend on the EW couplings and Z boson mass, η is a general pseudoscalar (loop-induced) coupling and β is the velocity. Far more sophisticated analyses techniques exist, often called "optimal observable" analyses [87], but are only for the terminally curious.

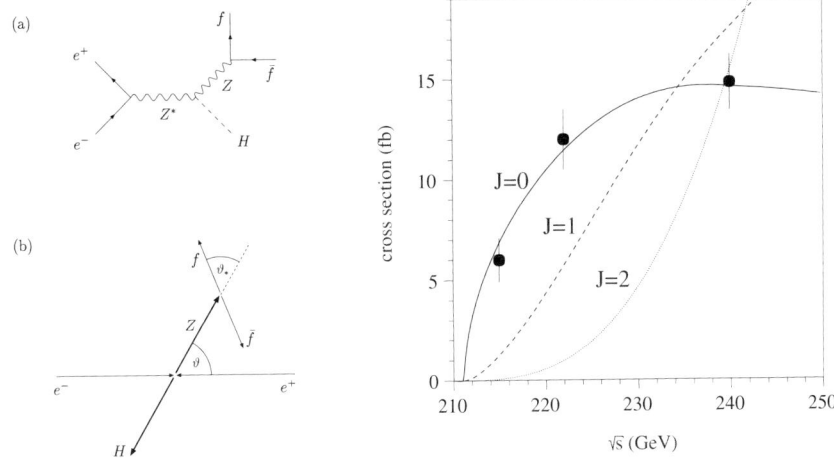

Fig. 8.38. Left: Feynman diagram for $e^+e^- \to ZH$ and schematic [88] showing the analyzing angles. Right: curves showing the threshold rate dependence for $J = 0, 1, 2$ states in this channel [71].

If one would have the liberty to perform a threshold scan of $Z\phi$ production at an ILC, distinguishing given-normality $J = 0, 1, 2$ states is straightforward due to their different β-dependence. For $J = 0$ it is linear, but for higher spin is higher-power in β [88]. The qualitative behavior is shown in the right panel of Fig. 8.38, complete with error bars for the SM Higgs case. However, while the physics is solid, experiments in the past have generally proved to be a horse race for highest energy, so there is no guarantee that one would have threshold scan data available. The angular distribution fortunately works at all energies.

8.3.3. Higgs couplings at LHC

Now to something much harder. It's commonly believed that LHC cannot measure Higgs couplings, only ratios of BRs [42]. This is incorrect, but requires a little explanation to understand why people previously believed in a limitation.

First, let me state that the LHC doesn't measure couplings or any other quantum number directly. It measures *rates*. (This is true for any particle physics experiment.) From those we extract various $\sigma_i \cdot \mathrm{BR}_j$ by removing detector, soft QCD and phase space effects, among other things, using Monte Carlo simulations based on known physics inputs.

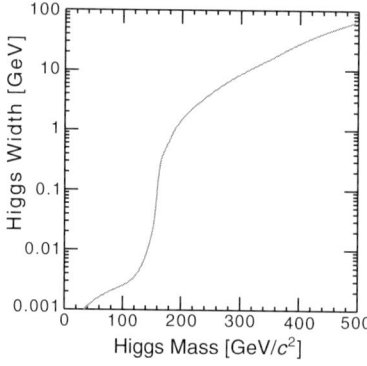 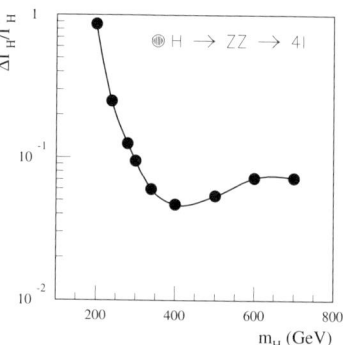

Fig. 8.39. Left: Standard Model Higgs total width as a function of M_H. Right: expected experimental precision on Γ_H at ATLAS using the $gg \to H \to ZZ \to 4\ell$ channel [42] (CMS similar).

Second, we note that for a light Higgs, which has a very small width (cf. Sec. 8.2.1.1), the Higgs production cross section is proportional to the partial width for Higgs decay to the initial state (the Narrow Width Approximation, NWA). That is, $\sigma_{gg \to H} \propto \Gamma_{H \to gg}$. Similarly, $\sigma_{\rm WBF} \propto \Gamma_{H \to W^+W^-}$. The student who has never seen this may easily derive it by recalling the definition of cross section and partial decay width – they share the same matrix elements and differ only by phase space factors[m]. Typically we abbreviate these partial widths with a subscript identifying the final state particle, thus we have Γ_g, Γ_γ, Γ_b, etc. Since a BR is just the partial decay width over the total width, we then write:

$$\left(\sigma_H \cdot {\rm BR}\right)_i \propto \left(\frac{\Gamma_p \Gamma_d}{\Gamma_H}\right)_i \qquad (8.19)$$

where Γ_p and Γ_d are the "production" and decay widths, respectively.

Third, count up the number of observables we have and measurements we can make. Assuming we have a decay channel for each possible Higgs decay (which we don't), we're still one short: Γ_H, the total width. Now, if the width is large enough, larger than detector resolution, we can measure it directly. Figure 8.39 shows that this can happen only for $M_H \gtrsim 230$ GeV or so [42], far above where EW precision data suggests we'll find the (SM) Higgs. Below this mass range, we have to think of something else.

---

[m]Well, slightly more than that in the case of WBF, but the argument holds after careful consideration.

In the SM, we know precisely what Γ_H is: the sum of all the partial widths. For the moment let's assume we have access to all possible decays or partial widths via production, ignore the super-rare decay modes to first- and second-generation fermions. This is a mild assumption, because if for some reason the muon or electron Yukawa were anywhere close to that of taus, where it might contribute to the total width, it would immediately be observable. The list of possible measurements we can form from accessible $(\sigma \cdot \mathrm{BR})_{i,exp}$ is:

$$X_\gamma, X_\tau, X_W, X_Z, Y_\gamma, Y_W, Y_Z, Z_b, Z_\gamma, Z_W \tag{8.20}$$

where X_i correspond to WBF channels, Y_i are inclusive Higgs production, and Z_i are top quark associated production[n] We could easily add measurements like X_μ, Y_e, etc. if we wanted, because measuring zero for any observable is still a measurement – it simply places a constraint on that combination of partial widths or couplings.

In the original implementation of this idea [89], the authors noted that the $t\bar{t}H$, $H \to b\bar{b}$ channel won't work, so there is no access at LHC to Γ_b. However, there is access to Γ_τ. In the SM, the b and τ Yukawa couplings are related by $r_b = \Gamma_b/\Gamma_\tau = 3c_{\mathrm{QCD}} m_b^2/m_\tau^2$, where c_{QCD} contains QCD higher-order corrections and phase space effects. Γ_W and Γ_Z are furthermore related by $SU(2)_L$, although we don't need to use it. Now write down the derived quantity

$$\tilde{\Gamma}_W = X_\tau(1+r_b) + X_W + X_Z + X_\gamma + \tilde{X}_g = \left(\sum \Gamma_i\right)\frac{\Gamma_W}{\Gamma_H} = (1-\epsilon)\Gamma_W \tag{8.21}$$

where $\tilde{X}_g$ is constructed from X_W, X_γ, Y_W and Y_γ. Although Γ_γ is an infinitesimal contribution to Γ_H, it is important as above, and it contains both the top quark Yukawa and W gauge-Higgs couplings. Our error is contained in ϵ and is typically small. This provides a good <u>lower</u> bound on Γ_W from data. The total width is then

$$\Gamma_H = \frac{\tilde{\Gamma}_W^2}{X_W} \tag{8.22}$$

and the error goes as $(1-\epsilon)^{-2}$. Assuming systematic uncertainties of 5% on WBF and 20% on inclusive production, this would achieve about a 10% measurement of Γ_W and 10 – 20% on the total width for $M_H < 200$ GeV.

---

[n]For this case, we actually use the Yukawa coupling squared (y_t^2) instead of Γ_t, because decays to top quarks is kinematically forbidden. But this is irrelevant for our argument.

Voilà! We have circumvented the naïve problem of not enough independent measurements. The astute observer should immediately protest, however, and rightly so. The result is achieved with a little too much confidence that the SM is correct. Not only does the trick rely on a very strong assumption about the b Yukawa coupling, but there could be funny business in the up-quark sector, giving a large partial width to e.g. charm quarks, which would not be observable either via production (too little initial-state charm, and anyhow unidentifiable) or decay (charm can't be efficiently tagged). Nevertheless, this was a useful exercise, because a much more rigorous, model-independent method is closely based on it.

The more sophisticated method is a powerful least-likelihood fit to data using a more accurate relation than Eq. 8.19 between data and theory [90]:

$$\sigma_H \cdot \mathrm{BR}(H \to xx) = \frac{\sigma_H^{\mathrm{SM}}}{\Gamma_p^{\mathrm{SM}}} \cdot \boxed{\frac{\Gamma_p \Gamma_d}{\Gamma_H}} \qquad (8.23)$$

where the partial widths in the box are the true values to be extracted from data, and the $(\sigma/\Gamma)_{\mathrm{SM}}$ ratio in front quantifies all effects shoved into Monte Carlo using SM values: phase space, QCD corrections, detector, etc. As before, the "sum" of all channels provides a solid lower bound on Γ_H, simply because some rate in each of a number of channels requires some minimum coupling. But these are found by a fit, rather than theory assumptions. It also properly takes into account all theory and experimental systematic and statistical uncertainties assigned to each channel. We then need only a firm upper bound on Γ_H and the fit then extracts *absolute* couplings (transformed from the partial widths). This bound comes from unitarity: the gauge-Higgs coupling can be depressed via mixing in any multi-doublet model, as well as any number of additional singlets, but it cannot exceed the SM value, which is strictly defined by unitarity. Thus $\Gamma_V \leq \Gamma_V^{\mathrm{SM}}$. (This bound is invalid in triplet models, but these have other characteristics which should make themselves apparent in experiment.) The WBF $H \to W^+W^-$ channel then provides an upper limit on Γ_H via its measurement of Γ_V^2/Γ_H.

The method can be further armored against BSM alterations by including the invisible Higgs channel, allowing additional loop contributions, and so on. Of course, the more possible deviations one allows, the larger the fit uncertainties become. We see this in the differences between the left and right panels of Fig. 8.40 [90]. It is obvious that LHC's weakness is lack of access to $H \to b\bar{b}$. Nevertheless, LHC can measure absolute Higgs couplings with useful constraints on BSM physics. This is especially

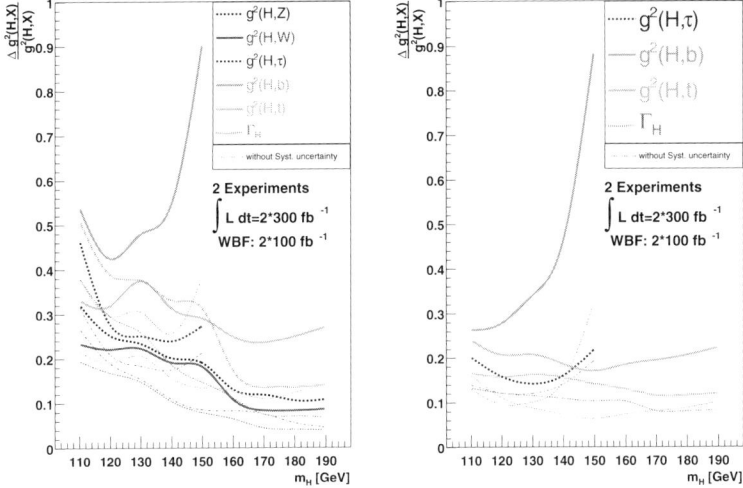

Fig. 8.40. Left: a least likelihood general fit on simulated LHC data, with no additional assumptions about the Higgs sector. Right: the fit assuming no new particles appear in Higgs loop-induced decays, and the gauge-Higgs coupling fixed exactly to the SM value. Figures from Ref. [90].

true for $M_H \gtrsim 150$ GeV, where LHC can achieve $\mathcal{O}(10\%)$ precision on the gauge-Higgs couplings and the total width.

The fit as implemented in Ref. [90] fixes M_H. This is a slight cheat, since for some M_H the BRs change quite rapidly, and a 1-2 GeV uncertainty can lead to a lot of slop in the coupling extraction. This is especially critical for the Higgs sector of the Minimal Supersymmetric Standard Model (MSSM). Eventually a fit to M_H will also have to be included, which will degrade measurement precision somewhat.

At the same time, there is cause for optimism. The results of Fig. 8.40 were based on very conservative, almost pessimistic assumptions: overly-large systematic errors, WBF not being possible at all at high-luminosity running, no minijet veto for WBF (cf. Sec. 8.2.3.3), and lack of progress in higher-order QCD calculations for signals and backgrounds. The reality is that significant progress has been made regarding QCD corrections, and we'll see one example shortly. Also, everyone knows that the minijet veto is a qualitatively correct aspect of the physics, we just can't accurately predict its impact. Early LHC data from Zjj production should take care of this. Furthermore, ATLAS and CMS experimentalists fully expect WBF to work at high-luminosity LHC running, they just don't have full simulation results for the probable efficiencies. Also, we may expect far better performance in the WBF $H \to \tau^+\tau^-$ channels as discussed in Sec. 8.2.3.4. Finally, if new physics exists up to a few TeV, it will be observable and we can take it into accounts in Higgs loop-induced decays.

Now to QCD corrections. Ref. [90] used large QCD uncertainties for $\sigma_{gg \to H}$ and Γ_g, 20% each, which is the correct NNLO uncertainty for each by itself. However, these two quantities appear as a ratio in our observables formula, Eq. 8.23. As pointed out in Ref. [91], most of these uncertainties drop out in the ratio. The reason for this is that the QCD corrections to the cross section and partial width are largely the same:

$$\Gamma \sim \alpha_s^2(\mu_R) C_1^2(\mu_R)[1 + \alpha_s(\mu_R)X_1 + ...] \qquad (8.24)$$
$$\sigma \sim \alpha_s^2(\mu_R) C_1^2(\mu_R)[1 + \alpha_s(\mu_R)Y_1 + ...] \qquad (8.25)$$

The correct uncertainty on the ratio is 5%, which will have an enormous impact on the fits of Fig. 8.40. We eagerly await new results from this and other improvements!

8.3.4. *Higgs couplings at an ILC*

Measuring Higgs couplings at an e^+e^- collider would be far more straightforward and rely on far fewer theoretical assumptions. Between that and being a colorless collision environment, it would also involve far fewer systematic uncertainties. I'll outline the basic idea.

In fixed-beam collisions it's possible to measure the *total* ZH production rate. To see this, we just apply a little relativistic kinematics, rewriting the invariant M_H^2:

$$M_H^2 = p_H^2 = (p_+ + p_- - p_Z)^2 = s + M_Z^2 - 2E_Z\sqrt{s} \qquad (8.26)$$

We see that observing the Higgs and measuring its total rate boils down to observing Z bosons via their extremely sharp dimuon peak and plotting this recoil mass. Figure 8.41 shows what the resulting event rate looks like in this distribution. The Higgs peak is clearly visible and sidebands allow one to subtract the SM background in the signal region. This captures all possible Higgs decays, even though that aren't taggable or even identifiable, simply by ignoring everything in the event except for the Z dimuons.

Simulations [71] suggest that the recoil mass technique would allow for about a 2.5% absolute measurement of the ZH rate. Since the cross section depends on the Z–Higgs coupling squared, the coupling uncertainty is then about a percent.

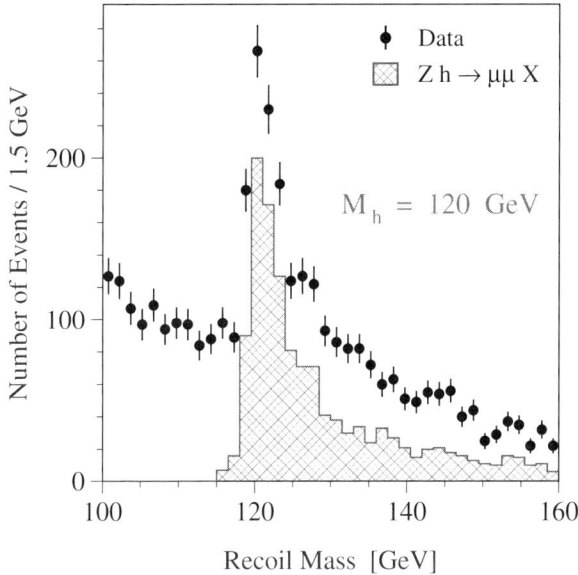

Fig. 8.41. Event rate of the recoil mass for $e^+e^- \to \mu^+\mu^- + X$ at a future high-energy linear collider. ZH production will fall into this sample, but the Higgs decays are ignored, thus capturing the total Higgs production rate. (Figure modified from Ref. [92] for a public talk by one of the authors.)

Getting from this one coupling and the total rate to any other coupling is formulaic:

1. In the total rate, measure the best branching ratios, whatever they may be. Depending on the mass and detector performance, that's likely one of $b\bar{b}$, $\gamma\gamma$ or W^+W^- decays.
2. Now look in WBF Higgs production[o] with the Higgs decaying to the same best final state. This yields the partial width Γ_W.
3. Calculate the total Higgs width as $\Gamma_W/BR(H \to W^+W^-)$.
4. Any other measured BR now gives that individual partial width, therefore the relevant coupling (or couplings for some loop-induced decays).

Table 8.2 enumerates the results of ILC simulation for select M_H [93]. (Clearly more thorough work should be done here.) There are a few noteworthy features. First, $H \to b\bar{b}$ would be accessible even as a rare BR at

---

[o]For a linear collider this is both $e^+e^- \to e^+e^- H$ and $e^+e^- \to \nu\bar{\nu}H$, since e and ν are distinguishable. Experimentally they become two different analyses.

Table 8.2. Estimated precision on various SM Higgs partial widths for a few select values of M_H, from measurements at a future e^+e^- collider [93].

M_H (GeV)	120	140	160	180	200	220
Decay	\multicolumn{6}{c}{Relative precision on Γ_i (%)}					
$b\bar{b}$	1.9	2.6	6.5	12.0	17.0	28.0
$c\bar{c}$	8.1	19.0				
$\tau^+\tau^-$	5.0	8.0				
gg	4.8	14.0				
W^+W^-	3.6	2.5	2.1			
ZZ				16.9		
$\gamma\gamma$	23.0					
$Z\gamma$		27.0				

larger M_H, due to the nearly QCD-free collision environment. Second, a weak measurement of $H \to c\bar{c}$ should be possible, for the same reason, and due to the superior b v. c resolution of the next generation of collider detectors. Third, $H \to jj$ is also accessible. This would be attributed to gg, which is a mild theoretical assumption. It is in principle sanity-checkable by the absence of an anomalous high-x Higgs production rate at LHC, which would come from sea or valence quarks and a non-SM coupling to lighter fermions (which would be difficult to accommodate theoretically, so not expected).

But what about the top Yukawa coupling? Its anticipated value of approximately one is curious enough to warrant special attention. A light Higgs can't decay to top quark pairs, so we'd have to rely on top quark associated production, as at LHC but without all the nasty QCD backgrounds. However, the event rate is far lower than at LHC and would require an 800 GeV machine collecting 1000 fb^{-1} [94], the planned lifetime of a next-generation second-stage machine (justifying my previous statement about the drive to go to maximum energy and sit there). One study combined expected LHC and ILC results [73, 95], and there are more recent results for ILC, summarized in Fig. 8.42 [96]. SLHC and an ILC would be complementary, granting superb coverage of M_H for a y_t measurement at the 10% level.

More sophisticated LC Higgs coupling analyses exist [87], but aren't often reviewed. They use a more complicated "optimal observables" (detailed kinematic shape information, for example) scheme. It's more powerful, but doesn't lend itself to the simplistic formulaic approach I just discussed.

I should emphasize that the results I reviewed are relevant only for the Standard Model. If the Higgs sector is non-minimal, or any new physics

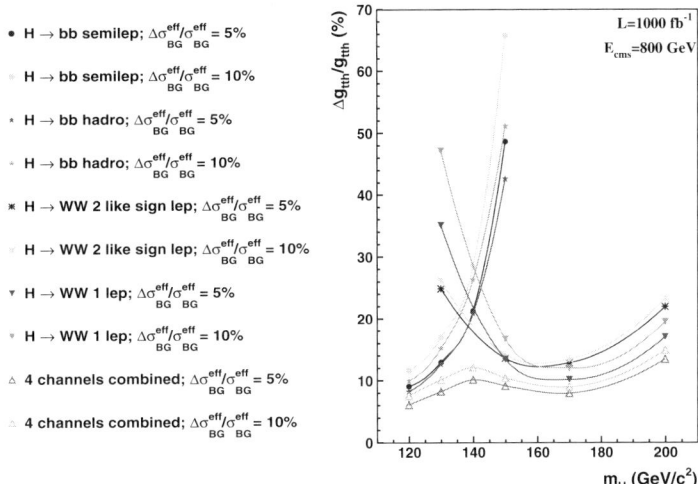

Fig. 8.42. Top Yukawa coupling measurement expectations for a future 800 GeV e^+e^- collider [96].

appears at the weak scale, it could result in altered couplings (and usually does; see Chapter 8.4). If they're suppressed, the event rate goes down, resulting in greater uncertainty. This is often glossed over or ignored in discussions of Higgs phenomenology, but is a potential reality and something we'd just have to lump. Nevertheless, it should be clear by now that an ILC would be a spectacular experiment for precision Higgs measurements.

8.3.5. *Higgs potential*

Finally we arrive at the most difficult Higgs property to test, the potential. This is the hallmark of spontaneous symmetry breaking, thus ranks at least as high in priority as finding Yukawa couplings proportional to fermion masses. To see what's involved, let's review the SM Higgs potential. The potential is normally written as:

$$V(\Phi) = \mu^2 \Phi^\dagger \Phi + \lambda (\Phi^\dagger \Phi)^2 \qquad (8.27)$$

where Φ is our $SU(2)_L$ complex doublet of scalar fields. The Higgs spontaneous symmetry-breaking mechanism is what happens to the Lagrangian when $\mu^2 < 0$ and the field's global minimum shifts to $v = \sqrt{-\mu^2/\lambda}$. We then expand $\Phi \to v + H(x)$ (ignoring the Goldstone modes which you learned about in Sally Dawson's lectures) where $H(x)$ is the radial excitation, the physical Higgs boson. The Higgs mass squared is then $2v^2\lambda$, and is the only free parameter, although constrained (weakly) by EW precision fits. The student performing this expansion will also notice HHH and $HHHH$ Lagrangian terms, which are self-interactions of the Higgs boson. The three- and four-point couplings are $-6v\lambda$ and -6λ, respectively[p].

To measure the potential is to measure these self-couplings and check their relation to the measured Higgs mass. Our phenomenological approach is to rewire the Higgs potential in terms of independent parameters and the Higgs candidate field η_H:

$$V(\eta_H) = \frac{1}{2} M_H^2 \eta_H^2 + \lambda v \eta_H^3 + \frac{1}{4} \tilde{\lambda} \eta_H^4 \qquad (8.28)$$

λ and $\tilde{\lambda}$ are now free parameters, which we measure from the direct production rate of HH and HHH events. This will ultimately be a voyage of frustration.

8.3.5.1. HH production at LHC

We begin with Higgs pairs at LHC. The dominant production mechanism is gluon fusion, $gg \to HH$ [97–99]. The Feynman diagrams are shown in Fig. 8.43. The first diagram is off-shell single Higgs production which split via the three-point self-coupling to a pair of on-shell Higgses, which then decay promptly. The second diagram is a box (four-point) loop contribution which involves only the top quark Yukawa coupling. Interestingly, the two diagrams interfere destructively and have a rather large cancellation. This means the rate is small [100], as shown in Fig. 8.44, making our life difficult with a small statistical sample. On the other hand, the destructive interference will turn out to be crucial to making constructive statements about the self-coupling λ.

The left panel of Fig. 8.44 tells us that we can expect $\mathcal{O}(10k)$ light Higgs pair events per detector over the expected 300 fb^{-1} lifetime of the first LHC run, and ten times that at SLHC. That sounds like a lot, but keep in mind that both Higgses have to decay to a final state we can observe, which will reduce the captured rate to something much smaller. Then we have to consider what backgrounds affect each candidate channel.

The right panel of Fig. 8.44 shows selected Higgs pair branching ratios. At low mass, decays to b pairs dominate, as expected, while for $M_H \gtrsim 135$ GeV mass it's W pairs. We can immediately discount the $4b$ final state

[p]Don't forget the identical-particle combinatorial factors.

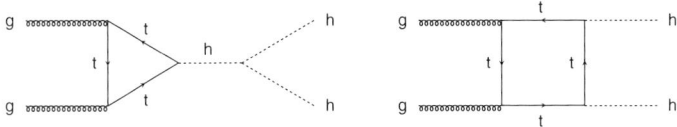

Fig. 8.43. Feynman diagrams for the dominant Higgs pair production rate at LHC, $gg \to HH$.

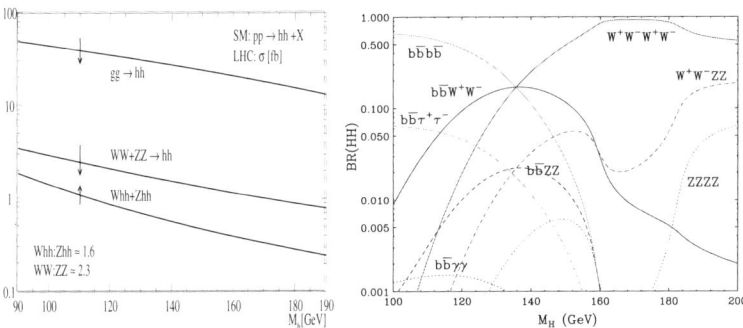

Fig. 8.44. Left: Higgs pair production cross sections at LHC as a function of M_H [100]. Arrows show the change of the cross section as λ is increase, and the tips are at one-half and twice the SM value. Right: Higgs pair branching ratios as a function of M_H, calculated using HDECAY [22].

as hopeless, based on what we already learned about QCD backgrounds – but $4W$ is promising for higher masses. The next-largest mode from those two is $b\bar{b}W^+W^-$, which unfortunately is the same final state as the far larger top quark pair cross section. A few minutes' investigation causes this to be discarded, even after trying various invariant mass constraints; b pair mass resolution is just not good enough. The $b\bar{b}\tau^+\tau^-$ mode has very low backgrounds, comparable to the signal, but suffers hugely from lack of statistics, due to low efficiency for subsequent tau decays. However, the rare decay mode $b\bar{b}\gamma\gamma$ is extremely clean and worth further consideration at low masses.

$HH \to W^+W^-W^+W^-$ at LHC

$HH \to W^+W^-W^+W^-$ has myriad decays, but for triggering purposes and to get away from QCD background sources of leptons (like top quarks) we need to select special multilepton final states [101]. The most likely

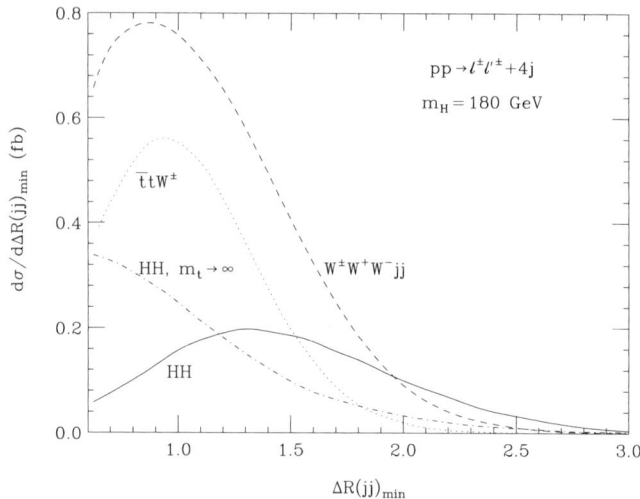

Fig. 8.45. Differential cross section as a function of the minimum jet pair lego plot separation for $\ell^+\ell^- + 4j$ at events at LHC. The solid curve is the correct distribution using exact matrix elements for HH, while the dash-dotted curve comes from effective-Lagrangian matrix elements where the top quark mass is taken to infinity. Figure taken from Ref. [101].

accessible channels are same-sign lepton pairs, $\ell^\pm\ell^\pm + 4j$, and three leptons, $\ell^+\ell^-\ell^\pm + 2j$, since the principal QCD SM backgrounds can't easily mimic them. Note that because of multiple neutrinos departing the detector unobserved, complete reconstruction is not possible. The principle backgrounds are $WWWjj$, $t\bar{t}W$, $t\bar{t}j$, $t\bar{t}Z/\gamma^*$ and $WZ + 4j$, but we also need to consider $t\bar{t}t\bar{t}$, $4W$, $W^+W^- + 4j$, W^+W^-Zjj as well as double parton scattering and overlapping events. The calculation of all of these is technical so I won't go into it, rather simply mention a few noteworthy points.

The first is a warning about using the $gg \to HH$ effective Lagrangian in practical calculations. It is still a mystery why the leading term in the $\sqrt{\hat{s}}/m_t$ expansion [97] should get the overall rate so close that of an exact calculation [98], but it does. Because of that, nobody has ever bothered to calculate higher-order terms in the effective Lagrangian expansion; in any case, the exact results are available, as well as NLO in QCD [99]. However, the leading terms in the expansion cancel too much close to threshold, yielding incorrect kinematics [101], as can be seen from Fig. 8.45. One should thus use only the exact matrix element results for practical $gg \to HH$ phenomenology.

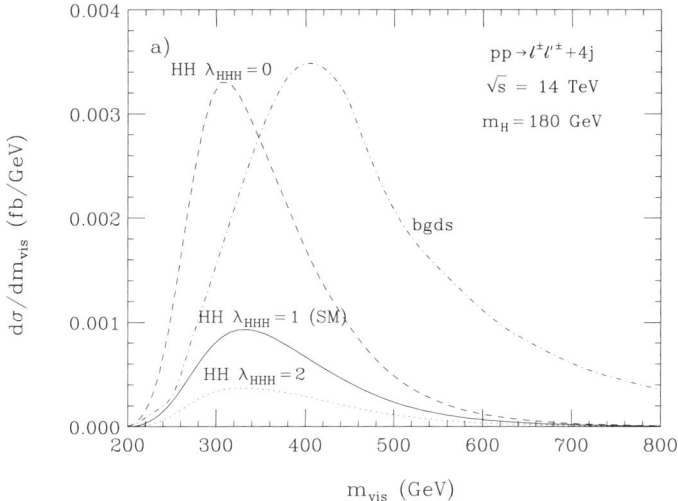

Fig. 8.46. Visible invariant mass distribution for same-sign dilepton plus four jet event at LHC [101]. All SM backgrounds are summed into one curve, while the $gg \to HH$ signal is shown separately, for the SM value of self-coupling λ, twice that value, and zero.

The second point is that our main systematic uncertainties will be our limited knowledge of the top quark Yukawa coupling, which drives the production rate, and the BR to W^+W^-, which drive the decay fraction. These must be known very precisely for any measurement to be useful.

We will need a discriminating observable to separate signal from background. We can speculate that nearly all the signal's kinematic information is encoded in the invariant mass of the visible final state particles, so let's construct a new variable, m_{vis}:

$$m_{vis}^2 = \left[\sum_i E_i\right]^2 - \left[\sum_i \mathbf{p}_i\right]^2 \qquad (8.29)$$

where i are all the leptons and jets in the event. We suspect a difference because the signal is a two-body process, which is threshold-like, while the backgrounds are multi-body processes which peak at much larger m_{vis} than the sum of their heavy resonances' masses.

Figure 8.46 displays the fruits of parameterizing our ignorance (or rather, the detector's). The separation between signal and background is exactly as expected: the signal peaks much lower, allowing a χ^2 fit to distinguish it from the backgrounds. But the plot also reveals a saving grace

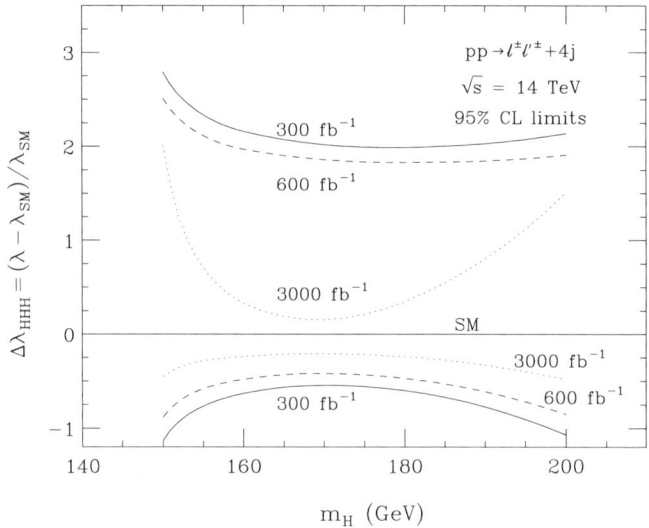

Fig. 8.47. 95% CL limits achievable at LHC on the shifted Higgs triple self-coupling (see text), $\triangle\lambda$, for LHC and SLHC expected luminosities [101].

in the destructive interference between triangle and box loop diagrams. If spontaneous symmetry breaking isn't the right description and there is no Higgs potential, then $\lambda = 0$ and the lack of destructive interference gives a wildly larger signal cross section, which is far easier to observe.

Figure 8.47 summarizes the results of Ref. [101]. It plots 95% CL limits on the shifted self-coupling, $\triangle\lambda = (\lambda - \lambda_{SM})/\lambda_{SM}$. This is somewhat easier to understand: zero is the SM, and -1 corresponds to no self-coupling, or no potential. For $M_H > 150$ GeV, the LHC can exclude $\lambda = 0$ at (for some M_H much greater than) 2σ with only the LHC. After SLHC running, this becomes a $20 - 30\%$ measurement, if other systematics are under control. Here, they're assumed to be smaller than the statistical uncertainty.

Another potential systematics issue is minimum bias, the presence of extra jets in an event which don't come from the primary hard scattering. Here, they could be confused with jets from the W bosons, causing a distortion of m_{vis}. ATLAS has investigated this and found it to not be a concern – the shape of m_{vis} for the signal remains largely unaltered [102].

$HH \to b\bar{b}\gamma\gamma$ at LHC

We've already ruled out as viable the vast majority of Higgs pair BRs for $M_H \lesssim 150$ GeV due to QCD backgrounds or too-small efficiencies.

Table 8.3. The major ID efficiencies and fake photon rejection factors at LHC. Note the two values for $P_{j\to\gamma}$, which represent the current uncertainty in detector capability for fake photon rejection. The true value won't be known until data is collected. See Ref. [103] for details.

	ϵ_γ	ϵ_μ	$P_{c\to b}$	$P_{j\to b}$	$P^{hi}_{j\to\gamma}$	$P^{lo}_{j\to\gamma}$
LHC	80%	90%	1/13	1/140	1/1600	1/2500
SLHC	80%	90%	1/13	1/23	1/1600	1/2500

However, the rare decay mode to $b\bar{b}\gamma\gamma$ is worth a closer look [103]. There are many backgrounds to consider, coming from b or c jets plus photons, or other jets which fake photons, just as in the single Higgs to photon pairs case. Table 8.3 highlights the major ID efficiencies and fake photon rejection factors at LHC and SLHC relevant for us. The backgrounds are all calculable at LO, but with significant uncertainties, probably a factor of two or more. However, that won't be a concern as we can identify distributions useful for measuring the background in the non-signal region. Note that with this channel we can completely reconstruct both Higgs bosons.

The background QCD uncertainties have a work-around. There are two angular distributions in the lego plot which look very different for the signal, principally because scalars decay isotropically and thus are uncorrelated, while the QCD backgrounds have spin correlations. The two distributions are shown in Fig. 8.48. The differences are rather dramatic (and even more so in 2-D distributions). Tevatron's experiments CDF and DØ have used such a pseudo-sideband analysis for some time to measure a background in a non-signal region to normalize their Monte Carlo tools, then extrapolating

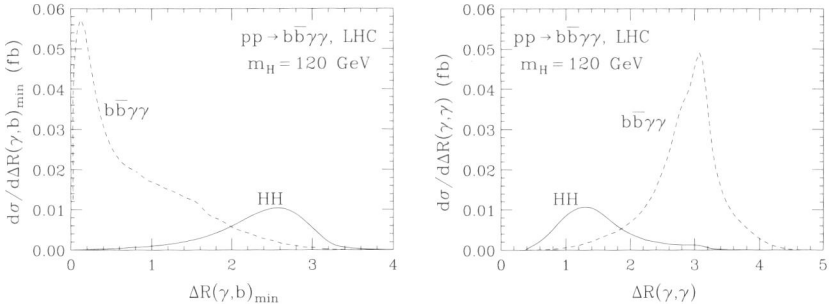

Fig. 8.48. Angular separations in the lego plot for b jets and photons in $gg \to HH \to b\bar{b}\gamma\gamma$ signal events and background at the LHC. Figures from Ref. [103].

Table 8.4. Expected event rates after ID efficiencies and all kinematic cuts for $b\bar{b}\gamma\gamma$ events at LHC (SLHC), two detectors and 600(6000) fb^{-1} of data [103]. LHC assumes only one b tag, while SLHC requires two. Note the increased fake rate at SLHC.

	HH	$b\bar{b}\gamma\gamma$	$c\bar{c}\gamma\gamma$	$b\bar{b}\gamma j$	$c\bar{c}\gamma j$	$jj\gamma\gamma$	$b\bar{b}jj$	$c\bar{c}jj$	γjjj	$jjjj$	$\sum$(bkg)	S/B
LHC	6	2	1	1	0	5	0	0	1	1	11	1/2
SLHC	21	6	0	4	0	6	1	0	1	1	20	1/1

to the signal region to perform a background subtraction. The technique is viable because QCD radiative corrections *in general* do not significantly alter angular distributions.

Table 8.4 summarizes the results of Ref. [103]. It gives event rates expected with 600(6000) fb^{-1} of data (two detectors) at LHC(SLHC). SLHC would not get ten times as many events because of lower efficiency of having to tag two b jets instead of only one, to overcome the low fake jet rejection rate in a high-luminosity environment. First, note that fake b jets or fake photons are the largest background: the measurement would be significantly hampered by detector limitations. Second, while the S/B ratio is excellent, the overall event rate is extremely small, definitely in the non-Gaussian statistics regime.

SLHC could make a useful statement about λ, ultimately achieving limits on $\Delta\lambda$ of about ± 0.5, but this is not such a strong statement. It could at best generally confirm the SM picture of spontaneous symmetry breaking and perhaps rule out wildly different scenarios, but would never be particularly satisfying. On the other hand, it's strong encouragement for ATLAS and CMS to push the envelope on tagging efficiency and fake rejection, especially for the detector upgrades necessary for SLHC. Doing studies like this well ahead of time is useful for this reason, our present case being a perfect example.

8.3.5.2. *HH production at an ILC*

While (S)LHC clearly has access to Higgs pair production and thus λ for $M_H > 150$ GeV, it would disappoint at lower masses. We should see if a future linear collider could also give a precision measurement for λ as it could for (most) other Higgs couplings.

For e^+e^- collisions below about 1 TeV, double Higgsstrahlung is the largest source of Higgs pairs. The Feynman diagrams appear in Fig. 8.49, while the cross sections as a function of M_H for 500 and 800 GeV collisions [104] are found in Fig. 8.50, which also shows the cross sections times

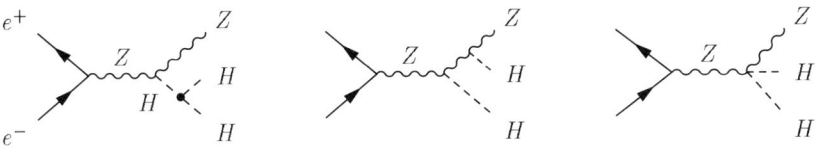

Fig. 8.49. Feynman diagrams for double Higgsstrahlung at a future linear collider, $e^+e^- \to HH$.

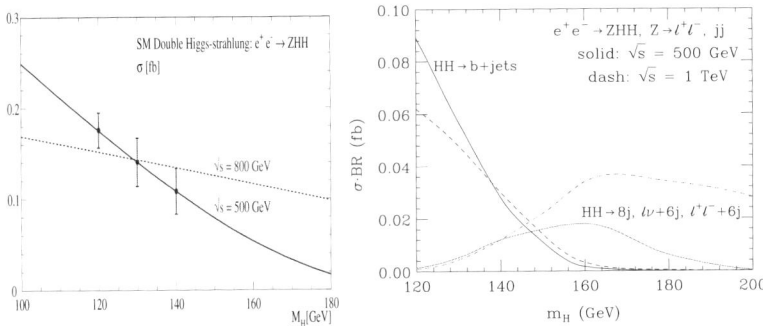

Fig. 8.50. Left: the double Higgsstrahlung cross section as a function of M_H for 500 and 800 GeV e^+e^- collisions [104]. Right: the cross section times BR at 500 GeV and 1 TeV e^+e^- collisions, for the dominant final state BRs as a function of M_H [105].

BRs for the dominant final states over the range of Higgs masses. Roughly, this corresponds to $4b$ and $4W$ final state. The former is very steeply falling with M_H, but the latter is much flatter over the 100–200 GeV mass region, suggesting broader access if at all visible.

The parton-level studies performed so far [105] are fairly encouraging. As shown in Fig. 8.51, an ILC could achieve about a $20-30\%$ measurement of λ over a broad mass range, with somewhat worse performance around $M_H \sim 140$ GeV, where the $b\bar{b}$ and W^+W^- BRs are roughly equal. Interestingly, for a lower Higgs mass, the analysis prefers lower machine energy, while the opposite is true at least to a small degree at higher mass. This is largely a phase space effect for the 3-body production mechanism. Also, SLHC is superior for $M_H \gtrsim 150$ GeV (largely due to better statistics), with an important caveat: controlling systematics in $gg \to HH \to 4W$ at LHC would require precision input from ILC for the Higgs couplings and BRs. This is an excellent example of synergy between experiments.

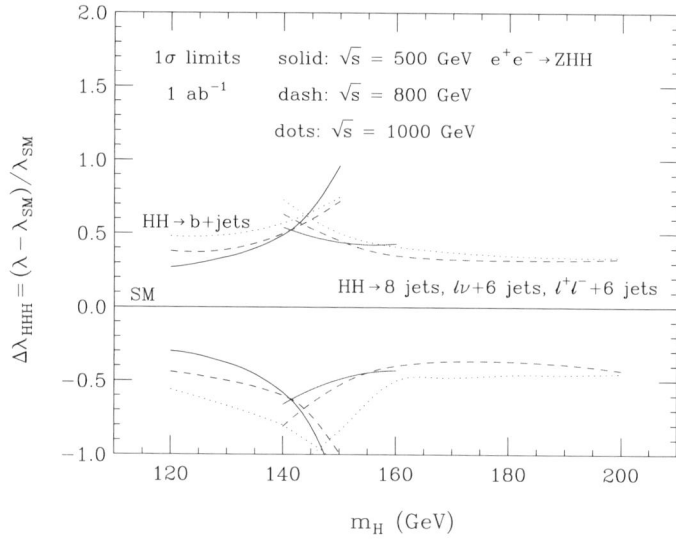

Fig. 8.51. Estimated achievable limits in the shifted self-coupling $\triangle\lambda$ (see Sec. 8.3.5.1) at future e^+e^- colliders of various energy, as a function of M_H [105].

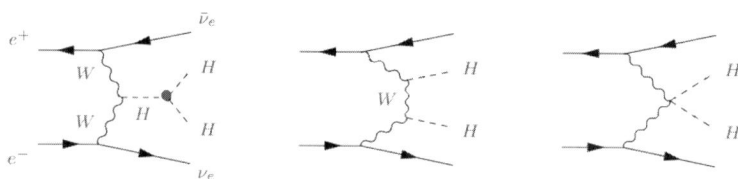

Fig. 8.52. Representative Feynman diagrams for the WBF process $e^+e^- \to \nu\bar{\nu}HH$.

Double Higgsstrahlung is not the only source of Higgs pairs at an e^+e^- collider, however. In fact, as the energy increases, WBF Higgs pair production becomes more and more important. Representative Feynman diagrams for $e^+e^- \to \nu\bar{\nu}HH$ are shown in Fig. 8.52. A preliminary analysis [106] for CLIC [107], a second-generation $1-5$ TeV e^+e^- collider collecting 5000 fb^{-1}, found rather interesting results, summarized graphically in Fig. 8.53. The principal finding is that no matter how high the collision energy goes, and regardless of Higgs mass, the precision on λ bottoms out at $10-15\%$. This is because the self-coupling has an s-channel suppression, and its contributions becomes washed out as by other diagrams as $\sqrt{s}$

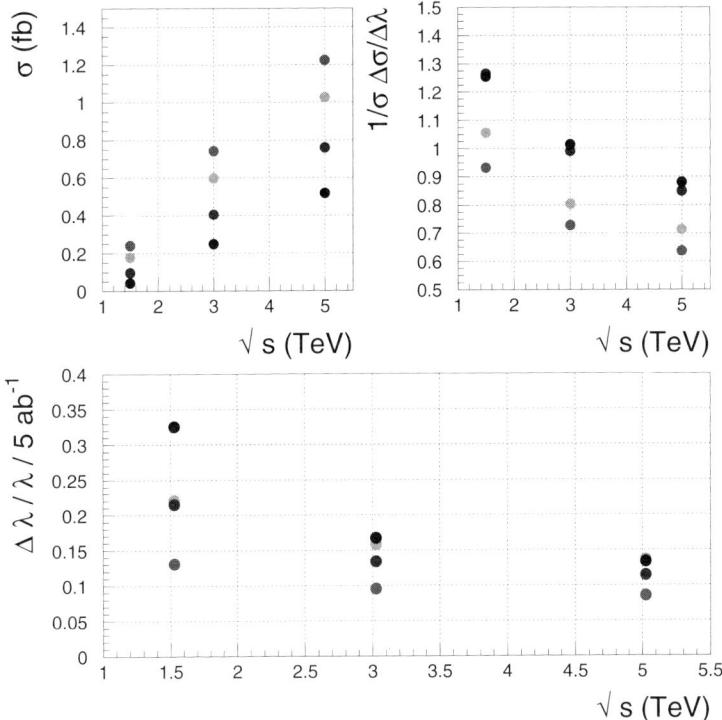

Fig. 8.53. The results of Ref. [106] for WBF HH production at CLIC, a second-generation multi-TeV e^+e^- collider. The plot labels are self-explanatory, while the colors are for various Higgs masses: 120 GeV in red, 140 GeV in blue, 180 GeV in green and 240 GeV in black.

increases. A corollary, though, is that CLIC could potentially achieve better precision than SLHC for larger M_H, although this may be marginal. Much more detailed work would be required for both SLHC and CLIC, as well as experience at LHC and SLHC to determine its true potential, to make conclusive statements.

8.3.5.3. *Electroweak corrections to* λ

One final word on the trilinear self-coupling λ: Ref. [108] calculated the leading 1-loop top quark EW corrections to $\lambda_{\rm SM}$. Their principal SM result is:

$$\lambda_{HHH}^{eff} = \frac{M_H^2}{2v^2}\left[1 - \frac{N_C}{3\pi^2}\frac{m_t^4}{v^2 M_H^2} + ...\right] \quad (8.30)$$

The correction is $-10\%(-4\%)$ for $M_H = 120(180)$ GeV, non-trivial for smaller Higgs masses, but those are excluded in the SM. This correction should obviously be taken into account in any future analysis, should the Higgs be found. But it should be clear that neither (S)LHC nor ILC will be sensitive to it. Even CLIC would have only marginal sensitivity, and then only for low M_H.

Non-minimal Higgs sectors and new physics effects can tell a very different story, however, as we'll see, coming up in Secs. 8.4.1 and 8.4.5.

8.3.5.4. *HHH production anywhere*

The trilinear self-coupling λ is only part of our phenomenological Higgs potential of Eq. 8.28, though. We also need to measure $\tilde{\lambda}$, the quartic self-coupling. In some sense this is equally important to measuring λ. Recall the structure of the Higgs potential: λ allows the global minimum to be away from zero, but a non-zero (and positive) $\tilde{\lambda}$ is required to keep the potential bounded from below. We can't really convince ourselves that the potential structure of Eq. 8.27 is the right picture without a measurement of both these ingredients. We've just seen that probing λ is extremely challenging. Just how difficult is this likely to be for $\tilde{\lambda}$?

For e^+e^- collisions we already know this is hopeless: the HHH rate is both too low and its dependence on $\tilde{\lambda}$ too weak [104]. However, the situation at (S)LHC was only very recently investigated [109, 110]. The authors calculated the $gg \to HHH$ cross section, which involves Feynman diagrams like those of Fig. 8.54. Note the appearance of numerous diagrams dependent on the trilinear self-coupling, in addition to diagrams dependent only on y_t.

The results of the study are shown in Fig. 8.55, *for a 200 TeV VLHC*. They're rather deflating because the cross section is miserably small. A challenge to the student: find a three-Higgs BR to a final state that could be observed at a VLHC, where the rate is not laughable. Good luck! In addition, the right panel shows that any variation of the trilinear coupling

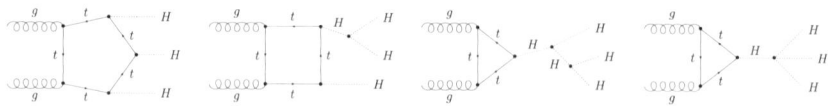

Fig. 8.54. Representative Feynman diagrams for $gg \to HHH$.

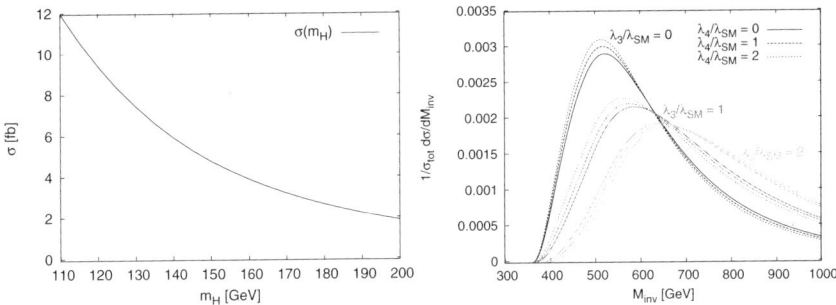

Fig. 8.55. Left: 200 TeV VLHC $gg \to HHH$ cross section as a function of M_H. Right: differential cross section as a function of M_{HHH} for three values each of λ and $\tilde{\lambda}$. Figures from Ref. [109].

λ completely swamps variation of the quartic $\tilde{\lambda}$, whose own variation is already infinitesimal.

In summary, it appears that we will likely never achieve a complete picture of the Higgs potential. This of course applies only to the Standard Model. Coming up in Chapter 8.4 we're going to see that for BSM physics the situation is even more discouraging.

8.4. Beyond-the-SM Higgs sectors

Now that we know how the Standard Model Higgs sector works – how it could be discovered and measured at LHC – it's natural to think about other possibilities for EWSB. The SM Higgs is elegant in its simplicity, but as you know from Sally Dawson's SM lectures, it's probably too minimal – nagging theoretical questions remain about Higgs mass stability, flavor (ignoring this is kind of a black eye), neutrino masses (another black eye), and so on. Because new physics that could explain dark matter is likely to also lie at the TeV scale, most model building makes an attempt to incorporate solutions to some of these other problems along with EWSB. The literature is vast, but let's try to roughly classify some of the major ideas to get a handle on the variations.

The broadest two categories of classes are weakly-coupled new physics which can be handled with perturbation theory, and strongly-coupled or "strong dynamics" models which are penetrable in some cases, others not. These include QCD-inspired theories like Technicolor [10, 11] (or more properly Extended [12, 13] or Walking [14] Technicolor, which can handle a top quark mass very different from the other quark masses) and Topcolorassisted Technicolor [15] ("TC2"), which incorporates additional weakly-coupled gauge structure. Strong dynamics assumes that some TeV-scale massive or heavier fermions' attraction became strong at low energy scales, eventually causing their condensation to mesonic states (Technipions, Technirho, Technieta, etc.), the neutral scalars of which can incite EWSB via their $SU(2)_L$ gauge interactions. Strong dynamics scenarios are beyond the scope of these lectures, however, so I leave it for the interested student to study the excellent review article of Ref. [111].

While strong dynamics theories are Higgsless in some sense, meaning no fundamental scalar fields, the terms is usually reserved for a new class of models where the EW symmetry is broken using boundary conditions on gauge boson wavefunctions propagating in finite extra dimensions (see e.g. Refs. [18, 112]). We'll also skip these.

There is far more theoretical effort expended on weakly-coupled EWSB, which is mostly variations on what we can add to the single Higgs doublet of the SM:

① 1HDM + invisible (high-scale) new physics, hidden from direct detection
② CP-conserving 2HDM: 4 types (minimal supersymmetry, MSSM, is Type II)
③ CP-violating 2HDM
④ Higgs singlet(s) (e.g. next-to-minimal supersymmetry, NMSSM)
⑤ Higgs triplets (often appear in Grand Unified Theories)
⑥ Little Higgs models: $SU(2)_L \times U(1)_Y$ is part of larger gauge and global group

The first item, new high-scale physics hidden from direct detection, sounds like a cheat. It actually involves an important aspect of phenomenology: effective Lagrangians from higher-dimensional operators. We'll come back to these in a moment. Two Higgs doublets instead of one is an idea with multiple sources. For instance, one doublet could give mass to the leptons and the other to the quarks, or one to the up-type fermions and the other to the down-type, etc. We'll return to these after effective operators. Additional Higgs singlets likewise have a variety of reasons for being written down, but usually it's just "we can do it, so we will". We'll skip these. Higgs triplets originated from natural appearance in left-right symmetric GUTs. They're a bit exotic and typically have issues with precision EW data, but are interesting in that they predict the existence of doubly-charged Higgs states $H^{\pm\pm}$, and a tree-level $H^{\pm}W^{\mp}Z$ coupling, which must be zero in most Higgs-doublet models. It would therefore stand out experimentally. For all these cases I don't have time to cover, the Higgs Hunter's Guide is the best place to start to learn more [6].

Little Higgs theories, on the other hand, are different in that they necessarily involve new scalar *and* gauge structure arising from an enlarged global symmetry from which the SM emerges, as well as additional matter content. Interestingly, in these models the Higgs looks very much like the SM Higgs, but with $\mathcal{O}(v^2/F^2)$ corrections, where F is typically a few TeV, parametrically 4π larger than the EW scale. The smallness of v^2/F^2 could make it very difficult to measure Little Higgs corrections to Higgs observables. These models are probably ultimately strongly-coupled at a scale $\Lambda \sim 4\pi F$, but this is an open question. If nature chose this course, the most interesting physics is the new gauge boson and matter fields that appears at a scale F. Refs. [19] provide nice overviews and simple explanations of the two primary Little Higgs mechanisms.

8.4.1. *Higher-dimensional operators*

The new physics responsible for dark matter, flavor, neutrino masses, etc., might very well be too massive to produce directly at colliders. This the dreaded SM-Higgs-only scenario, where LHC sees nothing new. It would really be an invitation to take a more rigorous look at all data – new physics effects might still appear as small deviations in precision observables.

The standard way of parameterizing this is to write down all the possible Lagrangian operators with the heavy fields integrated out which preserve $SU(3)_c \times SU(2)_L \times U(1)_Y$ gauge invariance. This was done over two decades ago for operators up to dimension six [86]. Although not often emphasized in today's phenomenology, I consider this paper a must-read for all students.

Let's begin by considering the possible operators involving only the SM Higgs doublet. There are two, of dimension six:

$$\mathcal{O}_1 = \frac{1}{2} \partial_\mu (\Phi^\dagger \Phi) \, \partial^\mu (\Phi^\dagger \Phi) \quad \& \quad \mathcal{O}_2 = -\frac{1}{3} (\Phi^\dagger \Phi)^3 \quad (8.31)$$

for the effective Lagrangian contribution

$$\mathcal{L}_{6D,\Phi} = \sum_{i=1}^{2} \frac{f_i}{\Lambda^2} \mathcal{O}_i \, , \quad f_i > 0 \quad (8.32)$$

Λ must be at least a couple TeV, otherwise we'd likely observe it directly at LHC. If you've somewhere seen an alternative effective theory for the

Higgs potential written as

$$V_{\text{eff}} = \sum_{n=0} \frac{\lambda_n}{\Lambda^{2n}} \left(|\Phi|^2 - \frac{v^2}{2}\right)^{2+n} \qquad (8.33)$$

the operators written above correspond to the $n = 1$ term in this expansion.

$\mathcal{O}_1$ modifies the Higgs kinetic term, while $\mathcal{O}_2$ modifies the EW vev, v:

$$\mathcal{L}_{\text{kin}} = \frac{1}{2}\partial_\mu \phi \partial^\mu \phi + \frac{1}{2} f_1 \frac{v^2}{\Lambda^2} \partial_\mu \phi \partial^\mu \phi, \quad \frac{v^2}{2} \approx \frac{v_0^2}{2}\left(1 - \frac{f_2}{4\lambda}\frac{v_0^2}{\Lambda^2}\right) \qquad (8.34)$$

where v is what G_F measures. We must also canonically normalize the physical Higgs field: $\phi = NH$ with $N = 1/(1 + f_1 \frac{v^2}{\Lambda^2})$.

This results in a number of alterations to masses and couplings [113]. First, the Higgs mass itself receives corrections from the expected value, given λ:

$$M_H^2 = 2\lambda v^2 \left(1 - f_1 \frac{v^2}{\Lambda^2} + \frac{f_2}{2\lambda}\frac{v^2}{\Lambda^2}\right) \qquad (8.35)$$

where the f_2 term is independent of λ. Next, Higgs gauge couplings receive v^2/Λ^2 shifts:

$$\frac{1}{2}g^2 v \left(1 - \frac{f_1}{2}\frac{v^2}{\Lambda^2}\right) H W_\mu^+ W^{-\mu} \qquad \frac{1}{4}g^2 \left(1 - f_1 \frac{v^2}{\Lambda^2}\right) HH W_\mu^+ W^{-\mu}$$

$$\frac{1}{2}\frac{g^2}{c_W} v \left(1 - \frac{f_1}{2}\frac{v^2}{\Lambda^2}\right) H Z_\mu Z^\mu \qquad \frac{1}{4}\frac{g^2}{c_W}\left(1 - f_1 \frac{v^2}{\Lambda^2}\right) HH Z_\mu Z^\mu \qquad (8.36)$$

Finally, the Higgs boson self-couplings are (phases vary with Feynman rule convention):

$$|\lambda_{3H}| = \frac{3m_H^2}{v}\left[\left(1 - \frac{f_1}{2}\frac{v^2}{\Lambda^2} + \frac{2f_2}{3}\frac{v^2}{M_H^2}\frac{v^2}{\Lambda^2}\right) + \frac{2f_1}{3M_H^2}\frac{v^2}{\Lambda^2}\sum_{i<j}^{3} p_i \cdot p_j\right] \qquad (8.37)$$

$$|\lambda_{4H}| = \frac{3m_H^2}{v^2}\left[\left(1 - f_1 \frac{v^2}{\Lambda^2} + 4f_2 \frac{v^2}{M_H^2}\frac{v^2}{\Lambda^2}\right) + \frac{2f_1}{3M_H^2}\frac{v^2}{\Lambda^2}\sum_{i<j}^{4} p_i \cdot p_j\right] \qquad (8.38)$$

Note that $\mathcal{O}_1$ and $\mathcal{O}_2$ both enter here, but more importantly there are momentum-dependent terms, which are typical of higher-dimensional operators. The effect of these terms would be anomalous high-p_T Higgses in pair production.

Only one phenomenological analysis exists for these effects, and only for precision experiments at a future ILC and CLIC [113]. In this study, measurements are expressed in terms of $a_i = f_i v^2/\Lambda^2$, since f_i and Λ

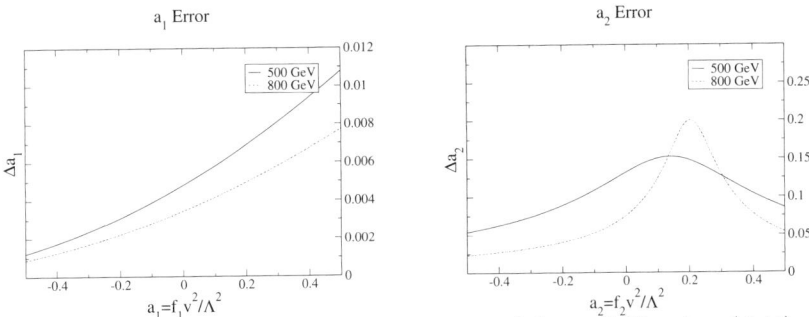

Fig. 8.56. Achievable uncertainty on measurements of the a_1 (left) and a_2 (right) coefficients of Eq. 8.31 (also see text) at a future ILC for 500 and 800 GeV running [113].

can't be easily separated from so few measurements. Higgsstrahlung, double Higgsstrahlung and WBF Higgs pair production together measure a combination of a_1 and a_2.

Figure 8.56 shows the expected achievable uncertainties (not limits!) on a_1 and a_2 at a future ILC. For $f_1 = 1$, this corresponds to a reach in Λ of about 4 TeV, possibly out of the reach of LHC depending on what might be directly produced. For $f_2 = 1$, however, this corresponds to only about $\Lambda \sim 0.8$ TeV, easily accessible at LHC. Put another way, an ILC could have access to new high-scale physics via altered Higgs–gauge boson couplings, but not via Higgs self-couplings. This is in line with what we'd come to expect, as HH production is much smaller. The shapes of the uncertainty curves in the figures depend on what values of the operator coefficients add to or subtract from the signal, with the added feature that the momentum dependence of the Higgs self-couplings that the $\mathcal{O}_1$ operator introduces changes to kinematic distributions.

In addition to the Higgs-only D6 operators, there are a handful of operators involving the Higgs and gauge boson fields together [86]:

$$O_{WW} = (\phi^\dagger \phi)[W^+_{\mu\nu} W^{-\mu\nu} + \tfrac{1}{2} W^3_{\mu\nu} W^{3\mu\nu}]$$
$$O_{BB} = (\phi^\dagger \phi) B_{\mu\nu} B^{\mu\nu}$$
$$O_{BW} = B^{\mu\nu}[(\phi^\dagger \sigma^3 \phi) W^3_{\mu\nu} + \sqrt{2}[(\phi^\dagger T^+ \phi) W^+_{\mu\nu} + (\phi^\dagger T^- \phi) W^-_{\mu\nu}]]$$
$$O_B = (D^\mu \phi)^\dagger (D^\nu \phi) B_{\mu\nu}$$
$$O_W = (D^\mu \phi)^\dagger [\sigma^3 (D^\nu \phi) W^3_{\mu\nu} + \sqrt{2}[T^+ (D^\nu \phi) W^+_{\mu\nu} + T^- (D^\nu \phi) W^-_{\mu\nu}]]$$
$$O_{\Phi,1} = (D_\mu \phi)^\dagger \phi \phi^\dagger (D^\mu \phi)$$

These induce momentum-dependent $HHVV$ vertices, so could be studied at an ILC or CLIC in the same manner as the Higgs-only couplings, as well as with rare Higgs decays [114], but in general they're highly constrained by EW precisions observables (S, ρ, g_{VVV}) [115]. Interestingly, it appears there has not been an update of the EW constraints on these operators since 1997 [116], although there are predictions for limits at an ILC [117]. There is, however, a new analysis for WBF Higgs at LHC includes the effects of some of these operators and finds that they would be encoded in the tagging jet azimuthal separation [118].

There is also a set of D6 operators involving the Higgs, fermion and gauge boson fields [86]:

$$\begin{aligned}
O_{d\phi} &= (\phi^\dagger \phi)(\bar{q}d\phi) \\
O_{\phi d} &= i(\phi^\dagger D_\mu \phi)(\bar{d}\gamma^\mu d) \\
O_{\bar{D}d} &= (D_\mu \bar{q}d)D^\mu \phi \\
O_{\phi q}^{(1)} &= i(\phi^\dagger D_\mu \phi)(\bar{q}\gamma^\mu q) \\
O_{\phi\phi} &= i(\phi^\dagger \epsilon D_\mu \phi)(\bar{u}\gamma^\mu d) \\
O_{dW} &= (\bar{q}\sigma^{\mu\nu}\sigma^i d)\phi W^i_{\mu\nu} \\
O_{\phi q}^{(3)} &= i(\phi^\dagger D_\mu \sigma^i \phi)(\bar{q}\gamma^\mu \sigma^i q) \\
O_{Dd} &= (\bar{q}D_\mu d)D^\mu \phi \quad O_{dB} = (\bar{q}\sigma^{\mu\nu}d)\phi B_{\mu\nu}
\end{aligned} \quad (8.39)$$

Some of these are constrained by precise LEP measurements of $Zb\bar{b}$, $\gamma b\bar{b}$ couplings, but not severely. They would give interesting rare Higgs decays like $H \to b\bar{b}Z, b\bar{b}\gamma$. Their phenomenology for LHC and even ILC is not really studied. Thus, I can't say to what scale they might be sensitive given a SM Higgs discovery with nothing else observed.

8.4.2. Two-Higgs doublet models (2HDMs)

The most-often studied extension to the SM Higgs sector is the two-Higgs doublet model (2HDM) [6, 119]. That is, we add one additional $SU(2)_L$ doublet. Both of the doublets acquire a vev. For now let's assume CP conservation and work with in the real-vev basis. Counting degrees of freedom, four per complex doublet, and knowing that three modes are "eaten" to give the $W^\pm$ and Z their masses, after SSB there must be five physical states. Two of them will necessarily be charged ($H^\pm$) regardless of how we assigned hypercharge to each doublet, leaving the other three neutral. Of those, two (h, H) will be CP-even and one will be CP-odd (A), the last of which won't couple to the weak bosons at tree level. The general 2HDM potential is quite messy [6, 120], so we'll not discuss it.

Recall the primary role of the Higgs sector: to restore unitarity to weak boson scattering. This requires the gauge coupling to WW to be exactly $\frac{1}{2}g_W^2 v$, where v is what we measure with G_F. In the amplitude, then, the coupling squared is $\frac{1}{4}g_W^4 v^2$. With two vevs, there is the automatic constraint $v_1^2 + v_2^2 \equiv v^2$ [121]. The ratio is $\tan\beta \equiv \frac{v_2}{v_1}$. The CP-even mass eigenstates, which couple to the weak bosons, thus boil down to simply mixing:

$$h = \sqrt{2}[-(\mathrm{Re}\phi_1^0 - v_1)\sin\alpha + (\mathrm{Re}\phi_2^0 - v_2)\cos\alpha] \quad (8.40)$$

$$H = \sqrt{2}[\ (\mathrm{Re}\phi_1^0 - v_1)\cos\alpha + (\mathrm{Re}\phi_2^0 - v_2)\sin\alpha] \quad (8.41)$$

where α is the angle which diagonalizes the 2×2 mixing matrix. The Higgs sector is typically defined by α, $\tan\beta$ and the potential parameters which govern the self-couplings. Some models are defined instead by M_A and M_Z.

Let's pause for a moment to reflect on what would happen if we introduced CP violation [119]. This is a well-motivated exercise since there isn't enough CP violation in the SM model to account for baryogenesis in the early universe. The most immediate impact is that h, H and A now mix. M_A is supposed to parameterize the pseudoscalar pole, but it's now mixed into three physical states, so it becomes ill-defined. Instead, we typically use the charged Higgs mass. It would be logical to use $M_{H^\pm}$ for CP-conserving scenarios as well, but this is one of those historical accidents that has too much momentum to change.

Regarding the fermions, we can apportion the two doublets in four general ways [6]:

I only Φ_2 couples to fermions

II Φ_1 couples to down-type, Φ_2 to up-type fermions

III Φ_1 couples to down quarks, Φ_2 to up quarks and down leptons

IV Φ_1 couples to quarks, Φ_2 to leptons

Types III and IV induce flavor-changing neutral currents (FCNCs), which are highly constrained, thus these models are not much studied any more. Types I and II are qualitatively different and worth a quick look at the differences in their couplings, shown in Table 8.5[q]. Because of which doublet gives the down-type fermions their masses, those Yukawa couplings to h and

---

[q]Note that various references use different phase conventions for the Lagrangian. The important distinction is the phase between Higgs couplings, and a reference SM coupling such as $ee\gamma$. I use positive terms in the covariant derivative and drop the overall superfluous factor of i typical of most Lagrangians.

Table 8.5. Fermion and gauge boson couplings in Type I (upper) and II (lower) 2HDMs.

Φ	$\dfrac{g_{\Phi u\bar{u}}}{g_f}$	$\dfrac{g_{\Phi d\bar{d}}}{g_f}$	$\dfrac{g_{\Phi VV}}{g_V}$	$\dfrac{g_{\Phi ZA}}{g_V}$
h	$-\dfrac{\cos\alpha}{\sin\beta}$	$-\dfrac{\cos\alpha}{\sin\beta}$	$\sin(\beta-\alpha)$	$-\dfrac{1}{2}i\cos(\beta-\alpha)$
H	$\dfrac{\sin\alpha}{\sin\beta}$	$\dfrac{\sin\alpha}{\sin\beta}$	$\cos(\beta-\alpha)$	$\dfrac{1}{2}i\sin(\beta-\alpha)$
A	$-i\gamma_5\cot\beta$	$i\gamma_5\cot\beta$	0	0
h	$-\dfrac{\cos\alpha}{\sin\beta}$	$\dfrac{\sin\alpha}{\cos\beta}$	$\sin(\beta-\alpha)$	$-\dfrac{1}{2}i\cos(\beta-\alpha)$
H	$\dfrac{\sin\alpha}{\sin\beta}$	$-\dfrac{\cos\alpha}{\cos\beta}$	$\cos(\beta-\alpha)$	$\dfrac{1}{2}i\sin(\beta-\alpha)$
A	$-i\gamma_5\cot\beta$	$-i\gamma_5\tan\beta$	0	0

H are swapped between models, with a phase factor from mixing. Similarly, the $Af\bar{f}$ coupling is inverted and changes sign: $\cot\beta \to -\tan\beta$. The gauge coupling for h and H, of course, are unaffected by the Yukawa couplings and are fixed to $\sin(\beta-\alpha)$ and $\cos(\beta-\alpha)$. (The sum of their squares in the amplitude must equal 1!)

The charged Higgs Yukawa couplings are slightly different yet. The left-handed coupling is proportional to the up-type Yukawa coupling, and the right-handed coupling the down-type Yukawa, *for an out-flowing H^-*. The reverse is true for an outflowing H^+. We have:

$$g_{H^-D\bar{U}} = \frac{g}{2\sqrt{2}M_W}\left[m_U\cot\beta(1+\gamma_5) - m_D\cot\beta(1-\gamma_5)\right] \quad (8.42)$$

$$g_{H^-D\bar{U}} = \frac{g}{2\sqrt{2}M_W}\left[m_U\cot\beta(1+\gamma_5) + m_D\tan\beta(1-\gamma_5)\right] \quad (8.43)$$

where H^- flows out, D is incoming and $\bar{U}$ is outgoing.

8.4.3. *Type II 2HDM in the MSSM*

At this point we should focus on the Type II 2HDM, because that's the one required to appear in the MSSM[r] (see Ref. [123] for a detailed description). Model I will have similar features, modulo the couplings swaps given in Table 8.5, so is understandable by analogy. We'll spend the remaining

---
[r]A superpotential can't be constructed from conjugate fields, else the supersymmetry transformations aren't preserved. For an excellent SUSY tutorial, see Ref. [122].

portion discussing only SUSY Higgs phenomenology, and specifically minimal SUSY, the MSSM. However, by the end it should be apparent that extended Higgs sectors may often be treated as variations on a theme, with much of the phenomenology based on the same collider signatures.

The MSSM imposes tree-level constraints on the Higgs potential which require the various λ to be gauge parameters (MSSM extensions add non-gauge terms). We'll come back to what the potential looks like in Sec. 8.4.5 and study its phenomenology, and for now simply examine the implication of this structure on the mass spectrum. Because we consider only the CP-conserving case here, we can get away with using M_A as an input. The others will be $\tan\beta$ as discussed before, the average top squark mass M_S, and an encoded trilinear mixing parameter for the top sector, X_t. This last one is important because of the large top Yukawa corrections the MSSM Higgs sector receives. The values 0 and $\sqrt{6}\,M_S$ are referred to as "no mixing" and "maximal mixing", because they extremize the loop corrections. The $h - H$ mixing angle is

$$\alpha = \frac{1}{2}\tan^{-1}\left[\tan 2\beta \frac{M_A^2 + M_Z^2}{M_A^2 - M_Z^2}\right], \qquad -\frac{\pi}{2} \leq \alpha \leq 0 \qquad (8.44)$$

to first order. The CP-even masses are given by:

$$M_{H,h}^2 = \frac{1}{2}\left(M_A^2 + M_Z^2 \pm \sqrt{(M_A^2 + M_Z^2)^2 + 4M_A^2 M_Z^2 \sin^2(2\beta)}\right)$$
$$+ \frac{3}{8\pi^2}\cos^2\alpha\, y_t^2 m_t^2 \left[\log\frac{M_S^2}{m_t^2} + \frac{X_t^2}{M_S^2}\left(1 - \frac{X_t^2}{12 M_S^2}\right)\right] \quad \text{for } M_h \text{ only} \qquad (8.45)$$

where the top Yukawa correction can be significant, a couple tens of GeV. The charged Higgs mass is rather more simple:

$$M_{H^\pm}^2 = M_A^2 + M_W^2 \qquad (8.46)$$

These equations exhibit the interesting property of h decoupling with increasing pseudoscalar mass: for large M_A the heavy states H, A and $H^\pm$ tend to be closely degenerate, and the light h has an asymptotic maximum mass which depends mostly on $\tan\beta$. We see this behavior, along with a plateau effect for M_h and M_H, in Fig. 8.57. There is always at least one CP-even Higgs boson in the mass region $90 \lesssim M_\phi \lesssim 145$ GeV, assuming perturbativity to high scales. For large M_A, toward the decoupling region, it is the lighter state, h, but at low M_A it is the heavier state, H. The transition region is sharper for larger $\tan\beta$.

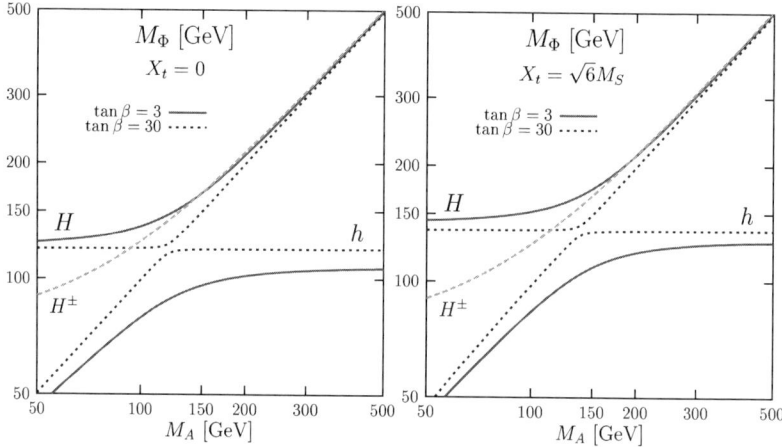

Fig. 8.57. MSSM Higgs boson masses as a function of pseudoscalar mass M_A and two choices of $\tan\beta$, for no (left) and maximal (right) mixing (X_t parameter; see text). Figures from Ref. [7].

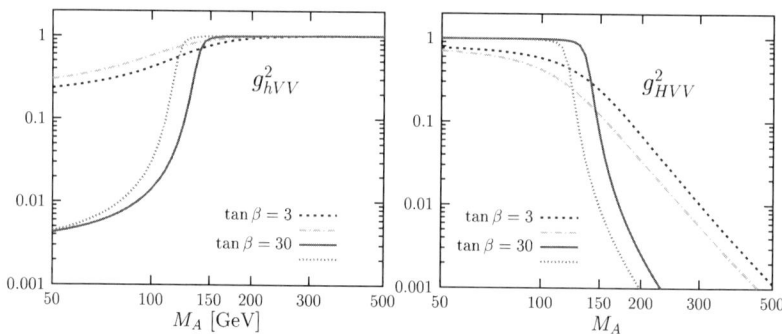

Fig. 8.58. MSSM CP-even Higgs boson couplings to the weak gauge bosons as a function of M_A and for two choices of $\tan\beta$, and for no mixing (darker colors) and maximal mixing (lighter colors). Figures from Ref. [7].

The mass spectrum is not the only feature to exhibit the decoupling and transition behavior, however. Both the gauge and Yukawa couplings do the same. The $VV\phi$ couplings are shown in Fig. 8.58. By comparison with Fig. 8.57, we easily see that when either h or H is in its plateau mass region, it holds most of the gauge coupling; $\sin(\beta-\alpha) \to 1$ or $\cos(\beta-\alpha) \to 1$. In the transition region, the two states share the gauge coupling, and both are of comparable importance in unitarity cancellation. As with the mass spectrum, and by now as anticipated, the transition region is sharper for larger $\tan\beta$. Hold these two figures in your mind, as they are going to play an extremely important phenomenological role shortly.

Using just trigonometry, let's rewrite the Yukawa couplings of Table 8.5 to see better how they depend on M_A and $\tan\beta$:

$$\begin{aligned}
g_{hu\bar{u}} &= -\frac{\cos\alpha}{\sin\beta}Y_u = -[\sin(\beta-\alpha)+\cot\beta\cos(\beta-\alpha)]Y_u \\
g_{hd\bar{d}} &= \frac{\sin\alpha}{\cos\beta}Y_d = -[\sin(\beta-\alpha)-\tan\beta\cos(\beta-\alpha)]Y_d \\
g_{Hu\bar{u}} &= -\frac{\sin\alpha}{\sin\beta}Y_u = -[\cos(\beta-\alpha)-\cot\beta\sin(\beta-\alpha)]Y_u \\
g_{Hd\bar{d}} &= -\frac{\cos\alpha}{\cos\beta}Y_d = -[\cos(\beta-\alpha)+\tan\beta\sin(\beta-\alpha)]Y_d
\end{aligned} \quad (8.47)$$

This is a far more convenient form, since $\tan\beta$ is an input and $\sin(\beta-\alpha)/\cos(\beta-\alpha)$ is the reduced h/H gauge coupling. These are both natural, convenient parameters to describe production cross sections and decay partial widths (thus branching ratios), rather than the CP-even mixing angle and $\sin\beta$ or $\cos\beta$, or their inverses. Check Fig. 8.59 to see if you agree.

These are the most salient features of the MSSM Higgs sector, sufficient to understand the bulk of MSSM Higgs phenomenology. For a more in-depth discussion, especially of why SUSY imposes these constraints, and for more detailed formulae, see Refs. [7, 123].

Now that we know the couplings, we can obtain cross sections for h and H production simply as correction factors to the SM channels of equal mass. There is no WBF or W/Z-associated pseudoscalar production, but there is both $gg \to A$ inclusive and top quark associated production, $t\bar{t}A$, which are easily obtained if one inserts the γ_5 factor into the loop derivation for $gg \to A$ [7]. The charged Higgs is a special case as there is no SM analogue; we'll discuss this in Sec. 8.4.4 in the context of searches. For the moment, let's examine the neutral states' branching ratios, just to get an idea of how they behave. It's easy to suffer plot overload about now, so don't try to absorb every last detail; focus on the general behavior, which you already should be able to guess from the couplings plots.

Figure 8.60 shows the BRs for the CP-even states h and H, cut off at the mass plateaus. They're basically what we would expect: both h and H behave like a SM Higgs of equal mass, except that the various couplings are dialed up or down. M_h can never be above ~ 145 GeV, so it almost never has a significant BR to gauge bosons. Because the fermionic partial widths

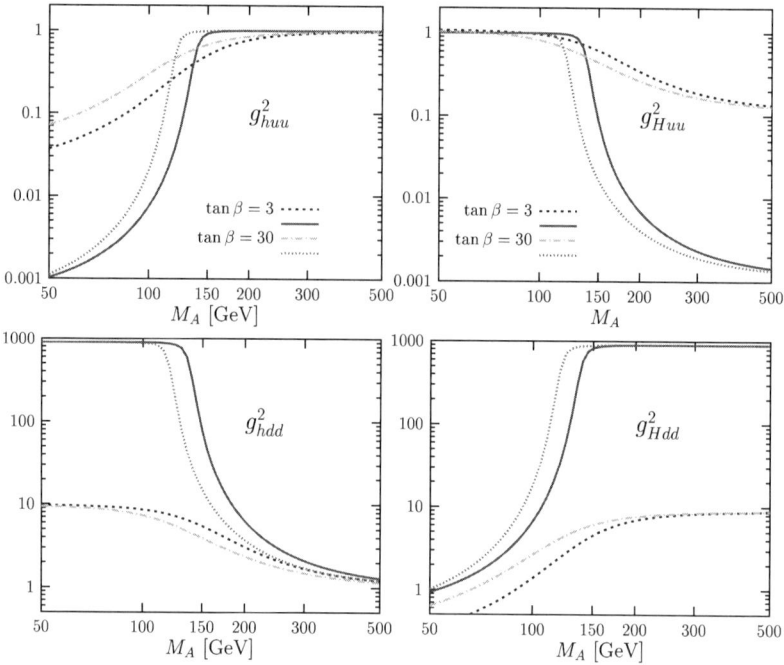

Fig. 8.59. MSSM CP-even Higgs boson couplings to fermions as a function of M_A and for two choices of $\tan\beta$, and for no mixing (darker colors) and maximal mixing (lighter colors). Figures from Ref. [7].

can be enhanced by a factor of $\tan^2\beta$, the rare modes like $\phi \to \gamma\gamma, gg$ tend to be suppressed (and the top quark loop can better cancel the W loop for some parameter choices, suppressing the partial width). The only new features are $H \to hh, AA$ decays, possible for limited parameter choices but making for interesting additional channels.

The pseudoscalar BRs behave similarly, as shown in Fig. 8.61. The new feature here is at small $\tan\beta$, where decays $A \to hZ$ are possible. But otherwise A prefers to decay $\sim 90\%$ to $b\bar{b}$ and $\sim 10\%$ to $\tau^+\tau^-$, unless it is heavy enough to produce top quark pairs. That dominates only at small $\tan\beta$ (large $\cot\beta$), where the up-type coupling dominates. At large $\tan\beta$, $b\bar{b}$ and $\tau^+\tau^-$ both still win by a considerable margin.

There are similar plots for $H^\pm$, but they're not particularly enlightening as its decay patterns are drastically simpler: as far as phenomenology is concerned, it's BR~ 1 to tb when kinematically accessible, $\tau\nu$ if lighter. For low $\tan\beta$ there is a rare BR to $hW^\pm$, but that is predicted to always be difficult to observe.

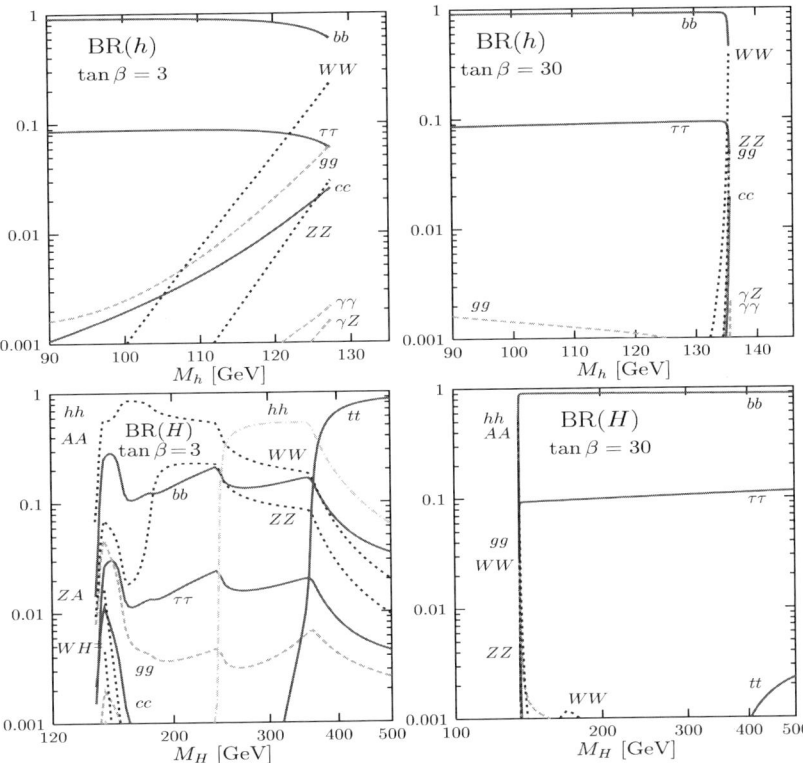

Fig. 8.60. MSSM CP-even Higgs boson branching ratios as a function of M_A for $\tan\beta = 3, 30$. Figures from Ref. [7].

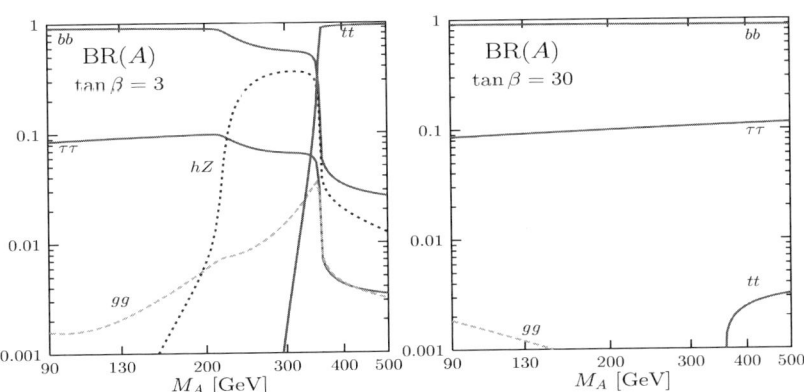

Fig. 8.61. MSSM CP-odd Higgs boson branching ratios as a function of M_A for two choices of $\tan\beta$. Figures from Ref. [7].

All Higgs bosons can decay to SUSY particle pairs if they're light enough, but this is not a very common occurrence across parameter space (especially since so much of it is ruled out already by LEP SUSY searches), so we'll bypass that discussion here.

8.4.4. *MSSM Higgs searches*

For MSSM Higgs searches past, we start again with LEP. It didn't find anything, but placed various limits. Let's begin with the charged Higgs search, because it's the simplest. This proceeded via H^+H^- pair production (the only mechanism accessible at LEP) and decay to $\tau\nu$ or cs, as there was never kinematic room for tb. Thus, the search had three channels: dual taus, mixed tau plus hadronic decays, and an all-hadronic mode [124]. Because the production mechanism depends on only gauge-fixed couplings, the MSSM charged Higgs search is usually presented as a more general 2HDM search, with limits presented in the $M_{H^\pm}$ v. BR($H^\pm \to \tau^\pm \nu$) plane. Fig. 8.62 summarizes the obtained limits. To translate the general search limits to the MSSM Higgs sector inputs, recall Eq. 8.46, $M_{H^\pm}^2 = M_A^2 + M_W^2$. The difficulty of this search was the low ID efficiency for taus and charm quarks. Unfortunately, there is no final combined limit, but each of the collaborations has published final independent limits [126–129]. Watch the LEP-Higgs web page for updates [130].

The basic neutral Higgs boson search channels are exactly the same as in the SM for each of h and H, to which we add $e^+e^- \to Z^* \to hA/HA$ production via the additional couplings of Table 8.5. Each of the four LEP collaborations presented a multitude of MSSM $h/H/A$ search limits, and there are combined LEP results with CP-conservation [131] and CP-violation (CPX) [131, 132]. However, one should be somewhat wary of what precisely is presented. The results are usually shown as shaded exclusion blobs in either M_A-$\tan\beta$ space (for a very specific set of additional assumptions) or M_{h_i}-$\tan\beta$ space, also given some assumptions. There are literally dozens of pages of exclusion plots, depending on what one chooses for the mixing parameter X_t, top quark mass (recall the strong M_h dependence on m_t), stop masses, μ, and so on. This is far too much to show here, because the exclusion contours change so much from assumption to assumption – it's impossible even to select a representative sample without misleading the uninitiated. See e.g. Ref. [133].

The curious student should flip through the plots in Refs. [131, 133] simply to get a feel for how wild this variation is. Observe how much the

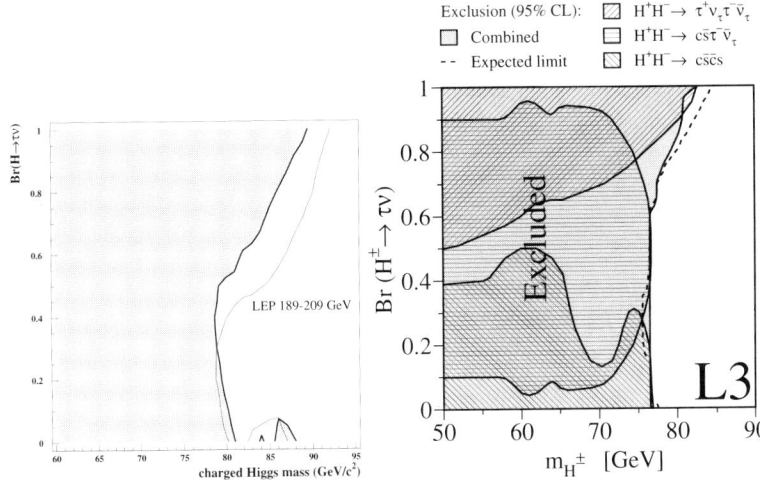

Fig. 8.62. Left: LEP preliminary combined-experiment charged Higgs search 95%CL limits (2001), from Ref. [125]. Right: L3 published limits from 2003, illustrating where each of the three decay channels discussed in the text contributes to the overall limit [128]. There is no final LEP combined limit, but judging from each of the individual limits [126–129], it does not change significantly from the preliminary results.

contours change depending on the top quark mass – it is obviously still fairly poorly measured, as far as fits to supersymmetry go. Note also that the plots are always logarithmic in $\tan\beta$, which compresses the unexcluded large-$\tan\beta$ region, making it appear that parameter space is vastly ruled out in many cases. This simply isn't true. Finally, I should comment that the "theoretically inaccessible" disallowed blobs are even more grossly misleading. All one has to do is move the stop masses up slightly and these retreat dramatically. Perhaps a more logical approach is the model-independent $h/H/A$ search of OPAL [134].

MSSM Higgs Searches at LHC are also mostly variants on the SM search channels, the exceptions being charged Higgses, rare (SUSY or Higgs pair) decay modes, and one new production channel, $b\bar{b}\phi$, which is important at large $\tan\beta$ where the coupling is enhanced to top-quark Yukawa strength. $t\bar{t}\phi$ rates tend to be about the same as the SM for equal mass, or slightly suppressed. WBF h or H rates can only be suppressed relative to the SM, due to the appearance of $\sin^2(\beta-\alpha)$ or $\cos^2(\beta-\alpha)$, respectively. Inclusive rates can change rather dramatically, however, because the b loop can be extremely important. Figure 8.63 shows the cross sections for $gg \to \phi$ as a

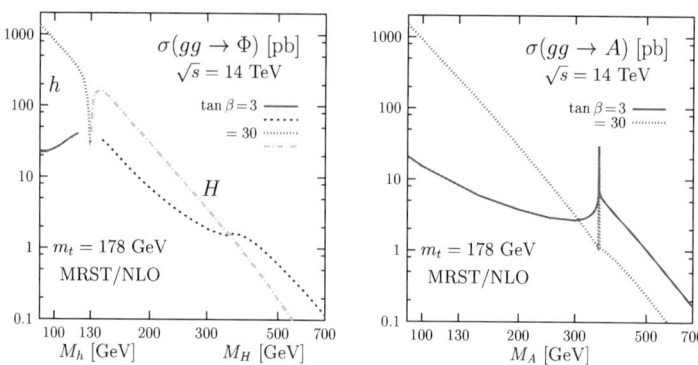

Fig. 8.63. Gluon fusion MSSM Higgs production cross sections at LHC for the CP-even states h and H (left) and the pseudoscalar A (right), for two values of $\tan\beta$. Figures from Ref. [7].

function of the physical masses, for small and large $\tan\beta$. These may be compared with the SM cross sections of Fig. 8.16.

Let's concentrate on the WBF modes, however, as they turn out to be the most interesting. Recall the plateau behavior of h and H masses as a function of M_A (cf. Fig. 8.57), and simultaneously the h and H gauge coupling behavior (cf. Fig. 8.58). The astute student will realize that this implies that WBF Higgs production in an accessible mass region probably always occurs at a good rate, somewhat suppressed but never much so. Figure 8.64 summarizes some of this previous information and goes on to show the cross section times BR to tau pairs (in the two accessible tau decay modes), also as a function of M_A [135]. Indeed, eyeballing the upper and lower rows, it appears that between h and H, there's always a signal in WBF. It may be slightly suppressed, but we know from SM WBF Higgs studies (cf. Sec. 8.2.3.4) that since so little data is required to make an observation, the signal could be suppressed by a factor of several and be detectable. The reason is that in the MSSM the h and H plateau mass ranges are in the "good" region of WBF Higgs observability. Actually, quite a large mass region is observable, but if the MSSM predicted Higgs masses closer to the Z pole, there could be trouble (but LEP would already have discovered such a Higgs).

This bit of luck forms the basis of the MSSM Higgs No-Lose Theorem: at least one of the CP-even Higgs states, h or H, is guaranteed to be observable in WBF at LHC [64, 135]. The original parton-level studies have since been confirmed with full ATLAS detector simulation, and actually im-

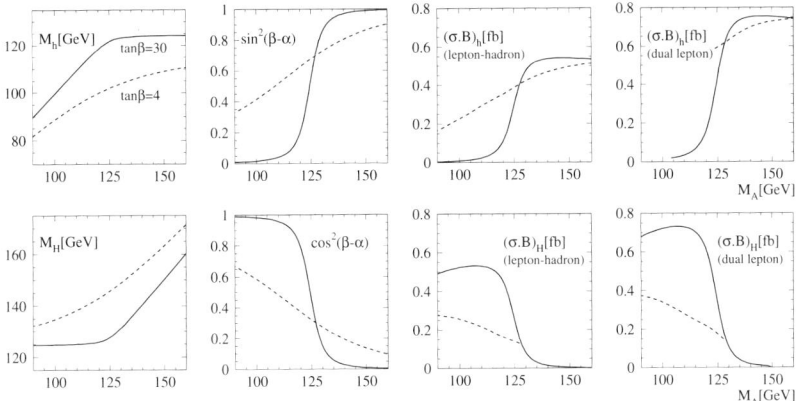

Fig. 8.64. From left to right, each plot as a function of M_A: h/H mass, gauge coupling suppression factor squared, WBF cross section times BR to taus to the lepton-hadron final state, and the same for the dual lepton mode. The upper (lower) row is for $h(H)$. Fig. from Ref. [64].

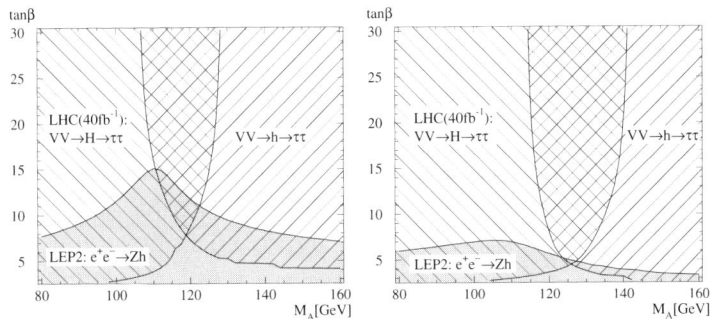

Fig. 8.65. MSSM parameter space coverage of WBF $h/H \to \tau^+\tau^-$ for the no-mixing ($A_t = 0$, left) and maximal mixing ($X_t = \sqrt{6}\,M_{SUSY}$, right) cases [135].

proved [136]. The parton-level coverage plots shown in Fig. 8.65, however, are simpler to grasp. Very little data would be required for discovery, and for some M_A it would be possible to observe both h and H simultaneously.

One caveat: the final state $\tau^+\tau^-$ is not always accessible![s] It's possible to zero out the MSSM down-type fermion coupling at tree level – an interesting exercise for the student. If this happens, $h/H \to \gamma\gamma$ and $h/H \to W^+W^-$ are "large" partial widths, so their BRs take up the

---
[s]There's always fine print...

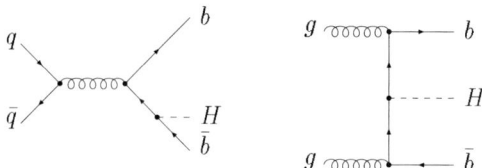

Fig. 8.66. Feynman diagrams for $gg \to b\bar{b}\phi$ production at LHC.

coverage slack [135, 136], saving the No-Lose Theorem. There's been some work on an NMSSM No-Lose Theorem [137–140], which extends the Higgs sector by a complex singlet [6]. The outlook for LHC is promising, but not obviously rock-solid.

The No-Lose Theorem is great for the CP-even states, but what about the other Higgses? I'll gloss over the bulk of searches, since they're mostly variants on the SM ones, and move on to the special case of heavy H/A (towards decoupling) and this new channel $b\bar{b}\phi$. The Feynman diagrams appear in Fig. 8.66. Recall that H has a $\tan\beta$ enhancement to down-type quarks in the decoupling region, and A always has this enhancement. We already know that means that H and A prefer to decay 90% of the time to $b\bar{b}$ and 10% to $\tau^+\tau^-$, but it would be impossible to observe either of those final states in inclusive production, and WBF production is zilch for H in the decoupling region. However, the LHC being essentially a gluon collider, the initial state can create high-energy b pairs, which can then Brem a Higgs, either H or A, which are essentially degenerate (but do not interfere due to the γ_5 coupling). Since the b jets are produced at high-p_T, the H/A must recoil against them, so it also produced with a transverse boost. It's decay products are then not back-to-back, allowing for tau pair reconstruction; $H/A \to \mu^+\mu^-$ may also be used, but is a rare mode. The final state is then $b\bar{b}\tau^+\tau^-$ (or $b\bar{b}\mu^+\mu^-$), which is taggable and distinguishable from mixed QCD-EW backgrounds because the tau pair invariant mass is in the several-hundred GeV region.

Figure 8.67 shows the cross section times BR to tau pairs for 300 GeV Higgs bosons as a function of $\tan\beta$, and also the CMS expected discovery reach for various final states in tau or muon pairs, with only 30 fb^{-1} of luminosity, or about 1/10 of the total LHC data expected. Coverage is not complete, because this mode doesn't produce enough rate at low $\tan\beta$ where there is little coupling enhancement, but is still a significant search tool. The mass resolution achievable for H and A using taus in this mode is

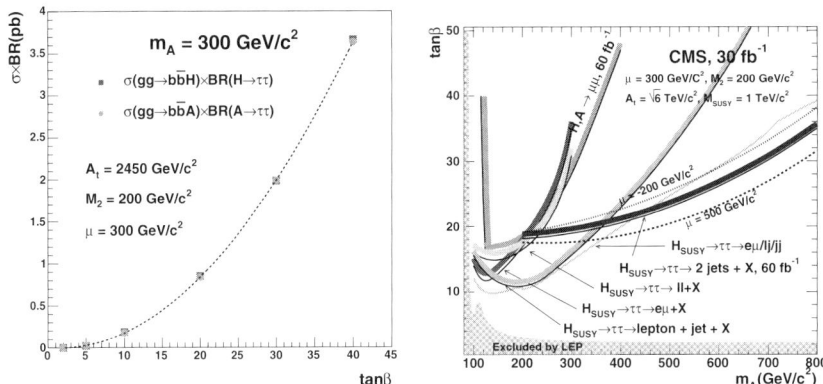

Fig. 8.67. Left: $b\bar{b}\phi$ production cross section at LHC times the BR to tau pairs, as a function of $\tan\beta$ for $M_A = 300$ GeV. Right: expected CMS reach using only 30 fb^{-1} of data for $b\bar{b}H/A \to \tau^+\tau^-, \mu^+\mu^-$ as a function of M_A. Figures from Ref. [141].

even pretty good, on the order of a couple tens of GeV, possibly better. Of course, if the decay to muons is accessible (at very large $\tan\beta$, then mass resolution would be on the order of a GeV.

This would determine M_A quite well, good enough for comparison with theory (at least at first), but what about the other major Higgs parameter, $\tan\beta$? The $b\bar{b}\phi$ production rate is directly proportional to $\tan^2\beta$, so we can measure it using the overall rate, with the mild (but not rock solid) assumption that the ratio of $b\bar{b}$ and $\tau^+\tau^-$ BRs is the ratio of the b and τ squared masses, i.e. that $BR(H/A \to \tau^+\tau^-) \sim 10\%$ [141]. The major sources of uncertainty are this assumption, the machine luminosity uncertainty of $5 - 10\%$, PDF uncertainties of probably about 5%, and higher-order QCD corrections to the production process of probably about 20% [142, 143].

Figure 8.68 shows the CMS expected uncertainty on $\tan\beta$ using this method, as a function of M_A and for 30 or 60 fb^{-1} of data. In general, $10 - 20\%$ appears achieveable. This is not spectacular, but would be a significant first step toward sorting out the new Higgs sector and presumably comparing to other SUSY discovery measurements. Clearly the higher-order QCD uncertainties dominate, which could probably be improved with better theoretical calculations over the next decade. This will be done if heavy Higgses are discovered.

Now, what about charged Higgs discovery? We know nothing about its phenomenology, because there is no SM analogue. All we do know is the

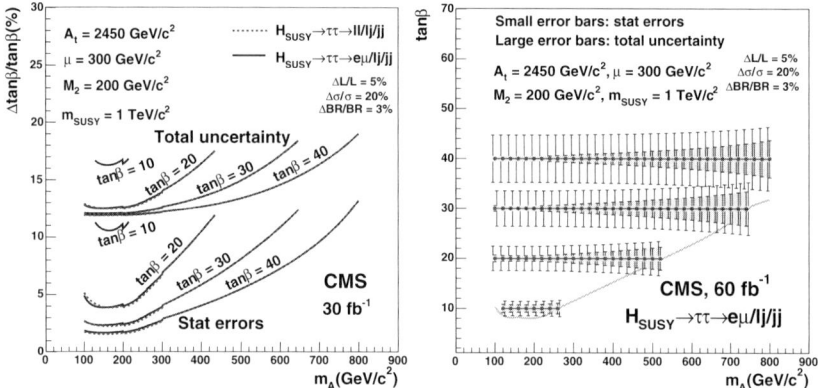

Fig. 8.68. CMS expected precision on $\tan\beta$ at LHC using $b\bar{b}\phi$ production as described in the text. Figures from Ref. [141].

very important fact that, *despite everything else we may see at Tevatron or LHC, the only way to prove the existence of two Higgs doublets is to directly observe the charged Higgs states.* I cannot emphasize this enough. For all we know, an extra neutral state might simply be the residue of an extra Higgs singlet; there could be more to the flavor sector that confuses us when we try to measure Yukawa couplings or $\tan\beta$. Thus, observing the $H^\pm$ states would be a huge qualitative step toward understanding what the Higgs sector is. How would this proceed experimentally?

At Tevatron there is very little energy available for direct charged Higgs production, since it must be produced in association with a top quark (large coupling), as shown in the Feynman diagrams of Fig. 8.69. However, if $M_{H^\pm}$ is small enough, the top quark can decay to $bH^\pm$ followed by

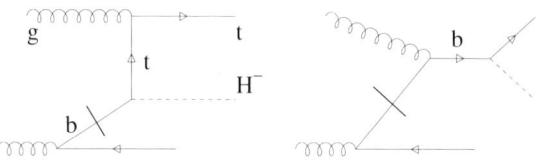

Fig. 8.69. Feynman diagrams for charged Higgs production at hadron colliders. The short line breaking the b quark propagator represents how the process may also be regarded as initiated by a b parton in the proton, rather than from gluon splitting to a b quark pair.

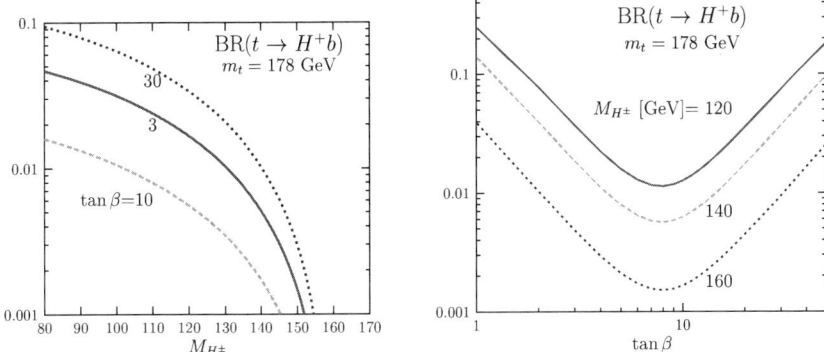

Fig. 8.70. Branching ratio for top quark to bottom quark plus charged Higgs boson, as a function of $M_{H^\pm}$ for a few select values of $\tan\beta$ (left) and as a function of $\tan\beta$ for $M_{H^\pm} = 120$ GeV (right). Figures from Ref. [7].

$H^\pm \to \tau\nu$ if $\tan\beta > 1$, and equally to bc and cs if $\tan\beta < 1$; if $M_{H^\pm} \gtrsim 120$ GeV, then the BR to $W^\pm b\bar{b}$ via a top quark loop becomes significant. Figure 8.70 shows the $t \to bH^\pm$ BR as a function of $M_{H^\pm}$ for a few select $\tan\beta$, and as a function of $\tan\beta$ for $M_{H^\pm} = 120$ GeV. At low $\tan\beta$, the partial width is driven mainly by the top quark Yukawa, while at large $\tan\beta$ it's primarily the bottom quark. Weakness of both Yukawas in the intermediate-$\tan\beta$ regime results in a comparatively reduced top quark partial width (recall Eqs. (8.42,8.43)). For fixed $M_{H^\pm}$, the partial width is symmetric in $\log(\tan\beta)$ about a minimum at $\tan\beta = \sqrt{m_t/m_b}$. Charged Higgs decays to $hW^\pm$ or $AW^\pm$ are generally disallowed in the MSSM from LEP mass limits on h and A.

The Tevatron search proceeds both as appearance (i.e. looking directly for $H^\pm$ in the top quark sample) and disappearance, or missing rate for top quark to $bW^\pm$. Figure 8.71 goes on to show the expected 95% CL limits in the $M_{H^\pm} - \tan\beta$ plane that Tevatron Run I achieved, and Run II might reach depending on how much data it ultimately records. The very slight change between 2 and 10 fb^{-1} reveals that the experiments there are statistics-limited, but not by a great margin.

LHC will search for $tH^\pm$ direct production (Fig. 8.69), covering the mass range $M_{H^\pm} > m_t$. Due to nasty QCD backgrounds, the tb decay will be inaccessible [145], leaving $\tau\nu$ with BR$\sim$ 10%. This is very difficult due to a subtlety of tau decays. Left-handed taus decay to soft leptons [146]. Since neutrinos are left-handed, helicity conservation in scalar decay means all taus are as well. We need a lepton to trigger the event, and it must come

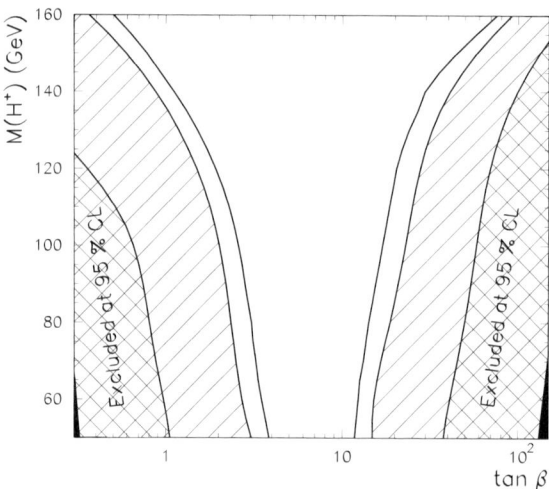

Fig. 8.71. Tevatron Run I 95% CL charged Higgs mass limits (double hatched lines) as a function of $\tan\beta$ from searches for top quark decays to bottom quark plus charged Higgs, and expected limits achievable in Run II (single hatched lines for 2 fb^{-1}, unhatched curves for 10 fb^{-1}). Figure from Ref. [144].

from $H^\pm$ instead of t, so that there is only one source of missing transverse momentum and we can fully reconstruct t, and $H^\pm$ transversely. Only a small fraction of the small rate could pass the necessary detector kinematic cuts to be recorded. This limits the search to large $\tan\beta$ or small $M_{H^\pm}$, where the production rate is largest. Fig. 8.72 shows ATLAS's expected transverse mass distributions for a fairly light and a heavy $H^\pm$.

Finally, we come to the overall picture of MSSM Higgs phenomenology at LHC. Primarily we're concerned with discovering all the states, but especially the charged Higgs as it's the key to confirming the existence of two Higgs doublets. That turns out to be extraordinarily difficult due to a combination of factors, from overwhelming QCD backgrounds to characteristics of left-handed tau decays. Figure 8.73 summarizes the reach for h, H, A and $H^\pm$ [70]. It's reassuring that the No-Lose Theorem holds and we're guaranteed to find at least one of the CP-even states, h or H. However, moderate $\tan\beta$ and the decoupling limit (large M_A) both present significant gaps in coverage to observe any of the additional states. This is especially more apparent once one realizes that the region below the solid black curve is already excluded by LEP, so those LHC access regions don't matter. The figure is from 2001 and needs updating – some significant positive changes exist – but the general picture remains.

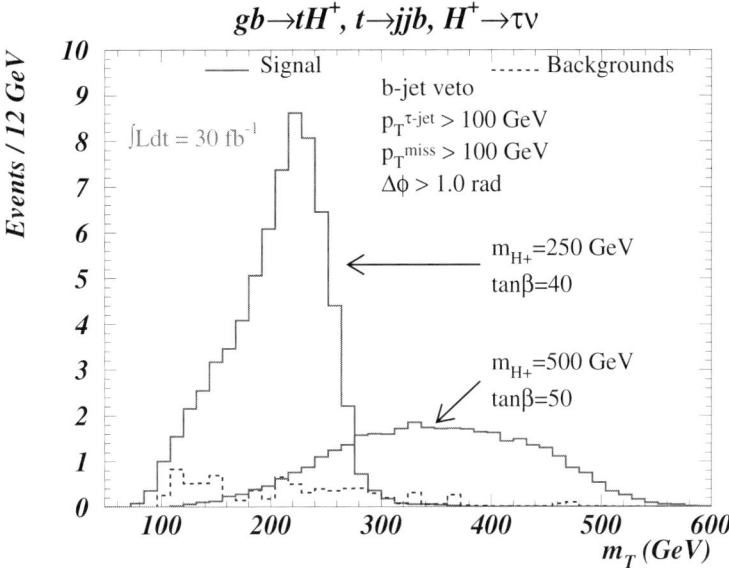

Fig. 8.72. Expected transverse mass distributions for light and heavy $H^\pm \to \tau\nu$ at ATLAS [147].

8.4.5. MSSM Higgs potential

I've touched on the bits of Higgs gauge and Yukawa couplings in the MSSM that are qualitatively different that the SM: M_A and $\tan\beta$. But we should look at self-couplings more closely, because in a general 2HDM (or the subset MSSM) they are radically different. First, because there are more Higgs bosons, there are more self-couplings – six for the neutral states alone, to be precise: λ_{hhh}, λ_{Hhh}, λ_{HHh}, λ_{HHH}, λ_{hAA}, λ_{HAA}. In the MSSM these are all equal to M_Z^2/v times various mixing angles (which aren't particularly enlightening so I don't show them) plus additional shifts from top quark Yukawa loop corrections. That is, they are all (mostly) gauge parameters. However, in the large-M_A decoupling limit which recovers the SM, $\lambda_{hhh} \to \lambda_{\rm SM}$.

If we discover SUSY, we'd start by assuming it's the MSSM. To measure the MSSM potential in that case, we'd have to observe at least six different Higgs pair production modes to measure the six self-couplings. (Note that I'm leaving out the possible self-couplings involving charged Higgses.) Inclusive Higgs pair production looks generally like it does in the SM, $gg \to \phi_1\phi_2$ via triangle and box loop diagrams as shown in Fig. 8.74, but the b quark loops become important and must be included.

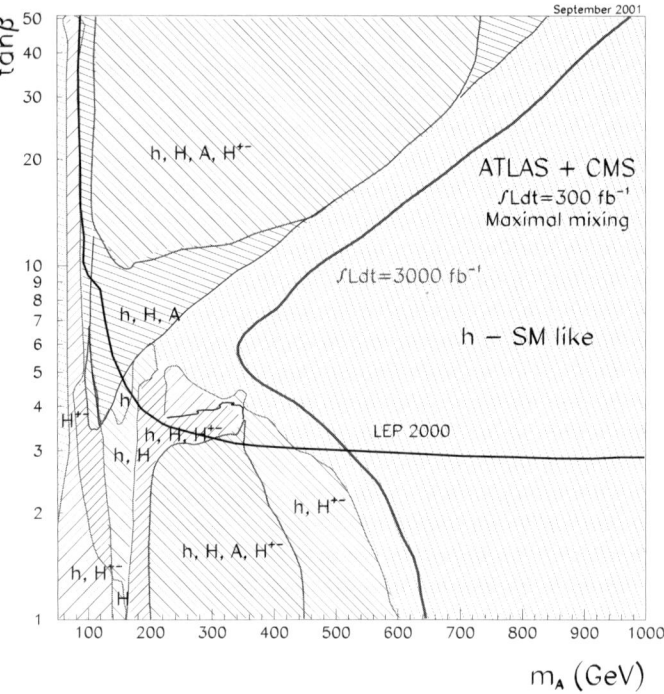

Fig. 8.73. Summary of MSSM Higgs boson discovery reaches at LHC (and extended to SLHC via the solid red line), combining ATLAS and CMS, in the $\tan\beta - M_A$ plane in the maximal mixing scenario. The reach is defined as 5σ discovery in at least one production and decay channel. Below the solid black curve is the region excluded by LEP. Figure from Ref. [70].

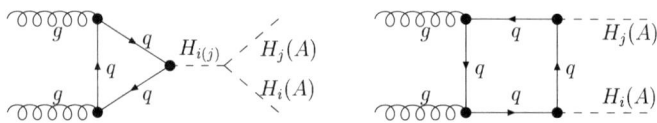

Fig. 8.74. Feynman diagrams for Higgs pair production in a 2HDM like the MSSM. The loops include both top and bottom quarks, and there are six possible processes (see text).

Unfortunately, the box diagram totally swamps the one containing the self-coupling we care about by a factor $\tan^2 \beta$, and in any case backgrounds from $H/A b\bar{b}$ production appear to be overwhelming [103]: very generally, LHC would not obtain any λ measurements at all. The one very limited

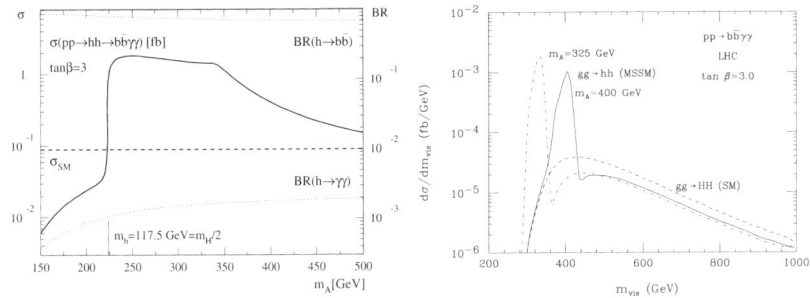

Fig. 8.75. Resonant MSSM Higgs pair production at LHC and decay to $b\bar{b}\gamma\gamma$ final states [103].

exception is that LHC could clearly observe Higgs pair production if it came from resonant heavy Higgs decay, $H/A \to hh$. An example peak is shown in Fig. 8.75. However, this would measure only a BR, at best, not an absolute coupling. Sadly, exactly the same situation exists for Higgs pairs at a future ILC [104].

8.5. Conclusions

The purpose of these lectures has not been to provide exhaustive coverage of all aspects of collider Higgs phenomenology. Rather, it's a solid introduction, focusing on the basics. This includes SM production and decay, mostly at LHC, where we're confident we could discovery a SM-like Higgs, and many non-SM-like variants. I focused on the most important channels which guarantee discovery, and especially in weak boson fusion (WBF) as those are the most powerful (best S/B, distinctive) search channels, covering the broadest range of Higgs mass. I emphasized that our understanding of LHC Higgs physics has changed dramatically from the days of the AT-LAS TDR, for example, which is now quite obsolete. However, ATLAS has produced a plethora of Notes and summaries of Notes to cover the changes, and CMS published a fresh TDR [48] in 2006 which covers the changes as well.

We now understand the LHC to be such a spectacular Higgs factory that not only can it discover any mass of SM-like Higgs boson, it can also do an impressive job of measuring all its quantum properties. Granted, Higgs couplings measurements won't be precision-level if the Higgs is light, as expected from EW precision data, but they would nonetheless be absolute couplings measurements. The LHC can even make significant steps toward measuring the SM Higgs potential, at least the Higgs trilinear self-coupling, although depending on M_h it may require precision gauge and Yukawa couplings input from a future e^+e^- collider (an ILC) to control the major systematic uncertainties. I also highlighted where an ILC could make improvements to the LHC's measurements, and where it would be vital to filling in gaps in LHC results.

The final third of the lectures discussed BSM Higgs sectors, but only the 2HDM MSSM Higgs sector in any detail. Many SM Higgs sector extensions are rather simple variants on SM phenomenology, involving factorizable changes in production and decay rates (couplings), mostly arising from mixing angles. This is not general, however, and there are plenty of "exotic" models – Higgs triplets, for example – which would be qualitatively different, but therefore simultaneously distinctive. The popular focus on the MSSM 2HDM is because of several other outstanding questions in particle physics, like dark matter or the theoretical dirty laundry of the SM Higgs sector, which strongly motivate the other new physics.

Students who wish to engage in Higgs phenomenology research should definitely take the time to expand their scope beyond the SM and the MSSM. Other extensions are equally well-motivated, such as Little Higgs, not to mention strong dynamics. But the two well-studied basic models I covered here give one a strong foundation for other BSM Higgs phenomenology by analogy. Happy Higgs hunting!

Acknowledgments

I would like to thank Sally Dawson and the TASI 2006 organizers for the opportunity to give these lectures, and for an extremely pleasant experience at the summer school. Gracious thanks also go to Dan Berdine, John Boersma, Fabio Maltoni, Tilman Plehn, Jürgen Reuter, and especially Steve Martin for proofreading contributions above and beyond the call of duty.

References

[1] B. W. Lee, C. Quigg and H. B. Thacker, Phys. Rev. Lett. **38**, 883 (1977); ibid., Phys. Rev. D **16**, 1519 (1977).
[2] T. Appelquist and M. S. Chanowitz, Phys. Rev. Lett. **59**, 2405 (1987) [Erratum-ibid. **60**, 1589 (1988)].
[3] F. Maltoni, J. M. Niczyporuk and S. Willenbrock, Phys. Rev. D **65**, 033004 (2002).

[4] D. A. Dicus and H. J. He, Phys. Rev. D **71**, 093009 (2005).
[5] D. A. Dicus and H. J. He, Phys. Rev. Lett. **94**, 221802 (2005).
[6] J. F. Gunion, H. E. Haber, G. L. Kane and S. Dawson, "THE HIGGS HUNTER'S GUIDE," SCIPP-89/13, Addison-Wesley, 1989.
[7] A. Djouadi, arXiv:hep-ph/0503172 and arXiv:hep-ph/0503173.
[8] M. Spira, Fortsch. Phys. **46**, 203 (1998) [arXiv:hep-ph/9705337].
[9] J. F. Gunion and H. E. Haber, Nucl. Phys. B **272**, 1 (1986) [Erratum-ibid. B **402**, 567 (1993)].
[10] S. Weinberg, Phys. Rev. D **13**, 974 (1976) and Phys. Rev. D **19**, 1277 (1979).
[11] L. Susskind, Phys. Rev. D **20**, 2619 (1979).
[12] S. Dimopoulos and L. Susskind, Nucl. Phys. B **155**, 237 (1979).
[13] E. Eichten and K. D. Lane, Phys. Lett. B **90**, 125 (1980).
[14] B. Holdom, Phys. Rev. D **24**, 1441 (1981).
[15] C. T. Hill, Phys. Lett. B **266**, 419 (1991) and Phys. Lett. B **345**, 483 (1995).
[16] C. T. Hill and E. H. Simmons, Phys. Rept. **381**, 235 (2003) [Erratum-ibid. **390**, 553 (2004)]; K. Lane, arXiv:hep-ph/0202255.
[17] N. Arkani-Hamed, A. G. Cohen and H. Georgi, Phys. Lett. B **513**, 232 (2001).
[18] E. H. Simmons, R. S. Chivukula, H. J. He, M. Kurachi and M. Tanabashi, AIP Conf. Proc. **857**, 34 (2006) [arXiv:hep-ph/0606019].
[19] For nice reviews of Little Higgs models, see e.g. :
M. Schmaltz and D. Tucker-Smith, arXiv:hep-ph/0502182;
M. Perelstein, Prog. Part. Nucl. Phys. **58**, 247 (2007).
[20] Z. Chacko, H. S. Goh and R. Harnik, Phys. Rev. Lett. **96**, 231802 (2006).
[21] W. M. Yao et al. [Particle Data Group], J. Phys. G **33**, 1 (2006).
[22] A. Djouadi, J. Kalinowski and M. Spira, Comput. Phys. Commun. **108**, 56 (1998).
[23] G. Pocsik and T. Torma, Z. Phys. C **6**, 1 (1980);
T. G. Rizzo, Phys. Rev. D **22**, 722 (1980);
W. Y. Keung and W. J. Marciano, Phys. Rev. D **30**, 248 (1984);
E. Gross, G. Wolf and B. A. Kniehl, Z. Phys. C **63**, 417 (1994) [Err.-ibid. C **66**, 321 (1995)].
[24] T. G. Rizzo, Phys. Rev. D **22**, 178 (1980) [Addendum-ibid. D **22**, 1824 (1980)].
[25] J. R. Ellis, M. K. Gaillard and D. V. Nanopoulos, Nucl. Phys. B **106**, 292 (1976);
M. B. Gavela, G. Girardi, C. Malleville and P. Sorba, Nucl. Phys. B **193**, 257 (1981).
[26] L. Lyons.
[27] E. Gross and A. Klier, arXiv:hep-ex/0211058.
[28] See the ALEPH event displays public web page for this and other pretty pictures:
http://aleph.web.cern.ch/aleph/ALPUB/seminar/wds/Welcome.html

[29] M. Carena et al. [Higgs Working Group Collaboration], arXiv:hep-ph/0010338.
[30] T. Hahn, S. Heinemeyer, F. Maltoni, G. Weiglein and S. Willenbrock, arXiv:hep-ph/0607308; see http://maltoni.web.cern.ch/maltoni/TeV4LHC/SM.html for references to all the specific latest calculations.
[31] U. Aglietti et al., arXiv:hep-ph/0612172.
[32] F. Gianotti, Phys. Rept. **403**, 379 (2004).
[33] L. Babukhadia et al. [CDF & D0 Working Group Members], Phys. Rev. D **66**, 010001 (2002).
[34] Analysis of V. Veszpremi, O. Gonzalez, D. Bortoletto, A. Garfinkel, S.-M. Wang [CDF], Public Note CDF-8842, 2006; see http://www-cdf.fnal.gov/~veszpv/ for the figure.
[35] M. Dittmar and H. K. Dreiner, Phys. Rev. D **55**, 167 (1997).
[36] T. Han and R. J. Zhang, Phys. Rev. Lett. **82**, 25 (1999);
T. Han, A. S. Turcot and R. J. Zhang, Phys. Rev. D **59**, 093001 (1999).
[37] D. L. Rainwater and D. Zeppenfeld, Phys. Rev. D **60**, 113004 (1999) [Erratum-ibid. D **61**, 099901 (2000)].
[38] G. Bernardi [D0 Collaboration], arXiv:hep-ex/0612044.
[39] R. V. Harlander and W. B. Kilgore, Phys. Rev. Lett. **88**, 201801 (2002);
C. Anastasiou and K. Melnikov, Nucl. Phys. B **646**, 220 (2002);
V. Ravindran, J. Smith and W. L. van Neerven, Nucl. Phys. B **665**, 325 (2003).
[40] N. Kauer, Phys. Rev. D **70**, 014020 (2004).
[41] T. Binoth, M. Ciccolini, N. Kauer and M. Kramer, JHEP **0503**, 065 (2005) and **0612**, 046 (2006).
[42] ATLAS TDR, report CERN/LHCC/99-15 (1999).
[43] E. Richter-Was and M. Sapinski, Acta Phys. Polon. B **30** (1999) 1001.
[44] V. Drollinger, T. Muller and D. Denegri, Phys. Rev. D **66**, 010001 (2002).
[45] S. Abdullin et al., Eur. Phys. J. C **39S2** (2005) 41.
[46] W. T. Giele, T. Matsuura, M. H. Seymour and B. R. Webber, FERMILAB-CONF-90-228-T,
published in Snowmass Summer Study 1990:0137-147.
[47] J. Cammin, Ph.D. Thesis [ATLAS], BONN-IR-2004-06
[48] CMS TDR, report CERN/LHCC/2006-001 (2006).
[49] K. Cranmer, B. Quayle, et al., ATL-PHYS-2004-034.
[50] K. Cranmer, private communication, publication forthcoming.
[51] G. L. Kane, G. D. Kribs, S. P. Martin and J. D. Wells, Phys. Rev. D **53**, 213 (1996)
[52] For the early history, applied to heavy Higgs bosons, see the first three citations of Ref. [53].
[53] D. L. Rainwater, R. Szalapski and D. Zeppenfeld, Phys. Rev. D **54**, 6680 (1996).
[54] J. Campbell, R. K. Ellis and D. L. Rainwater, Phys. Rev. D **68**, 094021 (2003).
[55] S. Asai et al., Eur. Phys. J. C **32S2**, 19 (2004).

[56] Y. L. Dokshitzer, V. A. Khoze and S. Troian, in *Proceedings of the 6th International Conference on Physics in Collisions*, p. 365, ed. M. Derrick (World Scientific, 1987);
J. D. Bjorken, Int. J. Mod. Phys. A **7**, 4189 (1992) and Phys. Rev. D **47**, 101 (1993).
[57] V. D. Barger, R. J. N. Phillips and D. Zeppenfeld, Phys. Lett. B **346**, 106 (1995).
[58] D. L. Rainwater, arXiv:hep-ph/9908378.
[59] V. D. Barger, K. m. Cheung, T. Han and R. J. N. Phillips, Phys. Rev. D **42**, 3052 (1990).
[60] V. Del Duca, W. Kilgore, C. Oleari, C. Schmidt and D. Zeppenfeld, Phys. Rev. Lett. **87**, 122001 (2001), Nucl. Phys. B **616**, 367 (2001) and Phys. Rev. D **67**, 073003 (2003).
[61] D. L. Rainwater and D. Zeppenfeld, JHEP **9712**, 005 (1997)
[62] V. Buscher and K. Jakobs, Int. J. Mod. Phys. A **20**, 2523 (2005)
[63] R. K. Ellis, I. Hinchliffe, M. Soldate and J. J. van der Bij, Nucl. Phys. B **297**, 221 (1988).
[64] T. Plehn, D. L. Rainwater and D. Zeppenfeld, Phys. Rev. D **61**, 093005 (2000).
[65] K. Cranmer, private communication.
[66] N. Kauer and D. Zeppenfeld, Phys. Rev. D **65**, 014021 (2002);
N. Kauer, Phys. Rev. D **67**, 054013 (2003) and Phys. Rev. D **70**, 014020 (2004).
[67] S. Dittmaier, P. Uwer, and S. Weinzierl, in preparation.
[68] F. Maltoni, D. L. Rainwater and S. Willenbrock, Phys. Rev. D **66**, 034022 (2002).
[69] V. Kostioukhine, J. Leveque, A. Rozanov, and J.B. de Vivie, ATL-PHYS-2002-019.
[70] F. Gianotti *et al.*[SLHC report], Eur. Phys. J. C **39**, 293 (2004).
[71] J. A. Aguilar-Saavedra *et al.* [ECFA/DESY LC Phys. Wrk. Grp.], arXiv:hep-ph/0106315.
[72] J. F. Gunion, H. E. Haber and R. Van Kooten, arXiv:hep-ph/0301023.
[73] G. Weiglein *et al.* [LHC/LC Study Group], Phys. Rept. **426**, 47 (2006).
[74] C. P. Burgess, J. Matias and M. Pospelov, Int. J. Mod. Phys. A **17**, 1841 (2002).
[75] M. Spira, private communication.
[76] S. P. Martin, arXiv:hep-ph/0701051.
[77] T. Hahn *et al.*, arXiv:hep-ph/0611373.
[78] V. Drollinger and A. Sopczak, Phys. Rev. D **66**, 010001 (2002).
[79] G. Velo and D. Zwanziger, Phys. Rev. **186**, 1337 (1969) and Phys. Rev. **188**, 2218 (1969);
B. Schroer, R. Seiler and J. A. Swieca, Phys. Rev. D **2** (1970) 2927.
[80] C. N. Yang, Phys. Rev. **77**, 242 (1950);
L.D.Landau, Dokl. Akad. Nawk., USSR 60, 207-209 (1948).
[81] E. Accomando *et al.*, arXiv:hep-ph/0608079.
[82] J. R. Dell'Aquila and C. A. Nelson, Phys. Rev. D **33**, 93 (1986).

[83] C. P. Buszello, I. Fleck, P. Marquard and J. J. van der Bij, Eur. Phys. J. C **32**, 209 (2004).
[84] S. Y. Choi, D. J. Miller, M. M. Muhlleitner and P. M. Zerwas, Phys. Lett. B **553**, 61 (2003).
[85] T. Plehn, D. L. Rainwater and D. Zeppenfeld, Phys. Rev. Lett. **88**, 051801 (2002).
[86] W. Buchmuller and D. Wyler, Nucl. Phys. B **268**, 621 (1986).
[87] K. Hagiwara, S. Ishihara, J. Kamoshita and B. A. Kniehl, Eur. Phys. J. C **14**, 457 (2000)
[88] D. J. Miller *et al.*, Phys. Lett. B **505**, 149 (2001).
[89] D. Zeppenfeld, R. Kinnunen, A. Nikitenko, E. Richter-Was, Phys. Rev. D **62**, 013009 (2000).
[90] M. Duhrssen *et al.*, Phys. Rev. D **70**, 113009 (2004).
[91] C. Anastasiou, K. Melnikov and F. Petriello, Phys. Rev. D **72**, 097302 (2005).
[92] P. Garcia-Abia, W. Lohmann and A. Raspereza, LC-PHSM-2000-062 *Prepared for 5th International Linear Collider Workshop, Fermilab, Batavia, Illinois, 24-28 Oct. 2000*.
[93] T. Abe *et al.* [American Linear Collider Working Group], in *Proc. of the APS/DPF/DPB Summer Study on the Future of Particle Physics (Snowmass 2001)* ed. N. Graf, arXiv:hep-ex/0106056.
[94] A. Juste and G. Merino, arXiv:hep-ph/9910301.
[95] K. Desch and M. Schumacher, Eur. Phys. J. C **46**, 527 (2006).
[96] A. Gay, arXiv:hep-ph/0604034.
[97] E. W. N. Glover and J. J. van der Bij, Nucl. Phys. B **309**, 282 (1988).
[98] T. Plehn, M. Spira and P. M. Zerwas, Nucl. Phys. B **479**, 46 (1996) [Erratum-ibid. B **531**, 655 (1998)].
[99] S. Dawson, S. Dittmaier and M. Spira, Phys. Rev. D **58**, 115012 (1998).
[100] A. Djouadi, W. Kilian, M. Muhlleitner and P. M. Zerwas, Eur. Phys. J. C **10**, 45 (1999).
[101] U. Baur, T. Plehn and D. L. Rainwater, Phys. Rev. Lett. **89**, 151801 (2002) and
Phys. Rev. D **67**, 033003 (2003).
[102] A. Dahlhoff, private communication.
[103] U. Baur, T. Plehn and D. L. Rainwater, Phys. Rev. D **69**, 053004 (2004).
[104] A. Djouadi, W. Kilian, M. Muhlleitner and P. M. Zerwas, Eur. Phys. J. C **10**, 27 (1999).
[105] U. Baur, T. Plehn and D. L. Rainwater, Phys. Rev. D **69**, 053004 (2004).
[106] A. De Roeck, *In the Proceedings of 32nd SLAC Summer Institute on Particle Physics (SSI 2004): Natures Greatest Puzzles, Menlo Park, California, 2-13 Aug 2004, pp FRT002*.
[107] E. Accomando *et al.* [CLIC Physics Working Group], arXiv:hep-ph/0412251.
[108] S. Kanemura, S. Kiyoura, Y. Okada, E. Senaha and C. P. Yuan, Phys. Lett. B **558**, 157 (2003).
[109] T. Plehn and M. Rauch, Phys. Rev. D **72**, 053008 (2005).

[110] T. Binoth, S. Karg, N. Kauer and R. Ruckl, Phys. Rev. D **74**, 113008 (2006).
[111] C. T. Hill and E. H. Simmons, Phys. Rept. **381**, 235 (2003) [Erratum-ibid. **390**, 553 (2004)]
[112] G. Cacciapaglia, C. Csaki, C. Grojean and J. Terning, eConf **C040802**, FRT004 (2004)
[Czech. J. Phys. **55**, B613 (2005)].
[113] V. Barger, T. Han, P. Langacker, B. McElrath and P. Zerwas, Phys. Rev. D **67**, 115001 (2003).
[114] K. Hagiwara, R. Szalapski and D. Zeppenfeld, Phys. Lett. B **318**, 155 (1993).
[115] K. Hagiwara, S. Matsumoto and R. Szalapski, Phys. Lett. B **357**, 411 (1995).
[116] R. Szalapski, Phys. Rev. D **57**, 5519 (1998).
[117] M. Beyer et al., Eur. Phys. J. C **48**, 353 (2006).
[118] V. Hankele, G. Klamke, D. Zeppenfeld and T. Figy, Phys. Rev. D **74**, 095001 (2006).
[119] T. D. Lee, Phys. Rev. D **8**, 1226 (1973).
[120] One of the early 2HDMs to be written down was in the Ph.D. dissertation of C. T. Hill (Caltech, 1977), but is unpublished. It may contain the first discussion of the 2HDM potential. Some of those results may be found in a much later publication:
C. T. Hill, C. N. Leung and S. Rao, Nucl. Phys. B **262**, 517 (1985).
[121] H. Huffel and G. Pocsik, Z. Phys. C **8**, 13 (1981).
[122] S. P. Martin, arXiv:hep-ph/9709356.
[123] J. F. Gunion and H. E. Haber, Nucl. Phys. B **272**, 1 (1986) [Erratum-ibid. B **402**, 567 (1993)] and Nucl. Phys. B **278**, 449 (1986).
[124] A. N. Okpara, Ph. D. dissertation, http://www.ub.uni-heidelberg.de/archiv/1873.
[125] [LEP Higgs Working Group for Higgs boson searches], arXiv:hep-ex/0107031.
[126] A. Heister et al. [ALEPH Collaboration], Phys. Lett. B **543**, 1 (2002).
[127] J. Abdallah et al. [DELPHI Collaboration], Eur. Phys. J. C **34**, 399 (2004).
[128] P. Achard et al. [L3 Collaboration], Phys. Lett. B **575**, 208 (2003).
[129] D. Horvath [OPAL Collaboration], Nucl. Phys. A **721**, 453 (2003).
[130] http://lephiggs.web.cern.ch/LEPHIGGS/www/Welcome.html
[131] S. Schael et al. [ALEPH Collaboration], Eur. Phys. J. C **47**, 547 (2006); see citations therein for the individual collaborations' results.
[132] P. Bechtle [LEP Collaboration], PoS **HEP2005**, 325 (2006).
[133] A. Sopczak [ALEPH Collaboration], arXiv:hep-ph/0602136.
[134] G. Abbiendi et al. [OPAL Collaboration], Eur. Phys. J. C **40**, 317 (2005).
[135] T. Plehn, D. L. Rainwater and D. Zeppenfeld, Phys. Lett. B **454**, 297 (1999).
[136] M. Schumacher, arXiv:hep-ph/0410112.
[137] U. Ellwanger, J. F. Gunion and C. Hugonie, arXiv:hep-ph/0111179.

[138] U. Ellwanger, J. F. Gunion, C. Hugonie and S. Moretti, arXiv:hep-ph/0305109.
[139] U. Ellwanger, J. F. Gunion and C. Hugonie, JHEP **0507**, 041 (2005).
[140] S. Moretti, S. Munir and P. Poulose, Phys. Lett. B **644**, 241 (2007).
[141] R. Kinnunen *et al.*, Eur. Phys. J. C **40N5**, 23 (2005); and references therein.
[142] S. Dittmaier, M. Kramer and M. Spira, Phys. Rev. D **70**, 074010 (2004).
[143] S. Dawson, C. B. Jackson, L. Reina and D. Wackeroth, Mod. Phys. Lett. A **21**, 89 (2006).
[144] D. Chakraborty, J. Konigsberg and D. Rainwater, Ann. Rev. Nucl. Part. Sci. **53**, 301 (2003).
[145] K. A. Assamagan *et al.* [Higgs Working Group Collaboration], arXiv:hep-ph/0406152.
[146] K. Hagiwara, A. D. Martin and D. Zeppenfeld, Phys. Lett. B **235**, 198 (1990).
[147] K. A. Assamagan, Y. Coadou and A. Deandrea, Eur. Phys. J. direct C **4**, 9 (2002).

Chapter 9

Z′ Phenomenology and the LHC

Thomas G. Rizzo

Stanford Linear Accelerator Center,
2575 Sand Hill Rd., Menlo Park, CA, 94025,
rizzo@slac.stanford.edu

A brief pedagogical overview of the phenomenology of Z′ gauge bosons is presented. Such particles can arise in various electroweak extensions of the Standard Model (SM). We provide a quick survey of a number of Z′ models, review the current constraints on the possible properties of a Z′ and explore in detail how the LHC may discover and help elucidate the nature of these new particles. We provide an overview of the Z′ studies that have been performed by both ATLAS and CMS. The role of the ILC in determining Z′ properties is also discussed.

9.1. Introduction: What is a Z′ and What is It Not?

To an experimenter, a Z′ is a resonance, which is more massive than the SM Z, observed in the Drell-Yan process $pp(p\bar{p}) \to l^+l^- + X$, where $l=e, \mu$ and, sometimes, τ, at the LHC(or the Tevatron). To a theorist, the production mechanism itself tells us that this new particle is neutral, colorless and self-adjoint, *i.e.*, it is its own antiparticle. However, such a new state could still be interpreted in many different ways. We may classify these possibilities according to the spin of the excitation, *e.g.*, a spin-0 $\tilde{\nu}$ in R-parity violating SUSY,[1] a spin-2 Kaluza-Klein(KK) excitation of the graviton as in the Randall-Sundrum(RS) model,[2,3] or even a spin-1 KK excitation of a SM gauge boson from some extra dimensional model.[4,5] Another possibility for the spin-1 case is that this particle is the carrier of a new force, a new neutral gauge boson arising from an extension of the SM gauge group, *i.e.*, a true Z′, which will be our subject below.[6] Given this discussion it is already clear that once a new Z′-like resonance is discovered it will first be necessary to measure its spin as quickly as possible to have some idea what kind of new physics we are dealing with. As will be discussed below this can be done rather easily with only a few hundred events by measuring the dilepton angular distribution in the reconstructed Z' rest frame. Thus, a Z' is a neutral, colorless, self-adjoint, spin-1 gauge boson that is a carrier of a new force.[a]

Once found to be a Z', the next goal of the experimenter will be to determine as well as possible the couplings of this new state to the particles (mainly fermions) of the SM, *i.e.*, to *identify* which Z' it is. As we will see there are a huge number of models which predict the existence of a Z'.[6,8] Is this new particle one of those or is it something completely new? How does it fit into a larger theoretical framework?

9.2. Z' Basics

If our goal is to determine the Z' couplings to SM fermions, the first question one might ask is 'How many fermionic couplings does a Z' have?' Since the Z' is a color singlet its couplings are color-diagonal. Thus (allowing for the possibility of light Dirac neutrinos), in general the Z' will have 24 distinct couplings-one for each of the two-component SM fields: $u_{L_i}, d_{L_i}, \nu_{L_i}, e_{L_i} + (L \to R)$ with $i = 1 - 3$ labeling the three generations. (Of course, exotic fermions not present in the SM can also occur but we will ignore these for the moment.) For such a generic Z' these couplings are *non-universal*, *i.e.*, family-dependent and this can result in dangerous flavor changing neutral currents(FCNC) in low-energy processes. The constraints on such beasts are known to be quite strong from both $K-\bar{K}$ and $B_{d,s}-\bar{B}_{d,s}$ mixing[9] as well as from a large number of other low-energy processes. There FCNC are generated by fermion mixing which is needed to diagonalize the corresponding fermion mass matrix. As an example, consider schematically the Z' coupling to left-handed down-type quarks in the weak basis, *i.e.*, $\bar{d}^0_{L_i} \eta_i d^0_{L_i} Z'$, with η_i being a set of coupling parameters whose different values would represent the generational-dependent couplings. For simplicity, now let $\eta_{1,2} = a$ and $\eta_3 = b$ and make the unitary transformation to the physical, mass eigenstate basis, $d^0_{L_i} = U_{ij} d_{L_j}$. Some algebra leads to FCNC couplings of the type $\sim (b-a)\bar{d}_{L_i} U^\dagger_{i3} U_{3j} d_{L_j} Z'$. Given the existing experimental constraints, since we expect these mixing matrix elements to be of order those in the CKM matrix and a, b to be O(1), the Z' mass must be huge, ~ 100 TeV or more, and outside the reach of the LHC. Thus un-

---
[a]Distinguishing a Z' from a spin-1 KK excitation is a difficult subject beyond the scope of the present discussion.[7]

less there is some special mechanism acting to suppress FCNC it is highly likely that a Z' which is light enough to be observed at the LHC will have *generation-independent* couplings, *i.e.*, now the number of couplings is reduced: $24 \to 8$ (or 7 if neutrinos are Majorana fields and the RH neutrinos are extremely heavy).

Further constraints on the number of independent couplings arise from several sources. First, consider the generator or 'charge' to which the Z' couples, T'. Within any given model the group theory nature of T' will be known so that one may ask if $[T', T_i] = 0$, with T_i being the usual SM weak isospin generators of $SU(2)_L$. If the answer is in the affirmative, then all members of any SM representation can be labeled by a common eigenvalue of T'. This means that u_L and d_L, *i.e.*, $Q^T = (u, d)_L$, as well as ν_L and e_L, *i.e.*, $L^T = (\nu, e)_L$ (and dropping generation labels), will have identical Z' couplings so that the number of independent couplings is now reduced from $8 \to 6 (7 \to 5)$. As we will see, this is a rather common occurrence in the case of garden-variety Z' which originate from extended GUT groups[6] such as $SO(10)$ or E_6. Clearly, models which do not satisfy these conditions lead to Z' couplings which are at least partially proportional to the diagonal SM isospin generator itself, *i.e.*, $T' = aT_3$.

In UV completed theories a further constraint on the Z' couplings arises from the requirement of anomaly cancellation. Anomalies can arise from one-loop fermionic triangle graphs with three external gauge boson legs; recall that fermions of opposite chirality contribute with opposite signs to the relevant 'VVA' parts of such graphs. In the SM, the known fermions automatically lead to anomaly cancellation in a generation independent way when the external gauge fields are those of the SM. The existence of the Z', together with gauge invariance and the existence of gravity, tells us that there are 6 new graphs that must also vanish to make the theory renormalizable thus leading to 6 more constraints on the couplings of the Z'. For example, the graph with an external Z' and 2 gluons tells us that the sum over the colored fermion's eigenvalues of T' must vanish. We can write these 6 constraints as (remembering to flip signs for RH fields)

$$\sum_{color triplets, i} T'_i = \sum_{isodoublets, i} T'_i = 0$$

$$\sum_i Y_i^2 T'_i = \sum_i Y_i T'^2_i = 0 \quad (9.1)$$

$$\sum_i T'^3_i = \sum_i T'_i = 0,$$

where here we are summing over various fermion representations. These 6 constraints can be quite restrictive, e.g., if $T' \neq aT_{3L} + bY$, then even in the simplest Z' model, ν_R (not present in the SM!) must exist to allow for anomaly cancellation. More generally, one finds that the existence of new gauge bosons will also require the existence of other new, vector-like (with respect to the SM gauge group) fermions to cancel anomalies, something which happens automatically in the case of extended GUT groups. It is natural in such scenarios that the masses of these new fermions are comparable to that of the Z' itself so that they may also occur as decay products of the Z' thus modifying the various Z' branching fractions. If these modes are present then there are more coupling parameters to be determined.

9.3. Z-Z' Mixing

In a general theory the Z' and the SM Z are not true mass eigenstates due to mixing; in principle, this mixing can arise from two different mechanisms.

In the case where the new gauge group G is a simple new $U(1)'$, the most general set of $SU(2)_L \times U(1)_Y \times U(1)'$ kinetic terms in the original weak basis (here denoted by tilded fields) is

$$\mathcal{L}_K = -\frac{1}{4}W^a_{\mu\nu}W_a^{\mu\nu} - \frac{1}{4}\tilde{B}_{\mu\nu}\tilde{B}^{\mu\nu} - \frac{1}{4}\tilde{Z}'_{\mu\nu}\tilde{Z}'^{\mu\nu} - \frac{\sin\chi}{2}\tilde{Z}'_{\mu\nu}\tilde{B}^{\mu\nu}, \quad (9.2)$$

where $\sin\chi$ is a parameter. Here W^a_μ is the usual $SU(2)_L$ gauge field while $\tilde{B}_\mu, \tilde{Z}_\mu$ are those for $U(1)_Y$ and $U(1)'$, respectively. Such gauge kinetic mixing terms can be induced (if not already present) at the one-loop level if $Tr(T'Y) \neq 0$. Note that if G were a nonabelian group then no such mixed terms would be allowed by gauge invariance. In this basis the fermion couplings to the gauge fields can be schematically written as $\bar{f}(g_L T_a W^a + g_Y Y \tilde{B} + \tilde{g}_{Z'} T' \tilde{Z}') f$. To go to the physical basis, we make the linear transformations $\tilde{B} \to B - \tan\chi Z'$ and $\tilde{Z}' \to Z'/\cos\chi$ which diagonalizes $\mathcal{L}_K$ and leads to the modified fermion couplings $\bar{f}[g_L T_a W^a + g_Y Y B + g_{Z'}(T' + \delta Y)Z'] f$ where $g_{Z'} = \tilde{g}_{Z'}/\cos\chi$ and $\delta = -g_Y \tan\chi/g_{Z'}$. Here we see that the Z' picks up an additional coupling proportional to the usual weak hypercharge. $\delta \neq 0$ symbolizes this gauge kinetic mixing[10] and provides a window for its experimental observation. In a GUT framework, being a running parameter, $\delta(M_{GUT}) = 0$, but can it can become non-zero via RGE running at lower mass scales if the low energy sector contains matter in incomplete GUT representations. In most models[10] where this happens, $|\delta(\sim \text{TeV})| \leq 1/2$.

Z-Z' mixing can also occur through the conventional Higgs-induced SSB mechanism (*i.e.*, mass mixing) if the usual Higgs doublet(s), H_i (with vevs v_{D_i}), are *not* singlets under the new gauge group G. In general, the breaking of G requires the introduction of SM singlet Higgs fields, S_j (with vevs v_{S_j}). These singlet vevs should be about an order of magnitude larger than the typical doublet vevs since a Z' has not yet been observed. As usual the Higgs kinetic terms will generate the W, Z and Z' masses which for the neutral fields look like

$$\sum_i \left[\left(\frac{g_L}{c_w} T_{3L} Z + g_{Z'} T' Z'\right) v_{D_i}\right]^2 + \sum_j [g_{Z'} T' v_{S_j} Z']^2 , \qquad (9.3)$$

where $c_w = \cos\theta_W$. (Note that the massless photon has already been 'removed' from this discussion.) The square of the first term in the first sum produces the square of the usual SM Z boson mass term, $\sim M_Z^2 Z^2$. The square of the last term in this sum plus the square of the second sum produces the corresponding Z' mass term, $\sim M_{Z'}^2 Z'^2$. However, the ZZ' interference piece in the first sum leads to Z-Z' mixing provided $T' H_i \neq 0$ for at least one i; note that the scale of this cross term is set by the doublet vevs and hence is of order $\sim M_Z^2$.

This analysis can be summarized by noting that the interaction above actually generates a mass (squared) matrix in the ZZ' basis:

$$\mathcal{M}^2 = \begin{pmatrix} M_Z^2 & \beta M_Z^2 \\ \beta M_Z^2 & M_Z'^2 \end{pmatrix} . \qquad (9.4)$$

Note that the symmetry breaking dependent parameter β,

$$\beta = \frac{4 c_w g_{Z'}}{g_L} \left[\sum_i T_{3L_i} T'_i v_{D_i}^2\right] \bigg/ \sum_i v_{D_i}^2 , \qquad (9.5)$$

can be argued to be O(1) or less on rather general grounds. Since this matrix is real, the diagonalization of $\mathcal{M}^2$ proceeds via a simple rotation through a mixing angle ϕ, *i.e.*, by writing $Z = Z_1 \cos\phi - Z_2 \sin\phi$, etc, which yields the mass eigenstates $Z_{1,2}$ with masses $M_{1,2}$; given present data we may expect $r = M_1^2/M_2^2 \leq 0.01 - 0.02$. $Z_1 \simeq Z$ is the state presently produced at colliders, *i.e.*, $M_1 = 91.1875 \pm 0.0021$ GeV, and thus we might also expect that ϕ must be quite small for the SM to work as well as it does. Defining $\rho = M_Z^2/M_1^2$, with M_Z being the would-be mass of the Z if no mixing occurred, we can approximate

$$\phi = -\beta r [1 + (1+\beta^2) r + O(r^2)] \qquad (9.6)$$
$$\delta\rho = \beta^2 r [1 + (1+2\beta^2) r + O(r^2)] ,$$

where $\delta\rho = \rho - 1$, so that β determines the sign of ϕ. We thus expect that both $\delta\rho, |\phi| < 10^{-2}$. In fact, if we are *not* dealing with issues associated with precision measurements[11] then Z-Z' mixing is expected to be so small that it can be safely neglected.

It is important to note that non-zero mixing modifies the predicted SM Z couplings to $\frac{g_L}{c_w}(T_{3L} - x_W Q)c_\phi + g_{Z'}T's_\phi$, where $x_W = \sin^2\theta_W$, which can lead to many important effects. For example, the partial width for $Z_1 \to f\bar{f}$ to lowest order (*i.e.*, apart from phase space, QCD and QED radiative corrections) is now given by

$$\Gamma(Z_1 \to f\bar{f}) = N_c \frac{\rho G_F M_1^3 (v_{eff}^2 + a_{eff}^2)}{6\sqrt{2}\pi}, \qquad (9.7)$$

where N_c is a color factor, ρ is given above and

$$v_{eff} = (T_{3L} - 2x_W Q)c_\phi + \frac{g_{Z'}}{g_L/(2c_w)}(T'_L + T'_R)s_\phi \qquad (9.8)$$

$$a_{eff} = T_{3L}c_\phi + \frac{g_{Z'}}{g_L/(2c_w)}(T'_L - T'_R)s_\phi,$$

and where $T'_{L,R}$ are the eigenvalues of T' for $f_{L,R}$. Other effects that can occur include decay modes such as $Z_2 \to W^+W^-, Z_1 H_i$, where H_i is a light Higgs, which are now induced via mixing. If T' has no T_3 component this is the only way such decays can occur at tree level. In the case of the $Z_2 \to W^+W^-$ mode, an interesting cancellation occurs: the partial width scales as $s_\phi^2 (M_2/M_W)^4$, where the second factor follows from the Goldstone Boson Equivalence Theorem.[12] However, since $s_\phi \simeq -\beta r$ and $r = M_1^2/M_2^2 \simeq M_Z^2/M_2^2$, we find instead that the partial width goes as $\sim \beta^2$ without any additional mass enhancement or suppression factors. The tiny mixing angle induced by small r has been offset by the large M_2/M_W ratio! In specific models, one finds that this small Z-Z' mixing leads to $Z_2 \to W^+W^-$ partial widths which can be comparable to other decay modes. Of course, $Z_2 \to W^+W^-$ can be also be induced at the one-loop level but there the amplitude will be suppressed by the corresponding loop factor as well as possible small mass ratios.

9.4. Some Sample Z' Models

There are many (hundreds of) models on the market which predict a Z' falling into two rather broad categories depending on whether or not they arise in a GUT scenario. The list below is *only* meant to be representative and is very far from exhaustive and I beg pardon if your favorite model is not represented.

The two most popular GUT scenarios are the Left Right Symmetric Model (LRM)[13] and those that come from E_6 grand unification.[6]

(i) In the E_6 case one imagines a symmetry breaking pattern $E_6 \to SO(10) \times U(1)_\psi \to SU(5) \times U(1)_\chi \times U(1)_\psi$. Then $SU(5)$ breaks to the SM and only one linear combination $G = U(1)_\theta = c_\theta U(1)_\psi - s_\theta U(1)_\chi$ remains light at the TeV scale. θ is treated as a free parameter[b] and the particular values $\theta = 0$, $-90°$, $\sin^{-1}\sqrt{(3/8)} \simeq 37.76°$ and $-\sin^{-1}\sqrt{(5/8)} \simeq -52.24°$, correspond to 'special' models called ψ, χ, η and I, respectively. These models are sometimes referred to in the literature as effective rank-5 models (ER5M). In this case, neglecting possible kinetic mixing,

$$g_{Z'}T' = \lambda \frac{g_L}{c_w} \sqrt{\frac{5x_W}{3}} \left(\frac{Q_\psi c_\theta}{2\sqrt{6}} - \frac{Q_\chi s_\theta}{2\sqrt{10}} \right), \tag{9.9}$$

where $\lambda \simeq 1$ arises from RGE evolution. The parameters $Q_{\psi,\chi}$ originate from the embeddings of the SM fermions into the fundamental **27** representation of E_6. A detailed list of their values can be found in the second paper in[6] with an abbreviated version given in the Table below in LH field notation. Note that this is the *standard* form for this embedding and there are other possibilities.[6] These other choices can be recovered by a shift in the parameter θ. Note further that in addition to the SM fermions plus the RH neutrino, E_6 predicts, per generation, an additional neutral singlet, S^c, along with an electric charge $Q = -1/3$, color triplet, vector-like isosinglet, h, and a color singlet, vector-like isodoublet whose top member has $Q = 0$, H (along with their conjugate fields). These exotic fermions with masses comparable to the Z' cancel the anomalies in the theory and can lead to interesting new phenomenology[6] but we will generally ignore them in our discussion below. In many cases these states are quite heavy and thus will not participate in Z' decays.

(ii) The LRM, based on the low-energy gauge group $SU(2)_L \times SU(2)_R \times U(1)_{B-L}$, can arise from an $SO(10)$ or E_6 GUT. Unlike the case of ER5M, not only is there a Z' but there is also a new charged $W_R^\pm$ gauge boson since here $G = SU(2)$. In general $\kappa = g_R/g_L \neq 1$ is a free parameter but must be $> x_W/(1-x_W)$ for the existence of real gauge couplings. On occasions, the parameter $\alpha_{LR} = \sqrt{c_w^2 \kappa^2/x_W^2 - 1}$ is also often used. In this case we

---

[b]The reader should be aware that there are several different definitions of this mixing angle in the literature, *i.e.*, $Z' = Z_\chi \cos\beta + z_\psi \sin\beta$ occurs quite commonly.

Table 9.1. Quantum numbers for various SM and exotic fermions in LH notation in E_6 models

Representation	Q_ψ	Q_χ
Q	1	-1
L	1	3
u^c	1	-1
d^c	1	3
e^c	1	-1
ν^c	1	-5
H	-2	-2
H^c	-2	2
h	-2	2
h^c	-2	-2
S^c	4	0

find that

$$g_{Z'}T' = \frac{g_L}{c_w}[\kappa^2 - (1+\kappa^2)x_W]^{-1/2}[x_W T_{3L} + \kappa^2(1-x_W)T_{3R} - x_W Q]. \quad (9.10)$$

The mass ratio of the W' and Z' is given by

$$\frac{M_{Z'}^2}{M_{W'}^2} = \frac{\kappa^2(1-x_W)\rho_R}{\kappa^2(1-x_W) - x_W} > 1, \quad (9.11)$$

with the values $\rho_R = 1(2)$ depending upon whether $SU(2)_R$ is broken by either Higgs doublets (or by triplets). The existence of a $W' = W_R$ with the correct mass ratio to the Z' provides a good test of this model. Note that due to the LR symmetry we need not introduce additional fermions in this model to cancel anomalies although right-handed neutrinos are present automatically. In the E_6 case a variant of this model[14] can be constructed by altering the embeddings of the SM and exotic fermions into the ordinary **10** and **5** representations (called the Alternative LRM, *i.e.*, ALRM).

(*iii*) The Z' in the Little Higgs scenario[15] provides the best non-GUT example. The new particles in these models, *i.e.*, new gauge bosons, fermions and Higgs, are necessary to remove at one-loop the quadratic divergence of the SM Higgs mass and their natures are dictated by the detailed group structure of the particular model. This greatly restricts the possible couplings of such states. With a W' which is essentially degenerate in mass with the Z', the Z' is found to couple like $g_{Z'}T' = (g_L/2)T_{3L}\cot\theta_H$, with θ_H another mixing parameter.

(iv) Another non-GUT example[17] is based on the group $SU(2)_l \times SU(2)_h \times U(1)_Y$ with l, h referring to 'light' and 'heavy'. The first 2 generations couple to $SU(2)_l$ while the third couples to $SU(2)_h$. In this case the Z' and W' are again found to be degenerate and the Z' couples to $g_{Z'}T' = g_L[\cot\Phi T_{3l} - \tan\Phi T_{3h}]$ with Φ another mixing angle. Such a model is a good example of where the Z' couplings are generation dependent.

(v) A final example is a Z' that has couplings which are exactly the same as those of the SM Z (SSM), but is just heavier. This is not a real model but is very commonly used as a 'standard candle' in experimental Z' searches. A more realistic variant of this model is one in which a Z' has *no* couplings to SM fermions in the weak basis but the couplings are then induced in the mass eigenstate basis Z-Z' via mixing. In this case the relevant couplings of the Z' are those of the SM Z but scaled down by a factor of $\sin\phi$.

A nice way to consider rather broad classes of Z' models has recently been described by Carena et al..[18] In this approach one first augments the SM fermion spectrum by adding to it a pair of vector-like (with respect to the SM) fermions, one transforming like L and the other like d^c; this is essentially what happens in the E_6 GUT model. The authors then look for families of models that satisfy the six anomaly constraints with generation-independent couplings. Such an analysis yields several sets of 1-parameter solutions for the generator T' but leaves the coupling $g_{Z'}$ free. The simplest such solution is $T' = B - xL$, with x a free par meter. Some other solutions include $T' = Q + xu_R$ (i.e., $T'(Q) = 1/3$ and $T'(u_R) = x/3$ and all others fixed by anomaly cancellation), $T' = d_R - xu_R$ and $T' = 10 + x\bar{5}$, where '10' and $\bar{5}$ refer to $SU(5)$ GUT assignments.

9.5. What Do We Know Now? Present Z' Constraints

Z' searches are of two kinds: indirect and direct. Important constraints arise from both sources at the present moment though this is likely to change radically in the near future.

9.5.1. *Indirect Z' Searches*

In this case one looks for deviations from the SM that might be associated with the existence of a Z'; this usually involves precision electroweak measurements at, below and *above* the Z-pole. The cross section and forward backward asymmetry, A_{FB}, measurements at LEPII take place at high center of mass energies which are still (far) below the actual Z' mass.

Since such constraints are indirect, one can generalize from the case of a new Z' and consider a more encompassing framework based on contact interactions.[19] Here one 'integrates out' the new physics (since we assume we are at energies below which the new physics is directly manifest) and express its influence via higher-dimensional (usually dim-6) operators. For example, in the dim-6 case, for the process $e^+e^- \to \bar{f}f$, we can consider an effective Lagrangian of the form[19]

$$\mathcal{L} = \mathcal{L}_{SM} + \frac{4\pi}{\Lambda^2(1+\delta_{ef})} \sum_{ij=L,R} \eta_{ij}^f (\bar{e}_i \gamma_\mu e_i)(\bar{f}_j \gamma^\mu f_j), \quad (9.12)$$

where Λ is called 'the compositeness scale' for historic reasons, δ_{ef} takes care of the statistics in the case of Bhabha scattering, and the η's are chirality structure coefficients which are of order unity. The exchange of many new states can be described in this way and can be analyzed simultaneously. The corresponding parameter bounds can then be interpreted within your favorite model. This prescription can be used for data at all energies as long as these energies are far below Λ.

Z-pole measurements mainly restrict the Z-Z' mixing angle as they are sensitive to small mixing-induced deviations in the SM couplings and not to the Z' mass. LEP and SLD have made very precise measurements of these couplings which can be compared to SM predictions including radiative corrections.[11] An example of this is found in Fig. 9.1 where we see the experimental results for the leptonic partial width of the Z as well as $\sin^2\theta_{lepton}$ in comparison with the corresponding SM predictions. Deviations in $\sin^2\theta_{lepton}$ are particularly sensitive to shifts in the Z couplings due to non-zero values of ϕ. Semiquantitatively these measurements strongly suggest that $|\phi| \leq a\ few\ 10^{-3}$, at most, in most Z' models assuming a light Higgs. Performing a global fit to the full electroweak data set, as given, e.g., by the LEPEWWG[11] gives comparable constraints.[8]

Above the Z pole, LEPII data provides strong constraints on Z' couplings and masses but are generally insensitive to small Z-Z' mixing. Writing the couplings as $\sum_i \bar{f}\gamma_\mu(v_{f_i} - a_{f_i}\gamma_5)f Z_i^\mu$ for $i=\gamma, Z, Z'$, the differential cross section for $e^+e^- \to \bar{f}f$ when $m_f = 0$ is just

$$\frac{d\sigma}{dz} = \frac{N_c}{32\pi s} \sum_{i,j} P_{ij}[B_{ij}(1+z^2) + 2C_{ij}z], \quad (9.13)$$

where

$$\begin{aligned} B_{ij} &= (v_i v_j + a_i a_j)_e (v_i v_j + a_i a_j)_f \\ C_{ij} &= (v_i a_j + a_i v_j)_e (v_i a_j + a_i v_j)_f, \end{aligned} \quad (9.14)$$

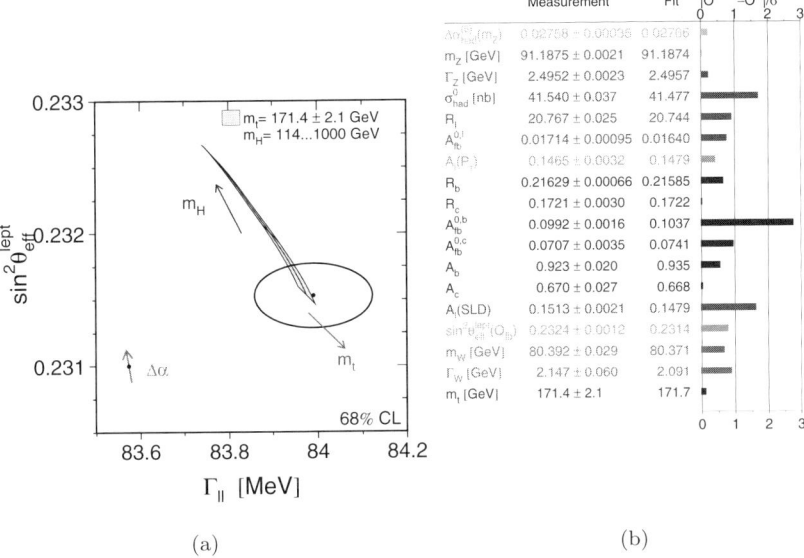

Fig. 9.1. Summer 2006 results from the LEPEWWG. (a) Fit for the Z leptonic partial width and $\sin^2\theta_{lepton}$ in comparison to the SM prediction in the yellow band. (b) Comparison of a number of electroweak measurements with their SM fitted values.

and

$$P_{ij} = s^2 \frac{(s-M_i^2)(s-M_j^2) + \Gamma_i\Gamma_j M_i M_j}{[(s-M_i^2)^2 + \Gamma_i^2 M_i^2][i \to j]}, \qquad (9.15)$$

with $\sqrt{s}$ the collision energy, Γ_i being the total widths of the exchanged particles and $z = \cos\theta$, the scattering angle in the CM frame. A_{FB} for any final state fermion f is then given by the ratio of integrals

$$A_{FB}^f = \left[\frac{\int_0^1 dz \frac{d\sigma}{dz} - \int_{-1}^0 dz \frac{d\sigma}{dz}}{'' + ''}\right]. \qquad (9.16)$$

If the $e^\pm$ beams are polarized (as at the ILC but not at LEP) one can also define the left-right polarization asymmetry, A_{LR}^f; to this end we let

$$\begin{aligned}B_{ij} &\to B_{ij} + \xi(v_i a_j + a_i v_j)_e (v_i v_j + a_i a_j)_f \\ C_{ij} &\to C_{ij} + \xi(v_i v_j + a_i a_j)_e (v_i a_j + a_i v_j)_f,\end{aligned} \qquad (9.17)$$

and then form the ratio

$$A_{LR}^f(z) = P\left[\frac{d\sigma(\xi=+1) - d\sigma(\xi=-1)}{'' + ''}\right], \quad (9.18)$$

where P is the effective beam polarization.

For a given Z' mass and couplings the deviations from the SM can then be calculated and compared with data; since no obvious deviations from the SM were observed, LEPII[11] places 95% CL lower bounds on Z' masses of 673(481, 434, 804, 1787) GeV for the $\chi(\psi, \eta, \mathrm{LRM}(\kappa=1), \mathrm{SSM})$ models assuming $\lambda = 1$. Note that since we are far away from the Z' pole these results are not sensitive to any particular assumed values for the Z' width as long as it is not too large.

The process $e^+e^- \to W^+W^-$ can also be sensitive to the existence of a Z', in particular, in the case where there is some substantial Z-Z' mixing.[20] The main reason for this is the well-known gauge cancellations among the SM amplitudes that maintains unitarity for this process as the center of mass energy increases. The introduction of a Z' with Z-Z' mixing induces tiny shifts in the W couplings that modifies these cancellations to some extent and unitarity is not completely restored until energies beyond the Z' mass are exceeded. As shown by the first authors in Ref. 20, the leading effects from Z-Z' mixing can be expressed in terms of two s−dependent anomalous couplings for the $WW\gamma$ and WWZ vertices, i.e., $g_{WW\gamma} = e(1 + \delta_\gamma)$ and $g_{WWZ} = e(\cot\theta_W + \delta_Z)$ and inserting them into the SM amplitude expressions. The parameters $\delta_{\gamma,Z}$ are sensitive to the Z' mass, its leptonic couplings, as well as the Z-Z' mixing angle. In principle, the constraints on anomalous couplings from precision measurements can be used to bound the Z' parameters in a model dependent way. However, the current data from LEPII[11] is not precise enough to get meaningful bounds. More precise data will, of course, be obtained at both the LHC and ILC.

The measurement of the W mass itself can also provides a constraint on $\delta\rho$ since the predicted W mass is altered by the fact that $M_Z \neq M_{Z_1}$. Some algebra shows that the resulting mass shift is expected to be $\delta M_W = 57.6 \frac{\delta\rho}{10^{-3}}$ MeV. Given that M_W is within $\simeq 30$ MeV of the predicted SM value and the current size of theory uncertainties,[21] strongly suggests that $\delta\rho \leq a \; few \; 10^{-3}$ assuming a light Higgs. This is evidence of small r and/or β if a Z' is actually present.

Below the Z pole many low energy experiments are sensitive to a Z'. Here we give only two examples: (i) The E-158 Polarized Moller scattering experiment[22] essentially measures A_{LR} which is proportional to a coupling combination $\sim -1/2 + 2x_{eff}$ where $x_{eff} = x_W +$ 'new physics'. Here x_W is the running value of $\sin^2\theta_W$ at low Q^2 which is reliable calculable. For a Z' (assuming no mixing) the 'new physics' piece is just $\frac{-1}{\sqrt{2}G_F}\frac{g_{Z'}^2}{M_{Z'}^2}v'_e a'_e$, which can be determined in your favorite model. Given the data,[22] $x_{eff} - x_W = 0.0016 \pm 0.0014$, one finds, e.g., that $M_{Z_\chi} \geq 960\lambda$ GeV at 90% CL. (ii) Atomic Parity Violation(APV) in heavy atoms measures the effective parity violating interaction between electrons and the nucleus and is parameterized via the 'weak charge', Q_W, which is again calculable in your favorite model:

$$Q_W = -4\sum_i \frac{M_Z^2}{M_{Z_i}^2} a_{e_i}[v_{u_i}(2Z+N) + v_{d_i}(2N+Z)], \qquad (9.19)$$

$= -N + Z(1 - 4x_W)+$ a Z' piece, in the limit of no mixing; here the sum extends over all neutral gauge bosons. The possible shift, ΔQ_W, from the SM prediction then constrains Z' parameters. The highest precision measurements from Cs^{133} yield[23] $\Delta Q_W = 0.45 \pm 0.48$ which then imply (at 95% CL) $M_{Z_\chi} > 1.05\lambda$ TeV and $M_{Z_{LRM}} > 0.67$ TeV for $\kappa = 1$. Note that though both these measurements take place at very low energies, their relative cleanliness and high precision allows us to probe TeV scale Z' masses. Figure 9.2 shows the predicted value of the running $\sin^2\theta_W$[25] together with the experimental results obtained from E-158, APV and NuTeV.[24] The apparent $\sim 3\sigma$ deviation in the NuTeV result remains controversial but is at the moment usually ascribed to our lack of detailed knowledge of, e.g., the strange quark parton densities and not to new physics.

9.5.2. *Direct Z' Searches*

In this case, we rely on the Drell-Yan process at the Tevatron as mentioned above. The present lack of any signal with an integrated luminosity approaching ~ 1 fb^{-1} allows one to place a model-dependent lower bound on the mass of any Z'. The process $p\bar{p} \to l^+l^- + X$ at leading order arises from the parton-level subprocess $q\bar{q} \to l^+l^-$ which is quite similar to the $e^+e^- \to f\bar{f}$ reaction discussed above. The cross section for the inclusive process is described by 4 variables: the collider CM energy, $\sqrt{s}$, the invariant mass of the lepton pair, M, the scattering angle between the q and the l^-, θ^*, and the lepton rapidity in the lab frame, y, which depends on its energy (E) and longitudinal momentum(p_z): $y = \frac{1}{2}\log\left[\frac{E+p_z}{E-p_z}\right]$. For a massless particle, this is the same as the pseudo-rapidity, η. With these variables the triple differential cross section for the Drell-Yan process is

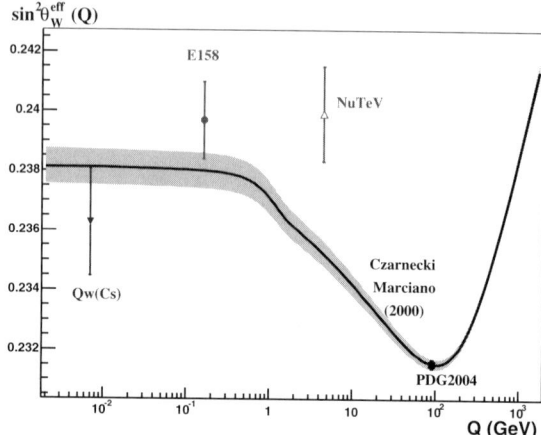

Fig. 9.2. A comparison by E-158 of the predictions for the running value of $\sin^2\theta_W$ with the results of several experiments as discussed in the text.

given by $(z = \cos\theta^*)$

$$\frac{d\sigma}{dM\,dy\,dz} = \frac{K(M)}{48\pi M^3} \sum_q \left[S_q G_q^+ (1+z^2) + 2 A_q G_q^- z \right], \tag{9.20}$$

where K is a numerical factor that accounts for NLO and NNLO QCD corrections[26] as well as leading electroweak corrections[28] and is roughly of order $\simeq 1.3$ for suitably defined couplings,

$$G_q^\pm = x_a x_b [q(x_a, M^2)\bar{q}(x_b, M^2) \pm q(x_b, M^2)\bar{q}(x_a, M^2)], \tag{9.21}$$

are products of the appropriate parton distribution functions (PDFs), with $x_{a,b} = M e^{\pm y}/\sqrt{s}$ being the relevant momentum fractions, which are evaluated at the scale M^2 and

$$\begin{aligned} S_q &= \sum_{ij} P_{ij}(s \to M^2) B_{ij}(f \to q) \\ A_q &= \sum_{ij} P_{ij}(s \to M^2) C_{ij}(f \to q), \end{aligned} \tag{9.22}$$

with B, C and P as given above. In order to get precise limits (and to measure Z' properties once discovered as we will see later), the NNLO QCD corrections play an important role[26] as do the leading order electroweak radiative corrections.[28] Apart from the machine luminosity errors the largest uncertainty in the above cross section is due to the PDFs. For $M \lesssim 1$ TeV or so these errors are of order $\simeq 5\%$[30] but grow somewhat bigger for larger invariant masses: $\sim 15(25)\%^{31}$ for $M = 3(5)$ TeV. As a point of comparison the corrected SM predictions for the W and Z production cross sections at the Tevatron are seen to agree with the data from both CDF and D0 at the level a few percent.[32]

It is somewhat more useful to perform some of the integrals above in order to make direct comparison with experimental data. To this end we define (for our LHC discussion below)

$$\frac{d\sigma^{\pm}}{dM\,dy} = \left[\int_0^{z_0} \pm \int_{-z_0}^0\right] \frac{d\sigma}{dM\,dy\,dz}, \quad (9.23)$$

and subsequently

$$\frac{d\sigma^{\pm}}{dM} = \left[\int_{y_{min}}^{Y} \pm \int_{-Y}^{-y_{min}}\right] \frac{d\sigma^{\pm}}{dM\,dy}. \quad (9.24)$$

Here Y is cut representing the edge of the central detector acceptance($\simeq 1.1$ for the Tevatron detectors and $\simeq 2.5$ for those at the LHC) with $z_0 = min[\tanh(Y-|y|), 1]$ being the corresponding angular cut. y_{min} is a possible cut employed to define the Z' boost direction which we will return to below. As in the case of e^+e^- collisions above, one can define an $A_{FB}(M) = d\sigma^-/d\sigma^+$.

A Z', being a weakly interacting beast, generally has a rather narrow width to mass ratio, i.e., $\Gamma^2_{Z'}/M^2_{Z'} \ll 1$; e.g., in the case of the SM Z this ratio is $\simeq 10^{-3}$. This being the case, almost the entire Z' event rate comes from a rather narrow window of M values: $M \simeq M_{Z'} \pm 2\Gamma_{Z'}$, or so. In this limit we can approximate the resonance as a δ-function in M and drop all of the SM contributions to the sums above. In this case, pieces of the P_{ij} that go as, e.g., $M^4/|(M^2 - M^2_{Z'}) + iM_{Z'}\Gamma_{Z'}|^2$ can be replaced by $\frac{\pi}{2}\delta(M - M_{Z'})\frac{M^2_{Z'}}{\Gamma_{Z'}}$, up to $\Gamma^2_{Z'}/M^2_{Z'}$ corrections, so that integrals over M can be performed analytically (since the integral over the PDFs is now just a constant factor). In such a limit, the contribution to the cross section for l^+l^- production from the Z' is just $\sigma_{Z'}B(Z' \to l^+l^-)$ with $\sigma_{Z'}$ being the integrated value of the cross section at $M = M_{Z'}$, i.e., at the Z' peak, and B being the leptonic branching fraction of the Z'. This is called the Narrow Width Approximation (NWA). In a similar way, A_{FB} on the Z' pole in the NWA is just the ratio $d\sigma^-/d\sigma^+$ evaluated at $M_{Z'}$; note that this ratio does *not* depend upon what decay modes (other than leptonic) that the Z' might have. Also note that in the NWA, the continuum Drell-Yan background makes no contribution to the event rate. This is a drawback of the NWA

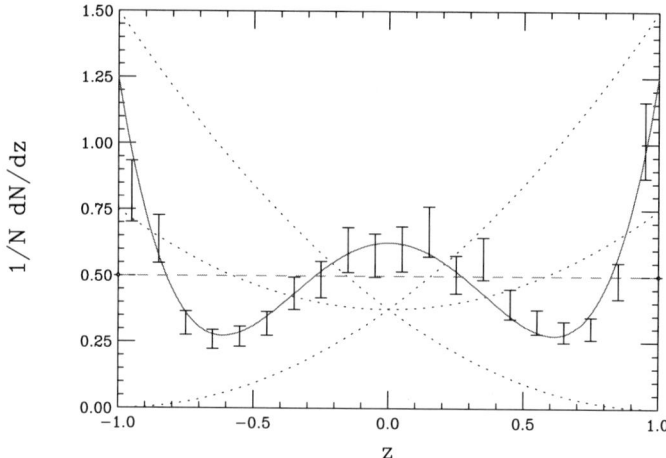

Fig. 9.3. Normalized leptonic angular distribution predicted from the decay of particles with different spin produced in $q\bar{q}$ annihilation. The dashed(solid,dotted) curves are for spin-0(2,1). The generated data corresponds to 1000 events in the spin-2 case.

since it is sometimes important to know the height of the Z' peak relative to this continuum to ascertain the Z' signal significance.

It is evident from the above cross section expressions that the Z' (as well as γ and Z) induced Drell-Yan cross section involves only terms with a particular angular dependence due to the spin-1 nature of the exchanged particles. In the NWA on the Z' pole itself the leptonic angular distribution is seen to behave as $\sim 1+z^2+8A_{FB}z/3$, which is typical of a spin-1 particle. If the Z' had not been a Z' but, say, a $\tilde{\nu}$ in an R-parity violating SUSY model[1] which is spin-0, then the angular distribution on the peak would have been z-independent, i.e., flat(with, of course, $A_{FB} = 0$). This is quite different than the ordinary Z' case. If the Z' had instead been an RS graviton[2] with spin-2, then the $q\bar{q} \to l^+l^-$ part of the cross section would behave as $\sim 1 - 3z^2 + 4z^4$, while the $gg \to l^+l^-$ part would go as $\sim 1-z^4$, both parts also yielding $A_{FB} = 0$. These distributions are also quite distinctive. Figure 9.3 shows an example of these (normalized) distributions and demonstrates that with less than a few hundred events they are very easily distinguishable. Thus the Z' spin should be well established without much of any ambiguity given sufficient luminosity.

An important lesson from the NWA is that the signal rate for a Z' depends upon B, the Z' leptonic branching fraction. Usually in calculating B one assumes that the Z' decays only to SM fields. Given the possible

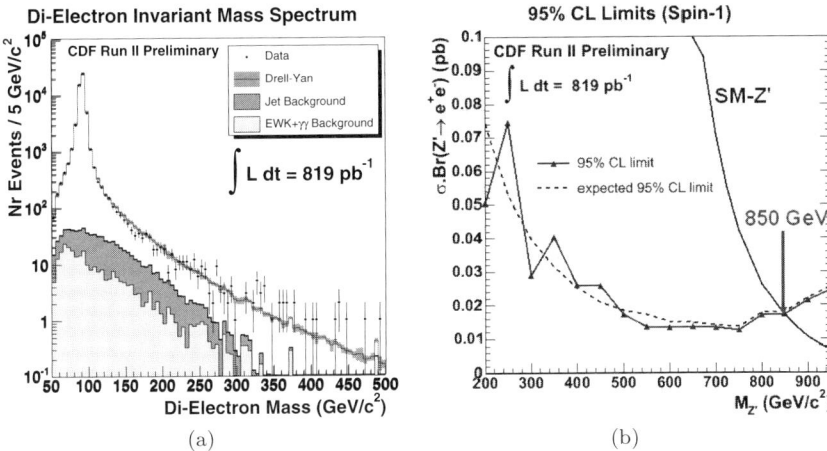

Fig. 9.4. (a) The Drell-Yan distribution as seen by CDF. (b) CDF cross section lower bound in comparison to the predictions for the Z' in the SSM.

existence of SUSY as well as the additional fermions needed in extended electroweak models to cancel anomalies this assumption may be wrong. Clearly Z' decays to these other states would decrease the value of B making the Z' more difficult to observe experimentally.

At the Tevatron only lower bounds on the mass of a Z' exist. These bounds are obtained by determining the 95% CL upper bound on the production cross section for lepton pairs that can arise from new physics as a function of $M(=M_{Z'})$. (Note that this has a slight dependence on the assumption that we are looking for a Z' due to the finite acceptance of the detector.) Then, for any given Z' model one can calculate $\sigma_{Z'}B(Z' \to l^+l^-)$ as a function of $M_{Z'}$ and see at what value of $M_{Z'}$ the two curves cross. At present the best limit comes from CDF although comparable limits are also obtained by D0.[34] The left panel in Fig. 9.4 shows the latest (summer 2006) Drell-Yan spectrum from CDF; the right panel shows the corresponding cross section upper bound and the falling prediction for the Z' cross section in the SSM. Here we see that the lower bound is found to be 850 GeV *assuming* that only SM fermions participate in the Z' decay. For other models an analogous set of theory curves can be drawn and the associated limits obtained.

Figure 9.5 shows the resulting constraints (from a different CDF analysis[35] with a lower integrated luminosity but also employing the A_{FB} observable above the mass of the SM Z) on a number of the models discussed

$E_6 Z'$ Model	Z_χ	Z_ψ	Z_η	Z_I	Z_N	Z_{sec}
Exp. limit (GeV/c^2)	735	725	745	650	710	675
Obs. limit (GeV/c^2)	740	725	745	650	710	680

(a)

Littlest Higgs Z'	$\cot\theta_H=0.3$	$\cot\theta_H=0.5$	$\cot\theta_H=0.7$	$\cot\theta_H=1.0$
Exp. $M_{Z'_H}$ limit (GeV/c^2)	625	765	835	910
Obs. $M_{Z'_H}$ limit (GeV/c^2)	625	760	830	900

(b)

Fig. 9.5. Experimental lower bounds from CDF on a number of Z' models: (a) E_6 models (b) Little Higgs models.

above all assuming Z' decays to SM particles only and no Z-Z' mixing. Looking at these results we see that the Tevatron bounds are generally superior to those from LEPII and are approaching the best that the other precision measurements can do. These bounds would degrade somewhat if we allowed the Z' to have additional decay modes; for example, if B were reduced by a factor of 2 then the resulting search reach would be reduced by 50-100 GeV depending on the model.

The Tevatron will, of course, be continuing to accumulate luminosity for several more years possibly reaching as high as 8 fb^{-1} per experiment. Assuming no signal is found this will increase the Z' search reach lower bound somewhat, $\sim 20\%$, as is shown in Fig. 9.6 from[36]. At this point the

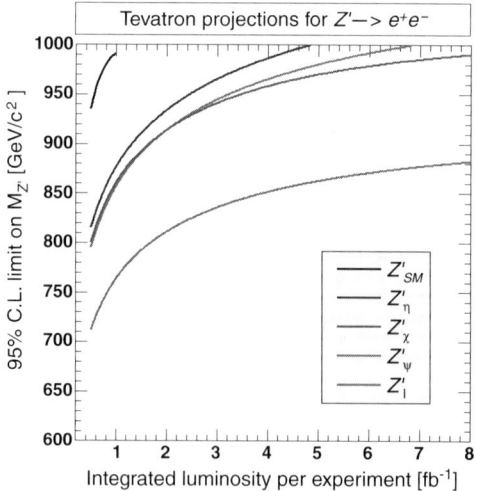

Fig. 9.6. Extrapolation of the Z' reach for a number of different models at the Tevatron as the integrated luminosity increases. Results from CDF and D0 are combined.

search reach at the Tevatron peters out due to the rapidly falling parton densities leaving the mass range above ~ 1 TeV for the LHC to explore.

9.6. The LHC: Z' Discovery and Identification

The search for a Z' at the LHC would proceed in the same manner as at the Tevatron. In fact, since the Z' has such a clean (*i.e.*, dilepton) signal and a sizable cross section it could be one of the first new physics signatures to be observed at the LHC even at relatively low integrated luminosities.[37–39] Figure 9.7 shows both the theoretical anticipated 95% CL lower bound and the 5σ discovery reach for several different Z' models at the LHC for a single leptonic channel as the integrated luminosity is increased; these results are mirrored in detectors studies.[40] Here we see that with only $10-20\ pb^{-1}$ the LHC detectors will clean up any of the low mass region left by the Tevatron below 1 TeV and may actually discover a 1 TeV Z' with luminosities in the $30-100\ pb^{-1}$ range! In terms of discovery, however, to get out to the $\sim 4-5$ TeV mass range will requite $\sim 100\ fb^{-1}$ of luminosity. At such luminosities, the 95% CL bound exceeds the 5σ discovery reach by about 700 GeV. In these plots, we have again assumed that the Z' leptonic branching fraction is determined by decays only to SM fermions. Reducing B by a factor of 2 could reduce these reaches by $\simeq 10\%$ which is not a large effect.

The Z' peak at the LHC should be relatively easy to spot since the SM backgrounds are well understood as shown[38,41] in Fig. 9.8 for a number of

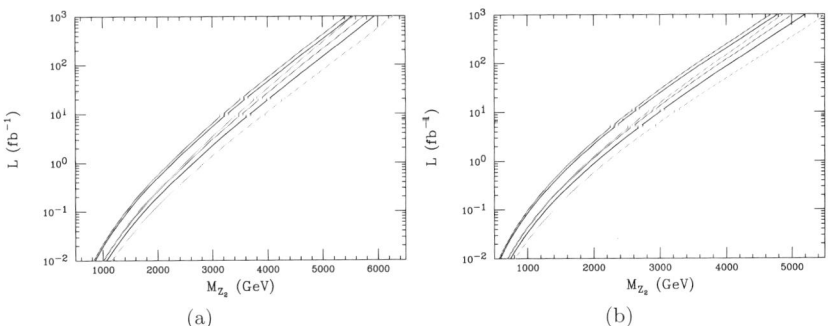

Fig. 9.7. (a) 95% CL lower bound and (b) 5σ discovery reach for a Z' as a function of the integrated luminosity at the LHC for ψ(red), χ(green), η(blue), the LRM with $\kappa = 1$(magenta), the SSM(cyan) and the ALRM(black). Decays to only SM fermions is assumed.

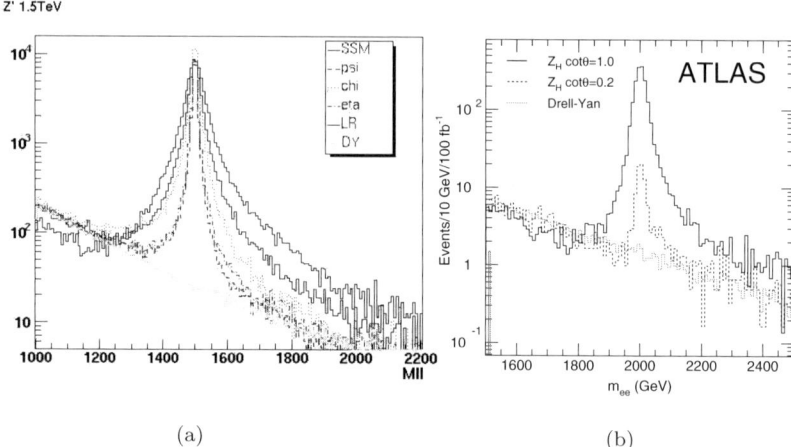

Fig. 9.8. Resonance shapes for a number of Z' models as seen by ATLAS assuming $M_{Z'} = 1.5$ TeV. The continuum is the SM Drell-Yan background.

different Z' models. The one problem that may arise is for the case where the Z' width, $\Gamma_{Z'}$, is far smaller than the experimental dilepton pair mass resolution, δM. Typically in most models, $\Gamma_{Z'}/M_{Z'}$ is of order $\simeq 0.01$ which is comparable to dilepton pair mass resolution, $\delta M/M$, for both ATLAS[42] and CMS.[43] If, however, $\Gamma_{Z'}/M_{Z'} \ll \delta M/M$, then the Z' resonance is smeared out due to the resolution and the cross section peak is reduced by roughly a factor of $\sim \Gamma_{Z'}/\delta M$ making the state difficult to observe. This could happen, e.g., if the Z' (before mixing with the SM Z) had no couplings to SM fields.[44]

Given the huge mass reach of the LHC it is important to entertain the question of how to 'identify' a particular Z' model once such a particle is found. This goes beyond just being able to tell the Z' of Model A from the Z' from model B. As alluded to in the introduction, if a Z'-like object is discovered, the first step will be to determine its spin. Based on the theoretical discussion above this would seem to be rather straightforward and studies of this issue have been performed by both ATLAS[45] and CMS.[46] Generally, one finds that discriminating a spin-1 or spin-2 object from one of spin-0 requires several times more events than does discriminating spin-2 from spin-1. The requirement of a few hundred events, however, somewhat limits the mass range over which such an analysis can be performed. If a particular Z' model has an LHC search reach of 4 TeV, then only for masses below $\simeq 2.5 - 3$ TeV will there be the statistics necessary to perform a

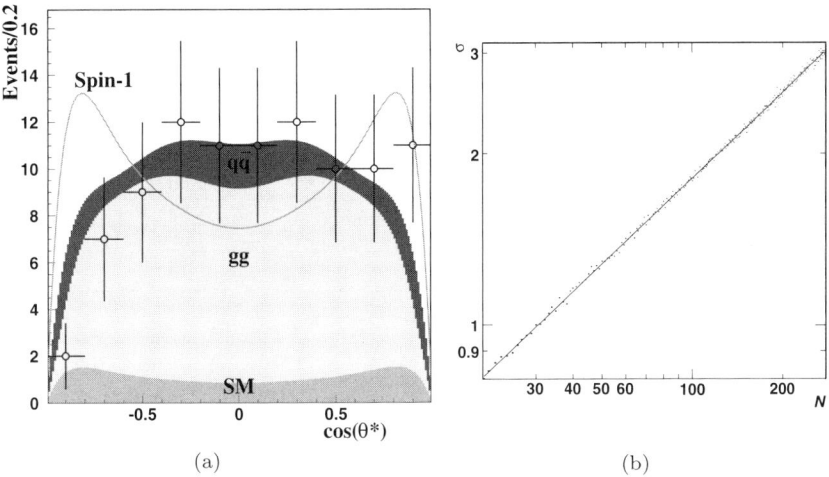

Fig. 9.9. (a) The theoretical predictions for 1.5 TeV SSM Z′ and RS graviton resonance shapes at ATLAS in comparison to the graviton signal data. (b) Differentiation, in σ, of spin-1 and spin-2 resonances at CMS as a function of the number of events assuming a 1.5 TeV mass.

reliable spin determination. Figure 9.9 shows two sample results from this spin analysis. For the ATLAS study in the left panel[45] the lepton angular distribution for a weakly coupled 1.5 TeV KK RS graviton is compared with the expectation for a SSM Z′ of identical mass assuming a luminosity of 100 fb^{-1}. Here one clearly sees the obvious difference and the spin-2 nature of the resonance. In the right panel[46] the results of a CMS analysis is presented with the distinction of a 1.5 TeV Z′ and a KK graviton again being considered. Here one asks for the number of events (N) necessary to distinguish the two cases, at a fixed number of standard deviations, σ, which is seen to grow as (as it should) with $\sqrt{N}$. For example, a 3σ separation is seen to require $\simeq 300$ events.

Once we know that we indeed have a spin-1 object, we next need to 'identify' it, i.e., uniquely determine its couplings to the various SM fermions. (Note that almost all LHC experimental analyses up to now have primarily focused on being able to distinguish models and not on actual coupling extractions.) We would like to be able to do this in as model-independent a way as possible, e.g., we should not assume that the Z′ decays only to SM fields. Clearly this task will require many more events than a simple discovery or even a spin determination and will probably be difficult for a Z′ with a mass much greater than $\simeq 2 - 2.5$ TeV unless

Table 9.2. Results on σ_{ll} and $\sigma_{ll} \times \Gamma_{Z'}$ for all studied models from ATLAS. Here one compares the input values from the generator with the reconstructed values obtained after full detector simulation.

		σ_{ll}^{gen}(fb)	σ_{ll}^{rec}(fb)	$\sigma_{ll}^{rec} \times \Gamma_{rec}$ (fb.GeV)
	SSM	78.4±0.8	78.5±1.8	3550±137
	ψ	22.6±0.3	22.7±0.6	166±15
$M = 1.5\,\text{TeV}$	χ	47.5±0.6	48.4±1.3	800±47
	η	26.2±0.3	24.6±0.6	212±16
	LR	50.8±0.6	51.1±1.3	1495±72
$M = 4\,\text{TeV}$	SSM	0.16±0.002	0.16±0.004	19±1
	KK	2.2±0.07	2.2±0.12	331±35

integrated luminosities significantly in excess of 100 fb^{-1} are achieved (as may occur at the LHC upgrade.[47]) Some of the required information can be obtained using the dilepton (i.e., e^+e^- and/or $\mu^+\mu^-$) discovery channel but to obtain more information the examination of additional channels will also be necessary.

In the dilepton mode, three obvious observables present themselves: (i) the cross section, σ_{ll}, on and below the Z' peak (it is generally very small above the peak), (ii) the corresponding values of A_{FB} and (iii) the width, $\Gamma_{Z'}$, of the Z' from resonance peak shape measurements. Recall that while A_{FB} is B insensitive, both σ_{ll} and $\Gamma_{Z'}$ are individually sensitive to what we assume about the leptonic branching fraction, B, so that they cannot be used independently. In the NWA, however, one sees that the product of the peak cross section and the Z' width, $\sigma_{ll}\Gamma_{Z'}$, is *independent* of B. (Due to smearing and finite width effects, one really needs to take the product of $d\sigma^+/dM$, integrated around the peak and $\Gamma_{Z'}$.) Table 9.2 from an ATLAS study[48] demonstrates that the product $\sigma_{ll}\Gamma_{Z'}$ can be reliably determined at the LHC in full simulation, reproducing well the original input generator value.

Let us now consider the quantity A_{FB}. At the theory level, the angle θ^* employed above is defined to be that between the incoming q and the outgoing l^-. Experimentally, though the lepton can be charge signed with relative ease, it is not immediately obvious in which direction the initial quark is going, i.e., to determine which proton it came from. However, since the q valence distributions are 'harder' (i.e., have higher average momentum fractions) than the 'softer' $\bar{q}$ sea partons, it is likely[49] that the Z' boost direction will be that of the original q. Of course, this is not *always* true so that making this assumption dilutes the true value of A_{FB} as does, e.g., additional gluon radiation. For the Z' to be boosted, the leptons in the

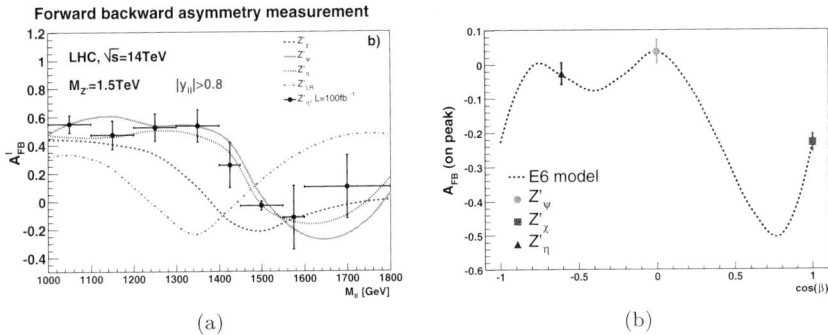

Fig. 9.10. (a) A_{FB} near a 1.5 TeV Z' in a number of models. (b) On-peak differentiation of E_6 models using A_{FB} showing statistical errors for a 1.5 TeV Z'.

final state need to have (significant) rapidity, hence the lower bound in the integration of the cross section expression above. Clearly, a full analysis needs to take these and other experimental issues into account.

The left panel of Fig. 9.10 shows[50] A_{FB} as a function of M in the region near a 1.5 TeV Z' for E_6 model η in comparison with the predictions of several other models. Here we see several features, the first being that the errors on A_{FB} are rather large except on the Z' pole itself due to relatively low statistics even with large integrated luminosities of 100 fb^{-1}; this is particularly true above the resonance. Second, it is clear that A_{FB} both on and off the peak does show some reasonable model sensitivity as was hoped. From the right panel[50] of Fig. 9.10 it is clear that the various special case models of the E_6 family are distinguishable. This is confirmed by more detailed studies performed by both ATLAS[48] and CMS.[51] Figure 9.11 from CMS[51] shows how measurements of the on-peak A_{FB} can be used to distinguish models with reasonable confidence given sufficient statistics (and in the absence of several systematic effects). Table 9.3 from the ATLAS study[48] shows that the original input generator value of the on-peak A_{FB} can be reasonably well reproduced with a full detector simulation, taking dilution and other effects into account.

If a large enough on-peak data sample is available, examining A_{FB} as a function of the lepton rapidity[52] can provide additional coupling information. The reason for this is that u and d quarks have different x distributions so that the weight of $u\bar{u}$ and $d\bar{d}$ induced Z' events changes as the rapidity varies. No detector level studies of this have yet been performed.

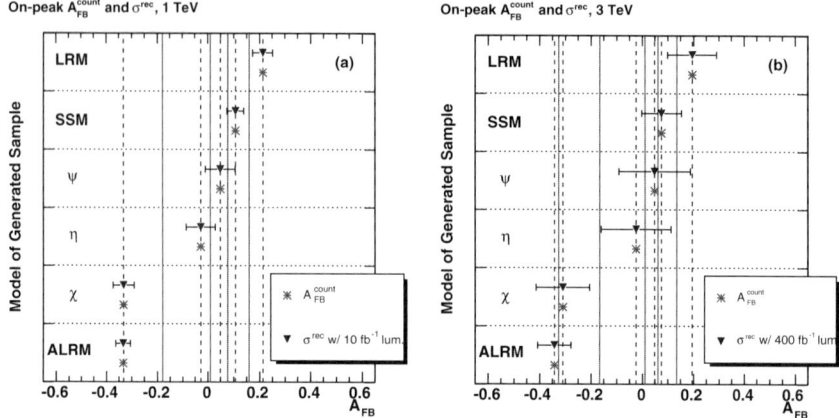

Fig. 9.11. CMS analysis of Z' model differentiation employing A_{FB} assuming $M_{Z'} = 1$ or 3 TeV.

Table 9.3. Measured on-peak A_{FB} for all studied models in the central mass bin from ATLAS. Here the raw value obtained before dilution corrections is labeled as 'Observed'.

Model	$\int \mathcal{L}(fb^{-1})$	Generation	Observed	Corrected
1.5 TeV				
SSM	100	$+0.088 \pm 0.013$	$+0.060 \pm 0.022$	$+0.108 \pm 0.027$
χ	100	-0.386 ± 0.013	-0.144 ± 0.025	-0.361 ± 0.030
η	100	-0.112 ± 0.019	-0.067 ± 0.032	-0.204 ± 0.039
η	300	-0.090 ± 0.011	-0.050 ± 0.018	-0.120 ± 0.022
ψ	100	$+0.008 \pm 0.020$	-0.056 ± 0.033	-0.079 ± 0.042
ψ	300	$+0.010 \pm 0.011$	-0.019 ± 0.019	-0.011 ± 0.024
LR	100	$+0.177 \pm 0.016$	$+0.100 \pm 0.026$	$+0.186 \pm 0.032$
4 TeV				
SSM	10000	$+0.057 \pm 0.023$	-0.001 ± 0.040	$+0.078 \pm 0.051$
KK	500	$+0.491 \pm 0.028$	$+0.189 \pm 0.057$	$+0.457 \pm 0.073$

Off-peak measurements of A_{FB} are also useful although in this case systematics are more important; as shown in the ATLAS study,[48] whose results are shown in Table 9.4, it is more difficult to reproduce the input generator value of this quantity than in the on-peak case.

There are, of course, other observables that one may try to use in the dilepton channel but they are somewhat more subtle. The first possibility[50] is to reconstruct the Z' rapidity distribution from the dilepton final

Table 9.4. Measured off peak, $0.8 < M < 1.4$ TeV, A_{FB} for all studied models from ATLAS using the same nomenclature as above.

Model	$\int \mathcal{L}(fb^{-1})$	Generation	Observed	Corrected
1.5 TeV				
SSM	100	$+0.077 \pm 0.025$	$+0.086 \pm 0.038$	$+0.171 \pm 0.045$
χ	100	$+0.440 \pm 0.019$	$+0.180 \pm 0.032$	$+0.354 \pm 0.039$
η	100	$+0.593 \pm 0.016$	$+0.257 \pm 0.033$	$+0.561 \pm 0.039$
ψ	100	$+0.673 \pm 0.012$	$+0.294 \pm 0.033$	$+0.568 \pm 0.039$
LR	100	$+0.303 \pm 0.022$	$+0.189 \pm 0.033$	$+0.327 \pm 0.040$

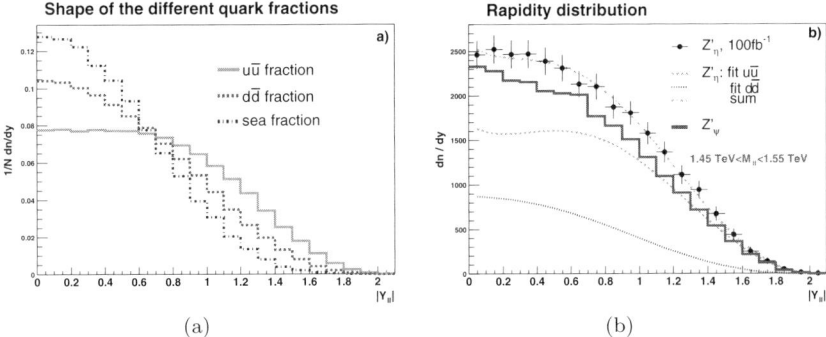

Fig. 9.12. (a) Rapidity distributions for different $q\bar{q}$ induced events. (b) Rapidity distribution differentiation of Z' models.

state. The left panel of Fig. 9.12 reminds us that the Z' rapidity distribution produced by only $u\bar{u}$, $d\bar{d}$ or sea quarks would have a different shape. The particular Z' couplings to quarks induce different weights in these three distributions and so one may hope to distinguish models in this way. An example of this is shown in the right panel of Fig. 9.12. The first analysis[50] of this type considered the quantity $R_{u\bar{u}}$, the fraction of Z' events originating from $u\bar{u}$, as an observable; a similar variable $R_{d\bar{d}}$ can also be constructed. Figure 9.13 from a preliminary ATLAS analysis[53] compares the values of these two parameters extracted via full reconstruction for a 1.5 TeV Z'; here we see that reasonable agreement with the input values of the generator are obtained although the statistical power is not very good. Knowing both $R_{d\bar{d},u\bar{u}}$ and the ratio of the $d\bar{d}$ and $u\bar{u}$ parton densities fairly precisely, one can turn these measurements into a determination of the coupling ratio $(v_u^{'2} + a_u^{'2})/(v_d^{'2} + a_d^{'2})$.

Model	Generation level Fitted values (%)		Reconstruction level Fitted values (%)	
	Prop(Z'←dd)	Prop(Z'←uu)	Prop(Z'←dd)	Prop(Z'←uu)
SSM	41.±10.	52.±12.	22.±16.	60.±16.
χ	62.±12.	29.±14.	79.±17.	17.±19.
η	23.±13.	75.±14.	33.±6.	67.±8.
ψ	36.±12.	61.±13.	32.±15.	62.±17.
LR	57.±4.	43.±14.	53.±13.	46.±15.

Fig. 9.13. Comparison of $R_{q\bar{q}}$ values determined at the generator level and after detector simulation by ATLAS.

A second possibility is to construct the rapidity ratio[54] in the region near the Z' pole:

$$R = \frac{\int_{-y_1}^{y_1} \frac{d\sigma}{dy} dy}{\left[\int_{y_1}^{Y} + \int_{-Y}^{-y_1} \frac{d\sigma}{dy} dy\right]}. \tag{9.25}$$

Here y_1 is some suitable chosen rapidity value $\simeq 1$. R essentially measures the ratio of the cross section in the central region to that in the forward region and is again sensitive to the ratio of u and d quark couplings to the Z'. A detector level study of this observable has yet to be performed.

In addition to the e^+e^- and $\mu^+\mu^-$ discovery channel final states, one might also consider other possibilities, the simplest being $\tau^+\tau^-$. Assuming universality, this channel does not provide anything new unless one can measure the polarization of the τ's, P_τ, on or very near the Z' peak.[55] The statistics for making this measurement can be rather good as the rate for this process is only smaller than that of the discovery mode by the τ pair reconstruction efficiency. In the NWA, $P_\tau = 2v'_e a'_e / (v'^2_e + a'^2_e)$, assuming universality, so that the ratio of v'_e/a'_e can be determined uniquely. Figure 9.14 shows, for purposes of demonstration, the value of P_τ in the E_6 model case where we see that it covers its fully allowed range.

A first pass theoretical study[55] suggests that $\delta P_\tau \simeq 1.5/\sqrt{N}$, with N here being the number of reconstructed τ events. Even for a reconstruction efficiency of 3%, with $M_{Z'}$ not too large $\sim 1-1.5$ TeV, the high luminosity of the LHC should be able to tell us P_τ at the ± 0.05 level. It would be

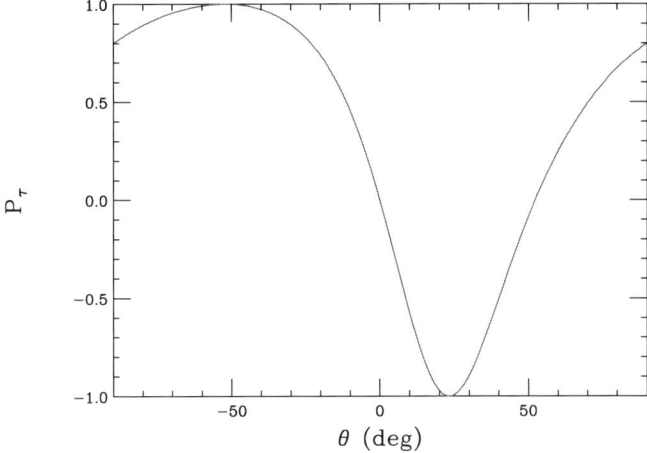

Fig. 9.14. τ polarization asymmetry for a Z' in E_6 models in the NWA.

very good to see a detector study for this observable in the near future to see how well the LHC can really do in this case.

Once we go beyond the dileptons, the next possibility one can imagine is light quark jets from which one might hope to get a handle on the Z' couplings to quarks. The possibility of new physics producing an observable dijet peak at the LHC has been studied in detail by CMS[56]; the essential results are shown in Fig. 9.15. Here we see that for resonances which are color non-singlets, *i.e.*, those which have QCD-like couplings, the rates are sufficiently large as to allow these resonances to be seen above the dijet background. However, for weakly produced particles, such as the SSM Z' shown here, the backgrounds are far too large to allow observation of these decays. Thus it is very unlikely that the dijet channel will provide us with any information on Z' couplings at the LHC.

Another possibility is to consider the heavy flavor decay modes, *i.e.*, $Z' \to b\bar{b}$ or $t\bar{t}$. Unfortunately, these modes are difficult to observe so that it will be quite unlikely that we will obtain coupling information from them. ATLAS[57] has performed a study of the possibility of observing these modes within the Little Higgs Model context for a Z' in the 1-2 TeV mass range. Figure 9.16 from the ATLAS study demonstrates how difficult observing these decays may really be due to the very large SM backgrounds. It is thus unlikely that these modes will provide any important information except in very special cases.

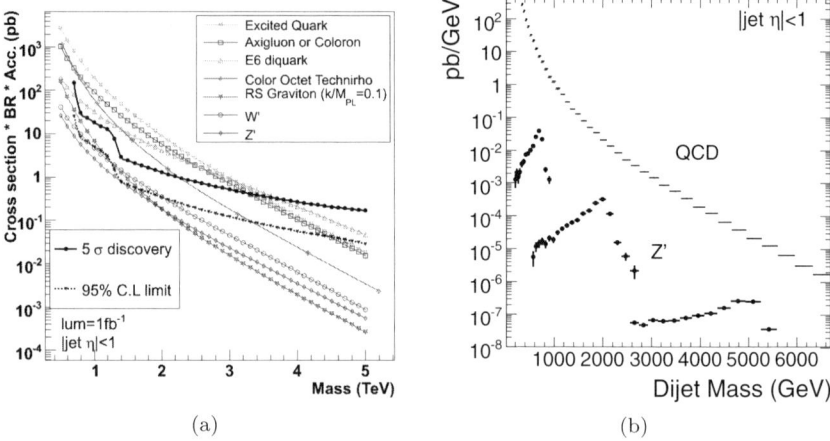

Fig. 9.15. (a) Dijet resonance discovery reach at CMS in comparison to the predictions for a number of models. (b) SSM Z' dijet signal for various masses in comparison with the SM background.

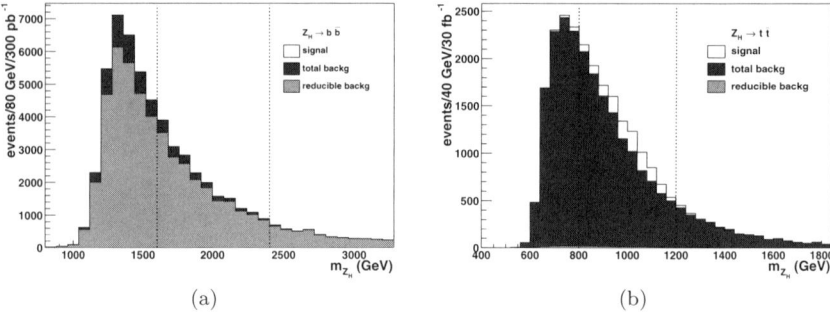

Fig. 9.16. Search for heavy flavor decays of the Z' in the Little Higgs model by ATLAS. $\cot\theta_H = 1$ has been assumed. Z' $\to b\bar{b}$ assuming $M_{Z'} = 2$ TeV and a luminosity of 300 fb^{-1}(a) and $t\bar{t}$(b) for $M_{Z'} = 1$ TeV and a luminosity of 30 fb^{-1}.

Another possible 2-body channel is Z' $\to W^+W^-$, which can occur at a reasonable rate through Z-Z' mixing as discussed above. Clearly the rate for this mode is very highly model dependent. ATLAS[58] has made a preliminary analysis of this mode in the $jjl\nu$ final state taking the Z' to be that of the SSM(for its fermionic couplings) and assuming a large integrated luminosity of 300 fb^{-1}. The mixing parameter β was taken to be unity in the

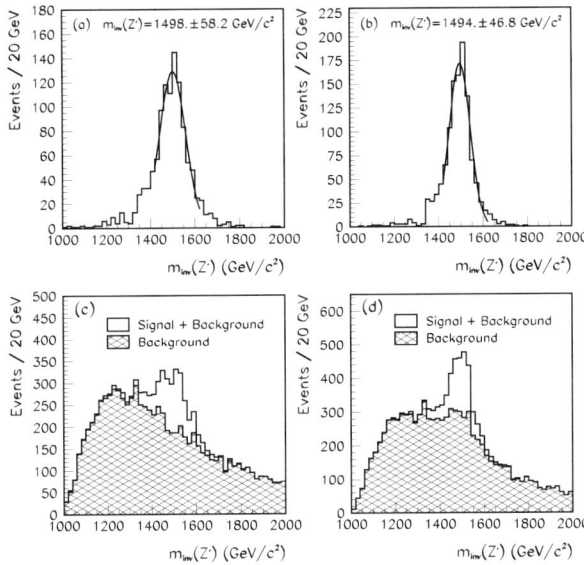

Fig. 9.17. Results of two ATLAS analyses showing the $Z' \to WW$ signal above SM backgrounds and Z' mass reconstruction in this channel for the SSM model assuming $M_{Z'} = 1.5$ TeV and $\beta = 1$.

calculations. The authors of this analysis found that a Z' in the mass range below $\simeq 2.2$ TeV could be observed in this channel given these assumptions. An example is shown in Fig. 9.17 where we clearly see the reconstructed Z' above the SM background. With a full detailed background study an estimate could likely be made of the relevant branching fraction in comparison to that of the discovery mode. This would give important information on the nature of the Z' coupling structure. More study of this mode is needed.

A parallel study was performed by ATLAS[41] for the $Z' \to ZH$ mode which also occurs through mixing as discussed above; this mixing occurs naturally in the Little Higgs model in the absence of T-parity. The results are shown in Fig. 9.18. Here we see that there is a respectable signal over background and the relevant coupling information should be obtainable provided the Z' is not too heavy.

Some rare decays of the Z' may be useful in obtaining coupling information provided the Z' is not too massive. Consider the ratios of Z' partial widths[54,59–61]

$$r_{ff'V} = \frac{\Gamma(Z' \to ff'V)}{\Gamma(Z' \to l^+l^-)}, \quad (9.26)$$

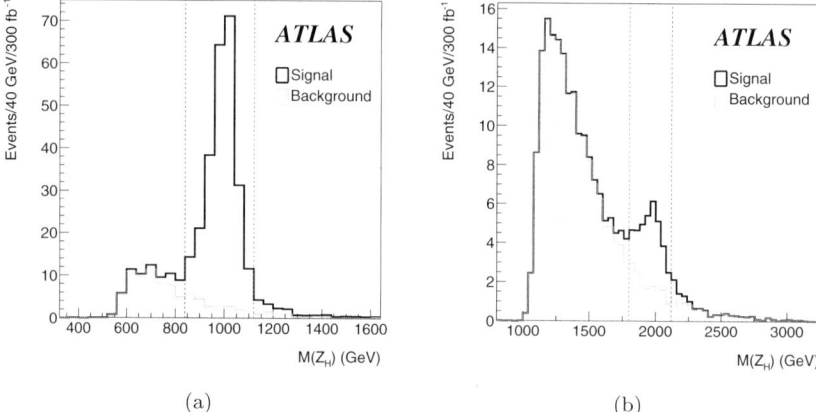

Fig. 9.18. Search study for the decay $Z' \to ZH$ by ATLAS in the Little Higgs model assuming $\cot\theta_H = 0.5$ for the $l^+l^-b\bar{b}$ mode assuming $M_{Z'} = 1$ (a) or 2(b) TeV.

where $V = Z, W$ and $ff' = l^+l^-, l^\pm\nu, \nu\bar{\nu}$, appropriately. The two $\Gamma(Z' \to ff'Z)$ (with $f = l, \nu$) partial widths originate from the bremsstrahlung of a SM Z off of either the f or $\bar{f}$ legs and are rather to imagine. Numerically, one finds that for the case $f = l$, little sensitivity to the Z' couplings is obtained so it is not usually considered. Assuming that the SM ν's couple in a left-handed way to the Z', it is clear that $r_{\nu\nu Z} = K_Z v_\nu^{'2}/(v_e^{'2} + a_e^{'2})$, where K_Z is a constant, model-independent factor for any given Z' mass. The signal for this decay is a (reconstructed) Z plus missing p_T with a Jacobean peak at the Z' mass.

$r_{l\nu W}$, on the otherhand, is more interesting; not only can the W be produced as a brem but it can also arise directly if a WWZ' coupling exists. As we saw above this can happen if Z-Z' mixing occurs *or* it can happen if T' is proportional to T_{3L}. If there is no mixing and if T' has no T_{3L} component then one finds the simple relation $r_{l\nu W} = K_W v_\nu^{'2}/(v_e^{'2} + a_e^{'2})$, with K_W another constant factor. Note that now $r_{l\nu W}$ and $r_{\nu\nu Z}$ are proportional to one another and, since T' and T_{3L} commute, one also has $v_e' + a_e' = v_\nu' + a_\nu' = 2v_\nu'$ so that both $r_{l\nu W}$ and $r_{\nu\nu Z}$ are *bounded*, i.e., $0 \leq r_{l\nu W} \leq K_W/2$ and $0 \leq r_{\nu\nu Z} \leq K_Z/2$. Thus, e.g., in E_6 models a short analysis shows that the allowed region in the $r_{l\nu W}, r_{\nu\nu Z}$ plane will be a straight line beginning at the origin and ending at $K_W/2, K_Z/2$. Other common models will lie on this line, such as the LRM and ALRM cases, but some others, e.g., the SSM, will lie elsewhere in this plane signaling the fact that T' contains a

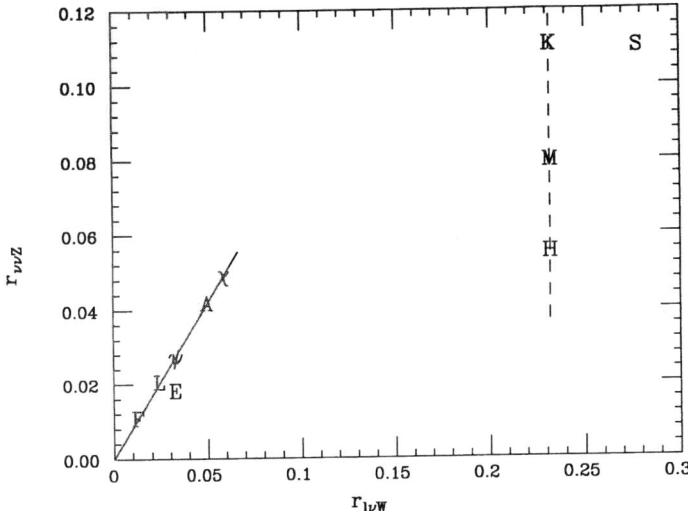

Fig. 9.19. Predictions for the rare decay mode ratios for a number of different models assuming a 1 TeV Z′: 'L' is the LRM with $\kappa = 1$, 'S'=SSM, 'A'=ALRM, *etc.* The solid line is the E_6 case.

T_{3L} component. Figure 9.19 from[61] shows a plot of these parameters for a large number of models, the solid line being the just discussed E_6 case and 'S' the SSM result.

While the coupling information provided by these ratios is very useful, the Z′ event rates necessary to extract them are quite high in most cases due to their small relative branching fractions. For a Z′ much more massive than 1-2 TeV the statistical power of these observables will be lost.

A different way to get at the Z′ couplings is to produce it in association with another SM gauge boson, *i.e.*, a photon[62] or a $W^{\pm}$,Z,[63] with the Z′ decaying to dileptons as usual. Taking the ratio of this cross section to that in the discovery channel, we can define the ratios

$$R_{Z'V} = \frac{\sigma(q\bar{q} \to Z'V)B(Z' \to l^+l^-)}{\sigma(q\bar{q} \to Z')B(Z' \to l^+l^-)}, \quad (9.27)$$

in the NWA with $V = \gamma, W^{\pm}$, or Z. (For the case $V = g$ there is little coupling sensitivity.[62]) Note that B trivially cancels in this ratio but it remains important for determining statistics. The appearance of an extra particle V in the final state re-weights the combination of couplings which appears in the cross section so that one can get a handle on the vector and axial-vector couplings of the initial u's and d's to the Z′. For example,

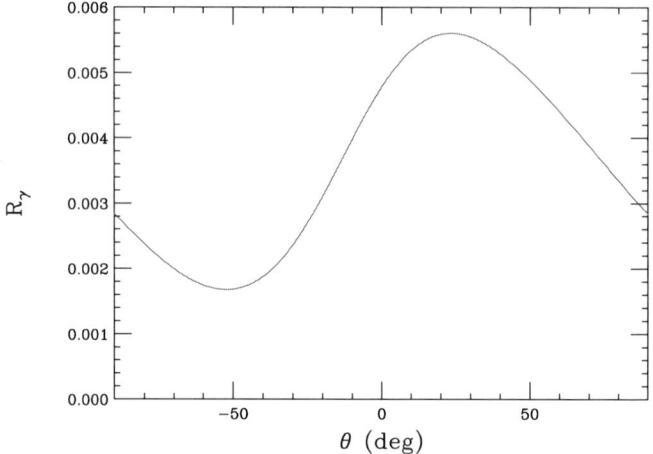

Fig. 9.20. R_γ in E_6 models for a 1 TeV Z' employing a cut $p_T^\gamma > 50$ GeV.

in the simple case of $V = \gamma$, the associated parton level $q\bar{q} \to Z'\gamma$ cross section is proportional to $\sum_i Q_i^2(v_i'^2 + a_i'^2)$ while the simple Z' cross section is proportional to $\sum_i (v_i'^2 + a_i'^2)$. Similarly, for the case $V = W$, the cross section is found to be proportional to $\sum_i (v_i' + a_i')^2$. Tagging the additional V, when $V \neq \gamma$, may require paying the price of leptonic branching fractions for the W and Z, which is a substantial rate penalty, although an analysis has not yet been performed. For the case of $V = \gamma$, a hard p_T cut on the γ will be required but otherwise the signature is very clean. All the ratios $R_{Z'V}$ are of order a few $\times 10^{-3}$ (or smaller once branching fractions are included) for a Z' mass of 1 TeV and (with fixed cuts) tend to grow with increasing $M_{Z'}$. For example, for a 1 TeV Z' in the E_6 model, the cross section times leptonic branching fraction for the $Z'\gamma$ final state varies in the range 0.65-1.6 fb, depending upon the parameter θ, assuming a photon p_T cut of 50 GeV. R_γ for this case is shown in Fig. 9.20. Generically, with 100 fb^{-1} of luminosity these ratios might be determined at the level of $\simeq 10\%$ for the $M_{Z'} = 1$ TeV case but the quality of the measurement will fall rapidly as $M_{Z'}$ increased due to quickly falling statistics. For much larger masses these ratios are no longer useful. It is possible that the Tevatron will tell us whether such light masses are already excluded.

It is clear from the above discussion that there are many tools available at the LHC for Z' identification. However, many of these are only applicable if the Z' is relatively light. Even if all these observables are available it still

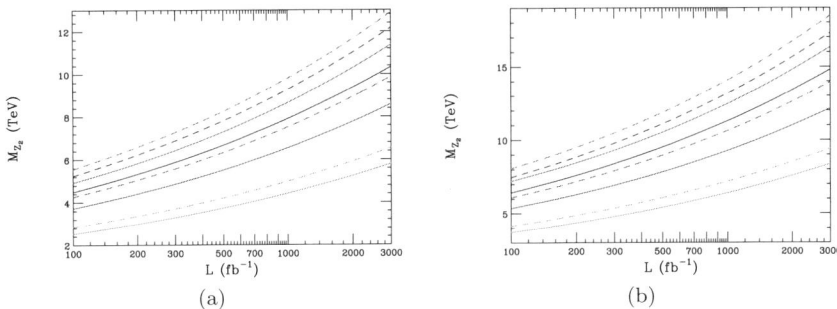

Fig. 9.21. Z' search reach at a $\sqrt{s}=0.5$ TeV(a) or 1 TeV(b) ILC as a function of the integrated luminosity without(solid) or with(dashed) 60% positron beam polarization for models ψ(green), χ(red), SSM(magenta) and LRM with $\kappa = 1$(blue).

remains unclear as to whether or not the complete set of Z' couplings can be extracted from the data with any reliability. A detailed analysis of this situation has yet to be performed. We will probably need a Z' discovery before it is done.

9.7. ILC: What Comes Next

The ILC will begin running a decade or so after the turn on of the LHC. At that point perhaps as much as $\sim 1\ ab^{-1}$ or more of integrated luminosity will have been delivered by the LHC to both detectors. From our point of view, the role of the ILC would then be to either extend the Z' search reach (in an indirect manner) beyond that of the LHC or to help identify any Z' discovered at the LHC.[64]

Although the ILC will run at $\sqrt{s} = 0.5 - 1$ TeV, we know from our discussion of LEP Z' searches that the ILC will be sensitive to Z' with masses significantly larger than $\sqrt{s}$. Figure 9.21[65] shows the search reach for various Z' models assuming $\sqrt{s} = 0.5, 1$ TeV as a function of the integrated luminosity both with and without positron beam polarization. Recall that the various final states $e^+e^- \to f\bar{f}$, $f = e, \mu, \tau, c, b, t$ can all be used simultaneously to obtain high Z' mass sensitivity. The essential observables employed here are $d\sigma/dz$ and $A_{LR}(z)$, which is now available since the e^- beam is at least 80% polarized. One can also measure the polarization of τ's in the final state. This figure shows that the ILC will be sensitive to Z' masses in the range $(7-14)\sqrt{s}$ after a couple of years of design luminosity, the exact value depending on the particular Z' model. Thus we see that

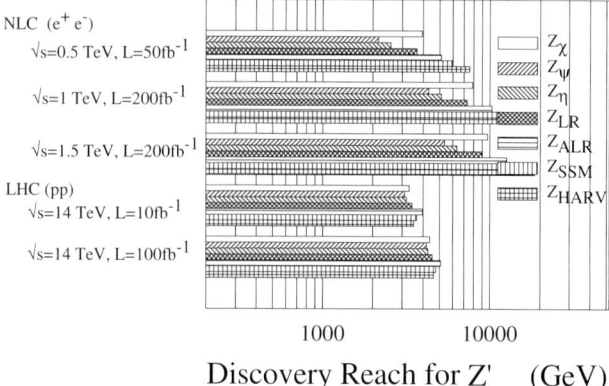

Fig. 9.22. A comparison of LHC direct and ILC indirect Z' search reaches.

it it relatively easy at the ILC to extend the Z' reach beyond the 5-6 TeV value anticipated at the LHC. Figure 9.22 from[66] shows a comparison of the direct Z' search reach at the LHC with the indirect reach at the ILC; note the very modest values assumed here for the ILC integrated luminosities. Here we see explicitly that the ILC has indirect Z' sensitivity beyond the direct reach of the LHC.

In the more optimistic situation where a Z' is discovered at the LHC, the ILC will be essential for Z' identification. As discussed above, it is unclear whether or not the LHC can fully determine the Z' couplings, especially if it were much more massive than $\simeq 1$ TeV.

Once a Z' is discovered at the LHC and its mass is determined, we can use the observed deviations in both $d\sigma/dz$ and $A_{LR}(z)$ at the ILC to determine the Z' couplings channel by channel. For example, assuming lepton universality (which we will already know is applicable from LHC data), we can examine the processes $e^+e^- \to l^+l^-$ using $M_{Z'}$ as an input and determine both v'_e and a'_e (up to a two-fold overall sign ambiguity); a measurement of τ polarization can also contribute in this channel. With this knowledge, we can go on to the $e^+e^- \to b\bar{b}$ channel and perform a simultaneous fit to $v'_{e,b}$ and $a'_{e,b}$; we could then go on to other channels such as $c\bar{c}$ and $t\bar{t}$. In this way *all* of the Z' couplings would be determined. An example of this is shown in Fig. 9.23 from[67] where we see the results of the Z' coupling determinations at the ILC in comparison with the predictions of a number of different models.

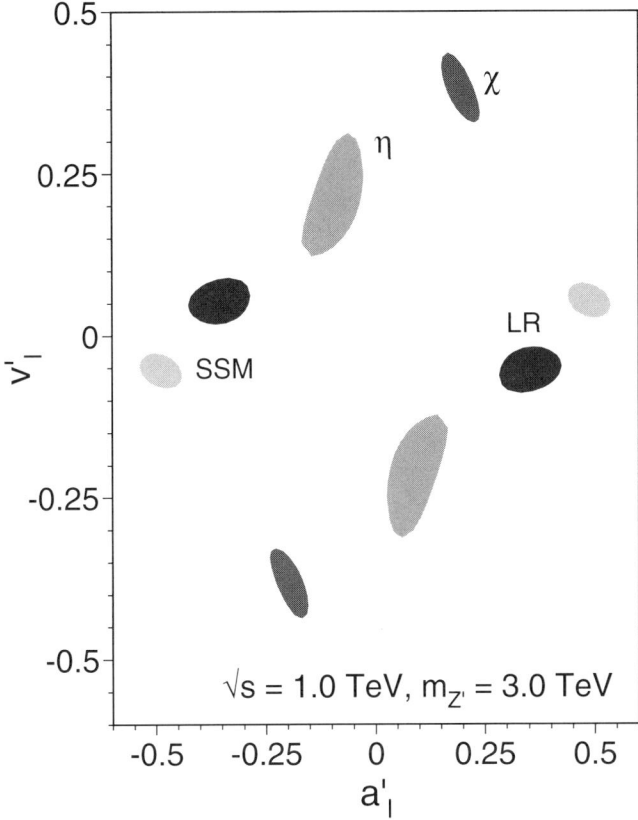

Fig. 9.23. The ability of the ILC to determine the Z′ leptonic couplings for a few representative models.

9.8. Summary

The LHC turns on at the end of next year and a reasonable integrated luminosity $\sim 1\ fb^{-1}$ will likely be accumulated in 2008 at $\sqrt{s} = 14$ TeV. The community-wide expectation is that new physics of some kind will be seen relatively 'soon' after this (once the detectors are sufficiently well understood and SM backgrounds are correctly ascertained). Many new physics scenarios predict the existence of a Z′ or Z′-like objects. It will then be up to the experimenters (with help from theorists!) to determine what these new states are and how they fit into a larger framework. In our discussion above, we have provided an overview of the tools which experiments at the LHC can employ to begin to address this problem. To complete this program will most likely require input from the ILC.

No matter what new physics is discovered at the LHC the times ahead should prove to be very exciting.

Acknowledgments

The author would like to thank G. Azuelos, D. Benchekroun, C. Berger, K. Burkett, R. Cousins, A. De Roeck, S. Godfrey, R. Harris, J. Hewett, F. Ledroit, L. March, D. Rousseau, S. Willocq, and M. Woods for their input in the preparation of these brief lecture notes. Work supported in part by the Department of Energy, Contract DE-AC02-76SF00515.

References

1. J. L. Hewett and T. G. Rizzo, arXiv:hep-ph/9809525; H. K. Dreiner, P. Richardson and M. H. Seymour, Phys. Rev. D **63**, 055008 (2001) [arXiv:hep-ph/0007228]; B. C. Allanach, M. Guchait and K. Sridhar, Phys. Lett. B **586**, 373 (2004) [arXiv:hep-ph/0311254].
2. L. Randall and R. Sundrum, Phys. Rev. Lett. **83**, 3370 (1999) [arXiv:hep-ph/9905221].
3. H. Davoudiasl, J. L. Hewett and T. G. Rizzo, Phys. Rev. Lett. **84**, 2080 (2000) [arXiv:hep-ph/9909255].
4. I. Antoniadis, Phys. Lett. B **246**, 377 (1990).
5. T. G. Rizzo and J. D. Wells, Phys. Rev. D **61**, 016007 (2000) [arXiv:hep-ph/9906234].
6. For classic reviews of Z' physics, see A. Leike, Phys. Rept. **317**, 143 (1999) [arXiv:hep-ph/9805494]; J. L. Hewett and T. G. Rizzo, Phys. Rept. **183**, 193 (1989); M. Cvetic and S. Godfrey, arXiv:hep-ph/9504216; T. G. Rizzo, "Extended gauge sectors at future colliders: Report of the new gauge boson subgroup," eConf **C960625**, NEW136 (1996) [arXiv:hep-ph/9612440];
7. T. G. Rizzo, JHEP **0306**, 021 (2003) [arXiv:hep-ph/0305077]; G. Azuelos and G. Polesello, Eur. Phys. J. C **39S2**, 1 (2005).
8. W. M. Yao et al. [Particle Data Group], J. Phys. G **33**, 1 (2006).
9. K. Cheung, C. W. Chiang, N. G. Deshpande and J. Jiang, "Constraints on flavor-changing Z' models by B/s mixing, Z' production, and arXiv:hep-ph/0604223.
10. K. S. Babu, C. F. Kolda and J. March-Russell, Phys. Rev. D **54**, 4635 (1996) [arXiv:hep-ph/9603212] and Phys. Rev. D **57**, 6788 (1998) [arXiv:hep-ph/9710441]; T. G. Rizzo, Phys. Rev. D **59**, 015020 (1999) [arXiv:hep-ph/9806397]; B. Holdom, Phys. Lett. **B166**, 196 (1986), Phys. Lett. **B259**, 329 (1991), Phys. Lett. **B339**, 114 (1994) and Phys. Lett. **B351**, 279 (1995); F. del Aguila, M. Cvetic and P. Langacker, Phys. Rev. **D52**, 37 (1995);

F. del Aguila, G. Coughan and M. Quiros, Nucl. Phys. **B307**, 633 (1988); F. del Aguila, M. Masip and M. Perez-Victoria, Nucl. Phys. **B456**, 531 (1995); K. Dienes, C. Kolda and J. March-Russell, Nucl. Phys. **B492**, 104 (1997).
11. Results from the LEPEWWG can be found at http://lepewwg.web.cern.ch/LEPEWWG/
12. B. W. Lee, C. Quigg and H. B. Thacker, Phys. Rev. Lett. **38**, 883 (1977).
13. For a classic review and original references, see R.N. Mohapatra, *Unification and Supersymmetry*, (Springer, New York,1986).
14. E. Ma, Phys. Rev. D **36**, 274 (1987).
15. N. Arkani-Hamed, A. G. Cohen and H. Georgi, Phys. Lett. B **513**, 232 (2001) [arXiv:hep-ph/0105239].
16. N. Arkani-Hamed, A. G. Cohen, E. Katz and A. E. Nelson, JHEP **0207**, 034 (2002) [arXiv:hep-ph/0206021].
17. K. R. Lynch, E. H. Simmons, M. Narain and S. Mrenna, Phys. Rev. D **63**, 035006 (2001) [arXiv:hep-ph/0007286].
18. M. Carena, A. Daleo, B. A. Dobrescu and T. M. P. Tait, Phys. Rev. D **70**, 093009 (2004) [arXiv:hep-ph/0408098].
19. E. Eichten, K. D. Lane and M. E. Peskin, Phys. Rev. Lett. **50**, 811 (1983).
20. The possible sensitivity of this reaction has been studied by a large number of authors; see, for example, A. A. Pankov and N. Paver, Phys. Lett. B **393**, 437 (1997) [arXiv:hep-ph/9610509], Phys. Rev. D **48**, 63 (1993), Phys. Lett. B **274**, 483 (1992), and Phys. Lett. B **272**, 425 (1991); R. Najima and S. Wakaizumi, Phys. Lett. B **184**, 410 (1987); P. Kalyniak and M. K. Sundaresan, Phys. Rev. D **35**, 75 (1987); S. Nandi and T. G. Rizzo, Phys. Rev. D **37**, 52 (1988).
21. F. Cossutti, Eur. Phys. J. C **44**, 383 (2005) [arXiv:hep-ph/0505232]. For the most recent status, see, S. Dittmaier, talk given at *Loopfest V*, SLAC, 19-21 June, 2006
22. P. L. Anthony *et al.* [SLAC E158 Collaboration], Phys. Rev. Lett. **95**, 081601 (2005) [arXiv:hep-ex/0504049].
23. J. S. M. Ginges and V. V. Flambaum, "Violations of fundamental symmetries in atoms and tests of unification Phys. Rept. **397**, 63 (2004) [arXiv:physics/0309054].
24. G. P. Zeller *et al.* [NuTeV Collaboration], Phys. Rev. Lett. **88**, 091802 (2002) [Erratum-ibid. **90**, 239902 (2003)] [arXiv:hep-ex/0110059].
25. A. Czarnecki and W. J. Marciano, Int. J. Mod. Phys. A **15**, 2365 (2000) [arXiv:hep-ph/0003049].
26. For a recent analysis and original references, see
27. K. Melnikov and F. Petriello, "Electroweak gauge boson production at hadron colliders through arXiv:hep-ph/0609070.
28. U. Baur and D. Wackeroth, Nucl. Phys. Proc. Suppl. **116**, 159 (2003) [arXiv:hep-ph/0211089];
29. V. A. Zykunov, arXiv:hep-ph/0509315.
30. J. Houston, talk given at the *Workshop on TeV Colliders*, Les Houches, France, 2-20 May 2005.
31. I. Belotelov *et al.*, CMS Note 2006/123.

32. C. Anastasiou, L. J. Dixon, K. Melnikov and F. Petriello, Phys. Rev. D **69**, 094008 (2004) [arXiv:hep-ph/0312266] and Phys. Rev. Lett. **91**, 182002 (2003) [arXiv:hep-ph/0306192];
33. A. D. Martin, R. G. Roberts, W. J. Stirling and R. S. Thorne, Eur. Phys. J. C **35**, 325 (2004) [arXiv:hep-ph/0308087] and Eur. Phys. J. C **28**, 455 (2003) [arXiv:hep-ph/0211080].
34. P. Savard, talk given at the *XXXIII International Conference on High Energy Physics(ICHEP06)*, Moscow, Russia, 26 July - 2 August, 2006; C. Ciobanu, talk given at the *40th Rencontres De Moriond On QCD And High Energy Hadronic Interactions*, La Thuile, Aosta Valley, Italy, 12-19 Mar 2005; K. Burkett, talk given at the *0th Rencontres De Moriond On Electroweak Interactions And Unified Theories*, La Thuile, Aosta Valley, Italy, 2-10 Mar 2005.
35. A. Abulencia *et al.* [CDF Collaboration], Phys. Rev. Lett. **96**, 211801 (2006) [arXiv:hep-ex/0602045].
36. See, for example, the analyses presented in http://www-cdf.fnal.gov/physics/projections/Zprime-CDF.html.
37. R. Alemany, talk given at *Beyond the Standard Model Physics at the LHC*, Cracow, Poland, 3-8 July 2006.
38. S. Willocq, talk given at the *XXXIII International Conference on High Energy Physics(ICHEP06)*, Moscow, Russia, 26 July - 2 August, 2006.
39. O.K. Baker, talk given at the *Third North American ATLAS Physics Workshop*, Boston, MA, 26-28 July, 2006.
40. R. Cousins, J. Mumford and V. Valuev, CMS Note 2006/062.
41. G. Azuelos *et al.*, Eur. Phys. J. C **39S2**, 13 (2005) [arXiv:hep-ph/0402037]; E. Roos, ATL-PHYS-CONF-2006-007.
42. ATLAS Detector and Physics Performance Technical Design Report, http://atlas.web.cern.ch/Atlas/GROUPS/PHYSICS/TDR/access.html.
43. CMS Physics Technical Design Report, https://cmsdoc.cern.ch/cms/cpt/tdr/.
44. See, for example, J. Kumar and J. D. Wells, arXiv:hep-th/0604203 and references therein. See also A. Freitas, Phys. Rev. D **70**, 015008 (2004) [arXiv:hep-ph/0403288].
45. B. C. Allanach, K. Odagiri, M. A. Parker and B. R. Webber, JHEP **0009**, 019 (2000) [arXiv:hep-ph/0006114].
46. R. Cousins, J. Mumford, J. Tucker and V. Valuev, JHEP **0511**, 046 (2005).
47. F. Gianotti *et al.*, "Physics potential and experimental challenges of the LHC luminosity Eur. Phys. J. C **39**, 293 (2005) [arXiv:hep-ph/0204087].
48. M. Schafer, F. Ledroit and B. Trocmé, ATL-PHYS-PUB-2005-010.
49. H. E. Haber, SLAC-PUB-3456 *Presented at 1984 Summer Study on the Design and Utilization of the Superconducting Super Collider, Snowmass, CO, Jun 23 - Jul 23, 1984*
50. M. Dittmar, A. S. Nicollerat and A. Djouadi, Phys. Lett. B **583**, 111 (2004) [arXiv:hep-ph/0307020].
51. R. Cousins, J. Mumford, J. Tucker and V. Valuev, CMS Note 2005/022.
52. J. L. Rosner, Phys. Rev. D **35**, 2244 (1987).

53. J. Morel and F. Ledroit, Talk given at *LPSC-Grenoble*, Grenoble, France, July 2005.
54. F. del Aguila, M. Cvetic and P. Langacker, Phys. Rev. D **48**, 969 (1993) [arXiv:hep-ph/9303299].
55. J. D. Anderson, M. H. Austern and R. N. Cahn, Phys. Rev. Lett. **69**, 25 (1992) and Phys. Rev. D **46**, 290 (1992).
56. K. Gumus, N. Akchurin, S. Esen and R.M. Harris, CMS Note 2006/070
57. S. González de la Hoz, L. March and E. Roos, ATL-PHYS-PUB-2006-003.
58. D. Benchekroun, C. Driouichi and A. Hoummada, Eur. Phys. J. directC **3**, N3 (2001).
59. T. G. Rizzo, Phys. Lett. B **192**, 125 (1987).
60. M. Cvetic and P. Langacker, Phys. Rev. D **46**, 14 (1992).
61. J. L. Hewett and T. G. Rizzo, Phys. Rev. D **47**, 4981 (1993) [arXiv:hep-ph/9206221].
62. T. G. Rizzo, Phys. Rev. D **47**, 956 (1993) [arXiv:hep-ph/9209207].
63. M. Cvetic and P. Langacker, Phys. Rev. D **46**, 4943 (1992) [Erratum-ibid. D **48**, 4484 (1993)] [arXiv:hep-ph/9207216].
64. G. Weiglein *et al.* [LHC/LC Study Group], arXiv:hep-ph/0410364.
65. T. G. Rizzo, arXiv:hep-ph/0303056. Such analyses have been performed by many authors; see, for example, F. Richard, arXiv:hep-ph/0303107 and and work by S. Riemann in J. A. Aguilar-Saavedra *et al.* [ECFA/DESY LC Physics Working Group], "TESLA Technical Design Report Part III: Physics at an e+e- Linear arXiv:hep-ph/0106315; S. Godfrey, P. Kalyniak and A. Tomkins, arXiv:hep-ph/0511335.
66. S. Godfrey, "Search limits for extra neutral gauge bosons at high energy lepton eConf **C960625** (1996) NEW138 [arXiv:hep-ph/9612384] and Phys. Rev. D **51**, 1402 (1995) [arXiv:hep-ph/9411237].
67. S. Riemann, LC-TH-2001-007, http://www.slac.stanford.edu/spires/find/hep/www?r=lc-th-2001-007, *In *2nd ECFA/DESY Study 1998-2001* 1451-1468*

Chapter 10

Neutrinoless Double Beta Decay

Petr Vogel

Kellogg Radiation Laboratory
Caltech, Pasadena, CA 91125, USA[*]

10.1. Introduction - fundamentals of $\beta\beta$ decay

In the recent past neutrino oscillation experiments have convincingly shown that neutrinos have a finite mass. However, in oscillation experiments only the differences of squares of the neutrino masses, $\Delta m^2 \equiv |m_2^2 - m_1^2|$, can be measured, and the results do not depend on the charge conjugation properties of neutrinos, i.e., whether they are Dirac or Majorana fermions. Nevertheless, a lower limit on the absolute value of the neutrino mass scale, $m_{scale} = \sqrt{|\Delta m^2|}$, has been established in this way. Its existence, in turn, is causing a renaissance of enthusiasm in the double beta decay community which is expected to reach and even exceed, in the next generation of experiments, the sensitivity corresponding to this mass scale. Below I review the current status of the double beta decay and the effort devoted to reach the required sensitivity, as well as various issues in theory (or phenomenology) concerning the relation of the $0\nu\beta\beta$ decay rate to the absolute neutrino mass scale and to the general problem of the Lepton Number Violation (LNV).

But before doing that I very briefly summarize the achievements of the neutrino oscillation searches and the role that the search for the neutrinoless double beta decay plays in the elucidation of the pattern of neutrino masses and mixing. In these introductory remarks I use the established terminology, some of which will be defined only later in the text.

There is a consensus that the measurement of atmospheric neutrinos by the SuperKamiokande collaboration[1] can be only interpreted as a conse-

---
[*]email: pxv@caltech.edu

quence of the nearly maximum mixing between ν_μ and ν_τ neutrinos, with the corresponding mass squared difference $|\Delta m^2_{atm}| \sim 2.4 \times 10^{-3}$ eV^2. This finding was confirmed by the K2K experiment[2] that uses accelerator ν_μ beam pointing towards the SuperKamiokande detector 250 km away, as well as by the very recent first result of the MINOS experiment located at the Sudan mine in Minnesota, 735 km away from the Fermilab.[3] Several large long-baseline experiments are being built to further elucidate this discovery, and determine the corresponding parameters more accurately.

At the same time the "solar neutrino puzzle", which has been with us for over thirty years since the pioneering chlorine experiment of Davis,[4] also reached the stage where the interpretation of the measurements in terms of oscillations between the ν_e and some combination of the active, i.e., ν_μ and ν_τ neutrinos, is inescapable. In particular, the juxtaposition of the results of the SNO experiment[5] and SuperKamiokande,[6] together with the earlier solar neutrino flux determination in the gallium experiments,[7,8] leads to that conclusion. The value of the corresponding oscillation parameters, however, remained uncertain, with several "solutions" possible, although the so-called Large Mixing Angle (LMA) solution with $\sin^2 2\theta_{sol} \sim 0.8$ and $\Delta m^2_{sol} \sim 10^{-4}$ eV^2 was preferred. A decisive confirmation of the "solar" oscillations was provided by the nuclear reactor experiment KamLAND[9,10] that demonstrated that the flux of the reactor $\bar{\nu}_e$ is reduced and its spectrum distorted at the average distance ~ 180 km from nuclear reactors.

The pattern of neutrino mixing is further simplified by the constraint due to the Chooz and Palo Verde reactor neutrino experiments[11,12] which lead to the conclusion that the third mixing angle, θ_{13}, is small, $\sin^2 2\theta_{13} \leq 0.1$. The two remaining possible neutrino mass patterns are illustrated in Fig. 10.1.

Altogether, clearly a *lower* limit for at least one of the neutrino masses, $\sqrt{\Delta m^2_{atm}} \simeq 0.05$ eV has been established. However, the oscillation experiments cannot determine the absolute magnitude of the masses and, in particular, cannot at this stage separate two rather different scenarios, the hierarchical pattern of neutrino masses in which $m \sim \sqrt{\Delta m^2}$ and the degenerate pattern in which $m \gg \sqrt{\Delta m^2}$. It is hoped that the search for the neutrinoless double beta decay, reviewed here, will help in foreseeable future in determining, or at least narrowing down, the absolute neutrino mass scale, and in deciding which of these two possibilities is applicable.

Moreover, the oscillation results do not tell us anything about the properties of neutrinos under charge conjugation. While the charged leptons are Dirac particles, distinct from their antiparticles, neutrinos may be the

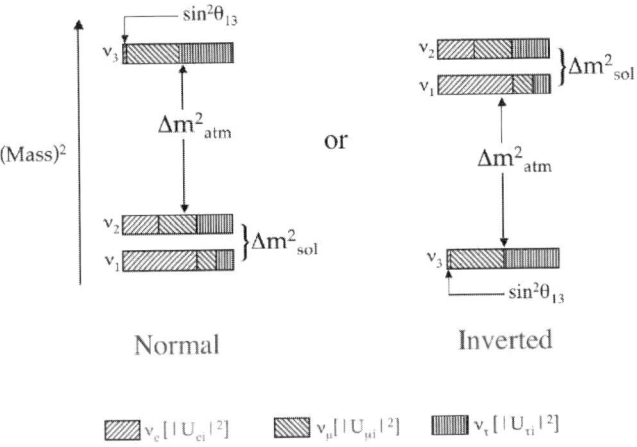

Fig. 10.1. Schematic illustration (mass intervals not to scale) of the decomposition of the neutrino mass eigenstates ν_i in terms of the flavor eigenstates. The two hierarchies cannot be, at this time, distinguished. The small admixture of ν_e into ν_3 is an upper limit.

ultimate neutral particles, as envisioned by Majorana, that are identical to their antiparticles. That fundamental distinction becomes important only for massive particles and becomes irrelevant in the massless limit. Neutrinoless double beta decay proceeds only when neutrinos are massive Majorana particles, hence its observation would resolve the question.

Double beta decay ($\beta\beta$) is a nuclear transition $(Z, A) \to (Z+2, A)$ in which two neutrons bound in a nucleus are simultaneously transformed into two protons plus two electrons (and possibly other light neutral particles). This transition is possible and potentially observable because nuclei with even Z and N are more bound than the odd-odd nuclei with the same $A = N + Z$. Analogous transition of two protons into two neutrons are also, in principle, possible in several nuclei, but phase space considerations give preference to the former mode.

An example is shown in Fig. 10.2. The situation shown there is not exceptional. There are eleven analogous cases (candidate nuclei) with the Q-value (i.e., the energy available to leptons) in excess of 2 MeV.

There are two basic modes of the $\beta\beta$ decay. In the two-neutrino mode ($2\nu\beta\beta$) there are two $\bar{\nu}_e$ emitted together with the two e^-. Lepton number is conserved and this mode is allowed in the standard model of electroweak interaction. It has been repeatedly observed by now in a number of cases and proceeds with a typical half-life of $\sim 10^{20}$ years. In contrast, in the

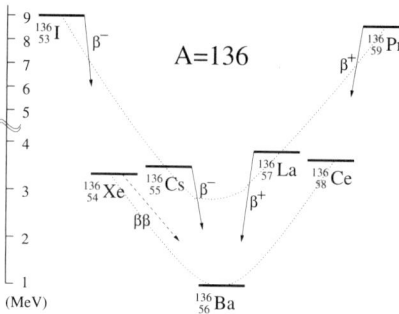

Fig. 10.2. Atomic masses of the isotopes with $A = 136$. Nuclei ^{136}Xe, ^{136}Ba and ^{136}Ce are stable against the ordinary β decay; hence they exist in nature. However, energy conservation alone allows the transition ^{136}Xe $\to$ ^{136}Ba $+ 2e^-$ (+ possibly other neutral light particles) and the analogous decay of ^{136}Ce with the positron emission.

neutrinoless mode ($0\nu\beta\beta$) only the $2e^-$ are emitted and nothing else. That mode clearly violates the law of lepton number conservation and is forbidden in the standard model. Hence, its observation would be a signal of a "new physics".

The two modes of the $\beta\beta$ decay have some common and some distinct features. The common features are:

- The leptons carry essentially all available energy. The nuclear recoil is negligible, $Q/Am_p \ll 1$.
- The transition involves the 0^+ ground state of the initial nucleus and (in almost all cases) the 0^+ ground state of the final nucleus. In few cases the transition to an excited 0^+ state in the final nucleus is energetically possible, but suppressed by the smaller phase space available. (But the $2\nu\beta\beta$ decay to the excited 0^+ state has been observed in few cases.)
- Both processes are of second order of weak interactions, $\sim G_F^4$, hence inherently slow. The phase space consideration alone (for the $2\nu\beta\beta$ mode $\sim Q^{11}$ and for the $0\nu\beta\beta$ mode $\sim Q^5$) give preference to the $0\nu\beta\beta$ which is, however, forbidden by the lepton number conservation.

The distinct features are:

- In the $2\nu\beta\beta$ mode the two neutrons undergoing the transition are uncorrelated (but decay simultaneously) while in the $0\nu\beta\beta$ the two neutrons are correlated.

- In the $2\nu\beta\beta$ mode the sum electron kinetic energy T_1+T_2 spectrum is continuous and peaked below $Q/2$. As $T_1+T_2 \to Q$ the spectrum approaches zero approximately like $(\Delta E/Q)^6$.
- On the other hand, in the $0\nu\beta\beta$ mode $T_1 + T_2 = Q$ smeared only by the detector resolution.

These last features allow one to separate the two modes experimentally by observing the sum electron spectrum with a good energy resolution, even if the corresponding decay rate for the $0\nu\beta\beta$ mode is much smaller than for the $2\nu\beta\beta$ mode. This is illustrated in Fig. 10.3 where the insert that includes the 0ν peak and the 2ν tail shows the situation for the rate ratio of $1:10^6$ corresponding to the most sensitive current experiments.

Various aspects, both theoretical and experimental, of the $\beta\beta$ decay have been reviewed many times. Here I quote just the more recent review articles,[13–16] earlier references can be found there.

In this introductory section let me make only few general remarks. The existence of the $0\nu\beta\beta$ decay would mean that on the elementary particle level a six fermion lepton number violating amplitude transforming two u quarks into two d quarks and two electrons is nonvanishing. As was first pointed out by Schechter and Valle[17] more than twenty years ago, this fact alone would guarantee that neutrinos are massive Majorana fermions (see

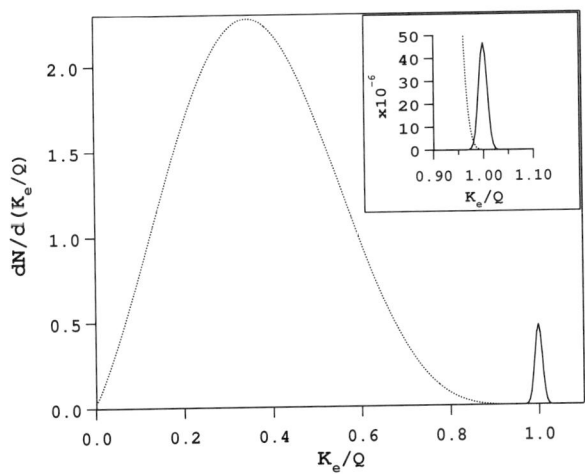

Fig. 10.3. Separating the $0\nu\beta\beta$ mode from the $2\nu\beta\beta$ by the shape of the sum electron spectrum (kinetic energy K_e of the two electrons), including the effect of the 2% resolution smearing. The assumed $2\nu/0\nu$ rate ratio is 10^2, and 10^6 in the insert.

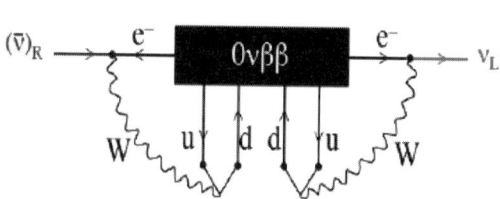

Fig. 10.4. By adding loops involving only standard weak interaction processes the $0\nu\beta\beta$ decay amplitude (the black box) implies the existence of the Majorana neutrino mass.

Fig. 10.4). This qualitative statement (or theorem), unfortunately, does not allow us to deduce the magnitude of the neutrino mass once the rate of the $0\nu\beta\beta$ decay have been determined.

There is no indication at the present time that neutrinos have nonstandard interactions, i.e. they seem to have only interactions carried by the W and Z bosons that are contained in the Standard Electroweak Model. All observed oscillation phenomena can be understood if one assumes that that neutrinos interact exactly the way the Standard Model prescribes, but are massive fermions forcing a generalization of the model. If we accept this, but in addition assume that neutrinos are Majorana particles, we can in fact relate the $0\nu\beta\beta$ decay rate to the quantity related to the absolute neutrino mass. With these caveats that relation can be expressed as

$$\frac{1}{T^{0\nu}_{1/2}} = G^{0\nu}(Q,Z)|M^{0\nu}|^2 \langle m_{\beta\beta}\rangle^2 \;, \tag{10.1}$$

where $G^{0\nu}(Q,Z)$ is a phase space factor that depends on the transition Q value and through the Coulomb effect on the emitted electrons on the nuclear charge Z and that can be easily and accurately calculated, $M^{0\nu}$ is the nuclear matrix element that can be evaluated in principle, although with a considerable uncertainty, and finally the quantity $\langle m_{\beta\beta}\rangle$ is the effective neutrino Majorana mass, representing the important particle physics ingredient of the process.

In turn, the effective mass $\langle m_{\beta\beta}\rangle$ is related to the mixing angles θ_{ij} that are determined or constrained by the oscillation experiments, to the absolute neutrino masses m_i of the mass eigenstates ν_i and to the as of now totally unknown additional parameters as fundamental as the mixing angles θ_{ij}, the so-called Majorana phase $\alpha(i)$,

$$\langle m_{\beta\beta}\rangle = |\Sigma_i |U_{ei}|^2 e^{i\alpha(i)} m_i| \ . \tag{10.2}$$

Here U_{ei} are the matrix elements of the first row of the neutrino mixing matrix.

It is straightforward to use the eq. (10.2) and the known neutrino oscillation results in order to relate $\langle m_{\beta\beta}\rangle$ to other neutrino mass dependent quantities. This is illustrated in Fig. 10.5. Traditionally such plot is made as in the left panel. However, the lightest neutrino mass m_{min} is not an observable quantity. For that reason the other two panels show the relation of $\langle m_{\beta\beta}\rangle$ to the sum of the neutrino masses $M = \Sigma m_i$ and also to $\langle m_\beta\rangle$ that represents the parameter that can be determined or constrained in ordinary β decay,

$$\langle m_\beta\rangle^2 = \Sigma_i |U_{ei}|^2 m_i^2 \ . \tag{10.3}$$

Several remarks are in order. First, the observation of the $0\nu\beta\beta$ decay and determination of $\langle m_{\beta\beta}\rangle$, even when combined with the knowledge of M and/or $\langle m_\beta\rangle$ does not allow, in general, to distinguish between the normal and inverted mass orderings. This is a consequence of the fact that the Majorana phases are unknown. In regions in Fig. 10.5 where the two hatched bands overlap it is clear that two solutions with the same $\langle m_{\beta\beta}\rangle$ and the same M (or the same $\langle m_\beta\rangle$) exist and cannot be distinguished.

On the other hand, obviously, if one can determine that $\langle m_{\beta\beta}\rangle \geq 0.1$ eV we would conclude that the mass pattern is degenerate. And in the so far hypothetical case that one could show that $\langle m_{\beta\beta}\rangle \leq 0.01$ - 0.02 eV but nonvanishing nevertheless the normal hierarchy would be established.[a]

It is worthwhile noting that if the inverted mass ordering is realized in nature, (and neutrinos are Majorana particles) the quantity $\langle m_{\beta\beta}\rangle$ is constrained from below by ~ 0.01 eV. This value is within reach of the next generation of experiments. Also, in principle, in the case of the normal hierarchy while all neutrinos could be massive Majorana particles it is still possible that $\langle m_{\beta\beta}\rangle = 0$. Such a situation, however, requires "fine tuning" or reflects a symmetry of some kind.

Let us remark that the $0\nu\beta\beta$ decay is not the only LNV process for which important experimental constraints exist. Examples of the other

---

[a]In that case also the $\langle m_\beta\rangle$ in the right panel would not represent the quatity directly related to the ordinary β decay. There are no realistic ideas, however, how to reach the corresponding sensitivity in ordinary β decay at this time.

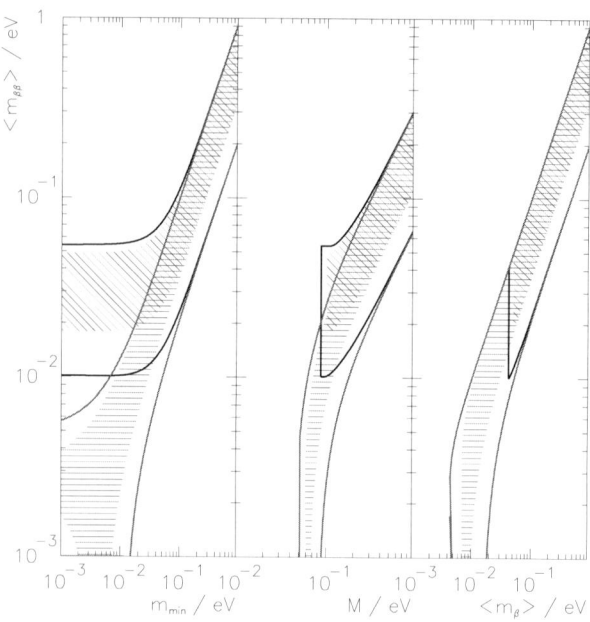

Fig. 10.5. The left panel shows the dependence of $\langle m_{\beta\beta} \rangle$ on the mass of the lightest neutrino $m_{\min}$, the middle one shows the relation between $\langle m_{\beta\beta} \rangle$ and the sum of neutrino masses $M = \Sigma m_i$ determined or constrained by the "observational cosmology", and the right one depicts the relation between $\langle m_{\beta\beta} \rangle$ and the effective mass $\langle m_\beta \rangle$ determined or constrained by the ordinary β decay. In all panels the width of the hatched area is due to the unknown Majorana phases and therefore irreducible. The solid lines indicate the allowed regions by taking into account the current uncertainties in the oscillation parameters; they will shrink as the accuracy improves. The two sets of curves correspond to the normal and inverted hierarchies, they merge above about $\langle m_{\beta\beta} \rangle \geq 0.1$ eV, where the degenerate mass pattern begins.

analogous processes are

$$\mu^- + (Z, A) \to e^+ + (Z-2, A); \text{ exp. branching ratio} \leq 10^{-12},$$
$$K^+ \to \mu^+ \mu^+ \pi^-; \text{ exp. branching ratio} \leq 3 \times 10^{-9},$$
$$\bar{\nu}_e \text{ emission from the Sun; exp. branching ratio} \leq 10^{-4}. \quad (10.4)$$

However, detailed analysis suggests that the study of the $0\nu\beta\beta$ decay is by far the most sensitive test of LNV. In simple terms, this is caused by the amount of tries one can make. A 100 kg $0\nu\beta\beta$ decay source contains $\sim 10^{27}$ nuclei. This can be contrasted with the possibilities of first producing muons or kaons, and then searching for the unusual decay channels. The Fermilab accelerators, for example, produce "a few" $\times 10^{20}$ protons on target per year in their beams and thus correspondingly smaller numbers of muons or kaons.

10.2. Mechanism of the $0\nu\beta\beta$ decay

It has been recognized long time ago that the relation between the $0\nu\beta\beta$ decay rate and the effective Majorana mass $\langle m_{\beta\beta} \rangle$ is to some extent problematic. The assumption leading to the eq. (10.1) is rather conservative, namely that there is an exchange of a virtual light, but massive, Majorana neutrino between the two nucleons undergoing the transition, and that these neutrinos interact by the standard left-handed weak currents. However, that is not the only possible mechanism. LNV interactions involving so far unobserved heavy ($\sim$ TeV) particles can lead to a comparable $0\nu\beta\beta$ decay rate. Thus, in the absence of additional information about the mechanism responsible for the $0\nu\beta\beta$ decay, one could not unambiguously infer $\langle m_{\beta\beta} \rangle$ from the $0\nu\beta\beta$ decay rate.

In general $0\nu\beta\beta$ decay can be generated by (i) light massive Majorana neutrino exchange or (ii) heavy particle exchange (see, e.g. Refs. 18,19), resulting from LNV dynamics at some scale Λ above the electroweak one. The relative size of heavy (A_H) versus light particle (A_L) exchange contributions to the decay amplitude can be crudely estimated as follows:[20]

$$A_L \sim G_F^2 \frac{\langle m_{\beta\beta} \rangle}{\langle k^2 \rangle}, \; A_H \sim G_F^2 \frac{M_W^4}{\Lambda^5}, \; \frac{A_H}{A_L} \sim \frac{M_W^4 \langle k^2 \rangle}{\Lambda^5 \langle m_{\beta\beta} \rangle}, \qquad (10.5)$$

where $\langle m_{\beta\beta} \rangle$ is the effective neutrino Majorana mass, $\langle k^2 \rangle \sim (50 \text{ MeV})^2$ is the typical light neutrino virtuality, and Λ is the heavy scale relevant to the LNV dynamics. Therefore, $A_H/A_L \sim O(1)$ for $\langle m_{\beta\beta} \rangle \sim 0.1 - 0.5$ eV and $\Lambda \sim 1$ TeV, and thus the LNV dynamics at the TeV scale leads to similar $0\nu\beta\beta$ decay rate as the exchange of light Majorana neutrinos with the effective mass $\langle m_{\beta\beta} \rangle \sim 0.1 - 0.5$ eV.

Obviously, the lifetime measurement by itself does not provide the means for determining the underlying mechanism. The spin-flip and non-flip exchange can be, in principle, distinguished by the measurement of the single-electron spectra or polarization (see e.g.,[21]). However, in most cases the mechanism of light Majorana neutrino exchange, and of heavy particle exchange cannot be separated by the observation of the emitted electrons. Thus one must look for other phenomenological consequences of the different mechanisms other than observables directly associated with $0\nu\beta\beta$. Here I discuss the suggestion[22] that under natural assumptions the presence of low scale LNV interactions also affects muon lepton flavor violating (LFV) processes, and in particular enhances the $\mu \to e$ conversion compared to the $\mu \to e\gamma$ decay.

The discussion is concerned mainly with the branching ratios $B_{\mu \to e\gamma} = \Gamma(\mu \to e\gamma)/\Gamma_\mu^{(0)}$ and $B_{\mu \to e} = \Gamma_{\text{conv}}/\Gamma_{\text{capt}}$, where $\mu \to e\gamma$ is normalized to the standard muon decay rate $\Gamma_\mu^{(0)} = (G_F^2 m_\mu^5)/(192\pi^3)$, while $\mu \to e$ conversion is normalized to the corresponding capture rate Γ_{capt}. The main diagnostic tool in the analysis is the ratio

$$\mathcal{R} = B_{\mu \to e}/B_{\mu \to e\gamma} , \qquad (10.6)$$

and the relevance of our observation relies on the potential for LFV discovery in the forthcoming experiments MEG[23] ($\mu \to e\gamma$) and MECO[24] ($\mu \to e$ conversion)[b] that plan to improve the current limits by several orders of magnitude.

It is useful to formulate the problem in terms of effective low energy interactions obtained after integrating out the heavy degrees of freedom that induce LNV and LFV dynamics. If the scales for both LNV and LFV are well above the weak scale, then one would not expect to observe any signal in the forthcoming LFV experiments, nor would the effects of heavy particle exchange enter $0\nu\beta\beta$ at an appreciable level. In this case, the only origin of a signal in $0\nu\beta\beta$ at the level of prospective experimental sensitivity would be the exchange of a light Majorana neutrino, leading to eq. (10.1), and allowing one to extract $\langle m_{\beta\beta} \rangle$ from the decay rate.

In general, however, the two scales may be distinct, as in SUSY-GUT[25] or SUSY see-saw[26] models. In these scenarios, both the Majorana neutrino mass as well as LFV effects are generated at the GUT scale. The effects of heavy Majorana neutrino exchange in $0\nu\beta\beta$ are, thus, highly suppressed. In contrast, the effects of GUT-scale LFV are transmitted to the TeV-scale by a soft SUSY-breaking sector without mass suppression via renormalization group running of the high-scale LFV couplings. Consequently, such scenarios could lead to observable effects in the upcoming LFV experiments but with an $\mathcal{O}(\alpha)$ suppression of the branching ratio $B_{\mu \to e}$ relative to $B_{\mu \to e\gamma}$ due to the exchange of a virtual photon in the conversion process rather than the emission of a real one, thus $\mathcal{R} \sim 10^{-(2-3)}$ in this case.

The case where the scales of LNV and LFV are both relatively low ($\sim$ TeV) is more subtle. This is the scenario which might lead to observable signals in LFV searches and at the same time generate ambiguities in

---
[b]Even though MECO experiment was recently cancelled, proposals for experiments with similar sensitivity exist elsewhere.

interpreting a positive signal in $0\nu\beta\beta$. Therefore, this is the case where one needs to develop some discriminating criteria.

Denoting the new physics scale by Λ, one has a LNV effective lagrangian of the form

$$\mathcal{L}_{0\nu\beta\beta} = \sum_i \frac{\tilde{c}_i}{\Lambda^5} \tilde{O}_i \qquad \tilde{O}_i = \bar{q}\Gamma_1 q \;\; \bar{q}\Gamma_2 q \;\bar{e}\Gamma_3 e^c \;, \tag{10.7}$$

where we have suppressed the flavor and Dirac structures (a complete list of the dimension nine operators $\tilde{O}_i$ can be found in Ref. 19).

For the LFV interactions, one has

$$\mathcal{L}_{\rm LFV} = \sum_i \frac{c_i}{\Lambda^2} O_i \;, \tag{10.8}$$

and a complete operator basis can be found in Refs. 27,28. The LFV operators relevant to our analysis are of the following type (along with their analogues with $L \leftrightarrow R$):

$$\begin{aligned}
O_{\sigma L} &= \frac{e}{(4\pi)^2} \overline{\ell_{iL}} \, \sigma_{\mu\nu} i\!\!\not{\!D} \, \ell_{jL} \, F^{\mu\nu} + \text{h.c.} \\
O_{\ell L} &= \overline{\ell_{iL}} \, \ell^c_{jL} \, \overline{\ell^c_{kL}} \, \ell_{mL} \\
O_{\ell q} &= \overline{\ell_i} \Gamma \ell_j \; \bar{q}\Gamma_q q \;.
\end{aligned} \tag{10.9}$$

Operators of the type O_σ are typically generated at one-loop level, hence our choice to explicitly display the loop factor $1/(4\pi)^2$. On the other hand, in a large class of models, operators of the type O_ℓ or $O_{\ell q}$ are generated by tree level exchange of heavy degrees of freedom. With the above choices, all non-zero c_i and $\tilde{c}_i$ are nominally of the same size, typically the product of two Yukawa-like couplings or gauge couplings (times flavor mixing matrices).

With the notation established above, the ratio $\mathcal{R}$ of the branching ratios $\mu \to e$ to $\mu \to e+\gamma$ can be written schematically as follows (neglecting flavor indices in the effective couplings and the term with $L \leftrightarrow R$):

$$\mathcal{R} = \frac{\Phi}{48\pi^2} \left| \lambda_1 \, e^2 c_{\sigma L} + e^2 \left(\lambda_2 c_{\ell L} + \lambda_3 c_{\ell q} \right) \log \frac{\Lambda^2}{m_\mu^2} \right.$$
$$\left. + \lambda_4 (4\pi)^2 c_{\ell q} + \ldots \right|^2 / \left[e^2 \left(|c_{\sigma L}|^2 + |c_{\sigma R}|^2 \right) \right] \;. \tag{10.10}$$

In the above formula $\lambda_{1,2,3,4}$ are numerical factors of $O(1)$, while the overall factor $\frac{\Phi}{48\pi^2}$ arises from phase space and overlap integrals of electron and muon wavefunctions in the nuclear field. For light nuclei $\Phi = (ZF_p^2)/(g_V^2 + 3g_A^2) \sim O(1)$ ($g_{V,A}$ are the vector and axial nucleon form factors at zero momentum transfer, while F_p is the nuclear form factor at $q^2 = -m_\mu^2$ [28]). The dots indicate subleading terms, not relevant for our discussion, such as loop-induced contributions to c_ℓ and $c_{\ell q}$ that are analytic in external masses and momenta. In contrast the logarithmically-enhanced loop contribution given by the second term in the numerator of $\mathcal{R}$ plays an essential role. This term arises whenever the operators $O_{\ell L,R}$ and/or $O_{\ell q}$ appear at tree-level in the effective theory and generate one-loop renormalization of $O_{\ell q}$[27] (see Fig. 10.6).

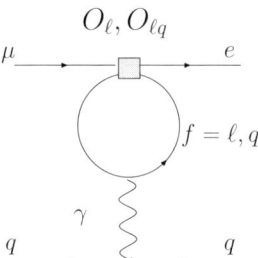

Fig. 10.6. Loop contributions to $\mu \to e$ conversion through insertion of operators O_ℓ or $O_{\ell q}$, generating the large logarithm.

The ingredients in eq. (10.10) lead to several observations: (i) In absence of tree-level $c_{\ell L}$ and $c_{\ell q}$, one obtains $\mathcal{R} \sim (\Phi \lambda_1^2 \alpha)/(12\pi) \sim 10^{-3} - 10^{-2}$, due to gauge coupling and phase space suppression. (ii) When present, the logarithmically enhanced contributions compensate for the gauge coupling and phase space suppression, leading to $\mathcal{R} \sim O(1)$. (iii) If present, the tree-level coupling $c_{\ell q}$ dominates the $\mu \to e$ rate leading to $\mathcal{R} \gg 1$.

Thus, we can formulate our main conclusions regarding the discriminating power of the ratio $\mathcal{R}$:

(1) Observation of both the LFV muon processes $\mu \to e$ and $\mu \to e\gamma$ with relative ratio $\mathcal{R} \sim 10^{-2}$ implies, under generic conditions, that $\Gamma_{0\nu\beta\beta} \sim \langle m_{\beta\beta} \rangle^2$. Hence the relation of the $0\nu\beta\beta$ lifetime to the absolute neutrino mass scale is straightforward.
(2) On the other hand, observation of LFV muon processes with relative ratio $\mathcal{R} \gg 10^{-2}$ could signal non-trivial LNV dynamics at the TeV scale, whose effect on $0\nu\beta\beta$ has to be analyzed on a case by case basis. Therefore, in this scenario no definite conclusion can be drawn based on LFV rates.

(3) Non-observation of LFV in muon processes in forthcoming experiments would imply either that the scale of non-trivial LFV and LNV is above a few TeV, and thus $\Gamma_{0\nu\beta\beta} \sim \langle m_{\beta\beta}\rangle^2$, or that any TeV-scale LNV is approximately flavor diagonal.

The above statements are illustrated using two explicit cases:[22] the minimal supersymmetric standard model (MSSM) with R-parity violation (RPV-SUSY) and the Left-Right Symmetric Model (LRSM).

RPV SUSY — If one does not impose R-parity conservation [$R = (-1)^{3(B-L)+2s}$], the MSSM superpotential includes, in addition to the standard Yukawa terms, lepton and baryon number violating interactions, compactly written as (see e.g.,[29])

$$W_{RPV} = \lambda_{ijk} L_i L_j E_k^c + \lambda'_{ijk} L_i Q_j D_k^c + \lambda''_{ijk} U_i^c D_j^c D_k^c + \mu'_i L_i H_u \,, \tag{10.11}$$

where L and Q represent lepton and quark doublet superfields, while E^c, U^c, D^c are lepton and quark singlet superfields. The simultaneous presence of λ' and λ'' couplings would lead to an unacceptably large proton decay rate (for SUSY mass scale $\Lambda_{SUSY} \sim$ TeV), so we focus on the case of $\lambda'' = 0$ and set $\mu' = 0$ without loss of generality. In such case, lepton number is violated by the remaining terms in W_{RPV}, leading to short distance contributions to $0\nu\beta\beta$ [e.g., Fig. 10.7(a)], with typical coefficients [cf. eq. (10.7)]

$$\frac{\tilde{c}_i}{\Lambda^5} \sim \frac{\pi\alpha_s}{m_{\tilde{g}}} \frac{\lambda'^2_{111}}{m_{\tilde{f}}^4} ; \frac{\pi\alpha_2}{m_\chi} \frac{\lambda'^2_{111}}{m_{\tilde{f}}^4} \,, \tag{10.12}$$

where α_s, α_2 represent the strong and weak gauge coupling constants, respectively. The RPV interactions also lead to lepton number conserving but lepton flavor violating operators [e.g. Fig. 10.7(b)], with coefficients [cf. eq. (10.8)]

$$\frac{c_\ell}{\Lambda^2} \sim \frac{\lambda_{i11}\lambda^*_{i21}}{m_{\tilde{\nu}_i}^2}, \frac{\lambda^*_{i11}\lambda_{i12}}{m_{\tilde{\nu}_i}^2} \,,$$
$$\frac{c_{\ell q}}{\Lambda^2} \sim \frac{\lambda'^*_{11i}\lambda'_{21i}}{m_{\tilde{d}_i}^2}, \frac{\lambda'^*_{1i1}\lambda'_{2i1}}{m_{\tilde{u}_i}^2} \,, \tag{10.13}$$
$$\frac{c_\sigma}{\Lambda^2} \sim \frac{\lambda\lambda^*}{m_{\tilde{\ell}}^2}, \frac{\lambda'\lambda'^*}{m_{\tilde{q}}^2} \,,$$

where the flavor combinations contributing to c_σ can be found in Ref. 30. Hence, for generic flavor structure of the couplings λ and λ' the underlying

LNV dynamics generate both short distance contributions to $0\nu\beta\beta$ and LFV contributions that lead to $\mathcal{R} \gg 10^{-2}$.

Existing limits on rare processes strongly constrain combinations of RPV couplings, assuming Λ_{SUSY} is between a few hundred GeV and ~ 1 TeV. Non-observation of LFV at future experiments MEG and MECO could be attributed either to a larger Λ_{SUSY} ($>$ few TeV) or to suppression of couplings that involve mixing among first and second generations. In the former scenario, the short distance contribution to $0\nu\beta\beta$ does not compete with the long distance one [see eq. (10.5)], so that $\Gamma_{0\nu\beta\beta} \sim \langle m_{\beta\beta}\rangle^2$. On the other hand, there is an exception to this "diagnostic tool". If the λ and λ' matrices are nearly flavor diagonal, the exchange of superpartners may still make non-negligible contributions to $0\nu\beta\beta$ without enhancing the ratio $\mathcal{R}$.

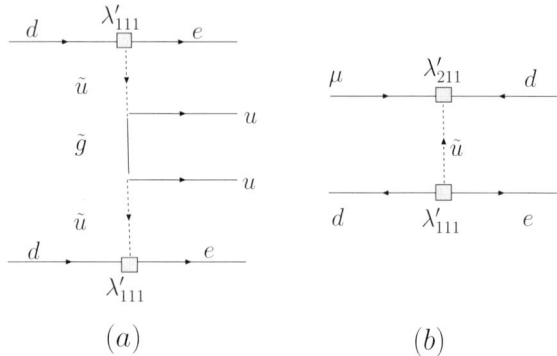

Fig. 10.7. Gluino exchange contribution to $0\nu\beta\beta$ (a), and typical tree-level contribution to $O_{\ell q}$ (b) in RPV SUSY.

LRSM — The LRSM provides a natural scenario for introducing non-sterile, right-handed neutrinos and Majorana masses.[31] The corresponding electroweak gauge group $SU(2)_L \times SU(2)_R \times U(1)_{B-L}$, breaks down to $SU(2)_L \times U(1)_Y$ at the scale $\Lambda \geq \mathcal{O}(\text{TeV})$. The symmetry breaking is implemented through an extended Higgs sector, containing a bi-doublet Φ and two triplets $\Delta_{L,R}$, whose leptonic couplings generate both Majorana neutrino masses and LFV involving charged leptons:

$$\mathcal{L}_Y^{\text{lept}} = -\overline{L_L}^i \left(y_D^{ij} \Phi + \tilde{y}_D^{ij} \tilde{\Phi} \right) L_R^j$$
$$- \overline{(L_L)^c}^i y_M^{ij} \tilde{\Delta}_L L_L^j - \overline{(L_R)^c}^i y_M^{ij} \tilde{\Delta}_R L_R^j . \quad (10.14)$$

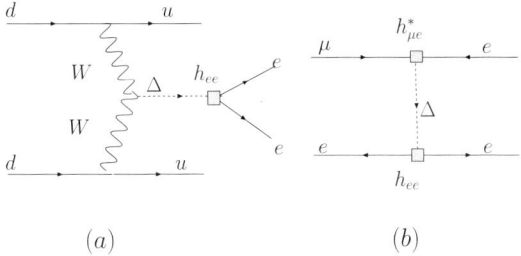

Fig. 10.8. Typical doubly charged Higgs contribution to $0\nu\beta\beta$ (a) and to O_ℓ (b) in the LRSM.

Here $\tilde{\Phi} = \sigma_2 \Phi^* \sigma_2$, $\tilde{\Delta}_{L,R} = i\sigma_2 \Delta_{L,R}$, and leptons belong to two isospin doublets $L^i_{L,R} = (\nu^i_{L,R}, \ell^i_{L,R})$. The gauge symmetry is broken through the VEVs $\langle \Delta^0_R \rangle = v_R$, $\langle \Delta^0_L \rangle = 0$, $\langle \Phi \rangle = \text{diag}(\kappa_1, \kappa_2)$. After diagonalization of the lepton mass matrices, LFV arises from both non-diagonal gauge interactions and the Higgs Yukawa couplings. In particular, the $\Delta_{L,R}$-lepton interactions are not suppressed by lepton masses and have the structure $\mathcal{L} \sim \Delta^{++}_{L,R} \overline{\ell^c_i} h_{ij} (1 \pm \gamma_5) \ell_j + \text{h.c.}$. The couplings h_{ij} are in general non-diagonal and related to the heavy neutrino mixing matrix.[32]

Short distance contributions to $0\nu\beta\beta$ arise from the exchange of both heavy νs and $\Delta_{L,R}$ (Fig. 10.8a), with

$$\frac{\tilde{c}_i}{\Lambda^5} \sim \frac{g_2^4}{M_{W_R}^4} \frac{1}{M_{\nu_R}} ; \frac{g_2^3}{M_{W_R}^3} \frac{h_{ee}}{M_\Delta^2}, \qquad (10.15)$$

where g_2 is the weak gauge coupling. LFV operators are also generated through non-diagonal gauge and Higgs vertices, with[32] (Fig. 10.8b)

$$\frac{c_\ell}{\Lambda^2} \sim \frac{h_{\mu i} h^*_{ie}}{m_\Delta^2} \qquad \frac{c_\sigma}{\Lambda^2} \sim \frac{(h^\dagger h)_{e\mu}}{M_{W_R}^2} \qquad i = e, \mu, \tau. \qquad (10.16)$$

Note that the Yukawa interactions needed for the Majorana neutrino mass necessarily imply the presence of LNV and LFV couplings h_{ij} and the corresponding LFV operator coefficients c_ℓ, leading to $\mathcal{R} \sim O(1)$. Again, non-observation of LFV in the next generation of experiments would typically push Λ into the multi-TeV range, thus implying a negligible short distance contribution to $0\nu\beta\beta$. As with RPV-SUSY, this conclusion can be evaded by assuming a specific flavor structure, namely y_M approximately diagonal or a nearly degenerate heavy neutrino spectrum.

In both of these phenomenologically viable models that incorporate LNV and LFV at low scale ($\sim$ TeV), one finds $\mathcal{R} \gg 10^{-2}$.[27,30,32] It is likely that the basic mechanism at work in these illustrative cases is generic: low scale LNV interactions ($\Delta L = \pm 1$ and/or $\Delta L = \pm 2$), which in general contribute to $0\nu\beta\beta$, also generate sizable contributions to $\mu \to e$ conversion, thus enhancing this process over $\mu \to e\gamma$.

In conclusion, the above considerations suggest that the ratio $\mathcal{R} = B_{\mu \to e}/B_{\mu \to e\gamma}$ of muon LFV processes will provide important insight about the mechanism of neutrinoless double beta decay and the use of this process to determine the absolute scale of neutrino mass. Assuming observation of LFV processes in forthcoming experiments, if $\mathcal{R} \sim 10^{-2}$ the mechanism of $0\nu\beta\beta$ is light Majorana neutrino exchange; if $\mathcal{R} \gg 10^{-2}$, there might be TeV scale LNV dynamics, and no definite conclusion on the mechanism of $0\nu\beta\beta$ can be drawn based only on LFV processes.

10.3. Overview of the experimental status of search for $\beta\beta$ decay

The field has a venerable history. The rate of the $2\nu\beta\beta$ decay was first estimated by Maria Goeppert-Mayer already in 1937 in her thesis work suggested by E. Wigner, basically correctly. Yet, first experimental observation in a laboratory experiment was achieved only in 1987, fifty years later. Why it took so long? As pointed out above, the typical half-life of the $2\nu\beta\beta$ decay is $\sim 10^{20}$ years. Yet, its "signature" is very similar to natural radioactivity, present to some extent everywhere, and governed by the half-life of $\sim 10^{10}$ years. So, background suppression is the main problem to overcome when one wants to study either of the $\beta\beta$ decay modes.

During the last two decades the $2\nu\beta\beta$ decay has been observed in "live" laboratory experiments in many nuclei, often by different groups and using different methods. That shows not only the ingenuity of the experimentalists who were able to overcome the background nemesis, but makes it possible at the same time to extract the corresponding 2ν nuclear matrix element from the measured decay rate. In the 2ν mode the half-life is given by

$$1/T_{1/2} = G^{2\nu}(Q,Z)|M^{2\nu}|^2 , \qquad (10.17)$$

where $G^{2\nu}(Q,Z)$ is an easily and accurately calculable phase space factor.

The resulting nuclear matrix elements $M^{2\nu}$, which have the dimension energy^{-1}, are plotted in Fig. 10.9. Note the pronounced shell dependence; the matrix element for ^{100}Mo is almost ten times larger than the one for ^{130}Te. Evaluation of these matrix elements, to be discussed below, is an

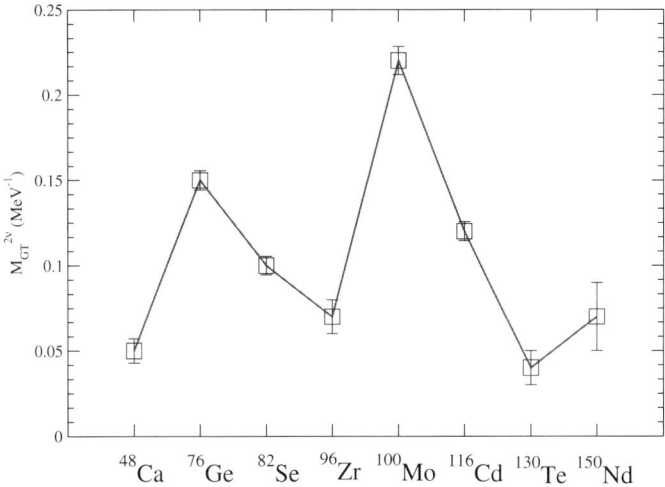

Fig. 10.9. Nuclear matrix elements for the $2\nu\beta\beta$ decay extracted from the measured half-lives.

important test for the nuclear theory models that aim at the determination of the analogous but different quantities for the 0ν neutrinoless mode.

The challenge of detecting the $0\nu\beta\beta$ decay is, at first blush, easier. Unlike the continuous $2\nu\beta\beta$ decay spectrum with a broad maximum at rather low energy where the background suppression is harder, the $0\nu\beta\beta$ decay spectrum is sharply peaked at the known Q value (see Fig. 10.3), at energies that are not immune to the background, but a bit less difficult to manage. However, as also indicated in Fig. 10.3, to obtain interesting results at the present time means to reach sensitivity to the 0ν half-lives that are $\sim 10^6$ times longer than the 2ν decay half-life of the same nucleus. So the requirements of background suppression are correspondingly even more severe.

The historical lessons are illustrated in Fig. 10.10 where the past limits on the $0\nu\beta\beta$ decay half-lives of various candidate nuclei are translated using the eq. (10.1) into the limits on the effective mass $\langle m_{\beta\beta} \rangle$. When plotted in the semi-log plot this figure represents the "Moore's law" of double beta decay, and indicates that, provided that the past trend will continue, the mass scale corresponding to Δm^2_{atm} will be reached in about 10 years. This is also the time scale of significant experiments these days. Indeed, as discussed further, preparations are on the way to reach this sensitivity

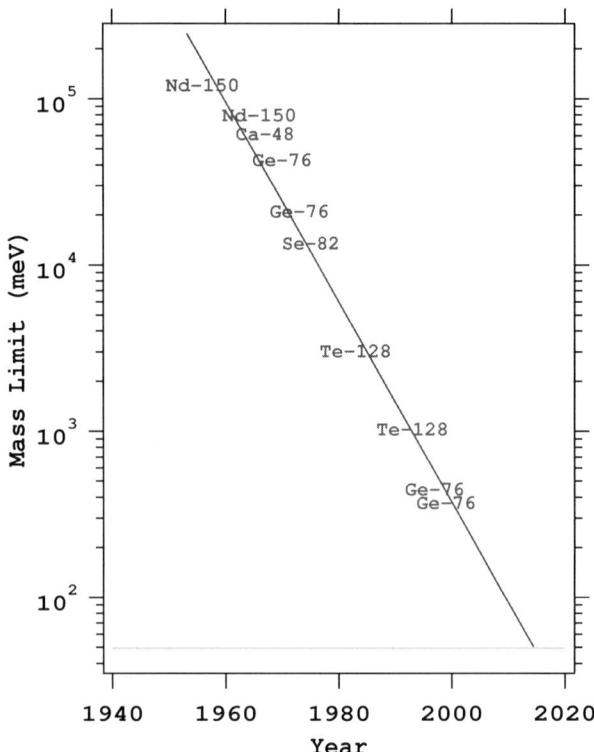

Fig. 10.10. The limit of the effective mass $\langle m_{\beta\beta} \rangle$ extracted from the experimental lower limits on the $0\nu\beta\beta$ decay half-life versus the corresponding year. The gray band near bottom indicates the $\sqrt{\Delta m^2_{atm}}$ value. Figure originally made by S. Elliott.

goal. Note that the figure was made using some assumed values of the corresponding nuclear matrix elements, without including their uncertainty. For such illustrative purposes they are, naturally, irrelevant.

The past search for the neutrinoless double beta decay, illustrated in Fig. 10.10, was driven by the then current technology and the resources of the individual experiments. The goal has been simply to reach sensitivity to longer and longer half-lives. The situation is different, however, now. The experimentalists at the present time can and do use the knowledge summarized in Fig. 10.5 to gauge the aim of their proposals. Based on that figure, the range of the mass parameter $\langle m_{\beta\beta} \rangle$ can be divided into three regions of interest.

- The degenerate mass region where all $m_i \gg \sqrt{\Delta m_{atm}^2}$. In that region $\langle m_{\beta\beta} \rangle \geq 0.1$ eV, corresponding crudely to the 0ν half-lives of 10^{26-27} years. To explore it (in a realistic time frame), ~ 100 kg of the decaying nucleus is needed. Several experiments aiming at such sensitivity are being built and should run and give results within the next 3-5 years. Moreover, this mass region (or a substantial part of it) will be explored, in a similar time frame, by the study of ordinary β decay (in particular of tritium) and by the observational cosmology. These techniques are independent on the Majorana nature of neutrinos. It is easy, but perhaps premature, to envision various scenarios depending on the possible outcome of these measurements.
- The so-called inverted hierarchy mass region where $20 < \langle m_{\beta\beta} \rangle < 100$ meV and the $0\nu\beta\beta$ half-lives are about 10^{27-28} years. (The name is to some extent a misnomer. In that interval one could encounter not only the inverted hierarchy but also a quasi-degenerate but normal neutrino mass ordering. Successful observation of the $0\nu\beta\beta$ decay will not be able to distinguish these possibilities, as I argued above. This is so not only due to the anticipated experimental accuracy, but more fundamentally due to the unknown Majorana phases.) To explore this mass region, $\sim$ ton size sources would be required. Proposals for the corresponding experiments exist, but none has been funded as yet, and presumably the real work will begin depending on the experience with the various ~ 100 kg size sources. Timeline for exploring this mass region is ~ 10 years.
- Normal mass hierarchy region where $\langle m_{\beta\beta} \rangle \leq$ 10-20 meV. To explore this mass region, ~ 100 ton sources would be required. There are no realistic proposals for experiments of this size at present.

Over the last two decades, the methodology for double beta decay experiments has improved considerably. Larger amounts of high-purity enriched parent isotopes, combined with careful selection of all surrounding materials and using deep-underground sites have lowered backgrounds and increased sensitivity. The most sensitive experiments to date use ^{76}Ge, ^{100}Mo, ^{116}Cd, ^{130}Te, and ^{136}Xe. For ^{76}Ge the lifetime limit reached impressive values exceeding 10^{25} years.[33,34] The experimental lifetime limits have been interpreted to yield effective neutrino mass limits typically a few eV and in ^{76}Ge as low as 0.3 - 1.0 eV (the spread reflects an estimate of the uncertainty in the nuclear matrix elements). The sum electron spectrum obtained in the Heidelberg-Moscow[33] experiment is shown in Fig. 10.11

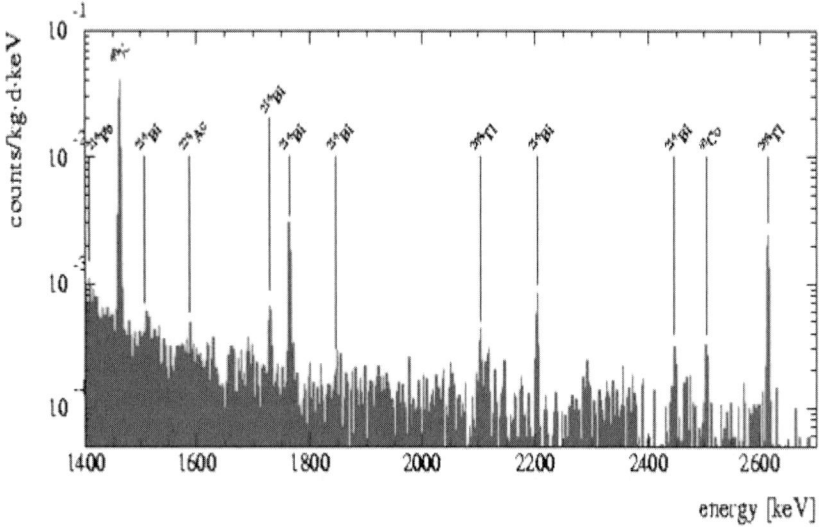

Fig. 10.11. The spectrum recorded in the Heidelberg-Moscow $\beta\beta$ decay experiment on ^{76}Ge. Identified γ lines are indicated.

over a broad energy range, and in Fig. 10.12 over a narrower range in the vicinity of the 0ν Q value of 2039 keV. Some residual natural radioactivity background lines are clearly visible in both figures, and no obvious peak at the 0ν expected position can be seen in Fig. 10.12.

Nevertheless, a subset of members of the Heidelberg-Moscow collaboration reanalyzed the data (and used additional information, e.g. the pulse-shape analysis and a different algorithm in the peak search) and claimed to observe a positive signal corresponding to the effective mass of $\langle m_{\beta\beta} \rangle = 0.39^{+0.17}_{-0.28}$ eV.[35] That report has been followed by a lively discussion. Clearly, such an extraordinary claim with its profound implications, requires extraordinary evidence. It is fair to say that a confirmation, both for the same ^{76}Ge parent nucleus, and better yet also in another nucleus with a different Q value, would be required for a consensus. In any case, if that claim is eventually confirmed, the degenerate mass scenario will be implicated, and eventual positive signal in the analysis of the tritium β decay and/or observational cosmology should be forthcoming. For the neutrinoless $\beta\beta$ decay the next generation of experiments, which use ~ 100 kg of decaying isotopes will, among other things, test this recent claim.

It is beyond the scope of these lecture notes to describe in detail the forthcoming $0\nu\beta\beta$ decay experiments. Rather detailed discussion of them

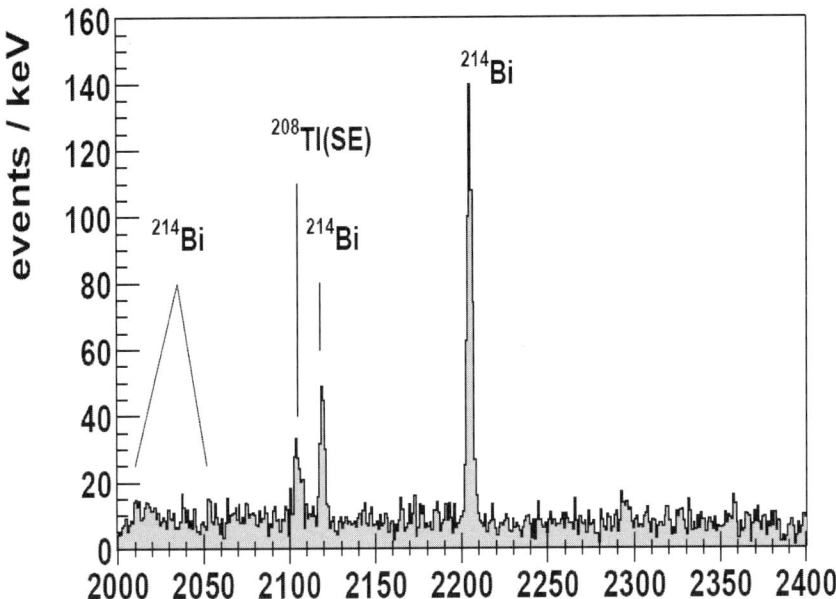

Fig. 10.12. Spectrum of the Heidelberg-Moscow experiment in the vicinity of the $0\nu\beta\beta$ decay value of 2039 keV.

can be found e.g. in Ref. 16. Also, the corresponding chapter of the APS neutrino study[36] has various details. Nevertheless, let me briefly comment on the most advanced of the forthcoming ~ 100 kg source experiments *CUORE, GERDA, EXO*, and *MAJORANA*. All of them are designed to explore all (or at least most) of the degenerate neutrino mass region $\langle m_{\beta\beta} \rangle \geq$ 0.1 eV. If their projected efficiencies and background projections are confirmed, all of them plan to consider scaling up the decaying mass to $\sim$ ton and extend their sensitivity to the "inverted hierarchy" region.

These experiments use different nuclei as a source, ^{76}Ge for *GERDA* and *MAJORANA*, ^{130}Te for *CUORE*, and ^{136}Xe for *EXO*. The requirement of radiopurity of the source material and surrounding auxiliary equipment is common to all of them, as is the placement of the experiment deep underground to shield against cosmic rays. The way the electrons are detected is, however, different. While the germanium detectors with their superb energy resolution have been used for the search of the $0\nu\beta\beta$ decay for a long time, the cryogenic detectors in *CUORE* use the temperature increase associated with an event in the very cold TeO$_2$ crystals, and in the *EXO* experiment a Time Projection Chamber (TPC) uses both scintillation and ionization to detect the events. The *EXO* experiment in its final form (still under development and very challenging) would use a positive identification of the final Ba^+ ion as an ultimate background rejection tool. These four experiments are in various stages of funding and staging. First results are expected in about 3 years, and substantial results within 3-5 years in all of them.

10.4. Nuclear matrix elements

It follows from eq. (10.1) that (i) values of the nuclear matrix elements $M^{0\nu}$ are needed in order to extract the effective neutrino mass from the measured $0\nu\beta\beta$ decay rate, and (ii) any uncertainty in $M^{0\nu}$ causes a corresponding and equally large uncertainty in the extracted $\langle m_{\beta\beta} \rangle$ value. Thus, the issue of an accurate evaluation of the nuclear matrix elements attracts considerable attention.

To see qualitatively where the problems are, let us consider the so-called closure approximation, i.e. a description in which the second order perturbation expression is approximated as

$$M^{0\nu} \equiv \langle \Psi_{final} | \hat{O}^{(0\nu)} | \Psi_{initial} \rangle . \qquad (10.18)$$

Now, the challenge is to use an appropriate many-body nuclear model to describe accurately the wave functions of the ground states of the initial and final nuclei, $|\Psi_{initial}\rangle$ and $|\Psi_{final}\rangle$, as well as the appropriate form of the effective transition operator $\hat{O}^{(0\nu)}$ that describes the transformation of two neutrons into two protons correlated by the neutrino propagator, and consistent with the approximations inherent to the nuclear model used.

Common to all methods is the description of the nucleus as a system of nucleons bound in the mean field and interacting by an effective residual interaction. The used methods differ as to the number of nucleon orbits (or shells and subshells) included in the calculations and the complexity of the configurations of the nucleons in these orbits. The two basic approaches used so far for the evaluation of the nuclear matrix elements for both the 2ν and 0ν $\beta\beta$ decay modes are the Quasiparticle Random Phase Approximation (QRPA) and the nuclear shell model (NSM). They are in some sense complementary; QRPA uses a larger set of orbits, but truncates heavily the included configurations, while NSM can include only a rather small set of orbits but includes essentially all possible configurations. NSM also can be tested in a considerable detail by comparing to the nuclear spectroscopy data; in QRPA such comparisons are much more limited.

For the 2ν decay one can relate the various factors entering the calculations to other observables (β strength functions, cross sections of the charge-exchange reactions, etc.), accessible to the experiment. The consistency of the evaluation can be tested in that way. Of course, as pointed out above (see Fig. 10.9) the nuclear matrix elements for this mode are known anyway. Both methods are capable of describing the 2ν matrix elements, at least qualitatively. These quantities, when expressed in natural units based on the sum rules, are very small. Hence their description depends on small components of the nuclear wave functions and is therefore challenging. In QRPA the agreement is achieved if the effective proton-neutron interaction coupling constant (usually called g_{pp}) is slightly (by $\sim$ 10 - 20 %) adjusted.

The theoretical description for the more interesting 0ν mode cannot use any known nuclear observables, since there are no observables directly related to the $M^{0\nu}$. It is therefore much less clear how to properly estimate the uncertainty associated with the calculated values of $M^{0\nu}$, and to judge their accuracy. Since the calculations using QRPA are much simpler, an overwhelming majority of the published calculations uses that method. There are suggestions to use the spread of these published values of $M^{0\nu}$ as a measure of uncertainty.[37] Following this, one would conclude that the uncertainty is quite large, a factor of three or as much as five. But that way of assigning the uncertainty is questionable. Using all or most of the published values of $M^{0\nu}$ means that one includes calculations of uneven quality. Some of them were devoted to the tests of various approximations, and concluded that they are not applicable. Some insist that other data, like the $M^{2\nu}$, are correctly reproduced, other do not pay any attention to such test. Also, different forms of the transition operator $\hat{O}^{0\nu}$ are used, in particular some works include approximately the effect of the short range nucleon-nucleon repulsion, while others neglect it.

In contrast, in Ref. 38 an assesment of uncertainties in the matrix elements $M^{0\nu}$ inherent in the QRPA was made, and it was concluded that with a consistent treatment the uncertainties are much less, perhaps only about 30% (see Fig. 10.13). That calculation uses the known 2ν matrix elements in order to adjust the interaction constant mentioned above. There is a lively debate in the nuclear structure theory community, beyond the scope of these lectures, about this conclusion.

It is of interest also to compare the resulting matrix elements of Rodin et al.[38] based on QRPA and its generalizations, and those of the avail-

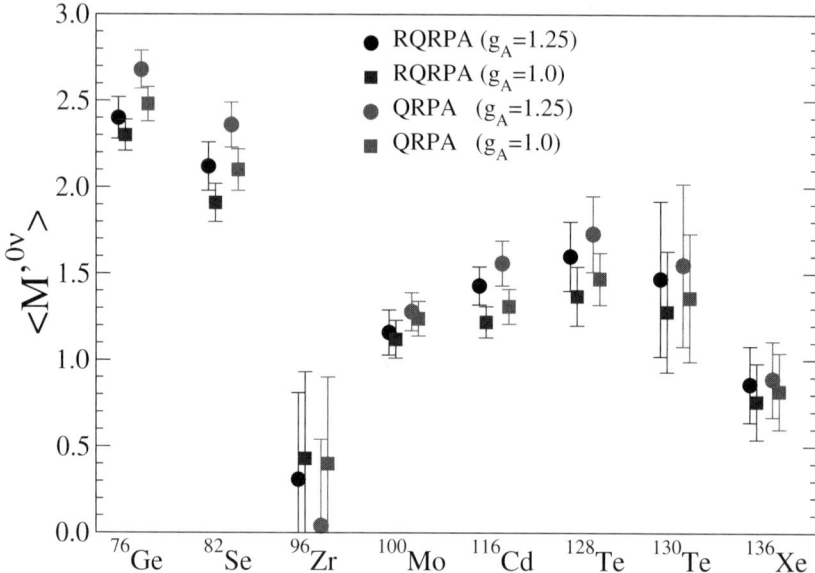

Fig. 10.13. Nuclear matrix elements and their variance for the indicated approximations (see Ref. 38).

able most recent NSM evaluation.[39] Note that the operators used in NSM evaluation do not include the induced nucleon currents that in QRPA reduce the matrix element by about 30%. The QRPA[38] and NSM[39] $M^{0\nu}$ are compared in Table 10.1. In the last column the NSM matrix elements are reduced by 30% to approximately account for the missing terms in the operator, and to make the comparison more meaningful. With this reduction, it seems that QRPA results are a bit larger in the lighter nuclei and a bit smaller in the heavier ones than the NSM results, but basically within the 30% uncertainty estimate. Once the NSM calculations for the intermediate mass nuclei ^{96}Zr, ^{100}Mo and ^{116}Cd become available, one can make a more meaningful comparison of the two methods.

When comparing the results shown in Table 10.1 as well as the results of other calculations (e.g. Refs. 40,41) with Fig. 10.9 it is important to notice a qualitative difference in the behaviour of the 2ν and 0ν matrix elements when going from one nucleus to another one. For 2ν the matrix elements change rapidly, but for the 0ν the variation is much more gentle (^{96}Zr is a notable exception, at least for QRPA). That feature, common to most calculations, if verified, would help tremendously in comparing the results or constraints from one nucleus to another one.

Table 10.1. Comparison of the calculated nuclear matrix elements $M^{0\nu}$ using the QRPA method[38] and the NSM.[39] In the last column the NSM values are reduced, divided by 1.3, to account approximately for the effects of the induced nucleon currents.

Nucleus	QRPA	NSM	NSM/1.3
^{76}Ge	2.3-2.4	2.35	1.80
^{82}Se	1.9-2.1	2.26	1.74
^{96}Zr	0.3-0.4		
^{100}Mo	1.1-1.2		
^{116}Cd	1.2-1.4		
^{130}Te	1.3	2.13	1.64
^{136}Xe	0.6-1.0	1.77	1.36

Once the nuclear matrix elements are fixed (by choosing your favorite set of results), they can be combined with the phase space factors (a complete list is available, e.g. in the monograph[42]) to obtain a half-life prediction for any value of the effective mass $\langle m_{\beta\beta} \rangle$. It turns out that for a fixed $\langle m_{\beta\beta} \rangle$ the half-lives of different candidate nuclei do not differ very much from each other (not more than by factors ~ 3 or so) and, for example, the boundary between the degenerate and inverted hierarchy mass regions corresponds to half-lives $\sim 10^{27}$ years. Thus, the next generation of experiments, discussed above, should reach this region using several candidate nuclei, making the corresponding conclusions less nuclear model dependent.

10.5. Neutrino magnetic moment and the distinction between Dirac and Majorana neutrinos

Neutrino mass and magnetic moments are intimately related. In the orthodox Standard Model neutrinos have a vanishing mass and magnetic moments vanish as well. However, in the minimally extended SM containing gauge-singlet right-handed neutrinos the magnetic moment μ_ν is nonvanishing and proportional to the neutrino mass, but unobservably small,[43]

$$\mu_\nu = \frac{3eG_F}{\sqrt{28}\pi^2} m_\nu = 3 \times 10^{-19} \mu_B \frac{m_\nu}{1 \text{ eV}} . \quad (10.19)$$

Here μ_B is the electron Bohr magneton, traditionally used as unit also for the neutrino magnetic moments. An experimental observation of a magnetic moment larger than that given in eq. (10.19) would be an uneqivocal indication of physics beyond the minimally extended Standard Model.

Laboratory searches for neutrino magnetic moments are typically based on the obsevation of the $\nu - e$ scattering. Nonvanishing μ_ν will be recognizable only if the corresponding electromagnetic scattering cross section is at least comparable to the well understood weak interaction cross section. The magnitude of μ_ν (diagonal in flavor or transitional) which can be probed in this way is then given by

$$\frac{|\mu_\nu|}{\mu_B} \equiv \frac{G_F m_e}{\sqrt{2}\pi\alpha}\sqrt{m_e T} \sim 10^{-10}\left(\frac{T}{m_e}\right)^{1/2}, \qquad (10.20)$$

where T is the electron recoil kinetic energy. Considering realistic values of T, it would be difficult to reach sensitivities below $\sim 10^{-11}\mu_B$. Present limits are about an order of magnitude larger than that.

Limits on μ_ν can also be obtained from bounds on the unobserved energy loss in astrophysical objects. For sufficiently large μ_ν the rate of plasmon decay into the $\nu\bar{\nu}$ pairs would conflict with such bounds. Since plasmons can also decay weakly into the $\nu\bar{\nu}$ pairs, the sensitivity of this probe is again limited by the size of the weak rate, leading to

$$\frac{|\mu_\nu|}{\mu_B} \equiv \frac{G_F m_e}{\sqrt{2}\pi\alpha}\hbar\omega_P, \qquad (10.21)$$

where ω_P is the plasmon frequency. Since usually $(\hbar\omega_P)^2 \ll m_e T$ that limit is stronger than that given in eq. (10.20). Current limits on μ_ν based on such considerations are $\sim 10^{-12}\mu_B$.

The interest in μ_ν and its relation to neutrino mass dates from $\sim$1990 when it was suggested that the chlorine data[4] on solar neutrinos show an anticorrelation between the neutrino flux and the solar activity characterized by the number of sunspots. A possible explanation was suggested in Ref. 44 where it was proposed that a magnetic moment $\mu_\nu \sim 10^{-(10-11)}\mu_B$ would cause a precession in solar magnetic field of the neutrinos emitted initially as left-handed ν_e into unobservable right-handed ones. Even though later analyses showed that the correlation with solar acivity does not exist, the possibility of a relatively large μ_ν accompanied by a small mass m_ν was widely discussed and various models accomplishing that were suggested.

If a magnetic moment is generated by physics beyond the Standard Model (SM) at an energy scale Λ, we can generically express its value as

$$\mu_\nu \sim \frac{eG}{\Lambda}, \qquad (10.22)$$

where e is the electric charge and G contains a combination of coupling constants and loop factors. Removing the photon from the diagram gives a contribution to the neutrino mass of order

$$m_\nu \sim G\Lambda. \qquad (10.23)$$

We thus arrive at the relationship

$$m_\nu \sim \frac{\Lambda^2}{2m_e} \frac{\mu_\nu}{\mu_B} \sim \frac{\mu_\nu}{10^{-18}\mu_B}[\Lambda(\text{TeV})]^2 \text{ eV}, \qquad (10.24)$$

which implies that it is difficult to simultaneously reconcile a small neutrino mass and a large magnetic moment.

This naïve restriction given in eq. (10.24) can be overcome via a careful choice for the new physics, e.g., by requiring certain additional symmetries.[45–50] Note, however, that these symmetries are typically broken by Standard Model interactions. For Dirac neutrinos such symmetry (under which the left-handed neutrino and antineutrino ν and ν^c transform as a doublet) is violated by SM gauge interactions. For Majorana neutrinos analogous symmetries are not broken by SM gauge interactions, but are instead violated by SM Yukawa interactions, provided that the charged lepton masses are generated via the standard mechanism through Yukawa couplings to the SM Higgs boson. This suggests that the relation between μ_ν and m_ν is different for Dirac and Majorana neutrinos. This distinction can be, at least in principle, exploited experimentally, as shown below.

Earlier, I have quoted the Ref. 17 (see Fig. 10.4) to stress that observation of the $0\nu\beta\beta$ decay would necessarily imply the existence of a novanishing neutrino Majorana mass. Analogous considerations can be applied in this case. By calculating neutrino magnetic moment contributions to m_ν generated by SM radiative corrections, one may obtain in this way general, "naturalness" upper limits on the size of neutrino magnetic moments by exploiting the experimental upper limits on the neutrino mass.

In the case of Dirac neutrinos, a magnetic moment term will generically induce a radiative correction to the neutrino mass of order[51]

$$m_\nu \sim \frac{\alpha}{16\pi} \frac{\Lambda^2}{m_e} \frac{\mu_\nu}{\mu_B} \sim \frac{\mu_\nu}{3 \times 10^{-15}\mu_B}[\Lambda(\text{TeV})]^2 \text{ eV}. \qquad (10.25)$$

Taking $\Lambda \simeq 1$ TeV and $m_\nu \leq 0.3$ eV, we obtain the limit $\mu_\nu \leq 10^{-15}\mu_B$ (and a more stringent one for larger Λ), which is several orders of magnitude more constraining than current experimental upper limits on μ_ν.

The case of Majorana neutrinos is more subtle, due to the relative flavor symmetries of m_ν and μ_ν respectively. For Majorana neutrinos the transition magnetic moments $[\mu_\nu]_{\alpha\beta}$ (the only possible ones) are antisymmetric in the flavor indices $\{\alpha, \beta\}$, while the mass terms $[m_\nu]_{\alpha\beta}$ are symmetric.

These different flavor symmetries play an important role in the limits, and are the origin of the difference between the magnetic moment constraints for Dirac and Majorana neutrinos.

It has been shown in Ref 52 that the constraints on Majorana neutrinos are significantly weaker than those for Dirac neutrinos,[51] as the different flavor symmetries of m_ν and μ_ν lead to a mass term which is suppressed only by charged lepton masses. This conclusion was reached by considering one-loop mixing of the magnetic moment and mass operators generated by Standard Model interactions. The authors of Ref. 52 found that if a magnetic moment arises through a coupling of the neutrinos to the neutral component of the $SU(2)_L$ gauge boson, the constraints for $\mu_{\tau e}$ and $\mu_{\tau \mu}$ are comparable to present experiment limits, while the constraint on $\mu_{e\mu}$ is significantly weaker. Thus, the analysis of Ref. 52 lead to a bound for the transition magnetic moment of Majorana neutrinos that is less stringent than present experimental limits.

Even more generally it was shown in Ref. 53 that two-loop matching of mass and magnetic moment operators implies stronger constraints than those obtained in[52] if the scale of the new physics $\Lambda \geq 10$ TeV. Moreover, these constraints apply to a magnetic moment generated by either the hypercharge or $SU(2)_L$ gauge boson. In arriving at these conclusions, the most general set of operators that contribute at lowest order to the mass and magnetic moments of Majorana neutrinos was constructed, and model independent constraints which link the two were obtained. Thus the results of Ref. 53 imply completely model independent naturalness bound that – for $\Lambda \geq 100$ TeV – is stronger than present experimental limits (even for the weakest constrained element $\mu_{e\mu}$). On the other hand, for sufficiently low values of the scale Λ the known small values of the neutrino masses do not constrain the magnitude of the transition magnetic moment μ_ν for Majorana neutrinos more than the present experimental limits. Thus, if these conditions are fulfilled, the discovery of μ_ν might be forthcoming any day.

The above result means that an experimental discovery of a magnetic moment near the present limits would signify that (i) neutrinos are Majorana fermions and (ii) new lepton number violating physics responsible for the generation of μ_ν arises at a scale Λ which is well below the see-saw scale. This would have, among other things, implications for the mechanism of the neutrinoless double beta decay and lepton flavor violation as discussed above and in Ref. 22.

10.6. Summary

In these lectures I discussed the status of double beta decay, its relation to the charge conjugation symmetry of neutrinos and to the problem of the lepton number conservation in general. I have shown that if one makes the minimum assumption that the light neutrinos familiar from the oscillation experiments which are interacting by the left-handed weak current are Majorana particles, then the rate of the $0\nu\beta\beta$ decay can be related to the absolute scale of the neutrino mass in a straightforward way.

On the other hand, it is also possible that the $0\nu\beta\beta$ decay is mediated by the exchange of heavy particles. I explained that if the corresponding mass scale of such hypothetical particles is ~ 1 TeV, the corresponding 0ν decay rate could be comparable to the decay rate associated with the exchange of a light neutrino. I further argued that the study of the lepton flavor violation involving $\mu \to e$ conversion and $\mu \to e + \gamma$ decay may be used as a "diagnostic tool" that could help to decide which of the possible mechanisms of the 0ν decay is dominant.

Further, I have shown that the the range of the effective masses $\langle m_{\beta\beta}\rangle$ can be roughly divided into three regions of interest, each corresponding to a different neutrino mass pattern. The region of $\langle m_{\beta\beta}\rangle \geq 0.1$ eV corresponds to the degenerate mass pattern. Its exploration is well advanced, and one can rather confidently expect that it will be explored by several $\beta\beta$ decay experiments in the next 3-5 years. This region of neutrino masses (or most of it) is also accessible to studies using the ordinary β decay and/or the observational cosmology. Thus, if the nature is kind enough to choose this mass pattern, we will have a multiple ways of exploring it.

The region of $0.01 \leq \langle m_{\beta\beta}\rangle \leq 0.1$ eV is often called the "inverted mass hierarchy" region. In fact, both the inverted and the quasi-degenerate but normal mass orderings are possible in this case, and experimentally indistinguishable. Realistic plans to explore this region using the $0\nu\beta\beta$ decay exist, but correspond to a longer time scale of about 10 years. They require much larger, $\sim$ ton size $\beta\beta$ sources and correspondingly even more stringent background suppression.

Finally, the region $\langle m_{\beta\beta}\rangle \leq 0.01$ eV corresponds to the normal hierarchy only. There are no realistic proposals at present to explore this mass region experimentally.

Intimately related to the extraction of $\langle m_{\beta\beta}\rangle$ from the decay rates is the problem of nuclear matrix elements. At present, there is no consensus among the nuclear theorists about their correct values, and the corresponding uncertainty. I argued that the uncertainty is less than some suggest, and that the closeness of the Quasiparticle Random Phase Approximation (QRPA) and Shell Model (NSM) results are encouraging. But this is still a problem that requires further improvements.

In the last part I discussed the neutrino magnetic moments. I have shown that using the Standard Model radiative correction one can calculate the contribution of the magnetic moment to the neutrino mass. That contribution, naturally, should not exceed the experimental upper limit on the neutrino mass. Using this procedure one can show that the magnetic moment of Dirac neutrinos cannot exceed about $10^{-15}\mu_B$, which is several orders of magnitudes less than the current experimental limits on μ_ν. On the other hand, due to the different symmetries of the magnetic moment and mass matrices for Majorana neutrinos, the corresponding constraints are much less restrictive, and do not exceed the current limits. Thus, a discovery of μ_ν near the present experimental limit would indicate that neutrinos are Majorana particles, and the corresponding new physics scale is well below the GUT scale.

Acknowlegment

The original results reported here were obtained in the joint and enjoyable work with a number of collaborators, Nicole Bell, Vincenzo Cirigliano, Steve Elliott, Amand Faessler, Michail Gorchtein, Andriy Kurylov, Gary Prezeau, Michael Ramsey-Musolf, Vadim Rodin, Fedor Šimkovic and Mark Wise. The work was supported in part under U.S. DOE contract DE-FG02-05ER41361.

References

1. T. Kajita and Y. Totsuka, *Rev. Mod. Phys.* **73**, 85–118 (2001); Y. Ashie *et al.*, *Phys. Rev.* **D71**, 112005 (2005).
2. M. H. Ahn, *et al.*, *Phys. Lett. B* **511**, 178–184 (2001); M. H. Ahn *et al.* hep-ex/0606032.
3. MINOS collaboration, hep-ex/0607088.
4. B. T. Cleveland *et al.*, *Astrophys. J.* **496**, 505 (1998).
5. Q. R. Ahmad *et al.*, *Phys. Rev. Lett.* **87**, 071301 (2001).
6. S. Fukuda *et al.*, *Phys. Rev. Lett.* **86**, 5651–5655, (2001); ibid **86**, 5656–5660 (2001).
7. W. Hampel *et al. Phys. Lett.* **B447**, 127 (1999).
8. J. N. Abdurashitov *et al. Phys. Rev.* **C60**, 055801 (1999).
9. K. Eguchi *et al. Phys. Rev. Lett.* **90**, 021802 (2003).

10. T. Araki et al. Phys. Rev. Lett. **94**, 081801 (2005).
11. M. Apollonio et al., Phys. Lett. B **466**, 415–430 (1999).
12. F. Boehm et al., Phys. Rev. **D64**, 112001 (2001).
13. A. Faessler and F. Šimkovic, F. Šimkovic, J. Phys. G **24**, 2139 (1998).
14. J. D. Vergados, Phys. Rep. **361**, 1 (2002).
15. S. R. Elliott and P. Vogel Ann. Rev. Nucl. Part. Sci. **52**, 115 (2002).
16. S. R. Elliott and J. Engel, J. Phys. G **30**, R183 (2004).
17. J. Schechter and J. Valle, Phys. Rev. **D25**, 2951 (1982).
18. R. N. Mohapatra, Phys. Rev. D **34**, 3457 (1986); J. D. Vergados, Phys. Lett. **B184**, 55 (1987); M. Hirsch, H. V. Klapdor-Kleingrothaus and S. G. Kovalenko, Phys. Rev. D **53**, 1329 (1996); M. Hirsch, H. V. Klapdor-Kleingrothaus, and O. Panella, Phys. Lett. **B374**, 7 (1996); A. Fässler, S. Kovalenko, F. Šimkovic and J. Schwieger, Phys. Rev. Lett. **78**, 183 (1997); H. Päs, M. Hirsch, H. V. Klapdor-Kleingrothaus and S. G. Kovalenko, Phys. Lett. **B498**, 35 (2001); F. Šimkovic and A. Fässler, Progr. Part. Nucl. Phys. **48**, 201 (2002).
19. G. Prezeau, M. Ramsey-Musolf and P. Vogel, Phys. Rev. D **68**, 034016 (2003).
20. R. N. Mohapatra, Nucl. Phys. Proc. Suppl. **77**, 376 (1999).
21. M. Doi, T. Kotani and E. Takasugi, Prog. Theor. Phys. Suppl. **83**, 1 (1985).
22. V. Cirigliano, A. Kurylov, M. J. Ramsey-Musolf and P. Vogel, Phys. Rev. Lett **93**, 231802 (2004).
23. G. Signorelli, "The Meg Experiment At Psi: Status And Prospects,"m J. Phys. G **29**, 2027 (2003); see also http://meg.web.psi.ch/docs/index.html.
24. J. L. Popp, NIM **A472**, 354 (2000); hep-ex/0101017.
25. R. Barbieri, L. J. Hall and A. Strumia, Nucl. Phys. B **445**, 219 (1995).
26. F. Borzumati and A. Masiero Phys. Rev. Lett. **57**, 961 (1986).
27. M. Raidal and A. Santamaria, Phys. Lett. B **421**, 250 (1998).
28. R. Kitano, M. Koike and Y. Okada, Phys. Rev. D **66**, 096002 (2002).
29. H. K. Dreiner, in 'Perspectives on Supersymmetry', Ed. by G.L. Kane, World Scientific, 462–479.
30. A. de Gouvea, S. Lola and K. Tobe, Phys. Rev. D **63**, 035004 (2001).
31. R. N. Mohapatra and G. Senjanovic, Phys. Rev. Lett. **44**, 912 (1980).
32. V. Cirigliano, A. Kurylov, M. J. Ramsey-Musolf and P. Vogel, Phys. Rev. **D70**, 075007 (2004).
33. H. V. Klapdor-Kleingrothaus et al. Eur. J. Phys. **A12**, 147 (2001).
34. C. E. Aalseth et al. Phys. Rev. **D65**, 092007 (2002).
35. H. V. Klapdor-Kleingrothaus, A. Dietz, I. V. Krivosheina and). Chvorets, Phys. Lett. **B586**, 198 (2004); Nucl. Inst. Meth **A522**, 371 (2004).
36. C. E. Aalseth et al., hep-ph/0412300.
37. J. N. Bahcall, H. Murayama and C. Pena-Garay, Phys. Rev. **D70**, 033012 (2004).
38. V. A. Rodin, Amand Faessler, F. Šimkovic and Petr Vogel, Phys. Rev. **C68**, 044302(2003); Nucl. Phys. **A766**, 107 (2006).
39. A. Poves, talk at NDM06, http://events.lal.in2p3.fr/conferences/NDM06/.
40. O. Civitarese and J. Suhonen, Nucl. Phys. **A729**, 867 (2003).
41. Aunola M. and Suhonen J., Nucl. Phys. **A643**, 207 (1998).

42. F. Boehm and P. Vogel, *Physics of Massive Neutrinos*, 2nd ed., Cambridge University Press, Cambridge, UK. 1992.
43. W. J. Marciano and A. I. Sanda, *Phys. Lett.* **B67**, 303 (1977); B. W. Lee and R. E. Shrock, *Phys. Rev.* **D16**, 1444 (1977); K. Fujikawa and R. E. Shrock, *Phys. Rev. Lett.* **45**, 963 (1980).
44. M. B. Voloshin, M. I. Vysotskij and L. B. Okun, *Soviet J. of Nucl. Phys.* **44**, 440 (1986).
45. M. B. Voloshin, *Soviet J. of Nucl. Phys.* **48**, 512 (1988).
46. R. Barbieri and R. N. Mohapatra, *Phys. Lett.* **B218**, 225 (1989.
47. H. Georgi and L. Randall, *Phys. Lett.* **B244**, 196 (1990).
48. W. Grimus and H. Neufeld, *Nucl. Phys.* **B351**, 115 (1991).
49. K. S. Babu and R. N. Mohapatra, *Phys. Rev. Lett.* **64**, 1705 (1990).
50. S. M. Barr, E. M. Freie and A. Zee, *Phys. Rev. Lett.* **65**, 2626 (1990).
51. N. F. Bell *et al.*, *Phys. Rev. Lett.* **95**, 151802 (2005).
52. S. Davidson, M. Gorbahn and A. Santamaria, *Phys. Lett.* **B626**, 151 (2005).
53. N. F. Bell *et al.*, Phys. Lett. to be published; hep-ph/0606248.

Chapter 11

Supersymmetry in Elementary Particle Physics

Michael E. Peskin

Stanford Linear Accelerator Center, Stanford University
2575 Sand Hill Road, Menlo Park, California 94025 USA

These lectures give a general introduction to supersymmetry, emphasizing its application to models of elementary particle physics at the 100 GeV energy scale. I discuss the following topics: the construction of supersymmetric Lagrangians with scalars, fermions, and gauge bosons, the structure and mass spectrum of the Minimal Supersymmetric Standard Model (MSSM), the measurement of the parameters of the MSSM at high-energy colliders, and the solutions that the MSSM gives to the problems of electroweak symmetry breaking and dark matter.

11.1. Introduction

11.1.1. Overview

It is an exciting time now in high-energy physics. For many years, ever since the Standard Model was established in the late 1970's, the next logical question in the search for the basic laws of physics has been that of the mechanism by which the weak interaction gauge symmetry is spontaneously broken. This seemed at the time the one important gap that kept the Standard Model from being a complete theory of the strong, weak, and electromagnetic interactions [1–3]. Thirty years later, after many precision experiments at high-energy e^+e^- and hadron colliders, this is still our situation. In the meantime, another important puzzle has been recognized, the fact that 80% of the mass in the universe is composed of 'dark matter', a particle species not included in the Standard Model. Both problems are likely to be solved by new fundamental interactions operating in the energy range of a few hundred GeV. Up to now, there is no evidence from particle physics for such new interactions. But, in the next few years, this situation should change dramatically. Beginning in 2008, the CERN Large Hadron Collider (LHC) should give us access to physics at energies well above 1 TeV and thus should probe the energy region responsible for electroweak symmetry breaking. Over a longer term, we can look forward to precision experiments in e^+e^- annihilation in this same energy region at the proposed International Linear Collider (ILC).

Given this expectation, it is important for all students of elementary particle physics to form concrete ideas of what new phenomena we might find as we explore this new energy region. Of course, we have no way of knowing exactly what we will find there. But this makes it all the more important to study the alternative theories that have been put forward and to understand their problems and virtues.

Many different models of new physics relevant to electroweak symmetry breaking are being discussed at this TASI school. Among these, supersymmetry has pride of place. Supersymmetry (or SUSY) provides an explicit realization of all of the aspects of new physics expected in the hundred GeV energy region. Because SUSY requires only weak interactions to build a realistic theory, it is possible in a model with SUSY to carry out explicit calculations and find the answers that the model gives to all relevant phenomenological questions.

In these lectures, I will give an introduction to supersymmetry as a context for building models of new physics associated with electroweak symmetry breaking. Here is an outline of the material: In Section 2, I will develop appropriate notation and then construct supersymmetric Lagrangians for scalar, spinor, and vector fields. In Section 3, I will define the canonical phenomenological model of supersymmetry, the Minimal Supersymmetric Standard Model (MSSM). I will discuss the quantum numbers of new particles in the MSSM and the connection of the MSSM to the idea of grand unification.

The remaining sections of these lectures will map out the phenomenology of the new particles and interactions expected in models of supersymmetry. I caution you that I will draw only those parts of the map that cover the simplest and most well-studied class of models. Supersymmetry has an enormous parameter space which contains many different scenarios for particle physics, more than I have room to cover here. I will at least try to indicate the possible branches in the path and give references that will help you follow some of the alternative routes.

With this restriction, the remaining sections will proceed as follows: In Section 4, I will compute the mass spectrum of the MSSM from its parameters. I will also discuss the parameters of the MSSM that characterize supersymmetry breaking. In Section 5, I will describe how the MSSM parameters will be measured at the LHC and the ILC. Finally, Section 6 will discuss the answers that supersymmetry gives to the two major questions posed at the beginning of this discussion, the origin of electroweak symmetry breaking, and the origin of cosmic dark matter.

Although I hope that these lectures will be useful to students in studying supersymmetry, there are many other excellent treatments of the subject available. A highly recommended introduction to SUSY is the 'Supersymmetry Primer' by Steve Martin [6]. An excellent presentation of the formalism of supersymmetry is given in the texbook of Wess and Bagger [7]. Supersymmetry has been reviewed at previous TASI schools by Bagger [8], Lykken [9], and Kane [10], among others. Very recently, three textbooks of phenomenological supersymmetry have appeared, by Drees, Godbole, and Roy [11], Binetruy [12], and Baer and Tata [13]. A fourth textbook, by Dreiner, Haber, and Martin [14], is expected soon.

It would be wonderful if all of these articles and books used the same conventions, but that is too much to expect. In these lectures, I will use my own, somewhat ideosyncratic conventions. These are explained in Section 2.1. Where possible, within the philosophy of that section, I have chosen conventions that agree with those of Martin's primer [6].

11.1.2. Motivation and Structure of Supersymmetry

If we propose supersymmetry as a model of electroweak symmetry breaking, we might begin by asking: What is the problem of electroweak symmetry breaking, and what are the alternatives for solving it?

Electroweak symmetry is spontaneously broken in the minimal form of the Standard Model, which I will refer to as the MSM. However, the explanation that the MSM gives for this phenomenon is not satisfactory. The sole source of symmetry breaking is a single elementary Higgs boson field. All mass of quarks, leptons, and gauge bosons arise from the couplings of those particles to the Higgs field.

To generate symmetry breaking, we postulate a potential for the Higgs field

$$V = \mu^2 |\varphi|^2 + \lambda |\varphi|^4 , \quad (11.1)$$

shown in Fig. 11.1. The assumption that $\mu^2 < 0$ is the complete explanation for electroweak symmetry breaking in the MSM. Since μ is a renormalizable coupling of this theory, the value of μ cannot be computed from first principles, and even its sign cannot be predicted.

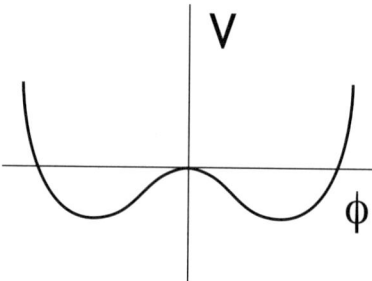

Fig. 11.1. The Standard Model Higgs potential (11.1).

In fact, this explanation has an even worse feature. The parameter μ^2 receives large additive radiative corrections from loop diagrams. For example, the two diagrams shown in Fig. 11.2 are ultraviolet divergent. Supplying a momentum cutoff Λ, the two diagrams contribute

$$\mu^2 = \mu^2_{\text{bare}} + \frac{\lambda}{8\pi^2}\Lambda^2 - \frac{3y_t^2}{8\pi^2}\Lambda^2 + \cdots \qquad (11.2)$$

If we view the MSM as an effective theory, Λ should be taken to be the largest momentum scale at which this theory is still valid. The presence of large additive corrections implies that the criterion $\mu^2 < 0$ is not a simple condition on the underlying parameters of the effective theory. The radiative corrections can easily change the sign of μ^2. Further, if we insist that the MSM has a large range of validity, the corrections become much larger than the desired result. To obtain the Higgs field vacuum expectation value required for the weak interactions, $|\mu|$ should be about 100 GeV. If we insist at the same time that the MSM is valid up to the Planck scale, $\Lambda \sim 10^{19}$ GeV, the formula (11.2) requires a cancellation between the bare value of μ and the radiative corrections in the first 36 decimal places. This problem has its own name, the 'gauge hierarchy problem'. But, to my mind, the absence of a logical explantion for electroweak symmetry breaking in the MSM is already problem enough.

How could we solve this problem? There are two different strategies. One is to look for new strong-couplings dynamics at an energy scale of 1 TeV or below. Then the Higgs field could be composite and its potential could be the result, for example, of pair condensation of fermion constituents. Higgs

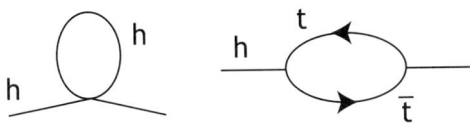

Fig. 11.2. Two Standard Model diagrams that give divergent corrections to the Higgs mass parameter μ^2.

actually proposed that his field was a phenomenological description of a fermion pair condensation mechanism similar to that in superconductivity [4]. Sometime later, Susskind [2] and Weinberg [3] proposed an explicit model of electroweak symmetry breaking by new strong interactions, called 'technicolor'.

Today, this approach is disfavored. Technicolor typically leads to flavor-changing neutral currents at an observable level, and also typically conflicts with the accurate agreement of precision electroweak theory with experiment. Specific models do evade these difficulties, but they are highly constrained [5].

The alternative is to postulate that the electroweak symmetry is broken by a weakly-coupled Higgs field, but that this field is part of a model in which the Higgs potential is computable. In particular, the Higgs mass term $\mu^2|\varphi|^2$ should be generated by well-defined physics within the model. A prerequisite for this is that the μ^2 term not receive additive radiative corrections. This requires that, at high energy, the appearance of a nonzero μ^2 in the Lagrangian should be forbidden by a symmetry of the theory.

There are three ways to arrange a symmetry that forbids the term $\mu^2|\varphi|^2$. We can postulate a symmetry that shifts φ

$$\delta\varphi = \epsilon v . \tag{11.3}$$

We can postulate a symmetry that connects φ to a gauge field, whose mass can then forbidden by gauge symmetry

$$\delta\varphi = \epsilon \cdot A . \tag{11.4}$$

We can postulate a symmetry that connects φ to a fermion field, whose mass can then be forbidden by a chiral symmetry.

$$\delta\varphi = \epsilon \cdot \psi . \tag{11.5}$$

The options (11.3) and (11.4) lead, respectively, to 'little Higgs' models [15–17] and to models with extra space dimensions [18,19]. The third option leads to supersymmetry. This is the route we will now follow.

The symmetry (11.5) looks quite innocent, but it is not. In quantum theory, a symmetry that links a boson with a fermion is generated by a conserved charge Q_α that carries spin-1/2

$$[Q_\alpha, \varphi] = \psi_\alpha \ , \qquad [Q_\alpha, H] = 0 \ . \tag{11.6}$$

Such a Q_α implies the existence of a conserved 4-vector charge R_m defined by

$$\{Q_\alpha, Q_\beta^\dagger\} = 2\gamma_{\alpha\beta}^m R_m \tag{11.7}$$

(It may not be obvious to you that there is no Lorentz scalar component in this anticommutator, but I will show this in Section 2.1.) The charge R_m is conserved, because both Q and $Q^\dagger$ commute with H. It is nonzero, as we can see by taking the expectation value of (11.7) in any state and setting $\alpha = \beta$

$$\langle A| \{Q_\alpha, Q_\alpha^\dagger\} |A\rangle = \langle A| Q_\alpha Q_\alpha^\dagger |A\rangle + \langle A| Q_\alpha Q_\alpha^\dagger |A\rangle$$
$$= \| Q_\alpha |A\rangle \|^2 + \| Q_\alpha^\dagger |A\rangle \|^2 \ . \tag{11.8}$$

This expression is non-negative; it can be zero only if Q_α and $Q_\alpha^\dagger$ annihilate every state in the theory.

However, in a relativistic quantum field theory, we do not have the freedom to introduce arbitrary charges that have nontrivial Lorentz transformation properties. Conservation of energy-momentum and angular momentum are already very constraining. For example, in two-body scattering, the scattering amplitude for fixed center of mass energy can only be a function of one variable, the center of mass scattering angle θ. If one adds a second conserved 4-vector charge, almost all values of θ will also be forbidden. Coleman and Mandula proved a stronger version of this statement: In a theory with an addtional conserved 4-vector charge, there can be no scattering at all, and so the theory is trivial [20].

If we would like to have (11.5) as an exact symmetry, then, the only possibility is to set $R_m = P_m$. That is, the square of the fermionic charge Q_α must be *the total energy-momentum of everything*. We started out trying to build a theory in which the fermionic charge acted only on the Higgs field. But now, it seems, the fermionic charge must act on every field in the theory. Everything—quarks, leptons, gauge bosons, even gravitons— must have partners under the symmetry generated by Q_α. Q_α is fermionic and carries spin $\frac{1}{2}$. Then every particle in the theory must have a partner with the opposite statistics and spin differing by $\frac{1}{2}$ unit.

The idea that the transformation (11.5) leads to a profound generalization of space-time symmetry was discovered independently several times in the early 1970's [22,23]. The 1974 paper by Wess and Zumino [24] which gave simple linear realizations of this algebra on multiplets of fields launched the detailed exploration of this symmetry and its application to particle physics.

The pursuit of (11.5) then necessarily leads us to introduce a very large number of new particles. This seems quite daunting. It might be a reason to choose one of the other paths, except that these also lead to new physics models of similarly high complexity. I encourage you to carry on with this line of analysis a bit longer. It will lead to a beautiful structure with many interesting implications for the theory of Nature.

11.2. Formalism of Supersymmetry

11.2.1. Fermions in 4 Dimensions

To work out the full consequences of (11.5), we will need to write this equation more precisely. To do this, we need to set up a formalism that describes relativistic fermions in four dimensions in the most general way. There is no general agreement on the best conventions to use, but every discussion of supersymmetry leans heavily on the particular choices made. I will give my choice of conventions in this section.

There are two basic spin-$\frac{1}{2}$ representations of the Lorentz group. Each is two-dimensional. The transformation laws are those of left- and right-handed Weyl (2-component) fermions,

$$\psi_L \to (1 - i\vec{\alpha} \cdot \vec{\sigma}/2 - \vec{\beta} \cdot \vec{\sigma}/2) \psi_L$$
$$\psi_R \to (1 - i\vec{\alpha} \cdot \vec{\sigma}/2 + \vec{\beta} \cdot \vec{\sigma}/2) \psi_R \,, \quad (11.9)$$

where $\vec{\alpha}$ is an infinitesimal rotation angle and $\vec{\beta}$ is an infinitesimal boost. The four-component spinor built from these ingredients, $\Psi = (\psi_L, \psi_R)$, is a Dirac fermion.

Define the matrix

$$c = -i\sigma^2 = \begin{pmatrix} 0 & -1 \\ 1 & 0 \end{pmatrix}. \quad (11.10)$$

This useful matrix satisfies $c^2 = -1$, $c^T = -c$. The combination

$$\psi_{1L}^T c \psi_{2L} = -\epsilon_{\alpha\beta} \psi_{1L\alpha} \psi_{2L\beta} \quad (11.11)$$

is the basic Lorentz invariant product of spinors. Many treatments of supersymmetry, for example, that in Wess and Bagger's book [7], represent c implicitly by raising and lowering of spinor indices. I will stick to this more prosaic approach.

Using the identity $\vec{\sigma}c = -c(\vec{\sigma})^T$, it is easy to show that the quantity $(-c\psi_L^*)$ transforms like ψ_R. So if we wish, we can replace every ψ_R by a ψ_L and write all fermions in the theory as left-handed Weyl fermions. With this notation, for example, we would call e_L^- and e_L^+ fermions and e_R^- and e_R^+ antifermions. This convention does not respect parity, but parity is not a symmetry of the Standard Model. The convention of representing all fermions in terms of left-handed Weyl fermions turns out to be very useful for not only for supersymmetry but also for other theories of physics beyond the Standard Model.

Applying this convention, a Dirac fermion takes the form

$$\Psi = \begin{pmatrix} \psi_{1L} \\ -c\psi_{2L}^* \end{pmatrix} \qquad (11.12)$$

Write the Dirac matrices in terms of 2×2 matrices as

$$\gamma^m = \begin{pmatrix} 0 & \sigma^m \\ \overline{\sigma}^m & 0 \end{pmatrix} \qquad (11.13)$$

with

$$\sigma^m = (1, \vec{\sigma})^m \qquad \overline{\sigma}^m = (1, -\vec{\sigma})^m \qquad c\sigma^m = (\overline{\sigma}^m)^T c \qquad (11.14)$$

Then the Dirac Lagrangian can be rewritten in the form

$$\begin{aligned} \mathcal{L} &= \overline{\Psi} i\gamma \cdot \partial \Psi - M\overline{\Psi}\Psi \\ &= \psi_{1L}^\dagger i\overline{\sigma} \cdot \partial \psi_{1L} + \psi_{2L}^\dagger i\overline{\sigma} \cdot \partial \psi_{2L} \\ &\quad - (m\psi_{1L}^T c \psi_{2L} - m^* \psi_{1L}^\dagger c \psi_{2L}^*) \,. \end{aligned} \qquad (11.15)$$

For the bilinears in the last line, we can use fermion anticommutation and the antisymmetry of c to show

$$\psi_{1L}^T c \psi_{2L} = +\psi_{2L}^T c \psi_{1L} \,. \qquad (11.16)$$

and, similarly,

$$(\psi_{1L}^T c \psi_{2L})^\dagger = \psi_{2L}^\dagger (-c) \psi_{1L}^* = -\psi_{1L}^\dagger c \psi_{2L}^* \,. \qquad (11.17)$$

The mass term looks odd, because it is fermion number violating. However, the definition of fermion number is that given in the previous paragraph. The fields ψ_{1L} and ψ_{2L} annihilate, respectively, e_L^- and e_L^+. So this mass term generates the conversion of e_L^- to e_R^-, which is precisely what we would expect a mass term to do.

If we write all fermions as left-handed Weyl fermions, the possibilities for fermion actions are highly restricted. The most general Lorentz-invariant free field Lagrangian takes the form

$$\mathcal{L} = \psi_k^\dagger i\overline{\sigma} \cdot \partial \psi_k - \frac{1}{2}(m_{jk}\psi_j^T c\psi_k - m_{jk}^*\psi_j^\dagger c\psi_k^*) \ . \qquad (11.18)$$

where j, k index the fermion fields. Here and in the rest of these lectures, I drop the subscript L. The matrix m_{jk} is a complex symmetric matrix. For a Dirac fermion,

$$m_{jk} = \begin{pmatrix} 0 & m \\ m & 0 \end{pmatrix}_{jk} \qquad (11.19)$$

as we have seen in (11.15). This matrix respects the charge

$$Q\psi_1 = +\psi_1 \ , \qquad Q\psi_2 = -\psi_2 \ , \qquad (11.20)$$

which is equivalent to the original Dirac fermion number. A Majorana fermion is described in the same formalism by the mass matrix

$$m_{jk} = m\delta_{jk} \ . \qquad (11.21)$$

The most general fermion mass is a mixture of Dirac and Majorana terms. We will meet such fermion masses in our study of supersymmetry. These more general mass matrices also occur in other new physics models and in models of the masses of neutrinos.

The SUSY charges are four-dimensional fermions. The minimum set of SUSY charges thus includes one Weyl fermion Q_α and its Hermitian conjugate $Q_\alpha^\dagger$. We can now analyze the anticommutator $\{Q_\alpha, Q_\beta^\dagger\}$. Since the indices belong to different Lorentz representations, this object does not contain a scalar. The indices transform as do the spinor indices of σ^m, and so we can rewrite (11.7) with $R^m = P^m$ as

$$\{Q_\alpha, Q_\beta^\dagger\} = 2\sigma_{\alpha\beta}^m P_m \ . \qquad (11.22)$$

It is possible to construct quantum field theories with larger supersymmetry algebras. These must include (11.22), and so the general form is [21]

$$\{Q_\alpha^i, Q_\beta^{\dagger j}\} = 2\sigma_{\alpha\beta}^m P_m \delta^{ij} \ , \qquad (11.23)$$

for $i, j = 1 \ldots N$. This relation can be supplemented by a nontrivial anticommutator

$$\{Q_\alpha^i, Q_\beta^j\} = 2\epsilon_{\alpha\beta}\mathcal{Q}^{ij} \qquad (11.24)$$

where the *central charge* Q^{ij} is antisymmetric in $[ij]$. Theories with $N > 4$ necessarily contain particles of spin greater than 1. Yang-Mills theory with $N = 4$ supersymmetry is an especially beautiful model with exact scale invariance and many other attractive formal properties [25]. In these lectures, however, I will restrict myself to the minimal case of $N = 1$ supersymmetry.

I will discuss supersymmetry transformations using the operation on fields

$$\delta_\xi \Phi = [\xi^T c Q + Q^\dagger c \xi^*, \Phi] \,. \tag{11.25}$$

Note that the operator δ_ξ contains pairs of anticommuting objects and so obeys commutation rather than anticommutation relations. The operator P_m acts on fields as the generator of translations, $P_m = i\partial_m$. Using this, we can rewrite (11.22) as

$$[\delta_\xi, \delta_\eta] = 2i \left(\xi^\dagger \overline{\sigma}^m \eta - \eta^\dagger \overline{\sigma}^m \xi \right) \partial_m \tag{11.26}$$

I will take this equation as the basic (anti-)commutation relation of supersymmetry. In the next two sections, I will construct some representations of this commutation relation on multiplets of fields.

11.2.2. Supersymmetric Lagrangians with Scalars and Fermions

The simplest representation of the supersymmetry algebra (11.26) directly generalizes the transformation (11.5) from which we derived the idea of supersymmetry. The full set of fields required includes a complex-valued boson field ϕ and a Weyl fermion field ψ. These fields create and destroy a scalar particle and its antiparticle, a left-handed massless fermion, and its right-handed antiparticle. Note that the particle content has an equal number of fermions and bosons. This particle content is called a *chiral supermultiplet*.

I will now write out the transformation laws for the fields corresponding to a chiral supermultiplet. It is convenient to add a second complex-valued boson field F that will have no associated particles. Such a field is called an *auxiliary field*. We can then write the transformations that generalize (11.5) as

$$\begin{aligned} \delta_\xi \phi &= \sqrt{2} \xi^T c \psi \\ \delta_\xi \psi &= \sqrt{2} i \sigma^n c \xi^* \partial_n \phi + \sqrt{2} F \xi \\ \delta_\xi F &= -\sqrt{2} i \xi^\dagger \overline{\sigma}^m \partial_m \psi \,. \end{aligned} \tag{11.27}$$

The conjugates of these transformations are

$$\delta_\xi \phi^* = -\sqrt{2}\psi^\dagger c\xi^*$$
$$\delta_\xi \psi^\dagger = \sqrt{2}i\xi^T c\sigma^n \partial_n \phi^* + \sqrt{2}\xi^\dagger F^*$$
$$\delta_\xi F^* = \sqrt{2}i\partial_m \psi^\dagger \overline{\sigma}^m \xi \ . \tag{11.28}$$

These latter transformations define the *antichiral supermultiplet*. I claim that the transformations (11.27) and (11.28), first, satisfy the fundamental commutation relation (11.26) and, second, leave a suitable Lagrangian invariant. Both properties are necessary, and both must be checked, in order for a set of transformations to generate a symmetry group of a field theory.

The transformation laws (11.27) seem complicated. You might wonder if there is a formalism that generates these relations automatically and manipulates them more easily than working with the three distinct component fields (ϕ, ψ, F). In the next section, I will introduce a formalism called *superspace* that makes it almost automatic to work with the chiral supermultiplet. However, the superspace description of the multiplet containing gauge fields is more complicated, and the difficulty of working with superspace becomes exponentially greater in theories that include gravity, higher dimensions, or $N > 1$ supersymmetry. At some stage, one must go back to components. I strongly recommend that you gain experience by working through the component field calculations described in these notes in full detail, however many large pieces of paper that might require.

To verify each of the two claims I have made for (11.27) requires a little calculation. Here is the check of the commutation relation applied to the field ϕ:

$$\begin{aligned}[\delta_\xi, \delta_\eta]\phi &= \delta_\xi(\sqrt{2}\eta^T c\psi) - (\xi \leftrightarrow \eta) \\ &= \sqrt{2}\eta^T c(\sqrt{2}i\sigma^n c\xi^* \partial_n \phi) - (\xi \leftrightarrow \eta) \\ &= -2i\eta^T (\overline{\sigma}^n)^T \xi^* \partial_n \phi - (\xi \leftrightarrow \eta) \\ &= 2i[\xi^\dagger \overline{\sigma}^n \eta - \eta^\dagger \overline{\sigma}^n \xi]\partial_n \phi \end{aligned} \tag{11.29}$$

The check of the commutation relation applied to F is equally straightforward. The check on ψ is a bit lengthier. It requires a *Fierz identity*, that is, a spinor index rearrangement identity. Specifically, we need

$$\eta_\alpha \xi_\beta^\dagger = -\frac{1}{2}(\xi^\dagger \overline{\sigma}_m \eta)\sigma^m_{\alpha\beta} \ , \tag{11.30}$$

which you can derive by writing out the four components explicitly. After some algebra that involves the use of this identity, you can see that the SUSY commutation relation applied to ψ also takes the correct form.

Next, I claim that the Lagrangian

$$\mathcal{L} = \partial^m \phi^* \partial_m \phi + \psi^\dagger i\overline{\sigma} \cdot \partial \psi + F^* F \tag{11.31}$$

is invariant to the transformation (11.27). I will assume that the Lagrangian (11.31) is integrated $\int d^4x$ and use integration by parts freely. Then

$$\begin{aligned}
\delta_\xi \mathcal{L} &= \partial^m \phi^* \partial_m (\sqrt{2}\xi^T c\psi) + (-\sqrt{2}\partial^m \psi^\dagger c\xi^*)\partial \phi \\
&\quad + \psi^\dagger i\overline{\sigma} \cdot \partial [\sqrt{2}i\sigma^n c\xi^* \partial_m \phi + \sqrt{2}\xi F] \\
&\quad + [\sqrt{2}i\partial_n \phi^* \xi^T c\sigma^n + \sqrt{2}\xi^\dagger F^*]i\overline{\sigma} \cdot \partial \psi \\
&\quad + F^*[-\sqrt{2}i\xi^\dagger \overline{\sigma}^m \partial_m \psi] + [\sqrt{2}i\partial_m \psi^\dagger \overline{\sigma}^m \xi] F \\
&= -\phi^* \sqrt{2}\xi^T c\partial^2 \psi + \sqrt{2}\partial_n \phi^* \xi^T c\sigma^n \overline{\sigma}^m \partial_n \partial_m \psi \\
&\quad + \sqrt{2}\psi^\dagger c\xi^* \partial^2 \phi - \sqrt{2}\psi^\dagger \overline{\sigma}^m \sigma^n c\xi^* \partial_m \partial_n \phi \\
&\quad + \sqrt{2}i\psi^\dagger \overline{\sigma}^m \partial_m F\xi + \sqrt{2}i\partial_m \psi^\dagger \overline{\sigma}^m \xi F \\
&\quad - \sqrt{2}i\xi^\dagger F^* \overline{\sigma}^m \partial_m \psi + \sqrt{2}i F^* \xi^\dagger \overline{\sigma}^m \partial_m \psi \\
&= 0 \; . \tag{11.32}
\end{aligned}$$

In the final expression, the four lines cancel line by line. In the first two lines, the cancellation is made by using the identity $(\overline{\sigma} \cdot \partial)(\sigma \cdot \partial) = \partial^2$.

So far, our supersymmetry Lagrangian is just a massless free field theory. However, it is possible to add rather general interactions that respect the symmetry. Let $W(\phi)$ be an analytic function of ϕ, that is, a function that depends on ϕ but not on ϕ^*. Let

$$\mathcal{L}_W = F \frac{\partial W}{\partial \phi} - \frac{1}{2}\psi^T c\psi \frac{\partial^2 W}{\partial \phi^2} \tag{11.33}$$

I claim that $\mathcal{L}_W$ is invariant to (11.27). Then we can add $(\mathcal{L}_W + \mathcal{L}_W^\dagger)$ to the free field Lagrangian to introduce interactions into the theory. The function W is called the *superpotential*.

We can readily check that $\mathcal{L}_W$ is indeed invariant:

$$\begin{aligned}
\delta_\xi \mathcal{L}_W &= F \frac{\partial^2 W}{\partial \phi^2} (\sqrt{2}\xi^T c\psi) - \sqrt{2} F\xi^T c\psi \frac{\partial^2 W}{\partial \phi^2} \\
&\quad - \sqrt{2}i\xi^\dagger \overline{\sigma}^m \partial_m \psi \frac{\partial W}{\partial \phi} - \psi^T c\sqrt{2}i\sigma^n c\xi^* \partial_n \phi \frac{\partial^2 W}{\partial \phi^2} \\
&\quad - \psi^T c\psi \frac{\partial^3 W}{\partial \phi^3} \sqrt{2}\xi^T c\psi \; . \tag{11.34}
\end{aligned}$$

The second line rearranges to

$$-\sqrt{2}i\xi^\dagger \overline{\sigma}\left(\partial_n\psi \frac{\partial W}{\partial \phi} + \psi\partial_n\phi\frac{\partial^2 W}{\partial \phi^2}\right), \tag{11.35}$$

which is a total derivative. The third line is proportional to $\psi_\alpha\psi_\beta\psi_\gamma$, which vanishes by fermion antisymmetry since the spinor indices take only two values. Thus it is true that

$$\delta_\xi \mathcal{L}_W = 0. \tag{11.36}$$

The proofs of invariance that I have just given generalize straightforwardly to systems of several chiral supermultiplets. The requirement on the superpotential is that it should be an analytic function of the complex scalar fields ϕ_k. Then the following Lagrangian is supersymmetric:

$$\mathcal{L} = \partial^m \phi_k^* \partial_m \phi_k + \psi_k^\dagger i\overline{\sigma}\cdot\partial\psi_k + F_k^* F_k + \mathcal{L}_W + \mathcal{L}_W^\dagger, \tag{11.37}$$

where

$$\mathcal{L}_W = F_k \frac{\partial W}{\partial \phi_k} - \frac{1}{2}\psi_j^T c\psi_k \frac{\partial^2 W}{\partial \phi_j \partial \phi_k}. \tag{11.38}$$

In this Lagrangian, the fields F_k are Lagrange multipliers. They obey the constraint equations

$$F_k^* = -\frac{\partial W}{\partial \phi_k}. \tag{11.39}$$

Using these equations to eliminate the F_k, we find an interacting theory with the fields ϕ_k and ψ_k, a Yukawa coupling term proportional to the second derivative of W, as given in (11.38), and the potential energy

$$V_F = \sum_k \left|\frac{\partial W}{\partial \phi_k}\right|^2. \tag{11.40}$$

I will refer to V_F as the *F-term potential*. Later we will meet a second contribution V_D, the *D-term potential*. These two terms, both obtained by integrating out auxiliary fields, make up the classical potential energy of a general supersymmetric field theory of scalar, fermion, and gauge fields.

The simplest example of the F-term potential appears in the theory with one chiral supermultiplet and the superpotential $W = \frac{1}{2}m\phi^2$. The constraint equation for F is [27]

$$F^* = -m\phi. \tag{11.41}$$

After eliminating F, we find the Lagrangian

$$\mathcal{L} = \partial^n \phi^* \partial_n \phi - |m|^2 \phi^* \phi + \psi^\dagger i\overline{\sigma}\cdot\partial\psi - \frac{1}{2}(m\psi^T c\psi - m^* \psi^\dagger c\psi^*) \qquad (11.42)$$

This is a theory of two free scalar bosons of mass $|m|$ and a free Majorana fermion with the same mass $|m|$. The Majorana fermion has two spin states, so the number of boson and fermion physical states is equal, as required.

The form of the expression (11.40) implies that $V_F \geq 0$, and that $V_F = 0$ only if all $F_k = 0$. This constraint on the potential energy follows from a deeper consideration about supersymmetry. Go back to the anticommutation relation (11.22), evaluate it for $\alpha = \beta$, and take the vacuum expectation value. This gives

$$\langle 0|\{Q_\alpha, Q_\alpha^\dagger\}|0\rangle = \langle 0|(H-P^3)|0\rangle = \langle 0|H|0\rangle \;, \qquad (11.43)$$

since the vacuum expectation value of P^3 vanishes by rotational invariance. Below (11.7), I argued that the left-hand side of this equation is greater than or equal to zero. It is equal to zero if and only if

$$Q_\alpha |0\rangle = Q_\alpha^\dagger |0\rangle = 0 \qquad (11.44)$$

The formulae (11.44) give the criterion than the vacuum is invariant under supersymmetry. If this relation is not obeyed, supersymmetry is spontaneously broken. Taking the vacuum expectation value of the transformation law for the chiral representation, we find

$$\langle 0|[\xi^T cQ + Q^\dagger c\xi^*, \psi_k]|0\rangle = \langle 0|\sqrt{2}i\sigma^n \xi^* \partial_n \phi_k + \xi F_k |0\rangle$$
$$= \xi \langle 0|F_k|0\rangle \;. \qquad (11.45)$$

In the last line I have used the fact that the vacuum expectation value of $\phi(x)$ is translation invariant, so its derivative vanishes. The left-hand side of (11.45) vanishes if the vacuum state is invariant under supersymmetry.

The results of the previous paragraph can be summarized in the following way: If supersymmetry is a manifest symmetry of a quantum field theory,

$$\langle 0|H|0\rangle = 0 \;, \text{ and } \langle 0|F_k|0\rangle = 0 \qquad (11.46)$$

for every F field of a chiral multiplet. In complete generality,

$$\langle 0|H|0\rangle \geq 0 \;. \qquad (11.47)$$

The case where $\langle H \rangle$ is positive and nonzero corresponds to spontaneously broken supersymmetry. If the theory has a state satisfying (11.44), this is necesssarily the state in the theory with lowest energy. Thus, supersymmetry can be spontaneously broken only if a supersymmetric vacuum state does not exist[a]

For the moment, we will work with theories that preserve supersymmetry. I will give examples of theories with spontaneous supersymmetry breaking in Section 3.5.

The results we have just derived are exact consequences of the commutation relations of supersymmetry. It must then be true that the vacuum energy of a supersymmetric theory must vanish in perturbation theory. This is already nontrivial for the free theory (11.42). But it is correct. The positive zero point energy of the boson field exactly cancels the negative zero point energy of the fermion field. With some effort, one can show the cancellation also for the leading-order diagrams in an interacting theory. Zumino proved that this cancellation is completely general [29].

I would like to show you another type of cancellation that is also seen in perturbation theory in models with chiral fields. Consider the model with one chiral field and superpotential

$$W = \frac{\lambda}{3}\phi^3 . \tag{11.48}$$

After eliminating F, the Lagrangian becomes

$$\mathcal{L} = \partial \phi^* \partial_m \phi + \psi^\dagger i \overline{\sigma} \cdot \partial \psi - \lambda(\phi \psi^T c \psi - \phi^* \psi^\dagger c \psi^*) - \lambda^2 |\phi|^4 . \tag{11.49}$$

The vertices of this theory are shown in Fig. 11.3(a).

From our experience in (11.2), we might expect to find an adddtive radiative correction to the scalar mass. The corrections to the fermion and scalar mass terms are given by the diagrams in Fig. 11.3(b). Actually, there are no diagrams that correct the fermion mass; you can check that there it is not possible to match the arrows appropriately. For the scalar mass correction, the two diagrams shown contribute

$$-4i\lambda^2 \int \frac{d^4p}{(2\pi)^4} \frac{i}{p^2} + \frac{1}{2}(-2i\lambda)(+2i\lambda) \int \frac{d^4p}{(2\pi)^4} \text{tr}\left[\frac{i\sigma \cdot p}{p^2} c \frac{i\sigma^T \cdot (-p)}{p^2} c\right] \tag{11.50}$$

Using $\sigma \cdot p \overline{\sigma} \cdot p = p^2$ in the second term and then taking the trace, we see that these two contributions cancel precisely. In this way, supersymmetry really does control radiative corrections to the Higgs mass, following the logic that we presented in Section 1.2.

---
[a]It is possible that a supersymmetric vacuum state might exist but that a higher-energy vacuum state might be metastable. A model built on this metastable state would show spontaneous breaking of supersymmetry [26].

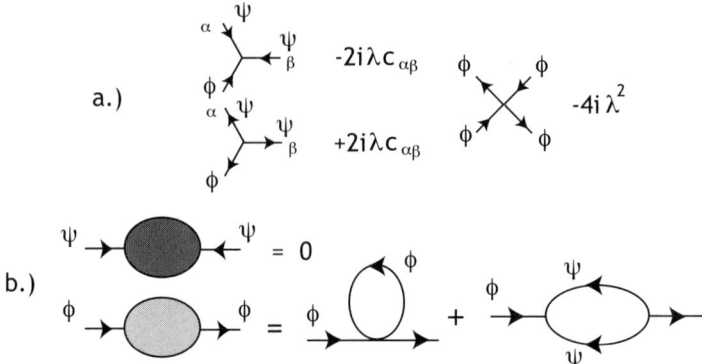

Fig. 11.3. Perturbation theory for the supersymmetric model (11.49): (a) vertices of the model; (b) corrections to the fermion and scalar masses.

In fact, it can be shown quite generally that not only the mass term but the whole superpotential W receives no additive radiative corrections in any order of perturbation theory [30]. For example, the one-loop corrections to quartic terms in the Lagrangian cancel in a simple way that is indicated in Fig. 11.4. The field strength renormalization of chiral fields can be nonzero, so the form of W can be changed by radiative corrections by the rescaling of fields. Examples are known in which W receives additive radiative corrections from nonperturbative effects [31].

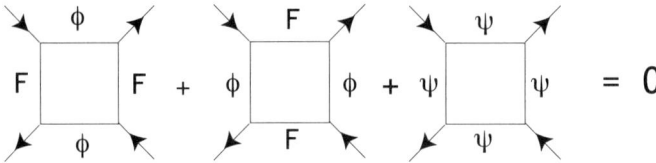

Fig. 11.4. Scheme of cancellations of one-loop corrections to the F-term potential.

11.2.3. Superspace

Because the commutation relations of supersymmetry include the generators of translations, supersymmetry is a space-time symmetry. It is an attractive idea that supersymmetry is the natural set of translations on a generalized space-time with commuting and anticommuting coordinates. In this section, I will introduce the appropriate generalization of space-time and use it to re-derive some of the results of Section 2.2.

Consider, then, a space with four ordinary space-time coordinates x^μ and four anticommuting coordinates $(\theta_\alpha, \bar\theta_\alpha)$. I will take the coordinates θ_α to transform as 2-component Weyl spinors; the $\bar\theta_\alpha$ are the complex conjugates of the θ_α. This is *superspace*. A *superfield* is a function of these superspace coordinates: $\Phi(x, \theta, \bar\theta)$.

It is tempting to define supersymmetry transformations as translations $\theta \to \theta + \xi$. However, this does not work. These transformations commute, $[\delta_\xi, \delta_\eta] = 0$, and we have seen in Section 1.2 that this implies that the S-matrix of the resulting field theory must be trivial. To construct a set of transformations with the correct commutation relations, we must write

$$\delta_\xi \Phi = \mathcal{Q}_\xi \Phi \,, \tag{11.51}$$

where

$$\mathcal{Q}_\xi = \left(-\frac{\partial}{\partial \theta} - i\bar\theta \bar\sigma^m \partial_m\right)\xi + \xi^\dagger \left(\frac{\partial}{\partial \bar\theta} + i\bar\sigma^m \theta \partial_m\right) \,. \tag{11.52}$$

This is a translation of the fermionic coordinates $(\theta, \bar\theta)$ plus a translation of the ordinary space-time coordinates proportional to θ, $\bar\theta$. It is straightforward to show that these operators satisfy

$$[\mathcal{Q}_\xi, \mathcal{Q}_\eta] = -2i \left(\xi^\dagger \bar\sigma^m \eta - \eta^\dagger \bar\sigma^m \xi\right) \partial_m \,. \tag{11.53}$$

Despite the fact that this equation has an extra minus sign on the right-hand side with respect to (11.26), it is the relation that we want. (The difference is similar to that between active and passive transformations.) Combined with the decomposition of the superfield that I will introduce below, this relation will allow us to derive the chiral supermultiplet transformation laws (11.27).

Toward this goal, we need one more ingredient. Define the superspace derivatives

$$D_\alpha = \frac{\partial}{\partial \theta_\alpha} - i(\bar\theta \bar\sigma^m)_\alpha \partial_m \qquad \overline{D}_\alpha = -\frac{\partial}{\partial \bar\theta_\alpha} + i(\sigma^m \theta)_\alpha \partial_m \,, \tag{11.54}$$

such that $(D_\alpha \Phi)^\dagger = \overline{D}_\alpha \Phi^\dagger$. These operators commute with $\mathcal{Q}_\xi$:

$$[D_\alpha, \mathcal{Q}_\xi] = 0 \qquad [\overline{D}_\alpha, \mathcal{Q}_\xi] = 0 \,. \tag{11.55}$$

Thus, we can constrain Φ by the equation

$$D_\alpha \Phi = 0 \quad \text{or} \quad \overline{D}_\alpha \Phi = 0 \,, \tag{11.56}$$

and these constraints are consistent with supersymmetry. What we have just shown is that the general superfield $\Phi(x,\theta,\overline{\theta})$ is a *reducible* representation of supersymmetry. It can be decomposed into a direct sum of three smaller representations, one constrained by the first of the relations (11.56), one constrained by the second of these relations, and the third containing whatever is left over in Φ when these pieces are removed.

Let us begin with the constraint $\overline{D}_\alpha \Phi = 0$. The solution of this equation can be written

$$\Phi(x,\theta,\overline{\theta}) = \Phi(x + i\theta\overline{\sigma}^m\overline{\theta}, \theta) , \qquad (11.57)$$

that is, this solution is parametrized by a general function of x and θ. Since θ is a two-component anticommuting object, this general function of x and θ can be represented as

$$\Phi(x,\theta) = \phi(x) + \sqrt{2}\theta^T c\psi(x) + \theta^T c\theta F(x) . \qquad (11.58)$$

The field content of this expression is exactly that of the chiral supermultiplet. The supersymmetry transformation of this field should be

$$\delta_\xi \Phi = \mathcal{Q}_\xi \Phi(x + i\theta\overline{\sigma}^m\overline{\theta}, \theta) . \qquad (11.59)$$

It is straightforward to compute the right-hand side of (11.59) in terms of θ, $\overline{\theta}$, and the component fields of (11.58). The coefficients of powers of θ are precisely the supersymmetry variations given in (11.27). Thus a superfield satisfying

$$\overline{D}_\alpha \Phi = 0 \qquad (11.60)$$

is equivalent to a chiral supermultiplet, and the transformation (11.59) gives the supersymmetry transformation of this multiplet. A superfield satisfying (11.60) is called a *chiral superfield*. Similarly, a superfield satisfying

$$D_\alpha \Phi = 0 \qquad (11.61)$$

is called an *antichiral superfield*. This superfield has a component field decomposition (ϕ^*, ψ^*, F^*), on which $\mathcal{Q}_\xi$ induces the transformation (11.28). I will describe the remaining content of the general superfield Φ in Section 2.5.

A Lagrangian on Minkowski space is integrated over d^4x. A superspace Lagrangian should be also be integrated over the θ coordinates. Integration over fermionic coordinates is defined to be proportional to the coefficient of the highest power of θ. I will define integration over superspace coordinates by the formulae

$$\int d^2\theta\, 1 = \int d^2\theta\, \theta_\alpha = 0 \qquad \int d^2\theta (\theta^T c\theta) = 1 \qquad (11.62)$$

and their conjugates. To use these formulae, expand the superfields in powers of θ and pick out the terms proportional to $(\theta^T c\theta)$. Then, if Φ is a chiral superfield constrained by (11.60) and $W(\Phi)$ is an analytic function of Φ,

$$\int d^2\theta\, \Phi(x,\theta) = F(x)$$

$$\int d^2\theta\, W(\Phi) = F(x)\frac{\partial W}{\partial \phi} - \frac{1}{2}\psi^T c\psi \frac{\partial^2 W}{\partial^2 \phi}\,, \qquad (11.63)$$

where, in the second line, W on the right-hand side is evaluated with $\Phi = \phi(x)$. With somewhat more effort, one can show

$$\int d^2\theta \int d^2\overline{\theta}\, \Phi^\dagger \Phi = \partial^m \phi^* \partial_m \phi + \psi^\dagger i\overline{\sigma}\cdot\partial\psi + F^* F\,. \qquad (11.64)$$

These formulae produce the invariant Lagrangians of chiral supermultiplets from a superspace point of view. The most general Lagrangian of chiral superfields Φ_k takes the form

$$\mathcal{L} = \int d^4\theta\, K(\Phi, \Phi^\dagger) + \int d^2\theta\, W(\Phi) + \int d^2\overline{\theta}\, (W(\Phi))^\dagger\,, \qquad (11.65)$$

where $W(\Phi)$ is an analytic function of complex superfields and $K(\Phi, \Phi^\dagger)$ is a general real-valued function of the superfields. The Lagrangian (11.37) is generated from this expression by taking $K(\Phi, \Phi^\dagger) = \Phi_k^\dagger \Phi_k$. The most general *renormalizable* Lagrangian of chiral supermultiplets is obtained by taking K to be of this simple form and taking W to be a polynomial of degree at most 3.

Because the integral $d^2\theta$ exposes the Lagrange multiplier F in (11.58), I will refer to a term with this superspace integral as an *F-term*. For similar reasons that will become concrete in the next section, I will call a term with a $d^4\theta$ integral a *D-term*.

In the remainder of these lectures, I will restrict myself to discussing renormalizable supersymmetric theories. But, still, it is interesting to ask what theories we obtain when we take more general forms for K. The Lagrangian for ϕ turns out to be a nonlinear sigma model for which the target space is a complex manifold with the metric [32]

$$g_{m\bar{n}} = \frac{\partial^2}{\partial \Phi^m \partial \Phi^{\dagger \bar{n}}} K(\Phi, \Phi^\dagger) \tag{11.66}$$

A complex manifold whose metric is derived from a potential in this way is called a *Kähler manifold*. The function K is the *Kähler potential*. It is remarkable that, wherever in ordinary quantum field theory we find a general structure from real analysis, the supersymmetric version of the theory has a corresponding complex analytic structure.

Now that we have a Lagrangian in superspace, it is possible to derive Feynman rules and compute Feynman diagrams in superspace. I do not have space here to discuss this formalism; it is discussed, for example, in [7] and [30]. I would like to state one important consequence of this formalism. It turns out that, barring some special circumstances related to perturbation theory anomalies, these Feynman diagrams always generate corrections to the effective Lagrangian that are D-terms,

$$\int d^4\theta \, X(\Phi, \Phi^\dagger) \,. \tag{11.67}$$

The perturbation theory does not produce terms that are integrals $\int d^2\theta$. This leads to an elegant proof of the result cited at the end of the previous section that the superpotential is not renormalized at any order in perturbation theory [30].

11.2.4. Supersymmetric Lagrangians with Vector Fields

To construct a supersymmetric model that can include the Standard Model, we need to be able to write supersymmetric Lagrangians that include Yang-Mills vector fields. In this section, I will discuss how to do that.

To prepare for this discussion, let me present my notation for gauge fields in a general quantum field theory. The couplings of gauge bosons to matter are based on the covariant derivative, which I will write as

$$\mathcal{D}_m \phi = (\partial_m - igA_m^a t_R^a)\phi \tag{11.68}$$

for a field ϕ that belongs to the representation R of the gauge group G. In this formula, t_R^a are the representation matrices of the generators of G in the representation R. These obey

$$[t_R^a, t_R^b] = if^{abc} t_R^c \tag{11.69}$$

The coefficients f^{abc} are the *structure constants* of G. They are independent of R; essentially, their values define the multiplication laws of G. They can be taken to be totally antisymmetric.

The generators of G transform under G according to a representation called the *adjoint representation*. I will denote this representation by $R = G$. Its representation matrices are

$$(t_G^a)_{bc} = if^{bac} \tag{11.70}$$

These matrices satisfy (11.69) by virtue of the Jacobi identity. The covariant derivative on a field in the adjoint representation takes the form

$$\mathcal{D}_m \Phi^a = \partial_m \Phi^a + g f^{abc} A_m^b \Phi^c \tag{11.71}$$

The field strengths F_{mn}^a are defined from the covariant derivative (in any representation) by

$$[\mathcal{D}_m, \mathcal{D}_n] = -ig F_{mn}^a t_R^a . \tag{11.72}$$

This gives the familiar expression

$$F_{mn}^a = \partial_m A_n^a - \partial_n A_m^a + g f^{abc} A_m^b A_n^c . \tag{11.73}$$

Now we would like to construct a supersymmetry multiplet that contains the gauge field A_m^a. The fermion in the multiplet should differ in spin by $\frac{1}{2}$ unit. To write a renormalizable theory, we must take this to be a spin-$\frac{1}{2}$ Weyl fermion. I will then define the *vector supermultiplet*

$$(A_m^a, \lambda_\alpha^a, D^a) \tag{11.74}$$

including the gauge field, a Weyl fermion in the adjoint representation of the gauge group, and an auxililary real scalar field, also in the adjoint representation, that will have no independent particle content. The particle content of this multiplet is one massless vector boson, with two transverse polarization states, and one massless fermion and antifermion, for each generator of the gauge group. The fermion is often called a *gaugino*. The number of physical states is again equal between bosons and fermions.

The supersymmetry transformations for this multiplet are

$$\delta_\xi A^{am} = [\xi^\dagger \overline{\sigma}^m \lambda^a + \lambda^{\dagger a} \overline{\sigma}^m \xi]$$
$$\delta_\xi \lambda^a = [i\sigma^{mn} F_{mn}^a + D^a]\xi$$
$$\delta_\xi \lambda^{\dagger a} = \xi^\dagger [i\overline{\sigma}^{mn} F_{mn}^a + D^a]$$
$$\delta_\xi D^a = -i[\xi^\dagger \overline{\sigma}^m \mathcal{D}_m \lambda^a - \mathcal{D}_m \lambda^{\dagger a} \overline{\sigma}^m \xi] \tag{11.75}$$

where

$$\sigma^{mn} = \frac{1}{4}(\sigma^m \overline{\sigma}^n - \sigma^n \overline{\sigma}^m) . \tag{11.76}$$

I encourage you to verify that these tranformations obey the algebra

$$[\delta_\xi, \delta_\eta] = 2i \left(\xi^\dagger \overline{\sigma}^m \eta - \eta^\dagger \overline{\sigma}^m \xi\right) \partial_m + \delta_\alpha , \qquad (11.77)$$

where δ_α is a gauge tranformation with the gauge parameter

$$\alpha = -2i(\xi^\dagger \overline{\sigma}^m \eta - \eta^\dagger \overline{\sigma}^m \xi) A_m^a . \qquad (11.78)$$

Acting on λ^a, the extra term δ_α in (11.77) can be combined with the translation to produce the commutation relation

$$[\delta_\xi, \delta_\eta] \lambda^a = 2i \left(\xi^\dagger \overline{\sigma}^m \eta - \eta^\dagger \overline{\sigma}^m \xi\right) (\mathcal{D}_m \lambda)^a . \qquad (11.79)$$

This rearrangement applies also for the auxiliary field D^a and for any matter field that tranforms linearly under G. The gauge field A^{am} does not satisfy this last criterion; instead, we find

$$\begin{aligned}[\delta_\xi, \delta_\eta] A_m^a &= 2i(\xi^\dagger \overline{\sigma}^n \eta - \eta^\dagger \overline{\sigma}^n \xi)(\partial_n A_m^a - \mathcal{D}_m A_n) \\ &= 2i(\xi^\dagger \overline{\sigma}^n \eta - \eta^\dagger \overline{\sigma}^n \xi) F_{nm}^a \end{aligned} \qquad (11.80)$$

The proof that (11.75) satisfies the supersymmetry algebra is more tedious than for (11.29), but it is not actually difficult. For the transformation of λ^a we need both the Fierz identity (11.30) and the relation

$$\eta_\alpha \xi_\beta - (\xi \leftrightarrow \eta) = -(\xi^T c \sigma_{pq} \eta)(\sigma^{pq} c)_{\alpha\beta} . \qquad (11.81)$$

The matrices $\sigma^{pq} c$ and $c \overline{\sigma}^{pq}$ are symmetric in their spinor indices.

Again, the transformation laws leave a simple Lagrangian invariant. For the vector supermultiplet, this Lagrangian is that of the renormalizable Yang-Mills theory including the gaugino:

$$\mathcal{L}_F = -\frac{1}{4}(F_{mn}^a)^2 + \lambda^{\dagger a} i\overline{\sigma} \cdot \mathcal{D} \lambda^a + \frac{1}{2}(D^a)^2 \qquad (11.82)$$

The kinetic term for D^a contains no derivatives, so this field will be a Lagrange multiplier.

The vector supermultiplet can be coupled to matter particles in chiral supermultiplets. To do this, we must first modify the transformation laws of the chiral supermultiplet so that the commutators of supersymmetry transformations obey (11.77) or (11.79). The modified transformation laws are:

$$\begin{aligned}\delta_\xi \phi &= \sqrt{2} \xi^T c \psi \\ \delta_\xi \psi &= \sqrt{2} i \sigma^n c \xi^* \mathcal{D}_n \phi + \sqrt{2} F \xi \\ \delta_\xi F &= -\sqrt{2} i \xi^\dagger \overline{\sigma}^m \mathcal{D}_m \psi - 2g \xi^\dagger c \lambda^{a*} t^a \phi \end{aligned} \qquad (11.83)$$

In this formula, the chiral fields ϕ, ψ, F must belong to the same representation of G, with t^a a representation matrix in that representation. From the transformation laws, we can construct the Lagrangian. Start from (11.31), replace the derivatives by covariant derivatives, add terms to the Lagrangian involving the λ^a to cancel the supersymmetry variation of these terms, and then add terms involving D^a to cancel the remaining supersymmetry variation of the λ^a terms. The result is

$$\mathcal{L}_D = \mathcal{D}^m \phi^* \mathcal{D}_m \phi + \psi^\dagger i \overline{\sigma} \cdot \mathcal{D}\psi + F^* F \\ - \sqrt{2} g (\phi^* \lambda^{aT} t^a c \psi - \psi^\dagger c \lambda^{a*} t^a \phi) + g D^a \phi^a t^a \phi \ . \quad (11.84)$$

The proof that this Lagrangian is supersymmetric, $\delta_\xi \mathcal{L} = 0$, is completely straightforward, but it requires a very large sheet of paper.

The gauge invariance of the theory requires the superpotential Lagrangian $\mathcal{L}_W$ to be invariant under G as a global symmetry. Under this condition, $\mathcal{L}_W$, which contains no derivatives, is invariant under (11.83) without modification. The combination of $\mathcal{L}_F$, $\mathcal{L}_D$, and $\mathcal{L}_W$, with W a polynomial of degree at most 3, gives the most general renormalizable supersymmetric gauge theory.

As we did with the F field of the chiral multiplet, it is interesting to eliminate the Lagrange multiplier D^a. For the Lagrangian which is the sum of (11.82) and (11.84), the equation of motion for D^a is

$$D^a = -g \phi^* t^a \phi \ . \quad (11.85)$$

Eliminating D^a gives a second potential energy term proportional to $(D^a)^2$. This is the *D-term potential* promised below (11.40). I will write the result for a theory with several chiral multiplets:

$$V_D = \frac{1}{2} g^2 \left(\sum_k \phi_k^* t^a \phi_k \right)^2 . \quad (11.86)$$

As with the F-term potential, $V_D \geq 0$ and vanishes if and only if $D^a = 0$. It can be shown by an argument similar to (11.45) that

$$\langle 0 | D^a | 0 \rangle = 0 \quad (11.87)$$

unless supersymmetry is spontaneously broken.

It makes a nice illustration of this formalism to show how the Higgs mechanism works in supersymmetry. For definiteness, consider a supersymmetric gauge theory with the gauge group $U(1)$.

Introduce chiral supermultiplets ϕ_+, ϕ_-, and X, with charges $+1$, -1, and 0, respectively, and the superpotential

$$W = \lambda(\phi_+\phi_- - v^2)X . \tag{11.88}$$

The $F = 0$ equations are

$$F_X^* = (\phi_+\phi_- - v^2) = 0 \qquad F_\pm^* = \phi_\pm X = 0 . \tag{11.89}$$

To solve these equations, set

$$X = 0 \qquad \phi_+ = v/y \qquad \phi_- = vy , \tag{11.90}$$

where y is a complex-valued parameter. The $D = 0$ equation is

$$\phi_+^\dagger \phi_+ - \phi_-^\dagger \phi_- = 0 . \tag{11.91}$$

This implies $|y| = 1$. So y is a pure phase and can be removed by a $U(1)$ gauge transformation.

Now look at the pieces of the Lagrangian that give mass to gauge bosons, fermions, and scalars. The gauge field receives mass from the Higgs mechanism. To compute the mass, we can look at the scalar kinetic terms

$$\phi_+^\dagger(-\mathcal{D}^2)\phi_+ + \phi_-^\dagger(-\mathcal{D}^2)\phi_- = \cdots + \phi_+^\dagger(g^2 A^2)\phi_+ + \phi_-^\dagger(g^2 A^2)\phi_- . \tag{11.92}$$

Putting in the vacuum expectation values $\phi_+ = \phi_- = v$, we find

$$m^2 = 4g^2 v^2 \tag{11.93}$$

for the vector fields. The mode of the scalar field

$$\delta\phi_+ = \eta/\sqrt{2} \qquad \delta\phi_- = -\eta/\sqrt{2} , \tag{11.94}$$

with η real, receives a mass from the D-term potential energy

$$\frac{g^2}{2}(\phi_+^\dagger \phi_+ - \phi_-^\dagger \phi_-)^2 \tag{11.95}$$

Expanding to quadratic order in η, we see that η also receives the mass $m^2 = 4g^2 v^2$. The corresponding mode for η imaginary is the infinitesimal version of the phase rotation of y that we have already gauged away below (11.91). The mode of the fermion fields

$$\delta\psi_+ = \chi/\sqrt{2} \qquad \delta\psi_- = -\chi/\sqrt{2} \tag{11.96}$$

mixes with the gaugino through the term

$$-\sqrt{2}g(\phi_+^\dagger \lambda^T c\psi_+ - \phi_-^\dagger \lambda^T c\psi_-) + h.c. \tag{11.97}$$

Putting in the vacuum expectation values $\phi_+ = \phi_- = v$, we find a Dirac mass with the value

$$m = 2gv \tag{11.98}$$

In all, we find a massive vector boson, a massive real scalar, and a massive Dirac fermion, all with the mass $m = 2gv$. The system has four physical bosons and four physical fermions, all with the same mass, as supersymmetry requires.

11.2.5. The Vector Supermultiplet in Superspace

The vector supermultiplet has a quite simple representation in superspace. This multiplet turns out to be the answer to the question that we posed in our discussion of superspace in the previous section: When the chiral and antichiral components of a general superfield are removed, what is left over? To analyze this issue, I will write a Lagrangian containing a local symmetry that allows us to gauge away the chiral and antichiral components of this superfield. Let $V(x, \theta, \bar{\theta})$ be a real-valued superfield, acted on by a local gauge transformation in superspace

$$\delta V = -\frac{i}{g}(\Lambda - \Lambda^\dagger) \tag{11.99}$$

where Λ is a chiral superfield and $\Lambda^\dagger$ is its conjugate. Since Λ satisfies (11.60), its expansion in powers of θ contains

$$\Lambda(x, \theta, \bar{\theta}) = \Lambda(x + i\bar{\theta}\bar{\sigma}\theta, \theta) = \alpha(x) + \cdots + i\bar{\theta}\bar{\sigma}^m\theta\partial_m\alpha(x) + \cdots \tag{11.100}$$

The general superfield V contains a term[b]

$$V(x, \theta, \bar{\theta}) = \cdots + 2\bar{\theta}\bar{\sigma}^m\theta\, A_m(x) + \cdots \tag{11.101}$$

So the superfield V contains a space time vector field $A_m(x)$, and under (11.99), A_m transforms as

$$\delta A_m = \frac{1}{g}\partial_m(\operatorname{Re}\alpha) . \tag{11.102}$$

This is just what we would like for an Abelian gauge field. So we should accept (11.99) as the generalization of the Abelian gauge transformation to superspace.

The real-valued superfield transforming under (11.99) is called a *vector superfield*. To understand its structure, use the gauge transformation to

---
[b]The factor 2 in this equation is convenient but disagrees with some standard treatments, e.g., [7].

remove all components with powers of θ or $\bar\theta$ only. This choice is called *Wess-Zumino gauge* [33]. What remains after this gauge choice is

$$V(x,\theta,\bar\theta) = 2\bar\theta\bar\sigma^m\theta\, A_m(x) + 2\bar\theta^2\theta^T c\lambda - 2\theta^2\bar\theta^T c\lambda^* + \theta^2\bar\theta^2 D \ . \qquad (11.103)$$

This expression has exactly the field content of the Abelian vector supermultiplet (A_m, λ, D).

This gauge multiplet can be coupled to matter described by chiral superfields. For the moment, I will continue to discuss the Abelian gauge theory. For a chiral superfield Φ with charge Q, the gauge transformation

$$\delta\Phi = iQ\Lambda\Phi \qquad (11.104)$$

contains a standard Abelian gauge transformation with gauge parameter $\mathrm{Re}\,\alpha(x)$ and also preserves the chiral nature of Φ. Then the superspace Lagrangian

$$\int d^2\theta d^2\bar\theta \ \Phi^\dagger e^{gQV} \Phi \qquad (11.105)$$

is gauge-invariant. Using the representation (11.103) and the rules (11.62), it is straightforward to carry out the integrals explicitly and show that (11.105) reduces to (11.84), with $t^a = Q$ for this Abelian theory.

We still need to construct the pure gauge part of the Lagrangian. To do this, first note that, because a quantity antisymmetrized on three Weyl fermion indices vanishes,

$$\overline{D}_\alpha \overline{D}^2 X = 0 \qquad (11.106)$$

for any superfield X. Thus, acting with $\overline{D}^2$ makes any superfield a chiral superfield. The following is a chiral superfield that also has the property that its leading component is the gaugino field $\lambda(x)$:

$$W_\alpha = -\frac{1}{8}\overline{D}^2 (Dc)_\alpha V \ . \qquad (11.107)$$

Indeed, working this out in full detail, we find that $W_\alpha = W_\alpha(x + i\bar\theta\sigma\theta, \theta)$, with

$$W_\alpha(x,\theta) = \lambda_\alpha + [(i\sigma^{mn} F_{mn} + D)\theta]_\alpha + \theta^T c\theta \,[\partial_m \lambda^* i\bar\sigma^m c]_\alpha \ . \qquad (11.108)$$

The chiral superfield W_α is the superspace analogue of the electromagnetic field strength. The Lagrangian

$$\int d^2\theta \frac{1}{2} W^T cW \qquad (11.109)$$

reduces precisely to the Abelian version of (11.82). It is odd that the kinetic term for gauge fields is an F-term rather than a D-term. It turns out that this term can be renormalized by loop corrections as a consequence of the trace anomaly [34]. However, the restricted form of the correction has implications, both some simple ones that I will discuss later in Section 4.3 and and more profound implications discussed, for example, in [35,36].

I will simply quote the generalizations of these results to the non-Abelian case. The gauge transformation of a chiral superfield in the representation R of the gauge group is

$$\Phi \to e^{i\Lambda^a t^a} \Phi \qquad \Phi^\dagger \to \Phi^\dagger e^{-i\Lambda^{\dagger a} t^a} , \qquad (11.110)$$

where Λ^a is a chiral superfield in the adjoint representation of G and t^a is is the representation of the generators of G in the representation R. The gauge transformation of the vector superfield is

$$e^{gV^a t^a} \to e^{i\Lambda^{\dagger a} t^a} e^{gV^a t^a} e^{-i\Lambda^a t^a} \qquad (11.111)$$

Then the Lagrangian

$$\int d^2\theta d^2\bar\theta \, \Phi^\dagger e^{gV^a t^a} \Phi \qquad (11.112)$$

is locally gauge-invariant. Carrying out the integrals in the gauge (11.103) reduces this Lagrangian to (11.84).

The form of the field strength superfield is rather more complicated than in the Abelian case,

$$W_\alpha^a t^a = -\frac{1}{8g} \overline{D}^2 e^{-gV^a t^a} (Dc)_\alpha e^{gV^a t^a} \qquad (11.113)$$

In Wess-Zumino gauge, this formula does reduce to the non-Abelian version of (11.108),

$$W_\alpha^a(x,\theta) = \lambda_\alpha^a + [(i\sigma^{mn} F_{mn}^a + D^a)\theta]_\alpha + \theta^T c\theta \, [\mathcal{D}_m \lambda^{*a} i\overline{\sigma}^m c]_\alpha . \qquad (11.114)$$

Then the Lagrangian

$$\int d^2\theta \, \text{tr}[W^T c W] \qquad (11.115)$$

reduces neatly to (11.82).

The most general renormalizable supersymmetric Lagrangian can be built out of these ingredients. We need to put together the Lagrangian (11.115), plus a term (11.112) for each matter chiral superfield, plus a superpotential Lagrangian to represent the scalar field potential energy. These formulae can be generalized to the case of a nonlinear sigma model

11.2.6. R-Symmetry

The structure of the general superspace action for a renormalizable theory of scalar and fermion fields suggests that this theory has a natural continuous symmetry.

The superspace Lagrangian is

$$\mathcal{L} = \int d^2\theta \, \text{tr}[W^T cW] + \int d^4\theta \, \Phi^\dagger e^{gV \cdot t} \Phi + \int d^2\theta \, W(\Phi) + \int d^2\bar{\theta} \, (W(\Phi))^\dagger \,. \tag{11.116}$$

Consider first the case in which $W(\phi)$ contains only dimensionless parameters and is therefore a cubic polynomial in the scalar fields. Then $\mathcal{L}$ is invariant under the $U(1)$ symmetry

$$\Phi_k(x,\theta) \to e^{-i2\alpha/3}\Phi_k(x,e^{i\alpha}\theta) \,, \quad V^a(x,\theta,\bar\theta) \to V^a(x,e^{i\alpha}\theta, e^{-i\alpha}\bar\theta) \tag{11.117}$$

or, in components,

$$\phi_k \to e^{-i2\alpha/3}\phi_k \,, \quad \psi_k \to e^{i\alpha/3}\psi_k \,, \quad \lambda^a \to e^{-i\alpha}\lambda^a \,, \tag{11.118}$$

and the gauge fields are invariant. This transformation is called *R-symmetry*. Under R-symmetry, the charges of bosons and fermions differ by 1 unit, in such a way that that the gaugino and superpotential vertices have zero net charge.

Since all left-handed fermions have the same charge under (11.118), the R-symmetry will have an axial vector anomaly. It can be shown that the R-symmetry current (of dimension 3, spin 1) forms a supersymmetry multiplet together with the supersymmetry current (dimension $\frac{7}{2}$, spin $\frac{3}{2}$) and the energy-momentum tensor (dimension 4, spin 2) [37]. All three currents have perturbation-theory anomalies; the anomaly of the energy-momentum tensor is the trace anomaly, associated with the breaking of scalar invariance by coupling constant renormalization. The R-current anomaly is thus connected to the running of coupling constants and gives a useful formal approach to study this effect in supersymmetric models.

It is often possible to combine the transformation (11.117) with other apparent $U(1)$ symmetries of the theory to define a non-anomalous $U(1)$ R-symmetry. Under such a symmetry, we will have

$$\Phi_k(x,\theta) \to e^{-i\beta_k}\Phi_k(x,e^{i\alpha}\theta) \,, \quad \text{such that} \quad W(x,\theta) \to e^{2i\alpha}W(x,e^{i\alpha}\theta) \,. \tag{11.119}$$

Such symmetries also often arise in models in which the superpotential has dimensionful coefficients.

In models with extended, $N > 1$, supersymmetry, the R-symmetry group is also extended, to $SU(2)$ for $N = 2$ and to $SU(4)$ for $N = 4$ supersymmetry.

11.3. The Minimal Supersymmetric Standard Model

11.3.1. Particle Content of the Model

Now we have all of the ingredients to construct a supersymmetric generalization of the Standard Model. To begin, let us construct a version of the Standard Model with exact supersymmetry. To do this, we assign the vector fields in the Standard Model to vector supermultiplets and the matter fields of the Standard Model to chiral supermultiplets.

The vector supermultiplets correspond to the generators of $SU(3) \times SU(2) \times U(1)$. In these lectures, I will refer to the gauge bosons of these groups as A_m^a, W_m^a, and B_m, respectively. I will represent the Weyl fermion partners of these fields as $\widetilde{g}^a$, $\widetilde{w}^a$, $\widetilde{b}$. I will call these fields the *gluino*, *wino*, and *bino*, or, collectively, *gauginos*. In the later parts of these lectures, I will drop the tildes over the gaugino fields when they are not needed for clarity.

I will assign the quarks and leptons to be fermions in chiral superfields. I will use the convention presented in Section 1.3 of considering left-handed Weyl fermions as the basic particles and right-handed Weyl fermions as their antiparticles. In the Standard Model, the left-handed fields in a fermion generation have the quantum numbers

$$L = \begin{pmatrix} \nu \\ e \end{pmatrix} \qquad \overline{e} \qquad Q = \begin{pmatrix} u \\ d \end{pmatrix} \qquad \overline{u} \qquad \overline{d} \qquad (11.120)$$

The field $\overline{e}$ is the left-handed positron; the fields $\overline{u}$, $\overline{d}$ are the left-handed antiquarks. The right-handed Standard Model fermion fields are the conjugates of these fields. To make a generalization to supersymmetry, we will extend each of the fields in (11.120)—for each of the three generations—to a chiral supermultiplet. I will use the symbols

$$\widetilde{L} \qquad \widetilde{\overline{e}} \qquad \widetilde{Q} \qquad \widetilde{\overline{u}} \qquad \widetilde{\overline{d}} \qquad (11.121)$$

to represent both the supermultiplets and the scalar fields in these multiplets. Again, I will drop the tilde if it is unambiguous that I am referring to the scalar partner rather than the fermion. The scalar particles in these supermultiplets are called *sleptons* and *squarks*, collectively, *sfermions*.

What about the Higgs field? The Higgs field of the Standard Model should be identified with a complex scalar component of a chiral supermultiplet. But it is ambiguous what the quantum numbers of this multiplet should be. In the Standard Model, the Higgs field is a color singlet with $I = \frac{1}{2}$, but we can take the hypercharge of this field to be either $Y = +\frac{1}{2}$ or $Y = -\frac{1}{2}$, depending on whether we take the positive hypercharge field or its conjugate to be primary. In a supersymmetric model, the choice matters. The superpotential is an analytic function of superfields, so it can only contain the field, not the conjugate. Then different Higgs couplings will be allowed depending on the choice that we make.

The correct solution to this problem is to include *both* possibilities, That is, we include a Higgs supermultiplet with $Y = +\frac{1}{2}$ and a second Higgs supermultiplet with $Y = -\frac{1}{2}$. I will call the scalar components of these multiplets H_u and H_d, respectively:

$$H_u = \begin{pmatrix} H_u^+ \\ H_u^0 \end{pmatrix} \qquad H_d = \begin{pmatrix} H_d^0 \\ H_d^- \end{pmatrix} \qquad (11.122)$$

I will refer to the Weyl fermion components with these quantum numbers as $\widetilde{h}_u$, $\widetilde{h}_d$. These fields or particles are called *Higgsinos*.

I will argue below that it is necessary to include both Higgs fields in order to obtain all of the needed couplings in the superpotential. However, there is another argument. The axial vector anomaly of one $U(1)$ and two $SU(2)$ currents (Fig. 11.5) must vanish to maintain the gauge invariance of the model. In the Standard Model, the anomaly cancels nontrivially between the quarks and the leptons. In the supersymmetric generalization of the Standard Model, each Higgsino makes a nonzero contribution to this anomaly. These contributions cancel if we include a pair of Higgsinos with opposite hypercharge.

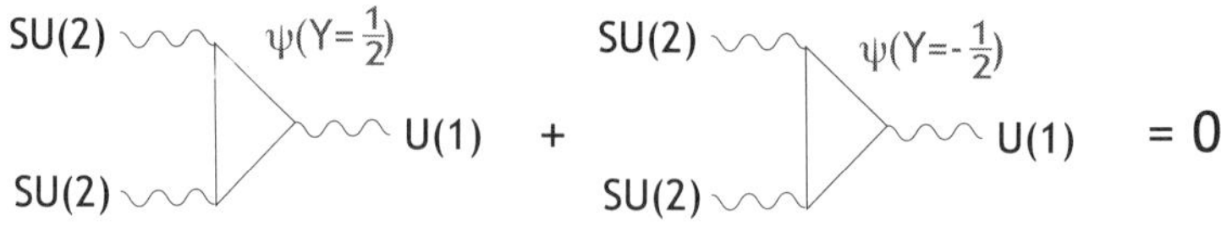

Fig. 11.5. The anomaly cancellation that requires two doublets of Higgs fields in the MSSM.

11.3.2. Grand Unification

Before writing the Lagrangian in detail, I would like to point out that there is an interesting conclusion that follows from the quantum number assignments of the new particles that we have introduced to make the Standard Model supersymmetric.

An attractive feature of the Standard Model is that the quarks and leptons of each generation fill out multiplets of the simple gauge group $SU(5)$. This suggests a very beautiful picture, called *grand unification*, in which $SU(5)$, or a group such as $SO(10)$ or E_6 for which this is a subgroup, is the fundamental gauge symmetry at very short distances. This unified symmetry will be spontaneously broken to the Standard Model gauge group $SU(3) \times SU(2) \times U(1)$.

For definiteness, I will examine the model in which the grand unified symmetry group is $SU(5)$. The generators of $SU(5)$ can be represented as 5×5 Hermitian matrices acting on the 5-dimensional vectors in the fundamental representation. To see how the Standard Model is embedded in $SU(5)$, it is convenient to write these matrices as blocks with 3 and 2 rows and columns. Then the Standard Model generators can be identified as

$$SU(3): \begin{pmatrix} t^a & \\ & 0 \end{pmatrix}; \quad SU(2): \begin{pmatrix} 0 & \\ & \sigma^a/2 \end{pmatrix}; \quad U(1): \sqrt{\frac{3}{5}} \begin{pmatrix} -\frac{1}{3}\mathbf{1} & \\ & \frac{1}{2}\mathbf{1} \end{pmatrix}. \quad (11.123)$$

In these expressions, t^a is an $SU(3)$ generator, $\sigma^a/2$ is an $SU(2)$ generator, and all of these matrices are normalized to $\text{tr}[T^A T^B] = \frac{1}{2}\delta^{AB}$. We should identify the last of these matrices with $\sqrt{3/5}\, Y$.

The symmetry-breaking can be caused by the vacuum expectation value of a Higgs field in the adjoint representation of $SU(5)$. The expectation value

$$\langle \Phi \rangle = V \cdot \begin{pmatrix} -\frac{1}{3}\mathbf{1} & \\ & \frac{1}{2}\mathbf{1} \end{pmatrix} \quad (11.124)$$

commutes with the generators in (11.123) and fails to commute with the off-diagonal generators. So this vacuum expectation value gives mass to the off-diagonal generators and breaks the gauge group to $SU(3) \times SU(2) \times U(1)$.

Matter fermions can be organized as left-handed Weyl fermions in the $SU(5)$ representations $\bar{5}$ and 10. The $\bar{5}$ is the conjugate of the fundamental representation of $SU(5)$; the 10 is the antisymmetric matrix with two 5 indices.

$$\bar{5}: \begin{pmatrix} \bar{d} \\ \bar{d} \\ \bar{d} \\ e \\ \nu \end{pmatrix}_L ; \quad 10: \begin{pmatrix} 0 & \bar{u} & \bar{u} & u & d \\ & 0 & \bar{u} & u & d \\ & & 0 & u & d \\ & & & 0 & \bar{e} \\ & & & & 0 \end{pmatrix}_L \quad (11.125)$$

It is straightforward to check that each entry listed has the quantum numbers assigned to that field in the Standard Model. To compute the hypercharges, we act on the $\bar{5}$ with (-1) times the hypercharge generator in (11.123), and we act on the 10 with the hypercharge generator on each index. This gives the standard results, for example, $Y = +\frac{1}{3}$ for the $\bar{d}$ and $Y = -\frac{1}{3} + \frac{1}{2} = \frac{1}{6}$ for u and d.

The $SU(5)$ covariant derivative is

$$\mathcal{D}_m = (\partial_m - ig_U A_m^A T^A) , \quad (11.126)$$

where g_U is the $SU(5)$ gauge coupling. There is only room for one value here. So this model predicts that the three Standard Model gauge couplings are related by

$$g_3 = g_2 = g_1 = g_U , \quad (11.127)$$

where

$$g_3 = g_s \qquad g_2 = g \qquad g_1 = \sqrt{\frac{5}{3}} g' . \quad (11.128)$$

Clearly, this prediction is not correct for the gauge couplings that we measure in particle physics.

However, there is a way to save this prediction. In quantum field theory, coupling constants are functions of length scale and change their values significantly from one scale to another by renormalization group evolution. It is possible that the values of g', g, and g_s that we measure could evolve at very short distances into values that obey (11.127).

I will now collect the formulae that we need to analyze this question. Let

$$\alpha_i = \frac{g_i^2}{4\pi} \quad (11.129)$$

for $i = 1, 2, 3$. The one-loop renormalization group equations for gauge couplings are

$$\frac{dg_i}{d\log Q} = -\frac{b_i}{(4\pi)^2} g_i^3 \quad \text{or} \quad \frac{d\alpha_i}{d\log Q} = -\frac{b_i}{(2\pi)} \alpha_i^2 . \quad (11.130)$$

For $U(1)$, the coefficient b_1 is

$$b_1 = -\frac{2}{3}\sum_f \frac{3}{5}Y_f^2 - \frac{1}{3}\sum_b \frac{3}{5}Y_b^2 , \tag{11.131}$$

where the two sums run over the multiplets of left-handed Weyl fermions and complex-valued bosons. The factors $\frac{3}{5}Y^2$ are the squares of the $U(1)$ charges defined by (11.123). For non-Abelian groups, the expressions for the b coefficients are

$$b = -\frac{11}{3}C_2(G) - \frac{2}{3}\sum_f C(r_f) - \frac{1}{3}\sum_b C(r_b) , \tag{11.132}$$

where $C_2(G)$ and $C(r)$ are the standard group theory coefficients. For $SU(N)$,

$$C_2(G) = C(G) = N , \quad C(N) = \frac{1}{2} . \tag{11.133}$$

The solution of the renormalization group equation (11.130) is

$$\alpha^{-1}(Q) = \alpha^{-1}(M) - \frac{b_i}{2\pi}\log\frac{Q}{M} . \tag{11.134}$$

Now consider the situation in which the three couplings g_i become equal at the mass scale M_U, the mass scale of $SU(5)$ symmetry breaking. Let α_U be the value of the α_i at this scale. Using (11.134), we can then determine the Standard Model couplings at any lower mass scale. The three $\alpha_i(Q)$ are determined by two parameters. We can eliminate those parameters and obtain the relation

$$\alpha_3^{-1} = (1+B)\alpha_2^{-1} - B\alpha_1^{-1} \tag{11.135}$$

where

$$B = \frac{b_3 - b_2}{b_2 - b_1} . \tag{11.136}$$

The values of the α_i are known very accurately at $Q = m_Z$ [38]:

$$\alpha_3^{-1} = 8.50 \pm 0.14 \quad \alpha_2^{-1} = 29.57 \pm 0.02 \quad \alpha_1^{-1} = 59.00 \pm 0.02 . \tag{11.137}$$

Inserting these values into (11.135), we find

$$B = 0.716 \pm 0.005 \pm 0.03 . \tag{11.138}$$

In this formula, the first error is that propagated from the errors in (11.137) and the second is my estimate of the systematic error from neglecting the two-loop renormalization group coefficients and other higher-order corrections.

We can compare the value of B in (11.138) to the values of (11.136) from different models. The hypothesis that the three Standard Model couplings unify is acceptable only if the gauge theory that describes physics between m_Z and M_U gives a value of B consistent with (11.138). The minimal Standard Model fails this test. The values of the b_i are

$$b_3 = 11 - \frac{4}{3}n_g$$

$$b_2 = \frac{22}{3} - \frac{4}{3}n_g - \frac{1}{6}n_h$$

$$b_1 = \phantom{\frac{22}{3}} - \frac{4}{3}n_g - \frac{1}{10}n_h \qquad (11.139)$$

where n_g is the number of generations and n_h is the number of Higgs doublets. Notice that n_g cancels out of (11.136). This is to be expected. The Standard Model fermions form complete representations of $SU(5)$, and so their renormalization effects cannot lead to differences among the three couplings. For the minimal case $n_h = 1$ we find $B = 0.53$. To obtain a value consistent with (11.138), we need $n_h = 6$.

We can redo this calculation in the minimal supersymmetric version of the Standard Model. First of all, we should rewrite (11.132) for a supersymmetric model with one vector supermultiplet, containing a vector and a Weyl fermion in the adjoint representation, and a set of chiral supermultiplets indexed by k, each with a Weyl fermion and a complex boson. Then (11.132) becomes

$$b_i = \frac{11}{3}C_2(G) - \frac{2}{3}C_2(G) - \left(\frac{2}{3} + \frac{1}{3}\right)\sum_k C(r_k)$$

$$= 3C_2(G) - \sum_k C(r_k) \qquad (11.140)$$

The formula (11.131) undergoes a similar rearrangement. Inserting the values of the $C(r_k)$ for the fields of the Standard Model, we find

$$b_3 = 9 - 2n_g$$

$$b_2 = 6 - 2n_g - \frac{1}{2}n_h$$

$$b_1 = - 2n_g - \frac{3}{10}n_h \qquad (11.141)$$

For the minimal Higgs content $n_h = 2$, this gives

$$B = \frac{5}{7} = 0.714 \qquad (11.142)$$

in excellent agreement with (11.138).

In Fig. 11.6, I show the unification relation pictorially. The three data points on the the left of the figure represent the measured values of the three couplings (11.137). Starting from the values of α_1 and α_2, we can integrate (11.130) up to the scale at which these two couplings converge. Then we can integrate the equation for α_3 back down to $Q = m_Z$ and see whether the result agrees with the measured value. The lower set of curves presents the result for the Standard Model with $n_h = 1$. The upper set of curves shows the result for the supersymmetric extension of the Standard Model with $n_h = 2$. This choice gives excellent agreement with the measured value of α_s.

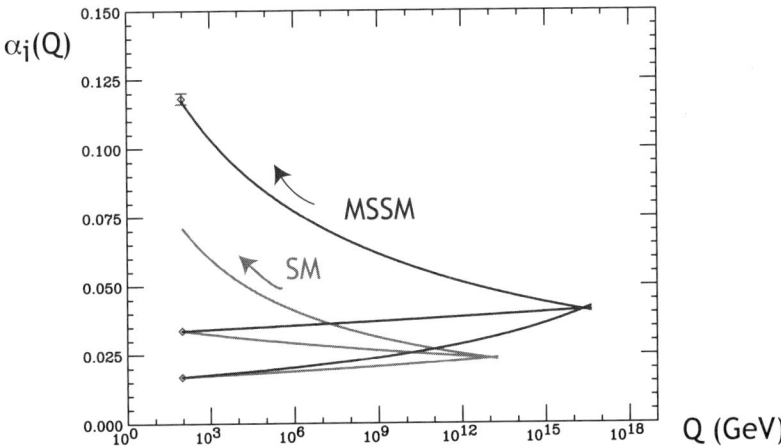

Fig. 11.6. Prediction of the $SU(3)$ gauge coupling α_s from the electroweak coupling constants using grand unification, in the Standard Model and in the MSSM.

Actually, I slightly overstate the case for supersymmetry by ignoring two-loop terms in the renormalization group equations, and also by integrating these equations all the way down to m_Z even though, from searches at high-energy colliders, most of the squarks and gluinos must be heavier than 300 GeV. A more accurate prediction of $\alpha_s(m_Z)$ from the electroweak coupling constants gives a slightly higher value, 0.13 instead of 0.12. However, these corrections could easily be compensated by similar corrections to the upper limit of the integration, following the details of the particle spectrum at the grand unification scale. For a more detailed formal analysis of these corrections, see [39], and for a recent evaluation of their effects, see [40]. It remains a remarkable fact that the minimal supersymmetric extension of the Standard Model is approximately compatible with grand unification 'out of the box', with no need for further model-building.

11.3.3. Construction of the Lagrangian

Now I would like to write the full Lagrangian of the minimal supersymmetric extension of the Standard Model, which I will henceforth call the MSSM.

The kinetic terms and gauge couplings of the MSSM Lagrangian are completely determined by supersymmetry, the choice of the gauge group $SU(3) \times SU(2) \times U(1)$, and the choice of the quantum numbers of the matter fields. The Lagrangian is a sum of terms of the forms (11.82) and (11.84). Up to this point, the only parameters that need to be introduced are the gauge couplings g_1, g_2, and g_3.

Next, we need a superpotential W. The superpotential is the source of nonlinear fermion-scalar interactions, so we should include the appropriate terms to generate the Higgs Yukawa couplings needed to give mass to the quarks and leptons. The appropriate choice is

$$W_Y = y_d^{ij}\overline{d}^i H_{d\alpha}\epsilon_{\alpha\beta}Q_\beta^j + y_e^{ij}\overline{e}^i H_{d\alpha}\epsilon_{\alpha\beta}L_\beta^j - y_u^{ij}\overline{u}^i H_{u\alpha}\epsilon_{\alpha\beta}Q_\beta^j \ . \quad (11.143)$$

The notation for the quark and lepton multiplets is that in (11.120); the indices $i, j = 1, 2, 3$ run over the three generations. The indices $\alpha, \beta = 1, 2$ run over SU(2) isospin indices. Notice that the first two terms require a Higgs field H_d with $Y = -\frac{1}{2}$, while the third term requires a Higgs field H_u with $Y = \frac{1}{2}$. If we leave out one of the Higgs multiplets, some quarks or leptons will be left massless. This is the second argument that requires two Higgs fields in the MSSM.

I have written (11.143) including the most general mixing between left- and right-handed quarks and leptons of different generations. However, as in the minimal Standard Model, we can remove most of this flavor mixing by appropriate field redefinitions. The coupling constants y_d, y_e, y_u are general 3×3 complex-valued matrices. Any such matrix can be diagonalized using two unitary transformations. Thus, we can write

$$y_d = W_d Y_d V_d^\dagger \qquad y_e = W_e Y_e V_e^\dagger \qquad y_u = W_u Y_u V_u^\dagger \ , \quad (11.144)$$

with W_a and V_a 3×3 unitary matrices and Y_a real, positive, and diagonal. The unitary transformations cancel out of the kinetic energy terms and gauge couplings in the Lagrangian, except that the W boson coupling to quarks is transformed

$$gu^\dagger \overline{\sigma}^m dW_m^+ \to gu^\dagger \overline{\sigma}^m (V_u^\dagger V_d) dW_m^+ \ . \tag{11.145}$$

From this equation, we can identify $(V_d^\dagger V_u) = V_{CKM}$, the Cabibbo-Kobayashi-Maskawa weak interaction mixing matrix. The Lagrangian term (11.143) thus introduces the remaining parameters of the Standard Model, the 9 quark and lepton masses (ignoring neutrino masses) and the 4 CKM mixing angles. The field redefinition (11.144) can also induce or shift a QCD theta parameter, so the MSSM, like the Standard Model, has a strong CP problem that requires an axion or another model-building solution [41].

There are several other terms that can be added to W. One possible contribution is a pure Higgs term

$$W_\mu = -\mu H_{d\alpha} \epsilon_{\alpha\beta} H_{u\beta} \ . \tag{11.146}$$

The parameter μ has the dimensions of mass, and consequently this *mu term* provides a supersymmetric contribution to the masses of the Higgs bosons. Because this term is in the superpotential, it does not receive additive raditive corrections. Even in a theory that includes grand unification and energies scale of the order of 10^{16} GeV, we can set the parameter μ to be of order 100 GeV without finding this choice affected by large quantum corrections. We will see in Section 4.2 that the mu term is needed for phenomenological reasons. If $\mu = 0$, a Higgsino state will be massless and should have been detected already in experiments. It is odd that a theory whose fundamental mass scale is the grand unification scale should require a parameter containing a weak interaction mass scale. I will present some models for the origin of this term in Section 3.5.

At this point, we have introduced two new parameters beyond those in the Standard Model. One is the value of μ. The other is the result of the fact that we have two Higgs doublets in the model. The ratio of the Higgs vacuum expectation values

$$\langle H_u \rangle / \langle H_d \rangle \equiv \tan \beta \tag{11.147}$$

will appear in many of the detailed predictions of the MSSM.

There are still more superpotential terms that are consistent with the Standard Model gauge symmetry and quantum numbers. These are

$$W_{\not R} = \eta_1 \epsilon_{ijk} \overline{u}_i \overline{d}_j \overline{d}_k + \eta_2 \overline{d} \epsilon_{\alpha\beta} L_\alpha Q_\beta$$
$$+ \eta_3 \overline{e} \epsilon_{\alpha\beta} L_\alpha L_\beta + \eta_4 \epsilon_{\alpha\beta} L_\alpha H_{u\beta} \ . \tag{11.148}$$

Here i, j, k are color indices, α, β are isospin indices, and arbitrary generation mixing is also possible. These terms violate baryon and lepton number through operators with dimensionless coefficients. In constructing supersymmetric models, it is necessary either to forbid these terms by imposing appropriate discrete symmetries or to arrange by hand that some of the dangerous couplings are extremely small [42].

If baryon number B and lepton number L are conserved in a supersymmetric model, this model respects a discrete symmetry called R-parity,

$$R = (-1)^{3B+L+2J} \ . \tag{11.149}$$

Here $(3B)$ is quark number and J is the spin of the particle. This quantity is constructed so that $R = +1$ on the particles of the Standard Model (including the Higgs bosons) and $R = -1$ on their supersymmetry partners. R acts differently on particles of different spin in the same supermultiplet, so R-parity is a discrete subgroup of a continuous R-symmetry.

In a model with grand unification, there will be baryon number and lepton number violation, and so B and L cannot be used as fundamental symmetries. However, we can easily forbid most of the superpotential terms (11.148) by introducing a discrete symmetry that distinguishes the field H_d from the lepton doublets L_i. A similar strategy can be used to forbid the first, 3-quark, term. With these additional discrete symmetries, the MSSM, including all other terms considered up to this point, will conserve R-parity.

11.3.4. The Lightest Supersymmetric Particle

If R-parity is conserved, the lightest supersymmetric particle will be absolutely stable. This conclusion has an important implication for the relation of supersymmetry to cosmology. If a supersymmetric particle is stable for a time longer than the age of the universe, and if this particle is electrically neutral, that particle is a good candidate for the cosmic dark matter. In Sections 6.3 and 6.4, I will discuss in some detail the properties of models in which the lightest Standard Model superpartner is the dark matter particle.

However, this is not the only possibility. Over times much longer than those of particle physics experiments—minutes, years, or billions of years— we need to consider the possibility that the lightest Standard Model superpartner will decay to a particle with only couplngs of gravitational strength. Complete supersymmetric models of Nature must include a superpartner of the graviton, a spin-$\frac{3}{2}$ particle called the *gravitino*. In a model with exact supersymmetry, the gravitino will be massless, but in a model with spontaneously broken supersymmetry, the gravitino acquires a mass through an analogue of the Higgs mechanism. If the supersymmetry breaking is induced by one dominant F-term, the value of this mass is [43]

$$m_{3/2} = \frac{8\pi}{3} \frac{\langle F \rangle}{m_{\text{Pl}}} \ . \tag{11.150}$$

This expression is of the same order of magnitude as the expressions for Standard Model superpartner masses that I will give in Section 3.6. In string theory and other unified models, there may be additional Standard Model singlet fields with couplings of gravitation strength, called *moduli*, that might also be light enough that long-lived Standard Model superpartners could decay to them.

Supersymmetric models with R-parity conservation and dark matter, then, divide into two classes, according to the identity of the lightest supersymmetric particle—the LSP. On one hand, the LSP could be a Standard Model superpartner. Cosmology requires that this particle is neutral. Several candidates are available, including the fermionic partners of the photon, Z^0, and neutral Higgs bosons and the scalar partner of one of the neutrinos. In all cases, these particles will be weakly interacting; when they are produced at high-energy colliders, they should not make signals in a particle detector. On the other hand, the LSP could be the gravitino or another particle with only gravitational couplings. In that case, the lightest Standard Model superpartner could be a charged particle. Whether this particle is visible or neutral and weakly interacting, its decay should be included in the phenomenology of the model.

11.3.5. Models of Supersymmetry Breaking

There is still one important effect that is missing in our construction of the MSSM. The terms that we have written so far preserve exact supersymmetry. A fully supersymmetric model would contain a massless fermionic partner of the photon and a charged scalar particle with the mass of the electron. These particles manifestly do not exist. So if we wish to build a model of Nature with supersymmetry as a fundamental symmetry, we need to arrange that supersymmetry is spontaneously broken.

From the example of spontaneous symmetry breaking in the Standard Model, we would expect to do this by including in the MSSM a field whose vacuum expectation value leads to supersymmetry breaking. This is not as easy as it might seem. To explain why, I will first present some models of supersymmetry breaking.

The simplest model of supersymmetry breaking is the O'Raifeartaigh model [44], with three chiral supermultiplets ϕ_0, ϕ_1, ϕ_2 interacting through the superpotential

$$W = \lambda\phi_0 + m\phi_1\phi_2 + g\phi_0\phi_1^2 . \tag{11.151}$$

This superpotential implies the $F = 0$ conditions

$$0 = F_0^* = \lambda + g\phi_1^2$$
$$0 = F_1^* = m\phi_2 + 2g\phi_0\phi_1$$
$$0 = F_2^* = m\phi_1 \tag{11.152}$$

The first and third equations contradict one another. It is impossible to satisfy both conditions, and so there is no supersymmetric vacuum state. This fulfils the condition for spontaneous supersymmetry breaking that I presented in Section 2.2.

This mechanism of supersymmetry breaking has an unwanted corollary. Because one combination of the scalar fields appears in two different constraints in (11.152), there must be an orthogonal combination that does not appear at all. This means that the F-term potential V_F has a surface of degenerate vacuum states. To see this explicitly, pick a particular vacuum solution

$$\phi_0 = \phi_1 = \phi_2 = 0 . \tag{11.153}$$

and expand the potential V_F about this point. There are 6 real-valued boson fields with masses

$$0, \quad 0, \quad m, \quad m, \quad \sqrt{m^2 - 2\lambda g}, \quad \sqrt{m^2 + 2\lambda g} . \tag{11.154}$$

These six fields do not pair into complex-valued fields; that is already an indication that supersymmetry is broken. The fermion mass term in (11.38) gives one Dirac fermion mass m and leaves one Weyl fermion massless. This massless fermion is the Goldstone particle associate with spontaneous supersymmetry breaking.

A property of these masses is that the sum rule for fermion and boson masses

$$\text{str}[m^2] = \sum m_f^2 - \sum m_b^2 = 0 \tag{11.155}$$

remains valid even when supersymmetry is broken. This sum rule is the coefficient of the one-loop quadratic divergence in the vacuum energy. Since supersymmetry breaking does not affect the ultraviolet structure of the theory, this coefficient must cancel even if supersymmetry is spontaneously broken [45]. In fact, if Q is a conserved charge in the model, the sum rule is valid in each charge sector $Q = q$:

$$\text{str}_q[m^2] = 0 \ . \tag{11.156}$$

In the O'Raifeartaigh model, supersymmetry is spontaneously broken by a nonzero expectation value of an F term. It is also possible to break supersymmetry with a nonzero expectation value of a D term. The D-term potential V_D typically has zeros. For example, in an $SU(3)$ supersymmetric Yang-Mills theory,

$$V_D = \frac{1}{2}\left(\sum_3 \phi^\dagger t^a \phi - \sum_{\overline{3}} \overline{\phi} t^a \overline{\phi}^\dagger \right)^2 \tag{11.157}$$

and it is easy to find solutions in which the terms in parentheses sum to zero. However, it is not difficult to arrange a V_F such that the solutions of the $F = 0$ conditions do not coincide with the solutions of the $D = 0$ conditions. This leads to spontaneous symmetry breaking, again with the sum rule (11.156) valid at tree level.

Unfortunately, the sum rule (11.156) is a disaster for the prospect of finding a simple model of spontaneously broken supersymmetry that extends the Standard Model. For the charge sector of the d squarks, we would need all down-type squarks to have masses less than 5 GeV. For the charge sector of the charged leptons, we would need all sleptons to have masses less than 2 GeV.

11.3.6. Soft Supersymmetry Breaking

The solution to this problem is to construct models of spontaneously broken supersymmetry using a different strategy from the one that we use for electroweak symmetry breaking in the Standard Model. To break electroweak symmetry, we introduce a Higgs sector whose mass scale is the same as the scale of the fermion and gauge boson masses induced by the symmetry breaking. To break supersymmetry, however, we could introduce a new sector at a much higher mass scale, relying on a weak coupling of the new sector to the Standard Model particles to communicate the supersymmetry breaking terms. In principle, a weak gauge interaction could supply this coupling. However, the default connection is through gravity. Gravity and supergravity couple to all fields. It can be shown that supersymmetry breaking anywhere in Nature is communicated to all other sectors through supergravity couplings.

We are thus led to the following picture, which produces a phenomenologically reasonable supersymmetric extension of the Standard Model: We extend the Standard Model fields to supersymmetry multiplets in the manner described in Section 3.1. We also introduce a *hidden sector* with no direct coupling to quark, leptons, and Standard Model gauge bosons. Supersymmetry is spontaneously broken in this hidden sector. A weak interaction coupling the two sectors then induces a supersymmetry-breaking effective interaction for the Standard Model particles and their superpartners. If Λ is the mass scale of the hidden sector, the supersymmetry breaking mass terms induced for the Standard Model sector are of the order of

$$m \sim \frac{\langle F \rangle}{M} \sim \frac{\Lambda^2}{M} ; \qquad (11.158)$$

where M is the mass of the particle responsible for the weak connection between the two sectors. M is called the *messenger scale*. By default, the messenger is supergravity. Then $M = m_{\rm Pl}$ and $\Lambda \sim 10^{11}$ GeV. In this scenario, the superpartners acquire masses of the order of the parameter m in (11.158).

It remains true that the quarks, leptons, and gauge bosons cannot obtain mass until $SU(2) \times U(1)$ is broken. It is attractive to think that the symmetry-breaking terms that give mass to the superpartners cause $SU(2) \times U(1)$ to be spontaneously broken, at more or less the same scale. I will discuss a mechanism by which this can happen in Section 6.1. The weak interaction scale would then not be a fundamental scale in Nature, but rather one that arises dynamically from the hidden sector and its couplings.

The effective interaction that are generated by messenger exchange generally involve simple operators of low mass dimensions, to require the minimal number of powers of M in the denominator. These operators are *soft* perturbations of the theory, and so we say that the MSSM is completed by including *soft supersymmetry-breaking interactions*.

However, the supersymmetry-breaking terms induced in this model will not include all possible low-dimension operators. Since these interactions arise by coupling into a supersymmetry theory, they are formed by starting with a supersymmetric effective action and turning on F and D expectation values as spurions. Only a subset of the possible supersymmetry-breaking terms can be formed in this way [46]. By replacing a superfield Φ by

$\theta^T c\theta \langle F \rangle$, we can convert

$$\int d^4\theta \, K(\Phi, \phi) \to m^2 \phi^\dagger \phi$$

$$\int d^2\theta \, f(\Phi) W^T cW \to m\lambda^T c\lambda$$

$$\int d^2 \, W(\Phi, \phi) \to B\phi^2 + A\phi^3 \qquad (11.159)$$

However, as long as the ϕ theory is renormalizable, we cannot generate the terms

$$m\psi^T c\psi \,, \quad C\phi^* \phi^2 \,, \qquad (11.160)$$

by turning on expectation values for F and D fields. Thus, we cannot generate supersymmetry-breaking interactions that are mass terms for the fermion field of a chiral multiplet or non-holomorphic cubic terms for the scalar fields.

There is another difficulty with terms of the form (11.160). In models with Standard Model singlet scalar fields, which typically occur in concrete models, these two interactions can generate new quadratic divergences when they appear in loop diagrams [46].

Here is the most general supersymmetry-breaking effective Lagrangian that can be constructed following the rule just given that is consistent with the gauge symmetries of the Standard Model:

$$\mathcal{L}_{soft} = -M_f^2 |\widetilde{f}|^2 - \frac{1}{2} m_i \lambda_i^{Ta} c \lambda_i^a$$

$$- (A_d y_d \widetilde{\overline{d}} H_{d\alpha} \epsilon_{\alpha\beta} \widetilde{Q}_\beta + A_e y_e \widetilde{\overline{e}} H_{d\alpha} \epsilon_{\alpha\beta} \widetilde{L}_\beta$$

$$- A_u y_u \widetilde{\overline{u}} H_{u\alpha} \epsilon_{\alpha\beta} \widetilde{Q}_\beta - B\mu H_{d\alpha} \epsilon_{\alpha\beta} H_{u\beta}) - h.c. \qquad (11.161)$$

I have made the convention of scaling the A terms with the corresponding Yukawa couplings and scaling the B terms with μ. The parameters A and B then have the dimensions of mass and are expected to be of the order of m in (11.158).

For most of the rest of these lectures, I will represent the effects of the hidden sector and supersymmetry breaking simply by adding (11.161) to the supersymmetric Standard Model. I will then consider the MSSM to be defined by

$$\mathcal{L} = \mathcal{L}_F + \mathcal{L}_D + \mathcal{L}_W + \mathcal{L}_{soft} \qquad (11.162)$$

combining the pieces from (11.82), (11.84), (11.143), (11.146), and (11.161).

There are two problems with this story. The first is the μ term in the MSSM superpotential. This a supersymmetric term, and so μ can be arbitrarily large. To build a successful phenomenology of the MSSM, however, we need to have μ of the order of the weak scale. Ideally, μ should be parametrically equal to (11.158).

There are simple mechanisms that can solve this problem. A fundamental theory that leads to the renormalizable Standard Model at low energies can also contain higher-dimension operators suppressed by the high-energy mass scale. Associate this scale with the messenger scale. Then a supersymmetric higher-dimension operator in the superpotential

$$\int d^2\theta \, \frac{1}{M} S^2 H_d H_u \qquad (11.163)$$

leads to a μ term if S acquires a vacuum expectation value. If S is a hidden sector field, we could find [47]

$$\mu = \frac{\langle S^2 \rangle}{M} \sim \frac{\Lambda^2}{M} , \qquad (11.164)$$

A supersymmetric higher dimension contribution to the Kähler potential

$$\int d^4\theta \, \frac{1}{M} \Phi^\dagger H_d H_u \qquad (11.165)$$

leads to a μ term if Φ acquires a vacuum expectation value in its F term. If Φ is a hidden sector field, we could find [48]

$$\mu = \frac{\langle F_\Phi \rangle}{M} \sim \frac{\Lambda^2}{M} , \qquad (11.166)$$

In models with weak-coupling dynamics, higher-dimension operators are associated with the string or Planck scale; then, these mechanisms work most naturally if supergravity is the mediator. However, it is also possible to apply these strategies in models with strong-coupling dynamics in the hidden sector at an intermediate scale.

Generating the μ term typically requires breaking all continuous R-symmetries of the model. This is unfortunate, because an R-symmetry might be helpful phenomenologically, for example, to keep gaugino masses small while allowing sfermion masses to become large, or because it might be difficult to break an R-symmetry using a particular explicit mechanism of supersymmetry breaking. In this case, it is necessary to add Standard Model singlet fields to the MSSM to allow all gaugino and Higgsino fields to acquire nonzero masses. Models of this type are presented in [49,50].

The second problem involves the flavor structure of the soft supersymmetry breaking terms. In writing (11.161), I did not write flavor indices. In principle, these terms could have flavor-mixing that is arbitrary in structure and different from that in (11.143). Then the flavor-mixing would not be transformed away when (11.143) is put into canonical form. However, flavor-mixing from the soft supersymmetry breaking terms is highly constrained by experiment. Contributions such as the one shown in Fig. 11.7 give contributions to K^0, D^0, and B^0 mixing, and to $\tau \to \mu\gamma$ and $\mu \to e\gamma$, that can be large compared to the measured values or limits. Theories of the origin of the soft terms in models of supersymmetry breaking should address this problem. For example, the models of *gauge-mediated* [52] and *anomaly-mediated* [53,54] supersymmetry breaking induce soft terms that depend only on the $SU(2) \times U(1)$ quantum number and are therefore automatically diagonal in flavor. A quite different solution, based on a extension of the MSSM with a continuous R-symmetry, is presented in [51].

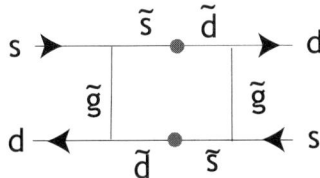

Fig. 11.7. A dangerous contribution to K-$\overline{K}$ mixing involving gluino exchange and flavor mixing in the squark mass matrix.

If I assume that the soft supersymmetry-breaking Lagrangian is diagonal in flavor but is otherwise arbitrary, it introduces 22 new parameters. With arbitrary flavor and CP violation, it introduces over 100 new parameters. This seems a large amount of parameter freedom. I feel that it is not correct, though, to think of these as new fundamental parameters in physics. The soft Lagrangian is computed from the physics of the hidden sector, and so we might expect that these parameters are related to one another as a part of a theory of supersymmetry breaking. Indeed, the values of these parameters are the essential data from which we will infer the properties of the hidden sector and its new high energy interactions.

If supersymmetry is discovered at the weak interaction scale, it will be a key problem to measure the coefficients in the soft Lagrangian and to understand their pattern and implications. Most of my discussion in the

11.4. The Mass Spectrum of the MSSM

11.4.1. Sfermion Masses

Our first task in this program is to ask how the parameters of the MSSM Lagrangian are reflected in the mass spectrum of the superparticles. The relation between the MSSM parameters and the particle masses is surprisingly complicated, even at the tree level. For each particle, we will need to collect all of the pieces of the Lagrangian (11.162) that can contribute to the mass term. Some of these will be direct mass contributions; others will contain Higgs fields and contribute to the masses when these fields obtain their vacuum expectation values. In this discussion, and in the remainder of these lectures, I will ignore all flavor-mixing.

Begin with the squark and slepton masses. For light quarks and leptons, we can ignore the fermion masses and Higgs couplings. Even with this simplification, though, there are two sources for the scalar masses. One is the soft mass term

$$\mathcal{L}_{soft} = -M_f^2 |\widetilde{f}|^2 \ . \tag{11.167}$$

The other comes from the D-term potential. The $SU(2)$ and $U(1)$ potentials contain the cross terms between the Higgs field and sfermion field contributions

$$V_D = \frac{g^2}{2} \cdot 2 \cdot (H_d^\dagger \frac{\sigma^3}{2} H_d + H_u^\dagger \frac{\sigma^3}{2} H_u) \cdot (\widetilde{f}^* t^3 \widetilde{f})$$
$$+ \frac{g'^2}{2} \cdot 2 \cdot (-\frac{1}{2} H_d^\dagger H_d + \frac{1}{2} H_u^\dagger H_u) \cdot (\widetilde{f}^* Y \widetilde{f}) \ . \tag{11.168}$$

To evalute this expression, we must insert the vacuum expectation values of the two Higgs fields. In terms of the angle β defined in (11.147), these are

$$\langle H_u \rangle = \begin{pmatrix} 0 \\ \frac{1}{\sqrt{2}} v \sin\beta \end{pmatrix} \qquad \langle H_d \rangle = \begin{pmatrix} \frac{1}{\sqrt{2}} v \cos\beta \\ 0 \end{pmatrix} , \tag{11.169}$$

where $v = 246$ GeV so that $m_W = gv/2$.

Inserting the Higgs vevs into the potential (11.168), we find

$$V_D = \tilde{f}^*[\frac{v^2}{4}(\cos^2\beta - \sin^2\beta)(g^2 I^3 - g'^2 Y)]\tilde{f}$$
$$= \tilde{f}^*[\frac{(g^2+g'^2)v^2}{4}\cos 2\beta\,(I^3 - s_w^2(I^3+Y))]\tilde{f}$$
$$= \tilde{f}^*[m_Z^2 \cos 2\beta(I^3 - s_w^2 Q)]\tilde{f}\,. \tag{11.170}$$

Then, if we define

$$\Delta_f = (I^3 - s_w^2 Q)\cos 2\beta\, m_Z^2\,, \tag{11.171}$$

the mass of a first- or second-generation sfermion takes the form

$$m_{\tilde{f}}^2 = M_f^2 + \Delta_f \tag{11.172}$$

when contributions proportional to fermion masses can be neglected. The D-term contribution can have interesting effects. For example, $SU(2)$ invariance of M_f^2 implies that

$$m^2(\tilde{e}) - m^2(\tilde{\nu}) = |\cos 2\beta|\, m_Z^2 > 0\,. \tag{11.173}$$

For some choices of parameters, the measurement of this mass difference is a good way to determine $\tan\beta$ [55].

For third-generation fermions, the contributions to the mass term from Yukawa couplings and from A terms can be important. For the $\tilde{b}$ and $\overline{\tilde{b}}$, these contributions come from the terms in the effective Lagrangian

$$|F_b|^2 + |F_{\overline{b}}|^2 = |y_b\langle H_d^0\rangle \tilde{b}|^2 + |y_b\overline{\tilde{b}}\langle H_d^0\rangle|^2 = m_b^2(|\tilde{b}|^2 + |\overline{\tilde{b}}|^2)$$
$$|F_{Hd}|^2 = (-\mu\langle H_d^0\rangle)^*(y_b\tilde{b}\overline{\tilde{b}}) + h.c. = -\mu m_b \tan\beta\,\tilde{b}\overline{\tilde{b}} + h.c.$$
$$-\mathcal{L}_{soft} = A_b y_b\langle H_d^0\rangle \tilde{b}\overline{\tilde{b}} = A_b m_b \tilde{b}\overline{\tilde{b}}\,. \tag{11.174}$$

In all, we find a mass matrix with mixing between the two scalar partners of the b quark,

$$\begin{pmatrix}\tilde{b}^* & \overline{\tilde{b}}\end{pmatrix}\mathcal{M}_b^2\begin{pmatrix}\tilde{b}\\ \overline{\tilde{b}}^*\end{pmatrix}, \tag{11.175}$$

with

$$\mathcal{M}_b^2 = \begin{pmatrix} M_b^2 + \Delta_b + m_b^2 & m_b(A_b - \mu\tan\beta) \\ m_b(A_b - \mu\tan\beta) & M_{\overline{b}}^2 + \Delta_{\overline{b}} + m_b^2 \end{pmatrix} \tag{11.176}$$

The mass matrix for $\tilde{\tau}$, $\overline{\tilde{\tau}}$ has the same structure. For $\tilde{t}$, $\overline{\tilde{t}}$, replace $\tan\beta$ by $\cot\beta$.

The mixing terms in the mass matrices of the third-generation sfermions often play an important role in the qualitative physics of the whole SUSY model. Because of the mixing, one sfermion eigenstate is pushed down in mass. This state is often the lightest squark or even the lightest superparticle in the theory.

11.4.2. Gaugino and Higgsino Masses

In a similar way, we can compute the mass terms for the gauginos and Higgsinos. Since the gauginos and Higgsino have the same quantum numbers after $SU(2) \times U(1)$ breaking, they will mix. We have seen in Section 2.4 that this mixing plays an essential role in the working of the Higgs mechanism in the limit where soft supersymmetry breaking terms are turned off.

The charged gauginos and Higgsinos receive mass from three sources. First, there is a soft SUSY breaking term

$$-\mathcal{L}_{soft} = m_2 \widetilde{w}^{-T} c \widetilde{w}^+ . \tag{11.177}$$

The μ superpotential term contributes

$$-\mathcal{L}_W = \mu \widetilde{h}_d^{-T} c \widetilde{h}_u^+ . \tag{11.178}$$

The gauge kinetic terms contribute

$$-\mathcal{L} = \sqrt{2} \frac{g}{\sqrt{2}} \left(\langle H_d^0 \rangle \, \widetilde{w}^{+T} c \widetilde{h}_d^- + \langle H_u^0 \rangle \, \widetilde{w}^{-T} c \widetilde{h}_u^+ \right) \tag{11.179}$$

Inserting the Higgs field vevs from (11.169), we find the mass term

$$\begin{pmatrix} \widetilde{w}^{-T} & \widetilde{h}_d^{-T} \end{pmatrix} c \, m_C \begin{pmatrix} \widetilde{w}^+ \\ \widetilde{h}_u^+ \end{pmatrix} , \tag{11.180}$$

with

$$m_C = \begin{pmatrix} m_2 & \sqrt{2} m_W \sin \beta \\ \sqrt{2} m_W \cos \beta & \mu \end{pmatrix} . \tag{11.181}$$

The mass matrix for neutral gauginos and Higgsinos also receives contributions from these three sources. In this case, all four of the states

$$(\widetilde{b}, \widetilde{w}^0, \widetilde{h}_d^0, \widetilde{h}_u^0) \tag{11.182}$$

have the same quantum numbers after $SU(2) \times U(1)$ breaking and can mix together. The mass matrix is

$$m_N = \begin{pmatrix} m_1 & 0 & -m_Z c_\beta s_w & m_Z s_\beta s_w \\ 0 & m_2 & m_Z c_\beta c_w & -m_Z s_\beta c_w \\ -m_Z c_\beta s_w & m_Z c_\beta c_w & 0 & -\mu \\ m_Z s_\beta s_w & -m_Z s_\beta c_w & -\mu & 0 \end{pmatrix}. \quad (11.183)$$

The mass eigenstates in these systems are referred to collectively as *charginos* and *neutralinos*. The matrix (11.183) is complex symmetric, so it can be diagonalized by a unitary matrix V_0,[c]

$$m_N = V_0^* D_N V_0^\dagger . \quad (11.184)$$

I will denote the neutralinos as $\widetilde{N}_i^0$, $i = 1, \ldots, 4$, in order of mass with $\widetilde{N}_1^0$ the lightest. Elsewhere in the literature, you will see these states called $\widetilde{\chi}_i^0$ or $\widetilde{Z}_i^0$. The mass eigenstates are related to the weak eigenstates by the transformation

$$\begin{pmatrix} \widetilde{b}^0 \\ \widetilde{w}^0 \\ \widetilde{h}_d^0 \\ \widetilde{h}_u^0 \end{pmatrix} = V_0 \begin{pmatrix} \widetilde{N}_1 \\ \widetilde{N}_2 \\ \widetilde{N}_3 \\ \widetilde{N}_4 \end{pmatrix} . \quad (11.185)$$

Note that the diagonal matrix D_N in (11.184) may have negative or complex-valued elements. If that is true, the physical fermion masses of the $\widetilde{N}_i$ are the absolute values of the corresponding elements of D_N. The phases will appear in the three-point couplings of the $\widetilde{N}_i$ and can lead to observable interference effects. Complex phases in D_N would provide a new source of CP violation.

The chargino mass matrix (11.181) is not symmetric, so in general it is diagonalized by two unitary matrices

$$m_C = V_-^* D_C V_+^\dagger . \quad (11.186)$$

I will denote the charginos as $\widetilde{C}_i^\pm$, $i = 1, 2$, in order of mass with $\widetilde{C}_1^\pm$ the lighter. Elsewhere in the literature, you will see these states called $\widetilde{\chi}_i^\pm$ or $\widetilde{W}_i^\pm$. The mass eigenstates are related to the weak eigenstates by the transformation

$$\begin{pmatrix} \widetilde{w}^+ \\ \widetilde{h}_u^+ \end{pmatrix} = V_+ \begin{pmatrix} \widetilde{C}_1^+ \\ \widetilde{C}_2^+ \end{pmatrix}, \quad \begin{pmatrix} \widetilde{w}^- \\ \widetilde{h}_u^- \end{pmatrix} = V_- \begin{pmatrix} \widetilde{C}_1^- \\ \widetilde{C}_2^- \end{pmatrix} . \quad (11.187)$$

---
[c]Note that this formula is different from that which diagonalizes a Hermitian matrix. A detailed discussion of the diagonalization of mass matrices appearing in SUSY can be found in the Appendix of [56].

It should be noted that μ are must be nonzero. If $\mu = 0$, the determinant of (11.183) vanishes and so the lightest neutralino must be massless. This neutralino will also have a large Higgsino content and thus an order-1 coupling to the Z^0. It is excluded by searches for an excess of invisible Z^0 decays and for $Z^0 \to \widetilde{N}_1 \widetilde{N}_2$. The condition $\mu = 0$ also implies that the lightest chargino has a mass below the current limit of about 100 GeV.

Often, one studies models for which m_1, m_2, and μ are all large compared to m_W and m_Z. The off-diagonal elements that mix the gaugino and Higgsino states are of the order of m_W and m_Z. Thus, if the scale of masses generated by the SUSY breaking terms is large, the mixing is small and the individual eigenstates are mainly gaugino or mainly Higgsino. However, there are two distinct cases. The first is the *gaugino region*, where $m_1, m_2 < |\mu|$. In this region of parameter space, the lightest states $\widetilde{N}_1$, $\widetilde{C}_1$ are mainly gaugino, while the heavy neutralinos and charginos are mainly Higgsino. In the *Higgsino region*, $m_1, m_2 > |\mu|$, the situation is reversed and $\widetilde{N}_1$, $\widetilde{C}_1$ are mainly Higgsino. In this case, the two lightest neutralinos are almost degenerate. In Fig. 11.8, I show the mass eigenvalues as a function of the mass matrix parameters along a line in the parameter space on which the $\widetilde{N}_1$ has a fixed mass of 100 GeV. As we will see in Section 6.4, the exact makeup of the lightest neutralino as a mixture of gaugino and Higgsino components is important to the study of supersymmetric dark matter.

To summarize this discussion, I present in Fig. 11.9 the complete spectrum of new particles in the MSSM at a representative point in its parameter space. Notice that the third-generation sfermions are split off from the others in each group. Note also that the parameter point chosen is in the gaugino region. The lightest superparticle is the $\widetilde{N}_1$. I will discuss the spectrum of Higgs bosons in Section 6.2.

11.4.3. Renormalization Group Evolution of MSSM Parameters

The spectrum shown in Fig. 11.9 appears to have been generated by assigning random values to the soft SUSY breaking parameters. But, actually, I generated this spectrum by making very simple assumptions about the relationships of the soft parameters, at a high energy scale. Specifically, I assumed that the soft SUSY breaking gaugino masses and (separately) the sfermion masses were equal at the scale of grand unification. The structure that you see in the figure is generated by the renormalization group

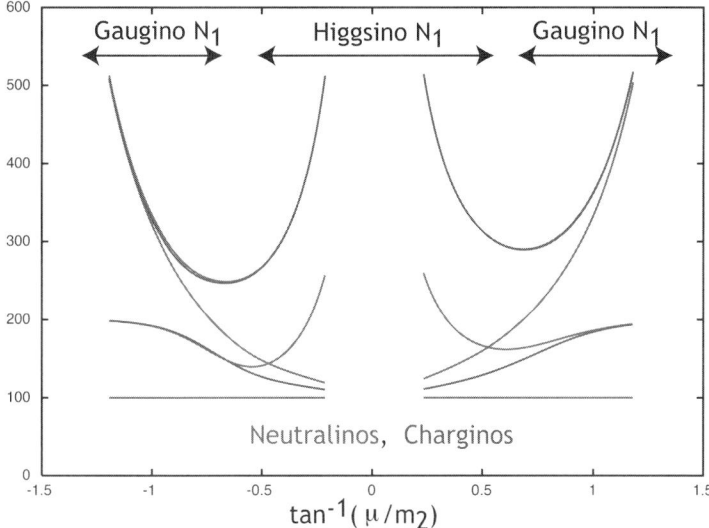

Fig. 11.8. Masses of the four neutralinos and two charginos along a line in the SUSY parameter space on which $m(\widetilde{N}_1^0) = 100$ GeV while the parameter μ moves from large negative to large positive values. The parameter m_1 is set to $m_1 = 0.5m_2$. Note the approximate degeneracies in the extreme limits of the gaugino and Higgsino regions.

evolution of these parameters from the grand unification scale to the weak scale.

The renomalization group (RG) evolution of soft parameters is likely to play a very important role in the interpretation of measurements of the SUSY particle masses. Essentially, after measuring these masses, it will be necessary to decode the results by running the effective mass parameters up to a higher energy at which their symmetries might become more apparent. The situation is very similar to that of the Standard Model coupling constants, where a renormalization group analysis told us that the apparently random values (11.137) for the coupling constants at the weak scale actually corresponds to a unification of couplings at a much higher scale.

In this section, I will write the most basic RG equations for the soft gaugino and sfermion masses. One further effect, which involves the Yukawa couplings and is important for the third generation, will be discussed later in Section 6.1.

The RG equation for the gaugino masses is especially simple. This

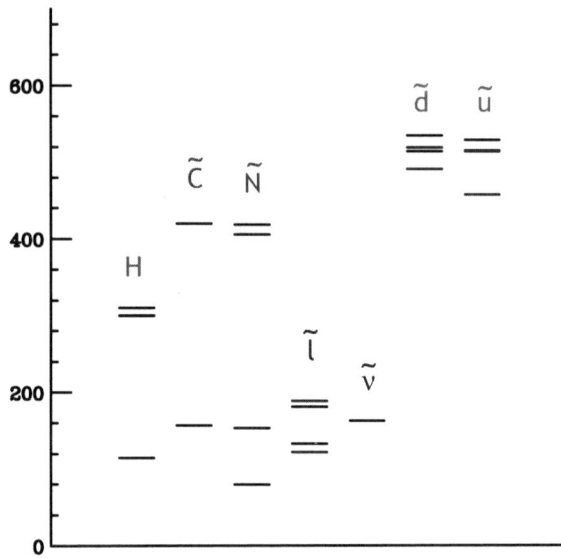

Fig. 11.9. Illustrative spectrum of supersymmetric particles. The columns contain, from the left, the Higgs bosons, the four neutralinos, the two charginos, the charged sleptons, the sneutrinos, the down squarks, and the up squarks. The gluino, not shown, is at about 800 GeV.

is because both the gaugino masses and the gauge couplings arise from the superpotential term (11.115), with the supersymmetry breaking terms arising as shown in (11.159). As I have already noted, this F-term receives a radiative correction proportional to the β function as a consequence of the trace anomaly [34,36]. The corrections are the same for the gauge boson field strength and the gaugino mass. Thus, if gaugino masses and couplings are generated at the scale M, they have the relation after RG running to the scale Q:

$$\frac{m_i(Q)}{m_i(M)} = \frac{\alpha_i(Q)}{\alpha_i(M)} . \qquad (11.188)$$

If the F term that generates the soft gaugino masses is an $SU(5)$ singlet, the soft gaugino masses will be grand-unified at M. Then, running down to the weak scale, they will have the relation

$$m_1 : m_2 : m_3 = \alpha_1 : \alpha_2 : \alpha_3 = 0.5 : 1 : 3.5 . \qquad (11.189)$$

This relation of soft gaugino masses is known as *gaugino unification*.

There are other models of the soft gaugino masses that also lead to gaugino unification. In *gauge-mediated SUSY breaking*, the dynamics responsible for SUSY breaking occurs at a scale much lower than the scale associated with mediation by supergravity. At this lower scale M_g (for example, 1000 TeV), some heavy particles with nontrivial $SU(3) \times SU(2) \times U(1)$ quantum numbers acquire masses from SUSY breaking. These fields then couple to gauginos and generate SUSY breaking masses for those particles through the diagram shown in Fig. 11.10(a). The heavy particles must fall into complete $SU(5)$ representations; otherwise, the coupling constant renormalization due to these particles between M_g and the grand unification scale would spoil the grand unification of the gauge couplings. Then the diagram in Fig. 11.10(a) generates soft gaugino masses proportional to $\alpha(M_g)$. Running these parameters down to the weak scale, we derive the relation (11.189) from this rather different mechanism.

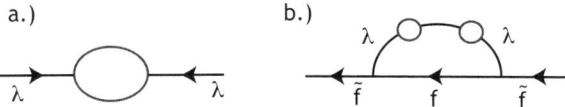

Fig. 11.10. Diagrams that generate the soft mass parameters in gauge mediated supersymmetry breaking: (a.) gaugino masses; (b.) sfermion masses.

Now let us turn to the RG running of soft scalar masses. In principle, there are two contributions, one from the RG rescaling of the soft mass term M_f^2 and one from RG evolution generating M_f^2 from the gaugino mass. The Feynman diagrams that contribute to the RG coefficients are shown in Fig. 11.11. The two one-loop diagrams proportional to M_f^2 cancel. The third diagram, involving the gaugino mass, gives the RG equation

$$\frac{dM_f^2}{d\log Q} = -\frac{2}{\pi} \sum_i \alpha_i(Q) C_2(r_i) m_i^2(Q) , \qquad (11.190)$$

with $i = 1, 2, 3$ and $C_2(r_i)$ the squared charge in the fermion representation r_i under the gauge group i. This equation leads to a positive contribution to M_f^2 as one runs the RG evolution from the messenger scale down to the weak scale. The effect is largest for squarks, for which the SUSY breaking mass is induced from the gluino mass.

As an example of this mechanism of mass generation, assume gaugino unification and assume that $M_f^2 = 0$ for all sfermions at the grand unifica-

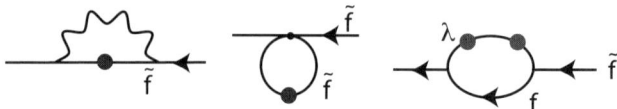

Fig. 11.11. Diagrams that generate the renormalization group evolution of the soft sfermion mass parameters M_f^2.

tion scale. Then the weak scale sfermion masses will be in the ratio

$$M(\widetilde{\overline{e}}) : M(\widetilde{e}) : M(\widetilde{\overline{d}}) : M(\widetilde{\overline{u}}) : M(\widetilde{d},\widetilde{u}) : m_2$$
$$= 0.5 : 0.9 : 3.09 : 3.10 : 3.24 : 1 \qquad (11.191)$$

This model of fermion mass generation is called *no-scale* SUSY breaking. It has the danger that the lightest stau mass eigenstate could be lighter than than the $\widetilde{N}_1$, leading to problems for dark matter. This problem can be avoided by RG running above the GUT scale [57]. Alternatively, it might actually be that the lightest Standard Model superpartner is a long-lived stau that eventually decays to a tau and a gravitino [58,59].

In gauge-mediated SUSY breaking, the diagram shown in Fig. 11.10(b) leads to the qualitatively similar but distinguishable formula

$$M_f^2 = 2 \sum_i \alpha_i^3(M) C_2(r_i) \cdot \left(\frac{m_2}{\alpha_2}\right)^2 . \qquad (11.192)$$

Each model of SUSY breaking leads to its own set of relations among the various soft SUSY breaking parameters. In general, the relations are predicted for the parameters defined at the messenger scale and must be evolved to the weak scale by RG running to be compared with experiment. Figure 11.12 shows four different sets of high-scale boundary conditions for the RG evolution, and the corresponding evolution to the weak scale. If we can measure the weak-scale values, we could try to undo the evolution and recognize the pattern. This will be a very interesting study for the era in which superparticles are observed at high energy colliders.

There are some features common to these spectra that are general features of the RG evolution of soft parameters:

(1) The pairs of sleptons $\widetilde{\overline{e}}$ and $\widetilde{e}$ can easily acquire a significant mass difference from RG evolution, and they might also have a different initial condition. It is important to measure the mass ratio $m(\widetilde{\overline{e}})/m(\widetilde{e})$ as a diagnostic of the scheme of SUSY breaking.

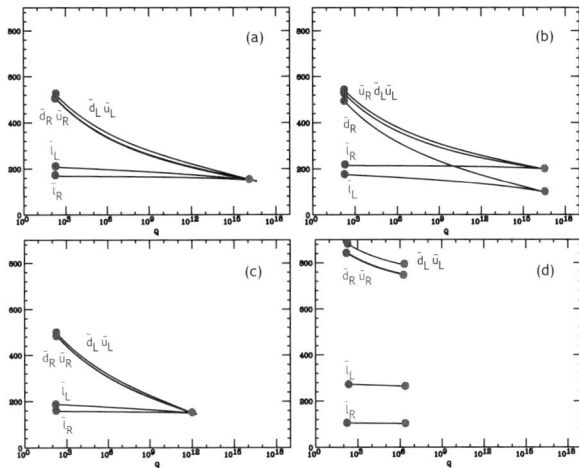

Fig. 11.12. Evolution of squark and slepton masses from the messenger scale down to the weak scale, for four different models of supersymmetry breaking: (a.) universal sfermion masses at the grand unification scale M_U; (b.) sfermion masses at M_U that depend on the $SU(5)$ representation; (c.) universal sfermion masses at an intermediate scale; (d.) gauge mediation from a sector of mass about 1000 TeV.

(2) Gaugino unification is a quantitative prediction of certain schemes of SUSY breaking. It is important to find out whether this relation is correct or not for the real spectrum of superparticles in Nature.

(3) When the RG effects on the squark masses dominate the values of M_f^2 from the initial condition, the various species of squark have almost the same mass and are much heavier than the sleptons. It is important to check whether most or all squarks appear at the same threshold.

11.5. The Measurement of Supersymmetry Parameters

11.5.1. Measurements of the SUSY Spectrum at the ILC

Now that we have discussed the physics that determines the form of the spectrum of superparticles, we turn to the question of how we would determine this spectrum experimentally. This is not as easy as it might seem. In this section, I will consider only models in which the dark matter particle is the $\widetilde{N}_1$, and all other SUSY particles decay to the $\widetilde{N}_1$. This neutral and weakly interacting particle would escape a collider detector unseen. Nevertheless, methods have been worked out not only to measure the masses of superparticles but also to determine mixing angles and other information needed to convert these masses to values of the underlying parameters of the MSSM Lagrangian.

Similar methods apply to other scenarios. For example, in models in which the neutralino decays to a particle with gravitational interactions, one would add that decay, if it is visible, to the analyses that I will present. It is possible in models of this type that the lightest Standard Model superpartner would be a charged slepton that is stable on the time scale of particle physics experiments. That scenario would produce very striking and characteristic events [58].

Most likely, this experimental study of the SUSY spectrum will begin in the next few years with the LHC experiments. However, at a hadron collider like the LHC, much of the kinematic information on superparticle production is missing and so special tricks are needed even to measure the spectrum. The study of supersymmetry should be much more straightforward at an e^+e^- collider such as the planned International Linear Collider (ILC). For this reason, I would like to begin my discussion of the experiments in this section by discussing SUSY spectrum measurements at e^+e^- colliders. More complete reviews of SUSY measurements at linear colliders can be found in [60,61].

I first discuss slepton pair production, beginning with the simplest process, $e^+e^- \to \tilde{\mu}^+\tilde{\mu}^-$ and considering successively the production of $\tilde{\tau}$ and $\tilde{e}$. Each step will bring in new complexities and will allow new measurements of the SUSY parameters.

The process $e^+e^- \to \tilde{\mu}^+\tilde{\mu}^-$, where $\tilde{\mu}$ is the partner of either the left- or right-handed μ, can be analyzed with the simple formulae for scalar particle-antiparticle production. The cross section for pair production from polarized initial electrons and positrons to final-state scalars with definite $SU(2) \times U(1)$ quantum numbers is given by

$$\frac{d\sigma}{d\cos\theta} = \frac{\pi\alpha^2}{2s}\beta^3 \sin^2\theta \, |f_{ab}|^2 \,, \qquad (11.193)$$

where

$$f_{ab} = 1 + \frac{(I_e^3 + s_w^2)(I_\mu^3 + s_w^2)}{c_w^2 s_w^2} \frac{s}{s - m_Z^2} \qquad (11.194)$$

and, in this expression, $I^3 = -\frac{1}{2}, 0$ for $a, b = L, R$. For the initial state, $a = L$ denotes the state $e_L^- e_R^+$ and $a = R$ denotes $e_R^- e_L^+$. For the final state, $b = L$ denotes the $\widetilde{\mu}$, $b = R$ the $\overline{\widetilde{\mu}}$. Notice that this cross section depends strongly on the polarization states:

$$\begin{aligned}
|f_{ab}|^2 &= 1.69 & e_R^- e_L^+ &\to \widetilde{\mu}^+ \overline{\widetilde{\mu}}^- \\
&= 0.42 & e_L^- e_R^+ &\to \widetilde{\mu}^+ \overline{\widetilde{\mu}}^- \\
&= 0.42 & e_R^- e_L^+ &\to \widetilde{\mu}^+ \widetilde{\mu}^- \\
&= 1.98 & e_L^- e_R^+ &\to \widetilde{\mu}^+ \widetilde{\mu}^-
\end{aligned} \qquad (11.195)$$

The angular distribution is characteristic of pair-production of a spin 0 particle; the normalization of the cross sections picks out the the correct set of $SU(2) \times U(1)$ quantum numbers.

If the smuon is light, its only kinematically allowed decay might be $\widetilde{\mu} \to \mu \widetilde{N}_1^0$. Even if the smuon is heavy, if the $\widetilde{N}_1$ is mainly gaugino, this decay should be important. As noted above, I am assuming that R-parity is conserved and that the $\widetilde{N}_1$ is the lightest particle in the superparticle spectrum. Then events with this decay on both sides will appear as

$$e^+ e^- \to \mu^+ \mu^- + (\text{missing } E \text{ and } p) \qquad (11.196)$$

The spectrum of the observed muons is very simple. Since the $\widetilde{\mu}$ has spin 0, it decays isotropically in its own rest frame. In $e^+ e^-$ production at a definite center of mass energy, the $\widetilde{\mu}$ is produced at a definite energy, and thus with a definite boost, in the lab. The boost of an isotropic distribution is a flat distribution in energy. So, the muon energy distribution should be flat, between endpoints determined by kinematics, as shown in the idealized Fig. 11.13.

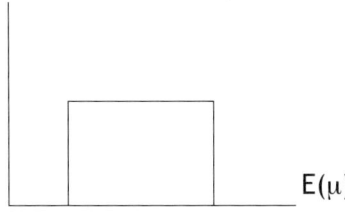

Fig. 11.13. Schematic energy distribution of final-state muons in $e^+ e^- \to \widetilde{\mu}^+ \widetilde{\mu}^-$.

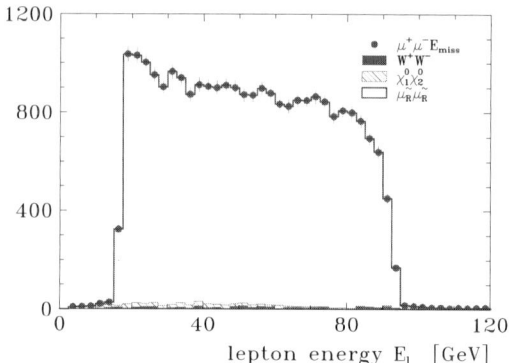

Fig. 11.14. Energy distribution of muons from $e^+e^- \to \widetilde{\mu}^-\widetilde{\mu}^+$ at the ILC, in a simulation by Blair and Martyn that includes realistic momentum resolution and beam effects [63].

The endpoint positions are simple functions of the mass of the $\widetilde{\mu}$ and the mass of the $\widetilde{N}_1$,

$$E_\pm = \gamma(1 \pm \beta)\, \frac{m^2(\widetilde{\mu}) - m^2(\widetilde{N}_1)}{2m(\widetilde{\mu})}\ , \qquad (11.197)$$

where $\gamma = E_{\rm CM}/2m(\widetilde{\mu})$, $\beta = (1 - 4m^2(\widetilde{\mu})/E_{\rm CM}^2)^{1/2}$. If we can identify both endpoint positions, we can solve for the two unknown masses. Figure 11.14 shows a simulation of the reconstructed smuon energy distribution from $\widetilde{\mu}$ pair production at the ILC [63]. The high-energy edges of the distributions are rounded because of initial-state radiation in the e^+e^- collision. The experimenters expect to be able to measure this effect and correct for it. Then they should obtain values of the smuon mass to an accuracy of about one hundred MeV, or one part per mil.

A similar analysis applies to $e^+e^- \to \widetilde{\tau}^+\widetilde{\tau}^-$, but there are several complications. First, for the τ system, mixing between the $\widetilde{\tau}$ and the $\widetilde{\overline{\tau}}$ might be important, especially if $\tan\beta$ is large. The production cross sections are affected directly by the mixing. For example, to compute the pair-production of the lighter $\widetilde{\tau}$ mass eigenstate from a polarized initial state, $e_R^- e_L^+ \to \widetilde{\tau}_1^- \widetilde{\tau}_1^+$, we must generalize (11.193) to

$$\frac{d\sigma}{d\cos\theta} = \frac{\pi\alpha^2}{2s}\beta^3 \sin^2\theta\, |f_{R1}|^2\ , \qquad (11.198)$$

where

$$f_{R1} = f_{RR} \cos^2 \theta_\tau + f_{RL} \sin^2 \theta_\tau \qquad (11.199)$$

and θ_τ is the mixing angle associated with the diagonalization of the $\tilde{\tau}$ case of (11.176).

Second, while the $\tilde{\tilde{\tau}}^-$ can decay to $\tau_R^- \tilde{b}$ through gauge couplings, this weak eigenstate can also decay to $\tau_L^- \tilde{h}_d$ through terms proportional to the Yukawa coupling. Both decay amplitudes contribute to the observable decay $\tilde{\tau}_1 \to \tau \tilde{N}_1^0$. With the $\tilde{\tau}$ mixing angle fixed from the measurement of the cross section, the τ polarization in $\tilde{\tau}$ decays can be used to determine the mixing angles in the diagonalization of the neutralino mass matrix (11.183) [62].

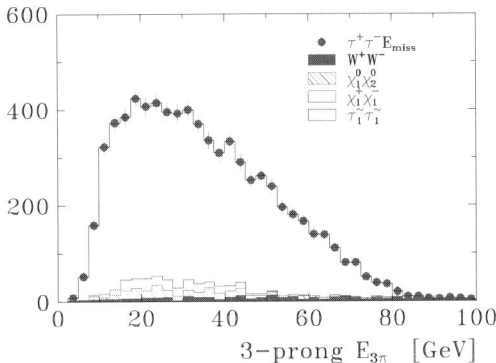

Fig. 11.15. Energy distribution of the three-pion system from $e^+e^- \to \tilde{\tau}_1^- \tilde{\tau}_1^+$ at the ILC, with a τ decay to 3π, in a simulation by Blair and Martyn that includes realistic momentum resolution and beam effects. [63].

In Fig. 11.15, I show the distribution of total visible energy in $\tilde{\tau} \to 3\pi + \nu + \tilde{N}_1^0$ at the ILC. Though there is no longer a sharp feature at the kinematic endpoint, it is still possible to accurately determine the $\tilde{\tau}$ mass by fitting the shape of this distribution.

The physics of $e^+e^- \to \tilde{e}^+\tilde{e}^-$ brings in further new features. In this case, there is a new Feynman diagram, involving t-channel neutralino exchange. The two diagrams contributing to the cross section for this process are shown in Fig. 11.16. The t-channel diagram turns out to be the more important one, dominating the s-channel gauge boson exchange and generating a large forward peak in selectron production. The cross section for

$e_R^- e_L^+ \to \tilde{e}^- \tilde{e}^+$ is given by another generalization of (11.193),

$$\frac{d\sigma}{d\cos\theta} = \frac{\pi\alpha^2}{2s} \beta^3 \sin^2\theta \, |\mathcal{F}_{RR}|^2 \,, \tag{11.200}$$

where

$$\mathcal{F}_{RR} = f_{RR} - \sum_i \left|\frac{V_{01i}}{c_w}\right|^2 \frac{s}{m_i^2 - t} \,, \tag{11.201}$$

with the sum running over neutralino mass eigenstates. The factor V_{01i} is a matrix element of the unitary matrix introduced in (11.184).

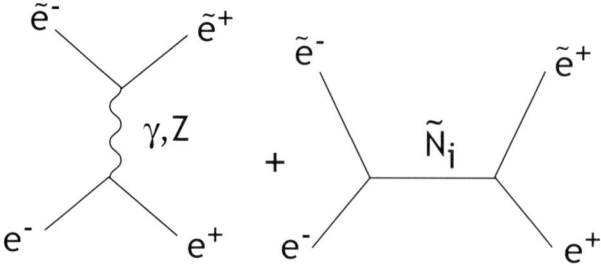

Fig. 11.16. Feynman diagrams contributing to $e^+ e^- \to \tilde{e}^- \tilde{e}^+$.

The t-channel diagram also allows new processes such as $e_L^- e_L^+ \to \tilde{e}^- \tilde{e}^+$. Note the correlation of the initial-state electron and position spins with the identities of the final-state selectrons. A complete set of polarized cross sections for selectron pair production in $e^+ e^-$ and $e^- e^-$ collisions can be found in [64].

The cross sections for chargino and neutralino pair production in $e^+ e^-$ collisions are somewhat more complicated, but still there are interesting things to say about these processes. Chargino pair production is given by the Feynman diagrams shown in Fig. 11.17. These diagrams are just the supersymmetric analogues of the diagrams for $e^+ e^- \to W^+ W^-$. As in that process, the most charcteristic final states are those with a hadronic decay on one side of the event and a leptonic decay on the other side, for example,

$$\tilde{C}_1^+ \to \ell^+ \nu \tilde{N}_1^0 \,, \quad \tilde{C}_1^- \to d\bar{u} \tilde{N}_1^0 \,. \tag{11.202}$$

A typical event of this kind is shown in Fig. 11.18.

The chargino and neutralino production cross sections have a strong dependence on the mixing angles in (11.184) and (11.186) and offer a number

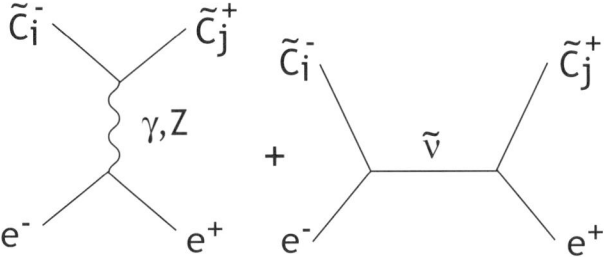

Fig. 11.17. Feynman diagrams contributing to $e^+e^- \to \widetilde{C}_i^- \widetilde{C}_j^+$.

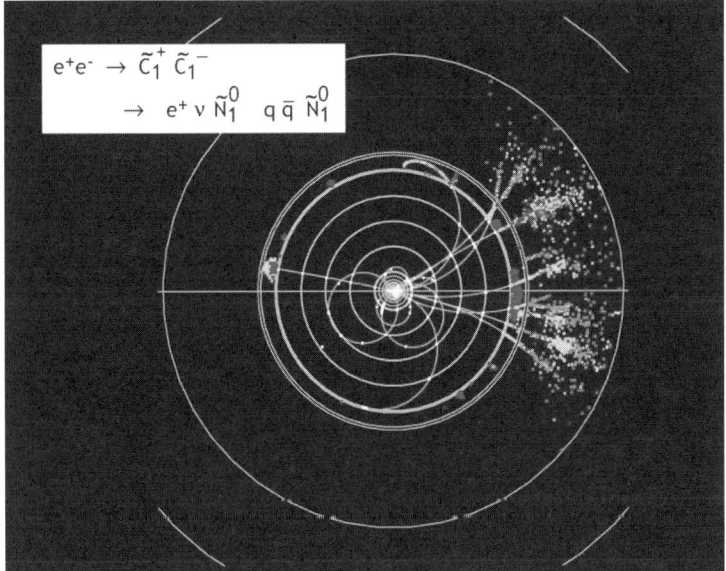

Fig. 11.18. A simulated chargino pair production event at the ILC [65].

of strategies for the determination of these mixing angles. Let me present one such strategy here. Consider the reaction from a polarized initial state $e_R^- e_L^+ \to \widetilde{C}_1^- \widetilde{C}_1^+$. Since we have an initial e_R^-, the t-channel diagram vanishes because the right-handed electron does not couple to the neutrino. Now simplify the s-channel diagram by considering the limit of high energies, $s \gg m_Z^2$. In this limit, it is a good approximation to work with weak gauge eigenstates (B^0, W^0) rather than the mass eigenstates (γ, Z^0). The weak eigenstate basis gives a nice simplification. The initial e_R^- couples only to B^0. But $\widetilde{w}^\pm$ couple only to W^0, so at high energy the s-channel diagram gets contributions only from the Higgsino components of the $\widetilde{C}_1^-$ and $\widetilde{C}_1^+$ eigenstates. If we go to still higher energies, $s \gg m(\widetilde{C}_1)^2$, there is a further simplification. The cross section for $\widetilde{h}_R^- \widetilde{h}_L^+$ production is forward-peaked, and the cross section for $\widetilde{h}_L^- \widetilde{h}_R^+$ production is backward-peaked. Then, the cross section for $e_R^- e_L^+ \to \widetilde{C}_1^- \widetilde{C}_1^+$ takes the form

$$\frac{d\sigma}{d\cos\theta} \sim \frac{\pi\alpha^2}{8c_w^2 s}\left[|V_{+21}|^4(1+\cos\theta)^2 + |V_{-21}|^2(1-\cos\theta)^2\right] . \quad (11.203)$$

In this limit, it is clear that we can read off both of the mixing angles in (11.186) from the shape of this cross section.

The use of high-energy limits simplified this analysis, but the sentivity of this cross section to the chargino mixing angles is not limited to high energy. Even relatively close to threshold, the polarized cross sections for chargino production depend strongly on the chargino mixing angles and can be used to determine their values. In Fig. 11.19, I show contours of constant cross section for $e_R^- e_L^+ \to \widetilde{C}_1^- \widetilde{C}_1^+$ in the (m_2, μ) plane (for $\tan\beta = 4$ and assuming gaugino unification) [66]. The value of this cross section is always a good measure of whether the SUSY parameters in Nature put us in the gaugino or the Higgsino region of Fig. 11.8.

11.5.2. Observation of SUSY at the LHC

Now we turn to supersymmetry production processes at the LHC. This subject, though more difficult, has immediate importance, since the LHC experiments are just about to begin.

The reactions that produce superparticles are typically much more complicated at hadron colliders than at lepton colliders. This is true for several reasons. High energy collisions of hadrons are intrinsically more complicated because the final states include the fragments of the initial hadrons that do not participate in the hard reaction. More importantly, the dominant reactions at hadron colliders are those that involve strongly interacting superparticles. This means that the primary particles are typically the heavier ones in the spectrum, which then decay in several steps. In addition, large backgrounds from QCD obscure the signatures of supersymmetric particle production in many channels.

Because of these difficulties, there is some question whether SUSY particle production can be observed at the LHC. However, as I will explain,

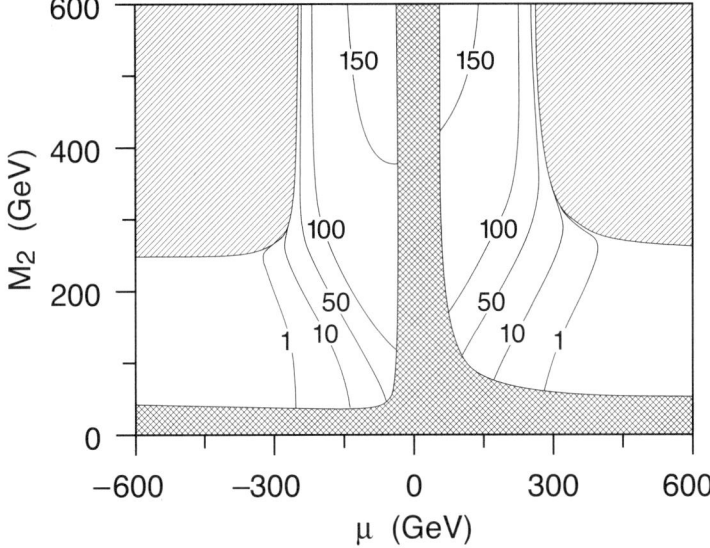

Fig. 11.19. Contours of constant cross section for the process $e_R^- e_L^+ \to C_1^- C_1^+$ (in fb, for $E_{\rm CM} = 500$ GeV), as a function of the underlying SUSY parameters [66]. The region shown is that in which the lightest chargino mass varies from 50 to 200 GeV. For fixed $\widetilde{C}_1^+$ mass, the cross section increases from zero to about 150 fb as we move from the gaugino region into the Higgsino region.

the signatures of supersymmetry are still expected to be striking and characteristic. It is not so clear, though, to what extent it is possible to measure the parameters of the SUSY Lagrangian, as I have described can be done from ILC experiments. This is an important study that still offers much room for new ideas.

The discovery of SUSY particles at the LHC and the measurement of SUSY parameters has been analyzed with simulations at a number of parameter points. Collections of interesting studies can be found in [63,67,68].

The dominant SUSY production processes at the LHC are

$$gg \to \widetilde{g}\widetilde{g} \, , \, \widetilde{q}\widetilde{q}^* \qquad gq \to \widetilde{g}\widetilde{q} \qquad (11.204)$$

These cross sections are large—tens of pb in typical cases. The values of numerous SUSY production cross sections at the LHC are shown in Fig. 11.20 [70].

We have seen that the squarks and gluinos are typically the heaviest

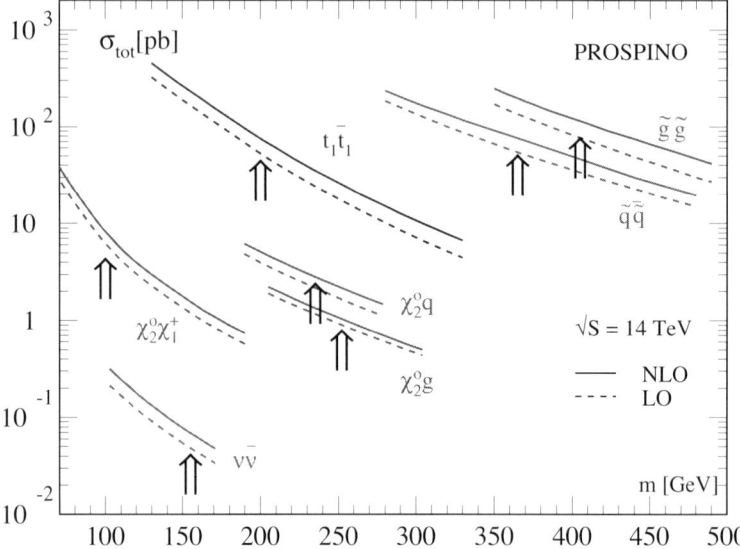

Fig. 11.20. Cross sections for the pair-production of supersymmetric particles at the LHC, from [70].

particles in the supersymmetry spectrum. The gluinos and squarks thus will decay to lighter superparticles. Some of these decays are simple, *e.g.*,

$$\widetilde{q} \to \overline{q}\widetilde{N}_1^0 \ . \tag{11.205}$$

However, other decays can lead to complex decay chains such as

$$\widetilde{q} \to q\widetilde{N}_2^0 \to q(\ell^+\ell^-)\widetilde{N}_1^0 \ , \qquad \widetilde{g} \to u\overline{d}\widetilde{C}_1^+ \to u\overline{d}W^+\widetilde{N}_1^0 \ . \tag{11.206}$$

With the assumptions that R-parity is conserved and that the N_1^0 is the LSP, all SUSY decay chains must end with the N_1^0, which is stable and very weakly interacting. SUSY production processes at hadron colliders then have unbalanced visible momentum, accompanied by multiple jets and, possibility, isolated leptons or W and Z bosons. Momentum balance along the beam direction cannot be checked at hadron colliders, because fragments of the initial hadrons exit along the beam directions, but an imbalance of transverse momentum will be visible and can be a characteristic signature of new physics. SUSY events contain this signature and the general large activity characteristic of heavy particle production. A simulated event of this type is shown in Fig. 11.21.

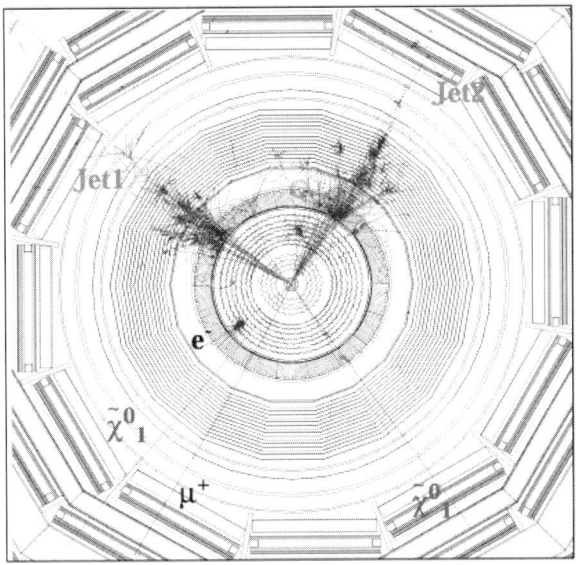

Fig. 11.21. Simulated SUSY particle production event in the CMS detector at the LHC [69].

Figure 11.22 shows a set of estimates given by Tovey and the ATLAS collaboration of the discovery potential for SUSY as a function of the LHC luminosity [71]. The most important backgrounds come from processes that are themselves relatively rare Standard Model reactions with heavy particle production,

$$pp \to (W, Z, t\bar{t}) + \text{jets} \ . \tag{11.207}$$

With some effort, we can experimentally normalize and control these backgrounds and reliably discovery SUSY production as a new physics process. In the figure, the contours for 5σ excesses of events above these backgrounds for various signatures of SUSY events are plotted as a function of the so-called 'mSUGRA' parameters. The SUSY models considered are defined as follows: Assume gaugino unification with a universal gaugino mass $m_{1/2}$ at the grand unification scale. Assume also that all scalar masses, including the Higgs boson mass parameters, are unified at the grand unification scale at the value m_0. Assume that the A parameter is universal at the grand unification scale; in the figures, the value $A = 0$ is used. Fix the value of $\tan \beta$ at the weak scale. Then it is possible to solve for μ and B, up to a sign, from the condition that electroweak symmetry is broken in such a way as to give the observed value of the Z^0 mass. (I will describe this calculation in Section 6.1.) This gives a 4-parameter subspace of the full 24-dimensional parameter space of the CP- and flavor-conserving MSSM, with the parameters

$$m_0 \ , \ m_{1/2} \ , \ A \ , \ \tan\beta \ , \ \text{sign}(\mu) \ . \tag{11.208}$$

This subspace is often used to express the results of phenomenological analyses of supersymmetry. In interpreting such results, one should remember that this choice of parameters is used for simplicity rather than being motivated by physics.

The figure shows contours below which the various signatures of supersymmetry significantly modify the Standard Model expectations. For clarity, the contours of constant squark and gluino mass are also plotted. The left-hand plot shows Tovey's results for the missing transverse momentum plus multijets signature at various levels of LHC integrated luminosity. It is remarkable that, in the models in which the squark or gluino mass is below 1 TeV, SUSY should be discoverable with a data sample equivalent to a small fraction of a year of running. The right-hand plot shows the contours for the discovery of a variety of SUSY signals, with up to three leptons plus jets plus missing transverse momentum, with roughly one year of data at the initial design luminosity. The signals are, as I have described, relatively robust with repect to uncertainties in the Standard Model backgrounds. This makes it very likely that, if SUSY is really present in Nature as the explanation of electroweak symmetry breaking, we will discover it at the LHC.

The general characteristics of SUSY events also allow us to estimate the SUSY mass scale in a relatively straightforward way. In Fig. 11.23, I show a correlation pointed out by Hinchliffe and collaborators [72] between the lighter of the squark and gluino masses and the variable

$$M_{eff} = \not{E}_T + \sum_1^4 E_{Ti} \tag{11.209}$$

given by the sum of the transverse momenta of the four highest E_T jets together with the value of the missing transverse momentum. The correlation applies reasonably well to mSUGRA models. In other models with smaller mass gaps between the squarks and the lightest neutralino, this relation can break down, but M_{eff} still measures the mass difference between the squark or gluino and the $\widetilde{N}_1^0$ [73]. Some more sophisticated techniques for

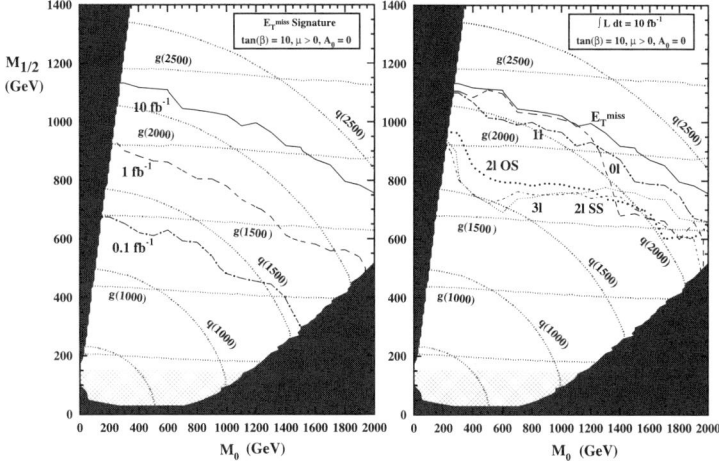

Fig. 11.22. Estimates by the ATLAS collaboration of the observability of various signatures of SUSY at the LHC. The plots refer to models with grand unification and universal sfermion and gaugino masses M_0 and $M_{1/2}$. The left-hand plot shows the region of this parameter space in which it is possible to detect the signature of missing E_T plus multiple jets at various levels of integrated luminosity. The right-hand plot shows the region of this parameter space in which it is possible to detect an excess of events with one or more leptons in addition to jets and missing E_T [71].

determining mass scales in SUSY models from global kinematic variables are described in [74].

11.5.3. Measurements of the SUSY Spectrum at the LHC

So far, I have only discussed the observation of the qualitative features of the SUSY model from global measures of the properties of events. Now I would like to give some examples of analyses in which specific details of the SUSY spectrum are measured with precision at the LHC. The examples that I will discuss involve the decay chain

$$\widetilde{q} \to q \widetilde{N}_2^0 \ , \ \widetilde{N}_2^0 \to \widetilde{N}_1^0 \ell^+ \ell^- \ , \qquad (11.210)$$

which is typically seen in models in which the gluino is heavier than the squarks and the LSP is gaugino-like.

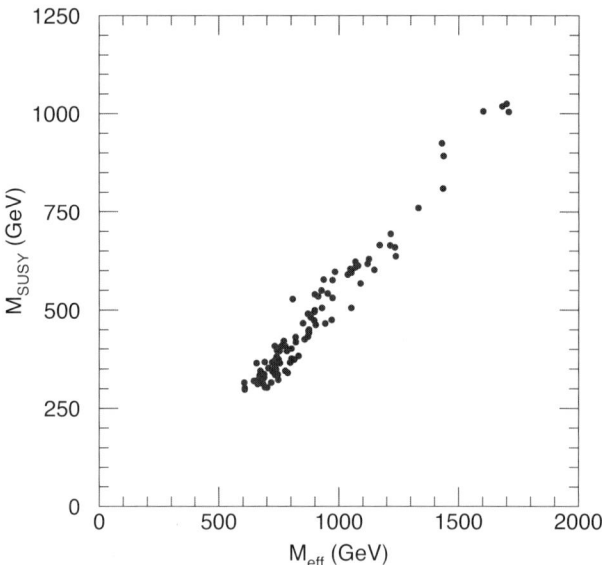

Fig. 11.23. Correlation between the value of the observable (11.209) and the lighter of the squark and gluino masses, from [72].

The decay of the N_2^0 can proceed by any of the mechanisms:

$$\widetilde{N}_2^0 \to \ell^\pm + \widetilde{\ell}^\mp \ , \ \widetilde{\ell}^\mp \to \ell^\mp \widetilde{N}_1^0$$
$$\widetilde{N}_2^0 \to \widetilde{N}_1^0 Z^0 \ , \ Z^0 \to \ell^+ \ell^-$$
$$\widetilde{N}_2^0 \to \widetilde{N}_1^0 Z^{0*} \ , \ Z^{0*} \to \ell^+ \ell^- \ . \quad (11.211)$$

The last line indicates a virtual Z^0, decaying off-shell. In a model with gaugino unification and heavy Higgsinos, $\widetilde{N}_2$ is mainly $\widetilde{w}^0$ and $\widetilde{N}_1$ is mainly $\widetilde{b}^0$. Then these modes are preferred in the order listed as long as they are kinematically allowed. If the slepton decay is allowed, this is the dominant model. Otherwise, the decay to $\widetilde{N}_1 Z^0$ or other open two-body decays dominate. If no two-body decays are open, the $\widetilde{N}_2$ must decay through three-body processes such as the last line of (11.211).

The decay to an on-shell Z^0 is hard to work with [75], but the other two cases can be explored in depth. It is useful to begin with the *Dalitz plot* associated with the 3-body $(\widetilde{N}_1, \ell^+, \ell^-)$ system. Let

$$x_0 = \frac{2E(\widetilde{N}_1)}{m(\widetilde{N}_2)} \ , \quad x_+ = \frac{2E(\ell^+)}{m(\widetilde{N}_2)} \ , \quad x_- = \frac{2E(\ell^-)}{m(\widetilde{N}_2)} \ , \quad (11.212)$$

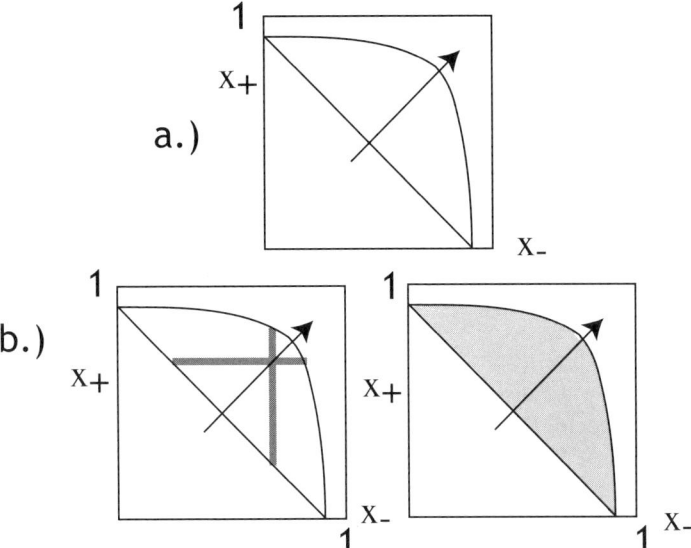

Fig. 11.24. The Dalitz plot describing 3-body neutralino decays, $\widetilde{N}_2^0 \to \widetilde{N}_1^0 \ell^+ \ell^-$.

where the energies are measured in the rest frame of the N_2. The three variables are related by

$$x_0 + x_1 + x_2 = 2 \ . \tag{11.213}$$

The three-body decay phase space is given by

$$\int d\Pi_3 = \frac{m^2(\widetilde{N}_2)}{128\pi^3} \int dx_+ \, dx_- \ ; \tag{11.214}$$

that is, phase space is flat in the variables (11.212). The basic kinematic identities involving the Dalitz plot variables are straightforward to work out, especially if we ignore the masses of the leptons. The kinematically allowed region is a wedge of the (x_+, x_-) plane bounded by the curves

$$x_+ + x_- = 1 - (m(\widetilde{N}_1)/m(\widetilde{N}_2))^2$$
$$(1-x_+)(1-x_-) = (m(\widetilde{N}_1)/m(\widetilde{N}_2))^2 \ , \tag{11.215}$$

as shown in Fig. 11.24(a). The invariant masses of two-body combinations are given in terms of the x_a by

$$\frac{m^2(\widetilde{N}_1 \ell^\pm)}{m^2(\widetilde{N}_2)} = (1 - x_\mp) \ , \quad \frac{m^2(\ell^+ \ell^-)}{m^2(\widetilde{N}_2)} = (1 - \frac{m(\widetilde{N}_1)^2}{m(\widetilde{N}_2)^2}) \ . \tag{11.216}$$

I am assuming that the $\widetilde{N}_1$ is stable and weakly interacting. In this case, the $\widetilde{N}_1$ will not be observed in the LHC experiments, and also the frame of the $\widetilde{N}_2$ cannot be readily determined. The only property of this system that is straightforward to measure is the two-body invariant mass $m(\ell^+\ell^-)$. So it is interesting to note that the distribution of this quantity distinguishes the first and third cases in (11.211), in the manner shown in Fig. 11.24(b). In the case of a two-body decay to an intermediate slepton, the decays populate two lines on the Dalitz plot, leading to a sharp discontinuity at the kinematic endpoint. In the case of a three-body decay, the events fill the whole Dalitz plot, producing a distribution with a slope at the endpoint. With a good understanding of the detector resolution in the dilepton invariant mass, these cases can be distinguished experimentally.

In the three-body case, the endpoint of the dilepton mass distribution is exactly

$$m(\widetilde{N}_2) - m(\widetilde{N}_1) \,, \tag{11.217}$$

so the observable mass distribution gives a precise measurement of this SUSY mass difference. The shape of the spectrum has more information. For example, for heavy slepton masses, the shape is distinctly different for gaugino-like or Higgsino-like neutralinos. Figure 11.25(a) shows the dilepton mass distribution for an mSUGRA parameter set for which the lightest two neutralinos are gaugino-like [72]. Figure 11.25(b) shows this distribution for a parameter set in which the two lightest neutralinos are Higgsino-like [73].

At the endpoint, the dilepton mass is maximal, and this requires that both the dilepton pair and the N_1 are at rest in the frame of the N_2. By measuring the four-vectors of the leptons, we would then know the N_1 and N_2 four-vectors, up to knowledge of the N_1 mass. It is possible to obtain this mass approximately from other measurements, for example, from the kinematics of $\widetilde{q}$ decays directly to N_1. With this information, we could determine the N_2 four-vector. Now the problem of missing momentum is solved. By adding observed jets to the N_2 four-vector, it is possible to find squarks as resonances [72]. Figure 11.26 shows the result of such an analysis for the SUSY parameter set of Fig. 11.25. The peak just below 300 GeV is a reconstructed $\widetilde{b}$ squark.

The two-body case of $\widetilde{N}_2$ decay is even nicer. In this case, we can see from the right-hand figure in Fig. 11.24(b) that the endpoint of the dilepton mass distribution is not located at the mass difference (11.217) but instead

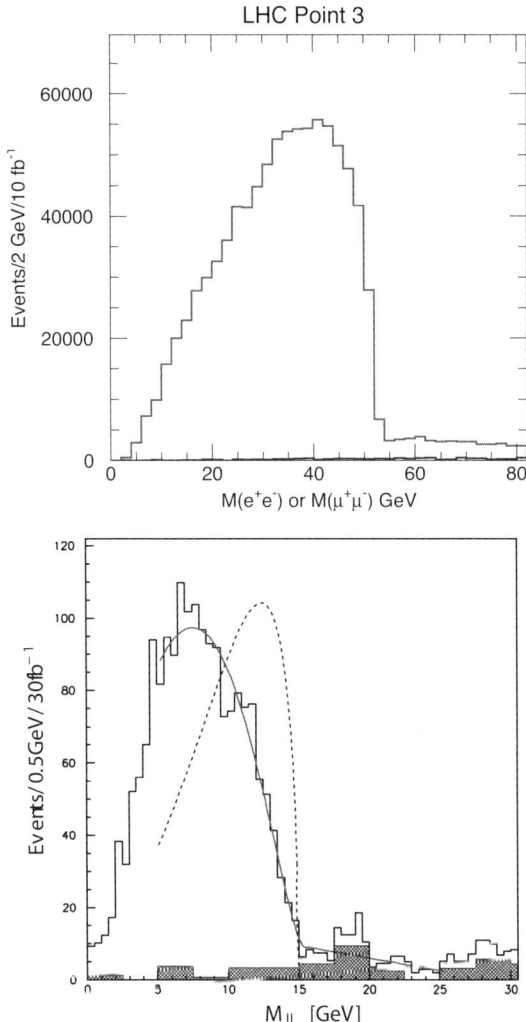

Fig. 11.25. Distribution of the dilepton invariant mass in two supersymmetry models with 3-body neutralino decays: (a.) a model with gaugino-like neutralinos [72], (b.) a model with Higgsino-like neutralinos [73]. In the second figure, the dashed curve indicates the $m(\ell^+\ell^-)$ spectrum expected for gaugino-like neutralinos with the same mass splitting.

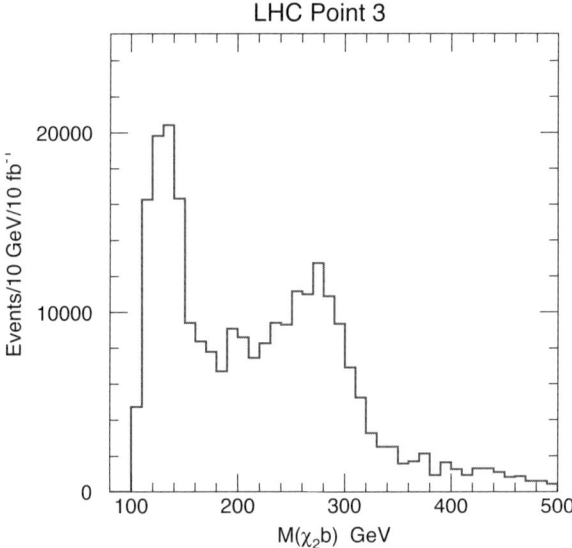

Fig. 11.26. Reconstruction of a squark in the model of Fig. 11.25(a) by combining a dilepton pair at the endpoint of the $m(\ell^+\ell^-)$ distribution, the $\widetilde{N}_1^0$ in the same frame with mass determined from kinematics, and a b-tagged quark jet.

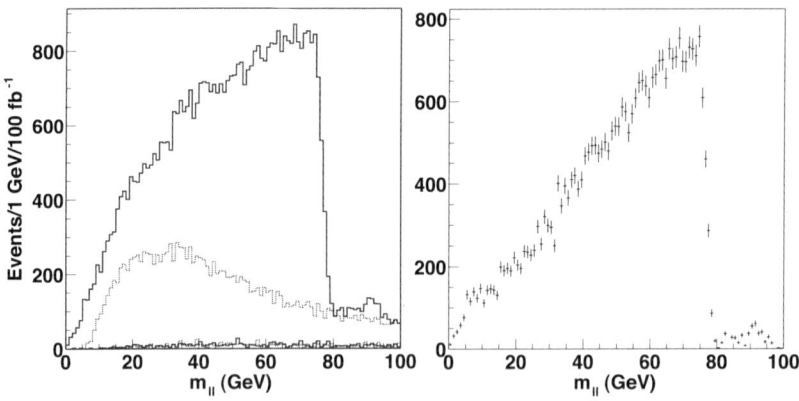

Fig. 11.27. Dilepton mass distribution in a model with two-body $\widetilde{N}_2$ decays, from [63]. The left-hand plot shows the dilepton mass distributions for opposite-sign same-flavor dileptons (solid) and for opposite-sign opposite-flavor dileptons (dashed). The lower histograms give the estimates of the Standard Model background. The right-hand plot shows the difference of the two distributions.

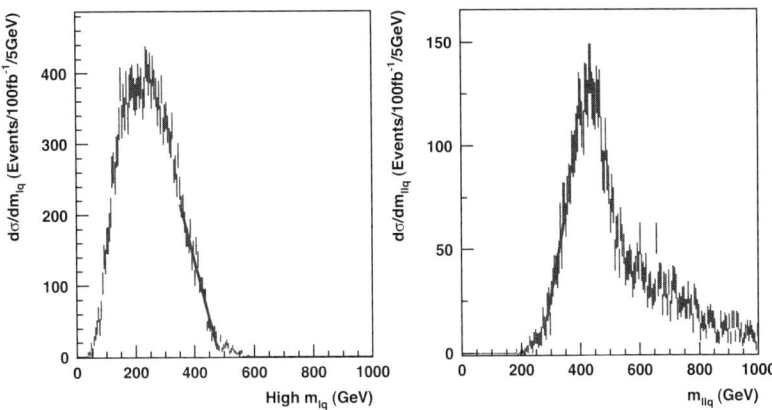

Fig. 11.28. Distributions of mass combinations of leptons and high-p_T jets showing kinematic endpoints in the analysis of [76]: (a.) the higher $m(q\ell)$ combination; (b.) the $m(q\ell^+\ell^-)$ distribution.

at the smaller value

$$m(\ell^+\ell^-) = m(\widetilde{N}_2)\sqrt{1 - \frac{m^2(\widetilde{\ell})}{m^2(\widetilde{N}_2)}}\sqrt{1 - \frac{m^2(\widetilde{N}_1)}{m^2(\widetilde{\ell})}}\ . \qquad (11.218)$$

Figure 11.27 shows an example of the dilepton spectrum from a SUSY parameter point in this region [63] The decay $\widetilde{q} \to q N_2$ is also a two-body decay, and there are similar kinematic relations for the upper and lower endpoints of the $(q\ell)$ and $(q\ell\ell)$ invariant mass distributions. These endpoints are likely to be visible in the collider data. Figure 11.28 shows two jet-lepton mass distributions from a similar analysis presented in [76]. In that analysis, it was possible to identify five well-measured kinematic endpoints, from which it was possible to solve (in an overdetermined way) for the four masses $m(N_1)$, $m(\ell)$, $m(N_2)$, $m(\widetilde{q})$.

There is one more case of an $\widetilde{N}_2 \to \widetilde{N}_1$ decay that should be mentioned. If two-body decays of $\widetilde{N}_2$ to sleptons are not kinematically allowed but the decay to $\widetilde{N}_1 h^0$ is permitted, this decay to a Higgs boson will be the dominant $\widetilde{N}_2$ decay. In this case, supersymmetry can provide a copious source of Higgs bosons. Figure 11.29 shows an analysis of a SUSY model in this parameter region [67]. Events with multijets and missing transverse energy are selected. In this sample, the mass distribution of two b-quark-tagged jets is shown. The signature of SUSY selects a sample of events in which the Higgs boson is visible in its dominant decay to $b\bar{b}$.

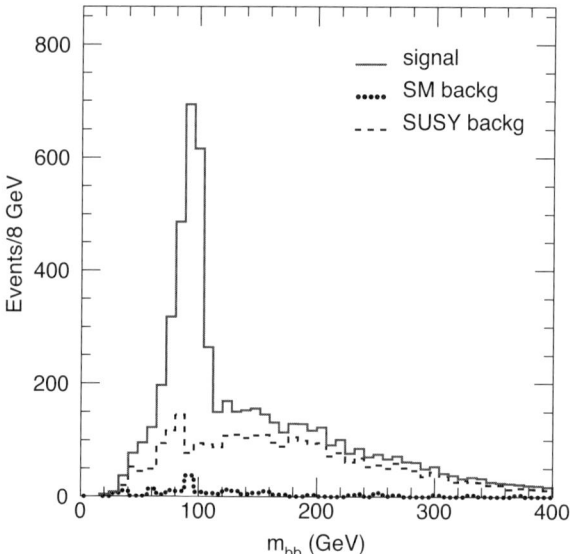

Fig. 11.29. The dijet mass distribution for 2 b-tagged jets at a point in the SUSY parameter space where the decay $\widetilde{N}_2^0 \to h^0 \widetilde{N}_1^0$ is dominant, from [67].

There is much more to say about the measurement of SUSY parameters at the LHC. Some more sophisticated sets of variables are introduced and applied in [76,77]. The question of measuring the spins of superparticles is discussed in [78–81]. And, we have not touched on alternative possibilities for the realization of SUSY, with R-parity violation or charged superparticles that are observed in the LHC experiments as stable particles. A broader overview of SUSY phenomenology at the LHC can be found in the references cited at the beginning of this section.

11.6. Electroweak Symmetry Breaking and Dark Matter in the MSSM

11.6.1. Electroweak Symmetry Breaking in the MSSM

In Section 1.2, I motivated the introduction of SUSY with the claim that SUSY could give an explanation of electroweak symmetry breaking, and for the presence of weakly interacting dark matter in the universe. Now that

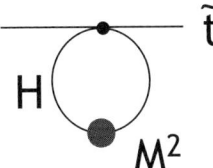

Fig. 11.30. Diagram contributing a term to the renormalization group equation for the soft mass parameter of $\tilde{t}$ proportional to the soft mass parameter for H_u.

we have a detailed understanding of the structure of the MSSM, it is time to come back and discuss these issues.

To present the mechanism of electroweak symmetry breaking in the MSSM, I need to add a term to one of the equations that I derived in Section 4.3. In (11.190), I presented the RG equation for the soft SUSY breaking scalar mass parameters, including renormalization effects from gauge interactions. I remarked that the contributions to this equation from Higgs Yukawa couplings are small for the scalars of the first and second generations. However, for the scalars of the third generation, these corrections can plan an important role.

The F-term interaction

$$\mathcal{L} = -\left|y_t H_u \cdot \tilde{t}\right|^2 \tag{11.219}$$

leads to a contribution to the RG equations for M_t, the mass parameter of $\tilde{t}$, proportional to $M_{H_u}^2$, from the diagram shown in Fig. 11.30. The value of the diagram is

$$-iy_t^2 \int \frac{d^4k}{(2\pi)^4} \frac{i}{k^2} (-iM_{H_u}^2) \frac{i}{k^2} = \frac{i}{(4\pi)^2} y_t^2 M_{H_u}^2 \log \Lambda^2 \ . \tag{11.220}$$

A scalar self-energy diagram is interpreted as $-i\delta m^2$, so this is a *negative* contribution to M_t^2. Each of the scalar fields $(H_u, \tilde{t}, \bar{\tilde{t}})$ gives a similar contribution that renormalizes the soft mass parameter of each of the others. For each correction, there is a counting factor from the number of color or $SU(2)$ degrees of freedom that run around the loop. There is also a correction to each of the scalar masses from the top quark A term. We must also remember that all of these terms add to the positive mass correction from the gaugino loops in Fig. 11.11, of which the gluino loop correction is the most important.

Taking all of these effects into account, we find for the RG equations of the soft mass parameters of H_u, t, and $\bar{t}$

$$\frac{dM_t^2}{d\log Q} = \frac{2}{(4\pi)^2} \cdot 1 \cdot y_t^2 [M_t^2 + M_{\bar{t}}^2 + M_{Hu}^2 + A_t^2] - \frac{8}{3\pi}\alpha_3 m_3^2 + \cdots$$

$$\frac{dM_{\bar{t}}^2}{d\log Q} = \frac{2}{(4\pi)^2} \cdot 2 \cdot y_t^2 [M_t^2 + M_{\bar{t}}^2 + M_{Hu}^2 + A_t^2] - \frac{8}{3\pi}\alpha_3 m_3^2 + \cdots$$

$$\frac{dM_{Hu}^2}{d\log Q} = \frac{2}{(4\pi)^2} \cdot 3 \cdot y_t^2 [M_t^2 + M_{\bar{t}}^2 + M_{Hu}^2 + A_t^2] m_3^2 + \cdots \qquad (11.221)$$

The structure is very interesting. The three scalar fields H_u, $\tilde{t}$, and $\tilde{\bar{t}}$ all receive negative corrections to their mass terms as these equations are integrated in the direction of decreasing $\log Q$. If any of these mass terms were to become negative, the corresponding field would have an instability to develop a vacuum expectation value, and the symmetry of the MSSM would be spontaneously broken. The symmetry-breaking we want is that associated with $\langle H_u \rangle \neq 0$. However, it seems equally possible that we could generate $\langle \tilde{\bar{t}} \rangle \neq 0$, which would break color $SU(3)$, or $\langle \tilde{t} \rangle \neq 0$, which would break both $SU(2)$ and $SU(3)$.

If the three mass parameters have similar values at a high mass scale, they race toward negative values according to (11.221). But H_u wins the race, and so the theory predicts the symmetry breaking pattern that is the one observed. In this way, the MSSM leads naturally to electroweak symmetry breaking and realizes the idea that electroweak symmetry breaking is connected to the large value of the top quark-Higgs coupling.

11.6.2. Higgs Boson Masses in the MSSM

Once we expect that $M_u^2 < 0$ at the weak scale, we can work out the details of the Higgs boson spectrum. First, we should write the potential for the Higgs fields H_u, H_d. As in the discussion of Sections 4.1 and 4.2, a number of terms need to be collected from the various pieces of the Lagrangian. The F terms contriubute

$$V_F = \mu^2 (H_u^{0*} H_u^0 + H_d^{0*} H_d^0) \qquad (11.222)$$

The D terms contribute

$$V_D = \frac{g^2 + g'^2}{8}(H_u^{0*} H_u^0 - H_d^{0*} H_d^0)^2 \qquad (11.223)$$

The soft SUSY breaking terms contribute

$$V_{soft} = M_{Hu}^2 H_u^{0*} H_u^0 + M_{Hd}^2 H_d^{0*} H_d^0 - (B\mu H_u^0 H_d^0 + h.c.) \qquad (11.224)$$

The sum of these terms gives the complete tree-level Higgs potential. Differentiating this potential with respect to H_u^0 and H_d^0, we obtain the equations that determine the Higgs field vacuum expectation values. If we write these equations with the parametrization of the vacuum expectation values given in (11.169), we find

$$\mu^2 + M_{Hu}^2 = B\mu \cot\beta + \frac{1}{2} m_Z^2 \cos 2\beta$$
$$\mu^2 + M_{Hd}^2 = B\mu \tan\beta - \frac{1}{2} m_Z^2 \cos 2\beta , \qquad (11.225)$$

where $m_Z^2 = (g^2 + g'^2)v^2/4$. This system of equations can be solved for μ to give

$$\mu^2 = \frac{M_{Hd}^2 - \tan^2\beta M_{Hu}^2}{\tan^2\beta - 1} - \frac{1}{2} m_Z^2 \qquad (11.226)$$

This is, for example, the way that we would determine μ in the mSUGRA parameter space described in Section 5.2.

It is interesting to turn this equation around and write it as an equation for m_Z in terms of the SUSY parameters,

$$m_Z^2 = 2 \frac{M_{Hd}^2 - \tan^2\beta M_{Hu}^2}{\tan^2\beta - 1} - 2\mu^2 . \qquad (11.227)$$

From this equation, a small value of m_Z would require a cancellation between the Higgs soft mass parameters and μ. The parameter μ sets the mass scale of the Higgsinos, and the Higgs soft mass parameters might be related to other masses of the SUSY scalar particles. Thus, if the masses of the charginos and neutralinos and, perhaps also, the sleptons are not close to m_Z, that disparity must be associated with an apparently unnatural cancellation between different SUSY parameters.

If we prohibit a delicate cancellation in (11.227), we put an upper bound on the SUSY partner masses. To avoid cancellations in more than two decimal places, μ must be less than 700 GeV. Similarly, we find bounds on the Higgs soft masses, and on the parameters that contribute to these masses through the RG equation. This consideration turns out to give a constraint on the gluino mass, $m_3 < 800$ GeV. Assuming gaugino universality, this becomes a condition $m_2 < 250$ GeV that restricts the chargino and neutralino masses. A variety of similar naturalness arguments that constrain the SUSY scale can be found in [82–84]. Though the logic is that of an estimate rather than a rigorous bound, this analysis strongly supports the idea that SUSY partners should be light enough to be discovered at the LHC and at the ILC.

Once we have the Higgs potential and the conditions for the Higgs vacuum expectation values, we can work out the masses of the Higgs bosons by expanding the potential around its minimum. A first step is to identify the combinations of Higgs fields that correspond to physical Higgs bosons. Look first at the charged Higgs bosons. There are two charged Higgs fields in the multiplets H_u, H_d. One linear combination of these fields is the Goldstone boson that is eaten by the W boson as it obtains mass through the Higgs mechanism. The orthogonal linear combination is a physical charged scalar field. If we decompose

$$H_u^+ = \cos\beta H^+ + \sin\beta G^+$$
$$H_d^- = \sin\beta H^- + \sin\beta G^- \qquad (11.228)$$

where $H^- = (H^+)^*$, $G^- = (G^+)^*$, and β is precisely the mixing angle in (11.169), it can be seen that $G^\pm$ are the Goldstone bosons and $H^\pm$ are the physical scalar states.

A similar analysis applies to the neutral components of H_u^0 and H_d^0. These are complex-valued fields. It is appropriate to decomposed them as

$$H_u^0 = \frac{1}{\sqrt{2}}(v\sin\beta + \sin\alpha H^0 + \cos\alpha h^0 + i\cos\beta A^0 + i\sin\beta G^0)$$
$$H_d^0 = \frac{1}{\sqrt{2}}(v\cos\beta + \cos\alpha H^0 - \sin\alpha h^0 + i\sin\beta A^0 - i\cos\beta G^0) \qquad (11.229)$$

The components H^0, h^0 are even under CP; the fields A^0, G^0 are odd under CP. The componet G^0 is the Goldstone boson eaten by the Z^0. The other three fields create physical scalar particles.

Having identified these fields, we can compute their masses. The formulae for the Higgs masses take an especially simple form when they are expressed in terms of the mass of the A^0. For the charged Higgs boson

$$m_{H+}^2 = m_A^2 + m_W^2 \; . \qquad (11.230)$$

For the CP-even scalars, one finds a mass matrix

$$\begin{pmatrix} m_A^2 \sin^2\beta + m_Z^2 \cos^2\beta & -(m_A^2 + m_Z^2)\sin\beta\cos\beta \\ -(m_A^2 + m_Z^2)\sin\beta\cos\beta & m_A^2 \cos^2\beta + m_Z^2 \sin^2\beta \end{pmatrix} \qquad (11.231)$$

The physical scalar masses m_h^2 and m_H^2 are the eigenvalues of this matrix, defined in such a way that $m_h^2 < m_H^2$. The angle α in (11.229) is the mixing angle that defines these eigenstates.

Taking the trace of (11.231), we find the relation

$$m_h^2 + m_H^2 = m_A^2 + m_Z^2 \ . \tag{11.232}$$

We can also obtain an upper bound on the lighter Higgs mass m_h^2 by taking the matrix element of (11.231) in the state $(\cos\beta, \sin\beta)$. The bound is a very strong one:

$$m_h^2 \leq m_Z^2 \cos^2\beta \ < m_Z^2 \ . \tag{11.233}$$

This seems inconsistent with lower bounds on the Higgs boson mass from LEP 2, which exclude $m_h < 114$ GeV for the Standard Model Higgs and for most scenarios of SUSY Higgs bosons [85].[d] However, the one-loop corrections to the tree-level result (11.231) give a significant positive correction

$$\delta m_h^2 = \frac{3}{\pi} \frac{m_t^4}{m_W^2} \sin^4\beta \log \frac{m_{\tilde{t}} m_{\tilde{\bar{t}}}}{m_t^2} \ . \tag{11.234}$$

This correction can move the mass of the h^0 up to about 130 GeV. The detailed summary of the radiative corrections to the h^0 mass in the MSSM is presented in [88]. A very clear and useful accounting of the major corrections can be found in [89].

It is possible to raise the mass of the h^0 by going outside the MSSM and adding additional $SU(2)$ singlet superfields to the model. However, this strategy is limited by a general constraint coming from grand unification. The requirement that the Higgs couplings do not become strong up to the grand unification scale limit the mass of the Higgs to about 200 GeV [90]. It is possible to raise the mass of the Higgs further only by enlarging the Standard Model gauge group or adding new thresholds that affect unification [91,92].

In the MSSM, we can easily have the situation in which $m_A \gg m_h$. In this limit, the couplings of the h^0 are very close to those of the Standard Model Higgs boson, and the H^0, A^0, and $H^\pm$ are almost degenerate. If $\tan\beta \gg 1$, the heavy neutral Higgs bosons decay dominantly to $b\bar{b}$ and $\tau^+\tau^-$.

Much more about the phenomenology of Higgs bosons in supersymmetry can be found in [93,94].

11.6.3. WIMP Model of Dark Matter

Now we turn to the second problem highlighted in the Introduction, the problem of dark matter in the universe. It has been known from many

---

[d]Some exceptional Higgs decay schemes that escape these bounds are considered in [86,87].

astrophysical measurements that the universe contains enormous amounts of invisible, weakly interacting matter. For an excellent review of the classic astrophysical evidence for this dark matter, see [95].

In the past few years, measurements of the cosmic microwave background have given a new source of evidence for dark matter. Since this data comes from an era in the early universe before the formation of any structure, it argues strongly that the invisible matter is not made of rocks or brown dwarfs but is actually a new, very weakly interacting form of matter. These measurements also determine quite accurately the overall amount of conventional and dark matter in the universe. Let ρ_b, ρ_N, and ρ_Λ be the large-scale energy densities of the universe from baryons, dark matter, and the energy of the vacuum. The data from the microwave background tells us that $\rho_b + \rho_N + \rho_\Lambda = \rho_c$, the 'closure density' corresponding in general relativity to a flat universe, to about 1% accuracy. If $\Omega_i = \rho_i/\rho_c$, the most recent data from the WMAP experiment and other sources gives [96,97]

$$\Omega_b = 0.042 \pm 0.003 \quad \Omega_N = 0.20 \pm 0.02 \quad \Omega_\Lambda = 0.74 \pm 0.02 \; . \qquad (11.235)$$

These results present a double mystery. We do not know what particle the dark matter is made of, and we do not have any theory that explains the observed magnitude of the vacuum energy or 'dark energy'.

I believe that supersymmetry will eventually play an essential role in solving the problem of dark energy. In ordinary quantum field theory, the value of the vacuum energy is quartically divergent, so the problem of computing the vacuum energy is not even well-posed. In supersymmetry, there is at least a well-defined zero of the energy associated with exact supersymmetry, which implies $\langle 0| H |0\rangle = 0$. Unfortunately, in most of today's models of supersymmetry, the vacuum energy is set by the SUSY breaking scale. This gives $\Lambda \sim (10^{11} \text{ GeV})^4$, about 80 orders of magnitude larger than the observed value of the vacuum energy. From this starting point, Λ must be fine-tuned to the scale of eV^4. This is an important problem that needs new insights which, however, I will not provide here.

On the other hand, supersymmetry offers a very definite solution to the problem of the origin of dark matter. We have already noted in Section 3.4 that it is straightforward to arrange that the lightest supersymmetric particle can be absolutely stable. If this particle were produced in the early universe, some density of this type of matter should still be present. In most, but not all, regions of parameter space, the lightest supersymmetric particle is neutral. Candidates include the lightest neutralino, the lightest sneutrino, and the gravitino. In the remainder of these lectures, I will concentrate on the case in which the lightest neutralino is the dark matter particle. For a discussion of the other candidates, see [98].

To begin our discussion, I would like to estimate the cosmic density of dark matter in a more general context. Let me make the following minimal assumptions about the nature of dark matter, that the dark matter particle is stable, neutral, and weakly interacting. To these properties, I would like to add one more, that dark matter particles can be created in pairs at sufficiently high temperature, and that, at some time in the early universe, dark matter particles were in thermal equilibrium. I will refer to a particle satisfying these assumptions as a 'weakly interacting massive particle' or WIMP. The assumption of thermal equilibrium is a strong one that is not satisfied even in many models of supersymmetric dark matter. For some exceptions, see [99,100]. However, let us see what implications follow from these assumptions.

The assumption that WIMPs were once in thermal equilibrium provides a definite initial condition from which to compute the current density of dark matter. In thermal equilibrium at temperature T, we have for the number density of dark matter particles

$$n_{eq} = \frac{g}{(2\pi)^{3/2}} (mT)^{3/2} e^{-m/T} \ . \tag{11.236}$$

where g is the number of spin degrees of freedom of the massive particle. As the universe expands, the temperature of the universe deccreases and the rate of WIMP pair production becomes very small. But the rate of dark matter pair annihilation also becomes small as the WIMPs separate from one another.

The expansion of the universe is governed by the Hubble constant $H = \dot{a}/a$, where a is the scale factor. Einstein's equations imply that

$$H^2 = \frac{8\pi}{3} \frac{\rho}{m_{\text{Pl}}^2} \ . \tag{11.237}$$

In a radiation-dominated universe where g_* is the number of relativistic degrees of freedom, $\rho = \pi^2 g_* T^4/30$. Then H is proportional to T^2. In a radiation-dominated universe, the temperature red-shifts as the universe expands, so that $T \sim a^{-1}$. Combining this relation with the equation $H = \dot{a}/a \sim T^2$, we find $t \sim T^{-2} \sim a^2$, that is, $a \sim t^{1/2}$ or $\dot{a}/a = 1/2t$. Setting this expression equal to the explict form of H in (11.237), we find a detailed formula for the time since the start of the radiation-dominated era for cooling to a temperature T,

$$t = \left(\frac{16\pi^3 g_*}{45}\right)^{-1/2} \frac{m_{\text{Pl}}}{T^2} . \tag{11.238}$$

The evolution of the WIMP density is described by the Boltzmann equation

$$\frac{dn}{dt} = -3Hn - \langle \sigma v \rangle (n^2 - n_{eq}^2) , \tag{11.239}$$

where H is the Hubble constant, σ is the $\widetilde{N}\widetilde{N}$ annihilation cross section—which appears thermally averaged with the relative velocity of colliding WIMPs—and n_{eq} is the equilibrium WIMP density (11.236). Assume, just for the sake of argument, that the temperature T is of the order of 100 GeV. At this temperature, the Hubble constant has the magnitude $H \sim 10^{-17} T$, so the expansion of the universe is very slow on the scale of typical elementary particle reactions. However, when T becomes less than the WIMP mass m, the WIMP density is exponentially suppressed and so the collision term in the Boltzmann equation is also very small. These two terms are of the same size at the *freezeout* temperature T_F satisfying

$$e^{-m/T_F} \sim \frac{1}{m_{\text{Pl}} m \langle \sigma v \rangle} . \tag{11.240}$$

At temperatures below T_F, we may neglect the production of WIMPs in particle collisions. The WIMP density is then determined by the expansion of the universe and the residual rate of WIMP pair annihilation. Maybe it is more appropriate to think of T_F as the temperature at which a WIMP density is frozen *in*. To determine the freezeout temperature, we take the logarithm of the right-hand side of (11.240). The result depends only on the order of magnitude of the annihilation cross section. For any interaction of electroweak strength,

$$\xi_F = T_F/m \sim 1/25 . \tag{11.241}$$

This physical picture suggests a way to estimate the cosmic density of WIMP dark matter. We can take as our initial condition the thermal density of dark matter at freezeout. We then integrate the Boltzmann equation, ignoring the term proportional to n_{eq}^2 associated with the production of WIMP pairs [102].

In analyzing the Boltzmann equation, it is useful normalize the particle density n of dark matter to the density of entropy s. Since the universe expands very slowly, this expansion is very close to adiabatic. Then entropy is conserved,

$$\frac{ds}{dt} = -3Hs \ . \tag{11.242}$$

In a radiation-dominated universe, $s = 2\pi^2 g_* T^3/45$. Now define

$$Y = \frac{n}{s} \ , \qquad \xi = \frac{T}{m} \ , \tag{11.243}$$

the latter as in (11.241). Using the expression (11.238), we can convert the evolution in time to an evolution in temperature or in ξ. Applying these changes of variables and dropping the n_{eq}^2 term, the Boltzmann equation (11.239) rearranges to the form

$$\frac{dY}{d\xi} = C \langle \sigma v \rangle Y^2 \ , \tag{11.244}$$

where

$$C = \left(\frac{\pi g_*}{45}\right)^{1/2} m m_{\text{Pl}} \ . \tag{11.245}$$

Let Y_F be the value of Y at $\xi = \xi_F$. If we assume that $\langle \sigma v \rangle$ is approximately constant, since we are at temperatures close to threshold, it is straightforward to integrate this equation to $\xi = 0$, corresponding to late times.

$$Y^{-1} = Y_F^{-1} + C\xi_F \langle \sigma v \rangle \ . \tag{11.246}$$

The second term typically dominates the first. Then we can put back the value of C in (11.245) and write the final answer in terms of the ratio of the mass density of dark matter to the closure density $\Omega_N = n m_N / \rho_c$. In this way, we find

$$\Omega_N = \frac{s_0}{\rho_c} \left(\frac{45}{\pi g_*}\right)^{1/2} \frac{1}{\xi_F m_{\text{Pl}} \langle \sigma v \rangle} \ , \tag{11.247}$$

where s_0 is the current entropy density of the universe. Turner and Scherrer observed that this formula gives a value of Ω_N that is usually within 10% of the result from exact integration of the Boltzmann equation [102]. If $\langle sigmav \rangle$ has a significant dependence on temperature, the derivation is still correct with the replacement

$$\xi \langle \sigma v \rangle \to \int_0^{\xi_f} d\xi \langle \sigma v \rangle (\xi) \tag{11.248}$$

in the denominator of the last term in (11.247).

This is a remarkable relation. Almost every factor in this relation is known from astrophysical measurements. The left-hand side is given by (11.235). On the right-hand side, the entropy density of the universe is dominated by the entropy of the microwave background photons and can be computed from the microwave background temperature. The closure density is known from the measurement of the Hubble constant and the observation that the universe is flat. The parameters g_* and ξ_F are relatively insensitive to the strength of the annihilation cross section, with values $g_* \sim 100$, $\xi_F \sim 1/25$. The mass of the WIMP does not appear explicitly in (11.247). We can then solve for $\langle \sigma v \rangle$. The result is

$$\langle \sigma v \rangle = 1 \text{ pb} . \tag{11.249}$$

This is the value of a typical electroweak cross section at energies of a few hundred GeV. If we convert this value to a mass M of an exchanged particle using the formula

$$\langle \sigma v \rangle = \frac{\pi \alpha^2}{8M^2} , \tag{11.250}$$

the value (11.249) corresponds to $M = 100$ GeV.

I consider this a truly remarkable result. From a purely astrophysical argument, relying on quite weak and general assumptions, we arrive at the conclusion that there must be new particles at the hundred GeV energy scale. It is probably not a concidence that this argument leads us back to the mass scale of electroweak symmetry breaking.

In our study of supersymmetry, we have found an argument from the physics of electroweak symmetry breaking that predicts the existence of dark matter. As I discussed at the beginning of these lectures, models that explain electroweak symmetry breaking are complex. They typically involve many new particles. It is easily arranged that the lightest of the new particles is neutral. In supersymmetry, there is a reason why the new particles are likely to carry a conserved quantum number (11.149). Other models of electroweak symmetry breaking, such as the extra dimensional and little Higgs models discussed in Section 1.2, have their own reasons to have a complex particle spectrum and discrete symmetries. Then these models lead in their own ways to WIMPs at the hundred GeV mass scale.

A slight extension of this argument adds more interest. In supersymmetry, the sector of new particles includes particles with QCD color. Since the top quark probably plays an essential role in the mechanism of electroweak symmetry breaking, it is very likely that, in any model, some of the new particles will carry color. If these particles have masses below 1 TeV, they have large (10 pb) pair-production cross sections at the LHC. These particles will then decay to the dark matter particle, producting complex events with several hard jets and missing transverse momentum. These mild assumptions thus lead to the conclusion, from any model that follows this general line of argument, that *we should expect exotic events with multiple jets and missing transverse momentum to appear with pb cross sections at the LHC*.

11.6.4. Dark Matter Annihilation in the MSSM

This argument of the previous section gives a very optimistic conclusion for the discovery of new physics at the LHC. However, we have already discussed that the first observation of supersymmetry or another model of new physics will only be the first step in a lengthy experimental program. Once we know that superparticles or other new particles exist, we will need to study them in detail to learn their detailed interactions and, eventually, to work out the underlying Lagrangian that governs their behavior. As we have already discussed in Section 3.5 and 4.3, this Lagrangian can give us a clue to the nature of the ultimate theory at very short distances.

The study of dark matter intersects this program in an interesting way. In principle, once we have discovered supersymmetric particles, we can try to measure their properties and see if these coincide with the properties required from astrophysical detections of dark matter. As we have seen in Section 5.3, the LHC experiments expect to measure the mass of the LSP to about 10% accuracy. These measurements can hopefully be compared to mass measurements at the 20% level that can be expected from astrophysical dark matter detection experiments [103,104]. We would also wish to find out whether the annihilation cross section $\langle \sigma v \rangle$ that is predicted from the supersymmetry parameters measured at colliders agrees with the value (11.249) required to predict the observed WIMP relic density. This comparison turns out to depend in a complex way on the parameters of the underlying supersymmetry theory.

To begin our discussion of the annihilation cross section, we can make a simple model of neutralino annihilation and see how well it works. We have seen in Section 4.3 that the right-handed sleptons are often the lightest charged particles in the supersymmetry spectrum. Consider, then, an idealized parameter set in which the neutralino is a pure bino and pair annihilation is dominated by the slepton exchange diagrams shown in Fig. 11.31. (Away from the pure bino case, there are also s-channel diagrams with Z^0,

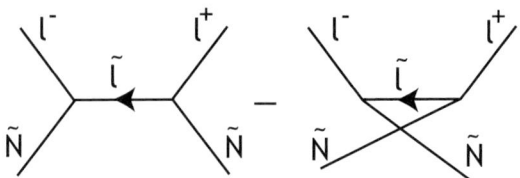

Fig. 11.31. Diagrams giving the simplest scheme of neutralino pair annihilation, leading to the annihilation cross section (11.251).

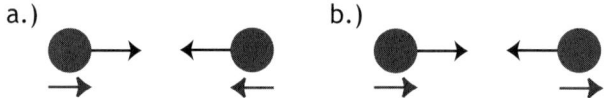

Fig. 11.32. Two possible spin configurations for neutralino annihilation: (a.) spin 0; (b.) spin 1. Because of Fermi statistics, the latter state does not exist in the S-wave.

h^0, H^0, A^0.) In this special limit, the annihilation cross section is given by

$$v\frac{d\sigma}{d\cos\theta} = \pi\alpha^2 m_N^2 \left|\frac{1}{c_w}\right|^2 \left|\frac{1}{m_{\tilde{\ell}}^2 - t} - \frac{1}{m_{\tilde{\ell}}^2 - u}\right|^2, \qquad (11.251)$$

where m_N is the $\widetilde{N}_1$ mass. The relative velocity v appears due to the flux factor in the cross section; this factor cancels in σv. I have ignored the lepton masses. This expression is of the order of (11.250) with $M \sim m_N$, except for one unfortunate feature: At threshold, $t = u$ and the cross section vanishes. This leads to a severe suppression, by a factor of

$$v^2 \cdot \left|\frac{m_N^2}{m_{\tilde{\ell}}^2 + m_N^2}\right|^4, \qquad (11.252)$$

which is at least of order $\xi_f/16$. So the relic density estimated in this simple way is too large by about a factor of 10.

There is an interesting physics explanation for the vanishing of this cross section at threshold [105]. Neutralinos are spin-$\frac{1}{2}$ fermions, and we might guess from this that, near threshold, they would annihilate in the S-wave either in a spin 0 or in a spin 1 state. The two spin configurations are shown in Fig. 11.32. However, because the neutralino is a Majorana fermion and therefore its own antiparticle, an S-wave state of two neutralinos must be antisymmetric in spin. Hence, the spin 1 S-wave state does not exist However, as we know from pion decay, a spin 0 state can convert to a pair of light leptons only with a helicity flip. Thus, there is an annihilation cross section from the spin 0 S-wave only when lepton masses are included, and even then with the suppression factor m_ℓ^2/m_N^2, which is 10^{-4} even for $\tau^+\tau^-$ final states.

To obtain a realistic value for the neutralino relic density, we have to bring in more complicated mechanisms of neutralino annihilation. These mechanisms are not difficult to find in various regions of the large supersymmetry paramet er space [106–108]. We need to look for annihilation processes that can proceed in the S-wave with full strength. Three possible mechanisms are shown in Fig. 11.33.

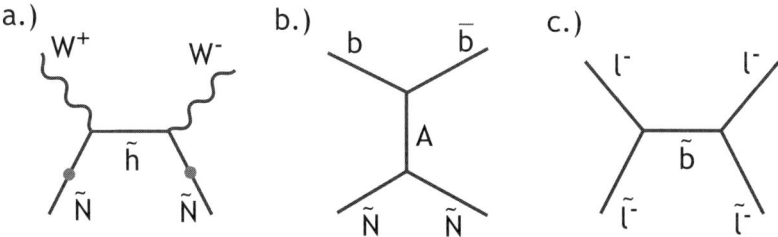

Fig. 11.33. Three mechanisms for obtaining a sufficiently large annihilation cross section to give the observed density of neutralino dark matter: (a.) gaugino-Higgsino mixing, opening the annihilation channels to W^+W^- and Z^0Z^0, (b.) resonance annihilation through the Higgs boson A^0, (c.) co-annihilation with another supersymmetric particle, here taken to be a $\widetilde{\ell}$.

Pairs of neutralinos can annihilate in the S-wave into vector bosons. The bino does not couple to W or Z pairs, but if the lightest neutralino has Higgsino or wino content, this reaction can be important. For charginos of mass about 200 GeV, this annihilation cross section can be 50 pb for a pure wino or Higgsino, so only a modest content of these states is needed to give a cross section of 1 pb.

The s-channel exchange of a Higgs boson can provide a mechanism for neutralino annihilation in the spin 0 S-wave. Because this state is CP-odd, it is the boson A^0 that is relevant here. If m_A is close to the neutralino threshold $2m_N$, the cross section has a resonant enhancement. Note that the $\widetilde{N}_1$ annihilation vertex to A arises as a Higgs-Higgsino-gaugino Yukawa term, so this vertex is nonzero only if $\widetilde{N}_1$ has both gaugino and Higgsino content. If $m_A = 2m_N$, the resonance enhancement is at full strength and the cross section can be as large as 50 pb. Thus, it is A boson masses about 20 GeV above or below the threshold that give the desired cross section (11.249).

The final mechanism shown in the figure is *coannihilation*. As we have discussed, the freezeout of the $\widetilde{N}_1$ occurs at a temperature given by $T/m_N \sim 1/25$. So if there is another particle in the supersymmetry spectrum that is within 4% of the $\widetilde{N}_1$ mass, this state will have a number density that remains in equilibrium with the number density of the $\widetilde{N}_1$. If this particle has S-wave annihilation reactions, those reactions can be the dominant mechanisms for the annihilation of supersymmetric particles. For a light slepton, the reactions

$$\widetilde{\ell}^- + \widetilde{N}_1^0 \to \ell^- + \gamma \qquad \widetilde{\ell}^- + \widetilde{\ell}^- \to \ell^- + \ell^- \qquad (11.253)$$

can give significant S-wave annihilation. In [106,109], the lighter stau is invoked as the coannihilating particle. In [110], the lighter top squark is

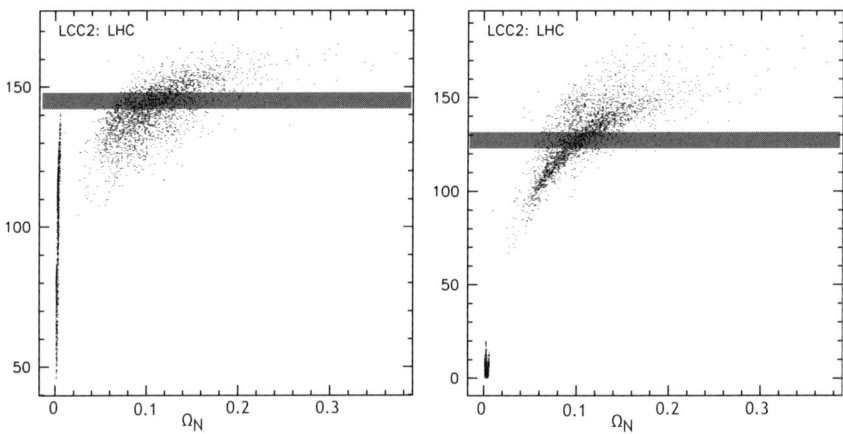

Fig. 11.34. Scatter plot of SUSY parameter points consistent with data from the LHC in the analysis of the parameter set LCC2 from [111]. The horizontal axis show the value of Ω_N at each parameter point. The vertical axes show polarized-beam cross sections measurable at the ILC, in fb: (a.) $\sigma(e_R^- e_L^+ \to \widetilde{C}_1^+ \widetilde{C}_1^-)$, (b.) $\sigma(e_R^- e_L^+ \to \widetilde{N}_2^0 \widetilde{N}_3^0)$. The colored bands show the $\pm 1\sigma$ region allowed after the ILC cross section measurements.

invoked as the coannihilating state. If the lightest neutralinos and charginos are Higgsino-like, chargino coannihilation can also be important.

It is, then, a complex matter to predict the neutralino relic density from microscopic physics. We will first need to learn what particles in the supersymmetry spectrum play the dominant role as particle exchanged in annihilation reactions or as coannihilating species. We will then need to measure the couplings and mixing angles of the important particles, since the dominant annihilation diagrams depend sensitively on these.

Some examples of how measurements at the LHC and ILC can accumulate the relevant information are described in [111]. Figure 11.34 shows a part of the analysis of this paper for a particular SUSY model in which the dominant annihilation reactions are $\widetilde{N}_1 \widetilde{N}_1 \rightarrow W^+W^-, Z^0 Z^0$. As a first step, the authors constructed numerous supersymmetry parameter sets that were consistent with the mass spectrum of this model as it would be measured at the LHC. These parameter sets included a variety of models in

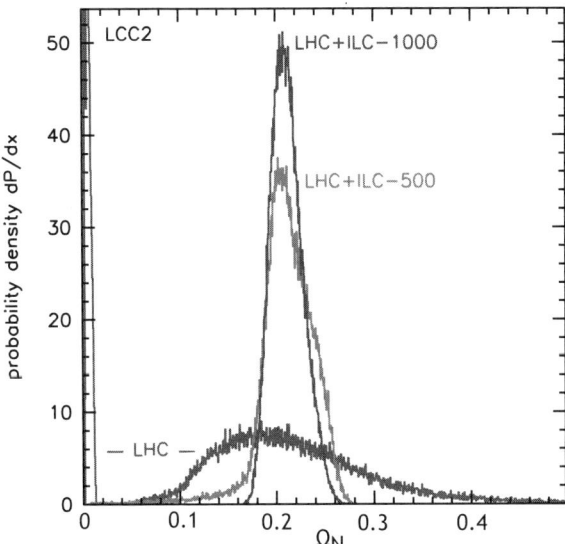

Fig. 11.35. Summary plot for the prediction of Ω_N from collider data for the SUSY parameter set LCC2 considered in [111]. The three curves show the likelihood distributions for the prediction of Ω_N using data from the LHC, the ILC at 500 GeV, and the ILC at 1000 GeV.

which the LSP was dominantly bino and wino. The figure shows scatter plots of the predictions of these models with ILC cross sections for neutralino and chargino pair production on the vertical axis and Ω_N on the horizontal axis. The two cross sections clearly separate the bino- and wino-like solutions. The second of these cross sections is the polarized reaction of chargino pair production for which the cross section is displayed in Fig. 11.19. The horizontal lines represent the accuracy of the measurements of these cross sections expected at the ILC. These measurements select the bino solution and also play an important role in fixing the bino-Higgsino mixing angle which is a crucial input to the annihilation cross sections. In Fig. 11.35, I show the distribution of predictions for Ω_N expected for this model, in the analysis of [111], from the data on SUSY particles that would be obtained from the LHC, from the ILC at a center-of-mass energy of 500 GeV, and from the ILC at a center-of-mass energy of 1000 GeV.

The similar summary plot for another of the models considered in [111] is shown in Fig. 11.36. The model considered in this analysis is one in which the neutralino relic density is set by stau coannihilation. In this

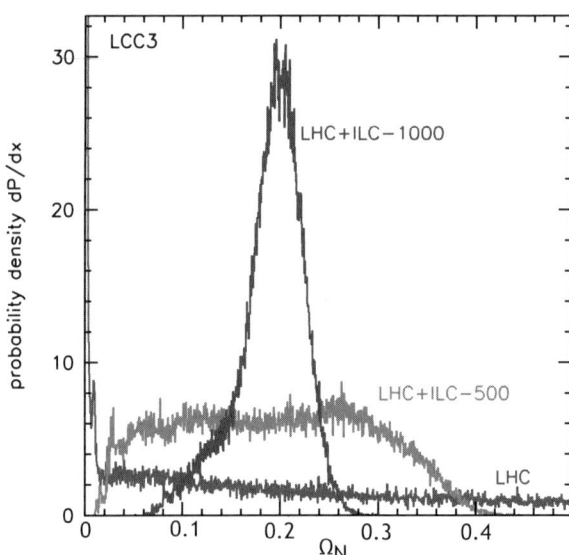

Fig. 11.36. Summary plot for the prediction of Ω_N from collider data for the SUSY parameter set LCC3 of [111]. The notation is as in Fig. 11.35.

model, the stau would be discovered at the LHC, and the stau-neutralino mass difference would be measured to about 10% accuracy at the 500 GeV ILC. However, the annihilation reactions also depend on mixing angles and on the value of $\tan\beta$. In this scenario, these are determined only by ILC measurements of some of the heavier states of the SUSY spectrum.

Collider measurements of the SUSY spectrum can also be used to constrain cross sections of the WIMP that are important for experiments that seek to detect dark matter, for example, the neutralino-proton cross section and the cross section for neutralino pair annihilation to gamma rays. If we can accurately predict these cross sections from collider data, the information about the SUSY spectrum that we learn from colliders will feed back into the astrophysics of dark matter. Some numerical examples that illustrate this are presented in [111].

11.7. Conclusions

In these lectures, I have given an overview of supersymmetry and its application to elementary particle physics. In the early sections of this review, I presented the formalism of SUSY and explained the rules for constructing supersymmetric Lagrangians. Our discussion then became more concrete, focusing on the mass spectrum of the MSSM and the properties of the particle states of the MSSM spectrum. This led us to a discussion of the experimental probes of this spectrum and the possibility of measurement of the parameters of the supersymmetric Lagrangian.

This possibility is now coming very near. As I have discussed in the last sections of this review, supersymmetry gives concrete answers to the major questions about elementary particle physics that we expect to be addressed at the hundred GeV scale — the questions of the origin of electroweak symmetry breaking and the identity of cosmic dark matter. In the next year, the LHC will begin to explore the physics of this mass scale. Supersymmetry is one candidate for what will be found. I hope that, after studying these lectures, you will agree that the picture provided by supersymmetry is highly plausible and even compelling.

Whatever explanations we will learn from the LHC data, our investigation of it will follow the general paradigm that I have described here. In successive stages, we will use data from the LHC and the ILC to learn the mass spectrum of new particles that are revealed at the LHC, to determine their quantum numbers and couplings, and to reconstruct their underlying

Lagrangian. On the basis of the detailed studies of this program that have been carried out for the MSSM, we have the expectation that we will be able to learn the underlying theory of the new particles and to test the specific explanations that this theory gives for the mysteries of the fundamental interactions.

Is supersymmetry just an attractive theory, or is it a part of the true description of elementary particles? We are about to find out.

Acknowledgements

I am grateful to Sally Dawson, Rabi Mohapatra and, especially, to K. T. Mahanthappa for organizing the 2006 TASI Summer School at which these lectures were presented. I thank Howard Haber and Thomas Dumitrescu for instructive comments on the manuscript.

References

[1] A. D. Linde, JETP Lett. **19**, 183 (1974) [Pisma Zh. Eksp. Teor. Fiz. **19**, 320 (1974)].
[2] L. Susskind, Phys. Rev. D **20**, 2619 (1979).
[3] S. Weinberg, Phys. Rev. D **19**, 1277 (1979).
[4] P. W. Higgs, Phys. Rev. Lett. **13**, 508 (1964).
[5] K. D. Lane, arXiv:hep-ph/9401324.
[6] S. P. Martin, arXiv:hep-ph/9709356.
[7] J. Wess and J. Bagger, *Supersymmetry and supergravity*. (Princeton U. Press, 1992).
[8] J. A. Bagger, arXiv:hep-ph/9604232.
[9] J. D. Lykken, arXiv:hep-th/9612114.
[10] G. L. Kane, arXiv:hep-ph/0202185.
[11] M. Drees, R. M. Godbole, and P. Roy, *Theory and Phenomenology of Sparticles*. (World Scientific, 2004).
[12] P. Binetruy, *Supersymmetry: Theory, Experiment, and Cosmology*. (Oxford U. Press, 2004).
[13] H. Baer and X. Tata, *Weak Scale Supersymmetry*. (Cambridge U. Press, 2006).
[14] H. Dreiner, H. E. Haber, and S. P. Martin, to appear.
[15] N. Arkani-Hamed, A. G. Cohen, E. Katz and A. E. Nelson, JHEP **0207**, 034 (2002) [arXiv:hep-ph/0206021].
[16] M. Schmaltz and D. Tucker-Smith, Ann. Rev. Nucl. Part. Sci. **55**, 229 (2005) [arXiv:hep-ph/0502182].
[17] For a pedagogical derivation of origin of the negative μ^2 in little Higgs

models, see M. Perelstein, M. E. Peskin and A. Pierce, Phys. Rev. D **69**, 075002 (2004) [arXiv:hep-ph/0310039].
[18] H. C. Cheng, B. A. Dobrescu and C. T. Hill, Nucl. Phys. B **573**, 597 (2000) [arXiv:hep-ph/9906327]; N. Arkani-Hamed, H. C. Cheng, B. A. Dobrescu and L. J. Hall, Phys. Rev. D **62**, 096006 (2000) [arXiv:hep-ph/0006238];
[19] C. Macesanu, Int. J. Mod. Phys. A **21**, 2259 (2006) [arXiv:hep-ph/0510418].
[20] S. R. Coleman and J. Mandula, Phys. Rev. **159** (1967) 1251.
[21] R. Haag, J. T. Lopuszanski and M. Sohnius, Nucl. Phys. B **88**, 257 (1975).
[22] Yu. A. Golfand and E. P. Likhtman, JETP Lett. **13** (1971) 323 [Pisma Zh. Eksp. Teor. Fiz. **13** (1971) 452].
[23] D. V. Volkov and V. P. Akulov, JETP Lett. **16** (1972) 438 [Pisma Zh. Eksp. Teor. Fiz. **16** (1972) 621].
[24] J. Wess and B. Zumino, Nucl. Phys. B **70** (1974) 39.
[25] For a taste, see N. Beisert, Comptes Rendus Physique **5**, 1039 (2004) [arXiv:hep-th/0409147].
[26] K. Intriligator, N. Seiberg and D. Shih, JHEP **0604**, 021 (2006) [arXiv:hep-th/0602239].
[27] In a common alternative notation, ϕ is written $-A^*$. Then this equation becomes $F = mA$, the *Newton-Witten equation*. See [28].
[28] V. Gates, *et al.*, in *Proceedings of the Workshop on Unified String Theories*, M. Green and D. Gross, eds. (World Scientific, 1986).
[29] B. Zumino, Nucl. Phys. B **89**, 535 (1975).
[30] M. T. Grisaru, W. Siegel and M. Rocek, Nucl. Phys. B **159**, 429 (1979).
[31] I. Affleck, M. Dine and N. Seiberg, Phys. Rev. Lett. **52**, 1677 (1984).
[32] B. Zumino, Phys. Lett. B **87**, 203 (1979).
[33] J. Wess and B. Zumino, Nucl. Phys. B **78**, 1 (1974).
[34] M. T. Grisaru, B. Milewski and D. Zanon, Nucl. Phys. B **266**, 589 (1986).
[35] V. A. Novikov, M. A. Shifman, A. I. Vainshtein and V. I. Zakharov, Nucl. Phys. B **229**, 381 (1983).
[36] N. Arkani-Hamed and H. Murayama, JHEP **0006**, 030 (2000) [arXiv:hep-th/9707133].
[37] S. Ferrara and B. Zumino, Nucl. Phys. B **87**, 207 (1975).
[38] I. Hinchliffe and J. Erler and P. Langacker, in W. M. Yao *et al.* [Particle Data Group], J. Phys. G **33**, 1 (2006).
[39] Y. Yamada, Z. Phys. C **60**, 83 (1993).
[40] M. L. Alciati, F. Feruglio, Y. Lin and A. Varagnolo, JHEP **0503**, 054 (2005) [arXiv:hep-ph/0501086].
[41] M. Dine, arXiv:hep-ph/0011376.
[42] S. Dimopoulos and L. J. Hall, Phys. Lett. B **207**, 210 (1988).
[43] S. Deser and B. Zumino, Phys. Rev. Lett. **38**, 1433 (1977).
[44] L. O'Raifeartaigh, Nucl. Phys. B **96**, 331 (1975).
[45] M. A. Luty, arXiv:hep-th/0509029.
[46] L. Girardello and M. T. Grisaru, Nucl. Phys. B **194**, 65 (1982).
[47] J. E. Kim and H. P. Nilles, Phys. Lett. B **138**, 150 (1984).
[48] G. F. Giudice and A. Masiero, Phys. Lett. B **206**, 480 (1988).
[49] L. J. Hall and L. Randall, Nucl. Phys. B **352**, 289 (1991).

[50] P. J. Fox, A. E. Nelson and N. Weiner, JHEP **0208**, 035 (2002) [arXiv:hep-ph/0206096]; A. E. Nelson, N. Rius, V. Sanz and M. Unsal, JHEP **0208**, 039 (2002) [arXiv:hep-ph/0206102].
[51] G. D. Kribs, E. Poppitz and N. Weiner, arXiv:0712.2039 [hep-ph].
[52] M. Dine, A. E. Nelson, Y. Nir and Y. Shirman, Phys. Rev. D **53**, 2658 (1996) [arXiv:hep-ph/9507378].
[53] G. F. Giudice, M. A. Luty, H. Murayama and R. Rattazzi, JHEP **9812**, 027 (1998) [arXiv:hep-ph/9810442].
[54] L. Randall and R. Sundrum, Nucl. Phys. B **557**, 79 (1999) [arXiv:hep-th/9810155].
[55] J. L. Feng and T. Moroi, Phys. Rev. D **56**, 5962 (1997) [arXiv:hep-ph/9612333].
[56] S. Y. Choi, H. E. Haber, J. Kalinowski and P. M. Zerwas, Nucl. Phys. B **778**, 85 (2007) [arXiv:hep-ph/0612218].
[57] M. Schmaltz and W. Skiba, Phys. Rev. D **62**, 095004 (2000) [arXiv:hep-ph/0004210].
[58] J. L. Feng and T. Moroi, Phys. Rev. D **58**, 035001 (1998) [arXiv:hep-ph/9712499].
[59] J. L. Feng, S. F. Su and F. Takayama, Phys. Rev. D **70**, 063514 (2004) [arXiv:hep-ph/0404198].
[60] T. Abe et al. [American Linear Collider Working Group], in *Proc. of the APS/DPF/DPB Summer Study on the Future of Particle Physics (Snowmass 2001)* ed. N. Graf [arXiv:hep-ex/0106056]
[61] J. L. Feng and M. M. Nojiri, arXiv:hep-ph/0210390.
[62] M. M. Nojiri, Phys. Rev. D **51**, 6281 (1995) [arXiv:hep-ph/9412374].
[63] G. Weiglein et al. [LHC/LC Study Group], Phys. Rept. **426**, 47 (2006) [arXiv:hep-ph/0410364].
[64] M. E. Peskin, Int. J. Mod. Phys. A **13**, 2299 (1998) [arXiv:hep-ph/9803279].
[65] I thank Norman Graf for providing this figure.
[66] J. L. Feng, M. E. Peskin, H. Murayama and X. R. Tata, Phys. Rev. D **52**, 1418 (1995) [arXiv:hep-ph/9502260].
[67] ATLAS Collaboration, *Detector and Physics Performance Technical Design Report*, vol.II. CERN/LHCC/99-14 (1999).
[68] A. Ball, M. Della Negra, A. Petrilli and L. Foa [CMS Collaboration], J. Phys. G **34**, 995 (2007).
[69] http://cmsinfo.cern.ch/outreach/CMSdetectorInfo/NewPhysics/
[70] W. Beenakker, R. Hopker, M. Spira and P. M. Zerwas, Nucl. Phys. B **492**, 51 (1997) [arXiv:hep-ph/9610490]; http://www.ph.ed.ac.uk/~tplehn/prospino/
[71] D. R. Tovey, Eur. Phys. J. direct C **4** (2002) N4.
[72] I. Hinchliffe, F. E. Paige, M. D. Shapiro, J. Soderqvist and W. Yao, Phys. Rev. D **55**, 5520 (1997) [arXiv:hep-ph/9610544].
[73] R. Kitano and Y. Nomura, Phys. Rev. D **73**, 095004 (2006) [arXiv:hep-ph/0602096].
[74] N. Arkani-Hamed, P. Schuster, N. Toro, J. Thaler, L. T. Wang, B. Knuteson and S. Mrenna, arXiv:hep-ph/0703088.

[75] Some models with on-shell gauge bosons in the final state of squark decays are analyzed in J. M. Butterworth, J. R. Ellis and A. R. Raklev, JHEP **0705**, 033 (2007) [arXiv:hep-ph/0702150].
[76] B. C. Allanach, C. G. Lester, M. A. Parker and B. R. Webber, JHEP **0009**, 004 (2000) [arXiv:hep-ph/0007009].
[77] C. G. Lester and D. J. Summers, Phys. Lett. B **463**, 99 (1999) [arXiv:hep-ph/9906349].
[78] A. J. Barr, Phys. Lett. B **596**, 205 (2004) [arXiv:hep-ph/0405052].
[79] T. Goto, K. Kawagoe and M. M. Nojiri, Phys. Rev. D **70**, 075016 (2004) [Erratum-ibid. D **71**, 059902 (2005)] [arXiv:hep-ph/0406317].
[80] J. M. Smillie and B. R. Webber, JHEP **0510**, 069 (2005); [arXiv:hep-ph/0507170]. C. Athanasiou, C. G. Lester, J. M. Smillie and B. R. Webber, JHEP **0608**, 055 (2006) [arXiv:hep-ph/0605286], arXiv:hep-ph/0606212.
[81] A. Alves, O. Eboli and T. Plehn, Phys. Rev. D **74**, 095010 (2006) [arXiv:hep-ph/0605067].
[82] J. R. Ellis, K. Enqvist, D. V. Nanopoulos and F. Zwirner, Mod. Phys. Lett. A **1**, 57 (1986).
[83] R. Barbieri and G. F. Giudice, Nucl. Phys. B **306**, 63 (1988).
[84] J. L. Feng, K. T. Matchev and T. Moroi, Phys. Rev. Lett. **84**, 2322 (2000) [arXiv:hep-ph/9908309]; Phys. Rev. D **61**, 075005 (2000) [arXiv:hep-ph/9909334].
[85] M. M. Kado and C. G. Tully, Ann. Rev. Nucl. Part. Sci. **52**, 65 (2002).
[86] R. Dermisek, J. F. Gunion and B. McElrath, Phys. Rev. D **76**, 051105 (2007) [arXiv:hep-ph/0612031].
[87] S. Chang and N. Weiner, arXiv:0710.4591 [hep-ph].
[88] G. Degrassi, S. Heinemeyer, W. Hollik, P. Slavich and G. Weiglein, Eur. Phys. J. C **28**, 133 (2003) [arXiv:hep-ph/0212020].
[89] H. E. Haber, R. Hempfling and A. H. Hoang, Z. Phys. C **75**, 539 (1997) [arXiv:hep-ph/9609331].
[90] N. Cabibbo, L. Maiani, G. Parisi and R. Petronzio, Nucl. Phys. B **158**, 295 (1979).
[91] P. Batra, A. Delgado, D. E. Kaplan and T. M. P. Tait, JHEP **0402**, 043 (2004) [arXiv:hep-ph/0309149].
[92] R. Harnik, G. D. Kribs, D. T. Larson and H. Murayama, Phys. Rev. D **70**, 015002 (2004) [arXiv:hep-ph/0311349].
[93] J. F. Gunion, H. E. Haber, G. Kane, and S. Dawson, *The Higgs Hunter's Guide.* (Addison-Wesley, 1990).
[94] M. S. Carena and H. E. Haber, Prog. Part. Nucl. Phys. **50**, 63 (2003) [arXiv:hep-ph/0208209].
[95] V. Trimble, Ann. Rev. Astron. Astrophys. **25**, 425 (1987).
[96] D. N. Spergel *et al.* [WMAP Collaboration], Astrophys. J. Suppl. **170**, 377 (2007) [arXiv:astro-ph/0603449].
[97] O. Lahav and A. R. Liddle, in W. M. Yao *et al.* [Particle Data Group], J. Phys. G **33**, 1 (2006).
[98] G. Bertone, D. Hooper and J. Silk, Phys. Rept. **405**, 279 (2005) [arXiv:hep-ph/0404175].

[99] T. Moroi and L. Randall, Nucl. Phys. B **570**, 455 (2000) [arXiv:hep-ph/9906527].
[100] R. Kitano and Y. Nomura, Phys. Lett. B **632**, 162 (2006) [arXiv:hep-ph/0509221].
[101] M. Ibe and R. Kitano, JHEP **0708**, 016 (2007) [arXiv:0705.3686 [hep-ph]].
[102] R. J. Scherrer and M. S. Turner, Phys. Rev. D **33**, 1585 (1986) [Erratum-ibid. D **34**, 3263 (1986)].
[103] E. A. Baltz, J. E. Taylor and L. L. Wai, arXiv:astro-ph/0610731.
[104] A. M. Green, JCAP **0708**, 022 (2007) [arXiv:hep-ph/0703217].
[105] H. Goldberg, Phys. Rev. Lett. **50**, 1419 (1983).
[106] J. R. Ellis, K. A. Olive, Y. Santoso and V. C. Spanos, Phys. Lett. B **565**, 176 (2003) [arXiv:hep-ph/0303043].
[107] J. Edsjo, M. Schelke, P. Ullio and P. Gondolo, JCAP **0304**, 001 (2003) [arXiv:hep-ph/0301106].
[108] H. Baer, A. Belyaev, T. Krupovnickas and X. Tata, JHEP **0402**, 007 (2004) [arXiv:hep-ph/0311351].
[109] R. Arnowitt, B. Dutta and Y. Santoso, Nucl. Phys. B **606**, 59 (2001) [arXiv:hep-ph/0102181].
[110] C. Balazs, M. S. Carena and C. E. M. Wagner, Phys. Rev. D **70**, 015007 (2004) [arXiv:hep-ph/0403224].
[111] E. A. Baltz, M. Battaglia, M. E. Peskin and T. Wizansky, Phys. Rev. D **74**, 103521 (2006) [arXiv:hep-ph/0602187].

TASI 2006 - Students Talks Schedule

Week 1: June 5 - June 9

Time	Monday	Tuesday	Wednesday	Thursday	Friday
**					N. Setzer and S. Spinner (U. Maryland) Predicting the Seesaw Scale Time: 3:45 – 4:25
**					Gabe Shaughnessy (U. Wisconsin) Higgs Sector in Singlet Extended MSSM Time: 4:30 – 5:00
**					
**					
**					
**					

Week 2: June 12 - June 16

Time	Monday	Tuesday	Wednesday	Thursday	Friday
7:15*	Delphine Perrodin (U. Arizona)	Alejandro Jenkins (Caltech) Limits On A Cosmic "Solid"	PICNIC Starting at 6:00pm	Rouven Essig (Rutgers) Direct Detection of Non-Chiral Dark Matter	Chiu Man Ho (U. Pittsburgh) Charged Lepton Mixing & Oscillation from Neutrino Mixing in the Early Universe *Time: 3:45 – 4:15
7:45*	Astrophysical Constraints on higher order theories of gravity.				
7:50*	Pearl Sandick (U. Minnesota)	Emel Gulez (Ohio State U.) Lattice Determination of form factors for Semi-leptonic B decays	PICNIC Starting at 6:00pm	Ken Hsieh (U. Maryland) Dark Matter in QED Models	Tommer Wizansky (Stanford U) Determination of Dark Matter Properties at High Energy Colliders *Time: 4:20 – 4:50
8:20*	Cosmological Supernovae: Neutrino Background & Gravitational Wave Signature.				

Week 3: June 19 - June 23

Time	Monday	Tuesday	Wednesday	Thursday	Friday
7:15	A. Bachri (Oklahoma State U.)	Yingchuan Li (U. Maryland)	Anibal Medina (U. Chicago)	Anupama Atre (U. Wisconsin)	FREE
–	Soft Leptogenesis in Left-Right Symmetry; RGE & SUSY Breaking effects.	SO(10) GUT Model, θ_{13} and Leptogenesis.	Soft Leptogenesis in Warped Extra Dimensions	Search for Lepton Number Violation	
7:45					
7:50	Natalia Shuhmaher (McGill University)	Jing Shu (U. Chicago)	Joachim Kopp (Tech U. Munich)	Haibo Yu (U. Maryland)	FREE
–	Undressing some of the hierarchies in Cosmology.	Phase Transition Baryogenesis.	Simulation of Neutrino Oscillation Experiments	Discrete Symmetry in Neutrino Physics and Beyond	
8:20					

Week 4: June 26 - June 30

Time	Monday	Tuesday	Wednesday	Thursday	Friday
**	John Mason (U Calif. Santa Cruz)	Andrew Noble (Cornell U.)	Guiyo Huang (U. Wisconsin)	Erin De Pree (William and Mary)	FREE
**	Enhanced Radiative Corrections to Higgs Couplings.	EW Constraints on the littlest Higgs model with T-Parity	Mass variables	Top pair production in Randall-Sundrum Models	
**	*Time: 2:00 – 2:30	*Time: 3:45 – 4:15	*Time: 7:15 – 7:45	*Time: 7:15 – 7:45	
**	Sky Bauman (U. Arizona)	Andrey Katz (Technion Israel)	Frank Tackmann (U. California Berkeley)	FREE	FREE
**	KK Masses & Couplings Radiative Corrections to Tree level relations.	Lorentz violation & Superpartner masses.	Flavor Physics & non perturbative effects in $B \to X_s \ell^+ \ell^-$ decays		
**	*Time: 2:40 – 3:10	*Time: 4:20 – 4:50	*Time: 7:50 – 8:20		

Notes: *Please note that Talks on Friday of the second week are taking place in the afternoon.